Productivity and Technology in the Metallurgical Industries

Productivity and Technology in the Metallurgical Industries

Proceedings of the International Symposium on Productivity and Technology in the Metallurgical Industries, co-sponsored by The Minerals, Metals & Materials Society and Gesellschaft Deutscher Metallhütten-und Bergleute, held in Cologne, West Germany, September 17-22, 1989.

Edited by

Dr. Michael Koch
Krupp Industrietechnik GmbH
West Germany

and

John C. Taylor
Jan H. Reimers and Associates
Ontario, Canada

A Publication of
TMS
Minerals • Metals • Materials

A Publication of The Minerals, Metals & Materials Society
420 Commonwealth Drive
Warrendale, Pennsylvania 15086
(412) 776-9024

The Minerals, Metals & Materials Society is not responsible for statements or opinions and absolved of liability due to misuse of information contained in this publication.

Printed in the United States of America
Library of Congress Catalog Number 89-61032
ISBN Number 0-87339-100-4

Authorization to photocopy items for internal or personal use, or the internal or personal use of specific clients, is granted by The Minerals, Metals & Materials Society for users registered with the Copyright Clearance Center (CCC) Transactional Reporting Service, provided that the base fee of $3.00 per copy is paid directly to Copyright Clearance Center, 27 Congress Street, Salem, Massachusetts 01970. For those organizations that have been granted a photocopy license by Copyright Clearance Center, a separate system of payment has been arranged.

© 1989

Preface

The Minerals, Metals & Materials Society (TMS) and Gesellschaft Deutscher Metallhütten-und Bergleute (GDMB) sponsored the International Symposium on "Productivity and Technology in the Metallurgical Industries." Members of the various committees of the Extraction and Processing Division (EPD) of TMS are pleased to join with their associates in GDMB in forming the Organizing Committee and supporting this Symposium.

The principal objective of the organizers is to provide a range of technical presentations covering a variety of ferrous and non-ferrous metals in order to provide an interchange of technical discussion between the various sectors of the metallurgical industry who traditionally meet separately. In addition, the economic, productivity and environmental concerns of the industry are addressed. To provide focus to the various technical sessions, eminent representatives from industry, research and academia were invited to provide plenary lectures and to participate in a Geo-Economic panel discussion at the end of the Symposium. The Organizing Committee particularly thanks these invited speakers for their contribution to the success of the meeting.

Unfortunately, the lead time required by our publisher dictates that manuscripts be submitted well in advance of the Symposium, thus preventing some of the latest developments, which may be presented, from appearing in print.

In order to allow the members of the Geo-Economic panel to comment on the most recent changes in the industry, their presentations have not been included in this publication. Their thoughtful comments and opinions, however, are greatly appreciated by all attendees. The Committee particularly wants to thank Dipl. Ing, Gerhard Berndt of Norddeutsche Affinerie, Dr. Takeshi Nagano of Mitsubishi Metal Corporation, and Dr. Robert E. Johnson of Phelps Dodge for their valuable contribution to the Geo-Economic discussion.

A second and important objective of the Symposium is to bring together leaders in the industry from both sides of the Atlantic and to provide a forum where mutual problems and concerns of the industry can be discussed on an informal basis. The extractive metallurgical industry is truly international, and becoming more so as concentrates, intermediate products and refined metals are traded extensively worldwide, and technology is transferred across national borders. Economic factors, in particular rates of exchange, the specifications for refined metals and environmental regulations in any one trading block have an impact on the industry as a whole. No longer can one group operate and survive in isolation. It is therefore timely that leading technical societies such as TMS, through its Extraction and Processing Division, and GDMB should take the initiative and provide a forum for industry representatives to meet and discuss their concerns.

The support provided to the Committee by industry and various institutions through sponsorship of the functions and in providing a wide range of technical presentations is gratefully acknowledged. The support and hospitality extended by Krupp Industrietechnik GmbH throughout, is particularly appreciated. It is the sponsorship and support of such companies which enable the technical societies to grow and serve the industry more effectively.

The Organizing Committee and all the attendees thank the Lord Mayor of Cologne, Norbert Burger, for the hospitality and friendly reception received during their stay in the city.

Finally, the Editors and the Organizing Committee express their appreciation to the authors, Session Chairmen, and the staff of TMS for their contributions to the success of the Symposium.

Dr. Michael Koch
Manager, Non Ferrous Metals
Krupp Industrietechnik GmbH
D-4100 Duisburg 14
West Germany

John C. Taylor
President
Jan H. Reimers and Associates Inc.
221 Lakeshore Road East
Oakville, Ontario, Canada L6J 1H7

Organizing Committee

GENERAL MEETING CHAIRMAN

Dr. Michael Koch
Krupp Industrietechnik GmbH

MEMBERS

John C. Taylor, O.S.M., P. Eng.
TMS Technical Director -
Extractive Metallurgy

James E. Hoffmann, P.E.
Jan H. Reimers & Associates USA Co.

Dr. Carlos A. Landolt
INCO Limited

Dr. Robert H. Maes
Metallurgie Hoboken-Overpelt

Dipl. Ing. Herbert Aly
Gesellschaft Deutscher Metallhütten und Bergleute (GDMB)

Ms. Barbara Kamperman
The Minerals, Metals & Materials Society (TMS)

Table of Contents

Preface v
Symposium Organizing Committee vii

Plenary Lectures

Non-Ferrous Metals—An Industrial/Economic Overview From a
European Viewpoint 3
   J. Castle, R. Maes, M. Leroy and H. Traulsen

Productivity and Technology Changes in the North American
Copper, Zinc and Nickel Extractive Industries 69
   F.G.T. Pickard and G.A. Crawford

Non-Ferrous Extractive Metallurgy: A Technical Overview 85
   J.G. Peacey

A Research Overview on Non-Ferrous Extractive Metallurgy
(Abstract Only) 103
   R. Winand

The Role of Mathematical Models in New Process Development 105
   J. Szekely

Energy for Metal Production in the 21st Century 145
   H.H. Kellogg

Materials for the Future (Abstract Only) 155
   Y. Farge

Pyrometallurgy

The Declining Influence of Economic Growth on Metal Demand 159
   G. Davis

The Mathematical Modelling of Non-Ferrous Blast Furnaces 187
   K.C. Wade, M. Cross and R. Smith

Treatment of Metallurgical Gases Now and Tomorrow 205
   K. Westerlund and R. Peippo

Carbothermic Reduction Kinetics of Tin Concentrate Pellets 215
   V.N. Misra

Dissolution Loss of Copper, Tin and Lead in FeO_n-SiO_2-CaO Slag 227
    Y. Takeda and A. Yazawa

Direct Smelting of Zinc 241
    S.K. Gupta and J.M. Floyd

Oxygen-Enhanced Fluid Bed Roasting 257
    D. Saha, J.S. Becker and L. Gluns

An Innovative Approach to Alloying Molten Metals with High Melting
Point Additions 281
    J. Schade, S.A. Argyropoulos and A. McLean

The Krupp HOM-TEC (Horizontal Oscillating Mould Technology) Caster
an Advanced Process for Horizontal Continuous Casting of Copper and
Copper Alloys 301
    E. Roller

Process Development of Electromagnetic Casting for Copper Base
Alloys (Abstract Only) 313
    J.C. Yarwood, P.C. Chatfield and B.G. Lewis

Copper Rod Casting Plant for Small Production Quantities 315
    E. Buch, K. Siebel and H. Berendes

Flash Smelting of Pyrite and Sulphur Recovery 341
    J.A. Asteljoki and T.P.T. Hanniala

Present and Future Trends of the Sintering-Blast Furnace Process
at MHO 357
    M. Van Camp and D. Crauwels

High Grade Matte Operation in the Mitsubishi Continuous Process
at Naoshima Smelter 365
    T. Shibasaki and K. Kanamori

Flash Smelting Gets a New Dimension: Contop-Continuous
Smelting and Top Blowing Now in Industrial Application
(Abstract Only) 379
    G. Melcher

Assessment of Copper/Zinc Smelter Modernization at Hudson Bay
Mining and Smelting (HBMS)—Flin Flon, Manitoba, Canada
(Abstract Only) 381
    W.J.S. Craigen, B. Barlin and B. Krysa

Restructuring Boliden's Rönnskär Smelter 399
    B. Lofkvist, P.-O. Lindgren and P.G. Broman

SO$_2$ Abatement and Productivity Programs at Inco's Copper
Cliff Smelter (Abstract Only) 413
    J. Blanco, H. Davies and P. Garritsen

Minimization of Heat Losses from Fuel Fired Heat Treatment
Equipment 415
    H.-O. Jochem

Optimal Fuel Consumption Strategy for Heating Furnaces 435
    M.A. Youssef and E.S. Geskin

Non-Ferrous Metallurgy: Decrease of Energy Consumption 459
    P. Paschen

Use of Oxy-Fuel Burners in Smelting Lead Battery Scrap in
Short Rotary Furnaces 473
    K.F. Lamm

Low Waste Technology for Reprocessing Battery Scrap 483
    K.F. Lamm and A.E. Melin

Processing of Lead-Acid-Battery-Scrap: The Varta-Process 495
    M. Koch and H. Niklas

"Contibat" Process for Treatment of Scrap Lead Batteries 501
    R. Fischer

Use of Electric Furnaces in Non-Ferrous Metallurgy 511
    C. König, G. Rath and T. Vlajcic

Treatment of Automotive Exhaust Catalysts 523
    V. Jung

Ferrochrome Production From Chromite Fines Via Coal-Based
Direct Reduction in Rotary Kilns and Subsequent Melting in
Electric Furnaces (Abstract Only) 541
    K.H. Ulrich and W. Janssen

Hydrometallurgy

The Disposal of Arsenical Solid Residues 545
    G.B. Harris and S. Monette

Nickel, Cobalt and Copper Bearing Synthetic Model Oxyhydroxides
of Manganese (IV)—Relevance in Manganese Nodule Processing 561
    R. Kumar, R.K. Ray and A.K. Biswas

The Hydrothermal Conversion of Jarosite-Type Compounds to
Hematite 587
    J.E. Dutrizac

Solvent Extraction of Precious Metals by S-Alkyl Derivatives of
Dehydrodithizone and Dithizone 613
    M. Grote, G. Pickert and A. Kettrup

Pressure Cyanidation: An Option to Treat Refractory Gold 627
    S.C. Girardi, O.M. Alruiz and J.P. Anfruns

Electrochemical Study of Silver Cementation on Lead in
Chloride Solution 639
    K.H. Kim, T. Kang and H.-J. Sohn

Recovery of Zinc from Blast Furnace Dust of Metallurgical
Company in Yugoslavia 651
    M. Stamatovic, S. Milosevic, N. Canic, L. Mihovilovic
    and T. Zivanovic

Extraction of Zinc From Complex Polymetallic Sulphide
Concentrates 659
    V.N. Misra

Relation Between the Leachability of Zinc-Concentrates in
Fe(III)-H_2SO_4 Media and Their Electric Conductivity......................................673
    X.Y. Xiong, M. Jacob-Dulière and J. Sterckx

Zinc Recovery by Solvent Extraction 695
    A. Selke and D. de Juan Garcia

Simulation of Continuous Leaching of Galena with $FeCl_3$ in
Hydrochloric Acid Solution 705
    C.-K. Lee, J.-B. Ryu, H.-J. Sohn and T. Kang

Effectiveness of Glue in Actual Tankhouse Electrolyte Controlled
by the Collamat®—Method 717
    B.E. Langner and P. Stantke

Column Leaching of Scrapped Petrochemical Catalysts 725
    S. Kelebek and P.A. Distin

An Engineering Approach to the Arsenic Problem in the Extraction
of Non-Ferrous Metals..735
    N.L. Piret and A.E. Melin

Improvements of Hydrometallurgical Processes by Application
of the Acid Retardation Procedure—Actual Situation and
Developments 815
    M. Gülbas, R. Kammel and H.-W. Lieber

Light and Other Metals

Advances in Application of Zirconium, Niobium & Titanium Alloys
Through Powder Metallurgy 829
    A.F. Condliff

VAW-Technology for the Alumina-Production According to the
Bayer-Process 843
    W. Arnswald

A New In-Line Aluminum Treatment System Using Nontoxic Gases
and a Gas-Permeable Vessel Bottom 855
    D. Saha, J.S. Becker and L. Gluns

Modernization of VAW-Potlines in the F.R.G. 879
    V. Sparwald

Lithium Extraction From Spodumene by Metallo-Thermic
Reduction 893
    E. Mast, R. Harris and J.M. Toguri

Mineral Processing and Feed Preparation

The Routine Use of Process Mineralogy in a Cost-Efficient
Gold-Silver Operation 909
    W. Baum and S.R. Gilbert

Stochastic Dynamic Simulation of Flotation Circuits
(Abstract Only) 921
    D. Yin, S. Li and W. Hu

Some Milling Practices and Technological Innovations on
Beneficiation of Mercury Sulfide Ores in China 923
    H.-K. Hu

Advances in Technology for Complex Custom Feed Material
Treatment at Noranda 929
    M. Chapados, M. Bédard and G. Kachaniwsky

The Development of a New Type of Industrial Vertical Ring
and Pulsating HGMS Separator 947
    X. Dahe, L. Shuyi, L. Yongzhi and C. Jin

Subject Index 953

Author Index 957

Plenary Lectures

NON-FERROUS METALS - AN INDUSTRIAL/ECONOMIC OVERVIEW

FROM A EUROPEAN VIEWPOINT

John Castle	Michel Leroy
RTZ Technical Services Bristol BS99 7YR United Kingdom	Pechiney - Aluval Developpement 38340 Voreppe France
Robert Maes	Heinrich Traulsen
Metallurgie Hoboken-Overpelt 2710 Hoboken Belgium	Norddeutsche Affinerie 2000 Hamburg F.R. Germany

ABSTRACT

The paper traces the technological and economic factors that have shaped the extractive non-ferrous metallurgical industry in Western Europe over recent years and comments on some future possible trends. Factors that have had a general worldwide influence on the industry such as : energy and raw materials costs and availability, environmental constraints, high cost of capital, changing capacity requirements and materials recycling and substitution are briefly introduced. Against this background the evolution in Europe of the extraction technologies of the individual metals Cu, Ni, Pb, Sn, Zn, Al, Mg and precious metals to the present "state of the art" and economic viability is discussed. Some common underlying trends that may influence the future development and growth of the industry are identified. Possible new technological developments in the extraction of the individual metals and their suitability for exploitation in Europe are discussed.

TABLE OF CONTENTS

1	GENERAL BACKGROUND

2	INDUSTRY RESTRUCTURING DURING THE EIGHTIES

3	LEAD

4	ZINC

5	COPPER

6	NICKEL

7	PRECIOUS METALS

8	ALUMINIUM

9	MAGNESIUM

10	TIN

11	SOME FUTURE TRENDS OF THE INTERNATIONAL AND EUROPEAN CONTEXT

1   GENERAL BACKGROUND

The importance of the world production of non-ferrous metals can be assessed by reference to Table 1, which gives the tonnages of the ten major metals and the corresponding revenue, estimated from current average prices. For comparison, the table also gives equivalent values for steel, which is about 2.5 times that of the total of all the non-ferrous metals.

Table 1   World Production of Metals and Corresponding Turn-over

	Production (tonnes)	Turn-over (G Ecu)	Classification
Ag	15.5 k	2.6	7
Al	15.5 M	35	1
Au	1.5 k	18	2
Cu	9.8 M	18	2
Mg	320 k	1.1	10
Ni	740 k	7	4
Pb	5.5 M	2.8	6
Pt	100	1.6	8
Sn	200 k	1.3	9
Zn	6.6 M	6.6	5
		94	
Fe	720 M	240	

1.1   The Metals Industry - Geographic Distribution

An estimate of the geographic distribution of the mine and metallurgical production of non-ferrous metals and their world consumption is given in Figure 1. This Figure gives average percentages for six metals (Al Cu Ni Pb Sn Zn) produced and consumed in the major regions of the world, namely: North America, Europe, the Centrally Planned Countries, the Far East, South America, Africa, Asia and Oceania. As a general rule the percentage consumptions are more or less the same for the different non-ferrous metals, since they relate to the level of development of the regions in question. The Figure also shows that the four "Northern" regions contribute about 90% of the world consumption, compared to about 10% for the four "Southern" regions.

An important portion of these metals is produced in the regions where they are consumed. This emphasises the importance that smelting and refining of non-ferrous metals has in the Northern regions, which on average contribute about 80% of the production of the metals consumed in the world.

If the geographic distribution of the mine production is considered, keeping in mind the pattern can differ markedly from the average for some metals because of the geographic distribution of their minerals, it may be observed that the North covers about 57% of the world's needs of non-ferrous minerals, while 43% originates from the Southern regions.

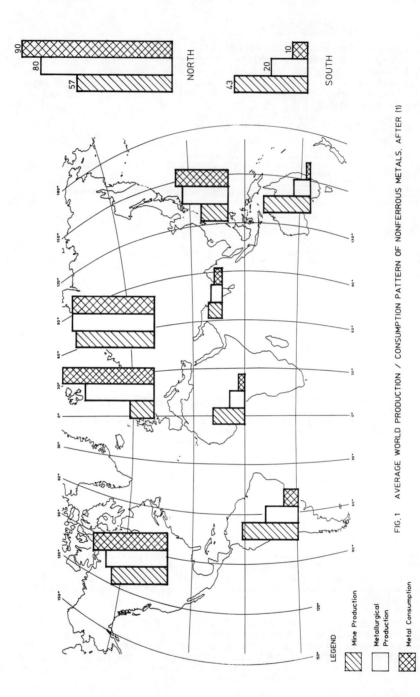

FIG. 1 AVERAGE WORLD PRODUCTION / CONSUMPTION PATTERN OF NONFERROUS METALS, AFTER (1)

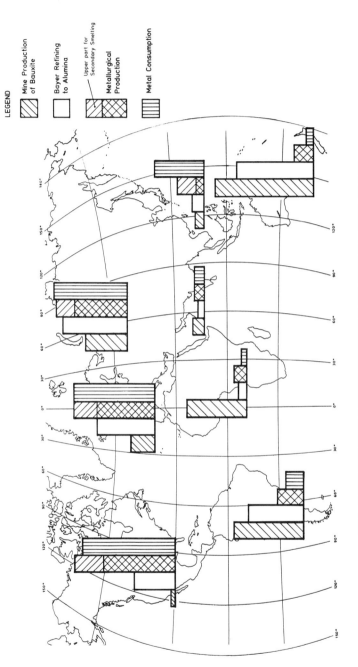

FIG. 2 WORLD PRODUCTION / CONSUMPTION PATTERN FOR ALUMINIUM AFTER (1) & (2)

This situation is less dramatic than the picture which is generally presented ie extreme dependence of the North upon the South for its minerals supply. In this respect Europe and the Far East are less favourably situated, although each on average still contributes about 8% to world needs. The situation of near self sufficiency of the Centrally Planned Countries is worth underlining.

The situation described above gives only general tendencies corresponding to an average of six metals. A more precise description would need to consider the situation metal by metal, when the specific geographic distribution of minerals and the extraction and consumption patterns would appear more accurately. In addition, most metals are produced and exchanged at intermediate stages of extraction (concentrates, mattes, blisters, bullions) which should be mentioned in the general picture. As illustration Figure 2 gives the world production/consumption pattern for aluminium from bauxite to metal.

1.2   The Metal Industry - Pre Oil Crisis

The historical evolution of the non-ferrous industry in the period since World War II is characterised by high growth at relatively constant rate until 1974, the year of the oil crisis, see Figure 3. The average annual

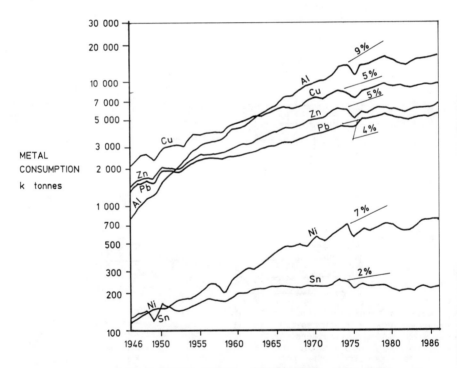

FIG.3 WORLD METAL CONSUMPTION 1946 - 1986

consumption growth rates at that time were approximately: 9% for Al, 7% for Ni, 5% for Cu and Zn, 4% for Pb and 2% for Sn. The exceptional growth of aluminium, whose consumption at least doubled every 8 years, is mainly due to the developments in transport, electrical power distribution and packaging industries. The growth of nickel may be related to its use in stainless steel and that of copper and zinc to the annual 4.8% growth rate which steel, a measure of industrial activity, realised during the same period. As regards lead and tin they lie below the average trend, which can be explained for lead by a marked tendency to substitution and for tin by the reduction of coating thicknesses.

As was emphasised by Strauss (3) in an interesting analysis of this situation, the geographic distribution of this post-war growth has been far from uniform amongst the developed countries. If, for instance the average % annual growth rate for the period 1950 - 1973 is analysed for both steel and the six non-ferrous metals, making the distinction between the USA and the other Market Economy Countries, the graph of Figure 4 is obtained. From this figure it is quite evident that the post-war restoration of Europe and Japan contributed a major part of the growth at that time.

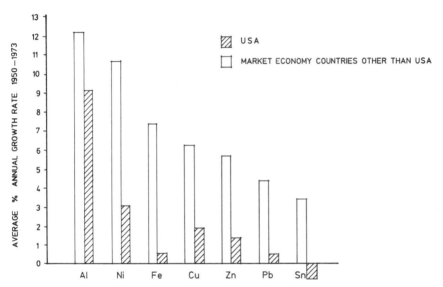

FIG. 4 AVERAGE PERCENT ANNUAL GROWTH RATE OF METAL CONSUMPTION 1950-1973 IN THE USA AND OTHER MARKET-ECONOMY COUNTRIES ; AFTER (3)

In the sixties, voices were raised to say that the growth rates which were observed could not be maintained over the long term due to exhaustion of resources. This culminated in 1972 with the Club of Rome which induced consumers' fears of eventual shortage and raised the spectre of zero growth. The OPEC action which tended to demonstrate by cartel formation the possibility of imminent exhaustion of energy resources, focused more attention on the vulnerability of the supply base of non-ferrous metals. With hindsight, the evolution in the seventies namely the slowing down of growth from 1974 may be described in a somewhat different way: the growth of the sixties could not last due to the fact that one of its essential contributors, the post war restoration, was disappearing. The OPEC action had the main effect of imposing a new cruising speed in the short-term.

2   INDUSTRY RESTRUCTURING DURING THE EIGHTIES

The evolution of the last fifteen years may be summarised as being that of the adjustment to a new growth rate and of the restoration of confidence which had been seriously unsettled. Restructuring, consolidation and rationalisation of our industry did not take place without serious social difficulties brought about by the need to assimilate production capacities installed during the previous high growth period. In Europe the effects were most strongly felt in sectors such as steel and zinc. In the meantime growth has resumed, although at rates lower than those experienced prior to 1975.

In parallel with these major effects related to the oil crisis, which have exerted a profound influence on the mineral and non-ferrous industry as well as on public opinion, our industry became increasingly aware of some disturbing aspects and weaknesses in its activity for example concern with regard to environmental protection. These different points will be discussed in the following paragraphs.

2.1   Consolidation/Rationalisation

The intensification of mine prospecting, which had already started in the sixties, led to the discovery and exploitation of many deposits which will contribute to supply in the future and the spectre of short-term exhaustion of resources is now firmly dismissed. This contributed to progressively diminish the cohesion of the various cartels tending to ration the market; the themes developed by the Club of Rome and the OPEC-type associations have now become considerably less significant. The supply uncertainties which they raised remain however a major concern. The non-ferrous industry responded to this by consolidating its position and its supply channels, by association of companies in mining joint ventures and by the constitution of industrial groups linking producers of minerals and metals more closely together. This is illustrated by some recent major company regroupings such as the participation of MIM, Metallgesellschaft and Teck Corporation in Cominco or the formation of Metaleurop by Preussag and Penarroya. It may be mentioned that during this period the participation of oil companies in the minerals industry and their subsequent withdrawal served to disturb this process of rationalisation.

The consolidation of the structures of our industry went hand in hand with a rationalisation of the production units. A major factor influencing this has been the escalating cost of capital and equipment. The ex-works cost of equipment rose by a factor of 2.0 between 1972 and 1980 and an overall factor of 2.6 from 1972 to date. This and the high cost of capital made it difficult to justify new green field projects even with the advantage of modern lower cost technology. The most attractive projects were technology retrofits and expansions of plants where existing infrastructure can contribute to reducing the total capital per annual tonne capacity. This situation is likely to continue for most non ferrous metals in Europe at least in the short term.

Tariff barriers either for imported metals or exported raw materials can also influence rationalisation and sometimes, because they can be varied or removed, distort the assessment of the potential of a metal production facility in a particular location.

2.2 Recycling

Minerals are finite and non-renewable and despite the present availability wasting them should be avoided. Metals by nature have the advantage they can be recycled, offering a solution to this concern, and recycling has developed considerably since the last world war. In the case of steel, in world terms one tonne of metal produced at present consists on average of half a tonne of recycled material; this corresponds to the total amount of steel which was brought into use fifteen years ago. It is thus estimated that the recycling rate is now close to 100%, if related to the metal at the beginning of its life cycle.

For non-ferrous metals the 1987 situation is summarised in Table 2, which gives the percentage of Al, Cu, Zn and Pb produced from recycling, related to the consumption in Europe, the USA and Japan, the three major regions where recycling actually takes place. As in the steel situation the tonnages of recycled Pb, Cu and Al correspond to the total amount of metal which was brought into use fifteen to twenty-five years ago. For Zn, recycling is less pronounced, due to the more diffuse uses of this metal, although it should be noticed that the tendency to up-grade Zn-containing residues and to incorporate the obtained metal concentrates in primary smelting circuits is developing.

In connection with recycling, primary and secondary sources are both important in providing the metal demand. However there is a growing source of metal that is recycled as a solution to environmental problems and authorities are progressively promoting the recycle of subeconomic metal containing residues by regulation of dumping and its cost.

Permissible dumping sites are limited and their cost increases particularly in developed countries. Recycle from these sources will play an increasing role in future as dumping costs rise.

2.3 Environment

Environment has been a concern of all times and industry has tended to cope with it according to prevailing standards and technical possibilities. From the seventies, the general awareness of environmental implications from metallurgical activity intensified however. Considerable efforts both in basic process design (eg reduced process gas volumes) and gas and liquid effluent control technology have been made since then and much retrofit capital spent. Modern acid plant technology can now control gaseous SO_2 emissions from sulphide smelting to high standards. Process and ventilation gas filtration with modern electrostatic and bag filters can control heavy metal emissions to low levels. This, combined with better ventilation, material handling techniques and changes in extraction technology have improved plant hygiene conditions. An illustration of the possible reduction is given for lead in air near the fence of a European lead smelter on Figure 5.

Table 2 1987 Recycling of Al, Cu, Zn and Pb, as Proportion of Consumption in Europe, USA and Japan, after (2), (4), (5) and (6)

		Europe		USA		Japan	
		Recycling	Consumption	Recycling	Consumption	Recycling	Consumption
Al	kTonne	1720	5770	1730	6270	970	2720
	% from recycling	30		28		36	
Cu	kTonne	1380	2680	1290	2130	600	1280
	% from recycling	51		60		47	
Zn	kTonne	610	2110	360	1340	170	830
	% from recycling	29		27		20	
Pb	kTonne	880	1610	780	1330	160	420
	% from recycling	55		59		38	

FIG. 5
EVOLUTION 1976 1987 OF THE CONCENTRATION OF LEAD IN AIR NEAR THE FENCE OF THE HOBOKEN SMELTER (7)

Similarly heavy metal emissions in liquid effluents can be controlled to acceptable standards by precipitation and modern solid/liquid separation techniques. Also SX, IX and bioaccumulation techniques are used in special applications.

A continuing concern of our industry is the further handling of its residual products. These may arise in the form of flue dust bleed streams, leach/purification residues, slags, etc. They can only be recirculated in the metallurgical circuits as far as they do not cause processing or product quality problems. They are preferably reintroduced in the economic circuit, possibly after transformation or adaptation following market requirements. According to their nature, they may be sold as commercial products or transferred to other companies for further treatment. If no market exists for them, they have to be produced in forms environmentally acceptable for disposal.

Environmental control often requires costly additional processes, and plants offering economy of scale are better situated to meet these costs than smaller operations. This is particularly true for some secondary plants where meeting environmental constraints will probably bring about regrouping to form fewer larger operations.

2.4 Technology Changes

Over the past decade many improvements have taken place in existing process technology and some new unit operations have been exploited. These innovations have been prompted by the need to reduce energy costs, improve productivity and meet ever more demanding environmental constraints.

In pyrometallurgy the opportunity for high intensity operation via two major types of reactor, namely flash/flame and bath smelting, has been exploited. Extensive use of oxygen enrichment is made. Hydrometallurgical developments include pressure leaching and biological leaching, use of selective solvent extractants, carbon adsorption and improved methods of solution iron control. Also extensive mechanisation has been installed along with improved larger electrode design and number of electrodes per cell to increase productivity in both electrowinning and electrorefining cells. All these technology changes progressively transformed the non-ferrous industry from labour intensive, as it was traditionally, to a capital intensive one; the quality of labour which it requires has also risen correspondingly.

There is a considerable amount of technology that has been developed to the pilot or semi commercial level which awaits a suitable economic climate and operator willing to take the risk to exploit. In a high cost environment, technical innovation to cut costs may but not always provide an adequate competitive edge at least for a limited period. Together with the necessity to maintain an industrial base for market and strategic reasons, it tends to counter a general movement towards transferring metallurgical activities to countries with low energy and labour costs.

2.5 Substitution

Finally, mention should be made of substitution, conservation in use, reduced consumption due to miniaturisation and competition by new materials as important factors conditioning our future. The non-ferrous

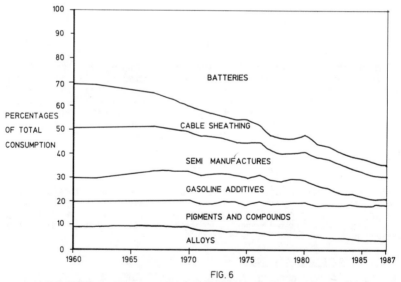

FIG. 6
PERCENTAGES OF TOTAL LEAD CONSUMPTION 1960-1987 IN SIX MAJOR INDUSTRIAL COUNTRIES (FRANCE GERMANY ITALY JAPAN UNITED KINGDOM UNITED STATES) AFTER (8)

industry can only have a limited impact on the advance of science and technology which is the driving force of this evolution and leads to greater or lesser consumption of the metals which it produces. In some cases, the evolution seems threatening, such as for the almost academic example of lead. As shown on Figure 6, usage of this metal since the last world war tends to progressively evolve towards the single market of automotive batteries. The substitution phenomenon should however neither be generalised nor over-emphasised as metals have their own specific properties which compete against possible substitutes. It has to some extent been allowed to occur by past lack of attention to market requirements by producers. New applications and markets are being developed and producers are providing an improved customer service with respect to purity, special cast shapes, alloy specifications and surface finish of ingots and semis. Substitution is also more often a long-term mechanism.

3   LEAD INDUSTRY

3.1   Location and Capacity versus Demand

The mine production of lead in Europe (EEC + EFTA countries) is about 300 000 tonnes per year, which corresponds more or less to 20% of the European consumption of the metal. With the addition of 50% produced by recycling, a total of about 70% of the needs is covered by internal supply.

Figure 7 gives a map of the main European smelters, including a dozen primary smelters and about 30 secondary smelters (those with a capacity lower than 10000 tonnes/year being generally omitted). Adding the primary and secondary capacity contributions of the plants, a total of 1.7Mt/y is obtained, which can be compared with the European consumption of lead, which is of the order of 1.5Mt/y. Most of the primary smelters operate on Pb or Pb/Zn concentrates, some of them produce lead in connection with a more complex smelting metallurgy, one of them, in the United Kingdom, limits its operations to the refining of bullion produced overseas.

Compared to the situation in 1960, the total Pb production capacity in Europe has approximately doubled. This was developed by different mechanisms, including some newcomers to the business. Primary smelter capacity increases since 1960 were achieved by increase in smelter size rather than their number; the introduction of the Imperial Smelting Process (ISP) also contributes now a significant part of the available capacity. There has also been conversion of several plants that were badly situated from the point of view of transport from primary to secondary smelting. Evolution towards greater production of the metal by recycling took place and the secondary industry has become increasingly important. Figure 6 indicated the changes in lead use over the period and the swing towards battery production; this now constitutes about 45% of the outlet of the metal in Europe as compared to 77% in the USA. The 1987 lead production of 50% from recycling in Europe compares with 65% in the USA and a 48% Western world average. Expansion of secondary smelter capacity since 1960 was brought about by increases in size and number of plants. Information, particularly on the smaller plants is more fragmentary than on primary smelters. Figure 8 gives a histogram of their European population classified by smelting capacity and provides a clear indication of the scatter in the sizes of the operations.

● PRIMARY SMELTER
▲ SECONDARY SMELTER

FIG. 7
LOCATION OF MAIN EUROPEAN PRIMARY AND SECONDARY LEAD SMELTERS

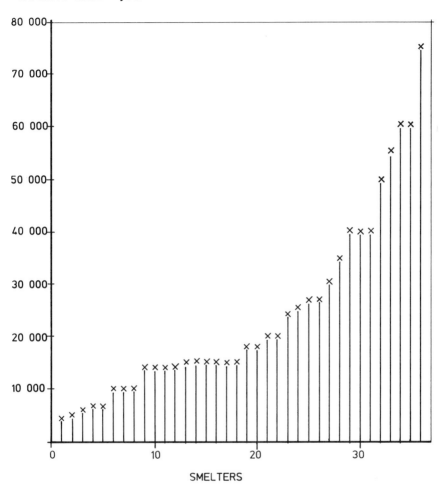

FIG.8
CLASSIFICATION OF EUROPEAN SECONDARY LEAD SMELTERS BY CAPACITY

3.2   Process Technology

The primary and secondary smelting technology is listed below.

(a)   Primary Smelting

The following primary lead smelting technology is presently operated or will soon be started in Europe:

- conventional sintering/blast furnace smelting;
- electric furnace smelting;
- ISP sintering/blast furnace smelting;
- direct flash smelting (Kivcet);
- direct bath smelting TBRC, QSL.

(b) Secondary Smelting

Most of the secondary Pb smelters operate in connection with battery producers. Traditionally, primary and secondary Pb metallurgies were not intermixed, and in most cases smelting processes are still kept distinct and, at most, some flows in the refining steps are combined. The smelting technology applied depends to a certain extent on the pretreatment of the main feed material scrap batteries. The following smelting methods are applicable:

- direct blast furnace smelting of as received whole batteries (Varta and Bergsoe processes);
- rotary furnace smelting of previously physically separated battery components;
- reverberatory (electric furnaces in USA) smelting of separated battery components.

The number of plants employing the various technologies and their respective total production capacities are given in Table 3. Together with the present situation, this table gives some assumptions (in brackets) on the probable evolution in the coming years. The table indicates that the primary extractive metallurgy of lead is diversified and in a state of change in Europe. With regard to lead refining both batch pyrometallurgical kettle and in a minority of cases Betts electrolytic technology is used in Europe.

3.3 Factors Influencing Evolution

The evolution of the European lead industry has been influenced by the following main factors:

Capacity

Due to environmental constraints on the more diffuse uses of the metal such as gasoline additives, market growth has continued to slow reaching more or less saturation. Some rationalisation of both primary and secondary capacity, which started in 1975, continues in the form of regroupings or association of mines and/or smelters.

Technology

Environmental control of emissions is a key aim of new primary smelting technology, and to this end the recently developed direct autogenous smelting processes using flash and bath smelting techniques will be considered 'state of the art' in future. These processes offer energy, investment and operating cost savings over conventional blast furnace technology. Their ability to directly treat lead sulphate residues from battery breaking will enable primary plants to treat increasing quantities of secondary lead. For Pb/Zn smelting and some complex Pb

Table 3   Smelting processes applied in European Pb smelters

Process	Number of plants	Production capacities (tonnes)
Primary smelters		
Sintering - blast furnace	8 (5)	520000 (320000)
Imperial Smelting	4	140000
Electric furnace	1 (2)	65000 (60000)
Kivcet	1	85000
TBRC	1	20000 (60000)
QSL	0 (1)	- (90000)
		830000 Total primary
Secondary smelters		
Blast furnace	7	100000
Rotary furnace	24	710000
Static furnace	5	60000
		870000 Total secondary
Total primary + secondary		1700000

feeds, sintering could remain an essential link of the processing operations. In secondary production the processes used remain traditional, the accent being principally placed on efficient environmental control.

Lead is an effective collector of minor metals and will be further used in this role in metallurgical flowsheets. Precious metals along with elements such as Sb, Sn, In, Te and Bi can be collected and recovered in the refining circuits. The recovery of all these metals entails however, extended separation processes, such as those provided by Harris refining which are not so traditional in lead metallurgy.

3.4 Some Future Trends of the Lead Industry in Europe

On the whole, the European lead industry is diversified and may be considered as adapted to needs. It incorporates a high proportion of modern technology. Formulating forecasts on its future evolution may seem hazardous, although trends which could significantly contribute to influencing its shape and appearance in the coming years are already clearly perceptible:

a) Lead is one of the metals whose extrapolation of supply and demand foreshadows possible negative growth. This will probably imply further rationalisation of the producing industry. Efforts will probably develop in the secondary lead industry to improve productivity and increase the size of smelting units, in order to reach economies of scale similar to those of primary smelters.

b) The tendency to use the metal in recyclable applications and to avoid it in diffuse ones will continue. The proportion of secondary smelting with regard to total production should consequently also continue to increase.

c) Where recycling is not profitable compared to primary smelting, authorities will tend to promote it by coercive measures or tax incentives, implying a growing public influence in industrial affairs.

d) Environmental constraints will keep weighing on the lead industry and will cause many of the traditional technology lead smelters to further improve their operation by taking significant environmental protection measures.

e) In primary smelting, particularly for those smelters for which sintering is not an essential link of the whole operation, the tendency to introduce the new direct smelting processes will continue. It will probably progressively take place over many years, the advantages claimed needing to be set against the considerable investments required.

f) Western world increases in zinc production are expected to drive byproduct lead production and could provide an overall lead concentrate surplus in the future.

4 ZINC INDUSTRY

4.1 Location and Capacity versus Demand

The European mine production, metal production and consumption for 1987 are summarised in Table 4 and compared to similar figures for other world regions. Locations of the primary zinc smelters are shown on Figure 9. From Table 4 it can be seen that Europe is responsible in western world terms for 19% mine, 40% metal production and 33% of consumption. Belgium, France, Germany, Italy, Netherlands are the main importers of concentrates along with Japan. Imports and exports of zinc metal, as with concentrates, reflect the differences between the domestic production and consumption of the individual countries. In Europe, West Germany and the UK are the two largest importers of refined metal, mainly from within the EEC/EFTA area. The main exporting European producers are Belgium, Finland, Netherlands, Norway and Spain.

The demand for zinc is closely tied to industrial activity and recent years saw record high world consumptions. Western world consumption by end use can be broken down as follows: protective coatings 35%, die casting 20%, brass making 20%, sheet zinc 10%, others including chemicals 15%. In Europe, use for brass is in a strong position, unlike the USA and developing countries. The main use namely galvanising is tied to activity in the construction, and more recently the car industry. More zinc alloys are now being produced for specific consumers of both die casting and galvanising metal (new long life coatings containing aluminium) thus requiring a closer link between the smelter and end use market.

Most zinc from secondary sources is reprocessed by direct melting/alloying or conversion to zinc chemicals. From an overall world wide viewpoint, secondary recovery of all types has averaged around 23% of total zinc consumption over the last decade. Compared to this an average of around 6% of primary metal can be assumed to come from smelting and refining secondary feed. A major source of secondary material is from galvanising drosses, much of which is directly recycled without smelting. Drosses from new alloy galvanising lines are proving more difficult to directly recycle. A growing source of secondary zinc for smelter feed, and one that Europe is in the technical forefront of treating is recovery from steel plant flue dusts. Increasing dumping regulation progressively promotes this source and steel plants are developing technology to produce an acceptable high zinc content bleed stream. Plants in Europe using the Waelz process to produce smelter feed from this material are in W Germany, Italy and Spain. The latter and most recent has attracted capital expenditure on new plant which would indicate that dumping costs have risen sufficiently to support the capital as well as the processing and transport costs of such an operation.

4.2 Process Technology

Leach electrowinning and ISP blast furnace smelting are the main technologies operated in Europe. The various plants, their capacities and some technical details are set out in Table 5. The major capacity is electrolytic with various leach residue treatments namely the jarosite, goethite and haematite processes differentiated mainly from each other by the amount and quality of Fe-residue which they produce; some pyrometallurgical treatments of leach residues are also in operation.

Table 4   1987 Western World Zinc Production and Consumption (k tonnes), after (9)

	Mine Production	Metal Production	Refined Metal Consumption
Austria	16	24	35
Belgium		285	163
Denmark (Greenland)	69		11
Finland	53	152	29
France	31	249	248
F R G	99	378	453
Greece	22		12
Ireland	177		1
Italy	41	256	249
Netherlands		205	46
Norway	23	113	19
Portugal		6	11
Spain	233	213	115
Sweden	214		33
UK	6	81	189
Total EEC + EFTA	984	1962	1630
Total Africa	304	213	173
Total Americas	2822	1443	1631
Total Asia	390	1006	1410
Total Oceania	695	313	104
Total Western World	5195	4937	4948

FIG. 9
LOCATION OF MAIN EUROPEAN PRIMARY ZINC SMELTERS

▲ Electrolytic   ● Imperial Smelting Process

Table 5   European Zinc Plants, after (1)

Electrolytic

Country	Plant	Capacity k tonnes	Residue Treatment
Austria	Gailitz	25	Fuming leach res
Belgium	Vieille Montagne Balen	180	Goethite
	MHO Overpelt	120	Goethite
Finland	Outokumpu Kokkola	160	Conversion
France	Vieille Montagne Auby	200	Goethite
FR Germany	Ruhr Zinc Datteln	135	Haematite
	Preussag Nordenham	120	Jarosite
Italy	Samim Porto Vesme	83	Goethite
	Pertusola Crotone	100	Fuming leach res
	Sameton Ponte Nossa	10	N/a
Netherlands	Budelco Budel	210	Jarosite
Norway	Norzinc Odda	120	Jarosite
Portugal	Quimigal Barreiro	11	Cinder to Fe prod
Spain	Asturiana	200	Jarosite
	Cartagena	60	Neutral leach only
	Bilbao	8	Cinder to Fe prod
	Total	1742	

Imperial Smelting Furnaces

Country	Plant	Capacity k tonnes	Refining
France	Penarroya Noyelles G	110	New Jersey
F R Germany	Berzelius Duisburg	85	New Jersey
Italy	Samim Porto Vesme	70	New Jersey
UK	AM&S Avonmouth	100	New Jersey
	Total	425	
	Grand Total	2167	

Europe also has a high proportion of blast furnace plants. To produce SHG metal a separate refluxer operation is required to treat ISP blast furnace metal.

4.3 Factors Influencing Evolution

The evolution of the European zinc industry has and will continue to be influenced by the following factors:

Capacity

Zinc is an example of metal whose metallurgical production in Europe exceeds consumption. Reasons for this are historical as in the past much of the zinc industry process development took place in Europe, a tradition that to a lesser extent prevails today.

Valid reasons for maintaining a zinc smelting industry at least in balance with consumption in Europe are:

o   the similar transport costs per unit of metal for concentrates and metal product due to the relatively high grade of concentrates (+50% Zn);

o   the high productivity level of the existing production plants;

o   the ability to dispose of acid with a slight financial contribution;

o   the need to tailor products to specific alloy markets;

o   the ability to recycle environmentally sensitive byproducts;

o   the high capital cost of replacement green field capacity.

It will have to be demonstrated in the future whether the present imbalance between consumption and available smelter capacity remains economically justified.

Technology

A look at the cost structure of the Electrolytic and ISP processes indicates that they are both energy (coke versus electric power) and labour intensive, the two factors contributing to around 70-80% of the direct operating costs. Thus most of the development achieved in European plants has been spurred by the need to stay competitive in a high energy and labour cost area with growing environmental pressures. Specific process applications developed according to differences such as: suitable types of feed material, required metal recoveries, acceptability of disposal of solid waste, grade of zinc product, byproduct contribution, useful heat recovery from roasting and ease of environmental control.

In the electrolytic process the development of large capacity fluid bed roasters with efficient heat recovery has provided single stream plants with considerable economies of scale. Continuous leach systems are in general use, with most of the neutral leach residues being treated hydrometallurgically, lead and silver being recovered in a concentrated form that is suitable for treatment in lead smelters. The Goethite and Haematite processes have been developed with the aim of producing a reduced quantity of iron residue that can be disposed of at minimum cost. Direct sulphide pressure leaching as an add on to electrolytic plants has been established and can treat high iron material and be used to redress a marginal short fall in roaster/acid capacity if economically attractive. Development of fluid bed cementation, automatic filters, and reliable instrumentation along with continuous/semi continuous operation has improved purification efficiency and solution purity, an essential when considering low current density operation in a modern tankhouse. New tankhouse technology has been developed to mechanise and automate operations. Capital and operating cost savings were achieved by increasing the size of cathode and the number of cathodes per cell. In order to reduce electrolytic power costs, some plants have adopted modulated operation at the expense of reduced capacity, making use of low tariffs during off peak times to deposit the major part of the zinc production.

For the lead/zinc blast furnace process the feed presently consists of many different types of complex primary and secondary materials. These lower cost feed materials make it possible for the custom IS plants to maintain a competitive position with conversion costs achieved by electrolytic plants and lead smelters treating clean separate concentrates. Sintering does not allow heat recovery from the roasting reactions as is possible with zinc fluid bed roasting or lead direct sulphide smelting. An increasing proportion of oxidised feed is prepared by briquetting. Integrated Waelz kiln/hot briquetting plants have been established and hot briquetted material constitutes over 30% of some furnace feeds. There are site specific flowsheet limits but byproduct recovery from complex feed is more comprehensive than for the electrolytic process and can contribute substantially to revenue. The blast furnace and zinc condensation system design have been improved over the years to give increased productivity. Today a furnace originally designed for 70000 t/y Zn can smelt 85000 t/y. Use of high blast preheat temperatures up to 1150°C, saves coke and increases smelting rate. Oxygen enrichment of blast has a similar effect to preheat at the expense of slightly increased coke consumption and is used at the 1-2% enrichment level. Reduction of coke costs by fuel oil, coal, coke fines injection has been tested (without condensation problems) but not adopted to date for general operation. Waste heat recovery from condenser lead to generate power is practiced in Japan but probably not justified in Europe. Recently there has been a need to remove Cd from GOB sold in Europe. This has resulted in all metal having to pass through refluxing prior to liquating a proportion as GOB product.

4.4 Some Future Trends of the Zinc Industry in Europe

Concerning feed material, some new indigenous sources of zinc concentrate may become available in Europe with the conversion of mines such as Aljustrel to concentrate production. There are considerable amounts of complex zinc material in Iberian pyrite belt ores, but it is yet to be seen if it is economically recoverable. As a result of growing environmental pressure, material supplies of ashes and flue dusts are likely to increase, particularly those tied to galvanising such as steelworks dusts.

It is clear from the above brief review of the 'state of the art' that a large proportion of the European zinc smelting capacity is at the forefront of the respective technologies. From figures in Tables 4 and 5 it would also seem that European plant utilization is at a high level. The imbalance however between production and consumption would point to smelter overcapacity that may or may not be economically sustainable in future. The severe competition this engenders, enables one to forecast that further capacity expansions will only take place as a result of rationalisation of a company's production facilities and that they could imply corresponding plant closures. The medium term is thus more likely to be one of consolidation/retrofitting of older plants if the economic climate permits.

Environmental pressure on the zinc industry will continue, whether plants operate following the pyro or hydro route. Due to the tonnages involved, the iron residues produced in the hydrometallurgical plants will remain a point of concern and most of the operators will have to consider, as far as feasible, their further use in the economic circuit or to improve the acceptability of their disposal.

Some technological improvements if economically justifiable are:

a   Electrolytic plants:

- o   some older plants may consider cell house redesign to benefit from larger cells and increased automation and mechanisation, modulated power operation may form part of this modernisation if power tariff structure makes it attractive;

- o   general improvements in solid/liquid separation technique such as automatic filtration could also be installed in leach and purification operations of older plants.

b   ISP blast furnace:

- o   continued move towards acquiring cheaper secondary materials for the blast furnace with subsequent increase in briquetted charge;

- o   use of high blast preheats and/or oxygen enrichment to increase blast furnace throughput;

- o   reduction in the use of expensive lump coke by injection of cheaper fossil fuels through the tuyeres.

In both processes there could be a move towards improving high value metal byproduct recoveries.

Long term development work is continuing to improve hydrometallurgical and pyrometallurgical zinc extraction technology. With regard to hydrometallurgy, pressure leach and pyro/hydro residue treatment development is continuing and early work on decreasing the anode potential in the electrowinning process is being undertaken in a European plant. New pyrometallurgical zinc processes under development mainly involve the use of tonnage oxygen and coal/coke fines as reductant in an intense flame/plasma or slag bath reactor configuration. Some of these processes designed for treatment of EAF dust will compete with the existing Waelz/ISP European approach described above. New pyro processes would depend on the use of a zinc or lead splash condenser, but in most cases would not have the benefit of efficient energy recovery and gas cleaning provided by the conventional ISP blast furnace shaft operation and constituting a key to efficient condensation. In the pursuit of increased smelting intensity in the conventional blast furnace, oxygen, fuel and charge injection via a conventional or possibly 'flame reactor type' tuyere may be worth considering. This combined with fluidised bed roasting of feed and briquetting of a portion of the charge could be a long term goal for the existing IS Process. Integration of the lead/zinc blast furnace process with other technologies such as slag fuming to improve zinc recovery and its use for treatment of high lead/zinc slag from an associated single stage direct lead smelting operation may come about in the long term.

5   COPPER INDUSTRY

5.1   Location and Capacity Versus Demand

Copper consumption in Europe has presently exceeded 2.6 million tonnes per year in accordance with Table 6. Some 140 000 annual tonnes originate from European mines and about 1.4 million tonnes are provided by the recycling of copper scrap. This means that approximately 45% of Europe's copper requirement has to be satisfied by imports in the form of concentrates, blister, cathodes and scrap.

The data given in Table 6 outline some further details of the current situation in the European copper industry, namely:

o   The mine production has decreased in recent years because of mine shutdowns and declining ore reserves. The start of a new mine in Portugal in 1989 (Neves/Corvo) will gradually improve the situation but not change the overall general tendency.

o   European based concentrate smelters contribute about 18% to the copper consumption, with half of the concentrates smelted being imported. Smelter capacities were significantly increased in the early seventies, mainly due to new facilities being installed in Spain, Germany and expansions in Sweden and Finland. Compared with 1960 primary smelter production has nearly tripled.

o   Electrolytic copper production is in the range of 1.1 to 1.2 million tonnes per year. The balance of the total European refined copper output of 1.32 million tonnes corresponds to that of fire-refined products.

o   Wire rod production accounted for more than 50% of the total consumption (Table 8). The remainder supplied the requirement for copper castings, sheets, tubes, copper alloy semis and copper salts.

o   Total scrap recovery in refined production and by direct use in the semis' industry is about 50%. Low grade secondary materials requiring processing by smelting and refining fluctuated around half a million tonnes a year over the last two decades as Table 6 shows. This indicates that nearly all European refineries are partly or completely supplied by copper scrap, including those plants operating in combination with concentrate smelters, a fact that underlines the importance of the role recycling plays in the European copper industry.

The major smelters and refineries are shown on the map on Figure 10. Concentrates are processed in smelters in Germany, Spain, Sweden and Finland, the operations being partly or completely based on imports. All other smelters base their raw material feed on non-metallic and/or

Table 6 Copper Production/Consumption in Europe (EEC and EFTA countries) in ktonnes, after (1) and (10)

	1950	1960	1970	1975	1980	1985	1987
Mine Production (1)	70.2	96.6	131.3	187.5	168.1	201.3	147.7
Smelter Production from Concentrates (10)	133.3	175.7	255	408.1	390	424.1	473.9
Refined Production (1)	634	938	1161.8	1157	1206.3	1317.2	1320.4
Copper Scrap Recovery in Refined Production (10)	190	240	534	405.6	459.5	566	516
(% of Refined Production)	(30)	(25.6)	(46)	(35)	(38.1)	(43)	(39)
Direct Use of Scrap (10)	330	590	731	571.8	817.3	893	868
Total Scrap Recovery	520	830	1265	977.4	1276.8	1459	1384
(% of Consumption)	(53)	(44)	(54.1)	(42.6)	(46.9)	(56.6)	(51.6)
Consumption (1)	982.2	1885.5	2366	2293.9	2723.5	2577.2	2681.9

- PRIMARY SMELTER
- ▲ SECONDARY SMELTER

FIG. 10
LOCATION OF MAIN EUROPEAN PRIMARY AND SECONDARY COPPER SMELTERS

metallic secondary materials. In addition, the concentrate smelters treat secondary materials in combination with primary feed or in separate facilities.

Figure 11 classifies primary and secondary smelters. The number of plants is plotted in relation to the capacity range. The majority of secondary smelters show capacities up to 50 000 t/y while primary smelters have outputs from 50 000 t/y upwards.

5.2 Process Technology

Table 7 indicates the main process type and the capacities of the major European primary and secondary copper smelters and refineries with copper as the main product.

a) Primary Smelting

Three out of the five concentrate smelters apply the Outokumpu flash smelting technology, utilising tonnage oxygen for increased specific throughput, better fuel economy and environmental control. Due to its specific conditions, a Swedish smelter applies partial roasting of concentrates in fluid bed roasters and matte smelting in an electric furnace. A smelter in Belgium is equipped to process complex concentrates in combination with secondary feed materials in shaft furnaces.

b) Secondary Smelting

Full line secondary smelters in Europe follow in general the conventional processing route where secondary copper materials are fed into the appropriate process step depending on grade, composition and nature (metallic/oxidic). Such concepts are based on shaft furnaces for oxidic low grade materials, converters, fire-refining furnaces, electrolytic refinery and in some cases "downstream" installations for cathode and No 1 scrap processing. Two smelters use TBRC giving the flexibility to treat a variety of secondary materials on a batchwise basis.

c) Electro refining

The majority of the electrolytic refineries operate with conventional DC tankhouse concepts. Periodic reversed current is practised in five operations while two plants apply permanent cathode approaches with a third facility under construction.

d) Electrowinning

The production of electrowon copper plays a minor role in Europe. In Norway up to 40 000 t/y electrowon cathodes are produced after nickel-copper matte treatment (11). A minor tonnage of electrowon cathode results at plants in Great Britain and Germany from intermediate products of tin, lead and precious metals operations and at a plant in Italy from secondary material processing.

5.3 Factors Influencing Evolution

Factors outlined in Section 2 such as: diversification of ownership (11), nationalisation in developing countries, new mining ventures in the 1960s, monetary and oil crises, sharply rising labour and capital costs,

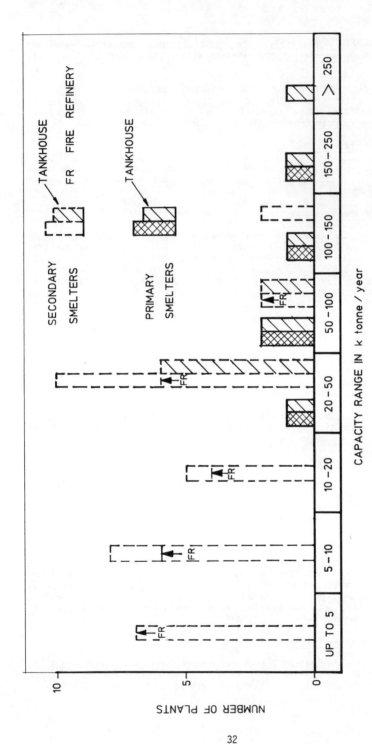

FIG.11
CLASSIFICATION OF EUROPEAN PRIMARY / SECONDARY COPPER SMELTERS & REFINERIES

Table 7: Main European Copper Smelters and Refineries (after 6 and 13)

Process	Number of Plants	Capacities k tonnes
Primary Smelters		
Blast furnaces	1	50
Roasting/electric furnace	1	95
Outokumpu flash furnaces	3	340
smelting facilities	5	485
Primary smelters with secondary material smelting facilities:		
All types	3	
Metallic copper scrap	2	
Primary smelters with refineries:		
Tankhouses	5	845
Secondary Smelters		
Blast furnaces	7	363
Scrap converter	1	45
Fire refining/direct use	24	534
TBRC	1	65
Secondary smelting facilities	33	1007
Tankhouses	10	509
Hydrometallurgical Plants		
Leach/electrowinning (intermediate product basis)	3	54
Total		
Smelting capacity	38	1492
Tankhouses capacity	15	1354
Hydrometallurgy	3	54

metal substitution and growing environmental constraints have had a considerable impact on the copper industry worldwide. Much of this change has now been absorbed and in the case of Europe evolution will take the form of consolidation of the existing to improve productivity and product diversity/quality against the following background:

Capacity/Raw Materials/Ownership

Several primary and secondary smelters have expanded capacities, shut down operations and/or went through several stages of plant modernisation and adjustment.

Ownership of some companies changed partly or completely. A similar regrouping could and can still be observed in the semis' industry; some of the remaining companies have expanded their operations upstream for direct scrap and blister processing, thus taking markets away from traditional producers.

The majority of concentrates supply will continue to be imported. Imported blister, cathodes and copper scrap are also significant raw material sources for European copper plants and reliability of supply is essential.

Technology

Substantial changes have taken place in the last two decades mainly in primary smelting and refining technology. The use of oxygen has become widespread and TBRC application for secondary smelting can be mentioned.

Today, the Outokumpu autogenous flash smelting using tonnage oxygen has in most cases substituted the conventional reverberatory furnace as the accepted 'state of the art' technology worldwide. Flash smelting results in substantial energy savings (consumption of 5.0 - 5.5 Gcal/t cathode compared to 7.5 - 11.0 for reverberatory and electric smelting (12)), the effect being even more significant when smelting to high matte grades. Furthermore, environmental control is eased because all process gases can be treated in gas cleaning and acid plants before being released to the atmosphere.

Two primary smelters in Spain and Germany have retrofitted and expanded their smelting facilities on the basis of the Outokumpu flash smelting process during the 1970's. Later, similar to Outokumpu's Harjavalta smelter, both operations were further improved by utilising tonnage oxygen produced by on site oxygen plants. Most of the smelters in Europe have gone over to high matte grades (around 60%) for better fuel economy, optimisation of converter operation and environmental control.

The Contimelt process, now in operation in Germany and Belgium in two smelting/refining complexes, has been developed specifically for the treatment of blister and scrap copper, substituting the conventional reverberatory furnaces for fire-refining.

Productivity improvement has prompted evaluation of modernisation or completely new construction of electrolytic refineries. Mechanisation and automation are the main features of such investments. Specifically tankhouses based on the Mount Isa technology using permanent cathode sheets have the potential of practising automation to a degree already established in the zinc industry.

Downstream Products

Some of the European primary and secondary copper smelters and refineries have adopted production of further downstream products, like continuously or discontinuously cast billets or cakes, and substituted wire bar production by wire rod from continuous cast rod mills. Hot milled rod from wire bar only had a 25% share of the rod market in 1985. The importance of wire rod may be illustrated by the fact that 50-60% of the copper is consumed by the cable industry. Table 8 gives the continuously cast rod capacity by country for 1985. Although some plant capacities have been adjusted since then, the table indicates that the installed capacity at that time was about 50% greater than actual production. This indicates the intensive competition on the wire rod market.

Table 8: 1985 Wirerod Capacity and Production in Western Europe (13)

Country	Capacity k tonnes	Production k tonnes
Belgium	453	277
Finland	25	25
France	314	276
Greece	92	24
Italy	343	225
Portugal	36	9
Spain	204	74
Sweden	90	54
UK	357	245
West Germany	520	402
Total	2434	1611

5.4 Some Future Trends of The Copper Industry in Europe

The European copper industry has achieved a high technological standard with a high level of productivity and compliance with environmental protection requirements. Increased pressure from tighter environmental legislation as well as heightened competition can however be forecast for the future and will challenge the industry for further improvements required to reduce operating costs. In any event it is not very likely that the actual situation in the future will change with regard to copper demand, supply from own resources and imports. Scrap recycling and consequently secondary copper processing will maintain its important role.

New technologies being proposed for primary materials like high intensity flash/flame smelting (Inco, FCR, Contop) or bath smelting (Mitsubishi Noranda, Isasmelt, Smelt in Melt) may offer answers to future requirements. In this context, availability of feed materials may have an impact on future decisions in regard to process selection or new developments.

Increased competition from other regions for the production of copper may be expected in the near future:

o   Low cost SX-Electrowon copper is gaining importance for example in North America and in main copper mining countries. The SX-cathode production actually amounts to about 10% of the western world's refined copper production and, as a result of disclosed expansion projects, is likely to double in the near future. Cathode quality constraints have been overcome, broadening the range of application for electrowon copper.

o   Although consumption in Europe increased steadily, its share of the western world's total consumption has levelled at about 35% within the last decade. This is due to the fact that in other areas like the Far East and Arabian countries, and including traditional European export markets, investments have been and are still being made to develop the local copper sector. Some substitution and reduced consumption through miniaturisation is also influencing copper demand worldwide.

o   Currently various new copper smelters and refinery projects or expansions are under investigation outside Europe mainly in copper mining countries. Some of them could be questioned in view of the high investment requirements and the overall copper market situation.

Predictions as to how these developments will influence the European copper industry are difficult, although Western Europe will probably maintain its position as a major copper consumer covering about half of its consumption from own supply. The position of the European refined metal producers is strengthened by their close market ties, together with the high productivity of their operations.

6.0   NICKEL INDUSTRY

6.1   Location and Capacity versus Demand

The western world's nickel reserves are estimated to be around 27 million tonnes, 36% being sulphide, the remainder oxide ores (14). The main portion of the total, secured world nickel reserves with more than 1% nickel are to be found in Cuba, Canada and the USSR. Although sulphide ores represent only 36% of the western secured reserves, they contribute for more than 60% to western world production (15). Nickel production from oxide ore to date is normally restricted to those deposits whose grade and composition result in cost competitive production compared to sulphide ore processing.

Furthermore, nickel is produced in the copper industry, most of the copper concentrates containing nickel in minor quantities (16). Secondary copper plants processing residues and scrap are a further source of nickel production. From these materials nickel is recovered as nickel salt or as alloy from alloy scrap.

In addition, scrap recycling plays an important role in the stainless steel industry. The percentage of scrap contained in stainless steel produced corresponds to approximately 60% (17).

As shown in Table 9, the European nickel consumption has steadily increased, with setbacks during phases of economic downswing. It has presently reached around 250 000 annual tonnes. Roughly 20 000 tonnes/y nickel in ore are recovered from European mines (8% of the consumption).

Table 9   Nickel Production/Consumption in Europe (EEC & EFTA Countries) in k tonnes, after (6) and (10)

	1960	1970	1980	1985	1987
Mine Production	2.1	13.7	20.9	23.3	20.4
Primary Metallurgical Production	78.2	99.4	93.5	95.9	110.6
Consumption	94.7	168.3	205.0	218.0	248.5

The primary metallurgical production is in the order of 110 000 annual tonnes, corresponding to 45% of the European consumption; of this, about 80 000 tonnes result from the refining of intermediate products (matte etc) imported from overseas (Canada, New Caledonia). These data illustrate that nickel consumption in Europe is greatly dependent on imports of refined metal or as ferro-nickel, in addition to intermediate products imported for further refining.

The major European primary nickel smelters and refineries are shown on the map of Figure 12. Important ore-based operations can be encountered in Greece based on oxide ore with ferro-nickel as the final product and in Finland from sulphide concentrate with electrowon nickel as the refined product. In France nickel matte imported from New Caledonia, in England nickel matte, metal and oxide from Canada and in Norway mattes coming from Canada and Botswana are refined. A plant in Austria is reported to treat residues and scrap producing ferro-nickel and oxide. In Belgium and Germany Ni-speiss is smelted from complex primary and secondary sources and pressure leached to produce commercial nickel salts.

The territorial pattern of consumption can be assumed for the western world as follows: Europe 39%, USA 28%, Japan 23%, others 10% (18). Western world consumption by end use can be broken down as follows: stainless steel 48%, nickel plating 10%, high nickel alloys 15%, alloy steels 9%, ferrous and non ferrous castings 8%, superalloys 5%, others 5%.

6.2   Process Technology

The principal processing steps applied by nickel operations for the two different ore types, oxide and sulphide, are outlined on Figure 13. The figure encompasses the variety of flowsheets applied in the various operations around the world. Specific raw materials dictate variations in basic parameters from plant to plant. The energy requirements for producing nickel from oxide and sulphide ore differ considerably. For sulphide ore figures around 22 kWh/kg Ni are quoted either for pyro or for hydro processes; depending on the ore composition and process selected oxide treatments can require 35-84 kWh/kg Ni (21).

Table 10 lists the processes applied in the main nickel operations in Europe. The total European capacity in accordance with the data reported amounts to about 175000 annual tonnes of nickel contained in the various types of products. The steps involved in the various treatment methods are as follows:

■ SMELTER

▲ REFINERY

FIG. 12
LOCATION OF MAIN EUROPEAN NICKEL SMELTERS AND REFINERIES

Table 10: Main European Primary Nickel Smelters/Refineries (EEC & EFTA) after (1)

Process	Raw material	No of plants	Capacities k tonnes
Primary Sulphides			
Flash smelting/ electrowinning	concentrate	1	18000
Sulphide intermediate products			
Leach/electrowinning	matte	2	70000
Vapour metallurgy	matte		
	nickel oxide	1	54000
Complex and Secondary Feed			
Electric furnace smelting	residues nickel oxide	1	approx 600
Pyro/hydrometallurgy	smelting to Ni-speiss	2	approx 3500
Oxide ore			
Prereduction/ Electric furnace/ converting	laterite type oxide	1	30000

Total capacity	176100
cathode nickel	88000
ferro-nickel	30000
others (nickel, oxide, powders, pellets, salts)	58100

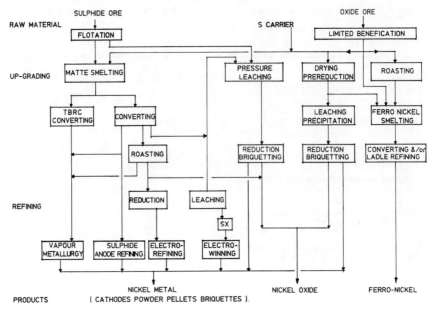

FIG. 13 WINNING OF PRIMARY NICKEL FROM SULPHIDE AND OXIDE ORES (15, 19, 20)

a) Sulphide Concentrate Smelting

The process flowsheet consists of: nickel concentrate flash smelting, matte conversion to white metal, nickel metal production from the white metal by a sulphuric acid treatment in a combination of atmospheric and pressure leach steps followed by solution purification and nickel electrowinning.

b) Matte Refining

The processed mattes may contain copper at various concentrations and the Cu-Ni separation is an essential feature of these refining operations. The flowsheets applied reflect the variety of options as briefly outlined on Figure 13:

o   Direct matte leaching/electrowinning using a chloride system for leaching and solvent extraction for cobalt and impurity elimination prior to electrowinning. This approach is used by two operators in France and Norway; it enables flexibility in regard to final product: by electrowinning nickel cathodes are obtained or by crystallisation of nickel chloride, pyrolysis and reduction, nickel metal granules can be produced (17, 21).

o   The flowsheet for a United Kingdom operation uses carbonylation for the refining of nickel oxide, impure metal or matte to produce pure nickel powders and pellets (21).

c) Oxide Smelting

The process flowsheet consists of : pre-reduction of the nickel and part of the iron, ferro-nickel smelting in electric furnaces, impurity elimination from ferro-nickel in converters and/or ladle refining.

6.3   Factors Influencing Evolution

Figure 14 illustrates how the world mine production of nickel evolved between 1950 and 1986; the changing geographic pattern of nickel mining is clearly demonstrated. Some decades ago similarly to some other non-ferrous metal sectors the nickel supply to the centres of consumption was controlled by a few major mining companies. In the 1960's the western world's nickel consumption suddenly increased dramatically with average annual growth rate of more than 14.8% over a period of about five years (23). Production could not follow this sharp rise in demand and consequently the world nickel market underwent its greatest supply crisis. Traditional nickel producers were caused to increase capacities or open up new plants whereas newcomers appeared in the business investing in operations, in Southern Africa, Central and South America, the Far East and Australia. In this context is should be mentioned that the two oil crises were followed by economic downswings which caused some high cost operations to be closed down.

In Europe efforts had been made during this period to expand existing mining operations in Finland and a new oxide ore based operation was opened in Greece. Therefore, by 1970 the nickel shortage was almost overcome. Nevertheless, the European nickel industry remains heavily dependent on imported primary metal or intermediate products and continuity of supply is essential.

6.4   Future Trends

It is significant with regard to treatment of imported intermediates that one important French and two Canadian producers have established refining operations decades ago in Europe and have modernised and/or streamlined them in recent years. This undoubtedly contributed to make Europe an area in the western world with a high share of nickel consumption, a situation which will hopefully continue for a some time.

Supply/demand considerations dominate the nickel industry and overshadow somewhat the process developments which took place during the last decades or which are in preparation for the future. As well in pyro as in hydrometallurgy considerable progress took place, as illustrated by the variety of processes schematically outlined on Figure 13. It may be anticipated that their optimisation will continue. The laterite treatment is desperately searching for improvements or alternatives which would reduce the economic margin from sulphide ore treatment. The development of reliable high capacity electric furnaces to smelt laterites to ferro-nickel should be mentioned as an important contribution in this context. Also further development of a sulphuric acid pressure leach process with counter-current treatment of the high magnesium ore fraction may offer energy savings compared with existing hydrometallurgical processes and allow a wider range of laterite feeds with higher specific acid consumptions to be treated. The processing of sulphides took advantage of the technological developments of autogenous copper smelting and its progress followed a parallel path. An important development now in consideration (24) is the direct smelting to metal,

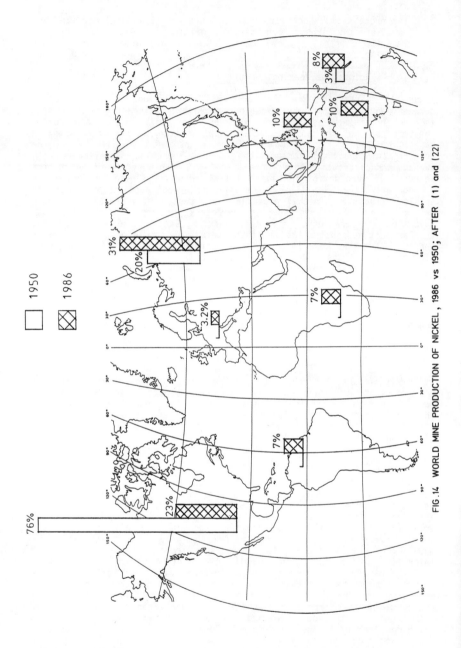

FIG.14 WORLD MINE PRODUCTION OF NICKEL, 1986 vs 1950; AFTER (1) and (22)

which would enable one to by-pass the matte converting step. And as a tradition, the nickel industry will probably continue to take good care of its markets, contributing to innovation and offering its services for any possible new development.

7 PRECIOUS METALS INDUSTRY

7.1 Location and Capacity

The mine production of PMs in Europe is rather limited, although not insignificant for some of them. For silver, the total extraction in EEC and EFTA countries is about 620 tonnes, which represents 5% of the world production; for gold, it is 11 tonnes or 0.7% of the world total.

Regarding the metallurgical production, only sketchy data are available, since production figures are generally kept confidential. Also recycling cannot be estimated quantitatively, although it certainly is important for most of the PMs. Silver is mainly recycled as scrap from jewellery, silverware and industrial applications; also many small recycling circuits for photographic material have been developed since the Hunt speculation. The contribution of recycled Ag is estimated at about 4000 tonnes on world scale (25) which represents about one third of the mine production. Gold is also mainly recycled as scrap from jewellery and industrial products; this contribution is estimated at 400 tonnes on world scale (also about one third of the mine production), of which 30 tonnes comes from Europe (26). Platinum group metals (PGMs) are recycled in important proportions, probably equivalent to the primary production; most of the massive applications are however toll refined and the corresponding metal does not actually reach the market. Recycled PGMs from products such as catalytic convertors or electronic components find their way back to the market, and this is for instance illustrated by autocatalysts, which contributed to an estimated secondary production of 3 tonnes of Pt at world scale in 1987 (27).

Consumption figures of PMs are better estimated; they are summarised in Table 11, which gives 1987 consumptions of the western world and distributions between major regions. Table 12 gives an estimate of the distribution of this consumption over the main application sectors.

Failing to accurately define the metallurgical production of PMs in Europe, we may nevertheless assert that it represents a very important industry sector, involving thousands of tonnes of silver, hundreds of tonnes of gold, tens of tonnes of platinum and palladium, and a few tonnes of rhodium.

Treatment plants fall into the following categories:

a) Primary non-ferrous smelting

   Here the recovery of PMs from trace quantities in feed materials takes place. Even concentrates where PMs are the primary product are taken by some smelters to complement the feed. As the case may be, these smelters stop their extraction of PMs at intermediate steps of realisation (Parkes alloy, tankhouse slimes, leach residue, dore alloy) or proceed to more advanced refining steps.

b) PM Refineries

   These plants grew from recycling activities of high grade products

Table 11: Consumption of Precious Metals (in tonnes); after (25), (26) and (27)

	Western World	North America	Europe	Japan	Others
Ag	13 400	4 100	4 400	2 700	2 700
Au	1 600	290	440	175	695
Pt	100	27	17	50	6
Pd	100	34	17	44	5
Rh	10	na	na	na	na
Ru	5	na	na	na	na
Ir	1	na	na	na	na

Table 12: PMs Consumption Pattern by Application (in % total); after (25), (26) and (27)

	Ag	Au	Pt	Pd	Rh	Ru	Ir
Jewellery	10	72	30	5	-	-	-
Coins, medals, bars	10	14	15	-	-	-	-
Electrical	20	8	5	50	5	53	-
Chemical	-	-	12	-	12	-	28
Photography	50	-	-	-	-	-	-
Dentistry	-	3	-	31	-	-	-
Autocatalysts	-	-	35	6	73	-	-
Chloralkali	-	-	-	-	-	36	13
Other	10	3	3	8	10	11	59
Total	100	100	100	100	100	100	100

(jewellery) or as complement to overseas smelting operations (South Africa, Canada). They often contain a small smelting circuit in their operations.

c) Scrap Treatment

This type of PMs plant is derived from the scrap collecting circuits, where smelting processes collecting PMs in base metal bullions were developed, possibly complemented by refining circuits.

Together they constitute the European PMs industry, with important operations, specifically involved with PMs, being located in United Kingdom, Germany, France, Belgium, Scandinavia, Italy, Switzerland, Austria. According to their flexibility and up and downstream diversification, these plants are more or less able to accept concentrates, by products and residues of other smelting operations from Europe or overseas, as well as the different materials collected for recycling.

7.2 Process Technology

Cupellation remains an important processing step for the recovery of PMs when Ag and Au predominate. It is normally followed by a Ag electrolysis of the dore alloy and a chemical separation of Au and PGMs from the slimes. For bullions and when PGMs predominate, there is a diversity of metallurgical and chemical treatments that can constitute the preferred processing route. At the smelting stages, mention should be made of high intensity smelting processes (TBRC, Plasma smelting) recently introduced as substitutes for the traditional dore furnaces or in connection with the recycling of refractory base PGM containing catalysts or other scrap, which constitute a growing market.

7.3 Factors Influencing Evolution

Feed Considerations

Development efforts of PM plants are concentrated on the possibility of treating increasingly diversified feed materials. Also in connection with recycling, there is a rapid evolution in available amounts and qualities of feed materials, especially in the catalyst sector. These diversifications imply increasing complexity in connection with the separation of the different PMs (Ag, Au, Pt, Pd, Rh, Ru, Ir, Os) and of the possibly associated impurities (Se, Te, Bi, Pb, Cu, As, Sb, as well as refractory base constituents such as SiO_2, Al_2O_3, C, etc) Adaptation of processes and the introduction of new separation techniques may therefore be required.

Technologies

As may be expected, energy connected considerations play a relatively minor role in technological developments in the PMs industry, since the scale of operation is always relatively small. Since it is a tradition, and indeed a necessity, to carefully control all effluents, environment considerations do neither constitute a major problem in this industry.

Major concerns are the efficiencies of recovery of the PMs, as well as the selectivity of their separation from each other. From this point of view, the PMs metallurgy may be compared to fine chemistry, requiring sophisticated techniques and processes. Improvements are continually introduced, even though innovation is only brought in carefully, paying attention not to disturb established practices.

The tendency in smelting operations is to extend the application of high intensity processes, in so far as they provide improved flexibility compared to traditional furnaces and that they contribute to shortening the hold up time of metal, another factor of prime importance in the processing of high value materials. In refining operations, there is a tendency to extend the application of solvent extraction in the separation and purification circuits.

7.4 Some Future Trends of the PMs Industry in Europe

The production, smelting and refining of the PMs is one of the strongest positions of the non-ferrous industry in Europe and it will probably remain so in the future. The main factors contributing to this are:

- o   a sustained demand for these metals, whose uses tend to intensify and to diversify;

- o   an increasing contribution of recycling to their production, ie raw materials which have not to be imported;

- o   a processing know how and a skill in refining techniques which have been developed by century old tradition;

- o   reliability in sampling, analysis and restitution - an essential factor in this business of expensive metals.

Concerning processes, the European PMs refineries will continue to consolidate their positions in a climate of growing markets and changing feed materials. The principal accents will remain on a rapid and reliable restitution of metals, flexibility with regard to varied feed materials and improved selectivities and recovery efficiencies. This may imply the introduction of new technologies to the extent that they can be integrated with established circuits; new processing circuits may be required but will generally be developed to complement the existing ones. Processes will thus remain, for a part, traditional whereas processing flow sheets will tend to increase in complexity, implying increased attention to control.

8   ALUMINIUM INDUSTRY

8.1   Location and Capacity versus Demand

The aluminium industry in general is characterised by the fact that ore quality and ore processing are basically the same worldwide:

- o   There is only one main ore, bauxite, which is well distributed over all tropical and equatorial regions, with no foreseeable shortage for centuries. The quality needed for pure alumina production must have a low silica content but high iron is acceptable, the reverse is required for alumina refractory production.

- o   The only industrial process used for concentrating bauxite to pure metallurgical alumina is the hydrometallurgical Bayer process. Some variations of this process can be observed depending upon bauxite crystallography, ie the nature of contained aluminium hydrate: Gibbsite (trihydrate), or Boehmite (gamma-monohydrate) are easier to crush and leach than Diaspore (alpha-monohydrate). To our knowledge, there is only one plant operating on Diaspore-containing bauxite: Aluminium de Grece.

o   Because of the exceptionally high stability of alumina, the only process for reducing aluminium oxide to primary aluminium metal is the electrolytic Hall-Heroult process.

The overall material balance of the production flowsheet and the corresponding energy consumptions are summarised on Figure 15. Actual production figures show only minor variations around the given values, whatever the geographic location of plants.

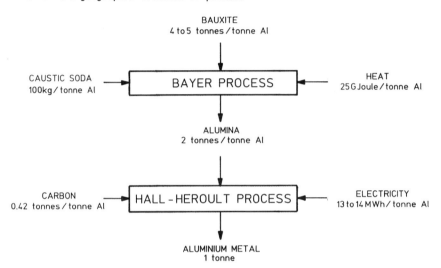

FIG. 15

FROM BAUXITE TO ALUMINIUM METAL

The existence of alumina plants in Europe can be explained mainly by an historically important (but decreasing) local extraction of bauxite. The high capital cost of greenfield alumina plant capacity (more than 1200$ per annual tonne) allow these "old" plants to survive as long as they have a competitive access to thermal energy and a competitive size. In some cases, as the Bayer process requires a large amount of concentrated caustic soda, there can be additional advantage for these plants located near chemical plants producing chlorine, and caustic soda as a by-product. Transport costs of raw materials hardly intervene since bauxite can be cheaply transported in high capacity mineral ships which benefit from return loads whereas alumina has to be transported in dedicated carriers 20-60 000 t.

Against this general background, the position of the European aluminium industry is outlined in Table 13 which gives the mine production of bauxite, the Bayer production of alumina and the metallurgical production of primary metal in the EEC and associated countries. The total production of primary metal, which reached 3.4 M tonnes in 1986, is close to the European consumption of primary metal of about 4 M tonnes. However the position considered by individual countries shows some large net importers or exporters. Recycling of aluminium is also important and is estimated at about 1.7 M tonnes, covering 30% of the total European consumption; as far as available, data on secondary production are given for some European countries in Table 14. The location of European primary and secondary smelters is shown on Figure 16.

8.2 Process Technology

Bayer Process

The Bayer process consists of a selective dissolution of aluminium hydrates in hot concentrated caustic soda, under pressure. Iron and titanium oxides remain undissolved, whereas silica combines with sodium aluminate to form an insoluble sodium aluminium silicate. These impurities are removed by filtration ("red mud"). The purified liquor containing dissolved sodium aluminate is then hydrolysed by addition of water and cooling. At atmospheric pressure, pure aluminium hydrate precipitates, is removed by filtration or decantation, and calcined to give non-hydrated alumina.

Many improvements have been made to the Bayer process particularly in the areas of energy saving and productivity. Thermal requirements have been decreased to 2 tonnes of steam per tonne of alumina, from 8 t/t in 1950. Productivity is linked to equipment and plant size. The latter has increased on the average from 200 000 tonnes Al_2O_3 per year in 1950, to a present 800 000 tonnes Al_2O_3 per year - with some giant plants like QAL or WAL in Australia, producing each more than 2.5 million tonnes per year.

Hall-Heroult Process

This process involving the electrolytic reduction of alumina dissolved in a cryolite melt was developed over a hundred years ago and has been improved systematically over the years. Two types of electrodes are used, namely Söderberg self baking anodes and pre-baked anodes. The latter, because of the higher energy and environmental control efficiencies achieved has become the 'state of the art'. The performance evolution is summarised on Figure 17. The individual cell current, which is in direct relation with productivity, increased to a present 280 kA per cell, whereas the energy consumption gradually decreased to about 13 000 kWh/tonne. As shown in Table 15 the best plants now operate with an anodic carbon efficiency of more than 80%, an energy efficiency exceeding 50% and a current efficiency of about 95%.

Table 13: 1986 Production of Bauxite, Alumina and Primary Metal in EEC + EFTA Countries, in k tonnes, after (1)

	Bauxite	Alumina	Primary Metal
Austria	-	-	92
France	1379	884	322
FR Germany	-	1560	764
Greece	2225	458	124
Iceland	-	-	80
Ireland	-	685	-
Italy	-	618	243
Netherlands	-	-	258
Norway	-	-	729
Spain	3	748	355
Sweden	-	-	77
Switzerland	-	-	81
United Kingdom	-	108	276
Total EEC + EFTA	3607	5061	3401

Table 14: 1986 Production of Secondary Aluminium, in k Tonnes, for some European Countries, after (1) and (28)

France	172
FR Germany	583
Italy	301
Netherlands	109
United Kingdom	116

- ● PRIMARY SMELTER
- ▲ SECONDARY SMELTER

FIG. 16
LOCATION OF **MAIN EUROPEAN PRIMARY AND SECONDARY ALUMINIUM SMELTERS**

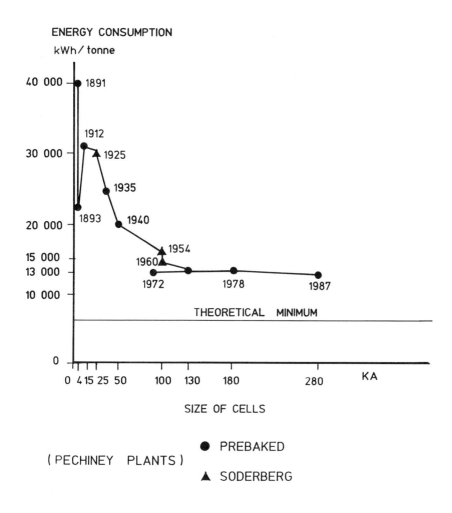

FIG. 17
PERFORMANCE EVOLUTION FOR INDUSTRIAL LINES

8.3 Factors Influencing Evolution

Capacity

The great impact of energy in the direct production costs of aluminium has brought about an important redistribution of production capacities and relocation of plants since 1973. Plants poorly located for low cost power supply closed down, particularly those operating on oil-fired power

Table 15: Present Efficiency of Hall-Heroult Process

	Theoretical minimum	Present modern plant	Efficiency %
Al_2O_3 consumption kg/tonne Al	1888	1920	98-99
Carbon consumption kg/tonne Al	333 (base CO_2)	400-420	80-83
Energy consumption kWh/tonne Al	7000 (liq metal)	13 000-13 500	52-54
Current efficiency %	100	95	95

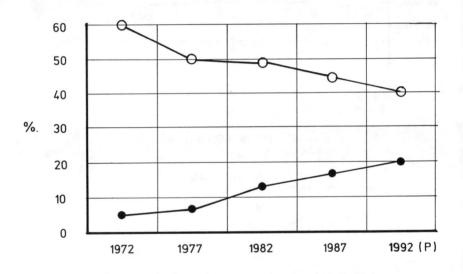

○ THE FIVE MAJORS

● PRODUCTION BY NON MEMBERS OF OECD

FIG 18   PRIMARY ALUMINIUM
REDISTRIBUTION OF WESTERN WORLD CAPACITIES

stations. New plants came on stream, based on abundant energy sources with low to moderate cost. Figure 18 illustrates this by showing how the share of the five major traditional producers decreased from 60% in 1972 to 43% in 1987. In parallel, the proportion of aluminium produced in non-OECD countries grew from 4% to 16% and is expected to reach 20% in 1992. Due to its high energy-dependence, aluminium is a metal for which high investment capital is still justified for the installation of greenfield plants where cheaper power supply is available. Energy may in fact be considered as one of the essential raw materials.

Integration downstream

In contrast with most other non-ferrous metals, traditional primary aluminium producers used to be much more vertically integrated not only upstream (bauxite, alumina and sometimes energy) but also downstream (mill products, extrusions). This integration downstream promoted a better collaboration with end-users, improvement of semis and subsequent growth of the market. A good example can be found in primary cast houses where vertical continuous casting of slabs appeared at the end of the 40's, continuous casting and rolling of rod at the end of the 50's, continuous casting and rolling of sheets at the beginning of the 70's. Direct casting of sheets is now under development.

Another example of improvements promoted by such close cooperation with end-users can be found in the ever-increasing purity of aluminium produced by the cells, in spite of the recycling of impurities from pollution control equipment. The normal quality produced by lines was 99.7% (grade A7) but will probably be in the range 99.8% (grade A8) in the nineties. Some plants are able to produce a significant amount with a purity better than 99.93%. The origin of this improvement can be found in at least four directions:

- o lower impurity level of alumina;

- o better process control allowing for smoother operation of cells and better magnetic balance limiting metal movements and cathode abrasion;

- o improvement of cell lining quality - which also induced a longer cathode life;

- o better cleaning of recycled anode butts and anode rods.

Technology

Vulnerable points of the Hall-Heroult process may be listed as follows:

- o the high investment cost of plants, due to the relatively low production per unit reactor area; this cost is as high as 3300 $ per annual tonne produced;

- o the high consumption of relatively expensive electric energy (typically 13000 to 16000 kWh/tonne);

- o the use of expensive consumable carbon anodes (typically 440 to 450 kg per tonne for pre-baked anodes).

Alternative routes have been investigated, at least at laboratory scale, in order to avoid these drawbacks; they are summarised on Figure 19 giving an overview of several possible attempts. Only a few of them were scaled up to significant pilot plants, amongst which two looked the most promising: the ALCAR process (carboreduction of alumina in electric arc furnace) and the ASP process (carbochlorination of alumina and fused electrolysis of the chloride). All had to be abandoned on economical rather than technical grounds and, to our knowledge, there is no new process under investigation at large pilot scale at the present day. There is, on the contrary, a strong belief that the Hall-Heroult process will remain the only industrial one for at least some decades.

Two technical versions of the Hall-Heroult process were progressively developed: the European approach, which is characterised by low current densities and high thermal efficiency, using insulating floury-type alumina; and the American approach, with higher current densities (higher production per unit area) using poorly insulating sandy-type alumina. Specific requirements for the feed material forced the aluminium producers to integrate into alumina production. Energy considerations have progressively favoured thermally efficient cells. This has brought about a move to a single technology used worldwide, which is similar to previous European practice. However for pollution and automated process control reasons a reactive alumina, similar to the original American sandy-type is required. As a consequence, alumina refineries are also being progressively retrofitted to produce the more reactive and insulating quality which will be used universally in the future. This allows disassociation of aluminium and alumina producers. The more reactive alumina offers the advantage of collecting and recycling fluorides contained in the cell fumes with an efficiency of more than 99%. Söderberg cells are less easily adapted to emission control and their share of total existing capacities has progressively decreased.

Many other improvements have been introduced during these last twenty years. Automated process control using microprocessors, computers and adaptive feed control strategies have allowed smooth operation "on the edge of the razor", near to the anode polarization limit, where current efficiency is at its maximum. Modifications of bath chemistry have allowed operation at lower temperatures and improved current efficiency (lowering bath temperature by $4°C$ improves current efficiency by 1%). The result is that the best industrial lines of the eighties operate with 95% current efficiency, to be compared to 90% for the best lines at the beginning of the seventies. This improvement is the most rapid ever experienced in the aluminium industry since the beginning of the century, but it is clear that further room for improvement is limited.

The situation of aluminium plants at the end of 1987 in the Western World is given on Figure 20:

    o   more than 40% of aluminium is produced in lines started after 1974; 3

    o   the newest generation of cells, with centre point feeding and pre-baked anodes, accounts for more than 50% of world production (including retrofitted lines);

    o   Söderberg cells, which accounted for more than half of the production in the sixties, represents now about 25% of the total;

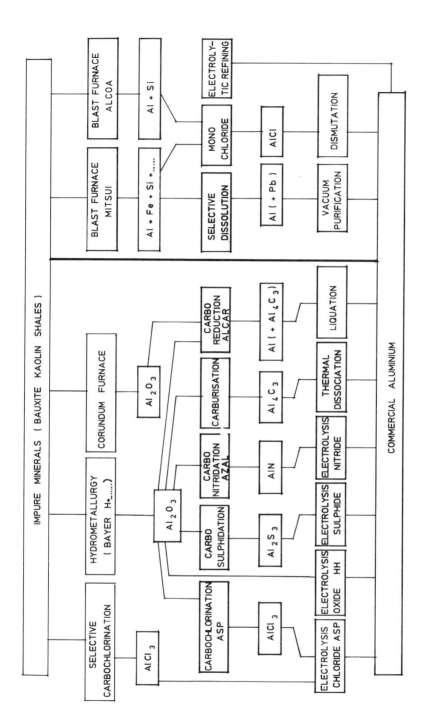

FIG.19 ALTERNATIVE PROCESSES

o   some older types "side feeding, pre-baked anodes", mainly installed before 1974, account for 20% of total capacity; they tend to be retrofitted to point feeding or may disappear.

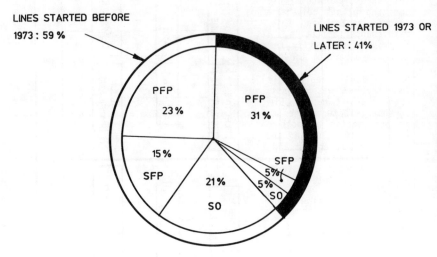

PFP   POINT FEEDING - PREBAKED
SFP   SIDE FEEDING - PREBAKED
SO   SÖDERBERG

FIG. 20. SITUATION OF ALUMINIUM PLANTS AT THE END OF 1987 IN THE WESTERN WORLD

8.4   Some Future Trends of the Aluminium Industry in Europe

The high growth rate experienced by the aluminium industry prior to the oil crisis was discussed in Section 2. During that period the production was controlled by a few "major" producers, who had to double capacities about every 8 years and used this opportunity to gradually improve the extraction process.

From the above considerations, it could however be concluded that the situation has radically changed since 1973 and that the European aluminium industry now progressively slips from the world stage, except for some regions where hydroelectric power is available in abundance. Other considerations must however also be taken into account. One of them is the industrial environment, which has already been mentioned in connection with the location of alumina plants. The availability of qualified labour combined with a transport, equipment and maintenance infrastructure provide evident advantages over smelting plants installed on sites with lower industrial development. Also the proximity of markets is essential for a metal whose uses mostly require severe quality tolerances, particularly concerning inclusions and purity control. Modern smelting plants rely more and more on a high degree of automation and process control and this in turn requires highly qualified labour and maintenance facilities. All these elements plus strategic considerations contribute to justify the continuation of a smelting activity, at least to cover a part of the total consumption.

Nevertheless cheaply negotiated power is the key to the survival of this industry, as was demonstrated recently with new investments decided in France and Canada. Elegant solutions can be found in partnership agreements between the smelter and the power supplier. These consist in an association of both partners where each brings his raw material - respectively alumina and energy, plus the corresponding production know-how - at agreed and flexible conditions, with participation of the power supplier in the profits of the smelter. It may be anticipated that such agreements will form the basis of more power contracts for existing plants and/or future aluminium smelter projects.

9 MAGNESIUM INDUSTRY

9.1 Location and Capacity Versus Demand

The European metal production and consumption for 1987 are summarised in Table 16 and compared to similar figures for other world regions. The primary magnesium industry is situated in France, Italy and Norway, the latter being the major producer and exporter of the metal. Some secondary production is also found in Austria and United Kingdom. Germany is the largest importer, without national production. On the whole, the production of the EEC and associated countries is approximately equivalent to the European consumption. The market however is international. On world basis, the largest producer is United States, where an important secondary production also takes place. The USSR and some other planned-economy countries may be ranked as other important producers and consumers. According to the table, the total recycled metal would amount to about 15% of the world consumption.

The main uses of the metal are as addition element to aluminium alloys and as magnesium alloys for die casting. They represent respectively about 55% and 15% of the total consumption. Important tonnages are also used for the desulphurization of pig iron (12%) and to a lesser extent for the magnesiothermic reduction of titanium and zirconium following the Kroll process.

9.2 Process Technology

Magnesium is extracted from dolomite ($MgCO_3.CaCO_3$), magnesite ($MgCO_3$), salt brines and ocean water. Deposits of magnesium rocks, minerals and brines are widely distributed and sea water is, of course, largely accessible. Most of the metal is produced by igneous electrolysis. Whatever the raw materials and the applied technologies may be, these processes are always based on the electrolysis of anhydrous $MgCl_2$. As an alternative, the metal can be produced by metallothermy combined with a distillation of the metal. In that case, the raw

Table 16: 1987 Magnesium Production and Consumption, in tonnes, after (6)

	Primary Production	Secondary Production	Metal Consumption
Austria	-	300	2 800
Belgium	-	-	3 600
France	13 600	-	11 000
FR Germany	-	-	23 800
Italy	7 600	-	5 100
Norway	56 900	-	7 000
Sweden	-	-	2 300
Switzerland	-	-	2 400
UK	-	600	5 100
Other	-	-	3 700
Total EEC & EFTA	78 100	900	66 800
Total Africa	-	-	2 200
Total Americas	136 900	46 600	153 800
Total Asia	8 500	10 300	35 800
Total Oceania	-	-	4 400
Total Western World	223 500	57 800	263 000
Planned Economy Countries	100 900	-	109 200
Total World	324 400	57 800	372 200

material is preferably calcined dolomite. Metals produced according to the two types of processes differ slightly in composition and have both their impurities, but most of their uses overlap, for instance for Al alloys and iron desulphurization, which cover 70% of the consumption.

For the igneous electrolysis of magnesium, anhydrous $MgCl_2$ can be prepared by different processes, according to the feed material. There are company traditions and processes specifically developed by Norsk Hydro, Dow, Amax, etc. In the electrolytic cells the electrolyte is a fused chloride mixture maintained at 700-750°C. Magnesium is removed as a liquid on top of the electrolyte bath, whereas chlorine is liberated at carbon anodes. In Europe, the electrolytic process is mainly applied in Norway, the magnesium source being brines and magnesite. In the United States, electrolytic extraction takes place from brines and sea water.

In the metallothermic process, calcined dolomite is heated with ferrosilicon at about 1100°C and magnesium is reduced whereas a silicate slag is formed. Since magnesium is the only volatile product of the reaction, it can be condensed directly from the gas phase. The Magnetherm process is practiced in electrically heated furnaces operating on a semi-continuous base. It is applied in France and in the United States. In the Pidgeon process, the charge is heated by transfer in retort type furnaces. This is still in application in Italy.

Data on process improvements in recent years are scarce. Contrary to aluminium and other non ferrous metals, the magnesium industry is rather partitioned. No transfer of technology takes place, although licencing of processes may occur.

9.3 Some Future Trends of the Magnesium Industry

The magnesium industry may be considered as mainly energy and market dependent, some of its characteristics being:

   o   the wide availability of magnesium containing raw materials;
   o   the high energy consumption of the extraction processes;
   o   the coexistence of two types of extraction processes.

The magnesium market is a moderately growing one, with expected growth rates of 2-3% per year. Consumption could increase more rapidly in Europe than in other areas, due to the development of new magnesium containing aluminium alloys as well as magnesium alloys for new applications. In this connection, technical breakthroughs could be realised in relation with corrosion resistance and mechanical properties by the production and use of higher purity magnesium and by the development of very fine grain alloys.

There has been a swing, in the past, from metallothermic to electrolytic processes, followed by a revival of interest in the direct reduction with ferrosilicon. This was again followed by important investments in electrolytic processes. The two types of processes are now in a state of coexistence, each being competitive according to local conditions and none being intrinsically more economic. It may be recalled in this connection that magnesium is essentially quoted in US dollars and that the metal is marketed worldwide, the main reference producers being American. The dollar quotation is thus also an important factor in the economic balance, a high dollar playing in favour if European producers.

Concerning energy requirements the magnesium industry in Europe is already established for a great part in regions where hydroelectric power is available and more particularly in Norway. Similarly to the aluminium industry there has been a recent tendency to develop production in areas where cheap energy is abundantly available such as Canada. This trend could amplify in the future, although investment is not only made in capacity but also in purity of the produced metal.

10   TIN INDUSTRY

10.1   Location and Capacity Versus Demand

The western world productions of tin concentrate for 1987 are given in Table 17. Approximate estimates of smelter production and refined metal consumption are also included in the table. There is very little production of tin concentrate in Europe (circa 4000 t/y from the UK) but consumption of the metal in the EEC and EFTA was 53 800 t in 1987 exceeding the 32 600 t and 36 700 t consumptions of Japan and the USA respectively.

Because of the 'cartel' like operation of the International Tin Agreements (Agreement between producing and consuming nations) tin prices doubled relative to other metals between 1973 and 1978 when most mining companies were facing the problems of the recession. This unreal situation culminated in the collapse of the Sixth Agreement and the tin crisis of 1985, since which time prices have fallen below their relative level of the early 1970s. This crisis caused the closure of some high cost mines and a necessary start to rationalisation and restructuring of the industry. This process is still taking place. The hardest hit mining countries were Bolivia and Malaysia where mine production fell dramatically. In general there has been a move away from the lower grade, smaller and hence higher cost mining operations. The country with the most influence on the tin price and production at present is Brazil. It is the worlds second largest and lowest cost producer and is not a member of the ATPC. Brazil has the potential to become the world's largest tin producer and has indicated that if necessary they will increase tin production to keep the metal competitive and avoid substitution.

In Europe the metallurgical production of tin covers about 40% of the consumption, the major producers being in United Kingdom, the Netherlands and Spain. Much of the tin feed to the smelters comes from secondary sources such as slags, flue dusts, drosses and residues. The smelters therefore form an extension of primary smelters flowsheets for recovery of tin values and in some cases associated base and precious metals. Lead refining via Harris technology produces a good grade tin intermediate for smelter feed in one instance. Flue dusts and slags from secondary copper operations are another source of tin.

The estimated western world average tin consumption by end use in 1987 was 35% tinplate, 32% solder, 10% chemicals and 23% other from a total production of 159 100 t. Most European countries use the majority of their tin intake (from 40 to 67%) for tinplate production; West Germany is an exception reportedly using only 14% for this application. There was a steep decline in the use of tin in the ten years after 1973. This was partly due to the high relative metal price during this time but mainly due to technical changes bringing about reduced consumption and substitution.

Table 17: 1987 Western World Tin Production and Consumption, in tonnes, after (6)

	Mine Production	Metal Production	Refined Metal Consumption
Austria	-	-	500
Belgium	-	200	1 400
Denmark	-	100	-
Finland	-	-	100
France	-	-	7 400
Germany	-	200	17 500
Greece	-	200	600
Italy	-	-	6 000
Netherlands	-	4 000	4 700
Norway	-	-	400
Portugal	100	100	800
Spain	100	1 700	3 300
Sweden	-	-	400
Switzerland	-	-	900
UK	4 000	17 000	9 800
Total ECC + EFTA	4 200	23 500	53 800
Other Europe	-	-	2 600
Total Africa	7 300	4 100	3 300
Total Americas	46 100	41 700	54 600
Total Asia	72 900	88 900	54 500
Total Oceania	7 700	900	1 300
Total Western World	138 200	159 100	170 100

10.2 Process Technology

The technology applied in the extractive metallurgy of tin can involve many diverse unit operations depending on the purity and types of trace elements contained in the feed. A brief summary of some of the available smelting and refining steps is as follows:

a) Primary Smelting

Clean high grade 60-70% Sn, low iron content concentrates are smelted with reductant in reverberatory and electric furnaces to produce metal and a slag containing around 15% Sn which is subsequently cleaned by resmelting on a separate campaign. Medium grade concentrates circa 50% Sn with increased iron and impurity contents produce larger quantities of recycle slag which is often for technical and/or economic reasons treated in a secondary circuit using lead washing or slag fuming for cleaning.

b) Low Grade Concentrates and Secondaries Smelting

Low grade concentrates 20-30% Sn and tin containing secondary materials such as slags, flue dusts and drosses are smelted using the following processes:

o conventional sintering/blast furnace smelting;
o short rotary furnace smelting;
o kiln smelting of zinc/tin/lead residues to produce solder and a zinc rich fume (Berzelius process);
o TBRC bath smelting.

These processes often smelt a solder charge. The presence of lead enables a low tin concentration to be maintained in the smelter slag. To reduce tin further these slags may along with primary smelter slags and some very low grade concentrates and residues be sulphide fumed in a submerged combustion fuming furnace. The TBRC technology claims to be able to smelt and subsequently fume the product slag in a two stage operation in the same furnace. Blast furnaces have also been used for tin fuming.

c) Refining

Tin is one of the most difficult metals to refine and separation from lead is a particular problem. Depending on the impurities to be removed the following refining techniques can be used:

o Molten metal dry refining using controlled temperature and chemical additions (S, Na, Mg, Ca) to remove unwanted elements such as Fe, Cu, As, Sb, Bi as drosses. Dross removal is sometimes made more efficient by use of a centrifuge, reducing costly recycle.

o Vacuum distillation at 1200°C is used to remove impurities such as Pb and Bi.

o Fractional crystallisation in a spiral crystalliser is used primarily for the removal of Pb and Bi.

o Alkaline electrolytic refining in electrolytes such as sodium stannate solution is used to deposit tin from impure high lead containing anodes (deposition from stannic solution).

o   Acid electrolytic refining is based on sulphuric acid with added phenol/cresol or both as antioxidants (deposition from stannous solution). Fluosilicic acid can also be used for electrorefining of tin, but its main application is for refining of solder.

Of the above refining processes, electrolytic techniques give the highest concentration of impurities per unit of tin in recycle materials and their application is essential if high levels of precious metals are present.

10.3   Factors Influencing Evolution

Capacity

The European tin smelting industry has developed in the past from a base of treating low to medium grade concentrates from such places as Bolivia along with secondary tin containing materials. Some smelters have installed medium to high grade feed smelter circuits with a view to expanding tin production. However there is considerable tin smelting overcapacity worldwide estimated at some 100% and there is also a tendency for the tin mining countries to construct their own smelters.

Despite the present low price, tin is finding difficulty in regaining tinplate market losses. However falls in developed countries have been offset by increases in developing countries. The growth of the electronics industry has made solder as important an end user as tinplate but this is not seen as a future growth area. The tin chemicals industry is a potential area for growth having benefitted from recent technical changes in the plastics industry (organotin stabilisers), however tins use in biocides has come under environmental pressure.

The above reduction in demand and general tin industry climate after the crisis made it very difficult for the European smelters to obtain raw materials, as mining operations have moved towards production of high grade materials and smelters competition favoured low cost operations close to sources of concentrate supply. This has caused industry restructuring and serious underutilisation of some of the medium to clean concentrate smelting capacity in Europe. Lack of low grade concentrate feed has affected the necessary balance of raw materials needed for smelting secondary feed.

Technology

In Europe, development and selection of processes has taken place on their ability to treat widely different chemical and physical raw materials. The main smelting techniques used at present are: blast furnaces, short rotary furnaces for low grade concentrates and secondaries and electric furnaces for cleaner higher grade concentrates.

The largest of the European plants situated in the UK has additional circuits for electrowinning of copper, electrorefining of lead and recovery of precious metals. This plant has the unique ability to electrorefine high grade tin from high lead content alloy. The refinery is the only alkaline one now operating and is the largest in the world either acid or alkaline. The plant also has a high capacity slag fuming plant. This combined with blast furnace and other metal circuits gives the plant flexibility to treat a wide range of residues.

Electrorefining of solder is also carried out on some European plants. Since the European smelters handle lead containing materials, they are therefore subject to stringent environmental regulations.

10.4 Some Future Trends of the Tin Industry in Europe

Considerable rationalisation of the tin industry has already taken place in Europe and production capacity has been reduced. It is considered that many of the adjustments required to put the tin industry on a firmer basis have been made and a more realistic situation is emerging.

In Europe some increase in tin mining capacity will take place with the realisation of the Neves Corvo project. The competitive position of European smelters to treat these concentrates has still to be demonstrated. However, this industry will continue as in the past to be based mainly on low grade material and secondaries. Raw material supply would therefore be limited and dependent on tin price; growth to previous capacity levels is improbable.

This specific situation of the European tin industry implies that the metallurgical flowsheets will need to remain very flexible particularly regarding efficient refining ability and byproduct recovery. Solder will be a major product from some plants. Metal purity will become increasingly important. And there is unlikely to be any great change in the existing technology as replacement costs would be high. However processes under development that combine bath smelting with fuming in a single vessel such as Sirosmelt may become attractive for retrofitting in the medium term. Also fractional crystallisation provides a cheap method of refining for some impurities and may be adopted to improve flowsheet flexibility.

Finally, environmental improvements will continue to be required in line with the lead smelting industry.

11 SOME FUTURE TRENDS OF THE INTERNATIONAL AND EUROPEAN CONTEXT

Summarising the evolution of the non-ferrous industry during the last fifteen years, it may be said that important changes took place in this period, during which our industry strengthened its position while gaining experience in its adaptation to change.

Whilst exercising prudence in forecasting possible future trends, the following evolution may be expected:

It is generally considered that the growth rate in the coming years will be largely shaped by the growing world population and by the advance of science and technology, which may either be in favour or against growth of some metals. The industrialisation of developing countries has also frequently been cited as a major possible growth factor, although it is now considered for many of these countries to be a long-term evolution with restraining cultural implications. The prevailing impression is that the combination of these effects will lead to a further average rise of metal consumption, although at a rate lower than the one prior to 1975. The trends which begin to appear in the right hand part of the curves in Figure 3 could well be those with which we will have to live in the coming years.

For some metals however, stagnation or reduction in consumption may actually lie ahead, with a corresponding zero or negative growth of the producing industry.

Improvements in extraction and processing technologies will go on, helping metals to remain competitive in face of possible substitutes. Rationalisations and restructuring of plants and enterprises will continue.

Confidence in supply sources will keep on recovering, while discovery of new supplies, even in excess of demand, may be expected. Initiation of new mineral projects could slow down however due to the slower expansion of markets for most metals.

Environment in its different aspects will remain an essential preoccupation of the industry in concert with authorities and public opinion.

In Europe, extractive metallurgy will remain a significant industry, although a progressive general shift takes place from primary smelting towards recycling and increased attention is paid to downstream operations. Continuation of smelting and refining activities, at least to cover a part of the total consumption, remains essential in connection with security of supply and close contact between producers and markets. Energy considerations are strong enough to attract some metal productions from the continent towards more favoured regions but availability of high quality labour, necessity of recycling, market requirements and capital cost tend to counter the movement. Environmental questions are resolutely tackled; environmental regulations are besides becoming universal. As opposed to other regions, Europe has maintained a high level of technological development spurred by the availability and complexity of raw materials, particularly with the increase of recycling.

As a complementary growth factor, the break-down of trade barriers in Europe is expected to promote economic growth in that continent and to increase trade with its various commercial partners. Similarly, although on a much longer term, there is renewed hope of tensions lessening between East and West, which could also result in improved international trade and consequently economic growth.

References

1. Annuaire MINEMET 1986, Minemet Holding, Paris

2. World Bureau of Metal Statistics, World Metal Statistics, vol. 41, September 1988

3. Strauss J.D., "Trouble in the Third Kingdom, The Minerals Industry in Transition", Mining Journal Books Limited, London, 1986

4. International Lead and Zinc Study Group, "Lead and Zinc Statistics", vol. 28, September 1988

5. Killington A.J, "Waste Management and Recycling", Meeting DG XI-Eurometaux, Cardiff, September 1987

6. Annuaire Minemet 1987, Minemet Holding, Paris

7. Belgian Ministry of Health and Environment, Institute of Hygiene and Epidemiology, Annual Report 1987

8. International Lead and Zinc Study Group, "Principal Uses of Lead and Zinc", London, 1985 and 1988

9. International Lead and Zinc Study Group

10. Metallstatistik, Metallgesellschaft AG, Frankfurt

11. A.Sutulov, "Copper at Crossroads", Intermet Publications, Santiago/Chile 1985

12. Traulsen H.R., Taylor J.C., George D.B., "Copper Smelting - an Overview", Journal of Metals, Volume 34, No 8 (1982), pp. 35-40

13. CRU Copper Studies, Volume 14, No 81, Feb 1987

14. Ph. Crowson: Minerals Handbook 1986-87 McMillan Publishers

15. C.M. Diaz et al: "A Review of Nickel Pyrometallurgical Operations", Journal of Metals, Vol. 40, No 9, (1988) P. 28-33

16 K.Emicke: "Nickel als Begleit- und Verunreinigungselement bei der Kupfergewinnung", Symposium Nickel, GDMB (1970)

17 Winnacker-Kuchler: Chemische Technologie, Band 4, Carl Hanser Verlag Munchen (1986), P. 389-407

18 W.Betteridge: "Nickel and its Alloys", Ellis Horwood Ltd, Chichester G.B. (1984)

19 P.Queneau, J. R. Boldt, jr.: "The Winning of Nickel", Longmans Canada Ltd, Toronto 1967

20 C.A.Landolt, J.C.Taylor: "Nickel Smelters - Survey Review", World Survey of Non-Ferrous Smelters, J.C.Taylor, H.R. Traulsen, eds. TMS (1987)

21 A.R.Burkin: "Extractive Metallurgy of Nickel", John Wiley and Sons, Chichester G.B. (1987)

22 Metallgesellschaft, "Tableaux Statistiques", 47e Publication 1950-1959, Frankfurt am Main 1960

23 H.Gruss: "Nickelversorgung der Bundesrepublik Deutschland", Symposium Nickel, GDMB (1970)

24 J.A. Asteljoki, H.J. Krogerus and T.S. Mäkinen, Recent pyrometallurgical process developments at Outokumpu, TMS paper Nr A89-26, Las Vegas, Feb 1989

25 Shearson Lehman Hutton Inc., London Metals Research Unit, Annual Review of the World Silver Industry, May 1988

26 Milling-Stanley G., Consolidated Gold Fields PLC, Gold 1988, May 1988

27 Robson G.G. and Smith F.J., Johnson Matthey PLC, Platinum 1988, May 1988

28 Aluminium recycling in Nederland, Recycling, August-September 1988, p.111

PRODUCTIVITY AND TECHNOLOGY CHANGES IN THE NORTH AMERICAN COPPER, ZINC AND

NICKEL EXTRACTIVE INDUSTRIES

Franklin G.T. Pickard and Gerald A. Crawford

Falconbridge Limited
P.O. Box 40
Commerce Court West
Toronto, Ontario
M5L 1B4
Canada

Crawford Associates
777 Meadow Wood Road
Mississauga, Ontario
L5J 2S7
Canada

Abstract

Factors contributing to improved productivity in the North American copper, zinc and nickel industries over the last ten years are examined with particular reference to corporate restructuring, labour rationalization and technology modernization. The industry has survived a wave of boom - bust - boom between 1979 and 1989 and is now more productive, more competitive and in better condition to face the next downturn than it was in 1981. Companies have decreased their debt, trimmed their manpower, modernized their plants and processes and now, with tight inventories and rising demand, are reaping welcome returns on the capital they have invested for the future.

Introduction

The worldwide recession of the early 1980's caught North American non-ferrous producers off-guard. As purveyors of primary metals to the world for generations, most of their mines and plants were relatively old with high labour costs. With the buoyant international expansion of the 1960's and 1970's, third world producers joined the party with new mines and plant, government-owned or subsidized operations and cheap labour.

Suddenly, North American suppliers were uncompetitive and bleeding to death. They had to become competitive in a hurry or quit. Some quit. This paper reviews what the survivors did to regain competitiveness. It largely ignores those factors over which individual corporations have no control such as exchange rates, interest rates, inflation and prices. It focuses instead on those changes by which companies exercise direct control over their competitiveness, including corporate restructuring, manpower rationalization and technology modernization.

Information for this review was drawn from many sources, as listed in the Bibliography, but primarily from annual reports of the U.S. Bureau of Mines. The USBM is in a class by itself in documenting statistics and other information regarding the mining and metals industries. Data are not always consistent from one source to another, however, particularly cost figures that are composites of several sub-elements and not always clearly or consistently defined. Some figures in this paper are expressed in ranges, therefore, usefully reflecting levels of magnitude and more importantly, trends.

In reviewing the copper, zinc and nickel industries in the United States and Canada within the last ten years, it is cumbersome to document every change that has occurred. Rather, selective information is presented which reflects the nature and extent of the changes that have been made both within these industries as a whole and also by some of the higher-profile corporations that represent them.

North American Overview

Few if any companies in North America escaped the impact of plummeting prices for copper, zinc and nickel in the early 1980's. For numerous reasons the nature and magnitude of the effects differed between individual companies and also between Canada and the U.S. but the response, while differing in degree, was qualitatively the same - to reduce cash costs as much and as fast as possible, staunch the flood of red ink and restore profitability. These goals were pursued by three principal devices:

i) minimizing debt and associated carrying charges by closing or selling losing or marginal mines, smelters, refineries or other non-core assets;

ii) minimizing payroll by pruning both production workers and office staff to maximize productivity; and

iii) investing capital in new plant and technologies to modernize or replace old or outmoded facilities, thereby displacing intolerably high operating charges with lower, long-term debt repayments.

If the amount documented about these metal industries in the U.S. and Canada is any measure of the seriousness of their situation, then the U.S. copper industry appears to have been hardest hit. It plays a larger role in the U.S. economy than zinc, and nickel is not really an issue. All three metals were also adversely affected in Canada although distinctions are more obscure because they occur commonly as by-products of one another in Canadian ores more than in the U.S., and allocation of costs is thus arbitrary and equivocal.

In any case, North American companies have risen to the challenge and slashed costs to competitive lows. Now, with increasing demand and prices, they are again reaping profits and returns on recent investments in modernization. For the time being at least, competitiveness has been restored. Problems and solutions of the 1980's are summarized below.

Copper in the United States

Copper is the most important of the non-ferrous metals in the U.S. and second only to steel in the contribution of primary metals to G.N.P. In 1980 the U.S. was still in first place internationally with more than 15% of world supply. Traditionally, it had benefited from numerous advantages including skilled labour and management, but suffered progressively from growing competitive pressures including low ore grades, depleting reserves, high labour rates, outdated and high cost facilities and technologies, and the additional costs of compliance with regulations respecting the environment and worker health and safety.

In addition, domestic demand and sales of copper were diminishing in the face of substitution in a maturing U.S. economy, thereby increasing competition for exports against growing foreign production. On top of this, with total cash costs averaging roughly $0.97 per pound, net cash costs were around $0.80 per pound or more after by-product credits. When copper prices fell below $0.70 per pound after 1981, the U.S. copper industry was clearly in a bad position and significant cost decreases were quickly required.

Shutdown and Sale of U.S. Copper Operations

To this end a number of losing or marginal operations were shut down temporarily or permanently. A partial list of such shutdowns is presented in Table I. A number of other operations were sold. Most were moved advantageously between copper companies, which decreased vendor debt and increased buyer assets. Included among these sales were those by oil companies which had diversified into the U.S. copper industry in the 1970's, and, with the sole exception of British Petroleum, abandoned it in the 1980's. A partial list of these sales is presented in Table II.

Table I. Partial List of U.S. Copper Operations Shut Down 1983-1987

Year	Operation	Location	Company
1983	Laurel Hill refinery	New York	Phelps Dodge
	White Pine refinery	Michigan	Copper Range
	Baltimore refinery	Ohio	Kennecott
1984	Ajo smelter	Arizona	Phelps Dodge
	Battle Mountain	Nevada	Duval
	EW refinery		
1985	Washington smelter	Washington	Asarco
	Twin Buttes mine & EW refinery	Arizona	Anamax
	Utah Copper Division	Utah	Kennecott
	Ajo mine	Arizona	Phelps Dodge
	Copperhill	Tennessee	Tennessee Chemical
	Tacoma smelter	Washington	Asarco
	Federated Metals	Texas	Asarco
	Garfield smelter & refinery	Utah	Kennecott
	Morenci smelter	Arizona	Phelps Dodge
	Port Nickel refinery	Louisiana	AMAX
	Three Rivers refinery	Michigan	United Technologies
1986	Carteret refinery	New Jersey	AMAX
1987	Douglas smelter	Arizona	Phelps Dodge

Table II. Partial List of U.S. Copper Operations Sold 1985 - 1987

Year	Operation	Location	Sold By	Bought By
1985	Amoco Minerals		Standard Oil, Indiana	Cyprus
	Butte properties	Montana	Anaconda	Washington Corp.
	Carr Fork property	Utah	Anaconda	Kennecott
	Pima Mine	Arizona	Cyprus	Asarco
	Sunnyside	Colorado	Standard Metals Corp.	Echo Bay
1986	Morenci (15%)	Arizona	Phelps Dodge	Sumitomo
	Ray Mines, smelter & EW refinery	Arizona	Sohio (Kennecott)	Asarco
	Chino Mines	New Mexico	Sohio	Phelps Dodge
	Sierrita Mine	Arizona	Duval	Cyprus
1987	Eisenhower	Arizona	Anamax Mining	Asarco
	Butte	Montana	Washington Corp.	Montana Mining Corp.
	Casa Grande	Arizona	Noranda	Cyprus

Profile of U.S. Copper Operations in the 1980's

Selected data illustrating changes by the U.S. copper industry in production, employment, productivity and costs during the 1980's are presented in Table III.

Table III. Profile of U.S. Copper in the 1980's

	1981	1982	1983	1984	1985	1986	1987
Copper Production, kilotonnes, total							
Mines	1538	1147	1038	1103	1106	1147	1256
Smelters	1300	1021	987	1183	1191	1196	1249
Refineries	1882	1533	1584	1490	1435	1479	1560
Employment, Production & Office Personnel, thousands							
	44.6	31.8	24.4	21.0	15.7	16.0	
Productivity, hours per tonne copper produced, total							
	57.9	51.9	44.8	36.0	29.0	27.7	29.3
Cash Costs, $ per pound copper produced, total							
Including by-product credits	0.79	0.77	0.69	0.65	0.58	0.56	0.53
Including recovery of capital and amortization	0.85-0.90			0.67		0.67	0.56-0.58

Production has dropped from 1981 highs and while it has regained some ground with rising prices in recent years, it is still significantly lower than at the beginning of the decade. Chile surpassed the U.S. as the world's largest producer in 1982 while the U.S. copper industry shrank and is unlikely to regain its former size.

Employment is barely a third of 1981 levels although it has recovered somewhat with the recent turnaround in production. Some of the decrease simply reflects the drop in production, and some is due to modernization as discussed below. The industry was also clearly over-employed and when the crunch came it was able to manage with much fewer people. With labour accounting for 35-40% of the cost of U.S. copper, manpower was a major factor in cost rationalization. To further assist the industry to survive, wages and benefits were trimmed by as much as 25% in contract negotiations in the mid 1980's.

Productivity has doubled and although it appears to be slipping somewhat with recently rising production, will likely remain close to current levels.

Cost data are at best imprecise and arguable but there is no doubt regarding levels of magnitude and trends. Costs were clearly $0.80 per pound or more in 1981 and have dropped below $0.60, an average decrease of at least a quarter and probably closer to a third. In any case, the industry can now survive the lowest prices that are likely to prevail in the foreseeable future. Beyond that is a question of continuing concern.

Technology Modernization

Technology improvements have been made in all sectors of the U.S. copper industry. Such changes are particularly appealing where savings in operating costs resulting from greater efficiencies and productivities more than offset carrying costs of capital required to make them. Some changes are necessary to comply with environmental or health and safety regulations. Numerous improvements have been implemented. Some of the more significant in each sector are briefly identified below.

Mining, Milling, Concentrating

- trucks of 170 tonnes or more in lieu of trains in open pits
- in-pit crushers and conveyor belts in lieu of trucks
- computerized control of trucks and shovels
- autogenous and semi-autogenous grinding
- large flotation cells for improved efficiency of scale, energy saving, simpler control and better performance
- column flotation in lieu of conventional cells
- automated process control in concentrators
- control loops in flotation operations for pH, slurry and reagent flow

Smelting

- traditional reverberatory furnaces being replaced by continuous smelting facilities with lower external fuel requirements, steady state operation and high-strength SO_2-bearing off-gases for acid fixation and environmental control

Smelter Technology	Copper Capacity, kilotonnes per year	
	1975	1987
Reverberatory, primary	1445	386
Reverberatory, secondary	208	281
Electric furnace	336	112
Inco flash	-	288
Noranda, modified	-	210
Outokumpu flash	-	160
	1989	1437

- table reflects displacement not only of old technology by new but also of total U.S. smelter capacity, shrinkage of more than 25%
- computers being applied for process control of smelting operations

Refining

- solvent extraction-electrowinning plants, (SX-EW); being installed increasingly; particularly advantageous with porphyry oxide ores common in the U.S.; lower capital cost and air pollution

Changes by Three High Profile U.S. Copper Producers

The information presented above provides an overall view of major changes that have been made by the U.S. copper industry as a whole to regain profitability and competitiveness in the world market. Actions taken by individual companies are exemplified by the following capsule review of Phelps Dodge, BP Minerals America and Asarco.

Phelps Dodge, PD

1981 - PD copper cost, $0.81/pound
1982 - $60 million interest on debt
1983 - lost $63 million when copper price dropped to $0.77/pound
1984 - planned to regain profitability at copper price of $0.65/pound
- PD share of U.S. mined copper about 20%
- wrote off losses, shut down high cost mines, closed 2 smelters, sold selected non-core assets
- $47.3 million SX-EW plant started at Tyrone, N.M.; total unit cost <$0.30/pound
- $37 million modernization at Hidalgo, N.M. including Outokumpu flash smelter and O_2 plant
1985 - 372 kt copper produced; surpassed Kennecott as #1 U.S. producer
- first U.S. copper producer to regain profitability
1986 - bought Chino, N.M. operation from Kennecott for $88 million
- installed SX-EW plant at Chino for 50 kt/year copper
- sold 15% interest in Morenci, AZ, to Sumitomo for $75 million
- bought 29 x 170 t trucks at Morenci for $22 million
- $90 million SX-EW plant at Morenci
- computerized truck dispatch at Morenci, Tyrone and Chino
- cut staff at Morenci by 45%
- sold various assets for $65 million
- sold share issue for $48 million
- $36 million interest on debt, down from $60 million in 1982
- record 401 kt copper output, nearly double 1983 production
- 44% decrease in production costs since 1981
1987 - 490 kt copper produced; nearly half of total U.S. output
- copper cost down to $0.55/pound from $0.81/pound in 1981
1988 - $275 million invested since 1984; another $275 million planned for 1989-1992

B.P. Minerals America, BPMA

1980 - Kennecott was #1 U.S. copper producer with 7,500 employees and productivity of 28 t copper/man year
1981 - Standard Oil of Ohio, Sohio, bought Kennecott; BP owned 25% of Sohio since 1960's
1984 - employment down to 4,300; productivity up more than 50% to 43 t copper/man year
1985 - completion of $350 million modernization at Chino including Inco flash smelter; cut costs by $0.31/pound copper
- shut down Bingham Canyon - world's largest man-made excavation; decision to embark on $400 million, 3 year modernization of Utah Copper Division
1986 - BP ownership of Sohio increased to 55%
- at Bingham Canyon trucks replaced by 8 km. conveyor system
- 3 fixed crushers replaced by one large mobile crusher
- >100 secondary crushing and grinding mills replaced by 3 x 10.4 m. diameter semi-autogenous grinding mills - largest in world;
- locomotives, ore cars and track replaced by 27 km. x 6" concentrate

                pipeline and 21 km. x 48" tailing pipeline
              - 2,000 flotation cells replaced by <100 larger cells including 33 x
                85 m³ cells--largest in world
              - wages and benefits decreased by 23%
       1987 - BP acquired all of Sohio for $7.8 billion
              - overall labour costs decreased by about 35%
              - Bingham Canyon lowest cost copper producer in North America and one
                of lowest in world
       1988 - employment down to 2,300; productivity up to 56 t copper/man year
              - unit costs down by 40% since 1980

Asarco

1981-1984
              - modernization at Mission mine, including truck fleet, flotation
                cells and computerized systems; increased output by nearly 50% and
                decreased costs by nearly 30%
              - Hayden smelter converted to flash technology to meet environmental
                regulations
              - pruned weaker operations
              - cut labour force by 21%, froze wages
              - renegotiated rail and truck contracts
1985 - closed Tacoma smelter in lieu of spending $150 million for environ-
        mental compliance
1986 - bought Ray mine from Kennecott for $72 million which doubled Asar-
        co's capacity
1987 - long-term debt decreased by 50%; excess cash in the bank

Copper in Canada

Canada is a major copper producer, just behind the U.S. in world ranking, but the industry is quite different from the U.S. in several significant respects and difficult to compare. Unlike the copper-only deposits in the U.S., for example, much of Canadian copper is mined from smaller, richer deposits of mixed nickel-copper and copper-zinc sulphides with significant by-product values including precious metals. With by-product credits that can approach or exceed the value of copper, determination of copper costs is arbitrary and equivocal at best. Also, the industry is dominated by a handful of major corporations with a modest number of mines, smelters and refineries, unlike the more pervasive nature of the U.S. industry.

With the high cost of underground mining and sulphide smelting, some data suggest that the cash cost of Canadian copper production in 1981 was substantially as high as in the U.S., roughly $0.96 per pound. With relatively high by-product credits, net cash costs are quoted variously between $0.50-$0.68, much lower than the $0.80 or more in the U.S. Thus while Canadian non-ferrous metal producers lost plenty of money in the early to mid 1980's, they were not as devastated as their U.S. counterparts. Whether for this or other reasons, net cash costs in Canada remain substantially unchanged at around $0.55 per pound of copper.

Employment was decreased by about 20% but wages actually rose 12%. In 1986 the average wage was about $16 per hour, more than 50% higher than the U.S. average of approximately $10 per hour. By cost-cutting measures and technology improvements similar to those identified earlier, there were real cost decreases in Canada of about $0.24 per pound but these appear to have been substantially offset by a similar decrease in by-product credits. Thus the Canadian industry in effect has drifted sideways and is probably no more competitive today than in 1981.

Production, on the other hand, has increased, unlike the U.S., and Canada has risen from 4th or 5th largest producer with about 8% of world output in 1981 to 3rd largest today with roughly 9-10%. Production data are summarized in Table IV.

Table IV. Canadian Copper Production in the 1980's, kilotonnes

	1981	1982	1983	1984	1985	1986	1987
Mines	691	612	640	713	738	698	767
Smelters		394	440	504	493	473	474
Refineries		354	525	539	533	526	525

It is noted that the increase in mine production occurred in spite of closing 160-170 kt of capacity because of low prices. Also, Canadian smelter production is roughly 40% less than mine output due to exports of concentrates, mainly to the U.S. and Japan.

Capital investments were made for expansion and new technology although the level of investment dropped from more than $500 million in 1981 to $200-$300 million per year thereafter. Since cash flow dropped even more, however, it is to the credit of Canadian copper producers that they were willing to make substantial commitments without any immediate return against the anticipated recovery to follow. They are presently being profitably rewarded.

While many changes occurred during the 1980's, two developments that had the greatest impact on the profile of the Canadian copper industry were the formation of Highland Valley Corp., B.C. by the Teck Group and the purchase and expansion of Kidd Creek, Ontario by Falconbridge Limited. Details of these developments are summarized below.

Highland Valley Corp.

Teck Corp. with its partners Metallgesellschaft, West Germany and MIM Holdings, Australia, acquired Cominco from Canadian Pacific for $135 million in late 1986. While gaining control of a $2 billion international metals empire, they also assumed more than $1 billion in debt. By selling assets and raising capital, debt was decreased to $375 million by late 1987, thereby saving over $40 million in interest payments which was used instead for a cost-cutting program.

The $62 million Highland Valley development is one example of that program. It connects Cominco's Valley copper mine with the Lornex mill by a 5 km. conveyor system. At 180 kt/year of copper it is the third largest mining operation in the world after Chuquicamata in Chile and Palabora in South Africa and one of the lowest cost. Highland Valley now supplies almost one-quarter of total mined copper in Canada.

Kidd Creek

North America's most modern copper smelter, based on the Mitsubishi continuous smelting process, was commissioned in 1981 with a capacity of 59 kt/year copper. A $60 million expansion to 90 kt/year was underway when Falconbridge acquired Kidd Creek Mines Limited from the Canada Development Corporation in 1986. The 50% expansion is now complete and when this is added to copper from Falconbridge's Sudbury nickel operations, it makes Falconbridge Canada's second largest producer of mined copper with nearly 20% of the total.

Major additions and improvements to the Kidd Creek copper smelter and refinery include:

- 250 t/d O_2 plant
- expansion of acid plant to 1,600 t/d
- additional jumbo electrorefining tanks, 22 m. x 4 m. x 1.5 m. which hold 16 cells with 89 electrodes each
- stainless steel blanks replacing copper starter sheets
- automated assembly of electrodes in cells, stripping of cathodes and baling for shipping

The other major players in Canadian copper are Inco, Noranda and Hudson Bay Mining and Smelting. About 40% of mine production comes from each of Ontario and B.C. with about 10% each from Manitoba and Quebec.

Zinc in Canada

Canada has been No.1 world producer of mined zinc since the 1960's and is No.2 or 3 supplier of refined zinc. About 80% of Canada's zinc occurs in sulphide ores associated with copper or lead and is therefore held largely by the same few organizations that also produce most of Canada's copper.

There are more than 20 producing zinc mines but two-thirds of total 1985 mine output of nearly 1,200 kt zinc was derived from the largest five at the time:

Rank	Mine	Location	Owner	Mined Zinc kt
1	Brunswick	New Brunswick	Noranda	237
2	Kidd Creek	Ontario	Falconbridge	220
3	Pine Point	NW Territories	Cominco	161
4	Polaris	NW Territories	Cominco	100
5	Sullivan	British Columbia	Cominco	76
				794

Smelting capacity in Canada in 1985 was around 700 kt zinc in four primary smelters:

Rank	Smelter	Location	Owner	Zinc Capacity kt
1	Trail	British Columbia	Cominco	272
2	Can. Electrolytic Zinc	Quebec	Noranda	227
3	Kidd Creek	Ontario	Falconbridge	130
4	Flin Flon	Manitoba	HBM&S	73
				702

Zinc production from mines and smelters in Canada from 1981-1987 is summarized in Table V.

Table V. Zinc Production in Canada 1981-1987, kilotonnes

	1981	1982	1983	1984	1985	1986	1987
Mines	1096	1036	1070	1207	1172	1291	1500
Smelters	618	512	617	632	692	571	611

The most notable fact is that Canadian production of mined zinc increased by nearly 50% during the period and Canada's share of world production of mined zinc increased from 17% in 1983 to 21% in 1987. The increase of 200 kt mined zinc in 1987 was due to the record output of 484 kt in concentrate from the Pine Point mine, the largest zinc mine in the country. Ore reserves are now depleted and the mine is being phased out.

Capital expenditures in the Canadian zinc industry were about $300 million in 1981 but dropped to around the $100 million per year level thereafter. Some of the major investments that have been announced or are being made include:

Year	Operation	Location	Owner	Investment $ millions	Zinc Capacity kt
1982	Polaris	NW Territories	Cominco	130	100
1983	Trail	British Columbia	Cominco	175	272
1988	Can.Electrolytic Zinc	Quebec	Noranda	120	230
1986-1991	Red Dog	Alaska	Cominco	420	750

Technology improvements have been made that are generally similar in nature to those applied to copper production with a couple of unique exceptions.

The Polaris mine, for example, is the most northerly in the world and thus technology has had to address severe problems associated with permafrost, extreme cold, freezing in storage, shipping difficulties. Similar problems must be resolved at the enormous Red Dog project presently under development in Alaska.

Cominco and Kidd Creek have the first two commercial pressure leaching operations for zinc based on Sherritt Gordon technology. The process has the significant advantage, among others, that sulphur reports in elemental form in the solid residues of the process and is therefore environmentally benign.

The Red Dog development in Alaska is unquestionably the most significant single undertaking by a Canadian zinc producer. It will be the largest zinc operation in the world and will underpin not only Cominco's future but also that of the Canadian zinc industry as a whole.

Cash costs of zinc in Canada have not changed dramatically in the 1980's but they have at least gone in the right direction. Average cost in 1981 was U.S. $0.36/lb. and in 1987 had dropped to $0.31/lb. This apparently modest improvement is perhaps more significant in view of the fact that Canadian mined zinc production increased by nearly one-half in the same period. Canada's competitive position in world zinc has not changed but it is secure and likely to remain so.

Zinc in the U.S.

U.S. zinc is an industry in decline. While world zinc production hit record highs from 1984 to 1987, U.S. output fell steadily to less than 4% of world production of mined zinc compared to nearly 6% in 1981. A number of aging U.S. mines and smelters were closed and venerable names including New Jersey Zinc and St. Joe Resources disappeared. There are several reasons for this unfortunate trend.

U.S. zinc ores have two basic disadvantages: they are relatively low-grade with only 3-4% zinc and have little by-product content such as lead and silver. Further, many mines are depleted and some smelters are obsolete. In 1981, less than half U.S. refining capacity was electrolytic, the technology of choice. Some of these have now been shut down and of the remainder, 80% are electrolytic. Labour and transportation costs are high, as is compliance with environmental regulations; instead of spending money to meet these regulations many old plants were shut down.

Some plant closings include:

Plant	Location	Company	Zinc Capacity kt
Corpus Christi refinery	Texas	Asarco	104
Bunker Hill mine & smelter	Indiana	Bunker Ltd.	103
Zinc dust plant	Oklahoma	Federated Metals (Asarco)	
Copperhill	Tennessee	Tennessee Chemical Co.	

The Copperhill plant was 140 years old!

U.S. production of mined and slab zinc is summarized in Table VI for 1981-1987.

Table VI. U.S. Zinc Production, 1981-1987, kilotonnes

	1981	1982	1983	1984	1985	1986	1987
Mined	343	326	295	277	252	216	233
Slab	397	302	305	331	334	316	343

While both mined and slab zinc production declined, slab zinc is higher, reflecting imports of zinc concentrates, largely from Canada. The U.S. uses about four times as much zinc as it produces and imports the difference.

Cash costs were decreased from about $0.43/lb. Zn in 1981 to around $0.38/lb. in 1987. While this is significant it is nevertheless modest and has not effectively changed U.S. competitiveness.

Nickel in Canada (North America)

Nickel in North America effectively means nickel in Canada and largely Inco and Falconbridge. During the 1980's the only two U.S. producers - AMAX and Hanna - closed down and Sherritt Gordon ran out of ore in Canada. Sherritt provides custom refining to primary producers including Inco and will also process some of the nickel concentrate from the new player on the Canadian nickel scene - the Namew Lake mine in Manitoba which is owned by Hudson Bay Mining and Smelting and Outokumpu. This is a state-of-the-art underground mining facility that is expected to produce about 9 kilotonnes per year of nickel in concentrate, roughly 5% of Canadian nickel output. Inco and Falconbridge are No.1 and 2 nickel producers in the Western World who together supply about one-third of the total from Canadian ores.

Inco and Falconbridge were both battered by low nickel prices in the early 1980's and responded in the same way as copper and zinc producers in North America - cutting costs by trimming debt and employment and investing in cost-saving technological improvements. The extent of these measures is reflected in Table VII, based on information drawn from annual reports.

Table VII. Collective Profile of Inco and Falconbridge, 1981-1987

1981	1982	1983	1984	1985	1986	1987	1988
Nickel produced from Canadian ores, kilotonnes							
158	85	128	175	170	168	200	201
Employment (Inco total + Falconbridge Sudbury), thousands							
35.7	29.5	28.3	25.3	23.6	22.7	20.9	21.0
Productivity, tonnes per employee							
4.4	2.9	4.5	6.9	7.2	7.4	9.6	9.6
Capital Investment, $ millions							
215	124	92	141	162	241	206	319
Debt, $ millions							
1821	1705	1660	1398	1213	2198	1602	1496

Production from Canada has been sustained except for the collapse in demand in 1982. The significant increase in 1987 and 1988 primarily reflects production from Inco's new open pit operation at Thompson, Manitoba. Total output in 1988 is 27% higher than 1981.

Employment has been decreased by more than 40% since 1981.

Productivity has more than doubled over the same period.

Capital investment was decreased in the early 1980's, like production, but has since been increased and in 1988 was 50% higher than in 1981.

Debt is gradually being diminished and in spite of the jump in 1986, reflecting the purchase of Kidd Creek by Falconbridge, is down almost 20% from 1981.

While neither company discloses production costs Inco has reported that unit costs in 1987 were down 33% from 1982 highs. Falconbridge has cut costs by about 8% per year for 4 years from 1985 to 1988, a similar overall decrease.

Technology improvements include computerization and large units for grinding and flotation, as for copper and zinc, but most focus on advances in mining technology.

Bulk mining is a major advance. Vertical Retreat Mining, VRM, has been developed by Inco and applied progressively through the 1980's. More than 90% of Inco's ore was mined by VRM in 1987. Other improvements associated with bulk mining include larger diameter drills, longer drill holes, larger explosive charges, double deck blasting, continuous loaders, conveyors and electric trucks. An interesting development being applied by Inco is a belt-bender to allow conveyors and ore to move around corners in the spirit of truly continuous mining.

The Crean Hill mine in the Sudbury Basin has been revamped as an all-electric mine. Inco expects to produce 3,000 tpd with 120 personnel this year instead of the 400 required when the mine was previously operated using conventional methods.

Continued cost-reductions at Falconbridge have been accomplished by increasing employee participation programs which include cost management seminars and simplified reward programs for ideas which generate cost reduction. More emphasis has been placed on grade control in mining to decrease the proportion of waste rock mined.

Conclusion

The 1980's have been a painful and costly period of salvation for survivors of the copper, zinc and nickel industries in North America. They are leaner and keener with lower debt, fewer employees, higher productivity, better technology and lower costs. Whether bigger or smaller they are better able to compete in an uncertain future against challenging circumstances. As the dust settles on the decade, the change that has been wrought by the will and wit of free enterprise to compete and survive is reflected in Table VIII.

The bottom line is that costs were decreased across-the-board and productivities increased, particularly for U.S. copper and Canadian nickel. Output increased in all cases in Canada but slipped in the U.S. The figures reflect a tremendous rationalization that has left the survivors in a stronger and more competitive position today than they were when the recession hit in 1981.

Table VIII. Trends in North American Copper, Zinc and Nickel Industries in the 1980's

	Canada			United States		
	1981	1987	Change, %	1981	1987	Change, %
Copper						
Cash cost, $/lb.	0.60	0.55	- 8	0.85	0.55	-35
Productivity, h/t	-	-	-	58	29	+50
Production, kt	690	770	+11	1575	1355	-14
World share, %	8	10	+ 2	15	13	- 2
World rank	4	3	+ 1	1	2	- 1
Zinc						
Cash cost, $/lb.	0.36	0.31	-14	0.43	0.38	-12
Production, kt	1095	1500	+37	343	233	-32
World share, %	17	21	+ 4	6	4	- 2
World rank	1	1	-	5	7	- 2
Nickel						
Cash cost	-	-	-33			
Productivity, t/employee	4.4	9.6	+53			
Production, kt	158	200	+27			
World share, %	23	33	+10			
World rank	1	1	-			

Bibliography

Copper

U.S. Industry

(1) Michael A.L. Cook, "Copper Smelting and Refining - the Cold Wind of Competition Begins to Blow," *Metal Bulletin*, International Copper Conference, Phoenix, December 4-7, 1988.

(2) K.E. Porter and Paul R. Thomas, "Competition Among World Copper Producers," *Engineering and Mining Journal*, November 1988, 38-44

(3) Alexander Sutulov, "Is Copper Demand Really Falling?," *Engineering and Mining Journal*, November 1988, 45-46.

(4) "Copper Technology and Competitiveness, Summary," Congress of the United States, Office of Technology Assessment, September 1988.

(5) Janice L. Jolly, "Competitiveness of the U.S. Copper Mining Industry," United States Bureau of Mines (USBM), *Minerals and Materials*, April-May, 1988.

(6) Martha M. Hamilton, "U.S. Copper Industry Is Lesson in Survival," *The Washington Post*, January 17, 1988.

(7) "Copper", USBM, *Bureau of Mines Minerals Yearbook*, 1987.

(8) William H. Dresher, "The Influence of Technology on the Copper Market," *Proceedings of International Symposium Copper 87*, Vol.1, Canadian Institute of Mining and Metallurgy and others, Chile.

(9) Nickolas J. Themelis and Phillip J. Mackey, "Copper Technology 1987 and Prospects for the Future," ibid.

(10) Leonard R. Judd, Remarks, The Metallurgical Society Extractive Metallurgy Luncheon, Denver, Colorado, February 25, 1987.

(11) The Competitiveness of American Metal Mining and Processing, *Congressional Report for the U.S. House of Representatives*, U.S. Government Printing Office, Washington, D.C., No.61-328, July, 1986.

(12) "Copper", USBM, *Bureau of Mines Minerals Yearbook*, 1986.

(13) ibid, 1985.

(14) "Copper", USBM, *Mineral Facts and Problems*, 1985.

U.S. Corporations

(1) "Phelps Dodge Expansion," *Mining Journal*, Vol. 310, No.7969, May 20, 1988, 406

(2) Sandra D. Atchison, "The Outlook for Copper is Bright as a New Penny," (Phelps Dodge) *Business Week*, July 13, 1987, 94.

(3) Carol L. Jordan, "Phelps Dodge Ups Productivity, Returns to Profitability," *American Metal Market*, March 16, 1987, 4.

(4) "B.P. Minerals Completes $400-million Modernization at Bingham Canyon," *Mining Engineering*, November 1988, 1017-1020.

(5) James Cook, "Lean, Mean and Prosperous," (Asarco Inc.), *Forbes*, December 28, 1987, 48.

Canadian Industry

(1) "Canadian Minerals and Metals Industry, Trends and Short-Term Outlook," *Energy Mines and Resources, Canada*, December 1988.

(2) "Annual Metals Review," *The Northern Miner Magazine*, November 1988, 45-59.

(3) Alan G. Green and M. Ann Green, "Productivity and Labour Costs in the Ontario Metal Mining Industry 1975 to 1985: An Update", Mineral Policy Background Paper No. 25, *Ontario Ministry of Northern Development and Mines*, September 1987.

(4) "The Mineral and Metal Policy of the Government of Canada", *Supply and Services Canada*, Cat. No. M37-37/1988.

Canadian Corporations

(1) William Annett, "The Cominco Offensive", *B.C. Business*, July 1988, 80.

(2) Patrick Whiteway, "A Quiet Giant", (Hudson Bay Mining & Smelting), *The Northern Miner Magazine*, November 1988, 30-42.

(3) Warren Rappleyea, "Noranda Upgrading CCR Unit", *American Metal Market*, September 26, 1988, 1.

(4) "The Scoop on Spending", *Canadian Mining Journal*, October 1988, 10-36.

(5) Marilyn Seals, "Output Up, Costs Down," (Kidd Creek), *Canadian Mining Journal*, October 1986.

Zinc

(1) "Zinc", USBM, *Bureau of Mines Minerals Yearbook*, 1987.

(2) ibid, 1986

(3) ibid, 1985.

(4) "Zinc, USBM, *Mineral Facts and Problems*, 1985.

Nickel

(1) Inco, *Annual Reports*, 1980's

(2) Falconbridge, *Annual Reports*, 1980's

(3) "Nickel", USBM, *Bureau of Mines Minerals Yearbook, 1987*

(4) "Nickel", USBM, *Mineral Facts and Problems*, 1985

NON-FERROUS EXTRACTIVE METALLURGY: A TECHNICAL OVERVIEW

John G. Peacey
Noranda Technology Centre

ABSTRACT

Non-ferrous metals processing is generally a mature capital intensive business and, as such, future technological developments are likely to be evolutionary rather than revolutionary. Nevertheless, new processes will still be developed when the need becomes sufficient, as evidenced by the new lead smelting processes developed primarily to meet environmental concerns. The current state-of-the-art in the production of non-ferrous metals will be reviewed, highlighting the most recent developments. The critical factors that will be the driving force for future technological change will be discussed. The most probable directions for future non-ferrous metal production processes will be outlined.

INTRODUCTION

Non-ferrous extractive metallurgy is a mature, capital intensive business and its basic processes have been used for many years. Nevertheless, technological improvements continue to be made in response to changing needs and new external pressures. In the seventies, it was the Energy Crisis and the need to reduce plant emissions. In the eighties, historically low metal prices forced most producers to become more flexible and innovative and to focus on cost competitiveness in order to survive. In the future, price cycles and currency fluctuations will continue and there will be increasing pressure to further reduce emissions and improve in-plant working conditions.

In this paper, I shall present a technical overview of the primary production of the four major non-ferrous metals: aluminium, copper, lead and zinc. I shall review the major trends in technology over the past 20 years and the current state-of-the-art in each of these metals, with particular emphasis on the experience at Noranda's own operations. Finally, I shall highlight promising developing technologies and outline possible directions for the future.

ALUMINUM

All of the world's primary aluminium metal is still produced using the Hall-Heroult process that was invented more than a century ago. Over the last twenty years considerable improvements have been made to increase productivity, reduce power requirements and fluoride emissions.

I shall illustrate the developments in aluminium cell technology with reference to our experience at Noranda Aluminum. In 1971, Noranda Aluminum started its first potline using Kaiser's P69 prebake cell, which was considered, at that time, the best available technology. The P69 cell was designed primarily for high productivity, operating at 150,000 amps and high current density to produce just over 1 tonne Al per pot-day. The P69 cell is a simple design with a centre-break and a long, roughly 60-minute alumina feed cycle. Because of this it uses a high-ratio bath and typically operates at about 88% current efficiency and 15.8 kWh/kg Al. When Noranda Aluminium decided to expand in 1980, the Alcoa 697 cell was selected as it then represented the most proven, state-of-the-art technology. This cell is designed to operate at 180,000 amps and 93% current efficiency, producing about 25% more aluminium per pot-day than the P69 cell using about 15% less power. The 697 cell has a much more sophisticated cell design, with very good balancing of the magnetic field giving a flat metal pad and allowing efficient operation at the higher amperage. Alumina is fed via point feeders that provide an almost continuous feed allowing close control of the alumina level in the bath and the use of low-ratio bath compositions, giving much higher current efficiencies.

Modern, higher amperage aluminium cells are designed almost entirely by computer. This is possible because of the extensive work on mathematical modelling of cell magnetics, dynamics and heat balances. The current state-of-the-art aluminium cell is the Pechiney 280kA cell[1] which is essentially a longer version of the proven 180 kA cell. The performance of the Pechiney 280 kA cell is compared with the Kaiser P69 and Alcoa 697 cells in Table I. It's very high current efficiency of over 95% is a result of Pechiney's excellent alumina feed control system. The 280 kA cell's main advantages are its higher productivity, low labour requirements and reduced capital cost at specific capacities.

It is difficult to foresee further significant improvements in conventional aluminium cell design, except perhaps for further increases in cell amperage. Some companies are already designing 350,000 amp cells[2], but the economic benefits seem marginal unless very large smelters are envisaged.

Apart from the alumina feed, the main costs in aluminium production are electrical energy (~30% of direct costs) and carbon anodes (~10%). The only significant way to lower electrical energy requirement is to reduce the anode-to-cathode voltage drop(ACD). Two future possibilities are:

(a) A wettable cathode design, using titanium diboride, to eliminate the aluminium metal pad and reduce the ACD from 4.5 cm to about 1.5 cm. This offers a potential reduction in energy requirements of about 20%([3]). Titanium diboride cathodes have been under development for many years and in this age of "materials" it is only a matter of time and effort before a practical design is developed.

(b) High lithium electrolytes to increase electrical conductivity. These could reduce energy requirements by about 15% and operate at lower temperatures but lithium make-up must be minimized and a highly-reactive alumina feed would be needed.

Production of carbon anodes is a significant operating and capital cost and also raises concerns regarding in-plant hygiene and long-term future supply. Current carbon consumption is in the range 0.40-0.45 kg/kg Al, close to the theoretical requirement of 0.34 kg/kg Al so there is little scope for significant reduction. Inert anodes, typically based on cermets([4]), have been investigated for many years and cost effective materials and designs may be on the horizon. The main potential benefits of an inert anode would be reduced capital and operating costs, and a sealed cell operation.

The inert anode and stable cathode could also open up the possibility of a multi-polar cell design, as suggested by Jarrett([5]). This Advanced Technology Reduction Cell concept, retrofitted into existing Hall-Heroult cells, could produce up to 5 times more aluminium per cell using 30% less energy than the current state-of-the art cell.

Over the years, considerable effort has been put into developing alternative aluminium production processes but none have proved competitive with the Hall-Heroult process and it is extremely unlikely that a more cost effective process exists. Although the Alcoa Smelting process (ASP) wasn't a commercial success, the ASP multipolar cell, producing aluminium from aluminium chloride, represented an outstanding technical achievement in fused salt technology. Based on operating a commercial-scale demonstration plant, LaCamera([6]) predicted that a single ASP cell was capable of producing over 30 tonnes per day of aluminium at a power consumption of only about 9 kWh/kg Al. This demonstrates the enormous productivity potential of this bipolar fused salt electrolysis cell. Although the ASP cell was not viable for aluminium, it offers exciting future potential in recovering lead, zinc and magnesium from their respective anhydrous chlorides.

The current state-of-the-art in aluminium cell technology appears close to the optimum, only futuristic developments in new anode and cathode materials offer the potential for significant improvements in energy and cost reduction. However, at the present time even such revolutionary improvements have less impact on an aluminium company's competitive position than relocating production to areas with low power costs.

COPPER

Unlike aluminium, copper is produced by many processes. The bulk of the world's primary copper is produced from sulphidic concentrates by matte smelting, converting to anode copper, followed by electrorefining to cathode copper. There are a large number of proven and potential options for copper smelting, most of which have been developed during the last twenty years. During this period, there were several major factors influencing copper smelting:

(a) The need to fix SO_2 emissions, following the lead set by the modern Japanese copper smelters built in the early seventies.

(b) The adoption of tonnage oxygen to increase productivity and decrease energy requirements, while generating a high-strength SO_2 gas for fixation.

(c) The production of higher matte grades to minimize energy requirements and reduce the load on converting.

The major modern copper smelting processes are compared in Table II. There is very little "real" difference in performance between the four processes. All are efficient, productive processes with high Cu recoveries and, for a typical 100,000 tpy "greenfield" copper smelter, each should offer similar capital and operating costs. Selection of a given process is based largely on subjective judgement, site specific factors and who offers the best financial/engineering package.

The Outokumpu flash smelting process is still the dominant copper smelting process. It is well-suited to smelting large tonnages of relatively clean concentrates and can operate with levels of oxygen enrichment close to those required for autogenous operation. It requires a fine, dry feed to ensure proper combustion but can only accept limited amounts of coarser revert-type material.

The Inco flash is unique in using 100% tonnage oxygen for smelting and, as a result, heat balance constraints limit the throughput and matte grade to about 1500 tpd at 55% Cu for typical chalcopyrite concentrates. This allows production of a very high strength SO_2 gas and a "discardable" slag, making the process attractive for modernizing copper smelters with existing converter acid plants.

The main advantage of the Noranda process is its simpler feed system and its flexibility in treating a wide-range of feeds, from high Pb, Zn-containing concentrates to scrap and reverts. At Noranda's Horne smelter, oxygen levels in the tuyeres up to 40% are used and productivity has been increased steadily to well over 2000 tpd, nearly three times the original design capacity. This level of oxygen seems to be close to the practical limit for conventional tuyeres at high blowing rates. The Mitsubishi process uses top-blown lances to inject oxygen and concentrates into the smelting furnace, which is interconnected with the converting and slag cleaning vessels via covered launders. Again, the productivity of the Mitsubishi process has been increased significantly using higher levels of oxygen enrichment, up to 45%, and it has been demonstrated recently that the process can accept some reverts ([7]).

In addition to these processes, new copper smelting processes are still being developed, notably NA's Flame cyclone reactor (FCR), KHD's Contop process and Mount Isa's Copper Isasmelt reactor. The first two are intense flash smelting processes generating very high smelting temperatures in compact reaction shafts, to enhance minor element volatilization. The effect of higher temperature on minor element distribution making a 70%Cu matte is shown in Table III([8]). At 1900 K, lead, zinc, arsenic and antimony are almost completely volatilized but, unfortunately, silver is also volatilized to a significant extent. These processes may find a niche treating "dirty" copper concentrates, if enough suitable feed is available, and also offer an interesting fit with an Imperial Smelting Furnace (ISF). The Contop process has recently been installed on reverberatory furnaces at Palabora([9]) and Chuciquemata to provide low-cost incremental smelting capacity.

The Copper Isasmelt([10]) process uses a single submerged lance in a vertical, cylindrical vessel and its main attraction is its simple construction and, presumably, low cost. All three processes are very interesting in their own right and each may find a specific niche, but the commercial justification for developing more copper smelting processes is sometimes difficult to understand given that the smelting vessel itself is generally a small fraction of the total smelter cost.

Operation at high matte grades means that the resulting slags must be cleaned to achieve acceptable copper recoveries. There are two proven options: a) slow cooling of the slag in ladles followed by milling and flotation to recover a slag concentrate for recycle; and b) reduction of slag in an electric furnace to produce a copper matte for recycle. At the present time, milling gives lower Cu levels in discard slag (~0.3%) and better minor element elimination, but most electric furnace slag cleaners are not operated very efficiently.

In converting copper mattes to blister copper, the Peirce-Smith converter still reigns supreme with little change over the past twenty years, except for better hoods and ventilation. It is a very efficient and flexible process for treating matte, with grades up to 65% Cu, and for smelting scrap and reverts. Its drawbacks are an intermittent, low-strength SO_2 gas stream and fugitive emissions during charging and tapping.

The modern alternatives to the Peirce-Smith converter for converting high-grade mattes are the Mitsubishi converter and Kennecott-Outokumpu's flash converting (SMOC) process[11]. Both processes are continuous, give high SO_2 gas strengths and have a long-enough campaign life to require only a single furnace for most potential applications. The Mitsubishi process uses a lime-slag, that has advantages for antimony and arsenic removal but not for lead. The flash converting process requires a finely-divided dry matte feed that can be prepared by granulation or solidification, crushing and grinding. A novel option recently evaluated by ER&S in Australia is leaching high-grade matte from a Noranda reactor followed by electrowinning[12]. Capital and operating costs were found to be slightly higher than converting and electrorefining. Although this option was not pursued, it could be attractive under different circumstances.

Direct production of copper from chalcopyrite concentrates in a single vessel is still a dream. This was achieved during the early operation of the Noranda process at the Horne smelter but it was discontinued for several reasons:

- High levels of impurities in reactor copper;
- Short tuyere line life;
- High recirculating load of copper from slag cleaning; and
- Higher productivity and lower costs making high-grade matte.

Since then, Outokumpu flash furnaces have been built to produce copper directly from low-iron concentrates. Mackinen[13] has also shown that direct production of high-sulphur copper has the lowest potential cost for a greenfield smelter. In the future, there will be increasing pressure to reduce in-plant fugitive emissions by eliminating ladle transfers of molten matte to converters. Continuous converting processes offer one potential solution but the most elegant one would still be direct copper production.

The future elimination of the Peirce-Smith converter will create a need to smelt scrap elsewhere. One option recently developed by Kennecott and Linde is the 'Smart' tuyere, a flexible triple tuyere allowing injection of different ratios of oxygen, natural gas and nitrogen into an anode furnace to either generate heat and melt scrap and/or oxidizing/reducing conditions to refine copper. If scrap melting is a major part of the business, the Contimelt process offers another option.

In the past, copper smelting developments have often closely-followed those in steelmaking. Ladle refining is a recent trend in steelmaking that could find application in copper smelters. Do we really need large poorly-mixed 300 ton rotary anode furnace with long holding and refining times? A ladle refining station would allow more flexibility and could reduce costs.

Having questioned the need for developing new copper smelting processes, I would like to suggest a couple more that are technically attractive. First is direct copper smelting with a lime flux, proposed by Inco[14], which may be especially attractive for high nickel and zinc concentrates. The second is to use the Pb Kivcet-concept using highly oxidizing conditions in the reaction shaft to oxidize most of the copper to oxide and then reduce it in a coke-layer in the smelting zone to lower the copper load in slag to more reasonable levels for subsequent cleaning.

Hydrometallurgical copper processes to replace smelting for sulphidic concentrates were the vogue in the early seventies but none were able to operate viably and most development efforts have been abandoned. It is now obvious that hydro processes cannot compete with large, modern copper smelters and refineries because of several drawbacks, including: sensitivity to concentrate feed; higher energy requirements; iron residue disposal and uncertainties about precious metal recoveries. However, hydrometallurgy has found its niche with oxide copper ores. Due to the development of better solvent extraction reagents and improvements in electrowinning technology, leach-SX-EW plants have sprung up in the Southern US, Zambia and Chile producing good-quality electrowon copper for well under 40¢US/lb.Cu.

Today's copper electrorefining process is basically the same as that practiced nearly a century ago apart from increased mechanization. The difference between a modern tankhouse and an old one are compared in Table IV. The key to high current efficiency operation and good cathode quality is precise anode-cathode geometry in the cell. Copper anodes must be of constant weight and thickness and hang vertically when placed in the cell. This can be achieved with conventional casting wheels using automatic, constant-weight spoons and good mould maintenance. The Rolls-Royce of anode casting systems is the Contilanode system, developed by MHO, and now performing well at White Pine and BICC. It is an expensive system but it can produce anodes with a high geometric consistency.

Traditionally, copper starter sheets are used as cathodes. Modern Japanese tankhouses, produce good quality embossed starter sheets and take great care in aligning electrodes in the cell to ensure accurate spacing to minimize shorts and obtain high current efficiency. Mt. Isa has introduced the permanent stainless steel cathode blank, which eliminates the production of starter sheets and gives a more geometrically consistent cathode but, again, at a very high cost, especially at current nickel prices. Nevertheless, most recent copper refinery projects have selected the Mt. Isa system, confirming it as the current state-of-the art technology.

Virtually, all copper is now sold as high-grade copper cathode with much stricter impurity specifications than previously applied to wirebars. Impurity control in copper smelters and refineries has, therefore, become more critical, especially for those treating custom material. The critical impurities are generally As, Bi, Sb, Pb and Se. Various techniques can be employed to control impurities in anode copper to satisfy refinery requirements, including: matte grade changes; dust bleed usually to a lead smelter, converter slag milling; overblowing and anode furnace fluxing. At the refinery, apart from anode composition, the important factors in making good quality cathodes are: good anode-cathode geometry, clean electrolyte, appropriate addition agents to produce good cathode morphology, and consistent electrolysis conditions with minimum shorts. At CCR, Noranda's copper refinery in Montreal, impurity levels in anodes have increased substantially over the last few years but, despite this, cathode quality has been improved significantly to meet the more demanding needs of today's market.

The current state-of-the art copper refinery would most probably comprise a Contilanode anode casting system, the Mt. Isa stainless steel blank and operation at a high current density (300-350 A/m^2) using periodic current reversal. Despite its high-technology components, this system still has its drawbacks, particularly its low productivity and large metal tie-up. Recent work by Winand[15] has shown that copper electrorefining is feasible at very high current densities, up to 2000 A/m^2, using high electrolyte velocities. This may be the direction for the copper refinery of the future.

LEAD

Lead smelting is undergoing a fundamental technical change from the traditional sinter-blast furnace process to the modern direct smelting processes. Three direct lead smelting processes: Kivcet, QSL and Outokumpu have been developed to meet stricter environmental regulations and reduce processing costs. Normally, new processes are developed by those with the greatest need, the primary producers, but these have been developed largely by engineering companies. I have been fortunate to have been closely involved in evaluating these processes for a future modernization of Brunswick Mining & Smelting's lead smelter in New Brunswick and I will discuss them from this viewpoint.

Brunswick's lead concentrate is unique in being low grade (34% Pb) and high in pyrite (22% Fe) and thus it presents a special challenge because of its high sulphur content and large slag make. The metallurgical performance of each process and its advantages/disadvantages on Brunswick concentrate are compared in Table V.

The Kivcet and Outokumpu processes both employ flash smelting followed by electric furnace slag reduction, but the philosophy and engineering of each process is very different. The Kivcet process uses a higher oxygen to concentrate ratio in the reaction shaft to ensure that all the lead is oxidized to PbO in the slag. The Outokumpu process uses a lower oxidation potential in the reaction shaft in order to produce part of the lead directly as bullion. As a result, the Kivcet process produces less lead fume and a lower sulphur bullion at the expense of higher oxygen requirements, and high PbO levels in slag. To reduce the load on the electric furnace, the Kivcet process maintains a coke layer in the smelting furnace that effectively reduces most of the PbO in slag before it enters the electric furnace. This considerably reduces the size of the electric furnace since the reduction of high-PbO slag droplets in a coke bed is much faster than continuous reduction by a floating coke layer in an electric furnace. To reduce the size of its electric furnace, Outokumpu uses a high power density and injects the coal reductant through lances. This ensures very good mixing in the furnace and increases the PbO reduction rate by a factor of 3 to 5 versus floating coke. Both processes tap bullion from the electric furnace.

The QSL process smelts as-received concentrates using Lee-Savard, high-pressure oxygen injectors shrouded with nitrogen to produce a high-PbO slag and some bullion directly. The slag underflows a refractory wall into the reduction zone, where it is reduced using coal and oxygen injected through special injectors. The use of an underflow wall, that physically separates the oxidation and reduction zones, was one of the keys to the successful development of the QSL process. It also allows tapping bullion from either the oxidation or the reduction zone. Bullion tapped from the oxidation zone is lower in sulphur, arsenic and antimony and this opens up the potential to eliminate more of the arsenic in slag. The QSL process has the highest lead fume make smelting Brunswick concentrates because of the high PbS activity in the oxidation zone slag. Despite the high fume make, no problems were experienced during pilot plant operations because of the vertical boiler uptake which ensured the off-gas was fully sulphated and cooled to below 700°C before entering the convection part of the boiler. This vertical uptake boiler design is also used successfully on the commercial Kivcet process in Sardinia.

Lead losses in slag are critical to Brunswick and all three processes have demonstrated that lead in discard slag can be reduced to the range 2 to 5% Pb, lower than that achieved in the present blast furnace operation. At this point in time, only the Samim Kivcet plant is operating commercially[16]. It is a very impressive operation and it has demonstrated that with good engineering design all the advantages of direct lead smelting (very good in plant hygiene, high-strength SO_2 gas, higher Pb recovery, and lower operating costs) can be realized. Everyone is now eagerly awaiting the start-up of Cominco's QSL plant, scheduled for late in 1989, to find out how it will perform commercially.

Copper removal from bullion is a concern for most lead smelters. Traditional kettle drossing followed by dross smelting with soda ash and coke is costly, dirty and produces a copper matte that nets a poor realization from copper smelters. Several companies have developed dross leaching processes to increase revenues by producing either copper sulphate or electrowon copper with a lead sulphate residue for recycle. Another approach is the continuous copper drossing furnace, developed by BHAS, but this is limited to bullions with Cu:S ratios less than 4 and/or low arsenic levels to avoid speiss problems. Also it still produces a high Pb-copper matte, with its correspondingly low value. There is a need for a better process to recover copper values from lead bullion. Mt. Isa have adapted the Isasmelt process to smelt copper dross in two stages, which potentially could produce a high-copper, low-lead matte. A novel idea is the direct electrorefining of copper-containing bullion, this has been tested at the lab-scale producing a good-quality refined lead and a copper-containing slime, which could then be treated by conventional copper slimes treatment methods.

After copper drossing, crude bullion is processed into refined lead either pyrometallurgically or by electrorefining. Most pyro-refineries use the traditional batch kettle refining process. Over the years BHAS in Australia has developed and now employs continuous pyro-refining processes for most steps, including sulphur drossing, softening, de-silverizing and de-bismuthizing. These continuous refining operations are most applicable to high-tonnage operations. Modern lead electrorefineries, such as those in Japan, only require about 150 kWh/tonne refined lead and have significant advantages over pyro-refineries including: a higher-purity lead product, improved working environment, better potential for by-product recovery, and lower operating costs. Future lead refinery projects are likely to use electrorefining, and the most attractive technology is the jumbo bipolar refining process successfully developed by Cominco[17].

Hydrometallurgical lead processes offer an attractive alternative for the future, if lead emissions and in-plant hygiene regulations are toughened even further. These processes, which have been successfully piloted, use chloride leaching followed by solution purification to produce a pure lead chloride solution for electrowinning or anhydrous lead chloride for fused salt electrolysis[18].

ZINC

Over 80% of the world's primary zinc metal production is produced by the roast-leach-electrowinning route. This process has seen many improvements over the past twenty years, and it has evolved into a very efficient process with high zinc recoveries and producing high-quality zinc metal. However, the process is still sensitive to impurities, requires considerable electric energy (~3300 kWh/tZn), has low by-product recoveries and revenues, and is facing growing problems with iron residue disposal.

Fluid bed roasters for zinc have increased in area and productivity. The largest units to-date have grate areas of about 120 m^2 and can process up to 800 tpd of concentrate, but they are sensitive to impurities, such as Pb, Cu, and pyrite, that cause agglomeration in the bed. MHO in Belgium has operated a pelletizing-fluid bed roasting process for many years, that can handle zinc feeds with very high Pb, Cu and chloride levels, and this type of operation could be applied to lower-grade zinc concentrates in the future.

Zinc leaching technology has advanced from batch neutral leach processes with relatively poor zinc extraction to continuous processes with over 98% zinc extraction. The big breakthrough was the development of the Jarosite process in the late sixties. Noranda's Canadian Electrolytic Zinc (CEZ) plant, which started up in 1963, converted to the batch ammonium jarosite process in 1971 increasing zinc extraction from 88% to 93%[19]. Later, in 1976, the leach was modified to use Outokumpu's continuous conversion process to further increase zinc extraction to over 98%. The main drawback to the jarosite process is the large volume of residue for ponding. Other iron precipitation processes, used commercially, are the Goethite process, developed by Vieille Montagne, and the hematite process used by Akita and Ruhr-Zinc. Goethite residues occupy 40% less volume than jarosite but still contain appreciable zinc. The hematite process requires autoclaves operating at 180°C with oxygen to precipitate iron as hematite. After drying, the hematite residue contains about 60% Fe and about 1% Zn and can be used in the cement and tile industry. However, the zinc level is still too high for use in ironmaking. The hematite process is also considerably more expensive to install and operate than the jarosite and goethite processes, but it currently offers the best method for iron disposal.

There are currently renewed efforts at developing other methods for iron disposal including: smelting of the neutral leach residue in a Contop or Flame cyclone reactor to recover a Zn-Pb-Ag fume and a disposable iron-containing slag, and solvent extraction of iron, with recovery of an iron oxide powder by spray roasting of the ferric chloride strip solution. All these processes, like the hematite process, involve major additional costs which can only be justified as a last resort to keep an operation from closing down because of iron disposal problems.

Direct pressure leaching of zinc sulphide concentrates, using the Sherritt Zinc process, was successfully commercialized in late 1980 at Cominco's Trail smelter[20]. This process has much lower capital costs than a roast-leach facility and fixes most of the sulphur as elemental sulphur. This technology is very attractive for treating high-pyrite zinc concentrates and for adding low cost incremental capacity to an existing electrolytic zinc plant. A grass-roots zinc plant using only pressure-leach technology has not yet been built, but may be realized in the future modernization of Hudson Bay's zinc plant in Manitoba.

In zinc electrowinning plants, the major improvements have been in increased mechanization of the tankhouse and the use of larger electrodes, Table VI. Vieille Montagne's Super Jumbo tankhouse[21] represents the current state-of-the-art in zinc electrowinning. This is an extremely productive and cost efficient technology and has recently been chosen by CEZ to replace its two existing cellhouses. Despite the enormous increases in zinc tankhouse productivity, little has been achieved commercially in reducing the high energy requirement. Several options to reduce the anode overvoltage have been investigated and of these the hydrogen depolarized anode appears to have the lowest energy requirement at about 1300 kWh/t Zn. It is being tested at Ruhr-Zinc[22] but it will only be attractive for locations where the cost of natural gas or hydrogen is significantly lower than the cost of electricity.

If electrowinning zinc from a sulphate electrolyte is to continue to be the dominant process for producing zinc in the future it must resolve its iron disposal problem, recover by-products more efficiently and increase its flexibility to treat secondary zinc feeds. An interesting option to replace roasting and iron precipitation would be high-temperature oxygen flash smelting of zinc concentrates and residues combined with slag fuming to volatilize most of the zinc and lead into a fume, recover copper and PM's as a matte and fix the iron as a slag. The zinc fume would be leached and halogens removed to acceptable levels by electrodialysis, as recently described by Boateng[23]. Since coatings are a major market for zinc, another interesting idea is to use the zinc fume directly for electrogalvanizing steel, thus eliminating the need for conventional electrowinning[24].

The major pyrometallurgical zinc process is the Imperial Smelting Furnace (ISF). This process has survived because it has found a niche for treating Zn-Pb bulk concentrates and secondary zinc feeds. Berzelius has been a leader in exploiting secondary feeds and in improving the viability of the ISF, developing the hot briquetting process and using higher blast preheat temperatures. The main drawbacks of the ISF are still: its sinter plant, its high coke requirement (about 1 tonne per tonne zinc produced), and its zinc metal quality. Although it is a remarkable process with the ability to recover zinc, lead, copper and precious metals. It is unlikely that future ISF plants will be built. An attractive alternative for treating Pb-Zn concentrates would be the Kivcet process with zinc recovery from the electric furnace gases via a lead splash condenser. The Kivcet electric furnace in Sardinia is very tight, with little air inleakage, and should be capable of producing high metallic zinc loadings suitable for zinc metal recovery.

There is still considerable interest in new pyrometallurgical zinc processes, especially in Australia and Japan. The CRA approach[25] is to oxygen smelt zinc concentrates to form a high zinc slag (up to 35% Zn) that is then reduced in a second stage by injecting coal and recovering the volatilized zinc in a lead-splash condenser. The Japanese process[26], which has been piloted in a 10 tpd Zn pilot plant, injects zinc calcine and coke with oxygen via lances into a slag bath and, again, recovers the volatilized zinc in a lead splash condenser. The key to all of these processes is to minimize dust carry-over to obtain efficient recovery of zinc metal in the lead splash condenser. The Japanese pilot plant obtained 74% zinc recovery and is projecting 85% for a commercial operation versus 92% currently achieved in commercial ISF plants. It was estimated that the injection smelting process would be 20% lower in capital cost and 30% lower in operating costs than a greenfield electrolytic zinc plant in Japan, where power costs are extremely high.

St. Joe is also evaluating zinc metal recovery from zinc calcine using its Flame Reactor Process[27], developed primarily for recovering zinc oxide values from electric arc furnace dust. Direct smelting of zinc concentrates with generation of zinc vapour in an SO_2-containing gas stream has been shown to be thermodynamically possible. The key question is can zinc metal be recovered efficiently from an SO_2 gas stream in a lead splash condenser? If it can, it makes high-temperature oxygen-flash smelting of zinc concentrates in a Flame-cyclone type reactor a possibility, which would be a most attractive process.

The main drawback to all pyrometallurgical zinc processes is still the zinc metal quality. Today's market is demanding purer zinc and, at some time in the not too distant future, all ISF zinc may have to be refined which will add greatly to costs. For a pyro zinc process to be successful a more efficient zinc refining process is needed, such as the continuous vacuum dezincing process propounded by Davey.

Zinc usually represents the main metal value in complex Zn-Cu-Pb-PM orebodies. Considerable work has been done on developing processes to treat bulk Zn-Cu-Pb-PM concentrates to increase metal recovery and revenues from complex orebodies compared with producing conventional selective concentrates. To date, none of these processes has been developed commercially, largely because an orebody complex enough and/or rich enough in metal values to justify a greenfield facility has not been found. In my opinion, it is more likely that, in the future, existing copper, lead and zinc smelters will continue to enhance their abilities to treat and recover larger quantities of by-product metals, making the economic justification for complex concentrate treatment processes even more difficult.

CONCLUDING REMARKS

I have attempted to give a personal overview of the developments over the past twenty years and current state-of-the-art in the extractive metallurgy of aluminium, copper, lead and zinc. Today's best technologies are highly productive and efficient. The potential for significant process improvements at this time appears relatively small, especially in view of the high cost of development. However, we will continue to see major improvements in process control and automation. Tomorrow's challenges will be to meet even stricter demands on emissions, in-plant working conditions and metal quality while keeping costs competitive with alternative materials. I am confident that extractive metallurgists will continue to demonstrate the ingenuity needed to enable the non-ferrous metals industry to successfully meet these challenges.

ACKNOWLEDGEMENTS

The author would like to thank Noranda Inc. for permission to publish this paper and his colleagues in Noranda's metallurgical operations and the Noranda Technology Centre for their assistance.

REFERENCES

1. Langon, B., and Varin, P., Aluminium Pechiney 280 kA Pots, Light Metals 1986, p. 343.

2. Abomelt, C., Echeverria, R., and Guzman, J., Development of the V-350 Venalum Cell, Light Metals 1989, p. 353.

3. Dorward, R.C., Energy Consumption of Aluminium Smelting Cells Containing Solid Wetted Cathodes, Journal of Applied Electrochemistry, 13 (1983), p. 569.

4. Baker, F.W., and Rolf, R.L., Hall Cell Operation with Inert Anodes, Light Metals 1986, p. 275.

5. Jarrett, N., United States Extractive Metallurgy - The 80's and Beyond, Metallurgical Transactions, 18B, 1987, p. 289.

6. LaCamera, A.F., Fluid Dynamics, A Key Factor in the Design of a Bipolar Electrochemical Cell for Aluminum Production, in Mathematical Modelling of Materials Processing Operations, edited by J. Szehely, AIME 1987, p. 671.

7. Shibasaki, T., Kanamori, K., and Ramio, S., Mitsubishi Process - Prospects to the Future and Adaptability to Varying Conditions in Metallurgical Processes for the Year 2000 and Beyond, ed. by H.Y. Sohn, AIME, 1989, p. 253.

8. Barin, I., Thermodynamics and Kinetics of Cyclone Smelting of Metal Sulphides, in Advances in Sulphide Smelting, AIME 1983, p. 257.

9. Smith, A., et al, Incremental Increase of Reverberatory Furnace Smelting Capacity at Palabora Using Contop Technology, TMS Paper No. A89-1, 1989.

10. Pritchard, J., Plazer, R., and Errington, W., Isasmelt Technology for Copper Smelting, TMS Paper No. A89-5, 1989.

11. Asteljoki, J., and Kyto, S., Minor Element Behaviour in Flash Converting, TMS Paper A86-57, 1986.

12. Cooper, P., Technologies for Low Cost Retrofitting of the ER & S Copper Smelter, op. cit., 7, p. 273.

13. Makinen, Y., et al, The Economics of Matte, White Metal and Blister Production in a Flash Smelter, paper presented at the First International SME-AIME Fall Meeting, Hawaii, September 1982.

14. Victorovich, G., et al, Direct Production of Copper, Journal of Metals, September 1987, p. 42.

15. Winand, R., and Harlet, P., High Current Density Copper Refining, in the Electrorefining and Winning of Copper, TMS, 1987, p. 239.

16. Perillo, A., et al, The Kivcet Lead Smelter at Portovesme - Commissioning and Operating Results, paper presented at the 111th AIME Annual Meeting, Pheonix, 1988.

17. Kerby, R., and Williams, R., The Cominco Bipolar Process for Lead Electrorefining, TMS Paper No. A84-15, 1984.

18. James, S., and Bounds, C., High Purity Lead From the Ferric Chloride Leaching of Complex Sulphides, in Complex Sulphides, ed. by A. Zunkel, AIME 1985, p. 441.

19. Rodier, D., Zinc Hydrometallurgical Practice at Canadian Electrolytic Zinc Ltd., Valleyfield, Canada. Paper presented at the 11th Annual C.I.M. Hydrometallurgical Meeting, Niagara Falls, Canada, October 1981.

20. Parker, E.G., McKay, D.R., and Salomon-de-Freidberg, H., Zinc Pressure Leaching at Cominco's Trail Operation, in Hydrometallurgy-Research, Development and Plant Practice, ed. by Osseo-Asare AIME, 1983, p. 927.

21. Ries, G., Four Years Operation at VM's Cell House in Balen, TMS Paper No. A84-13, 1984.

22. Juda, W., Ilan, A.B., Mager, K., and von Roepenack, A., Platinum Requirements in Zinc Electrowinning with Hydrogen Anodes, in Precious Metals 1986, ed. by U.V. Rao, IPMI 1986, p. 347.

23. Boateng, D.A.D., Removal of Chloride, Fluoride and Other Monovalent Ions from Zinc Sulphate Electrolyte by Electrodialysis, TMS Paper No. A89-18, 1989.

24. Srinivasan, V., et al, Evaluation of Electrogalvanized Deposits from Zinc Wastes, in Zinc '85, ed. by K. Tozawa, M.M.I.J., 1985, p. 749.

25. U.S. Patent No. 4,741,770, (to CRA Services Ltd., Melbourne, Australia) Zinc Smelting Process Using Oxidation Zone and Reduction Zone, 3 May 1988.

26. Goto, S., Nishikawa, M., and Fujikawa, M., Semi-Pilot Plant Test for Injection Smelting of Zinc Calcine, op. cit., 7, p. 301.

27. Pusateri, J.F., Bounds, C.O., and Lherbier, L.W., Zinc Recovery via the Flame Reactor Process, Journal of Metals, August 1988, p. 31.

Table I

Comparison of Aluminum Cell Performance

Parameter	Cell Type		
	Kaiser P-69	Alcoa 697	Pechiney 280kA
Design, amperage, kA	150	180	280
Cell voltage, V	4.65	4.20	4.15
Current density, A/cm^2	1.0	0.72	0.74
Bath ratio	1.4-1.5	1.1-1.2	1.1-1.2
Bath temperature, $^\circ$C	965	945	950
Alumina feeding	Centrebreak 60 min cycle	Point-feeders	Point-feeders
Current efficiency, %	88	93	95
Aluminum production, kg/pot-day	1060	1350	2150
Power requirement, kWh/kgAl	15.8	13.5	12.9
Pot life, yrs	4	5	6+

Table II

Comparison of the Major Copper Smelting Processes

Parameter	Outokumpu	Inco	Noranda	Mitsubishi
No. units installed	>20	3	4	2
Smelter type	Flash	Flash	Bath-tuyeres	Bath-lances
Feed type	Dry, fine	Dry, fine	As-received, coarse	Dry, fine
Maximum unit capacity, tpd conc.	up to 3000	~1500	>2000	~1500
Matte grade, % Cu	50-70	~55	55-75	65-75
Fuel type	Oil/gas	-	Coal/gas	Oil/gas
Fuel requirement, MBtu/t conc.	<0.5	-	0.5-1.0	<1.0
Slag cleaning unit	Mill or EF	-	Mill	EF
Cu in discard slag, %	0.3-0.6	0.6-1.0	0.3	0.6-1.0
Typical off-gas strength, %SO_2	15-50	~80	10-20	15-30
Typical campaign life, years	3-5	1-3	~1	~1

Table III

Theoretical Effect of Temperature on Impurity Distribution[8]

Impurity	Temperature	Impurity Distribution, %		
		Matte	Slag	Gas
Pb	1600 K	3	62	35
	1900 K	1	9	90
Zn	1600 K	10	70	20
	1900 K	2	8	90
Sb	1600 K	15	80	5
	1900 K	<5	10	85
As	1600 K	1	3	96
	1900 K	<1	-	>99
Ag	1600 K	99	1	0
	1900 K	79	1	20

Table IV

Comparison of Old Versus Modern Copper Refinery

	Old	Modern
Anode casting	Manual	Automatic, constant weight
Anode preparation	None	Lug milling Anode pressing Auto spacing
Starter sheets	Copper	Stainless steel
Current efficiency, %	~90	~95
Power, kWh/t Cu	350	320
Steam, t/t Cu	1.0	0.4
Anode scrap, %	20+	15
Productivity, man-hours/t Cu	2.4	1.0

Table V

Comparison of Direct Lead Smelting Processes for Brunswick Concentrate

Parameter	Outokumpu	QSL	Kivcet
Process type	Flash	Bath-injectors	Flash
Feed type	Dry, fine	As-received	Dry, fine
O_2 in smelting zone, %	60	85	95
O_2: concentrate ratio, Nm^3/t	250	280	350
Pb in primary slag, %	~20	30-35	~35
Pb fume make, t/t concentrate	0.4	0.6	~0.2
Off-gas strength, %SO_2	40	30	45
Slag reduction mode	Electric furnace with coal injection	Coal injection	Coke layer in smelting zone + electric furnace with floating coke
Reductant requirements, t/t concentrate	0.04	0.10	0.07
Pb in discard slag, %	2-4	3-5	~2
Bullion tapped from	EF	Smelting zone	EF
Composition, %S	~0.5	~0.5	0.1
%As	1.0	<0.2	1.0

Table VI

Evolution of Zinc Tankhouse Performance

Parameter	CEZ	Ruhr Zinc	VM Jumbo	Cominco	VM Super Jumbo
Cathode area, m^2	1.6	2.0	2.6	3.0	3.2
Cathodes per cell	44	40	44	50	100
Total area per cell, m^2	70	80	114	150	320
Current density, A/m^2	510	450	410	400	475
Cell productivity, tZn/day	1.1	1.0	1.2	1.6	4.0

A RESEARCH OVERVIEW ON NON-FERROUS EXTRACTIVE METALLURGY

R. Winand

Dir. du Service Metallurgie-Electrochimie
Faculté des Sciences Appliquées
Université Libre de Bruxelles, Belgium

Abstract

Instead of reviewing systematically what was done during the past decade in non-ferrous metals extractive metallurgy research metal by metal, the author will concentrate on main trends in a general perspective including economic factors.

Topics like raw materials availability, energy consumption, productivity, pyro- versus hydrometallurgy, process control, mathematical modelling, merging "new" metals and materials, will be emphasized. Where possible, the research needs will be pointed out for the future.

THE ROLE OF MATHEMATICAL MODELS IN NEW PROCESS DEVELOPMENT

Julian Szekely

Department of Materials Science and Engineering
Massachusetts Institute of Technology
Cambridge, MA 02139, USA

1. INTRODUCTION

In recent years there has been a major growth in interest in the use of mathematical models by the materials process industries. This interest is due to several factors, which will be briefly listed below:

- the dramatic improvement in the capabilities of computers coupled to the drastic reduction in the cost/benefit ratios and confident projections for at least comparable developments in the future.

- the development of "user friendly" software packages, which can greatly minimize the drudgery associated with computer program development.

- perhaps much more important is the growing computer literacy of the young graduates who enter the workforce and the rapidly accumulating experience with computer usage.

In the way of introduction it may be helpful to define what we mean by mathematical models and trace how the use of mathematical models or in general "theory" has evolved in new process development.

What is a mathematical model? (1,2,3)

Within the context of metals and materials processing a mathematical model may be defined as a set of differential or algebraic equations, which provide a quantitative description of a plant, a process or some aspect of the process. By quantitative description we mean a relationship between independent variables, such as system size, reactant feed rate, heating or cooling rate, power input and the like and dependent variables such as product yield and quality, temperature and velocity profiles, stress distribution and the like.

There are many types of models, including empirical relationships, equations deduced from statistical considerations and dynamic simulation models.

The present discussion will be restricted to <u>mechanistic models</u>, that is relationships which are based on physical and chemical laws, such as thermodynamic equilibria, Maxwell's equations of electrodynamics, the Navier-Stokes equations for fluid flow and the like.

<u>Brief History of Mathematical Modelling in the Materials Processing Field</u>

The history of modelling is perhaps best traced by examining the papers presented at a series of conferences devoted to this topic (4,5,6,7,8,9). It is of interest to examine these within three sets of contexts:

(a) topic

(b) methodology

(c) relationship to practice

(a) <u>topic</u>

In the late sixties heat flow was the principal topic, much reliance was placed on classical analytical techniques, since computer usage was in its infancy. At that time physical modelling (e.g. the use of water models and even electric analogues) was the preferred way of tackling many processing problems. By the early 1980's computer usage had become much more widely accepted and computational fluid mechanics was becoming a standard tool in the modeller's armory.

At recent meetings in the late 1980's (8,9) the increased sophistication of mathematical modelling efforts became clearly apparent. Not only did the problems studied include complex three-dimensional fluid flow and electromagnetic phenomena, but totally new subjects have also emerged, such as molecular dynamics and the use of artificial intelligence.

(b) <u>methodology</u>

The early modellers had to labor under quite serious handicaps, because of the very limited documented experience in the modelling field. Indeed it was only recently that the first textbook appeared, which was devoted to the modelling of metallurgical processes. The initial models were "hand built" from scratch, so they were either quite simple or took a very long time to develop. In more recent times it has become accepted that use must be made of computational subroutines and "packages" as part of the modelling sequence. This together with the fact that one can rely on modelling precendents has made model building a much speedier and much more cost effective exercise.

(c) <u>relationship to practice</u>

The most dramatic change in the posture of modelling has been in its relationship to practice. In the early days modelling was regarded as an esoteric activity that had only a marginal bearing on practice. Furthermore, in most cases the modellers were concerned with representing well established processes, such as ironmaking in the blast furnace, steelmaking kinetics and the like, and were often accused of "being able to explain what the practitioners already knew". At the present time there is a much closer relationship between the modellers and the operators; nonetheless much of the work is still centered on seeking an understanding of existing operations, such as the Hall Cell in aluminum production, ladle metallurgy systems and continuous casting in steel processing practice. In some of the more recent developments modelling is actually being used to tackle real process problems, such as the cracking of continuously cast blooms, the optimal design mold taper and the like.

More exciting is the new trend of using modelling as a complementary approach to experimental work, as is the case in near net shape casting (10) and the direct ironmaking technologies (11). An even more ambitious step in this direction would entail the use of mathematical models as the starting point in new process development. In the following we shall discuss some successes and failures of mathematical modelling in the light of current experience and then will illustrate the use and potential of mathematical models for new process development. Then we shall conclude with a brief listing of the key tasks ahead and comment on the types of problems modellers may be addressing in the next decade.

2. SOME MODELLING SUCCESSES AND FAILURES

Successful efforts in the modelling field may be divided into the following three categories:

2.1 Quantitative description of processes and their optimization

2.2 Qualitative process description

2.3 Evolution of methodology

2.1 Quantitative Process description

When we are dealing with purely physical processes, where the constitutive relationships are clearly established, such as conductive heat transfer or electromagnetic phenomena, exact relationships may be developed between the key process parameters. Indeed, some of the most spectacular successes of mathematical modelling to date may be found in this field.

Continuous casting of steel, sketched in Fig. 1, is one of the clearly dominant technologies in the production of billets, blooms or slabs. In all fairness this technology evolved without much help from modellers in the first instance. However, subsequently a great deal of useful modelling work has been done to describe heat transfer, solidification (12,13) and more recently crack formation and metal shrinkage upon solidification (14,15,16).

These problems are almost ideal subjects for the modeller, because the constitutive relationships (i.e. the transient heat conduction equation) are well established and the only areas of uncertainty are associated with the boundary conditions at the metal-mold interface and to a lesser extent, the relationship between stress and strain in the solidified shell.

In a series of elegant papers Brimacombe and Semarasekara addressed the issues associated with excessive thermal stresses and crack formation and provided guidelines for the optimal design of secondary cooling arrangements (14,15,16).

Of perhaps even greater interest is their recent work, which addresses the details of heat transfer in the mold and hence the definition of the optimal mold taper. Fig. 2, taken from a recent paper (17), shows a sketch of the series of resistances to heat transfer, including, from left to right, the convective heat transfer coefficient due to the cooling water, the mold wall, the air gap and the solidified shell. The practical point is that as the shell is formed, shrinkage will occur and thus an air gap is established. If one can partially compensate for this shrinkage by adjusting the taper of the mold wall, better heat transfer efficiency may be attained.

However, the shell formation will be markedly affected by the establishment of the air gap and vice versa. Using very precise mold heat flux measurements, such as depicted in Fig. 3, the authors were able to develop guidelines for optimal mold design. If the mold dimensions are not properly designed, the solid product may suffer significant distortion, as illustrated in Fig. 4.

This example is an excellent illustration of how mathematical modelling can provide quantitative answers to critical process problems. It also shows that actual in-plant experiments are often a key step in the refinement and implementation of a modelling effort. Last but not least, the experience gained in representing heat transfer phenomena in the molds of conventional continuous casters, can be extremely helpful in assessing the projected performance of novel continuous casting systems.

Many other success stories may be cited from the area of heat transfer; as one other example, through the intelligent use of mathematical modelling the design of reheating an annealing furnace can now be undertaken with a good degree of precision. Indeed one should state that mathematical modelling should be the logical choice in the rational design of heat transfer equipment in the metals processing field.

2.2 Qualitative Process Description

A common misconception among those not clearly associated with mathematical models is that a fully quantitative process description is the necessary or even desired output from a mathematical model. Far from it! As elegantly described by Evans of a recent symposium (18), "insight, rather than numbers" should be the desired objective in many cases.

Often the system may be too complex to be modelled exactly. The fully rigorous modelling of a Hall Cell, sketched in Fig. 5, or even that of an argon stirred ladle, sketched in Fig. 6, are still beyond our current capabilities.

As discussed in numerous recent publications (19,20,21), in the Hall Cell, where in essence aluminum is being produced by the electrolytic decomposition of alumina, we are concerned with a complex, three-dimensional circulation system, which involves three phases, a molten metal layer, a molten cryolite bath, in which the alumina is dissolved and carbon monoxide gas, which evolves at the electrode surface and is passed through the bath in the form of bubbles. The flow is driven by electromagnetic forces and also by the gas bubbles; this circulation plays a key role in distributing the reactant alumina and also in affecting the melt to wall heat transfer in the system. Since the reduction of the anode-cathode distance plays a key role in determining the overall energy efficiency of the system, the stability of the bath-melt interface is a parameter of major practical interest.

In spite of the massive research efforts that have been devoted to this subject by the various major aluminum companies and university-based researchers, a fully comprehensive representation of all the key phenomena in Hall Cells has yet to be achieved and possibly never will be accomplished. However, through the combination of computational and analytical tools very useful insights have been obtained into Hall Cell behavior, regarding the factors that govern bath circulation and perhaps more importantly the shape and stability of the bath-metal interface. As an example, Fig. 7 shows a typical computed bath-metal interface, which is far from being flat, while Fig, 8 shows the response of the interface height to an abrupt change in the operating conditions, such as an anode replacement.

Turning to ladle metallurgy, the secondary processing of molten steel, after the completion of the primary steelmaking step in an electric or basic oxygen furnace has become a very popular topic, because of the highly cost-effective nature of this operation. It is now an accepted fact that in the production of most steel grades, some form of a ladle treatment, that is additional refining (e.g. desulfurization, deoxidation, degassing), the adjustment of composition and temperature is an essential part of the operational sequence (22,23,24). In these systems some 50-200 tons molten steel is being held in a ladle, while reagents are being injected, vacuum may be applied and thermal energy may be supplied. It has been recognized quite early on that agitation plays a critical role in affecting the efficiency and the overall rate of these processes. Agitation may be provided either by injected gas streams (gas bubble driven recirculation) or by electromagnetic forces (induction stirring).

A great deal of work has been done in both these areas; due to the nature of the problem, electromagnetically driven flows are much more readily modelled in a rigorous manner (25,26) than gas plume driven circulation systems. Notwithstanding the important advances that have been made (27), in the latter case the definition of the plume boundaries still poses serious problems.

Fig. 9 (28) shows a comparison between the experimentally measured and the theoretically predicted velocities in a 4-ton inductively stirred ladle holding molten steel. The agreement between measurements and predictions is, of course gratifying, but of perhaps greater importance is the fact that both measurements and the theory predict a linear relationship between the applied current and the velocity. This finding is also supported by simple scaling laws (26).

Figs. 10 and 11 show computed results depicting the velocity fields and the maps of the turbulent kinetic energy in 6-ton melts agitated by an argon stream and by induction stirrers respectively (23). These plots are useful for the insight they provide. It is clearly seen that with induction stirring one can produce a relatively uniform velocity field and turbulent kinetic energy distribution; in contrast, with gas stirring high velocity and high turbulent kinetic energy are being generated in the gas plume and at the free surface, but the other portions of the melt are relatively quiescent. It follows that gas stirring is to be preferred if we wish to promote slag-metal reactions, while induction stirring would be more desirable from a technical standpoint if we need to provide uniform stirring throughout the melt.

Many other examples may be cited of cases where mathematical modelling has provided very useful qualitative insights into the behavior of systems. These range from injection metallurgy (24), to plasmas (28), welding (31), and the like.

2.3 Evolution of methodology

Perhaps the most important development in the whole field of mathematical modelling has been its transformation from a rather esoteric pursuit to a well established and well documented scientific-engineering endeavor (1,2,7,8).

There is now a well documented methodology of mathematical model development, a typical flowchart of which is given in Fig. 12. It is seen that the problem identification and formulation should ideally be followed by simple asymptotic scoping and scaling calculations and that the subsequent machine computation should be carried out in parallel with experimental work. These issues have been discussed in some detail in other publications (1,2,3), but here we should stress the following points:

  - ideally mathematical modelling should include both simple scoping and scaling calculations and machine computation

  - experimental and modelling work should be carried out in parallel, with very close interaction between the respective groups.

  - there is now an excellent body of knowledge accumulating in the mathematical modelling field so that prospective modellers can now be guided by precedents.

  - the cost of computational equipment has decreased dramatically in recent years. Very effective engineering workstations may now be purchased for sums that represent a fraction of the average annual paid to practicing engineers. It follows that modelling has emerged as a very cost-effective way of conducting research.

  - finally there has been a great deal of accumulated experience with the various software packages that are available for tackling metals processing problems; so in the majority of cases one would tend to use these packages at least in part in the model development. A partial listing of some of the more popular software packages has been is given in (2) and (3). Custom developing programs for essentially standard tasks would be very difficult to justify.

2.4 Failures in Modelling

As any other new scientific endeavor, the field of mathematical modelling is replete with failures, although for obvious reasons these are much less well documented than the success stories.

What do we mean by a failure? A modelling effort will be considered unsuccessful if it failed to attain the desired objectives, that is, an improved understanding of the system under study, or providing guidance for the implementation of a new technology. A perhaps stricter definition of failure, but appropriate under industrial conditions, would be the inefficient utilization of the resources devoted to the modelling project.

Even on acknowledging that research is an inherently risky enterprise, it is worthwhile to analyze some of the root causes that prevented the successful completion of mathematical modelling efforts:

- poor problem definition. At the beginning of any mathematical modelling project it is critical that we define the specific objectives and assess the limitations that may be posed by the poor understanding of the physics, chemistry and the constitutive relationships. If you do not understand the physics and the chemistry, modelling may be best restricted to a simple scoping exercise.

- poor coordination with the experimental part of the program. As we noted earlier, close interaction between the modellers and the experimentalists is a must throughout the lifetime of the project.

- inexperienced modellers, having poor analytical, computational or numerical skills and possibly inadequate tools at their disposal.

While the first entry can pose a serious, possibly insurmountable impediment (no one as yet suggested modelling cold fusion!) which has to be given very serious consideration, there is really no excuse for failures associated with poor communications and the use of inexperienced modellers. This field is now too mature for such occurrences!

3. MODELS IN NEW PROCESS DEVELOPMENT

Let us turn to the main theme of this discussion, that is, the use of mathematical models in new process development.

The most exciting role that one can envision for mathematical models is in new process development. The strong, growing pressures by the international marketplace are forcing all of us to develop new products and processes over a much shorter time period and at reduced cost. It should be clear that the survival of most major corporations will depend on their ability to achieve this. The intelligent use of mathematical modelling can take the following forms here:

3.1 Screening and refining new concepts

3.2 Actual development of new process concepts

3.1 Screening and refining of new process concepts

It is still quite widely held that the development of new technologies and new products is an essentially intuitive process and the "science and engineering component" has to be introduced in its refinement and actual physical realization.

When a process concept is proposed, often simple modelling calculations may provide a very effective screen to separate ideas that are feasible and unlikely to succeed. It is a well accepted axiom that in chemical process metallurgy the key thermodynamic relationships (at least the first law) will have to be obeyed. Indeed, "checking" the thermodynamic equilibria and the overall heat and mass balances has to be the first step in the evaluation of a new extraction or refining idea. (Volatile suboxides, unstable halides, certain intermetallic compounds can provide some interesting, at times overlooked complications.)

We should stress, however, the availability of an additional set of screening tools, which involve the rates of these processes, through the modelling of fluid flow, heat and mass transfer. Let us cite two conceptually simple examples.

In the plasma processing of materials, such as sketched in Fig. 13, we may now readily calculate, or at least estimate, the temperature profiles in the plasma plume, the trajectory of the solid particles, the plasma-particle heat transfer and hence the time-temperature trajectories of solid particles of a given size (30). On this basis we can readily assess the technical feasibility of melting or reacting particleds for specific reaction schemes. In short, for most conventional plasma systems it should be possible to melt most metallic particles in the 10-100 micron range; the melting of certain metal oxides, e.g. alumina, may be rather more difficult and the completion of reducing most metal oxides would be quite difficult to achieve.

Such calculations, which, of course would have to be carried out with some precision, should provide very helpful perspectives regarding the use of plasmas in certain branches of extractive metallurgy. (We note here that while the in-flight reduction of solid particles may be difficult, their reaction in a metal bath should be perfectly feasible.)

Another interesting, topical example is direct ironmaking. In one variant of this process, sketched in Fig. 14, the use of a converter is envisioned, with molten iron and slag as the principal operating media. Carbon, oxygen and (partially reduced) iron oxide would be fed into the vessel; oxide reduction would take place in the metal or at the slag-metal interface, giving up carbon monoxide, according to:

$$FeO + C = Fe + CO$$

While the oxidation of this carbon monoxide above the bath (termed post combustion) would provide the thermal energy needed to sustain the endothermic reduction process.

As per numerous calculations reported in the literature (31) this process is definitely feasible from the standpoint of thermodynamics. However, even if one can satisfy first law concerns, we have to recognize the need to meet the kinetic requirements, that is to verify, whether one can supply thermal energy from the post combustion region (i.e. above the bath) to the reduction region, the metal phase or the slag-metal interface.

Calculations by McRae have shown that for a given "reasonable converter configuration" one would need heat transfer rates in the region of 500 kW/m^2 from the post combustion zone to the slag-metal interface.

Very simple scoping calculations may be carried out, to test whether such heat transfer rate could be accomplished by thermal radiation. In considering an extreme case, postulating that the post combustion zone behaves like a black body, radiating at 1700° C, into an environment at absolute zero, we could achieve less than 20% of the figure cited by McRae. This simple calculation indicates that for this system the matter of heat transfer from the post combustion zone to the slag-metal region would need rather careful attention, preferably before an actual pilot plant is designed. One possibility of overcoming this problem would be to increase the effective heat transfer area by "splashing".

Many other examples may be cited, but it is hoped that the critical point has been made that simple modelling calculations, both thermodynamic and kinetic, should be carried out very early on in the lifetime of a project.

Let us now turn to a new, less explored area, that is the use of models in the development of new process concepts.

3.2 The use of models in the development of new process concepts

Up to the present mathematical models have been largely used to represent heat flow, fluid flow, chemical kinetics and mass transfer, deformation and the like, in systems that either exist or have been at least visualized conceptually. There is another, potentially more powerful side of modelling, which should allow us to invent new technologies, by addressing the inverse problem.

In other words, rather than calculating the velocity, temperature and concentration profiles in a metallurgical vessel, for a given set of inputs, we can define what the ideal flow pattern, temperature and concentration profiles would be and then devise the input parameters that would give us this desired end result. By addressing new process development this way, in a systematic manner, exciting new vistas may be opened up.

Let us illustrate this idea, using a concrete example, concerned with tundish design (32).

Tundishes represent an important component of all continuous casting systems. In steel production molten metal is poured from a ladle into a long trough, called a tundish, which then serves as a distributor of the molten metal stream into the various continuous casting strands. Typically the tundishes are some several meters long, about a meter deep and a meter wide, with a capacity ranging from a few tons to few tens of tons, providing a residence time of a few minutes for the metal stream. These tundishes perform several functions, which include being a buffer to even out the flow, as the ladle is being emptied, a means for "quieting down" the flow and encouraging the flotation of inclusion particles.

The initial tundish designs involved rectangular troughs, without any dams, weirs or baffles. The resultant flow fields, as shown in Fig. 15, lead to strong recirculatory flow patterns and markedly uneven flow. As shown by tracer tests such an arrangement resulted in short circuiting (by-passing) and hence in steel that retained many of the originally present inclusions.

A logical intermediate step was to introduce dams, weirs or baffles. As sketched in Fig. 16, these did provide a partial solution; however, the lee side of these baffles represented a stagnant zone, so that significant short circuiting was still inherent in these designs.

Now let us turn our attention to the inverse problem. The ideal flow field would be a parallel flow or plug flow arrangement, which would provide a uniform residence time and eliminate short circuiting. The use of mechanical baffles cannot achieve this, but the application of a magnetic field, applied in a direction perpendicular to the flow (Fig. 17), may produce this desired effect. In a physical sense the magnetic field will act to even out the flow, acting as a "brake" on faster moving melt streams. Fig. 18 shows the computed velocity field for "case b" when a magnetic field is being applied to one section of the tundish, while Fig. 19 depicts the same plot, when a field is being applied along the whole length of the tundish. It is seen that the latter case would provide the "ideal flow pattern". However, as seen in Fig 20, the residence time distribution that corresponds to "case b" would still be rather satisfactory, representing a major improvement over what may be achieved by the use of baffles or dams.

The tundish case cited here was a simple example, where by tackling the inverse problem a ready solution could be developed. Many other examples could be envisioned from the field of continuous casting, crystal growth, chemical vapor deposition and a whole range of materials processing operations, where new technologies could be developed in a similar manner. Near net shape casting, rheocasting and the production of composite materials may be ideal candidates for utilizing this approach. One should stress, however, that a good understanding of the interrelationships between the key process parameters is an essential prerequisite of formulating these inverse problems.

4. CONCLUDING REMARKS

A brief survey was attempted of mathematical modelling, as applied to materials processing operations. It was shown that mathematical modelling, which is maturing into an accepted scientific-engineering discipline, may play a very important role in both the optimization of existing operations and in the development of new technologies.

The availability of inexpensive computational hardware, the broad range of software packages and perhaps most important, the significant body of accumulated experience with modelling is making the more widespread use of modelling approaches an essential ingredient of the materials processing scene.

In closing, comments ought to be made regarding some key principles of modelling and also on the tasks that lie ahead.

In the development of modelling approaches it is critical that the modeller be in close contact with "the customers", that is, the people pursuing the experimental and the design aspects of the project. In this regard it is important to recognize that the modelling should begin at the very inception of the process.

The modeller should set clearly defined objectives and be fully aware of the limitations of the approaches to be adapted. If the chemistry or the physics of the process are poorly understood, the use of sophisticated software packages or supercomputers is unlikely to help - indeed, it may mask the real issues.

One of the most exciting opportunities that models offer is in their use in the initiation of new technological development, through the definition of the "inverse problems". This area is certain to grow in the near future.

The main impediments that remain regarding the more extensive use of models lie in our incomplete understanding of the chemistry and physics of the problem, or, in other words, the lack of constitutive relationships.

The development of a proper representation of multi-phase systems, or the non-newtonian behavior of melt-solid slurries, is far from a "mechanical task" and will require major new insights.

Another serious potential problem area is our lack of information on property values, such as viscosity, diffusivity, elastic modulus and the like, even for some quite commonly used materials. This is an area which will require a major commitment of experimental resources, to test and complement the interesting new predictive capabilities that are being provided through molecular dynamics.

Nevertheless, mathematical modelling is now established as a vigorous, growing discipline which offers important tools for tackling process development and process optimization problems.

REFERENCES

(1) J. Szekely, J.W. Evans, J.K. Brimacombe, The Mathematical and Physical Modelling of Primary Metals Processing Operations. Wiley, New York (1988).

(2) J. Szekely, The 1987 Extractive Metallurgy Lecture, Met. Trans. 19B, 525 (1988).

(3) J. Szekely, Ironmaking and Steelmaking, May, 1989 - in press.

(4) Mathematical Process Models in Iron and Steelmaking, London: The Metals Society (1969).

(5) Mathematical Process Models in Iron and Steelmaking, London: The Metals Society (1974).

(6) The Application of Mathematical and Physical Models in the Iron and Steel Industry, 3rd Process Technology Conference, Pittsburgh, PA. Warrendale, PA: Iron and Steel Society/AIME (1982).

(7) J. Szekely and W. Wahnsiedler, Mathematical Modelling Strategies in Materials Processing, New York: John Wiley, in press (1988).

(8) J. Szekely, L.B. Hales, H. Henein, N. Jarrett, K. Rajamani and I. Semarasekara, eds., Mathematical Modelling of Materials Processing Operations. The Metallurgical Society, Warrendale, PA (1987).

(9) Mathematical Models for Metals and Materials Applications. Preprint volume, for a Symposium held in Sutton Coldfield, The Metals Society, U.K., (1987).

(10) Y. Sahai, J.E. Battles, R.S. Carbonara and C.E. Mobley, eds. The Casting of Near Net Shape Products. The Metallurgical Society (1988).

(11) F. Oeters and A. Saatci, Proc. 5th International Iron and Steel Congress, vol. 6, p. 1021 (1986).

(12) A.W.D. Hills, J. Iron and Steel Inst., vol. 203, p. 18 (1965).

(13) E. Mizikar, Trans. Met. Soc. AIME, vol. 293, p. 1747 (1967).

(14) I.V. Samarasekera, J.K. Brimacombe and R. Bommaraju, ISS Trans., vol. 5, p. 79 (1984).

(15) R. Bommaraju, J.K. Brimacombe and I.V. Samaraseekera, ISS Trans., vol. 5, p. 95 (1984).

(16) I.V. Samarasekera, J.K. Brimacombe and R.W. Pugh, CIM, Montreal (1988).

(17) I.V. Semarasekara and J.K. Brimacombe, Proc. Philbrook Memorial Symposium, The Iron and Steel Society, p. 157 (1988).

(18) J.W. Evans, in J. Szekely et al, eds., Mathematical Modelling of Materials Processing Operations, p. 9, The Metallurgical Society (1987).

(19) W.E. Wahnsiedler, in J. Szekely et al., eds., Mathematical Modelling of Materials Processing Operations, p. 643, The Metallurgical Society (1987).

(20) S.D. Lympany, D.P. Ziegler and J.W. Evans, Light Metals, p. 507 (1983).

(21) V. Potocnik, Light Metals, p. 227 (1989).

(22) R.J. Fruehan, Ladle Metallurgy Principles and Practice, The Iron and Steel Society (1985).

(23) J. Szekely, G. Carlsson and L. Helle, Ladle Metallurgy, Springer, New York (1989).

(24) Proceedings of SCANINJECT IV, MEFOS, Lulea, Sweden (1986).

(25) S. Asai in H.Y. Sohn and E.S. Geskin, eds., Metallurgical Processes for the Year 2000 and Beyond, The Metallurgical Society (1989).

(26) J. Szekely, Fluid Flow Phenomena in Metals Processing, chap. 5. Academic Press, New York (1979).

(27) M.P. Schwartz, J.K. Wright and B.R. Baldock, in J. Szekely et el., eds., Mathematical Modelling of Materials Processing Operations, p. 565, The Metallurgical Society (1987).

(28) D. Apelian and J. Szekely, eds., Plasma Processing of Materials, North Holland, Amsterdam (1987).

(29) S. David, ed., Advances in Welding Science and Technology, The American Society of Metals (1986).

(30) N. El-Kaddah, J. McKelliget and J. Szekely, Met. Trans. 15B, 59 (1984).

(31) R.J. Fruehan, K. Ito and B. Ozturk, J. of Metals, November 1988, p. 83.

(32) J. Szekely and O.J. Ilegbusi, The Physical and Mathematical Modelling of Tundish Operations, Springer, New York (1988).

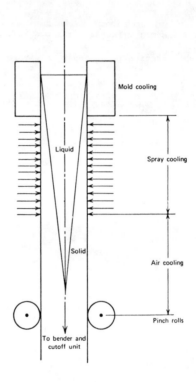

Fig. 1   Schematic sketch of a continuous casting system

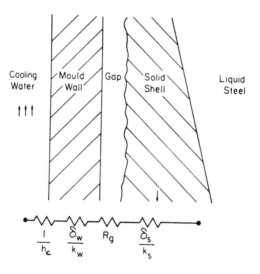

Fig. 2   Schematic sketch of the heat flow resistances in the mold of a continuous casting machine (17)

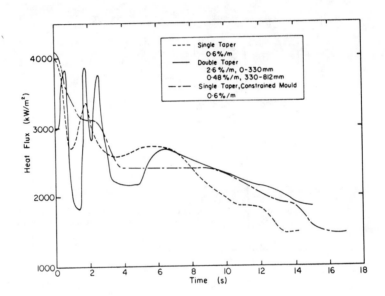

Fig. 3   Experimentally measured heat flux values as a function of residence time in the mold (17)

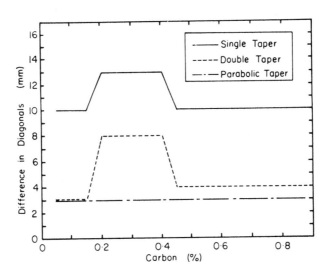

Fig. 4   Experimentally measured distorsion of the billet as a function of the carbon content and the mold design (17)

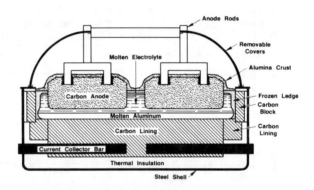

Fig. 5   Schematic sketch of a Hall Cell used for aluminum production

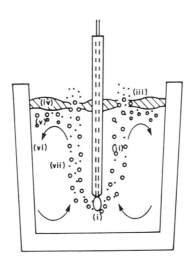

Fig. 6 Schematic sketch of an argon-stirred ladle

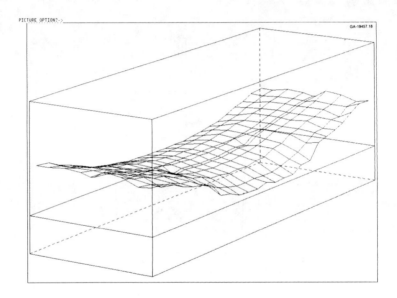

Fig. 7   Computed map of the bath-metal interface for a typical Hall Cell system (19)

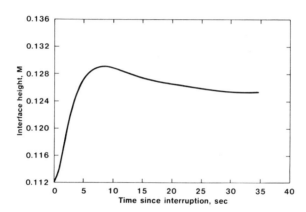

Fig. 8   The computed response of the position of a point at the bath-metal interface, as a function of time, in response to a step change in the operating conditions (19)

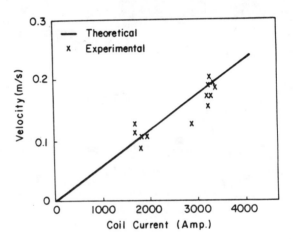

Fig. 9 Comparison of the experimentally measured and the theoretically predicted velocities in a 4 ton inductively stirred melt, showing the effect of the coil current. (23)

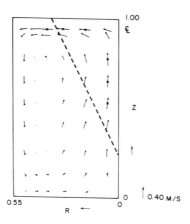

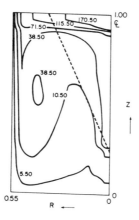

Fig. 10 Computed results showing the velocity field (upper plot) and the map of the turbulent kinetic energy (lower plot) for a six ton argon-stirred ladle. (23)

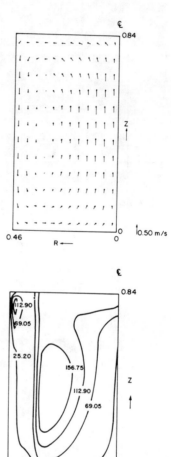

Fig. 11 Computed results showing the velocity field (upper plot) and the maps of the turbulent kinetic energy (lower plot) for an inductively-stirred 6 ton melt. (23)

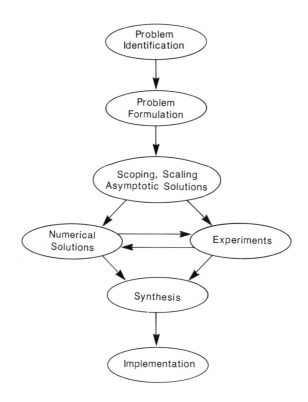

Fig. 12　Flow chart of a mathematical modelling project (2)

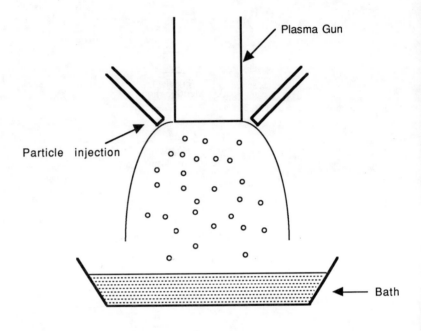

Fig. 13   Sketch of a plasma processing system

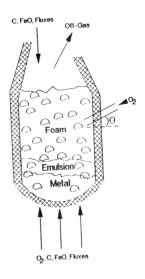

Fig. 14   Schematic sketch of a converter-based direct ironmaking system.

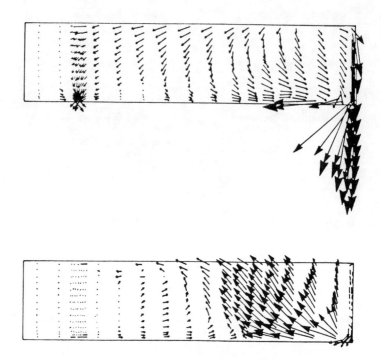

Fig. 15 The computed maps of the velocity vector for a typical tundish, in the absence of flow control devices. The upper plot corresponds to a plane near the inlet stream, while the lower plot represents a vertical plane close to the side wall.

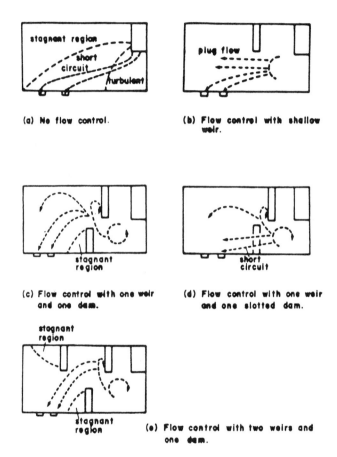

Fig. 16 Sketch of the effect of dams, weirs and baffles on the flow patterns in tundishes

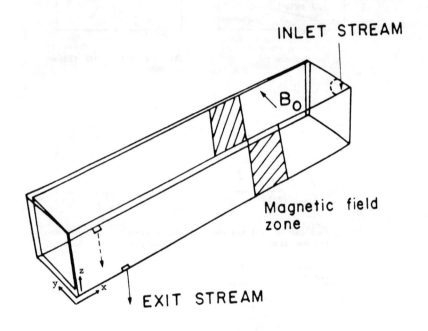

Fig. 17 Sketch of the application of a magnetic field across a tundish

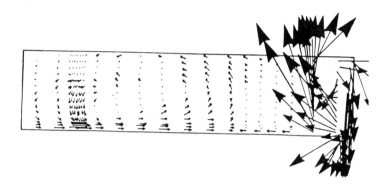

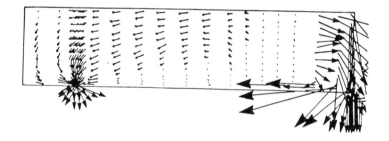

Fig. 18 The computed maps of the velocity vector for a typical tundish with a magnetic field imposed across a section near the inlet stream. The upper plot corresponds to a plane near the inlet stream, while the lower plot represents a vertical plane near the side wall.

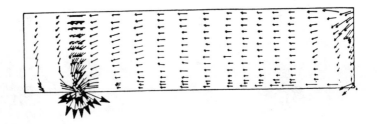

Fig. 19   The computed maps of the velocity vector for a typical tundish with a magnetic field imposed across the whole length. The upper plot corresponds to a plane near the inlet stream, while the lower plot represents a vertical plane near the side wall.

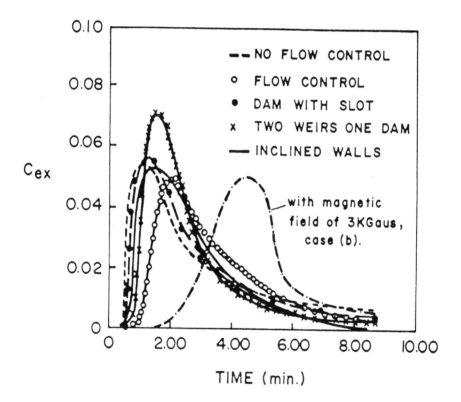

Fig. 20 The computed residence time distribution or C curves for various tundish flow control arrangements. Here case "b" corresponds to the velcity plots previously given in Fig. 18.

ENERGY FOR METAL PRODUCTION IN THE 21st CENTURY

Herbert H. Kellogg

Henry Krumb School of Mines
Columbia University, New York, N.Y. 10027

Abstract

The metallurgical industry, worldwide, has made considerable progress toward more energy-efficient processing since the oil crisis of the early 1970's. Some of the more important energy lessons and process improvements of the recent past are briefly reviewed.

Looking ahead, from now to the early years of the next century, rising energy prices seem inevitable as a result of declining petroleum resources, and governmental regulations aimed at controlling acid rain and the "greenhouse effect". To compete in this high-cost energy future the metallurgical industry must redouble its efforts to maximize the energy efficiency of existing processes, search diligently for new processes that are inherently more energy efficient, learn to use alternative energy sources, including the fuel value of waste materials, and adapt to a future where production of metal from scrap and waste grows in importance relative to primary metal production.

Energy Lessons Learned

Regardless of one's opinion of OPEC's policies, the oil crisis of the seventies served to awaken the world community to the critical role that energy can play in economic affairs. A brief review of some energy lessons from the past fifteen years will provide a background for speculating about future energy challenges.

The metallurgical community learned that primary metal production from ore in the ground is energy-intensive, more so if the ore is low-grade, more so if the metal-oxide is very stable, and more so if electric energy must be used rather than fuel. We also learned that secondary metal production from scrap is usually much less energy-intensive than primary production.

Table I illustrates some of these lessons about the energy-intensity of metal production. Here we see that copper produced from ore has 50 % greater energy-intensity (MFE) than zinc; this results from the very low grade of copper ore, which necessitates the mining and beneficiation of seven times as much copper ore in order to recover one ton of metal. Steel slabs require less than half the energy of zinc ingot; in this case the higher grade of the iron ore, and the fact that iron making requires very little electrical energy, contribute to the low MFE of steel. Aluminum produced from ore is far more energy intensive than the other primary metals. Although aluminum ore is relatively high-grade, it must be beneficiated by the energy-intensive Bayer process rather than by physical beneficiation methods used for other metals. Furthermore, the theoretical minimum free-energy required to reduce aluminum oxide is almost five-times greater than for iron oxide, and this inherent disadvantage is compounded by the need to supply that energy as electricity. Finally, we see that copper and aluminum recovered from scrap are far less energy-intensive than the primary metals; two features of scrap recycle are responsible for this difference -- the raw material treated is very high-grade, and, since it is already metallic, there is no need to supply the energy for reduction to the metallic state.

Prior to the oil crisis we lacked organized information of this kind simply because we failed to appreciate its importance. Now we gather and use such data to better set our priorities -- to focus attention on those metals that offer the greatest potential for energy saving from improved efficiency of current processes, from development of new processes, from increased secondary production relative to primary, and from substitution of less energy-intensive materials for end-use applications.

We also learned more about how process design influences energy-intensity, and the difficulty of meeting environmental regulations. In sulfide smelting, for example, we learned that sulfur and iron in concentrates are fuels at no cost to the process. When coupled with an intensive reactor of the flash or bath type and oxygen enrichment, oxidation of sulfur and iron supply most or all of the heat necessary for smelting. The process gas, enriched in sulfur dioxide, is ideally suited for fixation as acid, and "tail gas" from double-contact acid-making meets environmental regulations in most countries. The energy required to produce oxygen is less than the fuel energy saved, so that these new smelting processes both save energy and solve the environmental problem associated with reverberatory gases.

Table II illustrates these lessons on process design, using copper as an example. Compared to reverberatory smelting, which was the industry standard thirty years ago, the three newer processes shown in Table II have significantly reduced the overall consumption of energy resources, as measured by PFE, the process fuel equivalent; at the same time , the new processes produce much lower gas volumes that are much richer in sulfur dioxide, making possible almost complete recovery of sulfur as acid at an acceptable cost. Both flash smelting and the bath smelting used in the Mitsubishi process are intensive processes, as shown by their specific capacities, which are three to five times greater than the reverberatory. The rapid mass and heat transfer possible in intensive smelting make possible the effective use of oxygen, leading to reduction of the heat-loss associated with nitrogen from air, and increasing the concentration of sulfur dioxide in the process gas. The newer processes use far less fuel than the reverberatory, both because of the savings from use of oxygen, and because they burn more of the iron and sulfur from the concentrate, as shown by the higher matte-grades produced.

The foregoing examples from the metallurgical industry are but a sample of the energy-saving practices introduced in homes and industries throughout the industrialized world. We now know that reduced energy supply need not result in decreased industrial production or standard of living. Energy-

Table I. Energy Intensity of Common Metals

Metal product	Raw Material % M	MFE GJ/ton M	% of MFE as electric	Oxide Free Energy GJ/ton M
Steel slab cont. cast	ore 41% Fe	26.	13.	6.65
Zinc SHG ingot	ore 4.0% Zn	65.	91.	4.91
Copper wire-bar	ore 0.55% Cu	97.	65.	2.01
Copper wire-bar	No.2 scrap @97% Cu	20.	29.	--
Aluminum ingot	ore 25% Al	250.	67.	29.34
Aluminum alloy 380	Al scrap @96% Al	18.	15.	--

Notes to Table I

MFE, material fuel equivalent, sums the energy inputs for all process steps from mining to final product. Electric energy is charged at the average equivalent fuel rate of 11.1×10^6 J/kwhr.
 Steel slabs: based on ref.1, mining, beneficiation, blast furnace, basic-oxygen process, continuous slab casting.
 SHG Zinc: based on ref.1 for mining and beneficiation, ref.2 for roast, leach, electrolysis, melting, casting.
 Copper from ore: based on ref.3, mining, beneficiation, Outokumpu flash smelt with oxygen, electro-refining, melting, casting.
 Copper from scrap: based on ref.4, preparation, anode refining, electrolytic refining, melting, casting.
 Aluminum from ore: based on ref.1 for mining, drying, Bayer process, electrolysis, except electrolysis energy reduced to 15 kwhr/kg Al.
 Aluminum from scrap: based on ref.4, collection, sorting, scrap preparation, reverberatory melting, casting.

productivity, the amount of product or service that can be delivered per unit of energy used, has been increased since 1973, and can be further increased. This message has particular relevance to the United States, which lags Japan and many European countries in energy-productivity by almost a factor of two.

The New Energy Crisis

The record of the past fifteen years consists of facts that we all accept. In contrast, as I turn to my assigned topic --energy for the future -- facts are necessarily replaced by hypotheses, opinions and beliefs. The views of the future that I offer have been shaped by forty-seven years as a professional metallurgist. They are strongly held, but personal beliefs, and some are likely to be controversial.

The oil crisis of the seventies was just a sample of what we may face in the future. At least three portents suggest that more serious energy problems

Table II. Comparison of Copper Processes
(damp, 29.5% Cu concentrate to anode copper)

	Green Charge Reverb.	Outokumpu Flash	Inco Flash	Mitsubishi Continuous
Overall PFE GJ/ton Cu	25.6	16.4	14.2	17.2
Matte Grade % Cu	35.	62.	54.	65.
98 % O_2 used kg/ton Cu	0.	480.	790.	390.
Fuel used GJ/ton Cu	20.7	6.9	3.3	9.7
Tot. Gas Vol. Nm^3/ton Cu	17700.	5300.	3800.	5100.
% SO_2 in total gas	3.8	13.	20.	14.
Specific Cap. ton conc/d/m^3	0.81	3.5	2.3	4.3

Notes to Table II

a) **PFE**, Process Fuel Equivalent, sums the fuel equivalent of fuel, electricity, and supplies (oxygen, flux, etc.) used in the process, less a credit for surplus steam produced. For each process the operations included are drying (if used), smelting, converting, anode refining, acid production from all gas (except for the reverberatory gas, which is too low in sulfur dioxide for acid making), and fugitive gas collection.

b) The data are taken from ref.5, modified by addition of an estimate of the energy used for fugitive gas collection. Electric energy is counted at the average fuel rate of 11.1×10^6 J/kwhr.

lie ahead -- certainly by the early years of the next century. These are: <u>Acid rain, petroleum shortages and the "greenhouse effect"</u>. These problems bear a common theme -- they involve global, rather than national or local consequences. Deterioration of the global environment and depletion of natural resources are not issues that will go away, despite the wish of some elected governments and company managements. We, as pragmatic engineers, should lead, rather than drag our feet, in response to these global threats.

Acid Rain

This phenomenon has been studied by scientists, discussed in the media, and debated by politicians for the past twenty-five years. Now, there is growing realization that the time has come for political action to halt the progress of this devastating environmental problem. The culprits are the sulfur and nitrogen oxides that result from combustion of fuels -- particularly, high-sulfur coal and oil. In the atmosphere these gases are converted to sulfates and nitrates, and may be transported a thousand kilometers before falling as acid rain. The damage to forests, aquatic ecosystems, buildings, crops and human health are now well established. The phenomenon knows no international boundaries -- emissions from the U.S. midwest cause damage in eastern Canada, those from Great Britain affect Scandinavia.

In the past, metallurgical plants that smelted sulfide concentrates were the largest single point-sources of sulfur emissions, even though emissions from the total electric-utility industry were much larger than from the metallurgical industry. Today, the sulfide-smelting industry in most industrialized nations has learned to live with regulations that require sulfur capture, and no longer is a major contributor to acid-rain precursors, except for a few older plants that are not yet in compliance with emission standards.

Forthcoming regulations for control of acid-rain emissions will, however, affect the metallurgical industries indirectly by their effect on the cost of electric energy. Any of the many ways being studied for decrease in sulfur emissions from electric utility plants -- recovery from stack gases, fluid-bed combustion in the presence of limestone, reduction in the sulfur content of coal by ore-dressing methods, processing of coal into a cleaner fuel by a variety of methods, or substitution of nuclear or solar energy for coal -- will add significantly to the cost of electric energy. Modern industrial society, including the metallurgical industry, faces increased electrical prices as the cost of acid-rain control becomes a reality of the near future.

Petroleum Resources

For the past several years we have enjoyed a glut of relatively inexpensive oil on the world market, and many of us find it hard to believe another lesson we should have learned in the seventies -- <u>that geologic evidence and the world's enormous appetite for oil point to declining supply of this fuel in the early years of the next century</u>. New discoveries of petroleum resources will continue to be made, and a hundred or more years may pass before supplies dwindle to an insignificant amount, but it seems almost certain that discoveries, worldwide, will be unable to keep pace with our current use-rate. Furthermore, much of the newly discovered petroleum will be costly, both to discover and to produce, because of its location in hostile offshore and polar environments. Gradually declining supply and increasing prices for petroleum products seem inevitable.

Greenhouse Effect

Like acid rain, the "greenhouse effect" is an environmental problem resulting from too much human activity of a particular kind. In this case it is too much burning of fossil fuels, and too much destruction of forest cover, that combine to upset the global carbon cycle. The rise in concentration of atmospheric carbon dioxide over the past century is well documented, although there remain scientific questions regarding how and when this will alter world climate. Most atmospheric scientists believe that, if the accumulation of greenhouse gases in the atmosphere is unchecked, serious global warming and rising sea level will result, and warn that we must act now to avoid serious consequences twenty to thirty years hence.

It seems almost certain that our modern industrial civilization will find it necessary to curtail burning of fossil fuel in order to avoid a greenhouse crisis, and this will have a profound effect on the energy sources we use in our homes, our industries and, most important to us here, our metal production processes.

Energy for the 21st Century

How will the industrialized world meet its energy needs in the in the year 2000 and, at the same time, respond to these environmental and resource restrictions? Change from current energy sources will be slow, because the energy industry is huge and capital-intensive, and unable to change rapidly. The proportions of energy derived from petroleum, coal, water-power and nuclear fuel may remain substantially unchanged at the end on this century. What can be changed more rapidly is the amount of energy used to produce a unit of product or service, and I believe that improved energy efficiency -- or energy conservation, or improved energy productivity, or reduced energy demand, call it what you wish -- will be the first line of attack on future energy problems.

Energy Conservation

During the late nineteen seventies in the United States, energy efficiency was given a high priority, and remarkable gains were made after only a few years. New automobiles, refrigerators, and other appliances used less energy; added insulation and storm windows reduced fuel consumption in home heating; industry, including the metal industry, whole-heartedly adopted energy conservation -- after all, it saved money as well as energy. For the first time since the great depression the total energy used for all purposes in the U.S. failed to increase for several consecutive years. Unfortunately, the Government relaxed its standards for energy efficiency when petroleum became plentiful and cheaper in the 1980's, long before energy conservation had really taken root as an ethic.

Today, improved energy efficiency is once again recognized as an essential first step in meeting future energy problems. Energy productivity could be doubled in the U.S. and that country would only be where Japan is today. An energy-efficiency ethic should have particular appeal to this gathering because producing more from fewer resources, with less waste, is a fair statement of a credo for all engineers. To me it has particular appeal because it is an open-ended goal -- there will always be new and more efficient ways to do what we do today.

Essential as it is, energy efficiency alone will not solve the longer-range problem of dwindling petroleum resources, and it will only postpone the day when accumulation of carbon dioxide in the atmosphere becomes an inescapable problem. New energy sources, requiring long lead-time for deployment, are needed to solve these problems. From today's perspective the alternatives appear to be nuclear energy, the many forms of solar energy, tidal and geothermal energy.

Nuclear Energy

We have lived with an immature nuclear-energy industry for twenty-five years, and the experience has been far from comforting. Most of us would prefer to reside several hundred kilometers upwind from the nearest reactor or fuel-processing plant, and for good reasons. Before we commit ourselves to a vastly expanded nuclear industry, we should do what should have been done thirty years ago: perform the research and development that will tell us whether it is economically feasible to mine and process uranium, operate a nuclear reactor, reprocess spent fuel, and dispose of radioactive waste and obsolete facilities, all without irreparable contamination of the environment, and without danger of devastating accident. When all of these costs of generating nuclear electricity, in a manner free from seriously adverse environmental and safety hazards, are taken into account, this form of energy is likely to be no cheaper than alternative, and more benign, energy sources.

Solar Energy

If we had not been spoiled by an abundance of petroleum, and misled by the false promise of electricity "too cheap to meter" from nuclear energy, we might now have a sizeable solar-energy industry. Despite R&D expenditures on solar energy that are trifling in comparison with those spent on nuclear energy, we are approaching a time when various forms of solar energy might compete with conventional fuels. If we fund solar R&D at a level appropriate to its promise, solar energy should be able to supply a significant, and growing, fraction of our energy needs by the second decade of the 21st century.

Solar-electric, wind power, and alcohols from biomass appear to me the most promising forms of solar energy. Efficiencies as high as 30% in conversion of solar energy to electricity have recently been achieved from a "mechanically-stacked, multi-junction solar cell", in an experimental device built by Sandia Laboratories in New Mexico. "Farms" of wind turbines in California even now produce electricity at, or near to, competitive costs. Alcohol could supply our need for liquid fuel for transportation and firing of furnaces; because it is derived from biomass, its combustion would result merely in recirculation of carbon from biomass to energy production and, by photosynthesis, back to biomass.

Energy Efficiency

How should the metallurgical industry prepare itself to compete in the 21st century? If you believe my thesis of higher energy costs relative to other goods and services, then improved energy efficiency should be the first, and a lasting, goal of your research and development. This is not a new message -- the industry has been working seriously at saving energy for the past twenty years. But the effort needs to be redoubled. The opportunities abound. In hydrometallurgical plants there are still too many heated tanks, often outdoors, that lack adequate insulation. In plants using aqueous electrowinning, lead anodes are still the rule despite the energy saving possible if one could devise a "dimensionally-stable anode" of the kind that has revolutionized the chlor-alkali industry. Waste-heat recovery from slags and molten metal will prove more attractive as energy prices rise. Energy can be saved by more efficient electric motors that are properly sized to the task, by installation of improved sensing and measuring devices, and by countless "housekeeping" improvements.

New Processes

The foregoing energy-conservation measures, however, are only ways to improve the efficiency of current processes. The more important task is to seek entirely new process designs that are inherently more energy-efficient, and, at the same time, capable of processing lower quality raw materials than those we enjoy today. This will require costly, long-range R&D programs, staffed with well-trained and imaginative engineers. More than forty years of observation of the efforts of the metallurgical industry to carry out long-range process research have convinced me that, in many cases, the job is too costly for individual companies. Too often management withdraws support after three or four years, when ten years are needed; or the available personnel are too few or lack the needed skills; or the available facilities are inadequate. Pooling the resources of several companies in R&D consortia may be necessary to achieve the breakthroughs in process design that are needed for the next century. The metallurgical industry must change its image of an old-fashioned, smoke-stack industry by developing new processes that use the full range of modern technology.

Secondary Metal

The 21st century will bring an inevitable increase in production of secondary metal from scrap, relative to primary metal production. Scrap recycle not only saves huge amounts of energy, but conserves material resources, and helps to solve growing waste-disposal problems. Today we recycle most of the scrap generated in large quantities by the fabricating industry, and Table III shows that the total energy requirement -- for transporting, sorting, preparing and processing scrap to secondary metal -- is only a small fraction of that required for producing primary metal from ore. For aluminum, in particular, the energy saving from recycle of scrap has special significance -- not only is the energy for secondary metal about one-twentieth of that for primary metal, but we use more aluminum than any other metal except steel.

The values for three different grades of copper scrap, shown in Table III, illustrate the effect of scrap quality on the energy required for processing. No.1 wire scrap is uncontaminated copper that requires only remelting to produce high-quality wire bar. No.2 scrap, containing 1-3% of impurities, is melted, fire-refined and cast as anodes, then electro-refined to cathodes, and finally remelted and cast. Low-grade copper scrap, which may average about 25% copper, must first be smelted to black-copper and converted to blister before it undergoes the same treatment as No.2 scrap.

The unusually large energy requirement shown for titanium in Table III is misleading, in that most of this energy comes from the primary sponge titanium required to dilute impurities in the scrap to an acceptable level. There is, as yet, no economical means to "refine" impure titanium.

The improvement in scrap recycle must come from recovery of obsolete consumer-scrap, generated in small amounts in millions of households. Collecting, sorting and transporting such scrap will involve serious socio-economic problems that society must solve. Metallurgists will face the technological problem of how to process low-grade scrap into useful metal. Mankind in the twentieth century has been profligate in its use of both energy and materials; reclaiming obsolete scrap of all kinds will be a priority of the next century.

Waste Processing and "Tertiary" Metal

The solid and liquid wastes of modern civilization not only threaten the habitability of the earth, but disposal costs increase dramatically as governments impose regulations for "safe" disposal. Wastes have been aptly called "resources out of place", but the name "negative resources" better suggests that waste disposal costs money that subtracts from profitability. I foresee an increasingly important role of the metallurgical industry in relation to waste disposal.

For example, the steel industry is now faced with paying for the "safe" disposal of electric arc-furnace dust. A "raw material" such as this, containing 10-18% zinc, available in quantity, and bearing a premium, should provide enough financial incentive for the non-ferrous industry to devise an economical means of treatment to recover zinc, and perhaps other values, and to produce an environmentally "safe" residue. Metals that are recovered from waste having a negative value (i.e., where the waste-generating industry pays for treatment of the "raw material") have been termed "tertiary" metals, to differentiate them from "secondary" metal recovered from scrap, which bears a

Table III. Energy for Secondary Metals

Type of Scrap	Product	MFE ton M (note a)	% of Primary MFE (note b)
Aluminum can scrap	hot metal for can stock	10.	4.0
Aluminum, misc. scrap	alloy 380 ingot	18.	7.0
Copper, No.1 wire scrap	copper wire-bar	4.4	4.5
Copper, No.2 scrap	" " "	20.	21.
Copper, low-grade scrap	" " "	49.	51.
Lead, battery & misc.	soft & hard lead ingot	10.	32.
Nickel, superalloy	superalloy ingot	23.	14.
Steel, misc. scrap	steel ingot	10.	37.
Titanium 50% scrap + 50% primary sponge	double-melt ingot	302.	65.
Zinc, clean die-cast	die-cast ingot	3.0	4.6
Zinc, misc. scrap	zinc dust	23.	35.

Notes to Table III

a)   MFE, material fuel equivalent, includes energy for collection, transportation, sorting, preparation and processing to secondary metal. Values based on reference 4. Electric energy charged at the average equivalent fuel rate of 11.1×10^6 J/kwhr.

b)   This is the ratio (MFE for secondary metal) / (MFE for primary metal from ore)x100.

positive value. "Tertiary" metal recovery epitomizes the needs of the 21st century -- it solves a waste disposal problem, it conserves material resources, and it saves the energy required to produce an equivalent amount of primary metal.

Still another way that the smelting industry can interact with waste-generating industries is in the disposal of organic wastes -- oils, plastics, rubber, wood chips, etc -- where the smelting industry might substitute for incineration, and make effective use of the fuel value of these materials. We have an example of this in the use of shredded auto tires as supplemental fuel for copper-reverberatory furnaces in Japan a few years ago. Although the smelting industry might require special methods to deal with impurities in the waste, such as halogens, the waste-generating industry would pay the smelting industry to use these fuels so as to avoid costly incineration. The metal-smelting industry possesses unique know-how on high-temperature processing, and can best design processes that will, at the same time, incinerate organic waste and save an equivalent amount of primary fuel.

At this point my message regarding energy for the 21st century should be clear. It is a future that must stress conservation -- of energy, of resources and of the environment. There will be many other means than those I have cited that will contribute to conservation. It should be a busy and exciting time for engineers, with a panorama of new challenges. If the younger generation embraces such a conservation ethic, our grandchildren should enjoy a sustainable future.

References

1. "Energy Use Patterns in Metallurgical and Nonmetallic Mineral Processing, Phase 4, Energy Data and Flowsheets, High Priority Commodities", Battelle Columbus Laboratories, June 27, 1975, NTIS PB-245759.

2. H.H. Kellogg, "Energy Use in Zinc Extraction", 28-47 in Lead-Zinc-Tin '80, J.M. Cigan, T.S. Mackey, T.J. O'Keefe, eds. (The Metallurgical Society, Warrendale, PA, 1980)

3. C. Pitt, M.E. Wadsworth, "An Assessment of Energy Requirements in Proven and New Copper Processes", final report to U.S. Dept. Energy, Contract EM-78-S-07-1743, Univ. of Utah, Dec. 31, 1980.

4. C.L. Kusik, C.B. Kenahan, "Energy Use Patterns for Metal Recycling", U.S Bur. Mines, Inf. Cir. 8781, 1978.

5. H.H. Kellogg, "Energy Use in Sulfide Smelting of Copper", 373-415 in Extractive Metallurgy of Copper, Vol. 1, J.C. Yannopoulos, J.C. Agarwal, eds., (The Metallurgical Society, New York, NY, 1976)

MATERIALS FOR THE FUTURE

Yves Farge

Vice President
R&D Pechiney
Paris, France

Abstract

The technological advance of many industries is largely dependent on materials.

For modern components, the design begins with the definition of the functions of each part, followed by the selection of the most appropriate materials (or combination of materials) according to economic conditions.

In these new trends, the role of R and D and the evolution of the industry of materials will be considered, especially in Europe. The place of traditional materials and the difficulties in developing new advanced materials will be discussed. The relations between consumers and producers will also be examined.

Pyrometallurgy

THE DECLINING INFLUENCE OF ECONOMIC GROWTH ON METAL DEMAND

Graham Davis

Graduate School of Business
University of Cape Town
Private Bag
Rondebosch, 7700
Republic of South Africa

Abstract

Growth in Western World consumption of ferrous and non-ferrous metals has generally declined since the early 1960s, being largely negative in recent years. One of several factors that caused this fall in consumption was the falling growth rate of the economies of the major metal consuming countries. While mining industry leaders continue to blame poor demand for their commodities on poor economic growth, recent research has shown that the demand for raw materials has become uncoupled from economic growth. Applying this concept to metals, examination of the consumption of ten major metals over the past thirteen years reveals that economic growth is of minimal significance to metal consumption for several of the metals studied. This is due to a weak to unmeasurable correlation between metal consumption and economic growth, and a decreasing intensity of metal use. These findings have implications for anticipated future demand for metals and the strategic marketing position taken by mineral producers.

Introduction

Western World consumption of many metals has been growing at a declining annual rate since the early 1960s, and has been largely negative since the late 1970s (1,2). Carnegie (1), in a Group of Thirty report, states that one of several factors that has caused falling growth in demand for the main ferrous and non-ferrous metals is the low growth rate of the economies of the major metal consuming countries. His theory arises from an historical two-stage link between economic growth and raw material consumption. In the first stage, economic growth is linked to economic productivity, or the production of goods and services. There follows a secondary theoretical link between the production of goods and services and the consumption of raw materials used in their fabrication. Through the combination of these two relationships, a fall in economic growth has historically resulted in an indirect fall in demand for raw materials. It is this relationship to which Carnegie attributes the falling demand for metals.

Through a continued belief in this relationship, many mining industry leaders believe that the poor demand for most of their non-auriferous mineral products will improve with an upswing in their country's economy and the economies of their major trading partners. South African producers in particular look to foreign economic growth for revival of their industry, as export sales account for 85% of total South African mineral sales (3). Lower than expected growth in GNP for the developed nations is frequently stated in annual reports as the cause for continued poor demand for many of South Africa's minerals, with an economic upturn essential for an improvement in the markets (2). Carnegie (1) also promotes the view that an impending economic upturn will significantly improve the demand for metals, stating that mineral commodity markets over the next half decade seem likely to be influenced by modest economic growth in the industrialised and developing countries.

There are, however, a number of economists who believe that there has been a recent "uncoupling" of economic growth and the demand for raw materials (4,5). Drucker points out that within the past 10-15 years fundamental economic changes have occurred, one of which is the fact that the primary-products economy has become uncoupled from the industrial economy. As such, increasing productivity no longer implies primary resource demand. This uncoupling of economic growth and raw material demand, if real, has serious implications for South Africa's mining industry and other export oriented mineral-producing countries. In this paper, the extent of the relationship between economic growth and the consumption of ten ferrous and non-ferrous metals of importance to South Africa is assessed. The results are explored with a view to establishing the impact of the theory of uncoupling on the demand for metals in general.

Hypothesis

Previous research has shown that the function relating metal consumption to economic growth, or more specifically per capita metal consumption to per capita economic growth, changes with economic progress (6,7,8,9,10). A simple yet effective indicator of the ongoing relationship between per capita economic growth and per capita metal consumption is the intensity of metal use. Intensity of use is the amount of metal consumed per unit of GNP, and a decline in intensity of use with time is an indication of a decreasing requirement for metals in the production of goods and services. It is the hypothesis of this paper that, in accordance with Drucker's uncoupling theory, the influence of per capita economic growth on per capita metal consumption is now at a level of insignificance for several major non-auriferous metals. In other words, the decline in intensity of metal use and structural economic changes prevent the establishment of a measurable functional relationship between the current consumption of these metals and economic growth.

The Theoretical Link Between Economic Growth and Metal Demand

In an effort to explain variations in reported accounting profits, the relationship between economy-wide indexes and the firm's annual earnings was investigated by academics as early as 1967 (11). Lev, in 1980, identified that 21% of the sales variability for 573 U.S. firms from 1949 to 1973 was explained by changes in GNP (12). The corresponding figures for operating income and net income of the firm were 14% and 12%. Foster (13), using more recent figures, established that an economy variable explained 17% of the variability of net income for 315 U.S. firms from 1964 to 1983. There were significant interindustry differences in the importance of economic influences on net income, with an economy variable explaining 35% of the net income variability in the steel and blast furnace industry.

Mineral economists have done similar research into the link between economic indicators and the demand for metals. The relationship is apparently clear; metal demand at any point in time is affected by economic growth, usually indicated by GNP growth. Real GNP growth can be described in terms of three variables for analytical purposes: growth of the employed labour force, the trend of average hours worked per week, and the trend of output per man-hour worked ("productivity") (14). GNP is thus presumably a determining factor of the consumption of raw materials, such as metals, used in the production of "output" (goods and services). For example, a drop in real GNP is indicative of a drop in economic output and the production of goods and services, which should result in a drop in metal demand.

In illustration of this proposed link, average annual compound growth rates of real GNP are given in the Table I for the major metal consuming regions of the Western World, while the corresponding changes in Western World metal consumption are given in Table II (1). Metal consumption is a proxy for metal demand as stockpile changes, which create short-term differences between demand and consumption, are usually negligible in comparison to overall consumption.

Table I
Average Annual Compound Growth Rates of Real GNP (%)

	1960s	1970s	1979-1984
USA	5,0	3,5	2,4
Europe	5,3	3,3	1,2
Japan	15,0	6,8	4,5

Table II
Growth in Western World Metal Consumption
(% per annum)

	1960s	1970s	1979-1984
Aluminum	9,3	4,6	1,0
Copper	4,3	2,3	0,0
Lead	3,4	1,2	-1,6
Nickel	7,0	3,4	-1,3
Tin	1,4	-0,2	-3,2
Zinc	5,1	1,4	-1,0
Steel	5,8	1,0	-3,2

The proposed link between GNP growth and metal demand as indicated in Tables I and II appears to be valid. The present day utility of this theoretical link between metal demand and economic growth depends, however, on two assumptions. The first is that the direction of causality of the influence is from economic growth to metal demand. Secondly, the historical relationship is assumed to be currently applicable and measurable in the light of recent technological and economic progress.

Examining first the assumption of the relationship's causality, both national income and aggregate consumption are considered to be dependent (endogenous) variables in simultaneous equation modelling of national income, each affecting the other (see equations 1 and 2).

$$C_t = bY_t + u_t \quad (1)$$

$$Y_t = C_t + I_t + G_t \quad (2)$$

where;

C_t = aggregate consumption;
I_t = investment;
G_t = government spending;
Y_t = national income;
b   = marginal propensity to consume;

and u_t = random disturbance term.

When considering a disaggregate consumption variable such as metal consumption, the relevance of equation 2 must be determined. If it is in fact relevant, there will be circular causality, and the first assumption will be invalid.

The small influence of metal consumption on aggregate consumption in the major metal consuming nations tends to indicate that the main relationship between GNP and metal consumption is as stated in equation 1. For example, Ridker and Watson (9) point out that given the fact that the non-fuel minerals sectors constitute a small and diminishing portion of the U.S. economy (0.42%), it would take very large price increases in minerals before any measurable effects on aggregate economic growth would be observed. The ability of metal consumption to influence economic growth, as in equation 2, thus rests entirely on the scarcity of mineral resources. If mineral materials are in short supply it is possible for economic growth to be hindered and therefore to be dependent on mineral availability. Myers and Barnett (15) have shown that for the past 120 years there has been an absence of mineral scarcity in Europe and America, implying that the quantities of minerals supplied (and consumed) have not been a detriment to economic growth. The assumed singular causality, with economic growth as the independent variable as given in equation 1, is reasonable.

The second consideration for investigation, and the focus of this paper, is the present degree of influence of economic growth on the demand for metals. The historical representation of this function has assumed that metal consumption is some function of, inter alia, economic growth, as measured by GNP or other economic indicators;

$$\text{Metal Consumption} = f(\text{GNP}) \qquad (3)$$

This function can be divided into functions of both GNP per capita and population;

$$\text{Metal Consumption} = f(\text{GNP per capita, population}); \qquad (4)$$

where;

$$\text{GNP per capita} \times \text{population} = \text{GNP} \qquad (5)$$

It thus becomes evident that there are in essence two factors contained in the GNP function which may affect consumption: population growth, and growth in per capita economic productivity. The separate effects of these and other determinants on metal consumption has been researched elsewhere (9). This paper is specifically concerned with the influence of economic productivity growth on consumption, exclusive of population influences. As such, the influence of growth in _GNP per capita_ on _per capita_ metal consumption is examined in the balance of this paper.

Empirical Evidence of the Declining Influence
of Economic Growth on Metal Demand

Intensity of Use

A primary issue to be examined is the constancy of the intensity of metal use; that is, the degree of metal consumption per unit of economic output over time. This ratio analysis of metal consumption per unit of economic activity is useful for our purposes as it removes the population effect of crude GNP and metal consumption figures. A declining intensity of use over time indicates a diminishing relationship of per capita consumption to per capita economic output.

Drucker (4) states that the amount of raw materials needed for a given unit of economic output in the United States has been dropping for the entire century, except in wartime. A recent study by the International Monetary Fund calculates the decline as being at the rate of 1,25% per annum (compounded) since 1900. Thus, the amount of raw material needed for 1 unit of industrial production is now 2/5ths of what it was in 1900. Research by Carter (6) indicates that approximately 3% less metals (excluding aluminium) were required annually from 1947 to 1962 to produce a given amount of economic output in the U.S.. In Japan, every unit of industrial production in 1984 consumed only 60% of the raw materials consumed for the same amount of industrial production in 1973 (4).

While not explicitly stating so, these researchers were referring to a declining intensity of use with time. Examining actual consumption figures, Table III gives the average ratio of tonnage metal consumption per constant dollar of GNP for the top several consuming countries of the Organisation for Economic Cooperation and Development (OECD) (16). In all cases the intensity of use is decreasing with time, the reason for which is debated in academic circles (1,8,10,17,18). Research by Fisher (10) also shows a declining intensity of use for molybdenum.

These results support the original theory of declining intensity of metal use in developed countries as posited by Malenbaum (8) and Brooks and Andrews (7). The intensities of use of these metals in developing countries (China, India, Mexico), however, are also declining (16). This is contrary to the theory that as developing countries industrialise, their intensities of use increase as did those of the developed countries in the early 1900s (8). The common belief that a rise in intensity of metal use in developing countries will offset the fall in use in developed countries must be re-evaluated.

Table III
Metric Tons of Metal Consumed per
Million Constant (1975) $U.S. of GNP
(Average of Top Several Metal Consuming OECD Countries)

	1965	1970	1975	1980	1983
Copper	2.01	1.64	1.26	1.33	1.23
Lead	0.91	0.74	0.58	0.70	0.66
Nickel	0.32	0.24	0.19	0.20	0.17
Tin	0.07	0.05	0.04	0.03	0.03
Zinc	1.27	1.01	0.73	0.73	0.68

Rigorous forecasts of U.S. economic growth and metal consumption by Ridker and Watson (9) permit the estimation of U.S. intensity of use for these metals in the year 2000, given in Table IV (16). It is clear that based on these forecasts, the intensity of metal use will continue to decline in the future, though at a diminishing rate as it asymptotically approaches zero.

Table IV
Metric Tons of Metal Consumed per
Million Constant (1975) $U.S. of GNP
(United States)

	1965	1970	1975	1980	1983	2000*
Copper	1.54	1.35	0.90	1.01	0.93	0.57
Lead	0.64	0.65	0.53	0.59	0.59	0.44
Nickel	0.13	0.10	0.09	0.08	0.07	0.06
Tin	0.05	0.04	0.04	0.03	0.02	0.01
Zinc	1.03	0.78	0.54	0.44	0.40	0.28

*2000 estimate: based on 1983 data and forecasts of metal consumption, Ridker and Watson, 1980:141.

As stated in the hypothesis, it is of interest to determine whether the decline in intensity of metal use for the major metal consuming nations has reached the stage whereby there no longer exists a measurable relationship between economic growth and metal demand. The following sections explore this possibility on a statistical basis.

Econometric Metal Demand Modelling

The traditional econometric metal demand function, while composed of many determinants, is often expressed mathematically as;

$$D_t = f(PM_{t-s}, PS_{t-s}, A_t, T_t) \qquad (6)$$

where;

D_t = metal demand;
PM_{t-s} = metal prices with lag distribution;
PS_{t-s} = prices of substitute materials with lag distribution;
A_t = income or industrial activity;
T_t = time or another proxy for technological change.

The subscript "t" implies that these terms are usually observed over time, such as years, quarters, or months, with the "s" indicating the lag operator (18).

Examining only the link between metal consumption and economic activity, Fisher (10) states that this relationship is often explained as;

$$\ln(\text{Consumption}) = A + B \ln(\text{Industrial Production}) \qquad (7)$$

where "B" is an indicator of elasticity of consumption with respect to industrial production.

In his paper, Fisher warns against the assumption that the "B" value is constant with time, indicating that it is in fact decreasing due to declining intensity of use. Thus, a continuing decline in intensity of use as shown in Table III above precludes the use of temporal statistical analysis to determine the parameter coefficients of the above equation. At any given time t there will be a specific elasticity value linking consumption to economic growth. At time t+1, the elasticity will have decreased due to the declining intensity of metal use. Figure 1a illustrates this concept. Assuming that annual per capita economic growth oscillates by way of example between +2% and -2% in consecutive years, one might expect resultant metal consumption growth to move from A to B in the first year of analysis, B to C in the second, C to D in the third, and so on to year n as the elasticity value (slope) drops. For ease of presentation, straight lines connect the data points, while in actual fact the lines would be curved due to a continuously declining elasticity. The curvature of the lines would be as indicated in Figure 1b. A linear regression of these results over the entire time period would produce line XY (see Figure 1a). The resultant slope would be the average elasticity over the period, but would overestimate the elasticity value applicable at the end of the time series. At an advanced stage of decline in intensity of metal use, the average slope would not be significantly different from zero, indicating an absence of relationship between the two variables.

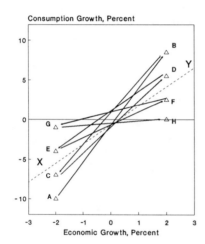 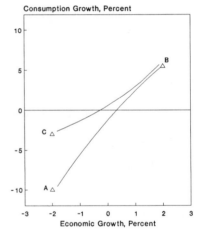

Figure 1a - Proposed temporal relationship between per capita metal consumption and per capita economic growth.

Figure 1b - Representation of a continuously declining elasticity of consumption.

A second possibility is that there would be breaks in the economic growth/metal consumption relationship due to increasing market instability and economic shocks (32). A resultant consumption pattern as illustrated in Figure 2 may result. In this case, the consumption elasticities would still drop with time, but regression results would be less significant due to the increased scatter of the data. A high frequency of nonconformities such as lines DE and FG may prevent a linear regression from being statistically established, again indicating an absence of relationship between the two variables. In this case however, a nonzero elasticity may periodically exist, but it would be frequently interrupted by shifts in the curve and would not be measurable.

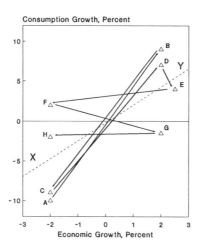

Figure 2 - Proposed temporal relationship between per capita metal consumption and economic growth with nonconformities present.

To test for these phenomena, a two stage process of investigation has been adopted. To determine whether there has in fact been any major structural change in historically derived metal demand/economic growth relationships, an existing pre-1970s model of steel consumption is tested for accuracy and applicability in a more recent time period. If changes in the consumption relationship have occurred, a second stage of investigation will examine the extent to which a current relationship between economic growth and metal consumption can be established.

A Test of an Existing Consumption Model for Current Applicability

The test model for consideration is a steel consumption model developed by Mo and Wang in 1970 (19). Steel consumption was chosen for analysis as it is a good proxy for the consumption of a number of metals whose demand is largely in the form of steel alloys. This model examines the annual per capita consumption of finished steel mill products in the United States from 1947 to 1965. The original model, given in equation 8, uses a linear function of two price variables and one income variable to describe aggregate steel consumption:

$$C_t = b_0 + b_1 PS_t + b_2 PNF_t + b_3 DI_t + u_t; \qquad (8)$$

where;

C_t = apparent per capita annual steel consumption at time t (10^{-2} short tons per capita);

PS_t = wholesale price index of finished steel mill products (adjusted by the wholesale price index for all commodities, 1957=100) at time t (1957-59 = 100);

PNF_t = wholesale price index of nonferrous metals at time t (1957-59 = 100);

DI_t = annual U.S. per capita disposable income at time t (10^2 dollars per capita);

u_t = random disturbance term;

and b_0, b_1, b_2, and b_3 are the coefficients to be estimated.

It was expected that the sign of b_1 would be negative, indicating that a rise in steel price would decrease consumption. b_2 was expected to be positive, as a rise in the price of a substitute product should increase the consumption of steel. The income variable, DI_t, was included in the equation to reflect the effect on steel consumption of a change in consumers' income, where at higher levels of income there would be a rising demand for goods and services. Thus, b_3 was expected to be positive.

The researchers also examined steel consumption technology as a factor affecting consumption, acknowledging that this is shifting the demand relationship of steel mill products. Changes in consumer tastes, induced usage changes resulting from research and development in steel and related commodity usages, changes in the mix of steel and related commodities, and other factors were included in a dummy technology variable. The difficulty in measuring this variable and its spurious derivation by the researchers, however, precludes its consideration in this analysis of the model.

Results of the ordinary least squares regression as derived by Mo and Wang are given in Table V (1947-65).

Table V
Regression Estimates for Equation 8
(Current price and income data)

Period	b_0	b_1	b_2	b_3	R^2	D-W
1947-65	31.32	-0.578	0.451	1.101	0.636	2.79
		(-4.084)	(3.948)	(1.792)		
1966-86	48.21	0.095	0.000	-0.196	0.521	1.70
		(0.487)	(0.006)	(-2.058)		

D-W = Durbin-Watson statistic.
Values in brackets are the coefficient t-statistics.

When a similar regression was run for data from 1966 to 1986 using Mo and Wang's regression variables, nonsense independent variable coefficients were produced (see Table V, 1966-86) (20,21,22). This is due to the fact that Mo and Wang used current nonferrous metal price and income data instead of real values, and the inflation effects in the post 1965 data invalidate the original model in the latter period. When the data from the original study was converted to real values using the wholesale price index to deflate the nonferrous metal price variable and the consumer price index to deflate the income variable, the original regression improves (see Table VI, 1947-65) (20,21,22). The R^2 of the regression increases, and the mean absolute percentage error (MAPE) is reduced from 6.07 to 5.83.

Table VI
Regression Estimates for Equation 8
(Real price and income data)

Period	b_0	b_1	b_2	b_3	R^2	D-W
1947-65	-3.68	-0.486	0.405	2.792	0.650	2.50
		(-3.966)	(3.400)	(2.767)		
1966-86	23.24	-0.527	0.341	1.610	0.565	1.82
		(-3.557)	(3.227)	(1.881)		

D-W = Durbin-Watson statistic.
Values in brackets are the coefficient t-statistics.

To test the validity of the improved 1947-65 model in the post 1965 period, data for the independent variables from 1966 to 1986 (real prices) was fitted into the model using the coefficients as presented in the first period in Table VI (1947-65). The resultant fitted consumption is mapped against actual consumption in Figure 3. It is obvious that the 1947-65 equation does not model the consumption of steel over this latter period well, with the MAPE jumping to 36.26 from 5.83 in the earlier period.

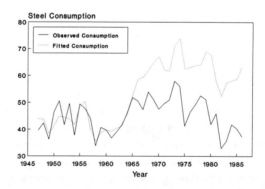

Figure 3 - Fitted vs. actual U.S. steel consumption (10^{-2} tons per capita), 1947-1986.

Considering the inability of the modelled equation derived from the earlier period data to explain steel consumption in the latter period, it may be concluded that either;

1. unmodelled factors had a larger influence on steel consumption in the latter period, or;

2. the relationship between the dependent and independent variables as specified changed dramatically in the latter period due to structural changes in the data.

When new independent variable coefficients were calculated using the 1966-86 real price data, reasonable regression results are obtained (see Table VI, 1966-86), though the coefficients change considerably from the first period. The lower t-statistic for the income variable should be noted. The fact that a reasonable regression was calculated for the latter period using the same variables as in the former period (R^2 = .565 versus .650, MAPE = 8.03 versus 5.83) tends to rule out the former explanation. The model _specification_ remains intact.

Piecewise linear regression (23) was used to investigate the proposal that the model error arises from a changing relationship between the dependent and independent variables. This technique tests parameter coefficients for a statistically significant change in slope when regressing the entire time series. Results of this test indicate that the income term (DI_t) of equation 8 had significant coefficient changes in the mid 1960s and, not surprisingly, in 1973. Regression results using both a simple regression on pooled data and the piecewise linear regression are given in Table VII (19,20,21,22). The piecewise regression is much improved on the simple regression, with the residual sum of squares dropping from 726 to 581.

Table VII
Regression Estimates for Equation 8
(Real price and income data, 1947-86)

	b_0	b_1	b_2	b_3	R^2	D-W
simple regression	23.62	-0.338 (-4.178)	0.357 (4.712)	0.742 (2.342)	0.535	1.78
piecewise regression		see equations 9, 10, and 11			0.628	2.01

D-W = Durbin-Watson statistic.
Values in brackets are the coefficient t-statistics.

The resultant period equations for the piecewise regression are given in equations 9, 10, and 11 below;

1947-64:

$$C_t = 4.071 - 0.405 PS_t + 0.366 PNF_t + 2.155 DI_t; \quad (9)$$
$$(-4.993) \quad (4.744) \quad (2.953)$$

1965-73:

$$C_t = 46.614 - 0.405 PS_t + 0.366 PNF_t + 0.920 DI_t; \quad (10)$$
$$(-4.993) \quad (4.744) \quad (0.047)$$

1974-86:

$$C_t = -17.072 - 0.405 PS_t + 0.366 PNF_t + 2.436 DI_t; \quad (11)$$
$$(-4.993) \quad (4.744) \quad (2.320)$$

with parameters as defined previously and t-statistics given in the brackets.

The income deflection in 1964 coincides with the start of an extended period of growth in the consumer price index in the United States, and no doubt a concomitant change in the structure of the economy and consumption and investment patterns (24). The 1973 deflection coincides with the first OPEC oil price increase. It is interesting to note that the changing income coefficients in equations 9, 10, and 11 indicate a ratchet-type link between economic growth and steel consumption. The income variable coefficient drops to zero during the period of highest economic growth (3.1% per annum from 1965 to 1973), yet is significantly different from zero during the two periods which experienced negative annual growth rates and lower average growth rates. An explanation for this phenomenon lies in the implications of changes in productivity on metal consumption. While a decline in productivity may logically reduce the consumption of metals, an increase in productivity is indicative of increasing investment in capital and technology. Productivity in itself implies improved output per unit of input, and the technological component can be assumed to further extend the degree of production possible from a given amount of material input. The ratchet effect would result in a slightly different transience in elasticity of consumption, as given in Figure 4. In this case, negative economic growth rates would result in an increasing slope or elasticity of consumption (line BC), as compared to Figure 1b where there was a continually decreasing elasticity.

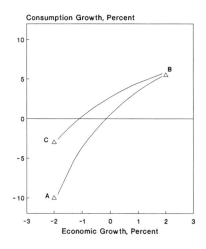

Figure 4 - The ratchet effect influence of economic growth on metal consumption.

We may conclude that the relationship between per capita steel consumption and real per capita income changed significantly in the mid 1960s and again in 1973 in the United States. The poor degree of fit of the inflation-adjusted pre 1965 model to actual post 1965 data is consistent with the theory that declining intensity of use and structural changes in the U.S. economy are creating changing metal consumption functions. Given the ties between steel consumption and the consumption of several associated alloy metals, an implication of this finding is the possible present day irrelevance of other metal consumption relationships and theories developed using pre 1974 data.

The Current Relationship Between Metal Consumption and Economic Growth

Having determined that the latest identifiable structural change to affect steel consumption appears to have taken place in the U.S. economy in 1973, the current (post 1973) economic growth/metal consumption relationship for several metals is now examined. Annual percentage changes in U.S. and Japanese per capita consumption of platinum, copper, iron ore and agglomerates, ferromanganese, ferrochromium, nickel, zinc, silver, tin, and lead are regressed against respective changes in per capita GNP and industrial production for the period 1974 to 1986 (1975 to 1986 for platinum). Consumption is defined as primary and secondary metal consumption, excluding recycled scrap. Consumption data collected for analysis was specifically chosen to avoid measuring apparent changes in consumption caused by stockpile movements. These ten metals have been selected for study as they represent approximately 70% of the total Rand value of South African export mineral sales, excluding gold, diamonds and coal (3). Japan and America together consume an average of 43% of Western World primary tonnage consumption of these metals and minerals (25,26,27), and are the two largest consumer markets susceptible to the uncoupling phenomenon. A single equation ordinary least squares linear regression of the annual percentage change in per capita consumption versus growth in per capita economic activity is used to test the significance of the relationship between economic growth and metal consumption.

Values of the variables under examination have been converted to yearly percentage change for three reasons. Firstly, univariate regressive modelling of time series data which contain trends often provide a statistical complication through serial correlation of the variables. First-differencing is a satisfactory technique for removing this autocorrelation (28). Secondly, it was desired to model the effect of economic growth on metal consumption, and not merely model the existing trend relationship between the two variables. Given the causality previously established in this paper, examining changes in consumption achieves this end. Thirdly, the coefficient of the independent variable in the resultant regression equation is the elasticity of consumption, a value which is desired for analysis. While logarithmic transformation of the variables, as with Fisher's equation, also satisfies the third criterion, it does not satisfy the two former criteria. As noted previously, the results of this regression will determine the average elasticity of consumption during the data period, and will not be an accurate reflection of the relationship in the most recent data period (1986).

Regression analysis of growth in per capita metal consumption and per capita economic growth is modelled as indicated in equation 12;

$$C_t = a_0 + a_1 EI_t + u_t; \qquad (12)$$

where;

C_t = percentage growth in per capita tonnage metal consumption during year t, USA or Japan;

EI_t = real percentage growth in per capita economic indicator. The economic indicators used in this study are GNP and the Industrial Production Index, adjusted to a per capita basis;

u_t = random disturbance term;

and a_0, a_1 are the parameters to be estimated. a_1 is the elasticity of consumption, and is expected to be non-negative.

The first relationship analysed is the effect of per capita U.S. GNP growth on per capita metal consumption. Results are given in Table VIII (25,26,27,29,30,31). The results of the regressions show that the correlation between metal consumption and economic growth is barely significant for lead, tin, and zinc, and not significant for platinum and silver. The average income elasticities of consumption (a_1) for the significant correlations are all significantly greater than zero, being between 3 and 7. Plots of the data using the method indicated in Figure 1a indicate an average estimated elasticity (slope) of 7 in the first half of the data series, dropping to near 3 in the more recent half. Several nonconformities in demand are evident for each metal, though in most cases these are not strong enough to prevent the derivation of a significant correlation coefficient. As an example of such a plot, a scatter graph of per capita growth in copper demand is given in Figure 5. The linear regression estimate is shown by the broken line.

Table VIII
Per Capita Metal Consumption Growth vs
Growth in Real GNP per Capita, 1974-86, USA

Metal	a_0	a_1	$t(a_1)$	R^2	D-W
Copper	-7.572	5.203	5.411#	0.745#	2.12
Chrome	-12.725	6.883	3.644#	0.570#	2.59
Iron ore	-13.350	4.418	4.405#	0.660#	2.94*
Lead	-4.516	3.808	2.069*	0.300*	2.07
Manganese	-16.333	5.239	5.522#	0.753#	1.99
Nickel	-8.750	3.214	3.244#	0.513#	2.24
Platinum	-1.541	3.802	1.628	0.228	2.22
Silver	-5.445	1.306	1.436	0.171	2.22
Tin	-6.257	2.511	2.441*	0.373*	2.14
Zinc	-6.455	3.167	2.645*	0.412*	2.15

D-W = Durbin-Watson statistic.
(*) indicates significance at the (one-sided) 5% level.
(#) indicates significance at the 1% level.

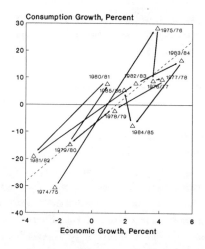

Figure 5 - Per capita growth in U.S. copper consumption vs. per capita growth in GNP, 1974-1986.

Since the elasticity in recent years is lower than the average elasticity as measured by the regression, significance of the t-statistic at the 1% level is probably the best indicator of a definite current relationship between economic growth and metal consumption. Thus, it is unlikely that any measurable relationship currently exists between per capita lead, platinum, silver, tin, or zinc consumption and economic growth as measured by GNP per capita.

Table IX gives the regression results for Japanese metal consumption. Data for the consumption of iron, manganese, and chrome in Japan were not available.

Table IX
Per Capita Metal Consumption Growth vs Growth in Real GNP per Capita, 1974-86, Japan

Metal	a_0	a_1	$t(a_1)$	R^2	D-W
Copper	-10.281	3.846	1.292	0.143	2.51
Lead	-18.897	7.278	1.407	0.165	2.33
Nickel	-27.414	8.716	1.691	0.222	2.74
Platinum	-17.212	4.228	0.802	0.067	1.91
Silver	1.848	0.695	0.231	0.005	2.07
Tin	-11.054	3.085	1.033	0.096	2.99*
Zinc	-15.242	4.765	1.524	0.188	2.62

D-W = Durbin-Watson statistic.
(*) indicates significance at the (one-sided) 5% level.
(#) indicates significance at the 1% level.

The Japanese results are remarkably worse than those for the U.S., with no significant correlation between metal consumption and per capita GNP growth. Examination of the original data in scatter plots reveals no discernible consumption pattern. Plots of the Japanese data in temporal sequence yield graphs similar to that illustrated in Figure 2, with the many nonconformities causing scattering of the data and low regression coefficients. As an example, Figure 6 shows a plot of per capita growth in zinc consumption and GNP. The regression estimate is shown by the broken line.

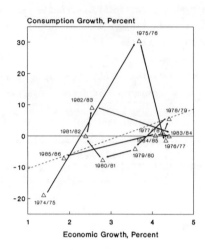

Figure 6 - Per capita growth in Japanese zinc consumption vs. per capita growth in GNP, 1974-1986.

Some mineral economists argue that the falling demand for metals is due to the changing composition of GNP as the developed countries move towards a more services-oriented and less metal intensive economy (8). In investigation of this claim, the same regressions are run using the per capita industrial production index as the economic indicator. The industrial production index is an indication of strictly industrial activity, such as mining, quarrying, manufacturing, and the production of electricity, gas and water, and specifically excludes services. The industrial production index for Japan, for example, accounts for the activity of 9 mining and 523 manufacturing concerns. Results of these regressions are given in Tables X and XI.

Table X
Per Capita Metal Consumption Growth vs
per Capita Growth in the Industrial Production Index,
1974-86, USA

Metal	a_0	a_1	$t(a_1)$	R^2	D-W
Copper	-2.992	2.385	6.691#	0.817#	2.21
Chrome	-6.739	3.202	4.258#	0.644#	2.98*
Iron ore	-9.270	1.902	4.202#	0.638#	3.25*
Lead	-1.340	1.859	2.438*	0.373*	1.90
Manganese	-11.498	2.257	5.195#	0.730#	2.64
Nickel	-6.020	1.536	3.969#	0.612#	2.95*
Platinum	2.060	1.467	1.255	0.149	1.89
Silver	-3.954	0.435	1.002	0.091	2.09
Tin	-4.295	1.310	3.362#	0.531#	2.27
Zinc	-3.727	1.489	3.011#	0.475#	2.11

D-W = Durbin-Watson statistic.
(*) indicates significance at the (one-sided) 5% level.
(#) indicates significance at the 1% level.

Table XI
Per Capita Metal Consumption Growth vs
per Capita Growth in the Industrial Production Index,
1974-86, Japan

Metal	a_0	a_1	$t(a_1)$	R^2	D-W
Copper	-0.391	1.032	2.134*	0.313*	1.97
Lead	-0.781	2.170	2.842#	0.447#	1.31
Nickel	-5.555	2.539	3.670#	0.574#	2.34
Platinum	-4.026	0.375	0.283	0.009	2.02
Silver	2.070	0.758	1.542	0.192	1.53
Tin	-4.533	1.343	3.535#	0.555#	2.46
Zinc	-3.534	1.477	3.502#	0.551#	2.02

D-W = Durbin-Watson statistic.
(*) indicates significance at the (one-sided) 5% level.
(#) indicates significance at the 1% level.

The regression results improve when the industrial production index is used as the economic indicator. This indicates that the inclusion of services in GNP tends to lessen the relationship between economic growth and metal consumption. The relationship between industrial activity and metal consumption is, however, still insignificant at the 1% level for several of the metals under study. The absence of relationship between economic growth and metal demand cannot therefore be totally explained by an increasingly services-oriented GNP. It would appear that declining intensity of metal use in industrial production, as highlighted by Drucker, is also occurring, resulting in limited instances of linkage between industrial production growth and metal demand. This finding is concurrent with a declining intensity of metal use in the industrially-oriented developing countries.

Conclusions and Implications of the Research

This paper has shown that, in the U.S. and Japan, there was at best a weak relationship between annual per capita economic growth and growth in per capita metal consumption for several major metals in the period 1974 to 1986. Significant correlations were absent in Japan for all metals analysed, and there was no relationship found for platinum and silver in the United States. Weak relationships were found for lead, tin and zinc metals in the U.S. during the sample period, but it is unlikely that these relationships existed at the end of the sample period due to apparently declining elasticities of consumption. A continuous decline in intensity of metal use in services and industrial production combined with structural economic changes were found to be the likely causes of these phenomena. Such structural changes were statistically shown to have twice changed U.S. consumption of steel in the period 1947 to 1986.

A declining intensity of metal use in developed and developing countries other than those examined in this paper permits the extrapolation of these findings to metal consumption in general. It may thus be concluded that a globally declining intensity of metal use is an ongoing phenomenon serving to minimise and in many cases eliminate the effect of economic growth on metal demand. This confirms that the theory of "uncoupling" posited by Drucker applies to metals. While the significant statistical correlations between U.S. economic growth and metal demand that were demonstrated for some metals would appear to refute this claim, further declines in intensity of metal use are anticipated to uncouple these correlations in the future. The advanced stage of uncoupling in technologically-advanced Japan is an indication that the existing relationships evident in the U.S. may be short-lived.

In the light of these conclusions the views of the South African mining industry must be reevaluated. The economic growth of trading partners cannot be seen as a measurable and identifiable factor that will revive export demand for metals in general, and any such linkage may in fact only be in effect during economic declines (the ratchet effect). The implications of these findings are that mineral producers will increasingly be required to actively engage in metal and non-metal market expansion activities if demand for their product is to be sustained. Many other commodity producing industries have faced flagging demand successfully through market creation and development. In the mining industry, some Western World mineral and metal producers have already begun to market their products through international market development organisations, following the success of Inco, Amax, Alcoa and Alcan in developing markets for their metal products (2). The foresight of those mineral producers who seek to expand existing markets and generate new markets for their products in defiance of falling metal demand is well placed, while those waiting for economic recovery to increase the demand for their product can expect disappointment.

REFERENCES

1. Sir Roderick H. Carnegie, *Outlook for Mineral Commodities* (New York, NY: Group of Thirty, 1986).

2. G. A. Davis, "Stimulating the Demand for South African Ferrous and Non-Ferrous Metals Through End-Use Marketing" (MBA thesis, University of Cape Town, 1987).

3. Minerals Bureau of South Africa, *South Africa's Mineral Industry 1986* (Braamfontein, South Africa: Department of Mineral and Energy Affairs, 1987).

4. Peter F. Drucker, *The Frontiers of Management: Where Tomorrow's Decisions Are Being Shaped Today* (London: Heinemann, 1986), 21-30.

5. Stuart Smith, "Canadian Mining Not in Hopeless State," *Northern Miner*, 72(50)(1987), B13-B15.

6. A. P. Carter, "Changes in the Structure of the American Economy, 1947 to 1958 and 1962," *The Review of Economics and Statistics*, 49 (1967), 207-224.

7. D. B. Brooks, and P. W. Andrews, "Mineral Resources, Economic Growth, and World Population," *Science*, 185(4145)(1974), 13-19.

8. W. Malenbaum, "Law of Demand for Minerals," *Proceedings of the Council of Economics*, 104th annual meeting (New York, NY: American Institute of Mining, Metallurgical, and Petroleum Engineers, 1975), 147-155.

9. R. G. Ridker, and W. D. Watson, *To Choose a Future* (Baltimore, MD: John Hopkins University Press, 1980).

10. J. F. C. Fisher, "The Analysis of Metal Markets and Future Prices," *Proceedings of the 13th Congress of the Council of Mining and Metallurgical Institutions*, vol. 5 (Singapore, 1986), 1-8.

11. P. Brown, and R. Ball, "Some Preliminary Findings on the Association between the Earnings of a Firm, Its Industry and the Economy," *Empirical Research in Accounting: Selected Studies, 1967*. Supplement to *Journal of Accounting Research*, 5 (1967), 78-80.

12. Baruch Lev, "On the Use of Index Models in Analytical Reviews by Auditors," *Journal of Accounting Research*, 18(2)(1980), 524-532.

13. G. Foster, *Financial Statement Analysis* (London: Prentice-Hall International, 1986), 196-200.

14. J. B. Cohen, E. D. Zinbarg, and A. Zeikel, *Investment Analysis and Portfolio Management*, 3rd edition (Homewood, Illinois: Irwin, 1977), 234-235.

15. John G. Myers, and Harold J. Barnett, "Minerals and Economic Growth," Economics of the Mineral Industries, 4th edition, ed. W. A. Vogely (New York, NY: American Institute of Mining, Metallurgical, and Petroleum Engineers, Inc., 1985), 3-17.

16. World Resources Institute and the International Institute for Environment and Development, World Resources 1986 (New York, NY: Basic Books Inc., 1986), 297-298.

17. Gary A. Campbell, "Theory of Mineral Demand," Economics of the Mineral Industries, 4th edition, ed. W. A. Vogely (New York, NY: American Institute of Mining, Metallurgical, and Petroleum Engineers, Inc., 1985), 176-179.

18. John E. Tilton, "The Metals," Economics of the Mineral Industries, 4th edition, ed. W. A. Vogely (New York, NY: American Institute of Mining, Metallurgical, and Petroleum Engineers, Inc., 1985), 383-415.

19. W. W. Mo, and K. Wang, A Quantitative Economic Analysis and Long-Run Projections of the Demand for Steel Mill Products (Washington, DC: U.S. Dept. of the Interior, Bureau of Mines Information Circular 8451, 1970).

20. U.S. Department of Commerce, Statistical Abstract of the United States, 1979, 100th ed. (Washington, DC: Government Printing Office, 1979).

21. U.S. Department of Commerce, Statistical Abstract of the United States, 1988, 108th ed. (Washington, DC: Government Printing Office, 1988).

22. U.S. Department of Commerce, Survey of Current Business, various issues, 1966-1987.

23. R. S. Pindyck, and D. L. Rubinfeld, Econometric Models and Economic Forecasts, 2nd edition (Tokyo: McGraw-Hill International Book Co., 1981).

24. William F. Sharpe, Investments, 3rd ed. (London: Prentice-Hall International, Inc., 1985), 241-249.

25. World Bureau of Metal Statistics, World Metal Statistics Year Book (New York, NY: World Bureau of Metal Statistics, 1984,1988).

26. Council for Mineral Technology (Mintek), private communication with author, Cape Town, 1988.

27. Johnson Matthey, Platinum 1987 (London: Johnson Matthey, 1987).

28. C. W. J. Granger, and P. Newbold, "Spurious Regressions in Econometrics," Journal of Econometrics, 2 (1974) 111-120.

29. American Bureau of Metal Statistics, Non-Ferrous Metal Data 1978 (New York, NY: American Bureau of Metal Statistics, 1979).

30. U.S. Bureau of Mines, *Minerals Yearbook*, various issues, 1974-86 (Washington, DC: U.S. Bureau of Mines, Dept. of the Interior).

31. International Monetary Fund, *International Financial Statistics*, various issues (Washington, DC: International Monetary Fund).

32. J. E. Tilton, and W. A. Vogely, ed., "Market Instability in the Metal Industries," Special Issue: *Materials and Society*, 5(3)(1981), 257-265.

THE MATHEMATICAL MODELLING OF NON-FERROUS BLAST FURNACES

K C Wade, M Cross and R Smith[*]

Centre for Numerical Modelling and Process Analysis
Thames Polytechnic
Wellington Street
London SE18 6PF

[*]Capper Pass Ltd
PO Box 6
North Ferriby
Hull HU14 3HD

ABSTRACT

The mathematical modelling of engineering applications is becoming highly developed. Many processes are now being examined in detail by using mathematical models resulting in major savings in operation and development costs. In the metallurgical industry much attention has been focussed on ferrous blast furnaces. It is the aim of this paper to convert the experience gained on such furnaces to non-ferrous furnaces. The general methodology for solving the governing differential equations is presented bearing in mind that the model is two-phase with additional source terms to account for the physics of the problem. These source terms include heat transfer and resistance to flow. The model also accommodates the change of state of the solid phase.

INTRODUCTION

The Problem Considered

With the operating costs of blast furnaces rising daily the need for mathematical models that can correctly predict the behaviour within the furnace are becoming increasingly important. Such models should be capable of predicting flow patterns and temperature distributions and have the ability to handle the complex chemical reactions that take place within the furnace.

Previous models of the lead blast furnace [1,2] have only considered the kinetics of the furnace where a uniform gas flow through the furnace is assumed. This approach is useful for post-operative analysis and for identifying trends, etc. They have not, however, proved reliable for detailed design criteria or as the basis of effective control schemes.

The situation can be compared with that previously used to model iron blast furnaces. A detailed knowledge of the gas flow distribution and the production of the cohesive zone [3] has led to a better understanding of the overall working of the furnace.

The present study is concerned with the prediction of the flow distribution and heat transfer in a lead blast furnace. Due to the slow transportation time of the solid a transient model is inappropriate and the settling of the solid removes the need for the effects of gravity to be included.

THEORETICAL MODEL

The Physical Problem

The basic features of tin/lead blast furnace, used at Capper Pass Ltd, are outlined in Figure 1. Air is injected through tuyeres on the side (14 in this case) while coke, sintered ore and unprepared charge are loaded in the top of the furnace. The coke, sintered ore and unprepared charge, come in various shapes and sizes, however, this study will assume that all solids are represented by spheres (by the use of a shape factor).

As cool air is blown up through the furnace it reacts with the coke (which is burnt and converted to the gaseous phase) raising its temperature and that of the surrounding sinter until the metal reduces and is collected at the base of the furnace (hearth). The molten metal is then tapped out. The study will not concern itself with the hearth of the furnace and will further assume that the solids flow out of the furnace as a molten metal slag product, through a tap hole, which is continuously open (at least in concept).

The Dependent and Independent Variables

The following are the dependent variables of the problem

- o gas and solid velocities
- o pressure
- o solid and gas concentrations
- o solid and gas enthalpies
- o three concentrations of the solid phase; viz coke, ore and molten metal/slag, (referred to hereafter as molten metal).

The independent variables include dimensions of the furnace, etc

The Partial Differential Equations

The equations are derived by considering the balance of fluxes over a control volume small enough to give the desired spatial distribution in the complete system, yet large enough to contain many solid particles, so that the average velocities and volume fractions are meaningful. The control volume can be regarded as containing a volume fraction, r_i, of the ith phase, so that if n phases exist altogether

$$\sum_{i=1}^{n} r_i = 1$$

Each phase is treated as a continuum in the control volume under consideration. The phases "share" the control volume.

In the following presentation of the differential equations, the dependent variables of the gaseous phase will be denoted by lower case letter and those of the solid phase by upper case. For variables other than main dependent variables (e.g. density, viscosity) the notation is subscript 1 for the gaseous phase and 2 for the solid phase.

In summary, the important assumptions upon which this flow model is based are listed below:

(1) each phase is a perceived as a continuum, so that derivations are uniquely defined,

(2) the phases are assumed interdispersed and coupled by appropriate interaction terms,

(3) all gases obey the ideal gas equation of state,

(4) all solids (coke, sintered ore and unprepared charge) are represented by one phase and thus one temperature, velocity, etc.,

(5) the density of the solid concentrations are assumed constant, although the bulk density of the solid phase is not,

(6) potential flow is assumed for the solid phase,

(7) the solid phase is unaffected by the gas pressure.

(a) The mass-conservation equations

Gas-phase mass equations

$$\frac{\partial}{\partial y}(\rho_1 rv) + \frac{\partial}{\partial x}(\rho_1 ru) = 0 \tag{1}$$

Particle-phase mass equations

$$\frac{\partial}{\partial y}(\rho_2 RV) + \frac{\partial}{\partial x}(\rho_2 RU) = 0 \tag{2}$$

The volumetric fractions are related by

$$r + R = 1 \tag{3}$$

(b) The conservation of momentum equations

Gas-phase momentum equations

u-velocity

$$\frac{\partial}{\partial y}(\rho_1 rvu) + \frac{\partial}{\partial x}(\rho_1 ru^2) + \frac{\partial}{\partial x}\left[r\mu_1 \frac{\partial u}{\partial x}\right] + \frac{\partial}{\partial y}\left[r\mu_1 \frac{\partial v}{\partial y}\right] = -r\left[\frac{\partial p}{\partial y} + \frac{\partial P}{\partial x}\right]$$
$$- f_2(u-U) \quad (4)$$

v-velocity

$$\frac{\partial}{\partial y}(\rho_1 rv^2) + \frac{\partial}{\partial x}(\rho_1 ruv) + \frac{\partial}{\partial y}\left[r\mu_1 \frac{\partial v}{\partial y}\right] + \frac{\partial}{\partial x}\left[r\mu_1 \frac{\partial u}{\partial x}\right] = -r\left[\frac{\partial p}{\partial y} + \frac{\partial p}{\partial x}\right]$$
$$- f_2(v-V) \quad (5)$$

where f_2 is the interphase friction term.

Particle-phase momentum equations

The solid phase is assumed to act under potential flow, i.e. ρ_2 = constant and μ_2 = 0, thus instead of solving the normal momentum equations for the particle-phase, stream functions are calculated. These being:

$$U = -\frac{\partial \Psi}{\partial y} \quad \text{and} \quad V = \frac{\partial \Psi}{\partial x} \quad (6)$$

Equation (6) can be substituted into the vorticity equation

$$\frac{\partial V}{\partial x} - \frac{\partial U}{\partial y} = 0 \quad (7)$$

and solved by the SOR method to obtain the stream functions, Ψ. The velocities U and V can then be calculated from (6).

(c) The conservation of enthalpy equations

Gas-phase enthalpy equations

$$\frac{\partial}{\partial y}(\rho_1 rvh) + \frac{\partial}{\partial x}(\rho_1 ruh) + \frac{\partial}{\partial x}\left[r\Gamma_h \frac{\partial h}{\partial x}\right] + \frac{\partial}{\partial y}\left[r\Gamma_h \frac{\partial h}{\partial y}\right] = hA(T_2 - T_1) \quad (8)$$

Particle-phase enthalpy equations

$$\frac{\partial}{\partial y}(\rho_2 RVH) + \frac{\partial}{\partial x}(\rho_2 RUH) + \frac{\partial}{\partial x}\left[R\Gamma_H \frac{\partial H}{\partial x}\right] + \frac{\partial}{\partial y}\left[R\Gamma_H \frac{\partial H}{\partial y}\right] = hA(T_1 - T_2) \quad (9)$$

(d) The conservation of species equations

Coke species (C)

$$\frac{\partial}{\partial y}(\rho_2 RVC) + \frac{\partial}{\partial x}(\rho_2 RUC) + \frac{\partial}{\partial x}\left[R\Gamma_C \frac{\partial C}{\partial x}\right] + \frac{\partial}{\partial y}\left[R\Gamma_C \frac{\partial C}{\partial y}\right] = 0 \quad (10)$$

Ore species (O)

$$\frac{\partial}{\partial y}(\rho_2 RVO) + \frac{\partial}{\partial x}(\rho_2 RUO) + \frac{\partial}{\partial x}\left[R\Gamma_o \frac{\partial O}{\partial x}\right] + \frac{\partial}{\partial y}\left[R\Gamma_o \frac{\partial O}{\partial y}\right] = 0 \qquad (11)$$

and C + O + M = 1 where M = molten metal.

(e) The continuity equation

$$\frac{\partial}{\partial y}(rv\rho_1) + \frac{\partial}{\partial x}(ru\rho_1) = 0 \qquad (12)$$

Normally for multi-phase flows a "joint" continuity equation would be employed, however, in this case the pressure is restricted to the first phase only. Hence a joint continuity equation is inappropriate.

Auxiliary Relations

Equation of State for the gas phase

The ideal gas law

$$P = \rho_1 \bar{R} T_1$$

is used as the equation of state for the gas phase, where $\bar{R} = 8314/32$ kg^{-1}k^{-1}, and T_1 is the temperature of the gas phase.

The Interphase Friction Coefficient

The expression used for the interphase friction force is the Ergun equation [4] to describe the pressure drop across a porous layer. This is, generalised to vectorial form to describe the gas flow through isotropic porous media [5] described as

$$-\nabla p = (f_1 + f_2 |\underline{V}|)\underline{V} \qquad (13)$$

where f_1 and f_2 represents the viscous and inertial resistances. Extending (13) to account for two-phase flow and taking the viscous effects as negligible due to a high enough Reynolds number in the furnace leads to

$$-\nabla p = f_2 |\underline{V} - \underline{v}|(\underline{V} - \underline{v}) \qquad (14)$$

where

$$f_2 = \frac{1\cdot 75(1-\epsilon)}{\epsilon(\bar{\Phi}_2 dp)} \rho_1$$

The solid particle density

Each of the concentrations of the 2nd phase, coke, ore and molten metal, can exhibit its own density and are all independent of the direct effect of temperature. The present model assumes that either coke and ore or coke and molten metal exists in a cell, thus the solid density can be calculated from either

$$\begin{array}{l} \rho_2 = \alpha \rho_{coke} + (1-\alpha)\rho_{ore} \quad \text{or} \\ \rho_2 = \alpha \rho_{coke} + (1-\alpha)\rho_{metal} \end{array} \qquad (15)$$

where α = either the coke:ore ratio or the coke:molten metal ratio.

For the purpose of the potential flow of the solids the density is assumed constant, and obtained via an average of the loading rates.

The Melting of the Ore

A simplistic approach to the melting of the ore has been taken. When the temperature of the solid phase increases above the melting temperature of the ore the diffusitivity of the ore changes from $\Gamma_o = \Gamma_c$ to $\Gamma_o = \alpha_o \Gamma_c$, i.e. simulating the change of state from ore to molten metal. This has the effect of allowing the molten metal to drip down to the base of the furnace and giving the model a better representation of the solids.

The Raceway Region

When the coke enters a region where combustion would occur, i.e. the tuyere combustion zone, the combustion of coke is modelled by setting the concentration of the coke to zero at that point.

FINITE-DOMAIN EQUATIONS

The Grid

The finite-domain versions of the governing differential equations are derived by integrating the above equation over a control volume. The grid arrangement employed is the standard staggered grid approach with the control volumes being different for the momentum and the scalar variables. The practice is conventional [6]. The grid employed is shown in Figure 1b. Clearly to accurately model the shape of the furnace partial cell blockages or body fitted coordinates will be required. However, the grid arrangement used serves to illustrate the furnace adequately.

The Equations

The result of integrating the equations is expressed in terms of the values of the variables pertaining to grid points. The practice for multi-phase flows is similar to that for single-phase flows [6] and is not presented here.

THE SOLUTION PROCEDURE

The solution procedure must determine sets of values, for all points of the grid for all phases; p, h, H, c_i, at central grid points and of u, U, v, V at velocity grid points. As all neighbouring grid cells affect the centre nodes an iterative method must be employed. The current trend for two-phase flow is to use the IPSA (<u>I</u>nter<u>p</u>hase <u>S</u>lip <u>A</u>lgorithm) procedure [7]. However for the case of the blast furnace with exhibits large pressure drops for the gas phase the linking of the pressure to both phases causes unrealistic solutions for the solid phase [3]. Thus a different approach has been taken in solving the finite-domain equations, an outline of this procedure is given below.

(1) Determine the boundary conditions for all variables.

(2) Solve the finite-domain equations for the stream functions by SOR and calculate U and V velocities from (6).

(3) Solve the finite-domain equation for the momentum at constant IX, by way of the tri-diagonal matrix algorithm (TDMA). The usual convention of solving two-phase flows by the partial elimination algorithm can not be

employed here as the code has been developed for true multi-phase problems.

(4) Solve the finite-domain equation for the enthalpy at constant IX.

(5) Compute the errors in the continuity equation and formulate the pressure-correction equation.

(6) Solve for the pressure corrections by TDMA and apply to the velocities (u and v) the resulting corrections, proportional to pressure, correction differences.

(7) Return to step 3 for the next IX step until the whole domain has been swept.

(8) Repeat 3-8 until the continuity errors computed at step 6 are sufficiently small.

The above procedure has been implemented into a multi-phase code environment CASCADE [8].

RESULTS

The following results are shown to give an idea of the flow patterns, temperature, etc and not accurate predictions hence no grid independent studies are shown.

The grid used was 15 × 10 which required about $4 \cdot 0 \times 10^{-3}$ cpu seconds per grid node/per sweep/per main variable, on an Apollo DN3000 to attain a converge solution. The above time includes of course the computation of the auxiliary variables.

The basic input data used is given summarised in Table 1. Where possible the data has been taken from the blast furnaces given in Figure 1(a).

Boundary Conditions

The boundary conditions for the lead furnace, with reference to Figure 1b, are set out below.

In the above model, an axis of symmetry is assumed at y = 0, defined as

$$\frac{\partial \Phi}{\partial y} = 0 \qquad (16)$$

where Φ represents the two velocity components of the gas phase and the enthalpy of both phases.

The gas inlet conditions are specified in the region of the tuyere as

$$\frac{\partial}{\partial y} (r\rho_1 v) = \beta$$

$$\frac{\partial^2 u}{\partial y^2} = 0 \qquad (17)$$

$$\frac{\partial^2 h}{\partial y^2} = 0$$

As no chemical reactions are included in this preliminary study, the incoming air is pre-heated to simulate the effect of chemical reactions near the tuyere region.

For the furnace wall the no-slip condition is assumed for both gas and solid phase, and adiabatic walls are assumed. The solid velocities are determined by potential flow involving the solution of stream functions, thus the boundary conditions are expressed in terms of these stream functions.

At the axis of symmetry the stream boundary condition is defined as

$$\psi(x,0) = 0 \tag{18}$$

The solid inlet is from the furnace top, thus the stream condition is

$$\psi(x_{top}, y) = U^{in} y \tag{19}$$

and the outlet via the tuyere on the furnace wall where the stream condition is set as

$$\psi(x, y_{wall}) = \begin{cases} U^{in} y_{wall}, & \text{above tuyere} \\ 0, & \text{at/below tuyere} \end{cases} \tag{20}$$

The base of the furnace is set by a no flow condition, i.e.

$$\psi(0,y) = 0 \tag{21}$$

To simulate the variable size of the solids as they enter the furnace a particle size distribution across the furnace has been provided for cases 2-6, with smaller particles located at the furnace walls.

Results for the Furnace Simulation

The results of the six cases considered are shown graphically in Figures 2 to 7.

Each figure shows five plots which indicate the following:

(a) gas velocity and temperature.
    The velocity field is relative to case 1 and the temperature contours are equally spaced and normalised with respect to the inlet temperature of the gas.

(b) gas pressure.
    The contours are equally spaced in the range 200 - 3000 N/m$_2$.

(c) solid velocity and temperature.
    The velocity field is relative to case 1 and the temperature contours are again equally spaced and normalised with respect to the inlet temperature of the solid. The shaded region represents the cohesive zone of the furnace.

(d) gas density.
    The contours are equally spaced in the range 0·146 - 0·863 kg/m^3.

(e) molten metal/slag product concentration.
    The contours are equally spaced in the range ·1 - 1.

The six cases considered vary the diffusivity of the solid (at the melting point of the ore) and the gas mass flow rates, as indicated in Table 2. Obviously the gas mass flow rates apply to the mass of gas injected into the furnace and not through one tuyere.

A closely packed contour field indicates that the variable under question is changing rapidly in that region, and where the contours are widely separated a small change is indicated.

The velocity vectors show the direction and magnitude of the resulting velocity components.

Allowing the particle size to vary across the top of the furnace has a marked effect on the gas distribution through the furnace, Figures 2-3. The smaller particles, near the furnace wall, are closely packed and hence the resistance to flow will be higher. This causes the gas to flow through the centre of the furnace.

Figure 2 also shows the effect of keeping the diffusivity of the solid the same all the way through the furnace, and the particle distribution across the furnace constant.

Clearly changing the gas flow rate, Figures 4-6, has a marked effect on the gas and solid temperature and the molten metal concentration. Due to the relatively small pressure drop across the bed, compared to the absolute pressure the density change is relatively unaffected. As the higher flow rate of the gas transports through the furnace the cohesive zone is pushed away from the tuyere area, (Figure 3), thus increasing the molten metal produced.

Not surprisingly changing the diffusivity of the ore only, Figure 7, has little effect on the gas or solid flow distribution. The concentration of liquid metal however is redistributed around the tuyere zone.

A general point of interest is the high pressure build up below the furnace cohesive zone. This is clearly noticed in Figures 4, 6 and 7, (cases 3, 5 and 6). This large pressure drop does not appear in the case of a large gas flow, presumably due to the cohesive zone being further up the bed.

Also the liquid concentrations are of general interest to engineers. If the liquid concentrations are significantly above the level of the tuyere the tuyeres may become blocked, indicated by a large pressure drop there. On the other hand the molten metal must be kept to a certain level to maintain the production rate.

CONCLUSIONS

The above model clearly serves to illustrate the behaviour of the gas and solid inside a non-ferrous blast furnace. However, to faithfully model the furnace the complex chemical reactions will have to be included and work is progressing in this direction.

These chemical reactions will be able to give a better representation of the heat release during combustion and hence the heat exchange between the two phases.

ACKNOWLEDGEMENTS

This paper is based on work funded jointly by Capper Pass Ltd, Hull and the Science and Engineering Research Council. The authors would further like to thank Capper Pass for providing all the data for the model.

REFERENCES

[1]   J T Chao, P J Dugdale, D R Morris and F R Steward, "Gas Composition, Temperature and Pressure Measurements in a lead blast furnace", Metall Trans B, 1978, Vol 9B, p 293.

[2]   J T Chao, "A Dynamic Simulation of a Lead Blast Furnace", Metall Trans B, 1981, Vol 12B, p 385.

[3]   K A Fenech, M Cross and V R Voller, "Numerical Modelling of the Cohesive Zone Formulation in the Iron Blast Furnace", PCH, Vol 9, No 1/2, 1987, pp 71-83.

[4]   S Ergun, (1952), "Fluid Flow through Packed Columns": Chemical engineering progress, Vol 48, No 2, pp 89-94.

[5]   V Stanek and J Szekely (1974), "Three Dimensional Flow of Fluids through non-uniform packed beds", AIChE Journal, Vol 20, No 5, pp 974-980.

[6]   S Patankar, (1982), Numerical Heat Transfer and Fluid Flow: McGraw-Hill.

[7]   D B Spalding, (1976), "The calculation of free-convection phenomena in gas-liquid mixtures", Proc ICHMT Semin, Dubrovnik, Yugoslavia, p 569.

[8]   M K Patel, K C Wade and M Cross, (1989), "The multi-phase approach" submitted to Numerical Methods in Laminar and Turbulent Flow Conference, Swansea 1989.

Coke density	1616 Kg/m^3
Ore density	1952 Kg/m^3
Gas flow rate	92·4 Kg/s
Solid flow rate	300 T/24 hr
Gas inlet temperature	1350°K
Solid inlet temperature	314°K
Melting temperature of the ore	1300°K
Softening temperature of the ore	1100°K
Solid specific heat	1·0467 KJ/Kg/K
Gas specific heat	1·1764 KJ/Kg/K
Heat transfer coefficient	1×10^{-1} $KJ/m^2/s/K$
Latent heat of melting	$1·84 \times 10^2$ KJ/Kg

Table 1:   Input data used for Case 1

Case	α_0	β kg/s	Figure
1	1	92·4	2
2	1	92·4	3
3	1×10^7	92·4	4
4	1×10^7	115·5	5
5	1×10^7	69·3	6
6	1×10^5	92·4	7

Table 2: α_0 and β for each case considered

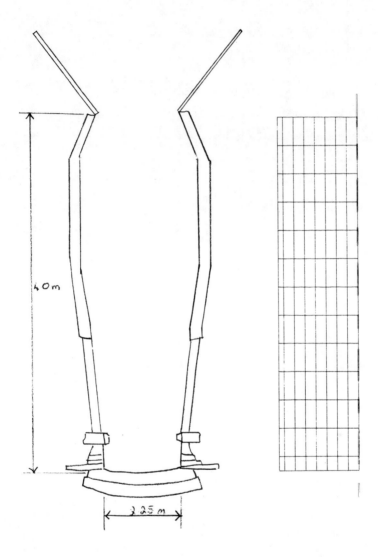

Figure 1: A section of the blast furnace and grid arrangement

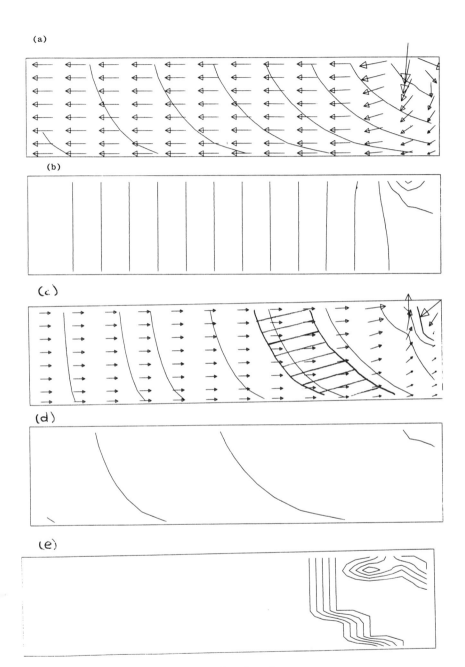

Figure 2: Case 1

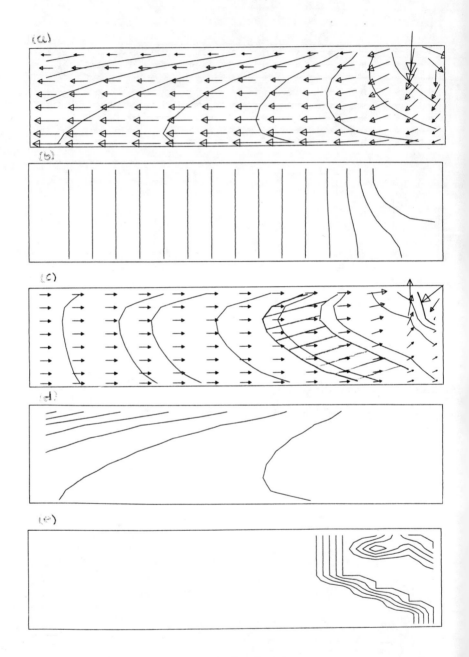

Figure 3: Case 2

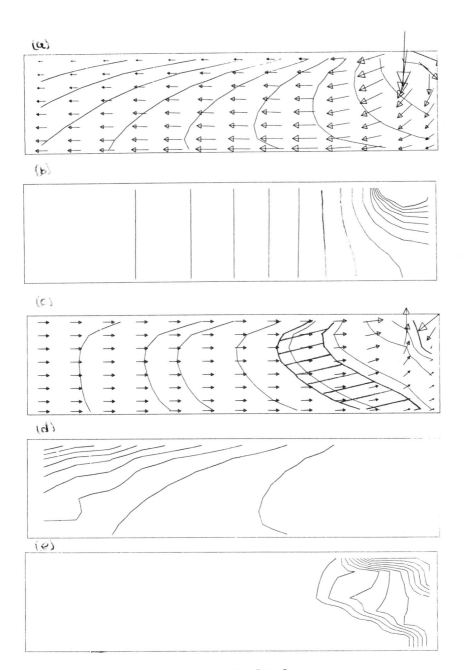

Figure 4: Case 3

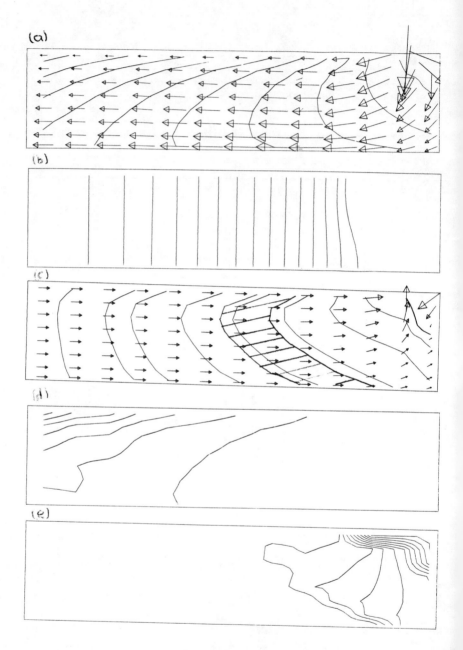

Figure 5: Case 4

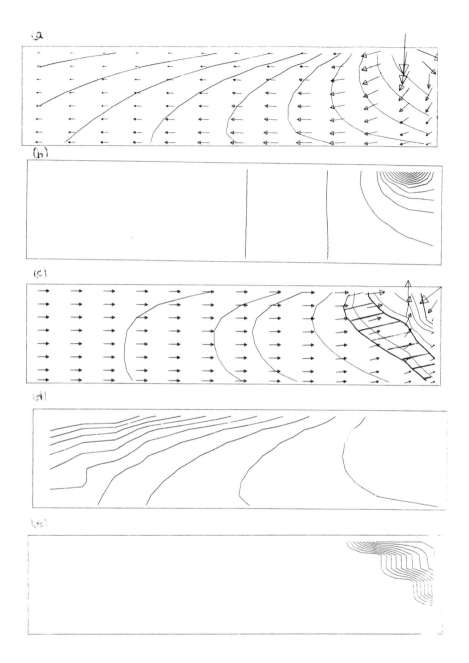

Figure 6: Case 5

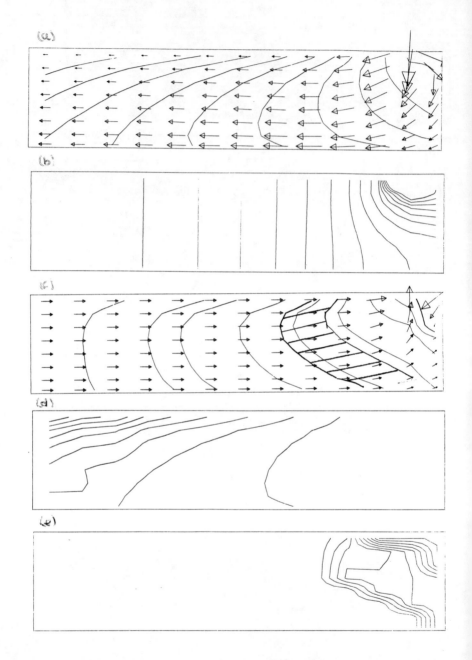

Figure 7: Case 6

TREATMENT OF METALLURGICAL GASES NOW AND TOMORROW

Mr Kurt Westerlund and Mr Rauno Peippo

A. Ahlstrom Corporation, Finland
Engineering Division

Kurt Westerlund
A. Ahlstrom Corp.
P.O.Box 5
00441 HELSINKI
FINLAND

Rauno Peippo
A. Ahlstrom Corp.
P.O.Box 184
78201 VARKAUS
FINLAND

TREATMENT OF METALLURGICAL GASES NOW AND TOMORROW

Gas treatment in metallurgical plants can generally be divided into the following functions:

* cooling and heat recovery
* dust precipitation
* elimination of gaseous substances that cannot be emitted in the atmosphere

The treatment of gases in metallurgical plants today account for a substantial amount of the conversion costs of the raw materials to metal. The investment cost of a gas handling line is 25 - 50 % of the total investment cost of a smelter. It is thus important to select a gas handling solution that will meet the future objectives of the smelter. Gas handling must be seen as an integrated part of process design which has implications on both performance, investment and operational costs. In the following we focus our discussion on the first function, gas cooling and heat recovery, its different technical solutions and implication on downstream equipment.

1. WASTE HEAT BOILERS (W.H.B.)

The cooling and recovery of heat from hot sticky gases in the non-ferrous industry has generally been done with forced circulation horizontal boilers cooling the gases, typically containing 200 g/nm3 molten particles, from 1250 °C to 400 °C. The A. Ahlstrom Corporation has supplied the metallurgical industry with waste heat boilers of which representative examples are:

* Outokumpu flash smelter boilers
* Kivcet flash smelter boilers
* Converter hoods and boilers
* Fluidized bed roster boilers for Zn, Co and Pyrites
* TBRC boilers

See Figure 1.

1.1 Heat transfer surface temperatures

The dusts depositing on the boiler tubes have a low sticking temperature. They are also good insulators and consequently the gas surface temperature of even thin dust layers raises to the sticking point of the dust. The dust can be knocked down easily at low temperatures when it is brittle. It is hence essential to keep the tube wall dust layers as thin as practically possible.

A lower limit on the tube temperature in SO2 containing gases is set by the corrosion occurring below the dew point for H2SO4 formed from the always present H2O and SO3. This limits the tube temperature for SO2 containing gases to a minimum of 250 °C corresponding to 40.6 bar saturated steam.

From a steam value point of view higher temperatures are generally preferred and in the case where H2SO4 corrosion is present this sets the lower limit.

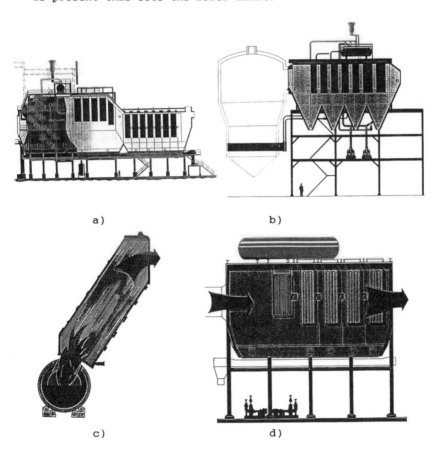

Figure 1. Waste Heat Boilers, a) flash smelter b) roster c) converter hood d) converter boiler

1.2 Horizontal W.H.B.

A typical Horizontal Waste Heat Boiler will consist of the following components:

- Waste Heat Boiler
  - steam drum
  - radiation chamber
  - convection chamber
  - dust hopper
  - spring hammers

- feed water treatment
- feed water tank
- feed water pumps
- water circulation pumps

Ahlstrom has supplied the majority of the Outokumpu flash smelters with boilers of this kind starting 1953.

Each Waste Heat Boiler has to be designed to meet properties of the gases entering properties. Generally the convection section is started at temperatures when the dust in the gas is not sticky. The sulphatisation of the dust in SO2 containing gases, that occurs when the dust is cooled, lowers the stickiness temperature of the dust. This makes it even more important to design the boiler so that the tubes and tube membrane walls can be kept clean and that the gases are cooled as fast as possible. The dust precipitated out of the boiler is generally recycled back to the process whereas more volatile components can be bled out in the final dust precipitation.

1.3 Vertical W.H.B.

A Vertical Waste Heat boiler offers the possibility of returning the dust collected on the tube walls directly into the furnace, thus eliminating the need of a hopper and simplifying dust handling. A Vertical Waste Heat boiler is easier to keep clean from dust than a Horizontal Boiler as it has no hopper. It is also more straightforward compact design.

Ahlstrom has delivered such boilers in the 80's, one example being the boiler on the Kivcet Pb-flash smelter in Italy on Sardinia. See figure 2.

In this vertical boiler 65 - 75 % of the dust emitted out of the furnace is returned directly back to the furnace.

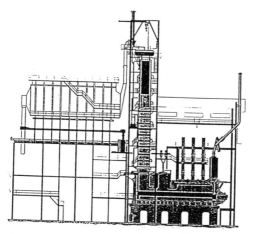

Figure 2. Vertical Waste Heat Boiler on the Kivcet Pb-flash smelter on Sardinia.

2.0 CLEANING METHODS

The importance of keeping the boiler tubes clean cannot be overemphasized. By process and boiler design the demand for cleaning can be kept within manageable limits. The dust build-ups in metallurgical boilers have been reduced by:

- gas recycling
- shot cleaning
- soot blowing with air or steam
- shaking
- building tube membrane wall radiation surfaces

As the mechanical impact methods for cleaning eroded the boiler tubes and where inefficient, Ahlstrom started to develop spring hammers to knock down dust in the early 70's. They have proven to be very efficient and non-damaging to the boiler. Spring hammers are now incorporated in all new Waste Heat Boilers designed by the company.

3.0 UTILIZATION OF RECOVERED HEAT

The utilization of the recovered heat has to be designed according to local needs and opportunities. Here are some examples:

- production of electricity
- drying of concentrate
- preheating of reaction air
- heating of other processes in the plant
- local heating in the plant and community

In some applications the super heater is not incorporated in the waste heat boiler as the higher surface temperatures of the super heater tubes might create sticky dust buildup on them. High temperature variations in the primary process is another reason for going to an external super heater that secures safe turbine operation.

4.0 FLUXFLOW

4.1 FLUXFLOW technology

FLUXFLOW is a registered trade mark of A. Ahlstrom Corporation and the system has been patented in over 10 countries. The FLUXFLOW system is based on the circulating fluid-bed technology. It is intended for heat recovery and precleaning of contaminated high temperature gases in the metallurgical industry, the steel industry and the chemical industry.

The operation principle of FLUXFLOW is shown in figure 3. The system comprises three main components:

* mixing chamber
* heat exchanger, water/steam or air
* separation of solids from the cooled gas (cyclone)

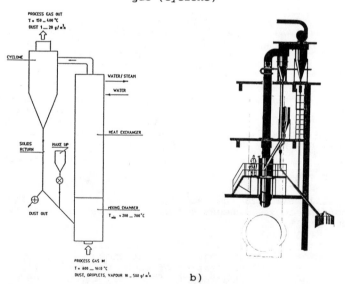

Figure 3. FLUXFLOW, a) the operation principle b) cement kiln alkali by pass application

In the mixing chamber the hot, uncleaned, process gas is rapidly "quenched" to the mixing temperature by the circulated cooled solids. The mixing temperature can be controlled to maintain it below the sticking point of the solids. From the mixing chamber the solids are entrained with the gas flow and transported as a suspension through the cooler. The heat is typically recovered as high pressure steam, hot water or as preheated air. In the primary separation stage, usually a cyclone, the solids are returned back to the mixing chamber.

In most FLUXFLOW applications the amount of solids in the primary circulation loop would be increased during operation by captured dust. To maintain the amount of solids in circulation at the desired level the system is equipped with a controllable outlet for dust. For start up a feeding system for make up solids is provided.

For final gas cleaning secondary cyclones have turned out to be effective and simple in applications where the dust amount is high and the dust is immediately returned to the process.

4.2 Features of FLUXFLOW

In general the FLUXFLOW technology offers the following benefits for heat recovery, gas cleaning and the total process:

* can handle dirty gases containing sticky particles or condensable material

* increases the particle size of the dust by collecting molten and evaporated material on the surface of the recycled particles

* high cooling rate provides the possibility to reach metastable equilibriums in order to avoid harmful reactions during cooling (e.g. sulphatisation)

* combined heat recovery and pretreatment of process feed is possible (preheating, prereduction)

  can be used as a chemical reactor integrated with the total process

* compact, vertical design economical installation

* self cleaning, eliminating the need for auxiliary cleaning systems

4.3 FLUXFLOW experience

Applications successfully tested	m³n/h	MW
Cu-flash smelting	600	0.25
Cu-Ni-slag reduction	600	0.25
PYROFLOW coal gasifier	3 600	1.00
Cu-matte smelting in an electric furnace	900	0.25
Pb-concentrate smelting in a top blown rotary converter	900	0.25

Full scale applications	m³n/h	MW
Cement kiln alkaline bypass (in continuous operation since late 1986)	2 200	1.00
Pb-concentrate smelting in a top blown rotary converter (on stream in spring 1989)	20 000	7.00

5.0 OBJECTIVES FOR CONCEPTUAL DESIGN

The features of the gas handling units discussed above can be characterized as follows:

	Horizontal W. H. B.	Vertical W. H. B.	FLUXFLOW W. H. B.
heat transfer coefficient W/m2C	moderate k	moderate k	high 2 to 4k
driving force temperature C	moderate t	moderate t	lower 0.5 to 0.6 t
heat transfer surface m2	high A	high A	lower 0.4 to 1.0 A
dust collection efficiency %	50	50	90
pressure drop	low	low	moderate
dust return	hopper conveyors	direct	direct or conveyed
need for heat transfer surface cleaning	yes	yes	no
cooling rate	low	low	very high
mass transfer solid/gas	low	low	very high

Treatment of metallurgical gases has to be seen as an integrated part of the total process design. The possibility of improved and new technology has to be evaluated with an open mind and on rational criteria.

The feasibility for different flow sheets can at a preliminary stage be evaluated based on material and energy balances, investment and operational costs.

CARBOTHERMIC REDUCTION KINETICS OF TIN CONCENTRATE PELLETS

V. N. Misra

Kalgoorlie Metallurgical Laboratory
Chemistry Centre, Department of Mines
PO Box 881, Kalgoorlie, 6430 Western Australia

Abstract

Tin metal is commercially produced by the carbothermic reduction of tin concentrate (SnO_2) in a furnace. The mechanism of tin oxide reduction reaction by carbon is not well understood, as comparatively few studies on the kinetics and mechanism of reduction of cassiterite have been carried out.

The present study was therefore carried out on the kinetics and mechanism of reduction of the naturally occurring tin concentrate pellets by charcoal powder. The influences of temperature, pellet size, time of reduction and the relative amount of carbon were ascertained. The results indicate that tin ore pellets are reduced directly to metallic tin without SnO formation as intermediate product and the overall rate is controlled by the oxidation of carbon by CO_2 and follows the diffusion controlled mechanism.

Introduction

Recently, there has been enormous interest in the study of the kinetics and mechanism of reduction of metal oxides. Much of the work published in the literature, however, belongs to the reduction of ferrous oxides. Hydrogen or ($H_2 + H_2O$), carbon monoxide or ($CO+CO_2$) and to some magnitude, solid carbonaceous material have been employed as reductant (1-5). On the other hand rather few studies have been reported on the carbothermic reduction of non-ferrous metal oxides (6-9).

The reduction of cassiterite by carbon is commonly expressed by:

$$SnO_2(s) + C_{(s)} = Sn(s, l) + CO_2(g) \qquad (1)$$

where s, l and g represent solid, liquid and gaseous phases respectively. The carbothermic reduction of solid SnO_2 can be conceived as the reaction occurring at the contact point between SnO_2 and carbon particles. After the production of a minute quantity of metallic tin, the reaction can progress only by the solid state diffusion of one reactant species to another through product layer. This reaction is excessively slow and complete reaction is almost unachievable by this mechanism. Thus, another scheme represented by following equations:

$$SnO_2(s) = 2CO(g) = Sn(s, l) + 2CO_2(g) \qquad (2)$$

$$CO_2(g) + C(s) = 2CO(g) \qquad (3)$$

is expected to be well founded for the carbothermic reduction of tin oxide.

A great majority of investigations quoted on the reduction of metal oxides with solid carbon involve the two stage mechanism. Eq. (3) progresses at a slower rate and therefore, is the rate controlling step in the overall reduction process. As will be seen afterwards, this has also be shown to be the case for the carbothermic reduction of tin oxides. Consequently, the reduction of cassiterite by carbon monoxide is close to equilibrium and its kinetics do not affect the overall rate. The oxidation of carbon by CO_2 has been comprehensively studied and the mechanism is reasonably well understood (10, 11). Turkdogan and Vinters (12) have studied the oxidation of various types of carbon in CO_2-CO mixtures. Rao and Jalan (13) studied the oxidation of graphite in CO_2 gas.

Studies prior to these two have also been published in the literature. The catalytic effect of metals on the oxidation of carbon has been investigated by several researchers. Misra and Standish (11) have reviewed the literature critically. Turkdogan and Vinters (14) examined the response of iron impregnated graphite to the oxidation in $CO-CO_2$ mixtures and also studied the oxidation of graphite impregnated with silver, copper, zinc, cobalt, nickel and iron in air.

The generation of the lower oxide of tin, SnO is thermodynamically unfavourable (9, 15, 16, 17). Platteeuw and Meyer (15), demonstrated experimentally that solid stannous oxide is unstable and dissociates into tin and tin dioxide at temperatures about 300°C. They also revealed that the pSnO over SnO_2/Sn mixtures is small at low temperatures and is less than 0.5 mmHg at low contents in $CO-CO_2$ mixtures.

The investigations reported here were undertaken to determine the nature and extent of the reactions occurring during the reduction of tin oxide ores by solid carbon. It was hoped that the information obtained regarding the characteristics, rate and mechanism of reduction of tin ore would be useful in the production of the reduction reactions that take place in the tins melting furnaces.

The rate and mechanism of reduction of tin concentrate pellets by charcoal was determined at temperatures ranging from 700 to 1,100°C, the reduction being carried out in a packed bed reactor in nitrogen atmospheres. The change in mass of the sample during reduction was measured and the products of reactions were examined by X-ray diffraction, chemical analysis and optical microscopy.

Experimental Details

Materials

The tin concentrate used throughout these tests was a high grade ($\approx$ 65% Sn). Tables I and II show the size distribution and XRF analysis of cassiterite sample. Mineralogical examination indicated that cassiterite was a major valuable mineral in the ore with minor associations of iron oxide, titanium oxide and wolframite. The predominant gangue mineral was quartz. The concentrate was ground to -75µm size and pelletized with 0.5% bentonite and varying proportions of charcoal powder (stoichiometric and 5, 10, 15 and 20% excess carbon) in the laboratory. The pellets were dried at 110°C for 2 hours.

Table I. Size Distribution of Cassiterite Sample

Size Range	% Weight Retained	% Weight Accumulated
+ 1.7mm	0.21	0.21
-1.7mm + 425µm	18.9	19.18
-425µm + 212µm	35.51	54.69
-212µm + 106µm	36.16	90.85
-106µm + 75µm	5.35	96.20
- 75µm	3.80	100.00

Table II. Cassiterite Sample Analysis by X-Ray Fluorescence Analysis

Compound	Percent	Element	Percent
SnO_2	83.12	Sn	65.51
NES LOI	0.75	-	-
Fe_2O_3	0.663	Fe	0.464
WO_3	0.56	W	0.444
Nb_2O_5	0.199	Nb	0.139
TiO_2	0.090	Ti	0.054
CuO	0.017	Cu	0.014
SiO_2	<0.01	Si	0.0046
NiO	<0.001	Ni	7.87×10^{-4}
SO_3	<0.001	S	4×10^{-4}
ZnO	<0.001	Zn	8.02×10^{-4}
ZnO_2	<0.001	Zn	7.4×10^{-4}
As_2O_3	140.22 (PPM)	As	85.5 (PPM)

Experimental Procedures

The kinetics of reduction were determined by means of the weight loss measurement techniques. The experiments were carried out in a packed bed reactor.

The composite pellets were placed in a recrystallised alumina crucible and was introduced into the hot zone of the furnace, previously set at the required temperature. During the period of experimental runs, a flow of nitrogen was maintained. The change in mass was continuously recorded and at the end of the run the reduced samples were rapidly removed from the hot zone of the furnace and cooled in air. The reduced sample was subjected to chemical analysis, to ascertain total tin, iron and carbon contents. This determination was necessary as a check on the calculation of the degree of reduction of the pellets. XRF and optical microscopy were used for the detection of changes in the mineralogy of tin oxides at various stages in reaction with charcoal. The percentage reduction is defined as the percentage removal of the available oxygen i.e. the oxygen present in the ore associated with tin and iron.

Results and Discussions

Figure 1 shows theoretical equilibrium ratio pCO_2/pCO as a function of temperature for SnO_2 reduction. This figure reveals that simultaneous equilibrium between tin oxide, metallic tin and carbon will take place at the temperature where curves for Eqs. (2) and (3) meet. In equilibrium with an atmosphere of $pCO_2 + pCO = 1$ atm., this is at about 630°C.

This implies that SnO₂ will be reduced by carbon at any temperature above 630°C, and the product gas mixture will have concentration between the values for Eqs. (2) and (3). It will be nearer to the value for the reaction which has the higher reaction rate.

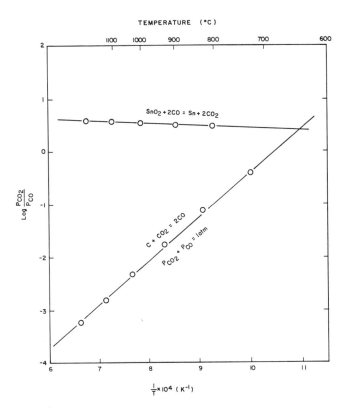

Figure 1: THEORECTICAL EQUILIBRIUM RATIO pCO_2/pCO AS A FUNCTION OF TEMPERATURE FOR SnO_2 REDUCTION

The effect of pertinent parameters such as temperature, various concentration of charcoal powder and size of the pellets on the overall reduction was studied. Figure 2 shows a typical weight loss data for the reduction of SnO_2 by charcoal powder at various temperatures (700-1,000°C).

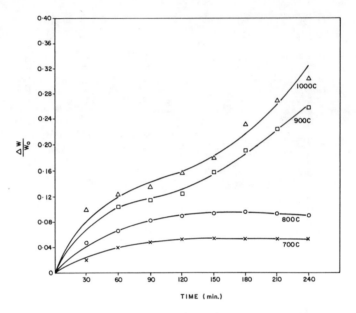

FIGURE 2 - EFFECT OF TEMPERATURE ON THE REDUCTION OF SnO_2 BY CHARCOAL (STOICHIOMETRIC AMOUNT)

The percentage reduction with time increased markedly as the temperature increased, and the extent of reduction for a given addition of charcoal increased significantly with temperature, as shown in Fig. 3. Charcoal additions above 10% (by mass) had very little effect on the extent of reduction at lower temperatures (700-800°C), but the effect of increased charcoal additions became more important at 1,000°C. In these diagrams ΔW is the weight loss of the sample and W_0 is the initial weight.

Of the various physical and chemical processes that are involved in the reduction of cassiterite intimately mixed with fine charcoal powder, the first to be considered is the initial formation of carbon monoxide. This may take place by one or more of the following mechanisms:

a) oxygen of the entrapped air combining with carbon particles;

b) carbon reacting with the oxygen released by the decomposition of tin oxide;

c) oxygen chemically adsorbed on the carbon surfaces liberated as carbon monoxide; and

d) direct reduction taking place at the points of contact between oxide and carbon particles.

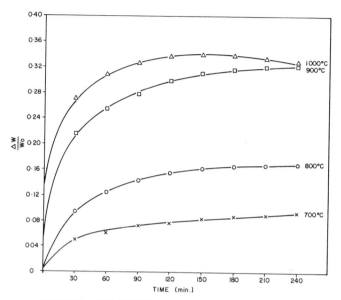

FIGURE 3 - THE EFFECT OF TEMPERATURE ON THE REDUCTION OF SnO_2 BY 5% EXCESS CARBON

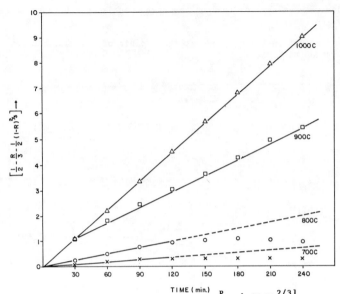

FIGURE 4 - PLOTS OF FUNCTION $[\frac{1}{2} - \frac{R}{3} - \frac{1}{2}(1-R)^{2/3}]$ VS TIME FOR THE REDUCTION OF SnO_2 BY CHARCOAL

Carbon monoxide thus generated readily reacts with cassiterite particles and reduction proceeds. This supports the assumption that the overall rate is controlled by the rate of oxidation of carbon.

The reduction samples were examined by X-ray analysis and the results confirmed that SnO was not produced as an intermediate compound during the reduction. At this stage the reduced pellets contained SnO_2, Sn and carbon. X-ray analysis after complete reduction showed the presence of mostly tin.

In order to understand the kinetic aspects of the reduction process, it is desirable to examine all possible rate controlling steps involved in the reduction of oxides. These may be classified into two categories:

a) Chemical reactions at a solid-gas or solid-solid interface.

b) Transport processes (both heat and mass) to or from the different interfaces.

For solid-solid and solid-gas reactions that are chemical, diffusion or mixed controlled, the following kinetic equations described the overall process. These equations, reproduced below, have been found to be very useful in correlating experimental data.

Chemical Control

$$r_o \, d_o \, [1-(1-R^{1/3})] = K_c \, t \qquad (4)$$

Diffusion Control

$$d_B r_o \, [1/2 - \frac{R}{3} - 1/2 \, (1-R)^{2/3}] = 3bDC_s t \qquad (5)$$

where r_o is the initial radius of the oxide particle, and R is the fraction reacted at time t, d_o is the density of oxide, D is diffusion coefficient, b is the number of moles of component B which react with component A, Cs is the concentration of reactant on the surface of the particle, d_B is the molar density of component B. Kc is rate constant for chemical controlled reactions.

The equations were derived on the common understanding that:

a) The reacting particles remain spherical during the course of reduction.

b) The reaction is diffusion-controlled and obeys the Fick's law.

c) One of the reactants diffuses into the particles of the other.

In deriving Eqs. (4) and (5), it is further assumed that the size of the particle remains unchanged during the reaction. In addition to these, Eq. (4) also assumes that diffusion takes place through a plane surface.

Figure 4 shows the data for reduction of cassiterite pellets plotted in terms of the function [1/2 - R/3 - 1/2 (1-R)2/3] against time, because this function was found to fit the data best.

After an initial time delay, the data followed a linear relationship up to approximately 95% reduction when plotted on this basis. Above 95% reduction the rate of reduction decreases. Furthermore, according to Eq. (5), the time to effect a given value of R should be proportional to r_o^2. For runs of equal values of effective diffusivity the time taken to reduce the pellets of different sizes to 60% reduction was found to be approximately proportional to r_o^2 and not proportional to r_o. Hence on the basis of Eq. (5), the data supports the hypothesis that reduction is controlled by gaseous diffusion through the product layer. The rate constant is a strong function of temperature, as shown in Fig. 5.

An activation energy of 58.4 Kcal/mol for charcoal was calculated in the temperature range 1,198 to 1,273K. Meaningful comparison of the energy of activation of this work, with the result of earlier work is difficult since there is no kinetic data on the carbothermic reduction of natural tin oxide available in the literature. The results of separate work indicates that particle size has little effect on the reaction rate. This could happen if the kinetics of the Boudouard reaction (C/CO_2 reaction) were much faster than the reduction of SnO_2 by CO.

Since this is not the case, the insensitivity of the reaction rate on the particle size of carbon is most likely to be due to the high porosity of the carbon particles, in which case particle size does not significantly change the surface area of the carbon.

The overall rate of reaction between SnO_2 and charcoal in this study is controlled by the oxidation of carbon by CO_2 and thus the measured rate reflects the kinetics of the latter.

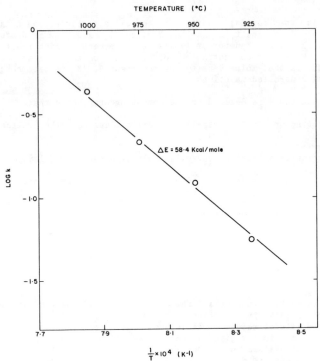

FIGURE 5 - ARRHENIUS PLOT FOR THE REDUCTION RATE USING CHARCOAL POWDER

Microscopic examination of the reduced samples indicated that the amount of metallic tin increased with increasing time and temperature of reduction. In some cases, tin layer were encapsulating the unreduced cassiterite particles. Table III shows the chemical analysis of various carbothermically reduced tin concentrate pellets respectively. As evident from Table III, the metallic tin content in the reduced pellets ranged between 74.6 and 91.6%. The degree of metallization was observed to be more than 95% in the pellets reduced at 1,000°C for 60 minutes. This result appears to be promising for the production of sponge tin from high grade tin concentrates as practiced in the case of iron ore pellets.

Table III. Chemical Analysis of the Reduced Tin Ore Pellets

Temperature °C	Reduction Time in min.	Amount of Carbon % Excess	Sn Content
700	60	5	67.7
800	60	5	71.7
900	60	5	90.9
1000	60	5	93.5
1100	60	5	91.6

Conclusions

The results of the present work clearly indicate that the reduction of cassiterite by charcoal proceeds by the way of gaseous intermediate CO and CO_2 and the rate is controlled by the oxidation of carbon. The rate of reaction of SnO_2C composite pellets depended on the amount of carbon present indicating the rate controlling reaction is associated with the carbon. An activation energy of 58.4 KCal.mole calculated for the reduction of SnO_2 with charcoal indicates that temperature has a pronounced effect on the reaction rate. Also the rate controlling factor is reactant diffusion through the product layer (Sn-layer). The reduced metallic tin pellets (sponge tin) could be used either as a charge in the tin smelting or as a source of tin in hydrometallurgical process of tin extraction.

References

1. Y. K. Rao, "The Kinetics of Reduction of Haematite by Carbon" Metall. Trans, 2, (1971), 1439-1447.

2. H. Y. Sohn, and P. C. Chanbal, "Kinetics of Iron Ore Reduction" Trans. Iron Steel Inst., Japan, 24 (5), (1985), 387-395.

3. R. J. Fruehan, "The Rate of Reduction of Iron Oxides by Carbon", Metall. Trans., 8B, (1977), 279-286.

4. R. J. Tien, and E. T. Turkdogan, "Mathematical Analysis of Reaction in Metal Oxide/Carbon Mixture", Metall. Trans., 8B, (1977), 305-313.

5. V. N. Misra, and P. R. Khangaonkar, "Recent Progress in the Understanding of Iron-Oxide Reduction". J. Sci. and Ind. Res., New Delhi, 34, (1976), 107-115.

6. I. J. Lin, and Y. K. Rao, "Reduction of Lead Oxide by Carbon", Trans. Inst. Min. Metall. 1984, Section C, 1(975), 76-82.

7. W. J. Rankin, "Reduction of Chromite by Graphite and Carbon Monoxide" Trans. Inst. Min. Metall. 1989, Section C, (1979), 107-113.

8. V. N. Misra, and P. R. Khangaonkar, "Hydrogen Reduction Kinetics of Manganese Ore Pellets", Trans. ind. Inst. Metals 28, (1975), 169-173.

9. P. A. Wright, Extractive Metallurgy of Tin (Elsevier Publishing Co. New York, 1982).

10. V. N. Misra, and N. B. Gray, "Reactivity of Brown Coal Charcoal to Carbon Dioxide", Presented at Aust/Japan Extractive Metall. Symp., Sydney, AusI.M.M. Publ. (1980), 419-427.

11. V. N. Misra, and N. Standish, "Reactivity of Carbon in Pyrometallurgical Process - A Review of Theoretical Analysis and Process Variables", Presented at the 13th Congress of C.M.M.R. Ed. L. E. Fielding, and A. R. Gordin, 4, (1986), 143-153.

12. E. T. Turkdogan, and J. V. Vinters, "Oxidation of Carbon in Co-CO Atmosphere", Carbon, 8, (1970), 39-53.

13. Y. K. Rao, and B. P. Jalan, "Oxidation of Graphite in Co Gas", Metall. Trans., 3, (1972), 2465-77.

14. E. T. Turkdogan, and J. V. Vinters, "Oxidation of Iron Impregnated Carbon in CO Gas", Carbon, 10, (1972), 97-111.

15. J. C. Platteeuw, and G. Meyer, "Thermodynamic Stabilities of Tin Oxides", Trans. Faraday Soc. 2, (1956), 1066-73.

16. U. J. Ibok, and R. Rawlings, "Effect of Cassiterite grain Size on it's Rate of Reduction by Hydrogen", Trans. Inst. Min. Metall., 91, Section C, (1982), 93-94.

17. I.d J. Bear, and R. J. T. Caney, "Selective Reduction of a Low-Grade Cassiterite Concentrate", Trans. Inst. Min. Metall. Section C, (1976), 139-146.

DISSOLUTION LOSS OF COPPER, TIN AND LEAD

IN FeO_n-SiO_2-CaO SLAG

Yoichi Takeda* and Akira Yazawa**

*Department of Metallurgy, Faculty of Engineering
Iwate University
Ueda 4-3-5, Morioka 020, Japan

**Research Institute of Mineral Dressing and Metallurgy
Tohoku University
Katahira 2-1-1, Sendai 980, Japan

Abstract

Equilibrium experiments between FeO_n-SiO_2-CaO slags and metals in magnesia crucibles have been carried out at 1573K. Solubilities of copper, tin and lead in FeO_n-SiO_2-CaO slags were determined as functions of oxygen potential and slag composition while the activity coefficients of $CuO_{0.5}$, SnO and PbO for the slag were derived from experimental data. The dissolution loss of valuable metals in slag were discussed based on the experimental data. Prominent features of the FeO_n-SiO_2-CaO system are formations of stable solution and compounds such as $2CaO \cdot SiO_2$ by reaction of basic CaO and the acidic SiO_2. Minimum solubilities of the metals were observed in slags saturated with $2CaO \cdot SiO_2$. The stable CaO-SiO_2 slag was immiscible with the Cu_2O slag and a phase diagram showing the miscibility gap was presented.

Introduction

It is important to minimize metal losses to slag in a smelting process to achieve maximum metal recovery from raw materials. The loss of a valuable metal to slag should be estimated to develop a new smelting system or to modify an existing system. Metal loss to slag is caused by mechanical entrapment and dissolution. It is not easy to estimate the amount of metal entrapped mechanically in slag from fundamental data, since its amount may be affected by the physical properties of slag, metal and gas and also by the smelting system. On the other hand, the amount of dissolution loss is estimated by multiplying solubility by slag volume, when solubility of the metal to the slag is known. Solubility data for the estimation of dissolution loss in practical smelting require a large number of equilibrium experiments because slags consisting of many components are formed which varies depending on the raw materials and smelting system. The FeO_n-SiO_2-CaO is a common slag in practical smelting. Several experimental data on equilibria between non-ferrous metals and the FeO_n-SiO_2-CaO slag of a particular composition have been published. In this study, a wide composition region of the slag in equilibrium with the metals was determined to know the outline of the solubilities.

Experimental

Procedure

FeO_n-SiO_2-CaO slag and copper, tin or lead in a magnesia crucible were kept at 1573K under argon stream for several hours. After equilibrating the slag with the metal, oxygen potentials in the metal phase were measured with the following oxygen concentration cell.

Pt/Ni-NiO(s)/ZrO_2-MgO/metal(1)-slag(1)/electrode/Pt

Oxygen partial pressure, pO_2 at 1573K is calculated as

$$\log pO_2 = 12.821E - 6.711 \qquad (1)$$

from the electromotive force, E/V of the cell and the standard free energy change for the formation of NiO(s)(1), where pO_2 = PO_2(Pa)/101325Pa O_2.
Then the slag was sampled by immersing a water cooled copper tube and liquid metal was sucked out with a silica tube. The compositions of the slag and metal were determined by chemical analysis.

Materials

The materials used in this study were all chemical reagent grade. To investigate the effect of oxygen partial pressure, seven master slags of A, B, C, D, a, c and d were prepared as illustrated in Figure 1. Each slag has different molar ratios of $n_{CaO}/(n_{CaO}+n_{SiO2})$ and $n_{FeOn}/(n_{CaO}+n_{SiO2}+n_{FeOn})$. As shown in the figure, the compositions of a, c and d are close to solid wustite saturation. The shadowed region on the bottom of the tetrahedron show homogeneous melt at 1573K. With increasing oxygen potential the liquidus line saturated

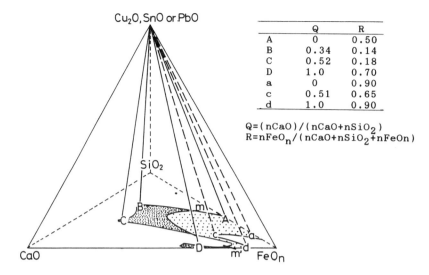

Figure 1 - Homogeneous liquid region for FeO_n-CaO-SiO_2 system at 1573K and composition of starting slag.

with iron oxide expands to the CaO-SiO_2 edge as represented by mm' and the homogeneous melt region is reduced because ferrous oxide is oxidized to ferric oxide having a high melting temperature. Oxygen potential in the sample was controlled by adding Cu_2O, SnO or PbO to the master slag. Large addition of Cu_2O, SnO or PbO resulted in high oxygen potential.

Results and Discussions

Metal phases

Copper, tin or lead melted with iron oxide bearing slags contain some amount of iron depending on the oxygen potential. The biggest solubilities of iron to the metals are observed during solid iron saturation. Lead is kept pure under iron saturation because the solubility of iron to lead is very small. That is the activity of lead, a_{Pb}, is kept almost unity at any oxygen potential in this experiment. The maximum iron solubility to copper saturated with iron is 8% and the activity coefficient of copper, γ_{Cu}, is unity (2). Therefore, activity of copper, a_{Cu}, varies from 1 to 0.9 depending on oxygen potential.

On the other hand, a large amount of iron is capable of dissolving in tin, which is one of the difficulty in practical tin production. Figure 2 shows the phase diagram of Fe-Sn system (b) (3) and activity coefficients of tin and iron at 1573K (a) estimated based on reference (3). The experimental conditions of the present study are represented by line ihh's and activity of tin, a_{Sn} varies from 1 to 0.45 depending on oxygen potential. Two liquid metal phases of tin rich and iron rich alloys were in coexistence with slag phase at some reducing conditions and activity of tin is 0.8 at this equilibrium.

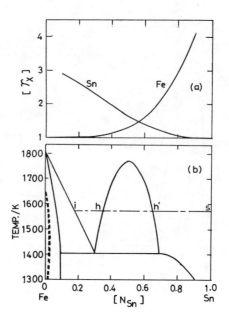

Figure 2 - Activity coefficients of Sn and Fe at 1573K (a) and phase diagram of Fe-Sn system (b) presented by Nunoue et al. (3).

Oxygen dissolves to the metals from the slag phase. The maximum oxygen solubility to copper saturated with its oxide is 4% at 1573K (4). The maximum oxygen solubility to tin is 0.2% (5). As no available data for oxygen solubility to lead have been reported at high temperature, it is estimated to be around 1% at 1573K based on reference (6).

Effects of oxygen partial pressure

Oxidic dissolution to slag of X having a valence of 2ν may be discussed thermodynamically on the basis of the following equations:

$$X(l) + \nu/2\ O_2 = XO_\nu(l), \qquad (2)$$

$$K = a_{XO_\nu}/(a_X\ p_{O_2}^{\nu/2}) \qquad (3)$$

and

$$(\%X) = K[\%X](n_T)p_{O_2}^{\nu/2}[\gamma_X]/(\gamma_{XO_\nu})[n_T] \qquad (4)$$

where, n_T is the total number of moles of constituents in 100g of metal or slag , and ( ) and [ ] are quantities in slag and metal, respectively. Equilibrium constant, K is derived from standard free energy change of formation, $\Delta G°_{XO_\nu}$ for the reaction (2) by equation (5)

$$K = \exp.(-\Delta G°_{XO_\nu}/RT) \qquad (5)$$

where R is the gas constant and T is absolute temperature. The following data (4,7,8) are available for the standard free energy changes of formation for $CuO_{0.5}(1)$, $SnO(1)$ and $PbO(1)$

$$\Delta G°_{CuO0.5(1)}/J = -242680 + 85.84T \qquad (6)$$

$$\Delta G°_{SnO(1)}/J = -274500 + 93.3T \qquad (7)$$

$$\Delta G°_{PbO(1)}/J = -190900 + 75.52T. \qquad (8)$$

The relationship between solubility of copper, tin or lead in various kind of slags and oxygen partial pressures are illustrated in Figures 3, 4 or 5, respectively. The dependence of copper solubility on oxygen partial pressure is smaller than that of tin or lead which is divalent in slag because copper dissolves as a monovalent oxide. It is expected that the slope in Figure 3 is around 1/4 for copper solubility and the slope in Figure 4 or 5 is 1/2 for tin or lead solubility from the equation (4). Minimum solubilities of the metals at a given oxygen partial pressure are observed in slag C which has a considerable amount of CaO and SiO_2 and a small amount of iron oxide. The composition of slag C is close to the liquidus composition saturated with dicalcium silicate. The minimum solubilities are thought to be due to the stable slag formed by neutralizing basic CaO with acidic SiO_2, which is well known from the study of FeO activity in the $CaO-SiO_2-FeO$ system.

Copper solubilities in the plain fayalite slag melted in silica crucibles (9) are also plotted in Figure 3 and are not different from that of slag A containing around 6% MgO. But solid magnetite precipitates at log pO_2 = -6 when a plain fayalite slag and copper are melted in a silica crucible. When melted in a magnesia crucible, solid magnetite did not precipitate as shown in Figure 3. Slags B and C containing considerable amounts of CaO and SiO_2 are immiscible with Cu_2O base slag under high oxygen partial pressure and $aCuO_{0.5}$ nearing unity, and this miscibility gap is discussed later.

From equation (4), the logarithmic plot between (%Sn) and oxygen partial pressure in Figure 4 will give the straight line with a slope of 1/2 if the metal phase is mostly tin and the activity coefficient remains constant. This is observed in Figure 4 for each slag composition at the region of moderate oxygen partial pressure, suggesting the divalent dissolution of tin in slag. The deviations from the slope of 1/2 at the region of lower oxygen partial pressure in Figure 4 are caused by metal composition varying from Sn to Sn-Fe alloy. On the contrary, those in the region of higher oxygen partial pressure may be ascribed to the variation of slag composition due to increasing content of SnO. In Figure 4, s at the extreme top right shows the equilibrium of plain SnO with Sn at 1573K, and h at lower oxygen partial pressure corresponds to the miscibility gap in the metal phase while i shows the equilibria saturated with solid iron, as suggested in Fe-Sn phase diagram of Figure 2. The solubility of tin to slag d free from silica at 1473K is also demonstrated in Figure 4. The difference in the solubility between 1473 and 1573K is mostly due to the difference in the equilibrium constant, K.

Figure 5 shows the solubility of lead to the slags at 1573K. Unfortunately, experimental data points are not plotted in the figure since all the research on this topic has not been

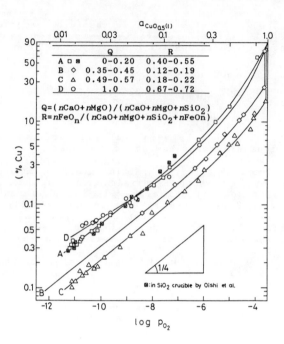

Figure 3 - Effect of oxygen partial pressure on copper solubility to slag with data for plain fayalite slag free from magnesia presented by Oishi et al.(9) at 1573K.

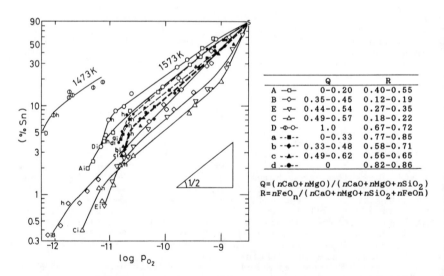

Figure 4 - Effect of oxygen partial pressure on tin solubility to slag at 1473K and 1573K.

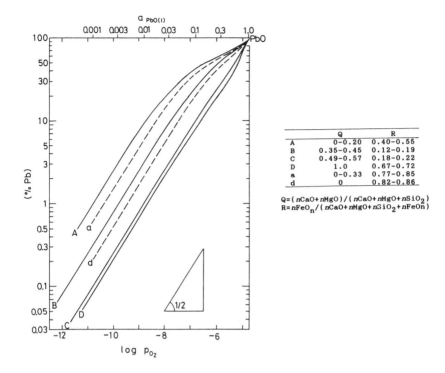

Figure 5 - Effect of oxygen partial pressure on lead solubility to slag at 1573K.

accomplished yet. The solubility of lead in FeO_n-CaO-SiO_2 slag at a restricted composition range has been presented by Taskinen et al. (10). The logarithmic relation between (%Pb) and oxygen partial pressure is also represented by a straight line with a slope of 1/2 below around 20% lead in slag, and this linear relation gives a constant activity coefficient of PbO as listed in Table I . As slag compositions vary to PbO with increase of oxygen partial pressure, activity coefficients and (n_T) change to 1 and $100/M_{PbO}$, respectively, where M_{PbO} is the molecular weight of PbO. Consequently, the lines in Figure 5 converge at the point marked by PbO.

Effect of slag composition on activity coefficients

The activity coefficients of $CuO_{0.5}(1)$, SnO(1) and PbO(1) derived from combining equation (4) with the experimental results in magnesia crucibles at 1573K are shown on a mole fraction basis as shown in Figures 6, 7 and 8. Since the activity coefficient of $CuO_{0.5}$ and not Cu_2O obeys Henry's law(11,12), the systems are considered as FeO-$FeO_{1.5}$-CaO-SiO_2-XO_y to derive the activity coefficients, where XO_y is $CuO_{0.5}$, SnO or PbO. The remaining part after deducting the concentration of $CuO_{0.5}$, SnO or PbO which is 10% at most was expanded to the FeO_n-(CaO+MgO)-SiO_2 system. The MgO concentration dissolved from the crucible in the equilibrium experiment between the slag and tin is illustrated in Figure 9. No magnesia was added to the initial slags. The magnesia content in FeO_n-CaO slags of D or d is around 1%, and a large solubility of magnesia is observed in slag D whose ratio of CaO to SiO_2 is around 2.

With regard to the physical properties of the FeO_n-CaO-SiO_2 system, slags rich in CaO and SiO_2 is viscous and has a rather high melting temperature. The addition of iron oxide is effective in decreasing viscosity and melting temperature. On the contrary, slags rich in iron oxide are apt to precipitate solid magnetite with an increase of oxygen partial pressure, especially in experiments with copper.

Activity coefficients of $CuO_{0.5}$ in the pseudo-binary slag

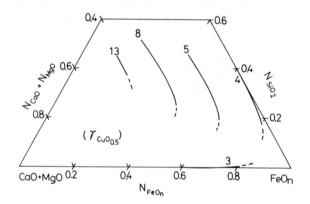

Figure 6 - Activity coefficient of $CuO_{0.5}(l)$ in FeO_n-CaO-SiO_2 system melted in magnesia crucible at 1573K.

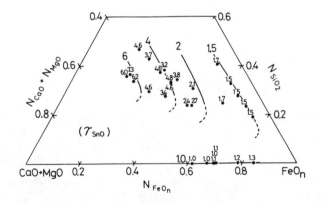

Figure 7 - Activity coefficient of $SnO(l)$ in FeO_n-CaO-SiO_2 system equilibrated with two metal phases of 80% Sn-Fe and 53% Sn-Fe in magnesia crucible at 1573K.

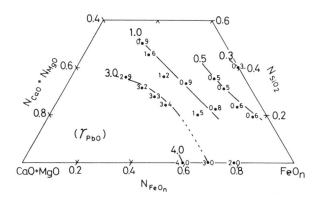

Figure 8 - Activity coefficient of PbO(l) in FeO_n-CaO-SiO_2 system melted in magnesia crucible at 1573K.

of the FeO_n-SiO_2 and FeO_n-CaO are 3 to 4. Addition of CaO to the FeO_n-SiO_2 slag or SiO_2 to the FeO_n-CaO slag makes the activity coefficient of $CuO_{0.5}$ or SnO bigger. The minimum activity coefficient is around 13 for $CuO_{0.5}$ and 5 for SnO, as observed in the slag which has a CaO content almost equal to that of SiO_2 and a little iron oxide. This slag composition is close to that saturated with dicalcium silicate. When a constituent U, which has a big affinity with constituent V, makes a stable solution UV, the solution tends to exclude constituent W, which does not have a big affinity with U and V. In an extreme case, a solution of UV is immiscible with a liquid of W as the Cu_2O-CaO-SiO_2-Fe_3O_4 system in Figure 11 or the FeO-CaO-P_2O_5 system (13). The activity coefficient is in inverse proportion to solubility. The stable slag formed by the reaction between basic CaO and acidic SiO_2 has a big activity coefficient of $CuO_{0.5}$ or SnO and a small solubility of Cu or Sn as oxide.

Table I. Activity coefficients of $CuO_{0.5}$(l), SnO(l) and PbO(l) in FeO_n-CaO-SiO_2 slag saturated with MgO.

	Q	R	$CuO_{0.5}$	SnO	PbO
A	0.10	0.50	4	1.8	0.3
B	0.40	0.15	8	5.0	1.0
C	0.55	0.20	13	6.2	3.0
D	1.0	0.70	3-4	1.2	4.0
a	0.10	0.80	4	1.8	0.5
c	0.55	0.60	5	2.5	1.0
d	1.0	0.84	3-4	1.2	2.0

$Q = (nCaO + nMgO)/(nCaO + nMgO + nSiO_2)$
$R = nFeO_n/(nCaO + nMgO + nSiO_2 + nFeOn)$

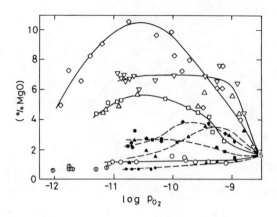

Figure 9 - Magnesia content in slag in equilibrium experiment between slag and tin as shown in Figure 4.

The activity coefficient of PbO(l) is demonstrated in Figure 8. PbO has a big affinity with SiO_2 and makes stable compounds such as $2PbO \cdot SiO_2$ and $PbO \cdot SiO_2$ (14). The activity coefficient decreases with an increase in the effective concentration of silica that will be related with the activity of SiO_2. The pseudo-binary FeO_n-SiO_2 slag with a small activity coefficient of PbO has a big solubility of lead as opposed to the FeO_n-CaO slag with no silica. If a slag contains a large amount of silica which is neutralized with lime, the activity coefficient of PbO in the slag is not small as shown in Figure 8. The degrees of interactions between CaO and SiO_2 and between PbO and SiO_2 may be the main governing factors of the activity coefficient of PbO in Figure 8. The activity coefficients of $CuO_{0.5}$(l), SnO(l) and PbO(l) are listed in Table I.

Relation between Sn content in slag and Fe content in metal.

The affinity of tin is close to that of iron, and the solubility of iron in tin is rather high as shown in Figure 2. A considerable amount of iron tends to be reduced along with tin in the reduction smelting. For this reason, tin smelting usually consists of two steps of weak and strong reduction. In the first weak reduction smelting, tin metal containing a few per cent of iron is in coexistence with slag dissolving 10 to 20 per cent SnO. As the second step, this slag is reduced strongly to recover dissolved tin as hardhead, Sn-Fe alloy containing around 40% Fe. For practical purposes, the tin content in slags obtained in the equilibrium experiments demonstrated in Figure 4 were rearranged against iron content in the metal as shown in Figure 10. i, h and h' in the figure have the same meaning as those in Figure 2. A, B, C, D, E, a, b, c, and d denoting types of slag in Figures 4 and 10 are the same. Naturally a slag with less iron oxide was equilibrated with a metal having less content of iron. A proper amount of iron oxide may be needed in tin smelting to get a good fluid

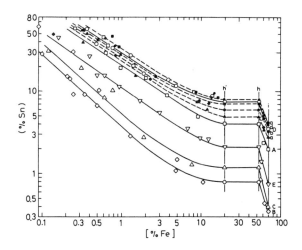

Figure 10 - Relations between Sn content in slag and Fe content in metal at 1573K.

slag. During the second step in tin smelting, it may be better to make a tin rich hardhead of 80% Sn-20% Fe because tin recovery from slag will not improve by making an iron rich hardhead of 50% Sn-50% Fe under the same oxygen partial pressure and activity of SnO judging from Figure 10.

Miscibility gap in the $Cu_2O-CaO-SiO_2-Fe_3O_4$ system.

The miscibility gap between calcium silicate slag and Cu_2O base slag co-existing with metallic copper at 1573K is illustrated in Figure 11. The top right figure shows the cross section of the miscibility gap through point x, Cu_2O and Fe_3O_4 in the bottom figure. This plane is close to the liquidus surface saturated with dicalcium silicate. Most constituents concentrate into one slag phase or the other, but iron oxide distributes equally to both slag phases especially at low concentration of iron oxide. Therefore, the miscibility gap at 7, 12 or 17 % Fe_3O_4 is illustrated in the figure. Calcium silicate with no iron oxide does not melt at 1573K. MgO content in the slag is 12 % at most and in the Cu_2O base slag is minimal. A miscibility gap was also observed between sodium silicate melt and Cu_2O (16). The miscibility gaps in both systems are caused by the formation of a stable melt by the reaction of acid and base mentioned above. The biggest miscibility gap will be observed in the silicate composition of $2CaO \cdot SiO_2$ judging from the similarity in phase diagrams between the $CaO-SiO_2$ and $NaO_{0.5}-SiO_2$ systems (17) if the sample of that composition is in the molten state.

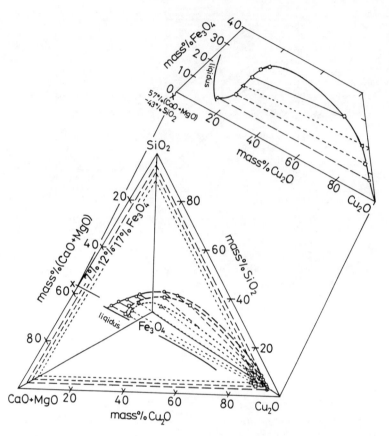

Figure 11 - Miscibility Gap in tetrahedron diagram of Cu_2O-SiO_2-$(CaO+MgO)$-Fe_3O_4 system with metallic copper at 1573K (15). On right top figure, miscibility gap on vertical plane through points x, Cu_2O and Fe_3O_4.

Conclusions

The best slag composition for smelting depends on several factors such as raw materials, energy requirement, circumstances of smelter etc. But one of the most important factors may be the solubility of the valuable metal. In this regard, the solubility data in the commonly used FeO_n-CaO-SiO_2 slag may be very useful in developing an effective slag for smelting operations. To clarify the effects of oxygen partial pressure and slag composition on the solubility of copper, tin or lead in slag, a number of equilibrium experiments between the metals and various slags of FeO_n-CaO-SiO_2 were carried out at 1573K. The activity coefficients of $CuO_{0.5}$, SnO and PbO were derived from the experimental data. The dependence of copper solubility in slag on oxygen partial pressure is smaller than that of tin or lead which is divalent in slag. Copper or tin solubilities are similar for binary fayalite (FeO_n-SiO_2) and ferrite (FeO_n-CaO) slags, but lower in ternary FeO_n-CaO-SiO_2 slag. The minimum solubilities of copper or tin are expected at the ratio of $CaO/SiO_2 = 2$ both experimentally and thermodynamically. The solubility of lead, having a big affinity for SiO_2 is affected by the activity of SiO_2. Increasing iron oxide content results in increasing solubilities of the metals in slag.

The miscibility gap in the Cu_2O-CaO-SiO_2-Fe_3O_4 was also revealed. The miscibility gap is caused by the formation of a stable melt reacted with basic CaO and acidic SiO_2.

References

1. N. Kemori, I. Katayama, and Z. Kozuka, "Measurements of standard molar Gibbs energies of formation of NiO, Cu_2O and CoO from solid and liquid metals and oxygen gas by an e.m.f. method at high temperatures," J.Chem.Thermodynamics, 11 (1979), 215-228.

2. Ralph Hultgren et al., Selected Values of the Thermodynamic Properties of Binary Alloys (Metals Park, Ohio: American Society for Metals, 1973).

3. S. Nunoue and E. Kato, "Mass Spectrometric Determination of the Miscibility Gap in the Liquid Fe-Sn System and the Activities of This System at $1550°C$ and $1600°C$," Tetu to Hagane, 73(7)(1987),868-875.

4. Y. Kayahara et al., "Thermodynamic Study of the Liquid Cu-O System," Transactions of the Japan Institution of Metals, 22(7)(1981),493-500.

5. U. Kuxmann and R. D. Lunen, "Untersuchungen im System Zinn -Zinn(IV)-Oxid im Temperaturbereich der Mischungslucke," Metall, 9(34)(1980), 821-828.

6. A. Taskinen, "Thermodynamics and Solubility of Oxygen in Liquid Lead," Scandinavian Journal of Metallurgy, 8 (1979), 185-190.

7. Y. Takeda et al., "Equilibria between Liquid tin and FeO-CaO -SiO_2 Slag," to be published in Materials Transactions, JIM.

8. Y. Takeda, unpublished data.

9. T. Oishi et al., "A Thermodynamic Study of Silica-Saturated Iron Silicate Slags in Equilibrium with Liquid Copper," Metallurgical Transactions B 14B(1983), 101-104.

10. A. Taskinen, L. M. Toivonen, and T. T. Talonen, "Thermodynamics of Slags in Direct Lead Smelting," Second International Symposium on Metallurgical Slags and Fluxes, ed. H. A. Fine and D. R. Gaskell (Pennsylvania: The Metallurgical Society of AIME, 1984), 741-756.

11. R. Altman and H. H. Kellogg, "Solubility of Copper in Silica -saturated Iron Silicate Slag," *Trans. Instn. Min. Metall,* 81(1972), C163-175.

12. Y. Takeda, S. Ishiwata, and A. Yazawa, "Distribution Equilibria of Minor Elements between Liquid Copper and Calcium Ferrite Slag," *Transactions of the Japan Institute of Metals,* 24(7)(1983), 518-528.

13. W.Oelsen and H.Maetz, "Zur Metallurgie des Thomasverfahrens," *Arch. Eisenhuttenw.,* 19(1948), 111-117.

14. U. Kuxmann and P. Fischer, "Beitrag zur Kenntnis der Zustandsdiagramme $PbO-Al_2O_3$, $PbO-CaO$ und $PbO-SiO_2$," *Erzmetall,* 27(11)(1974), 533-537.

15. Y. Takeda, "Miscibility Gap in $Cu_2O-CaO-SiO_2-Fe_3O_4$ Slag and Distribution of minor elements," to be published in *Materials Transactions, JIM.*

16. Y. Takeda et al., "Equilibria between Liquid Copper and Soda Slag," *Transactions of the Japan Institute of Metals,* 27(8) (1986), 608-615.

17. Y. Takeda and A. Yazawa, "Activity of FeO and Solubility of Copper in $NaO_{0.5}-SiO_2-FeO$ Slag Saturated with Solid Iron," *Transactions of the Japan Institute of Metals,* 29 (3)(1988), 224-235.

DIRECT SMELTING OF ZINC

Suresh K. Gupta and John M. Floyd

G.K. Williams Laboratory for Extractive Metallurgy Research
Department of Chemical Engineering
Melbourne University
Parkville, Victoria, Australia 3052

Abstract

A new idea is developed for the production of zinc from zinc and complex zinc concentrates. It is a two stage process. In the first stage a zinc oxide rich slag is produced by injecting zinc concentrates in iron silicate slag in an oxidizing atmosphere. The zinc oxide is reduced from the slag in the subsequent operation by injecting coke breeze. The vapours thus produced can be condensed in a lead splash condenser such as those used in the Imperial Smelting Process (ISP)

Introduction

Zinc is mainly produced by two major processes - electrolytic and Imperial smelting (blast furnace). At present the electrolytic process or roast-leach-electrowinning process account far more than 80% of the world zinc production. The blast furnace or ISP accounts for most of the remaining production.

The electrolytic process typically uses high grade zinc concentrates and an expensive form of energy (electric power). It also suffers from high energy requirements. The major problem is the disposal of solid waste jarosite, goethite or hematite residues. An overview of these processes has been given by Gordon (1).

The Imperial Smelting Process is currently the main pyrometallurgical process used for zinc recovery. This process is very well suited to the treatment of a wide range of relatively low grade and complex materials such as mixed zinc-lead-copper concentrates, residues and fumes. One of the major drawbacks of this process is that it utilizes a sinter plant for sulfur elimination from concentrates. It is difficult to prevent loss of lead containing dusts and vapours from this system and this leads to difficulty and expense in environment protection. It also entails regular and frequent mechanical and materials maintenance, large recycling loads of fine sinter, multiple handling of materials and strict control of composition of feed for good quality sinter. The production of sinter involves extra expenses. The blast furnace requires metallurgical coke as both a fuel and a reductant, which is also an expensive raw material.

There have been no major commercial developments in zinc smelting in the period since the first commercial application of the Imperial Smelting Process in 1960. In 1985 Goto (2,3) carried out pilot trials on a bath smelting process in which powdered coke, oxygen and zinc calcine were injected into a slag bath and zinc recovered using a lead splash condenser. However, recoveries of zinc were poor. Other processes have also been suggested on the basis of theoretical calculations, they are the the direct smelting (4,5) and roast-smelt (6) processes. No report has been published on the experimental testing of these processes as far as it is known. Warner (7,8,9) has suggested a new concept for treating complex concentrates or run of mine ore (Cu-Pb-Zn) incorporating the principles of liquid- liquid extraction and counter current extraction. The concept involves smelting complex concentrate under conditions such that the products are absorbed in a continuously circulating stream of high grade carrier matte. The matte is subjected to vacuum treatment to recover zinc and other volatile metals, then oxidized to produce blister copper and a high copper slag which is cleaned pyrometallurgically. Lead is recovered from the matte by counter current extraction with copper. The matte from slag cleaning is recycled to the melting unit. Pilot plant tests are in progress for this process at the University of Birmingham using a semi-pilot scale experimental rig.

In the light of evidence in the literature it may therefore be concluded that it is rather difficult to produce zinc directly from the zinc concentrates in a single process. Therefore, an intermediate process is required to treat zinc concentrates prior to extracting zinc metal. Such an intermediate process may be either an oxidizing roasting of concentrates in the solid state or the oxidation of zinc sulfide to zinc oxide and dissolution into a slag. Since roasting operations at higher lead contents have not been successful because of sintering and melting, the former approach is limited to concentrates with less than 5-6% lead. However, using the latter approach the limitation on lead content is removed, and in principle any concentrate may be used. Hence this concept is followed in the present investigation.

Brief Description of the Process

In the first stage zinc concentrate, flux and oxidizing gas (air or oxygen) are injected into the molten slag which is mainly iron silicate. The sulfides in the concentrate are oxidized and dissolved into the slag. The quartz is added to flux iron oxide formed upon oxidation of iron sulfide in the concentrate. A zinc oxide rich slag is thus produced. Some of the zinc vapourizes during the reactions of zinc sulfide. This is collected as ZnO fume in the bag filter.

The zinc rich slag produced in the first stage is used to produce zinc metal by the reduction of zinc oxide with carbon when coke breeze powder is injected into the furnace for this purpose. A splash condenser such as in operation in the ISP process can be used to condense zinc without reoxidation. The oxygen potential of the outcoming gas should be high enough not to reduce iron oxide in slag to metallic iron. Valuable elements such as gold, silver and others in the zinc concentrate are recovered in a separate matte or metal phase under the slag layer. For zinc concentrates which generate slags with greater than approximately 30% zinc there is a need to recycle or dilute the slag because of ZnO solubility limits in iron silicate slags at smelting temperatures of 1250-1350°C. A process has been patented by Ausmelt Pty Ltd to achieve this without solidification and remelting of slag.

Experimental Details

Materials

A zinc concentrate of Australian origin was used in this investigation. The analysis of the concentrate is given in Table I. The slag used was a mixture of two slags from commercial copper smelting and converting operations. The average analysis of this slag is given in Table II. Quartz was used as a fluxing agent. The reduction of zinc oxide in the slag was carried out by using coke breeze powder, its proximate analysis and size distribution is given in Table III.

Preparation of Pellets

The zinc concentrate powder was pelletized in the size range of 2-5 mm by rolling the powder in a porcelain jar. Black liquor (a waste product from paper manufacture containing approximately 55% sodium lignosulfonate) was used as a binding agent. The pellets were dried first at room temperature and then indurated at 120°C in an oven overnight.

Smelting of Concentrate

500 g of slag was melted in an alumino-silicate crucible under a cover of nitrogen gas at temperatures of between 1250 and 1300°C using an induction furnace. Different bore alumina lances, 3-6mm, were used for the injection of air into the slag depending upon the flow rate and concentrate feed rate. When the slag was heated to the required temperature the nitrogen gas cover was removed. The first charge of concentrate pellets with quartz in the proportion Fe/SiO_2 = 1.4 to flux iron oxide was fed to

Table I. Chemical Analysis of Zinc Concentrate

Compound	Wt.%
Zn	53
Pb	3.6
Cu	0.33
Fe	7.0
Cd	0.16
As	0.15
Sb	0.022
S	32.7
Mn	0.16
SiO_2	1.7
CaO	0.11
Ag	0.0078
Au	0.00014

Table II. Chemical Analyses of Slag in Weight Percent

Fe	SiO_2	CaO	MgO	Al_2O_3	Cu	Pb	Zn	As	Sb	S
43.8	29.0	5.08	0.32	2.33	2.868	0.473	0.179	0.02	0.008	0.374

Table III. Composition and Size Distribution of Coke Breeze

(a) Composition in Weight Percent

Fixed carbon	60.85
Ash	38.15
Moisture	1.00

(b) Size Distribution in Percent

210-104 micron	50
104-44 micron	50

the crucible and then predetermined quantity of air was injected into the slag bath for the conversion of sulfides to oxides and sulfur to sulfur dioxide. The air was calculated for the following reaction and is termed 100% stoichiometric:

$$2MS + 3O_2 = 2MO + 2SO_2$$

Experiments were carried out with air ratios equivalent to 100% and less than 100% stoichiometry. Pellets were fed into the crucible manually every minute at a predetermined rate over periods of up to 60 minutes. The progress of the reaction was followed by analyzing the slag samples for zinc content. In some samples sulfur was also analyzed. The slag samples were taken after every 10 minutes by inserting a steel rod and quenching into a water bath. The samples were analyzed by X-ray fluorescence (XRF). The outcoming gas from the furnace was allowed to pass through a bag filter to collect samples of the fume generated. The temperature of slag was measured continuously with the help of a Pt/Pt-13%Rh thermocouple immersed into the slag, and was controlled by controlling the power input of the furnace. The temperature variation was of ±10°C during the run. The experimental arrangement is shown in Fig. 1.

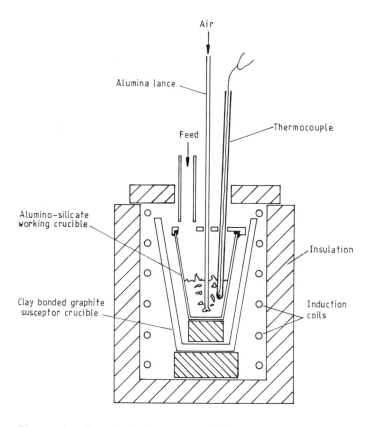

Figure 1 - Experimental arrangement for concentrate smelting.

Reduction of Zinc Oxide

A zinc rich slag produced as described above was cooled, crushed, mixed and was used to recover zinc by reducing with the coke breeze powder of composition and sizing shown in Table III to simulate the second stage. The same experimental set up was used as shown in Fig. 1. 500 g of slag was melted at 1300°C in the alumino-silicate crucible. The coke breeze powder was injected continuously into the slag bath with the help of an alumina lance of 2 mm internal diameter fitted with a powder dispenser. Nitrogen gas at 5 litres per minute was used as a carrier gas.

Condensation of Zinc Vapours

The zinc vapours were condensed in a lead splash condenser. About 5 kg of lead was melted at 450°C in a graphite crucible which was enclosed in a stainless steel chamber. The lead was melted with a gas burner and was splashed with the help of an impeller. The temperature of the lead was measured with a thermocouple. The graphite crucible was connected to an alumino-silicate crucible containing slag through an alumima tube which was well insulated with kaowool. All the joints were sealed thoroughly with refractory cement to prevent the infiltration of air.

About 500 g of zinc rich slag produced in stage one was melted in the alumino-silicate crucible at 1300°C under nitrogen cover using induction furnace. The nitrogen was injected through an alumina tube of 2 mm internal diameter; the same tube was used as a lance for the injection of coke breeze powder. After attaining the required temperature, the alumina lance with nitrogen blowing through it was lowered into the slag and the opening of the crucible around the lance was sealed with the refractory cement. The coke powder was injected into a slag using a powder dispenser at 0.75 g per minute for 60 minutes and the nitrogen flow rate was adjusted to 5 litres per minute. The stainless steel chamber was opened after the run and the samples were taken from the graphite crucible for analysis of zinc. The experimental set up is shown in Fig. 2.

Results and Discussion

Concentrate Smelting

The effects of quantity of air, temperature and feed rate were studied on the smelting of zinc concentrate. The results are discussed below:

Effect of Quantity of Air. The effect of using 90 and 100% stoichiometric air with a concentrate feed rate of 4 g per minute at 1300°C on the zinc content of slag is shown in Fig. 3. This clearly shows that a higher zinc content was obtained with 100% stoichiometric air, although the slag was very viscous after 45 minutes and the experiment was completed with difficulty.

The thermodynamic equilibrium calculations were carried out with the CSIRO-SGTE THERMODATA computer system, using a free energy-minimization program called CHEMIX (10). The calculation method is based on the equilibrium in a closed system. Therefore, the equilibrium calculations were carried out for every ten-minute period of the operations by assuming that all the concentrate to be fed and the air to be blown into a crucible during that period were charged in the crucible at the beginning of the period at a time, that the air stayed there until an equilibrium was established, and that the liberated gas left the crucible at the end of the

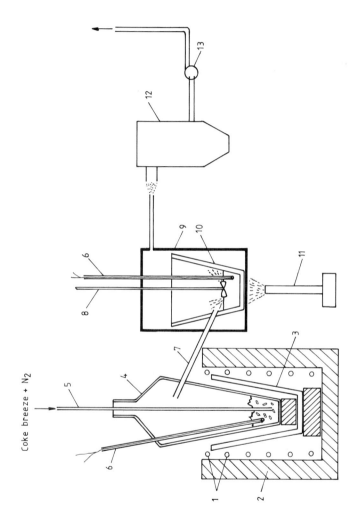

Figure 2 - Experimental set up for zinc condensation.
(1) induction coils; (2) insulation; (3) clay bonded graphite susceptor crucible; (4) alumino-silicate working crucible; (5) alumina lance; (6) thermocouples; (7) alumina tube; (8) impeller; (9) stainless steel chamber; (10) graphite crucible; (11) gas burner; (12) bag filter; and (13) fan.

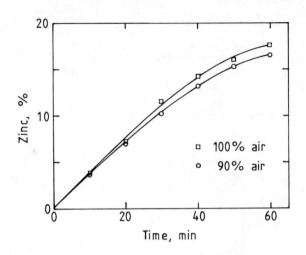

Figure 3 – The effect of quantity of air on the zinc content of slag at 1300°C and 4 g per minute feed rate.

period. By reiterating this step of calculation, the changes of factors with time in the process were simulated. The following activity coefficient data given by Goto (2) were used in the calculations:

ZnO	FeO	Fe_3O_4
1.2	0.3	5.0

The equilibrium zinc data at the feed rate of 4 g per minute with 90 and 100% air are compared with those of experimental values in Fig. 4. This figure shows that there is not much difference in the theoretical and experimental values of zinc with 100% air and therefore indicates that reactions are progressing almost at equilibrium. However, there are slight differences in two values with 90% air when the theoretical zinc content in slag is 0.5 - 1.75% higher than the experimental results [Fig. 4(b)]. This indicates slightly more zinc fuming at 90% oxidation than theoretical predictions. With 90% air due to insufficient oxygen in the slag some of the ZnS and ZnO would be dissociated in its components and zinc is lost by volatilization and results into lower zinc content of the slag. The theoretical calculations have shown that the oxygen potential of the slag was lower at 90% air, 0.12×10^{-5} atm, than at 100% air, 6×10^{-5} atm.

It has been stated before that the slag was very viscous with 100% air. This is because of higher magnetite content of the slag with 100% oxidation. Thermodynamic calculations also support this hypothesis. The calculated magnetite content in slag for 90 and 100% stoichiometric air is shown in Fig. 5. This figure shows greater levels of magnetite at 100% air but note that magnetite saturation is not predicted. It was decided to use 90% air in subsequent experiments to avoid the viscosity problems with 100% air.

Effect of Temperature. The effect of smelting temperature on the zinc content of slag is shown in Fig. 6. At a feed rate of 8 g per minute with

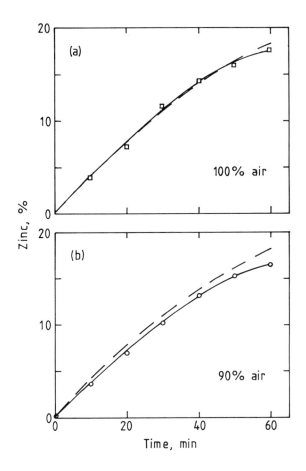

Figure 4 - The comparison of equilibrium and experimental zinc contents of slag at 1300°C and 4 g per minute feed rate. ———, experimental; ----------, equilibrium.

90% air slightly higher zinc levels in slag were obtained at 1250°C than at 1300°C. This would be expected because of the higher vapour pressure of zinc at 1300°C than at 1250°C and hence loss of more zinc at 1300°C. The slag was relatively more viscous at 1250°C than at 1300°C.

Effect of Feed Rate. The effect of concentrate feed rate of 4, 6, 8 and 10 g per minute with 90% air at 1300°C on the zinc contents of slag is shown in Fig. 7. The equilibrium zinc contents in the slag at each feed rate were calculated and are compared with those of experimental results in Fig. 8. It is quite evident from the figure that the zinc contents of the slag is not proportional to feed rate. More zinc is fumed at higher feed rates. This is presumably because the dissociation of ZnS is favoured. This is because of the longer time that it takes for dissolution of the

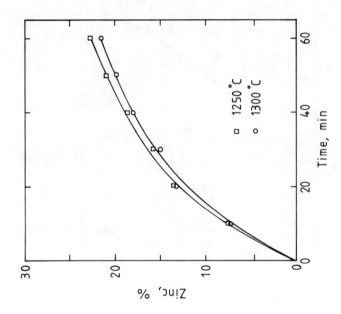

Figure 6 - The effect of temperature on the zinc content of slag at 8 g per minute feed rate and 90% stoichiometric air.

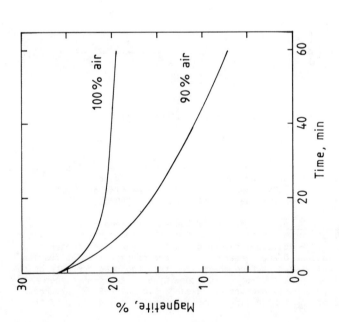

Figure 5 - Theoretical magnetite contents of the slag in a smelting run at 1300°C and 4 g per minute feed rate for 90 and 100% stoichiometric air.

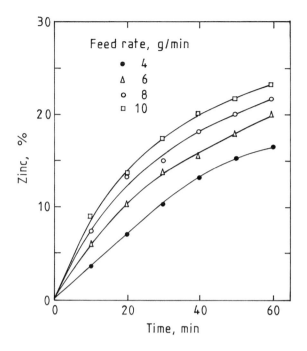

Figure 7 - The effect of feed rate on the zinc content of slag with 90% stoichiometric air at 1300°C.

greater quantities of feed pellets in the slag because of thermal limitations in the system.

A maximum of about 23% zinc was obtained in the slag in 60 minutes at a feed rate of 10 g per minute, which is not much less than the maximum solubility 35.9% ZnO (or 28.8% Zn) in a system ZnO - "FeO" - SiO_2 saturated with silica (11). However, the production of zinc oxide saturated slag would be detrimental in the present investigation because of the high viscosity of slags which would occur. Moreover, the initial slag used in this work was higher in "FeO" than occurs with silica saturation and also contains Al_2O_3 and other components (see Table II). The solubility of zinc oxide would be little different in this slag than reported by Umetsu et al. (11).

Some experiments were also carried out with slags lower in "FeO" and higher in CaO, but the trials were not successful because the system generated a foam causing the slag to run out of the crucible even at a very low concentrate feed rate (2 g per minute). The foamability was a function of temperature and type of oxidizing gas. The foaming was very vigorous as the temperature was increased from 1250 to 1300°C and when the oxidizing gas was changed from air to oxygen. This problem was overcome by the use of the slag shown in Table II.

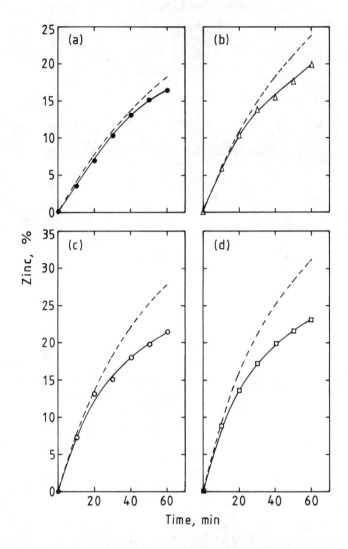

Figure 8 - The comparison of theoretical and experimental zinc contents of slag with 90% air at 1300°C and at the feed rate of (a) 4 g/min, (b) 6 g/min, (c) 8 g/min, and (d) 10 g/min. ─────, experimental; ─────, theoretical.

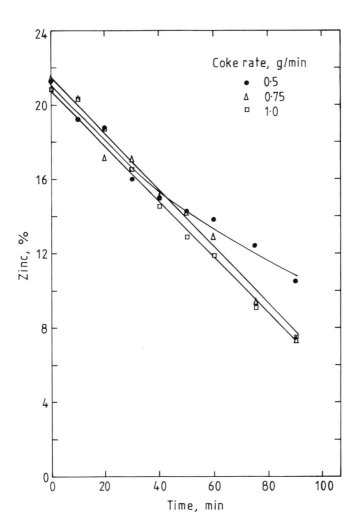

Figure 9 - The reduction of zinc oxide in slag with coke breeze powder at 1300°C.

Reduction of Zinc Oxide in Slag

The zinc oxide in the slag was reduced at 1300°C by injecting coke breeze powder entrained in nitrogen at a rate of 0.5, 0.75 and 1.0 g per minute into the slag for up to 90 minutes and the results are shown in Fig. 9. This figure indicates that the rate of zinc fuming is linear with the coke rate of 0.75 and 1.0 g per minute indicating zero order reaction. This is consistent with the previous published work (12-15). The rate of zinc fuming at the coke rate of 0.5 g per minute is comparable with those at 0.75 and 1.0 g per minute up to about 30 minutes. However, the rate of zinc fuming was slowed down beyond 30 minutes at the coke rate of 0.5 g per minute. The maximum zinc fuming rate was observed with 0.75 g per minute coke rate. The average rate of zinc fuming with the coke rate is given in the following table:

Coke rate, g/min	Zinc fuming rate, % Zn/min
0.5	0.12
0.75	0.16
1.0	0.15

The zinc fuming rates observed are much higher in the present investigation than the previous studies on zinc fuming from slags (12,13). Further, in most of the previous studies coal was used as reductant, which is more reactive than coke breeze. Coal was not used in the present work because zinc collection in a lead splash condenser is used. Goto (3) found that the hydrogen content of the fuel produced water vapour levels in the product gases which caused rapid oxidation of zinc metal on condensation and decreased the zinc recovery drastically in the splash condenser.

Condensation of Zinc

The zinc vapours produced by the reduction of zinc oxide in the slag by coke breeze were condensed in the lead splash condenser. The temperature of the out coming gas was held at 1250-1300°C at the exit of

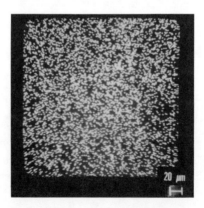

Figure 10 - SEM photograph showing the distribution of zinc in a sample from lead splash condenser.

the slag containing crucible, but the gas was rapidly cooled in the alumina tube which was connected to slag crucible with the lead crucible. Some of the zinc in the outcoming gas can readily be oxidized at the temperature of less than 1000°C before condensation in the splash condenser according to the reaction:

$$Zn_{(g)} + CO_{2(g)} = ZnO + CO_{(g)}$$

Some of the zinc oxide thus formed was deposited on the wall of the alumina tube and some carried to the lead splash condenser together with some coke particles. However, the dross formation in the splash condenser was very little. The oxidation of zinc vapour according to the above reaction could be prevented by increasing the temperature in the alumina tube. The aim of this part of the project was to show that the zinc could successfully be condensed in the lead splash condenser. A zinc rich lead sample was taken from top of the condenser and was analyzed qualitatively by XRF. The analysis has shown the presence of zinc in the sample. A photograph showing the distribution of zinc in the sample was taken by a scanning electron microscope and is given in Fig. 10.

Conclusion

The laboratory scale trials have evaluated the possibility of developing a new pyrometallurgical process for the direct smelting of zinc using gas and concentrate injection into a slag phase and reducing zinc from the slag phase. The zinc thus produced can successfully be collected in a lead splash condenser.

Acknowledgements

The authors gratefully acknowledge the financial support provided by the Commonwealth Government of Australia and BHP and CRA industries for this project.

References

1. A.R. Gordon: Zinc extraction and refining, Proc. Bull. Australas. Inst. Min. Metal., 291(6), 1986, pp. 57-69.

2. S. Goto: Thermodynamic consideration and basic test for new smelting process of zinc calcine, Metall. Review MMIJ, 1 (1), 1984, pp. 88-104.

3. S. Goto: Experimental work of injection smelting for new smelting process of zinc calcine, K. Tozawa (ed.), Zinc'85, Proc. Int. Symp. on Extractive Metallurgy of Zinc, Oct. 14-16, 1985, Tokyo, Japan, MMIJ, pp. 841-53.

4. A. Yazawa, T. Kiyomizu and S. Katoh: Direct zinc distillation by the high temperature oxidation method-thoretical study of its possiblity, Bull. Res. Inst. Min. Dress. Metall., Tohoku Univ., 33, 1977, pp. 59-65.

5. T.R.A. Davey and A.G. Turnbull: The direct smelting of zinc sulfide concentrate, Australia/Japan Ext. Met. Symp. 1980, Aus. Ins. Min.Met., pp 23-29.

6. H.H. Kellogg: Trends in non-ferrous pyrometallurgy. T.K. Tien and

J.F. Elliott (eds.), Metallurgical Treatises, TMS - AIME, Warrendale, Pa., 1981, pp. 199-38.

7. N.A. Warner: Direct smelting of zinc-lead ore, Trans. Inst. Min. Metall. Sec. C, 92, 1983, pp. C147-52.

8. N.A. Warner: Towards polymetallic sulfide smelting, A.D. Zunkel, R.S. Boorman, A.E. Morris and R.J. Wesely (eds.), Complex Sulfides, TMS-AIME, 1985, pp. 847-65.

9. T. Jones and N.A. Warner: Top blowing requirements for direct polymetallic smelting, Pyrometallurgy '87, Inst. Min. Metall., London, 1987, pp 605-26.

10. A.G. Turnbull and M.W. Wadsley: Thermodynamic modelling of metallurgical processes by the CSIRO-SGTE THERMODATA system, Extractive Metallurgy Symposium, Melbourne, Nov 12-14, 1984, Aus. Inst. Min. Met., pp. 79-85.

11. Y. Umetsu, T. Nishimura, K. Tozawa and K. Sasaki: Liquidus surfaces in a part of the systems $ZnO-PbO-SiO_2$ and $ZnO-"FeO"-SiO_2$, Australia/Japan Ext. Met. Symp. 1980, Aus. Inst. Min. Met., pp. 95-106.

12. B.W. Lightfoot, J.M. Bultitude-Paull and J.M. Floyd: Recovery of zinc and other valuable metals from slag dumps by use of Sirosmelt, Pyrometallurgy '87, Inst. Min. Metall, London, 1987, pp. 725-42.

13. G.G. Richards, J.K. Brimacombe and G.W. Toop: Kinetics of the zinc slag-fuming process, Met. Trans., 16B, 1985, pp. 513-27.

14. T. Lehner and R. Lindgren: Investigations of fluid flow phenomena in Boliden's zinc fuming plant, Can. Inst. Met., Annual Metting 1985, Vancouver, B.C.

15. R.M. Grant: The derivation of thermodynamic properties of slag from slag fuming plant data, Australia/Japan Ext. Met. Symp.1980, Aus. Inst. Min. Met., pp. 75-93.

OXYGEN-ENHANCED FLUID BED ROASTING

Debabrata Saha and J. Scott Becker, and L. Gluns

Air Products and Chemicals, Inc., 7201 Hamilton Boulevard,
Allentown, Pennsylvania 18195-1501 and Air Products, GmbH,
Klosterstr., D 4000 Dusseldorf 1, Germany

Abstract

The fluid bed roasting process is now widely employed for roasting zinc sulfide concentrate. In this process the feed is introduced from one side of the either round or rectangular reactor and the calcine discharges through an overflow on the opposite side. The residence time of the material in the fluidized bed, i.e., the roasting time, depends on the rate of feed and the air flow rate. The quantity of air blown into the reactor has a decisive effect on the condition of the bed. To increase the roasting capacity, some external source of oxidizing agent must be provided to the fluidized bed. While the current air-to-charge ratio must be closely maintained in order to satisfy the required fluidization velocity, the oxygen supply can be supplemented by injecting pure oxygen into the blower air while cutting back slightly on blower output. The level of oxygen injection must be determined by thermodynamic consideration, heat and mass balance, and the desired increment in throughput rate. This paper will describe both the process theory and the results of several on-stream applications of this technology. In particular, one plant has utilized 3% oxygen enrichment for five years and has realized a 21% production increase. A second plant utilizes the technology only during the summer months to make up for the loss of production rate due to low feed air density. Both initial test results and longer-term operating trends will be discussed as well as process limitations on the effective use of oxygen, with particular emphasis paid to handling the increased bed cooling which may be required.

Summary

Fluidized bed roasting accounts for most of the world's zinc production today. Recently, several U.S.-based zinc fluidized bed roaster operators have augmented their existing capacities with oxygen enrichment.

Fluid bed roasters face production increase limitations due to existing air blower capacities, requirements for fluidization characteristics of the bed, and downstream gas-handling capacities. Oxygen enrichment provides roaster operators with an opportunity to increase the productivity of their roasters in a very cost-effective way which can also provide additional revenue from increased sulfuric acid production. Additionally, oxygen enrichment can be utilized to improve calcine quality, resulting in a higher zinc recovery.

Air Products and Chemicals, Inc. (APCI) first demonstrated the oxygen enrichment technology for fluidized bed zinc roasters in April 1983. With 3% oxygen enrichment in a 13.5 short-ton-per-hour roaster, this producer achieved a 21% increase in calcine and sulfuric acid production. The quality of the calcine and the operating practices of the roaster and sulfuric acid plant were maintained essentially unchanged. At a second location, 3 to 3.5% enrichment provided a 13.6% production increase with a simultaneous reduction of 22.9% in sulfide sulfur in the calcine.

For a 3% enrichment practice, the oxygen consumption per month can vary between 10 to 17 million SCF per roaster. With credit from additional acid production, the net cost per extra ton of throughput is about $25 as compared to a conventional operating cost of $45 to $65 per ton. The total capital investment will typically be between $20,000 and $40,000 for one roaster. The technology is thus considered technically and economically attractive.

A number of fluid bed and multiple-hearth suspension zinc roasters are amenable to this application. Other nonferrous roasters, such as molybdenum, nickel, and lead, that use sulfide concentrates may also be technically and economically amenable to oxygen enrichment.

Introduction

In early 1983, the Metallurgy Group of Air Products' Research and Development Department started exploring the primary zinc industry to find new opportunities for increasing efficiency and production. A U.S.-based zinc producer approached Air Products around this time with the need to increase the capacity of their fluid bed zinc roaster by 20%. Upon investigation, and mass and heat balance calculations, it was determined that supplemental oxygen enrichment technology could meet this need. Starting in April of 1983, Air Products and this company demonstrated the technology in an in-plant trial in their roaster. Using 3% oxygen enrichment (i.e., 23.9% oxygen in the oxygen-enriched roaster air), the roaster capacity was increased by 21%. Subsequently, oxygen enrichment was adopted by a second U.S. plant where an increase in production and decrease in sulfide sulfur were simultaneously achieved.

Fluidized Bed Roasting Process

The fluid bed process is now widely employed for roasting zinc sulfide concentrate. The bed of powdered material is fluidized by blowing air through a dispersion plate having holes uniformly spaced over its area. Rising through the bed of material, the air brings the charge into a state resembling the boiling of a liquid. In the fluidized bed the surface of each particle is constantly blown with air, and therefore the roasting reactions proceed at a very high rate and to near completion.

Furnaces or reactors for the fluidized bed roasting process are either round or rectangular in shape. Figure 1 shows a typical circular reactor. The feed is introduced from one side of the reactor and the calcine discharges through an overflow on the opposite side. The residence time of the material in the fluidized bed, i.e., the roasting time, depends on the rate of feed and the air flow.

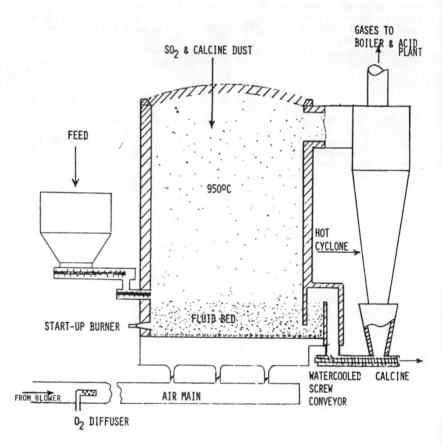

FLUIDIZED BED ZINC ROASTER

FIGURE 1

The quantity of air blown into the reactor has a decisive effect on the condition of the bed. With an excess of air, the particles pass into suspension and are carried out of the reactor. With too little air, fluidization is upset, and the bed settles down, resulting in a lower roasting efficiency.

No auxiliary heating is required for roasting sulfide concentrates by the fluidization process. The amount of heat generated by the rapid combustion of the sulfides is enough to maintain the requisite temperature in the reactor (940-980°C). In fact, external cooling is required. Cooling water is supplied by either bed coils built into the reactor at the level of the fluidized bed or by direct injection of water inside the reactor. The temperature can be controlled by varying the concentrate feed rate or the cooling water rate.

Some of the feed material inevitably goes into suspension and is carried off with the gas stream. However, it is collected in different stages of the gas cleaning plant before the roaster offgas is fed into the sulfuric acid plant. This material is then recycled back to the roaster.

Proposed Technique To Increase Throughput

To increase the current concentrate throughput rate (and consequently, the calcine production rate), an additional source of oxidizing agent must be provided in the fluidized bed. However, the current air rate from the blower must be maintained in order to satisfy the critical fluidization velocity. The bed must remain in the same fluidized state when the throughput rate is increased by oxygen usage. The air rate is supplemented by injecting pure oxygen into the air main through a diffuser, before the blower air reaches the plenum. This extra oxygen then reacts with the extra feed material to undergo the principal reaction, $ZnS + \frac{3}{2}O_2 \rightarrow ZnO + SO_2$. The higher oxygen concentration in the process air enhances the reaction rate of the above principal chemical reaction, thereby improving operating efficiency. The optimum level of oxygen injection can be determined by thermodynamic consideration, heat and mass balance, and the desired increment in throughput rate.

Theoretical Basis for A Throughput Increase

The oxidation of zinc sulfide (principal reaction for roasting zinc sulfide concentrate) is expressed as:

$$ZnS + \frac{3}{2} O_2 \rightarrow ZnO + SO_2 \tag{1}$$

$$\Delta H = -106 \text{ Kcal/mol ZnS} \quad (800 \text{ to } 1000°C) \tag{2}$$

The rate of oxidation of ZnS is directly proportional to the rate of diffusion of oxygen through the solid concentrate particles. The rate of diffusion of oxygen is written as:

$$-\frac{dN_{O_2}}{dt} = A \cdot k \cdot P_{O_2} \quad (3)$$

where A = Area of the Particle
        k = Overall Mass Transfer Coefficient
       P_{O_2} = Partial Pressure of Oxygen

The overall mass transfer coefficient depends upon the gas phase mass transfer coefficient, the effective diffusivity of oxygen through a solid zinc oxide layer, the extent of reaction, the intrinsic chemical rate coefficient, and the initial size of the sulfide particle.

In a fluidized bed, k and P_{O_2} are both functions of position in the bed. Usually, from a macroscopic point of view, the rate of oxygen consumption is written as:

$$F_{O_2}(\text{in}) - F_{O_2}(\text{out}) = A \cdot \bar{k} \cdot \bar{P}_{O_2} \quad (4)$$

where $\bar{k}$ and $\bar{P}_{O_2}$ are averaged values of k and P_{O_2}

Under the conventional situation (only blower air, no oxygen enrichment), P_{O_2} lies between 0.21 and 0.03 (assuming 3% O_2 in roaster offgas under existing air-based system). As an approximation, it will be assumed that $\bar{k}$ is independent of the level of oxygen enrichment. This approximation should be valid for low levels of oxygen enrichment (<5%) and in cases where the feed (concentrate) and product (calcine) characteristics are essentially unchanged with oxygen enrichment. In other words, an x% increase in the throughput rate requires an increase of x% in $\bar{P}_{O_2}$. The desired increase in $\bar{P}_{O_2}$ can be achieved by injecting oxygen by injecting oxygen at a specific flow rate into the wind main.

The use of oxygen-enriched air will, however, necessitate the removal of extra heat (from the exothermic reactions of increased feed materials) and closer temperature control in order to avoid localized temperature excursions and sintering. The amount of extra coolant (water, secondary feed materials, etc.) can be calculated via a heat balance. Varying the oxygen enrichment level can provide an excellent means for controlling bed temperature without interfering with the fluidized air flow.

Theoretical Projections

Using oxygen enrichment technology in a fluidized bed zinc roaster, the existing capacity of the roaster can be increased by 20% with a corresponding increase in sulfuric acid production. The net operating cost per extra ton of throughput is estimated at $25. The total investment for a one-roaster plant is approximately $40,000. These cost figures should justify the oxygen enrichment technology as a very cost-effective alternative to augment the capacity of an existing roaster.

The following predictions are from theoretical calculations:

Increased Throughput

O_2 Enrichment (%)	Throughput Increase (%)
1.0	6 - 8
2.0	13 - 16
3.0	20 - 23

Higher Coolant Rate (Water Injection)

O_2 Enrichment (%)	Coolant Rate Increase (%)
1.0	41
2.0	73
3.0	95

Increased Sulfuric Acid Production

If the dilution air flow rate is maintained, the SO_2% in the gas to the acid plant has been calculated to increase from 7.3 to 8.0% for a 20% throughput increase. The total gas flow to the acid plant is increased by 8% for the same throughput increase. Since both the SO_2 content (%) and the total SO_2 volume in the roaster offgas will increase, sulfuric acid production must be increased at the same rate as the throughput increment.

Increased Heat Recovery

More steam is generated from the waste heat boiler and water bed coils (if any) due to the extra heat generation from increased throughput. The increased steam generation may have a favorable impact on the overall economics of the plant.

Improved Control and Flexibility of Operations

The ability to adjust oxygen concentration in the roaster air provides an excellent means of controlling bed temperature without significantly interfering with the fluidizing air flow. The reactor response time should be much shorter when oxygen concentration is varied than when the feed mix is varied.

Operating Results, Plant "A"

The actual operating parameters from a zinc plant which has utilized the oxygen enrichment technology continuously for over five years are used below to illustrate the impact of oxygen enrichment.

Theoretical Projections

Basis: a) Air Flow Rate = 12,000 SCFM
       = 2,520 SCFM O_2, 9,480 SCFM N_2
    b) Concentrate Analysis = 58% Zn, 31% S, (Dry basis) and 5% Moisture
    c) Calcine Analysis = 70% Zn, 2.3% S
    d) 1 lb-mol of a Gas = 380 SCF

Case I: Air-Based Operation

Input:

Concentrate = 275 T/D ≡ 382 lbs/min ≡ 221.5 lbs Zn, 118 lbs S

Secondary Material = 45 T/D ≡ 63 lbs/min (86% ZnO) ≡ 43.1 lbs Zn

Moisture = 5% x 275 T/D ≡ 20.1 lbs/min   95%

Output:

Calcine = 272 T/D = 378 lbs/min ≡ 264.6 lbs Zn, 8.7 lbs S

Offgas = $\frac{(118-8.7) \text{ lbs S/min}}{32 \text{ lbs S/lb-mol}}$ x 380 SCF/lb-mol

= 1,298 SCFM SO_2 (11.0%)

(2520 − 1298 x $\frac{3 \text{ mol-}O_2}{2 \text{ mol-}SO_2}$) SCFM

= 573 SCFM O_2 (4.9%)

9,480 SCFM N_2 (80.5%)

$\frac{20.1 \text{ lbs } H_2O/\text{min}}{18 \text{ lbs } H_2O/\text{lb-mol}}$ x 380 SCF/lb-mol = 424 SCFM H_2O (3.6%)

11,775 SCFM Total (100%)

Case II: Oxygen-Enriched Operation

Goal → 10% Production Increase
To increase production by 10%, the partial pressure of oxygen input should also be increased by 10%.

So P_{O_2}(in) = 21% x 1.10 = 23.1%   (5)

Now, 0.231 (12,000 + F) = 0.21 (12,000) + F   (6)

Where: F = Oxygen Flow Rate, SCFM
       F = 327 SCFM

Hence,
Total Oxygen = (12,000 x 0.21 + 327 or 2,847 SCFM   (7)
Total Nitrogen = (12,000 x 0.79) or 9,480 SCFM   (8)

Input:

Concentrate = 382 lbs/min x 1.10 = 420 lbs/min ≡ 243.6 lbs Zn, 130.3 lbs S
Secondary Material = 63 lbs/min x 1.10 = 69 lbs/min ≡ 47.4 lbs Zn
Moisture = 20.1 lbs/min x 1.10 = 22.1 lbs/min

Output:

Calcine = 378 lbs/min x 1.10 = 415 lbs/min ≡ 291 lbs Zn, 9.5 lbs S

$$\text{Offgas} = \frac{(130.3 - 9.5)}{32 \text{ lbs S/lb-mol}} \text{ lbs S/min} \times 380 \text{ SCF/lb-mol}$$

= 1,435 SCFM SO_2 (11.9%)

$$(2847 - 1435) \times \frac{3 \text{ mol-}O_2}{2 \text{ mol-}SO_2} \text{ SCFM}$$

= 694 SCFM O_2 (5.7%)

9,480 SCFM N_2 (78.5%)

$$\frac{22.1 \text{ lbs } H_2O/\text{min}}{18 \text{ lbs } H_2O/\text{lb-mol}} \times 380 \text{ SCF/lb-mol}$$

= 467 SCFM H_2O (3.9%)

12,076 SCFM Total   (100%)

So P_{O_2}, out = 5.7%, which is 16% higher compared to the base figure of 4.9%.

The above calculations indicate: a) For a 10% production increase, oxygen enrichment of 2.1% is sufficient. b) The SO_2 content of the offgas will increase from 11.0% to 11.9%. Similarly, the O_2 content will change from 4.9% to 5.7%. The $\frac{SO_2}{O_2}$ ratio will be 2.08 with 2.1% enrichment.

c) The actual production increase with a 2.1% enrichment will lie somewhere between 10% and 16% (i.e., $\frac{5.7}{4.9} \times 100\%$) depending upon the relationship between $\bar{P}_{O_2}$ and the inlet and exit P_{O_2}.

Production Increase vs. Oxygen Usage. The following table shows the predicted oxygen requirement for different levels of production increase.

Table I. Base Throughput Rate: 320 Short Tons/Day

Production Increase (As % of Base)	Blower Rate (SCFM)	Oxygen Flow Rate (SCFM)	Enrichment Level (%)
Base	12,000	--	--
6	12,000	154	1.0
13	12,000	312	2.0
20	12,000	474	3.0

Actual Operating Results

A summary of plant data is tabulated below (Table II). The actual operating data and a summary report are included in Tables III and IV.

Table II.

	Oxygen Enrichment Level (%)			
	0%	1%	2%	3%
Throughput, Tons/Hour	13.44	14.28	15.43	16.36
Production Increase (%)	--	6.25	14.80	21.70
Water Injection, Gals/Min	5.23	7.77	9.03	10.36
Water Injection, Gals/Ton	23.41	32.66	35.04	37.94

Comparison Between Theoretical and Operating Data

Throughput Rate. Table IV indicates that the production increment achieved actually slightly exceeds the calculated values as follows:

Enrichment Level (%)	Theoretical Production Increase (%)	Actual Production Increase (%)
1	6	6.25
2	13	14.80
3	20	21.70

The positive deviation of production increase may be due to a higher oxygen utilization efficiency with oxygen enrichment. But this could not be proved due to the unavailability of the roaster offgas analysis.

Calcine Quality. The theoretical calculations for oxygen enrichment were made based on the given condition that the total sulfur in the calcine (2.3% S) should remain unchanged. The actual operating data indicates that the calcine quality does, in fact, meet this requirement. Consequently, no problems were encountered in treating the calcine in downstream operations.

TABLE 1II
OXYGEN ENRICHMENT IN FLUID BED ZINC ROASTER

9:04 THURSDAY, JANUARY 12, 1984

OBS	DATE	OXYGEN ENRICHMENT	THRUPUT, TONS PER HOUR	WATER, GALLONS PER MINUTE	WATER, GALLONS PER TON
1	04/01/83	0	13.66	.	.
2	04/02/83	0	13.83	.	.
3	04/03/83	0	14.12	.	.
4	04/04/83	0	13.33	.	.
5	04/05/83	0	12.58	.	.
6	04/06/83	0	14.70	.	.
7	04/07/83	0	13.29	4.80	21.66
8	04/08/83	0	13.50	.	.
9	04/09/83	0	13.41	.	.
10	04/10/83	0	12.08	.	.
11	04/11/83	0	13.33	.	.
12	04/12/83	0	14.45	.	.
13	04/13/83	0	12.95	.	.
14	04/15/83	0	13.12	.	.
15	04/16/83	0	13.50	.	.
16	04/17/83	0	13.25	.	.
17	04/18/83	1	14.29	7.35	30.86
18	04/19/83	1	14.26	8.00	33.68
19	04/20/83	1	14.29	7.96	33.45
20	04/21/83	2	16.25	9.66	35.69
21	04/22/83	2	14.68	9.42	38.49
22	04/23/83	2	15.17	9.46	36.93
23	04/24/83	2	13.67	7.22	31.67
24	04/25/83	2	15.60	8.89	34.19
25	04/26/83	2	15.95	9.60	36.11
26	04/27/83	3	15.44	10.73	41.70
27	04/28/83	3	17.78	10.00	33.75
28	04/29/83	3	16.40	9.88	36.15
29	04/30/83	3	16.25	10.45	38.59
30	05/01/83	3	16.29	8.67	31.92
31	05/02/83	3	16.25	9.99	36.88
32	05/03/83	3	16.03	10.98	41.09
33	05/05/83	0	13.59	5.67	25.02
34	05/06/83	0	13.50	5.52	24.54
35	05/07/83	2	16.67	8.96	32.24
36	05/08/83	3	16.30	10.84	39.91
37	05/09/83	0	17.06	10.42	36.67
38	05/10/83	0	13.06	5.13	23.59
39	05/11/83	0	13.60	5.04	22.24
40	05/12/83	3	16.68	10.93	39.30
41	05/13/83	3	16.41	11.38	41.57
42	05/14/83	3	15.91	9.84	37.00
43	05/15/83	3	16.62	10.60	38.75
44	05/18/83	3	16.16	.	.
45	05/19/83	3	16.16	.	.
46	05/20/83	3	16.12	.	.
47	05/21/83	3	16.12	.	.
48	05/22/83	3	16.54	.	.
49	05/23/83	3	16.04	.	.
50	05/24/83	3	16.41	.	.

TABLE IV
OXYGEN ENRICHMENT IN FLUID BED ZINC ROASTER

9:04 THURSDAY, JANUARY 12, 1984   2

VARIABLE	LABEL	N	MEAN	STANDARD DEVIATION	MINIMUM VALUE	MAXIMUM VALUE	STD ERROR OF MEAN	SUM	VARIANCE
					--OXYGEN ENRICHMENT % = 0--				
TPH	THRUPUT, TONS PER HOUR	20	13.44250000	0.58441041	12.08000000	14.70000000	0.13067814	268.850000	0.34153553
GPM	WATER, GALLONS PER MINUTE	5	5.23200000	0.35660903	4.80000000	5.67000000	0.15948041	26.160000	0.12717000
GPT	WATER, GALLONS PER TON	5	23.41000000	1.44332948	21.66000000	25.02000000	0.64547657	117.850000	2.08320000
					--OXYGEN ENRICHMENT % = 1--				
TPH	THRUPUT, TONS PER HOUR	3	14.28000000	0.01732051	14.26000000	14.29000000	0.01000000	42.840000	0.00030000
GPM	WATER, GALLONS PER MINUTE	3	7.77000000	0.36428011	7.35000000	8.00000000	0.21031722	23.310000	0.13270000
GPT	WATER, GALLONS PER TON	3	32.66333333	1.56596083	30.86000000	33.68000000	0.90410791	97.990000	2.45223333
					--OXYGEN ENRICHMENT % = 2--				
TPH	THRUPUT, TONS PER HOUR	7	15.42714286	1.01955406	13.67000000	16.67000000	0.38535521	187.990000	1.03949048
GPM	WATER, GALLONS PER MINUTE	7	9.03000000	0.85238880	7.22000000	9.66000000	0.32217268	63.210000	0.72656667
GPT	WATER, GALLONS PER TON	7	35.04571429	2.48197406	31.67000000	38.49000000	0.93809802	245.320000	6.16019524
					--OXYGEN ENRICHMENT % = 3--				
TPH	THRUPUT, TONS PER HOUR	20	16.36300000	0.46725514	15.44000000	17.78000000	0.10448143	327.260000	0.21832737
GPM	WATER, GALLONS PER MINUTE	13	10.36230769	0.69579875	8.67000000	11.38000000	0.19297985	134.710000	0.48413590
GPT	WATER, GALLONS PER TON	13	37.94461538	2.95333883	31.92000000	41.70000000	0.81910881	493.280000	8.72221026

Coolant Requirement. A heat and mass flow diagram for calcination can be presented as follows:

Basis: 1 Ton of concentrate Feed

```
1000 lbs—Oxygen———→ ┌─────────────┐ ──→ SO₂ ————1300 lbs; 0.43 x 10⁶ BTU
                    │  Heat of    │
                    │  Reaction   │
1625 lbs—Water (Liq.)→│            │──→ Water (Gas) 1625 lbs; 3.17 x 10⁶ BTU
                    │  4.1 x 10⁶  │
2000 lbs—Concentrate→│   BTU      │──→ Calcine————1700 lbs; 0.50 x 10⁶ BTU
                    └─────────────┘
```

Using the above diagram, the coolant requirement (water injection) can be calculated for different production rates (assuming 100% concentrate in the feed).

Base Data: Throughput Rate 13.44 Tons/Hr
Water Injection 5.23 GPM

O_2 Enrichment Level (%)	Actual Throughput Production Rate (T/Hr)	Calculated Water Injection Rate (GPM)	Actual Water Injection Rate (GPM)
0	13.44	--	5.23
1	14.28	7.97	7.77
2	15.43	11.72	9.03
3	16.36	14.75	10.36

The actual injection rate is lower compared to the calculated value. This is primarily due to the fact that the feed material contains 16% secondary materials which do not provide exothermic heat. Consequently, less water is required for temperature control compared to a 100% concentrate feed.

Model Development for Projecting Operating Results With Oxygen Enrichment

Applying the data from this installation, multiple regression analysis has been used as a tool to model the effect of oxygen enrichment in a fluid bed zinc roaster. The following model forms have been found to reasonably approximate the trends in the data (figs. 2-6). The dotted lines show the 90% confidence limits for long-term operation.

Figures 5 and 6 show the water flow rate requirements in gallons per ton and gallons per minute, respectively, as a function of throughput. The solid line represents the model quadratic equation for throughputs less than 16 TPH and a constant for throughputs over 16 TPH.

For each model, approximate 90% confidence limits for the long-term process average have been developed. At any point, the confidence limits are a statistical quantification of how well the underlying relationship has been determined. These confidence limits should only be considered reliable for substantially similar systems.

These graphs should not be extrapolated outside of the data ranges investigated. Figures 2, 3, and 4 are limited to 0% $\leq$ O_2 ENRICH $\leq$ 3% and Figures 5 and 6 are limited to 13 $\leq$ TPH $\leq$ 18 (tons per hour). Special care must also be taken when attempting to extend this model to a fluid bed zinc roaster of substantially different size than that used in this study.

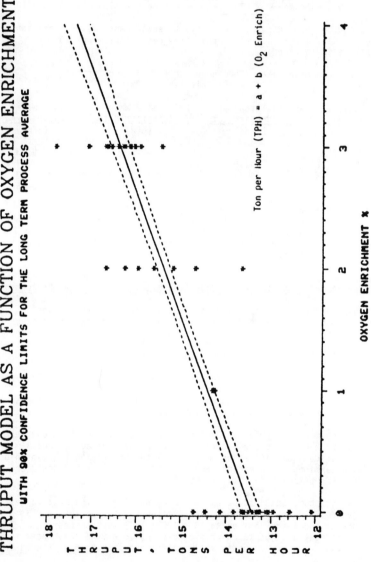

FIGURE 2

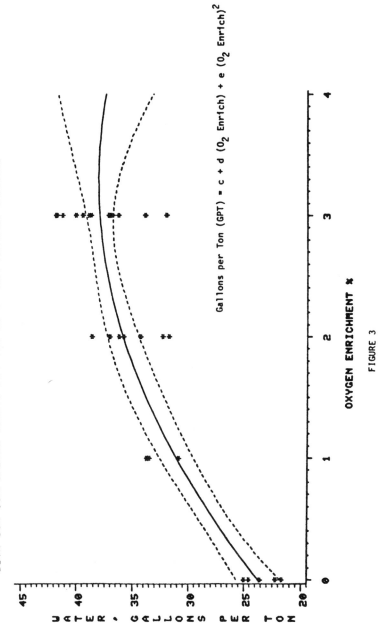

FIGURE 3

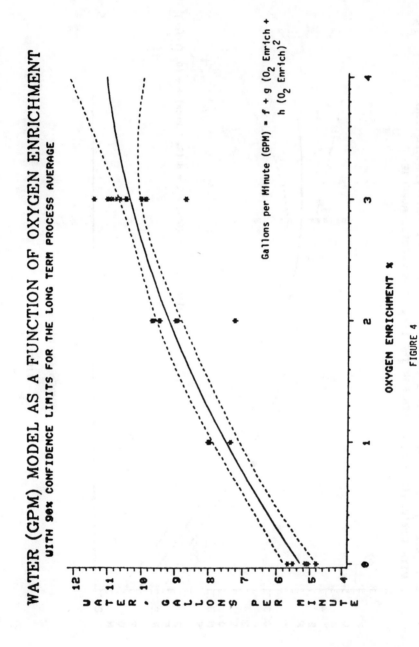

FIGURE 4

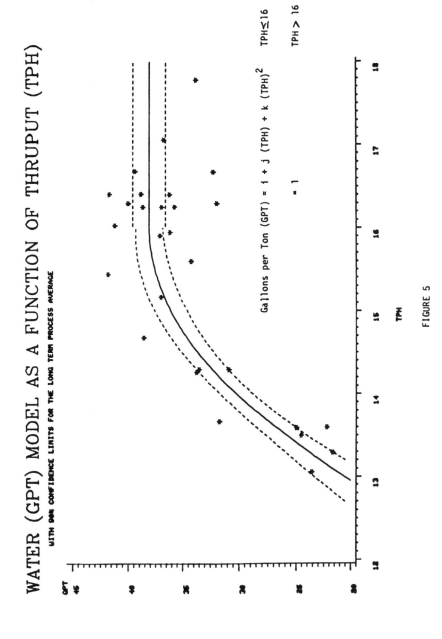

FIGURE 5

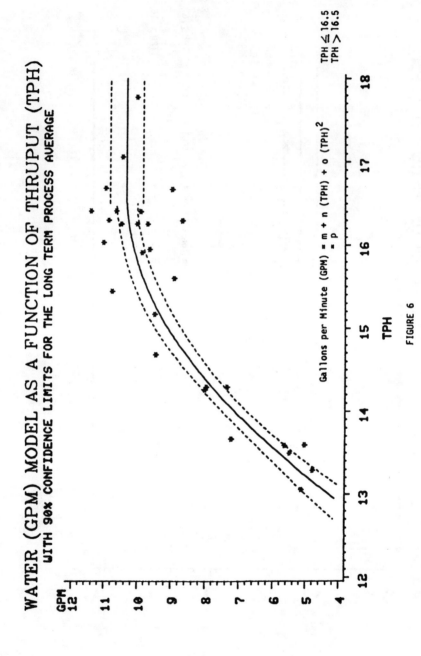

FIGURE 6

FIGURE 7

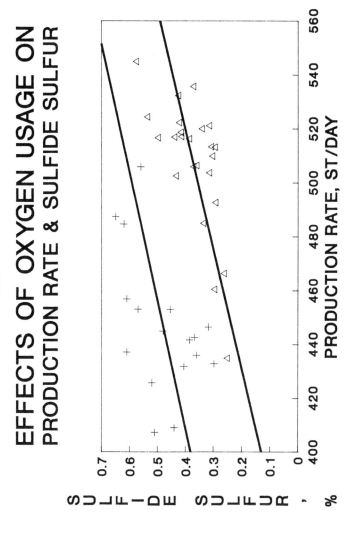

EFFECTS OF OXYGEN USAGE ON PRODUCTION RATE & SULFIDE SULFUR

ENRICHMENT LEVEL 3 TO 3.5%

Based upon the above regression analysis, the following predictions can be made with 90% confidence limits (CL) for the long-term average process parameters.

Throughput Rate

O_2 Enrichment (%)	Throughput Increase (%)		
	Predicted Value	Lower 90% CL For Mean	Upper 90% CL For Mean
0.5	3.6	2.3	5.0
1.0	7.8	6.2	8.4
1.5	10.9	9.8	11.9
2.0	14.5	13.4	15.6
2.5	18.2	16.9	19.5
3.0	21.8	20.3	23.4

Coolant (Water Injection) Rate

O_2 Enrichment (%)	Required Coolant Rate Increase (%)	
	Per Ton	Per Minute
0.5	16.8	21.8
1.0	30.0	41.3
1.5	42.1	58.3
2.0	50.6	73.1
2.5	56.3	85.4
3.0	59.4	95.4

This table is valid when the total coolant requirement is met by water injection only. If bed coils are used, an adjustment has to be made to the above table due to the different heat contents of steam at various temperatures and pressures. Steam tables will be helpful to perform the calculations for coolant requirements in these cases.

If the existing water injection system cannot handle the extra coolant requirement, the possibility of using bed coils and/or secondary materials (ZnO) should be explored.

Additional impacts of oxygen enrichment are summarized below:

Sulfuric Acid Production. The increase in sulfuric acid production is directly proportional to the increase in throughput rate. For example, if a 20% increase in throughput rate is achieved with a 3% oxygen enrichment system, the sulfuric acid production will also be augmented by 20%.

Roaster Offgas Volume and Gas Volume to Acid Plant. Both of these parameters will increase with oxygen enrichment when the current blower rate is maintained. If the gas-handling capacity limits the attainable production increase, the following alternatives may ease the problem:
a) Reduce the blower rate and make up with oxygen injection. b) Reduce the air leakage into the system. c) Substitute water injection partially by either increasing the bed coil capacity or injecting CO_2 in the roaster.

In summary, for an average-sized roaster (10 to 20 tons per hour) the gas handling plant will have an additional load of around 10% for a 3% oxygen enrichment practice.

Economics. The total capital cost for installing the oxygen enrichment system in this two-roaster plant was about $40,000. Using the credit for additional sulfuric acid production ($35-45 per ton of acid), the operating cost per incremental ton of throughput may range between $18 to $25.

This analysis has not taken into account the following additional benefits that may also be realized from oxygen enrichment:

- Increased value of calcine (compared to concentrate).
- Higher zinc yield due to the lower sulfide-sulfur content of the calcine.
- Lower maintenance cost (per ton of throughput) for downstream units (less nitrogen and dust carry-over).
- Quicker and smoother start-ups after shutdowns.

The net cost per incremental ton of throughput should compare favorably with the current operating cost per ton of throughput which is generally in the range of $45 to $65.

This installation demonstrated that the economics of this technology are very attractive if a production increase is desired. In particular, a production increase of up to 30% may be achieved more economically by oxygen enrichment compared to installation of an additional roaster.

Operating Results, Plant "B"

A second plant has utilized oxygen enrichment in their fluid bed zinc roaster intermittently over the past two years to balance their production demands, calcine quality, and other operating parameters. Operating data both with and without oxygen enrichment are presented in Tables V and VI. Standard operating conditions averaged 445.5 dry tons per day of concentrate with 0.48% sulfide sulfur in the calcine. Introduction of a 3 to 3.5% oxygen enrichment practice resulted in an increase in production to 506.3 dry tons per day of concentrate with only 0.375% sulfide sulfur in the calcine. Thus, both a 13.6% increase in production and a 22.9% reduction in sulfide sulfur in the calcine were achieved. With oxygen enrichment, sulfide sulfur in the calcine has been reported to be as low as 0.2% at this plant. Figure 7 shows the simultaneous improvements in production rate and sulfide sulfur content with oxygen enrichment. This plant has not encountered any operating problems with oxygen enrichment.

The choice between an increase in production rate and an improvement in product quality represents a classic operating trade-off in any production facility. Oxygen enrichment provides the capability to improve one or the other, or a partial improvement in both. Greater productivity could have been achieved in this case had the quality level been maintained at the pre-enrichment level. Conversely, even higher calcine quality could have been achieved but at the expense of a smaller increase in production rate. The choice must be made based upon a given company's needs and relative economics.

TABLE V
PLANT B BASE DATA
NO OXYGEN ENRICHMENT

		DATE	DST/DAY	% SS	CFM	% SS	REGRESSION CALCULATIONS -1.5 STD	+1.5 STD
1986:	JAN	20	487.51	0.650	23125	0.595	0.417	0.772
		21	453.01	0.570	23000	0.498	0.321	0.675
		23	484.69	0.620	23000	0.587	0.410	0.764
	MAR	29	425.64	0.520	22000	0.421	0.244	0.598
		30	407.30	0.510	22000	0.370	0.193	0.547
		31	437.22	0.610	22000	0.454	0.277	0.631
	OCT	04	456.92	0.610	22813	0.509	0.332	0.686
1987:	AUG	27	441.60	0.385	22593	0.466	0.289	0.643
		28	431.70	0.405	23313	0.438	0.261	0.615
		29	446.50	0.317	23229	0.480	0.303	0.657
		30	432.70	0.297	23283	0.441	0.264	0.618
	SEP	05	453.06	0.454	21908	0.498	0.321	0.675
		06	435.89	0.360	22104	0.450	0.273	0.627
		07	442.66	0.367	22417	0.469	0.292	0.646
		AVG	445.46	0.477	22628	0.477		
		STD	20.59	0.118	524	0.058		

X = DST/DAY
Y = % SS

REGRESSION OUTPUT:

Constant	-0.77100
Std Err of Y Est	0.111324
R Squared	0.238546
No. of Observations	14
Degrees of Freedom	12
X Coefficient(s)	0.002801
Std Err of Coef.	0.001444

DST = Dry Short Ton
SS = Sulfide Sulfur

TABLE VI
PLANT B RESULTS
WITH OXYGEN ENRICHMENT

		DST/DAY	% SS	SULFUR T/DAY	AIRFLOW SCFM	CAL. % SS	1.5 SIGMA LIMITS LOW	HIGH
3-3.5% Oxygen		FIRST REGRESSION						
SEPT '87	11	516.74	0.497	2.568	21063	0.3971	0.267	0.527
	12	524.44	0.535	2.806	20996	0.4156	0.286	0.545
	13	544.99	0.575	3.134	20875	0.4648	0.335	0.595
	17	466.36	0.261	1.217	22104	0.2764	0.146	0.406
	18	484.97	0.334	1.620	22313	0.3210	0.191	0.451
	19	460.43	0.295	1.358	21271	0.2622	0.132	0.392
	20	434.86	0.250	1.087	21000	0.2009	0.071	0.331
	25	518.76	0.415	2.153	22000	0.4020	0.272	0.532
	26	502.53	0.432	2.171	22000	0.3631	0.233	0.493
	27	492.68	0.291	1.434	22000	0.3395	0.210	0.469
	28	503.75	0.313	1.574	21875	0.3660	0.236	0.496
	29	509.90	0.302	1.540	22000	0.3807	0.251	0.511
	30	506.42	0.359	1.818	22000	0.3724	0.242	0.502
OCT '87	01	522.34	0.421	2.198	22000	0.4105	0.281	0.540
	02	516.95	0.438	2.262	22042	0.3976	0.268	0.528
	03	513.38	0.306	1.570	22000	0.3891	0.259	0.519
	04	521.20	0.313	1.629	22000	0.4078	0.278	0.538
	05	513.22	0.293	1.504	22000	0.3887	0.259	0.519
	31	520.06	0.339	1.764	21250	0.4051	0.275	0.535
NOV '87	01	532.32	0.424	2.258	20850	0.4344	0.305	0.564
	02	516.33	0.385	1.988	20263	0.3961	0.266	0.526
	03	517.15	0.413	2.133	21021	0.3981	0.268	0.528
Wt. Avg.:		506.35	0.375	1.899	21587	0.3722		
Std. Dev.:		24.627	0.0866	0.5069	561	0.0590		

Concentrate vs. S/S
Regression Output:

Constant	-0.84113
Std. Err of Y Est.	0.066493
R Squared	0.464217
No. of Observations	22
Degrees of Freedom	20
X Coefficient(s)	0.002396
Std. Err of Coef.	0.000575

Airflow vs. Concentrate
Regression Output:

Constant:	645.0987
Std. Err of Y Est.	25.55057
R Squared	0.021438
No. of Observations	22
Degrees of Freedom	20
X Coefficient(s)	-0.00642
Std. Err of Coef.	0.009709

DST = Dry Short Ton
SS = Sulfide Sulfur

Operating Results, Plant "C"

A recent oxygen enrichment test conducted at Plant C has demonstrated that: 1) Throughput rates of up to 24% above the roaster design capacity (117 tpd dry) can be attained while maintaining calcine quality at the pre-enrichment level. 2) With no increase in throughput rate (5.5 short wet tph), a 1.7% O_2 enrichment practice can lower sulfide sulfur from 1.35% to 0.6% in the calcine.

Conclusions

Both theoretical analysis and substantial operating data verify that oxygen enrichment of fluidized bed zinc roasters can provide both productivity increases and calcine quality improvements. There is substantial agreement between the projected and actual operating results, thus providing the ability to accurately project operating benefits and relative economics. Over many years of operation, oxygen enrichment has proved these benefits to be attainable without negative impact on the fluid bed roaster or other downstream operations.

An Innovative Approach to Alloying Molten Metals With High Melting Point Additions

Johannes Schade, Stavros A. Argyropoulos, and Alexander McLean

Ferrous Metallurgy Research Group
Department of Metallurgy and Materials Science
University of Toronto
184 College Street
Toronto, Ontario, Canada
M5S 1A4

Abstract

A microprocessor-based intelligent measurement system (IMS) has been used to develop a new class of powdered ferroalloys for use in post-furnace treatment vessels. The alloying elements of interest are those for which the melting range lies above the bath temperature, viz. ferroalloys based on chromium, niobium, molybdenum, vanadium, and tungsten. Using the heat liberated by an exothermic reaction, it is possible to greatly reduce the time required for assimilation of the alloy into the bath. A cored-wire technique has proven very successful to introduce the powders into iron and nickel melts. Assimilation rates and sample temperature history were monitored using the IMS. It will be demonstrated that silicon modified cored-wire additions exhibit markedly different and superior assimilation characteristics when compared to conventional lump additions and unmodified cored-wire additions. In steelmaking operations, the dramatic improvement in the time required for assimilation into a liquid bath renders these alloys ideal for trimming addition in the tundish of a continuous caster, where residence time restrictions demand a very rapid alloy assimilation period. In nickel-based melts, ferroalloying additions can now be made much closer to the actual casting process without the risk of decreasing recoveries of the alloying elements.

Introduction

The assimilation of solid additives into liquid melts depends to a great extent on the relative melting ranges of the solute addition and the solvent bath. If the melting range of the addition lies below that of the bath, the process may generally be classified as a melting process. If, on the other hand, the addition has a melting point which lies above that of the solvent melt, then in the case of mass transfer control mechanisms the assimilation is controlled by solute diffusion through a liquid phase boundary layer around the perimeter of the solid addition, and the assimilation process is termed dissolution. Melting or dissolution may be either endothermic or exothermic in nature, as in the cases of additions of tungsten or niobium to steel, respectively (1). In the case of endothermic additions, where heat needs to be supplied on a continuous basis to effect a thorough alloying process, dissolution rates are expected to be small (2). By contrast, the temperature inside exothermic additions may considerably exceed the bath temperature, and heat and mass transfer phenomena can in fact influence each other, resulting in a self-accelerating assimilation process. A more complete discussion of the solution kinetics of ferroalloys may be found in previous publications (1,3-5).

Economic considerations and assimilation kinetics are the reasons that it is rather uncommon to encounter addition of alloys in elemental form. In the case of the alloys considered in the present study, additions generally are made in form of ferroalloys, namely FeCr, FeMo, FeNb, FeV, and FeW. For these materials, dissolution tends to govern the assimilation process of the solid addition in both iron and nickel melts, the exception being the solution of FeCr in liquid steel, which is classified as a melting process (6). Conventionally, alloying is performed using either lump or briquette additions to the primary melting vessel or, more commonly, during post-furnace ladle metallurgical treatment. In steelmaking operations, cored-wire treatment in the ladle is often applied in preference to lump additions or powder injection (7). However, in order to allow for rapid post-furnace treatment, the assimilation of these ferroalloys must not be governed by the normal dissolution process; rather, an accelerated solution of the alloying element must be sought without compromising the recovery characteristics. The application of the principle of microexothermicity, which is discussed below, allows the alteration of the assimilation characteristics of the ferroalloys mentioned previously. The application of microexothermic cored-wire additions is expected to gain acceptance for the production of micro-alloyed steels since the improved chemical control obtained with tundish alloying will be reflected directly in enhanced material characteristics. Furthermore, such a technique provides a potential new technology for the non-ferrous industries, allowing the addition of alloy components to be performed just prior to casting.

The steelmaking tundish has traditionally been regarded as a transfer vessel between the batch ladle refining process stage and the continuous casting operations. The tundish itself is a continuous reactor which, if properly controlled, can be used for certain alloying purposes more effectively that additions to molten steel in the ladle. The advantages of using the tundish as a continuous metallurgical refining vessel were described by McLean (8) in the 65th Henry Marion Howe Memorial Lecture in a presentation entitled "The Turbulent Tundish - Contaminator or Refiner?". Metal alloying in the tundish is an innovative support technology with respect to achieving the final desired chemical composition, intended to allow trimming control of large heats as well as a tailoring of the chemical composition of individual slabs. A successful addition in the tundish of ferroalloys which normally require long assimilation times would be of great benefit to the steel industry in terms of meeting stringent customer specifications while minimizing the possible loss of yield due to remelting of semi-finished products which fall outside of customer specifications.

Currently practised techniques of adding the necessary alloying additions to the ladle can be somewhat erratic in terms of recovery levels, primarily due to the bulk volumes involved and the liquid steel temperature variability (6). While some microexothermic lump-form additions have been developed (3,9-12) and patented (13) previously, the use of these alloys was restricted to primary alloying additions in the ladle. The use of the tundish as a reaction vessel in which the chemical composition of the final product can be trimmed has been recognized for some time. However, the realization of this potential has been slowed by the impediment that alloys are not available commercially in a form that allows for adequate assimilation in the tundish, given that residence times rarely exceed the order of 60 to 90 seconds. This is an especially critical restraint for some ferroalloys such as FeMo, FeW, or FeNb which have melting ranges that exceed the temperature of the steel bath.

In present non-ferrous pyrometallurgical practice, alloying elements such as niobium, tungsten, or vanadium are added to the nickel charge in either the primary melting vessel or during a refining process such as the AOD, and alloying is always a batch charging process. Since there are only very few cases where recovery problems are encountered with regard to the above high melting point elements in non-ferrous melts, very little work has been performed using post-furnace wire feeding. While the present work will focus on the application of microexothermicity to the design and testing of cored-wire additions to iron- and nickel-base melts, the concept can be extended to other non-ferrous melts, especially to those where some type of continuous casting process is practised. A typical example of a potential application would be the addition of chromium to copper-base alloys, where chromium recovery has been found to be problematic at times (14), prior to casting sheets using the Mitsubishi process. In the case of nickel-base melts, limited success was achieved (14) when aluminum and calcium additions were fed in wire form to Alloy 800, which contains iron, nickel, and chromium.

Theory : Principles of Microexothermicity

The phenomena of microexothermicity and macroexothermicity were first introduced by Argyropoulos (10) in 1984 in conjunction with the solution behaviour of ferroalloys in liquid steel. The mass transfer of a solid ferroalloy addition into liquid steel can be characterized as either a melting or a dissolution process. These are illustrated schematically in Fig. 1. In all cases, a solid steel shell freezes around the cold addition upon immersion and eventually melts back. Which of the six mass transfer routes shown is followed depends on the particular thermophysical characteristics of an alloying addition and its thermodynamic interaction with the solvent bath. "Class I" additions such a ferrochromium (FeCr), which have a melting range which lies below that of the bath, will follow either route 1, 2, or 3, depending on the thermal conductivity of the alloy, bath superheat, and prevalent fluid flow conditions. The factors governing melting additions have been discussed in detail elsewhere (3,10).

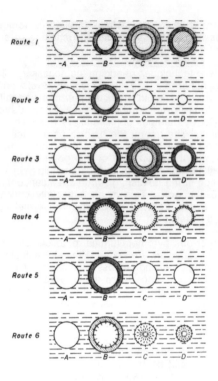

Fig. 1 - Assimilation Routes of Various Additions into Liquid Steel (3)

Routes 4 and 5 in Fig. 1 illustrate the two dissolution sequences that can be distinguished for conventional ferroalloys which have a melting range above that of the melt, *i.e.* "Class II" ferroalloys. Alloys which react exothermically when dissolving into the solvent bath, such as silicon additions to steel, follow the path indicated by route 4. The mixing of such an alloy is accompanied by a release of heat due to the reaction at the interface between the ferroalloy and the bath, and macroexothermicity is observed (10). For alloy additions such as ferrotungsten, where no exothermic reaction is observed (in liquid steel), the dissolution process is extremely slow (2) and follows route 5. The actual dissolution rate depends on a variety of factors that have been addressed in earlier publications (9,10).

The solution characteristics of ferroalloys must be dramatically improved if rapid alloying is to be successful, which applies especially to "Class II" ferroalloy additions to the steelmaking tundish. The underlying principle of rapidly assimilating ferroalloys is that heat is released by means of the formation of an intermetallic compound. For the current family of ferroalloys, the valuable alloy components (niobium, vanadium, tungsten, molybdenum, or chromium) all form at least one congruently melting intermetallic compound with silicon (15). The niobium-silicon phase diagram is shown as an example in Fig. 2.

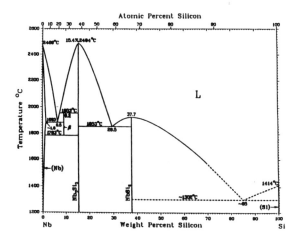

Fig. 2 - Binary Phase Diagram : Nb - Si (15)

In general, the reaction is of the following form:

$$x A + y B = A_xB_y + \text{heat} \tag{1}$$

where component **A** can be either the valuable alloying element or the iron base of the ferroalloy, and **B** is the modifying agent or the alloying element. Most commonly, component **B** is the modifying agent, silicon in the case of the alloys in the present study. It was shown earlier (10) that reactions such as the one above proceed much more rapidly if the alloying components are in powder form rather than the industry-standard lump form, and the term *microexothermicity* was coined to characterize the heat released in reactions involving powder additions. The enhanced reaction rate between powders as opposed to alloying lump- or plate-shaped pieces has also been observed in a recent study of the sintering behaviour of iron-silicon alloys (16), where it was found that a much enhanced reaction took place in the case where the powders were allowed to react.

The reaction products in the case of "Class II" ferroalloys are generally solid at post-furnace treatment temperatures. In order to obtain rapid assimilation rates and high alloy recovery, however, a liquid interaction between the alloy and the bath is desirable. To liquefy the reaction product, two endothermic heat terms are needed. The first enthalpic term, ΔH_e, represents the heat required to raise the temperature of the reaction product from the reaction temperature to its melting point. At the melting point of this compound, the latent heat of fusion (or melting), termed ΔH_m, must be supplied to effect the solid-liquid transition. The total enthalpy in the formation of a liquid intermetallic is thus given by

$$\Delta H_{tot} = \Delta H_f + \Delta H_e + \Delta H_m \qquad (2)$$

The reaction will release heat if the exothermic term associated with the formation of an intermetallic compound, ΔH_f, is sufficient to overcome the endothermic heat requirements ($\Delta H_e + \Delta H_m$). In order to illustrate this, consider that a reaction occurs in a silicon-modified ferroniobium cored-wire addition in which the solid powders react to form the liquid reaction product Nb_5Si_3, which has a melting temperature of 2484 °C :

$$5\,Nb_{(s)} + 3\,Si_{(s)} = Nb_5Si_{3(l)} \qquad (3)$$

For this reaction, calculations were made using programs available on the F*A*C*T thermodynamic computer databank (17). For the case of reaction (3), the overall enthalpy balance for a reaction between solid niobium and solid silicon yielded a value for ΔH_{tot} of -180.93 kJ/mol, which implies that this is a highly exothermic process.

Clearly, then, the heat released in the formation of the Nb_5Si_3 intermetallic (15.4 wt% Si, Fig. 2) is sufficient to raise it from a typical secondary steelmaking temperature, say 1600 °C, to its melting point of 2484 °C, and to provide further enthalpic energy to melt it, assuming that the process is adiabatic. An estimation of the latent heat of fusion term was made using a rule formulated by Kubaschewski and Alcock (18), which gives the value of the latent heat of fusion (in J/mol) as $9.6 \cdot T_m$ to $14.6 \cdot T_m$, where T_m is the absolute melting point of the intermetallic compound. In the calculations for the present study, which are summarized in Table I, the greater of these two value was used to calculate the total enthalpy of the reaction. For the case of the silicides denoted by an asterisk, the thermodynamic calculations show that exothermicity is observed only if the temperature exceeds the melting point of silicon (1414 °C).

Table I - Microexothermic Heat Release of Various Intermetallic Compounds

Intermetallic Compound	T_m (K)	ΔH_{tot} (kJ/mol)
FeSi	1683	- 57.08
VSi_2 *	1950	- 71.11
Cr_3Si	2043	- 49.85
Mo_5Si_3	2453	-129.01
Nb_5Si_3	2757	-180.93
W_5Si_3 *	2797	- 5.68

Table II - Characteristics of Ferroalloy Powders

Ferroalloy Powder	Elemental * Analysis (wt%)		Screen * Analysis	Liquidus ** Temperature	Bulk ** Density (kg/m³)
FeCr Batch no. A 4843	Cr C Si S P Fe	66.16 6.07 0.97 0.069 0.017 balance	70 % passing 200 mesh	1500 °C 1773 K	3850
FeMo Batch no. A 4711	Mo C Si S P Cu Al Fe	61.60 0.034 0.90 0.35 0.08 0.32 0.045 balance	76 % passing 200 mesh	1900 °C 2173 K	5135
FeNb (FeCb) Batch no. G 1777	Nb C Si S P Fe	64.50 0.25 1.95 0.011 0.063 balance	80 % passing 200 mesh	1580 °C 1853 K	4810
FeV Batch no. G 1915	V Si Al C S Fe	57.82 3.10 0.39 0.13 0.053 balance	85 % passing 200 mesh	1595 °C 1868 K	3840
FeW Batch no. G 7101	W C Si S P Fe	78.10 0.044 0.17 0.045 0.023 balance	90 % passing 200 mesh	2100 °C 2373 K	N/A
Silicon Batch no. G1871	Si	99.8 nominal	84% passing 200 mesh	1414 °C 1687 K	1360

* Provided by Shieldalloy Metallurgical Corporation, Newfield, N.J.
** Reference (19)

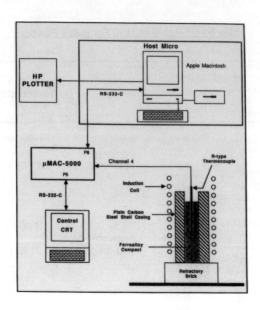

Fig. 3 - Schematic Illustration of the IMS in Thermal History Tests

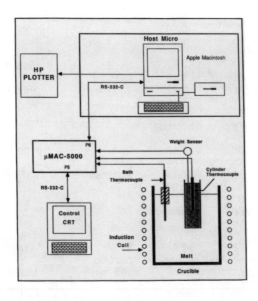

Fig. 4 - Schematic Illustration of the IMS in Alloy Assimilation Tests

Experimental Considerations

The experimental program consisted of two parts. The first part involved the heating of modified and unmodified samples using a 10 kHz - 30 kW TOCCO induction furnace to determine the effect of silicon additions on the temperature response of the material. Both unmodified (commercially available grades) and silicon-modified ferroalloys were compacted into 125 mm long mild steel casings with an outer diameter of 28.6 mm ($9/8$") at a pressure of about 110 MPa on a uniaxial press. Table II lists the chemical compositions of the powders used. After pressing the powder into the steel casing, a 3.2 mm ($1/8$") hole was drilled into the centre for the insertion of an R-type thermocouple. The steel shell thickness of 7.1 mm ($9/32$") in these induction testing samples was chosen such that it exceeded the penetration depth of the magnetic field into the mild steel, and thus inductive effects did not interfere with the microexothermic phenomenon. The analog outputs generated by the R-type thermocouple were evaluated using an intelligent measurement system (IMS).

The second part of the experimentation involved the measurement of the dissolution characteristics of these ferroalloys in liquid steel and nickel. For this purpose, melts of 70 kg of Armco iron or Inco pellets were prepared using a 1 kHz induction furnace. Each steel melt was deoxidized using commercial purity silicon lumps followed by aluminum wire. The ferroalloy samples were prepared in a manner analogous to that described above, but in order to simulate the cored-wire aspect, the steel-shell wall thickness was reduced to 1.6 mm ($1/16$"). With the exception of wall-thickness, the other dimensions of the sample for assimilation studies were identical to those of the induction heating sample.

Figures 3 and 4 schematically illustrate the experimental setup for the induction heating and alloy assimilation tests, respectively. In these experiments, real-time temperature data from both the centre of the compact cored-wire specimen and from the steel melt were obtained at regular intervals of 250 ms over the entire measuring period, again using R-type thermocouples. The force exerted by the sample on a load cell was recorded simultaneously. In the alloy assimilation studies, the force history of the cored-wire sample was monitored using a load cell with a maximum capacity of 11.36 kg. This mass measurement device was attached to a motorized lever powered by a dc-motor and could be raised or lowered to any desired position. In this way the system functioned as a wire-feeding facility. All data were stored on micro-floppy diskettes using the Apple Macintosh host computer that is part of the IMS. Details pertaining to hardware and software requirements and functions for data acquisition with the IMS are given elsewhere (6).

Measurement Aspects

The load cell registers the net downward force exerted by the specimen on a weight sensor. The strain gauge in the load cell, actuated by a high precision constant voltage dc-source, measures the difference between the gravitational force (F_G) exerted by the specimen and the buoyancy force (F_B) experienced by the sample upon immersion in a liquid bath. The buoyancy force is equivalent to the displaced mass, and the process can be expressed mathematically as follows:

$$F_{NDF} = F_G - F_B \qquad (4)$$

where F_{NDF} defines the net downward force, which is the measurement that the load cell registers. Dividing the cylinder volume into two segments, namely an immersed portion with volume V_{imm} and a portion above the bath with volume V_{out}, and denoting the density of the solid addition as ρ_S and the density of the liquid bath as ρ_L, this equation becomes

$$F_{NDF} = [V_{out} \cdot \rho_S + V_{imm}(\rho_S - \rho_L)] g \qquad (5)$$

where g is the gravitational constant. Differentiation of the above equation with respect to time, under the assumption that the density difference is constant, yields

$$dF/dt = (dV_{imm}/dt)(\rho_S - \rho_L) g \qquad (6)$$

Since only material in the liquid bath will be dissolved or melted, the volume portion above the bath is constant. Consequently, the derivative of V_{out} with respect to time is zero.

Equation (6) can be used to predict the shape of the curve of net downward force with respect to time, provided that the values of ρ_S and ρ_L are known. Since the change in immersed volume will always be a negative quantity during dissolution, the slope of the resulting line is effectively determined by the difference between the densities of the solid and liquid components. Thus, for a ferroalloy with a density which is less that of the bath, the net downward force will actually increase with time during the dissolution period, since dF/dt is a positive quantity. Conversely, if the density of the addition exceeds that of the solvent bath, the value of F_{NDF} will decrease since dF/dt < 0.

Results and Discussion

(A) Thermal History Tests

In order to quantify the effect of silicon modifications on the behaviour of cored-wire ferroalloys, a series of induction heating tests were performed to investigate the thermal history of the compacts. Tests were conducted under two types of power conditions : (i) under constant power input (4.2 kW) throughout the thermal history of the sample, and (ii) increasing the power to 7.2 kW after exceeding the magnetic transition temperature, at which the electrical properties of the carbon steel casing change. The additional power input allowed for higher sample temperatures than could be attained in the constant power tests.

The induction heating technique, sample configuration, and the data obtained for FeCr, FeMo, and FeNb compacts under non-constant power input conditions have been discussed previously (6). The thermal histories of non-modified and silicon-modified compacts based on FeV and FeW have now also bee investigated. The modifying effect of silicon on the thermal history of the compacts is illustrated in Fig. 5. The numbers next to the curves denote the quantity of silicon added for modifying purposes in weight percent, the unmodified compact being given the designation "0".

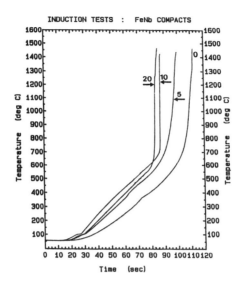

Fig. 5 - Typical Data for Induction Heating Tests Conducted on Modified and Unmodified FeNb Compacts. The numbers next to each curve denote the quantity of silicon used to modify the compacts (in wt%).

The data show that modifications of 5 to 20 wt% significantly accelerate the thermal response within the compact. The thermal history data for the other ferroalloys show virtually identical trends. In general, the time required to attain a certain elevated temperature tends to decrease as the amount of silicon used to modify the ferroalloy is increased. For instance, in the case of silicon-modified ferroniobium compacts, the data in Fig. 5 show that the time required to attain a temperature of 1400 °C decreases from 112.5 seconds for an unmodified FeNb sample to 97.5 s when 5 wt% silicon is added, and is further reduced to 86 s and 82.5 s with additions of 10 and 20 wt% silicon, respectively.

While the time improvements may not appear to be significant at first glance, a second fact must be considered in evaluating the influence of the silicon modifications on the thermal history of the ferroalloy. The addition of sufficient quantities of silicon alters the high temperature characteristics of the compacts, and the high temperature regime is characterized by a melting process well below steelmaking temperatures. A liquid breakout is observed at the bottom of the compact while the steel shell casing used to contain the powder compact is still solid. This breakout serves as an unambiguous indicator that the microexothermic reaction (between powder particles to form a liquid in a process that is essentially adiabatic in nature) does in fact occur. Sectioning of the samples revealed that in the case of unmodified compacts only powder particles persisted following the induction heating tests. In the case of the silicon-modified alloys the core of the sample had liquefied while the solid steel shell had remained relatively intact. These observations indicate that, after the addition of silicon as a modifying agent, a phase was formed which liquefies below the melting point of steel. Constant power tests have shown that a slope change occurs in the thermal history curve of modified samples in the temperature range of 1100 °C to 1180 °C (6), which corresponds closely to the temperature of a eutectic in the Fe-Si binary system (15).

At the time corresponding to this change of slope, a liquid breakout was observed in all silicon-modified compact specimens, irrespective of the principal component of the ferroalloy base. This breakout is an important visual confirmation of the microexothermic heat release and lends much credence to the thermodynamic calculations, which clearly show that the modified alloy would assimilate into a melt as a liquid phase. The corrosive attack of the low melting eutectic phase on the thermocouple is so severe that the thermocouple is actually destroyed in this process. This accounts for the fact that analog signals in thermal history tests generally tend to become erratic above 1400 to 1500 °C. By contrast, the thermal history curves of non-modified compacts do not exhibit a similar sharp temperature rise and, as expected, no liquid formation was observed.

If the silicon content in the final cast product is not critical, then the user can specify any degree of modification to suit his operation and process. However, in order to utilize modified cored-wire as a trimming addition in either the ladle or the tundish steelmaking stage, the amount of silicon used for modification purposes must be carefully considered since silicon is not necessarily a desirable element in the final product. For steelmaking applications, the data indicate that an "ideal" modification quantity lies in the range of 3 to 10 wt% Si, depending on the thermal characteristics of the unmodified ferroalloy. Calculations performed for "Class II" ferroalloys in the present study have shown that for each one weight percent of the valuable component added (*i.e.* Mo or Nb), the additional pick-up of silicon is of the order of 75 ppm per wt% silicon used to modify the compact.

While this figure may initially be regarded as unacceptably high, it must be emphasized that the primary role of these microexothermic additions is to trim the final chemical composition, specifically for the control of the valuable microalloys such as niobium or vanadium in HSLA steels. In order to illustrate this, suppose that a trimming operation using a microexothermic cored-wire modified with 5 wt% silicon is aimed at increasing the niobium content of an HSLA steel, such as *e.g.* ASTM A737, from 0.032 % in the ladle to a customer specification of 0.038 % niobium. The calculations show that some 22 ppm (0.0022%) of silicon would be added to the steel by the microexothermic reaction, assuming that the recovery of the silicon is 100%.

(B) Assimilation Characteristics in Liquid Steel and Liquid Nickel

Figure 6 illustrates a typical output from an alloy solution test in liquid steel using a cored-wire compact modified with 5 wt% silicon. In these tests, samples weighing 500 to 600 g were dissolved or melted in 70 kg of liquid steel. The IMS was responsible for tracking three analog inputs simultaneously, namely the bath and cylinder temperatures, read off the left hand vertical axis, as well as the net downward force, read off the right hand axis. The temperature at the centre of the specimen is represented by line 1, while line 3 characterizes the bath temperature, which remained constant (1610 °C) during the entire experiment. The uncertainty associated with the temperature measurements was of the order of ±10 C°. The temperature measurements were confirmed using a dip-tip temperature probe. The course of the net downward force registered by the load cell is given by line 2. This convention is used in all subsequent figures.

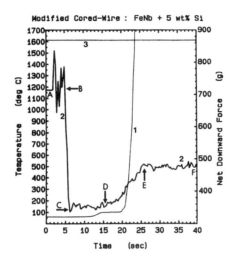

Fig. 6 - Typical Experimental Results for the Dissolution of a Silicon-Modified Compact Immersed into Liquid Steel at 1600 °C.
Curve 1 shows the temperature at the centre of the ferroalloy compact. Curve 2 denotes the net downward force exerted on the load cell by the specimen. Curve 3 reflects the temperature of the liquid steel. *Note*: this convention applies to all subsequent figures.

For the experiment shown in Fig. 6, the combined mass of the cored-wire sample and the sample holder was 720 g, defined by point **A** on line 2. At this stage, the only force acting on the load cell is the gravitational force exerted by the sample and the holder. The data acquisition process is started while the sample is held immobile above the steel bath. The wire-feeding facility is then used to lower the sample down towards the steel bath, and the vibrational instability in this phase is evident in segment **AB**. At point **B**, denoted by the arrow, the cored-wire specimen physically touches the top of the bath surface, and thereafter a buoyancy component is introduced. At point **C**, the sample has been lowered further to the so-called operating position, at which it is held until the experiment is terminated. The effect of buoyancy on the net downward force registered by the load cell is illustrated by the difference between points **A** and **C**. Accordingly, the values of F_G and F_B are 720 g and 390 g, respectively, resulting in a value of 330 g for F_{NDF} at point **C**.

Upon immersion, a solid steel shell freezes around the addition. This shell eventually melts back, and at point **D** has totally re-liquefied. It may be observed that the increase in gravitational pull due to the solid steel shell freezing around the compact is generally negated by an increased buoyancy experienced by the sample; therefore, the net downward force registered by the load cell remains approximately constant. This period was of the order of up to 10 seconds for the cored-wire test specimens, and depends on factors such as the thermal conductivity of the ferroalloy and the bath superheat (6).

The actual assimilation of the alloy into the bath takes place in the segment identified as **DE**. This process occurs over a period of about 10 s, finishing at about the 26th second. From the shape of the curve, it may be deduced that the density of the ferroalloy is actually less than that of the steel bath, which agrees well with the bulk density value for FeNb reported in literature (19). Note that this segment corresponds to the time period where a rapid temperature increase is observed in line 1. The temperature at the centre of the cored-wire rises from about 150 °C to 1700 °C within some 2.5 s. These curves attest to the system's ability to monitor real-time events, thus allowing the observation of phenomena which occur beneath the bath surface. At point **E**, assimilation is complete, and only the portion of the sample above the bath remains. It was alluded to earlier that the value of F_{NDF} does not change thereafter since V_{out} is constant with respect to time. The gravitational component is again the only force acting on the load cell, which registers an approximately constant force of just under 500 g. The results of this experiment thus indicate that some 220 g of ferroniobium were dissolved in the liquid steel bath over a period of about 20 s, which includes the steel shell freezing and remelting phase.

The assimilation characteristics of ferroalloys into liquid baths under static and dynamic conditions have been discussed thoroughly in literature (1,3-6). The induction power was left on during the course of the experiments that are described below, and therefore dynamic conditions and forced convection prevailed. Static experiments were performed after the power to the furnace had been turned down, and the data obtained showed that solution rates are enhanced by up to 60% with induction stirring, which agrees well with published data (3,4). The same principles applied to the investigation of the assimilation of modified FeMo, FeNb, and FeCr cored-wires in liquid steel (6) were applied to the study of the solution behaviour of FeV and FeW.

Figure 7 illustrates the dissolution behaviour of a ferrovanadium lump with an initial mass of 1378 g in liquid steel. The lump material analyzed 42.49 wt%V, 6.65 wt%Si, 0.37 wt%C, and 0.29 wt%N. In spite of the significant silicon content of this material no exothermic reaction was observed. This is due to the fact that the silicon in the ferroalloys, introduced in the production process, has already formed compounds, such that the highly exothermic heat of compound formation no longer is available to accelerate the reaction. Alloying of this lump therefore is controlled by mass transfer kinetics at the interface, as well as some chemical reaction control due to the heat of mixing. In comparison to a silicon-modified cored-wire, the dissolution process of a lump is rather slow, and was not completed at the end of the measurement period of 120 s. Based on earlier research by Argyropoulos and Guthrie (3), the dissolution time of this particular lump was estimated to be 800 s. The assimilation of lumps into a bath of molten metal is often characterized by breakage of sections of the lump. This is reflected by well-defined breaks in the curve of net downward force, as indicated in Fig. 7.

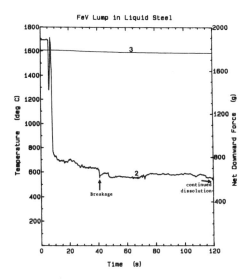

Fig. 7 - Dissolution Data for a FeV Lump in Liquid Steel at 1605 °C under Forced Convection

Cored-wire operations owe much of their success to the fact that materials with a high surface area to volume ratio (*viz.* powder) can be introduced into a melt rapidly and at precise locations. The data published previously (6) for commercial non-modified cored-wire ferroalloys show that significant improvements have been achieved with cored-wires in comparison to lump additions, but not to a level where they can be used for late trimming additions. The addition of silicon as a modifying agent, however, reduces the assimilation period required by as much as 80 percent. The transition in the assimilation behaviour from a dissolution process to a melting process, due to the microexothermic phenomenon, is characterized by smooth curves of net downward force.

Figures 8 and 9 illustrate the assimilation behaviour at typical steelmaking temperatures of two modified "Class II" alloys, ferrovanadium and ferrotungsten, respectively. The beginning and end of the assimilation period are identified by the solid triangles (normal and inverted, respectively) marked on curve 2. A direct comparison of the present data with the data due to Argyropoulos and Guthrie (3) and Benda (20) shows that the assimilation period for identical quantities of FeV can be reduced from about 150 s with a lump to about 30 s with a cored-wire containing five percent by mass of silicon as a modifying agent. This can be reduced by a further 20 % by increasing the silicon content of the cored-wire to 10 %. In the case of FeW cored-wire, the improvement is even more significant. A modification of 2.5 wt% silicon leads to a reduction in the time required for complete dissolution from about 300 s with a lump (3) to less than 20 s with a

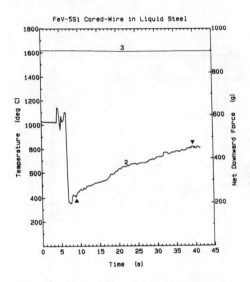

Fig. 8 - Assimilation of a 5 %Si Modified FeV Compact at 1610 °C with Bath Stirring. The solid triangles on curve 2 denote the beginning and end of assimilation. *Note* : this convention applies to all subsequent figures.

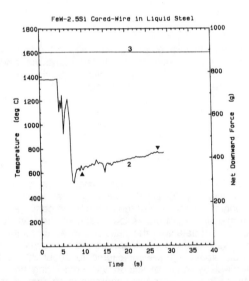

Fig. 9 - Assimilation Data for a 2.5%Si Modified FeW Compact in Liquid Steel at 1605 °C with Bath Stirring

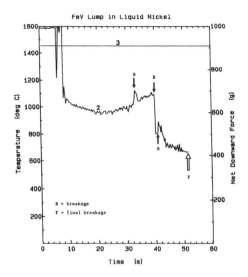

Fig. 10 - Dissolution Data for a FeV Lump in Liquid Nickel at 1460 °C under Forced Convection

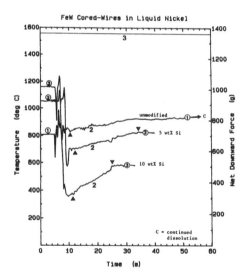

Fig. 11 - Assimilation Data for Various FeW Compacts in Liquid Nickel at 1555 °C with Bath Stirring

modified cored-wire addition. However, increasing the silicon-content in the FeW cored-wire to 7.5 and 10 wt% did not result in notable improvements.

Very little has been published in literature on the dissolution of vanadium and tungsten in iron-base melts. Benda (20) has analyzed the solution mechanisms and kinetics of various ferroalloys, but only comments on the kinetics of "Class I" additions such as FeV-36 or FeSiMn. Bungardt and coworkers (2) have reported on the macroscopic solution kinetics of ferroalloy additions to baths subjected to forced convection. However, since their study was restricted to pure binary Fe-Mo and Fe-W systems, and since their melt size was comparatively small (260 g solvent bath), the results cannot be readily applied to the present work. For example, Bungardt *et al.* report an assimilation period of more than 10 minutes for an alloying content of 17 wt% W at 1635 °C (2). Clearly, this does not compare favourably with the present data, where an addition of about 1 wt% W was alloyed within 20 s, even at lower bath temperatures.

From a review of extant literature, it would appear that no work has been published on the dissolution characteristics of post-furnace FeV and FeW additions to liquid nickel. Representative data from the present study are illustrated in Figs. 10 and 11. The breakage observed in the case of lump additions to liquid steel was again observed in liquid nickel. After three breakages (B, Fig. 10), the FeV lump broke off at the bath/air interface just after the 50th second of the experiment (F). The lump that dropped into the bath was recovered and weighed, and a dissolution of some 20 % was observed. By contrast, FeV and FeW modified cored-wire additions exhibited improved assimilation and recovery characteristics. It was further observed that a modified FeV cored-wire assimilated about 25 % more rapidly than a non-modified FeV cored-wire, even though the bath superheat in the latter case was almost 100 degrees higher than for the modified sample.

Figure 11 illustrates the improvement in the assimilation rate of FeW in liquid nickel at 1560 °C. Whereas more than 60 seconds were required to dissolve the non-modified FeW sample (identified by the circled "1"), additions of 5 wt% silicon (circled "2") and 10 wt% silicon (circled "3") allowed for assimilation within 25 seconds. Again the effect of increasing the silicon level in the cored-wire resulted in an improved assimilation period. On a percentage basis the improvement registered with 10 wt% silicon in comparison to a lump addition is about 95 %. The rapid assimilation rates shown in this figure serve to illustrate the concept that late stage additions and compositional adjustment of high melting point alloys to nickel-base melts are feasible, by virtue of the microexothermic phenomenon. Research is currently being performed to assess recovery characteristics of microexothermic cored-wire ferroalloys, and the results will be presented in a future publication.

Conclusions

Microexothermicity plays a large role in altering the assimilation of high melting point alloying additions such as molybdenum, niobium, chromium, vanadium, and tungsten from a dissolution process to a melting process. This allows for greatly enhanced assimilation rates and greater bath homogeneity once alloying has occurred. It has been shown that a microexothermic reaction is also beneficial in introducing some of these alloys to nickel-base melts. The addition of 5 wt% silicon as a modifying agent to ferroalloys has been found to effect a significant decrease in the time required to alloy, as well as resulting in a liquid addition which allows for more rapid homogenization than a diffusing solid. The rapid assimilation into the liquid bath, which is usually of the order of 15 to 20 seconds for Class II ferroalloys, is made possible by the release of heat from the reaction forming an intermetallic silicide compound.

Both small scale induction tests and pilot stage trials in 70 kg liquid iron and nickel melts were performed, and it was shown that the microexothermic additions based on FeMo, FeNb, FeCr, FeV, and FeW exhibit a markedly altered and improved assimilation due to the formation of a liquid phase. This rapid alloying behaviour renders these modified ferroalloys ideal for wire-feeding into the ladle and tundish as trimming additions. These alloys represent the first possibility to perform tundish trimming operations with microalloying components as they circumvent the restrictions imposed by residence times in such vessels. In addition, the feasibility of employing modified cored-wire ferroalloys for post-furnace compositional adjustments in nickel-base melts has been demonstrated.

Acknowledgements

The authors gratefully acknowledge the financial support of the Natural Sciences and Engineering Research Council of Canada (NSERC) for this work. Further, the help of Mr. R. Nicolic in furnace preparation and operation is gratefully noted. Our gratitude also goes to the Shieldalloy Corporation for supplying the ferroalloy powders used in preparing the compacts, the International Nickel Company for providing nickel pellets as melt material, as well as the Electronite Company for their donation of rapid temperature measurement equipment.

References

1. P.G. Sismanis and S.A. Argyropoulos, "Modelling of Exothermic Dissolution", Can. Metall. Q., **27**(2) (1988), 123-133.

2. K. Bungardt et al., "Die Aufloesung von Molybdaen und Wolfram in Eisenschmelze als Beitrag zum Vorgang der Aufloesung von festen in fluessigen Stoffen", DEW Techn. Ber., **9**(3) (1969), 407-438.

3. S.A. Argyropoulos and R.I.L. Guthrie, "Dissolution Kinetics of Ferroalloys in Steelmaking" (Proc. Steelmaking Conf., ISS-AIME, Vol. **65**, 1982), 156-167.

4. P.G. Sismanis and S.A. Argyropoulos, "The Dissolution of Niobium, Boron, and Zirconium Ferroalloys in Liquid Steel and Liquid Iron" (Proc. Electric Furnace Conf., ISS-AIME, Vol. **45**, 1987), 35-48.

5. P.G. Sismanis and S.A. Argyropoulos, "Studies on Recovery of Alloy Additions" (Proc. Int. Symp. on Ladle Steelmaking and Ladle Furnaces, R.I.L. Guthrie and F. Wheeler, eds., CIM, 1988), 47-57.

6. J. Schade, S.A. Argyropoulos, and A. McLean, "Cored-Wire Microexothermic Alloys for Tundish Metallurgy" (Proc. Electric Furnace Conf., ISS-AIME, Vol. **46**, 1988), in press.

7. E. Vachiery, "Advent of Cored-Wire Injection for In-Ladle Metallurgy" (Proc. First International Ca-Treatment Symposium, Glasgow, Scotland, June 30, 1988), 1-2.

8. A. McLean,"The Turbulent Tundish - Contaminator or Refiner ?" (Proc. Steelmaking Conf., ISS-AIME, Vol. **71**, 1988), 3-23.

9. S.A. Argyropoulos, "Dissolution Characteristics of Ferroalloys in Liquid Steel" (Proc. Electric Furnace Conf., ISS-AIME, Vol. **41**, 1983), 81-93.

10. S.A. Argyropoulos, "The Effect of Microexothermicity and Macroexothermicity on the Dissolution of Ferroalloys" (Proc. Electric Furnace Conf., ISS-AIME, Vol. **42**, 1984), 133-148.

11. P.G. Sismanis and S.A. Argyropoulos, "The Dissolution of Microexothermic Alloying Additions in Cast-Iron" (Proc. Electric Furnace Conf., ISS-AIME, Vol. **43**, 1985), 39-55.

12. P.G. Sismanis and S.A. Argyropoulos, "The Effect of Oxygen on the Recovery and Dissolution of Ferroalloys" (Proc. Steelmaking Conf., ISS-AIME, Vol. **69**, 1986), 315-326.

13. S.A. Argyropoulos and P.D. Deeley, "Exothermic Alloy for Addition of Alloying Elements to Steel", United States Patent, Patent No. 4,472,196, Sept. 1984.

14. J.J. deBarbadillo, private communication with author, Inco Alloys International, August 15th, 1988.

15. T.B. Massalski, ed., Binary Alloy Phase Diagrams (Metals Park, OH : American Society for Metals, 1986), 864, 1084, 1632, 1696, 2061 and 2063.

16. J.A. Lund, Y. Tanaka, and X. Qu, "Flux-Enhancement of Alloying in Iron-3% Silicon Compacts", Int. J. Powder Met., **24**(4) (1988), 301-313.

17. W.T. Thompson, A.D. Pelton, and C.W. Bale, F*A*C*T Manual (Montreal, Quebec : McGill University Press, Montreal, 1984).

18. O. Kubaschewski and C.B. Alcock, Metallurgical Thermochemistry (Toronto, Ontario : Pergamon Press, 1979), 181-187.

19. P.D. Deeley, K.J.A. Kundig, and H.R. Spendelow, Jr., Ferroalloys and Alloying Additives Handbook (New York, NY : Metallurg Corp., 1981), 115-122.

20. M. Benda, "Kinetik und Mechanismus der Aufloesung von Ferrolegierungen im fluessigen Stahl", Neue Huette, **29**(4) (1984) 130-133.

THE KRUPP HOM-TEC (HORIZONTAL OSCILLATING MOULD TECHNOLOGY) CASTER AN ADVANCED PROCESS FOR HORIZONTAL CONTINUOUS CASTING OF COPPER AND COPPER ALLOYS

Dr. Erling Roller
Managing Director
KRUPP Technica GmbH
P.O. Box 1128, 8707 Veitshöchheim, Bavaria, F.R.G.

Abstract

Ten years ago, KRUPP Technica GmbH started a research programme for a new horizontal continuous casting process in cooperation with a Swiss brass manufacturer: The KRUPP HOM-TEC Caster.Today, this technology has found wide industrial acceptance for the production of copper and brass billets, copper and copper alloy rods - and even for special steel billets. Unlike all other horizontal continuous casting processes, the Krupp HOM-TEC Caster uses an oscillating mould comparable to vertical casting and withdraws the cast strand with linear speed. This paper describes plant set-up details, casting parameters and improvements in productivity and product quality with HOM-TEC Casters for copper and copper alloys presently in operation.

The Development of Horizontal Continuous Casting with an Oscillating Mould.

Horizontal continuous casting using an oscillating mould and constant casting speed is by no means a new idea.

As early as the beginning of this century a patent application by a Swedish engineer(1) contained a description of every conceivable form of movement of the mould and strand of a horizontal continuous caster. One of the versions described was the simultaneous oscillation of both tundish and mould. It took around 70 years, however, before the idea becomes reality on an industrial scale in the form of the "HOM-TEC" Caster - "HOM-TEC" standing for "horizontal oscillating mould technology".

Around the beginning of the 70s, horizontal continuous casting of nonferrous metals began to gain acceptance in industry. The casters used, however, had moulds of exclusively stationary design. To achieve relative movement between the strand and the mould the strand is withdrawn from the mould with an intermittent movement.

The advantages of this horizontal casting technique compared with vertical casting were obvious:
- low foundation costs
- small space requirement
- good access to the caster components
- lower personnel costs with multi-strand casting
- high yield due to continuous casting compared with
  commonly practised semi-continuous vertical casting.

Horizontal casting with an oscillating strand had, however, several disadvantages compared with vertical casting, particularly for casting extrusion billets from copper and copper alloys: the casting rate per strand was only around 50% of that achieved with vertical casting. The horizontal position of the mould also led to eccentric solidification of the strand which can cause problems, for example, when tubes are formed by extrusion.

The HOM-TEC process was developed to avoid these difficulties and to combine the advantages of vertical and horizontal casting.

Based on the experience of around 50 vertical continuous casters with oscillating moulds for copper and copper alloys since the 50s, the logical step was to turn these casters round 90°, so to speak.

A first pilot plant was tested in a NF-metal works in Switzerland in the middle of the 70s.

This first stage in the development process ended with the successful production of small brass extrusion billets with a diameter of 140 mm at a casting capacity of around 2.5 t/h (Fig. 1). The billet showed already largely concentric solidification. The first production caster for rods suitable for cold rolling from copper alloys, largely special bronzes (Fig. 2), was also developed in this works. During the next stage in 1981/82 it was succeeded in casting copper extrusion billets horizontally(2) for the first time, in cooperation with a copper refinery in England.

These developments encouraged Krupp Stahl AG to make use of the HOM-TEC process. The company needed a casting process for

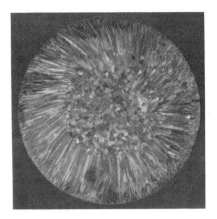

Figure 1 - Macro structure of a brass billet.

Figure 2 - Production of 8 copper alloy rods with a diameter of 19 mm.

crack-sensitive special steels which still had to be cast in ingots as they could not be cast without cracks in the available bow-type caster. This large-scale development project continued for several years. It resulted in the HOM-TEC caster at Krupp Stahl AG's Siegen works(3) - a fully integrated part of the company's steel production.

Since then two further HOM-TEC casters for brass billets have been put into operation. The installation of four additional casters will shortly be completed.

To date a total of 11 horizontal continuous caster of this type have been built (Fig. 3).

Technology of the HOM-TEC Caster. The design of these casters can be seen from a nonferrous caster for copper or brass extrusion billets (Fig. 4).

An induction furnace with an attached forehearth is used as a holding furnace, as is often employed for vertical casters.

The molten metal is fed into an inductively heated tundish through a dosing system with a discharge pipe. The mould is built onto the side of this tundish. Assembled in this way, the system can be made to oscillate with the aid of a servo-hydraulic high-precision cylinder (Fig. 5). The amplitude can be varied between 0 and 5 mm; the oscillating frequency is around 20 to 400 oscillations per minute. When the casting is interrupted the holding furnace is tipped back and the tundish can be moved away from the stationary mould on the joint sled guidance, providing access to the entrance of the mould and allowing, for example, the dummy bar to be sealed.

A secondary cooling line, comprising several rings with spray nozzles, directly follows the mould.

HOM-TEC-Casters presently in operation or under construction:

Steel billets	3
Copper billets	2
Brass billets	5
Copper alloy rod	1
Total	11

Figure 3 - References HOM-TEC Caster.

Figure 4 - HOM-TEC - Caster for copper billets.

The strand or strands are withdrawn from the mould at a constant speed. The speed of the guide rolls is infinitely adjustable.

Figure 6 shows the design of the tundish. It is a horizontal inductor with two or three channels, depending on the width of the caster.

Molten metal is fed from the holding furnace into the left-hand opening of the inductively heated tundish from the holding furnace. The metal is super-heated in the inductor channels and flows in the direction of the mould entrance. The heating capacity of the inductors varies between 120 and 300 kW, depending on the size of the tundish.

The inductor is controlled by a thermo couple installed near the mould entrance so that the casting temperature can be kept within very narrow tolerances.

The metal in the tundish is levelled by an automatic control system so that the growing strand shell can be pressed against the mould wall with sufficient metallostatic pressure.

If molten metal is charged into the casting vessel in batches from one or several melting furnaces, a tundish of pressurized design is employed (Fig. 7). This tundish has three chambers. The filling chamber and the chamber toward the mould are open, while the pressure is built up in the central chamber with inert gas such as nitrogen. This compensates the drop of the bath level in the mould-side chamber between charges. In this way the tundish can be charged in batches of up to five tonnes without a change in the metallostatic pressure above the moulds.

Unlike other channel-type induction furnaces the tundish can be completely emptied at the end of the casting sequence, or when the mould has to be changed, by removing a plug; the tundish can be cooled down to room temperature and heated up again when next required.

Primary solidification of the strand takes place in the usual way by indirect cooling in the mould. A mould of conical design is chosen, if the alloy to be cast permits.

Figure 5 - View of precision oscillation cylinder.

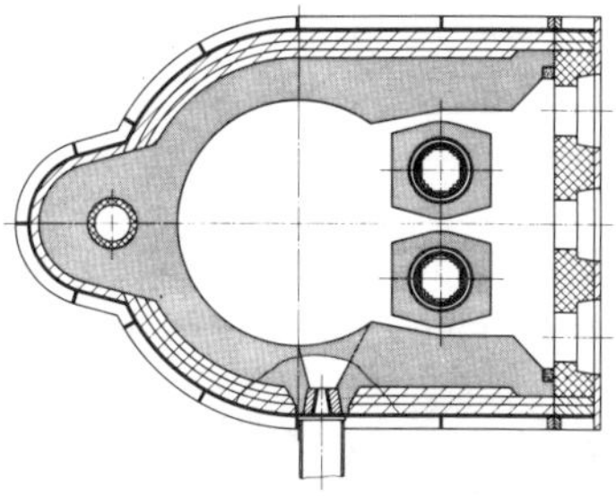

Figure 6 - Schematic view of the inductively heated tundish of NF-HOM-TEC Casters.

A mould of conical shape, as is often used in high performance vertical casting, has particular advantages for horizontal casting. With billet moulds of cylindrical design, the strand shrinks off the top of the mould as soon as it has formed a firm shell and rests by gravity on the bottom of the mould over its whole length.

To avoid a strand break-out on the upper side when it emerges from the mould, the casting speed has to be lowered, which results in eccentric solidification (Fig. 8).

With a conical mould, however, the tapering of the mould matches the shrinkage of the strand so that nearly all of the mould length is effective as a cooling surface over the whole circumference of the strand.

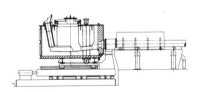

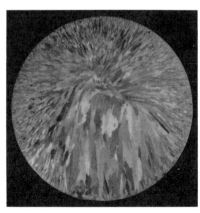

Figure 7 - Cross section of a NF-casting machine with pressurized tundish vessel

Figure 8 - Brass billet with eccentric Solidification.

The secondary cooling system directly following the mould is not only used to cool down the strand further as it emerges, but is also used to influence the final solidification of the strand.

The result of this interplay of indirect mould cooling and direct secondary cooling is shown in Figure 9. This is a macro-section of a copper extrusion billet. As one can see, a concentric solidifcation comparable to vertical casting has taken place.

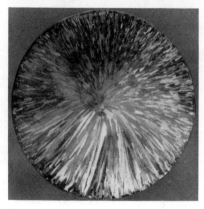
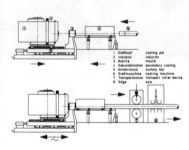

Figure 9 - Macro structure of a copper billet (HOM-TEC caster).

Figure 10 - Operational NF-HOM-TEC caster.

The next figure (10) shows the functions of the caster. In the upper part of the picture the inductively heated tundish, mould and secondary cooling system, forming three units, have been moved apart so that the dummy bar can be introduced into the mould and sealed. In the lower part of the picture the tundish, mould and secondary cooling units have been moved together on their joint sled and connected to each other.The casting furnace is lowered, so that the line is ready to cast.

As the strands are withdrawn from the mould at a constant speed, a withdrawal system of relatively simple design can be used. All strands run on a joint lower roll which is driven by a DC motor and a gear unit which has virtually no play. The upper rolls are pressed individually on the strand by a hydraulic system. The steel rings connected to the upper roll are to stop individual strands. If the upper roll is raised the steel ring lifts the strand off the driven lower roll so that the casting process can be interrupted for that strand. The downstream high-performance cold circular saw moves synchronously with the casting speed and is able to saw up to three strands one after the other. It is nevertheless able to saw the strands directly to length for the extrusion press, which for modern extrusion presses is in the region of 600 - 1000 mm.

Operating results. The following figures show the HOM-TEC Caster for copper billets (Fig. 11) and brass billets (Fig. 12) in production.

Figure 11 - HOM-TEC caster for copper billets.

Figure 12 - HOM-TEC caster with 3 strands for leaded brass.

If one compares the average casting speeds in production for the different profiles (Fig. 13) one can see that they are around 30-50% higher than was previously possible with horizontal casting, approaching speeds which used only to be associated with high-performance vertical continuous casting plants.

As, furthermore, similar concentric solidification of the billet can be achieved and no inclusion problems with vertical mould covers arise with closed-system horizontal casting, the advantages of horizontal and vertical continuous casting processes are ideally combined in the HOM-TEC casting process.

With respect to the casting of rods from copper alloys such as special bronzes, the casting rate depends less on the cooling capacity of the mould system, but rather on the material properties required for further cold rolling or drawing. It is important that the strand is always in contact with the mould wall, as considerable sweating of tin and other segregating components can only be suppressed in this way. The casting rate for rods of this type, with a diameter of approximately 18 to 20 mm, is around 250 - 350 mm/min. If conductivity copper rods are cast on the same casting line the casting speed can be increased to around 1000 mm/min.

Sequence of movements of the strand and mould - breakring and negative strip. Unlike most industrial horizontal casting processes, the HOM-TEC casting process for nonferrous metals employs a so-called "breakring". For horizontal continuous casting of steel, however a breakring is always employed, irrespective of the casting process. This breakring has number of functions and characteristics:

- it effects thermal separation between the heated tundish and the water-cooled mould
- strand solidification takes place in a defined manner directly at the interface between breakring and mould

- the breakring must be a good insulator to prevent primary solidification taking place inside the ring
- the breakring must not be penetrated by the cast metal
- on the other hand, high thermal shock resistance is required as the ring would otherwise crack on heating or cooling.
- finally, in nonferrous casting plants where the mould is lined with graphite the breakring prevents this graphite from oxidizing and thus becoming ineffective.

In view of the requirements for the breakring material - some of which even contradict each other - the search for suitable ceramic materials was difficult. Today we use aluminium titanate ($AlTiO_5$) for casting brasses, and boron nitride, or a mixture of boron nitride and silicon nitride, for steels. For pure copper a patented ring system comprising an inner graphite ring, an intermediate ring of ceramic material and an outer ring of heat-resistant steel has proved to be a good solution.

It is no exaggeration to say that the success of modern horizontal continuous casting was only made possible by the availability of suitable ceramic materials.

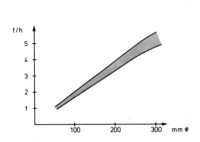

Figure 13 - HOM-TEC caster casting rates for copper and brass extrusion billets per strand.

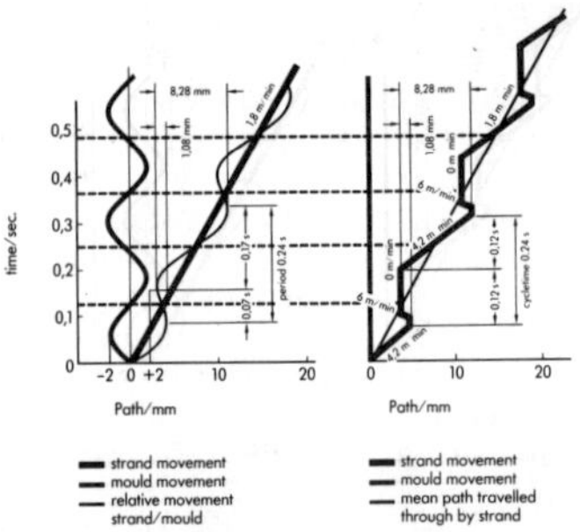

Figure 14 - Strand and mould movement with different horizontal continuous casting processes.

Let us now turn to the differing sequence of movements of strand and mould, looking at how the two different processes with an oscillating mould or oscillating strand compare with each other (Fig.14) The left-hand side of the picture shows the sequence of movements of a HOM-TEC caster with an oscillating mould, and the right-hand side the intermittent strand withdrawal process with a stationary mould. The numerical values selected here refer to the continuous casting of steel, but the movements can be applied without difficulty to the casting of copper and copper alloys. As one can see, the cycle has to be completed within a fraction of a second. With the intermittent withdrawal process such short cycle times are difficult to achieve and require complex machinery and control systems; one of the major problems is precise transfer of the movement to the strand.

The left half of the picture shows how simple it is to create an equal degree of relative movement between strand and mould when an oscillating mould is used. The mould movement represents simple harmonic motion, while the strand is moved with a simple linear conveying movement - in other words the same movements and laws apply as for vertical or bow-type casting. The only difference between horizontal continuous casting with an osciallting mould and vertical continuous casting with an oscillating mould is that the so-called "negative strip" part of the movement is fully converted into compression of the strand shell in the casting direction.

This sinoidal mould movement can be set without difficulty to a very high frequency, by contrast to the intermittent motion process in the right-hand side of the picture. HOM-TEC casters for copper operate at frequencies of up to seven cycles per second. A particular advantage of this mould oscillation is that the negative strip for the strand increments can be set with a high degree of precision and reproducibility (Fig.15). If the distance over which the strand is compressed (= negative strip) is too large , this can lead to scales on the cast surface. In addition there is a danger of damaging the breakring, which serves as an artificial meniscus. If the negative strip is too small the witness mark "B" is insufficiently welded. If strand compression is lost occasionally or periodically this results in casting defects or even strand break-offs .

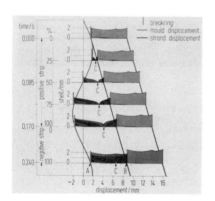
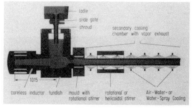

Figure 15 - Strand movement and shell formation - HOM-TEC - process.

Figure 16 - Tundish-mould arrangement with stirrers and secondary cooling zone.

HOM-TEC casters for stainless steel billets. There are certain differences between HOM-TEC casters for stainless steel billets (Fig. 16) and those for nonferrrous metals. The tundish has a flanged, inductively heated crucible with a heating capacity of 400 kW instead of an integrated channel type inductor. The mould is made of a copper-beryllium alloy. Graphite can not be used in the mould as steel and its alloys, such as nickel for example, would transform the graphite into carbides within a very short time, thus destroying it. A first electromagnetic stirrer is installed inside the mould which ensures a fine-grained sub-surface and high heat dissipation. A second electromagnetic stirrer is installed behind the mould which influences the solidification of the core and enables a fine-grained, dense structure in the core. The results of this stirring action are shown in Figure 17.

Behind the mould, instead of the usual water spray cooling system a water/air mist cooling system is used (Fig. 18). This cooling system allows sensitive, high-alloyed tool steels to be produced without cracking. The following figures 19 and 20 show a HOM-TEC steel caster at Krupp Stahl AG.

17

18

Figure 17 - Macro structure of stainless steel billet without stirring

and electromagnetically stirred.

Figure 18 - Water/air mist cooling system

Figure 19 - HOM-TEC caster at Krupp Stahl AG's Siegen works.

19

Future Prospects. Experience gained in the continuous casting of steel will be able to be made use of to a certain extent in the further installation of nonferrous casters. We are thinking here above all of HOM-TEC casters for large leaded brass extrusion billets. Leaded brass is a relatively brittle alloy with a long solidification interval. Unlike copper, for example, the liquid core extends a considerable distance beyond the mould. We therefore envisage building large brass casters in the future with air/water mist cooling systems and an electromagnetic strand stirrer to increase the casting speed whilst assuring a homogeneous internal structure. As you can see, even a highly developed casting process still leaves room for improvements - and that is what we are working on.

Figure 20 - General view on HOM-TEC steel caster (Krupp Stahl AG, Siegen works).

References

1. A.H. Pehrson, "Verfahren zur Erleichterung des Herausziehens eines gegossenen Metallstranges aus einem Mundstück oder einem Kühlrohr", Patentschrift Nr. 174791,(1905).

2. P.J. Clark, E. Roller, "Betrieb einer horizontalen Kupferbolzen-Stranggießanlage bei IMI Refiners Limited, UK", (Paper presented at the Deutsche Gesellschaft für Metallkunde Symposium Stranggießen , Bad Nauheim, 1985).

3. R. Heinke et al.,"Das Krupp-Horizontalstranggießverfahren mit oszillierender Kokille und linearem Strangabzug",Stahl und Eisen 12/13 (1984)590-594.

PROCESS DEVELOPMENT OF ELECTROMAGNETIC

CASTING FOR COPPER BASE ALLOYS

John C. Yarwood, Phillip C. Chatfield, Brian G. Lewis

Olin Metals Research Laboratory
91 Shelton Avenue
New Haven, CT 06511, U.S.A.

Abstract

Electromagnetic casting (EMC) is now a widely applied technology in the aluminum industry. This paper chronicles the development of the process from bench to industrial scale for the casting of copper base process ingots. The principles involved in extending this technique to copper base revolve around density, surface tension and temperature, all of which are considerably more challenging for copper. Critical to scale-up are EM containment and mold level control, molten metal distribution and thermal management and their interaction with alloy properties; surface tension, solidification characteristics, oxidation potential and cracking propensity. In addition to developments in these areas, transfer to production operation also necessitated full automation of what is by present industry standards a sophisticated casting technique. Emphasis is given here to the automation and computer management of the containment and level controls during the critical start-up sequence of the EM process. Finally, a review of test results is given in terms of an economic analysis for plant introduction of EM casting.

COPPER ROD CASTING PLANT

FOR SMALL PRODUCTION QUANTITIES

Dr. E. Buch, K. Siebel and H. Berendes
Rolling Mills Div.,
Krupp Industrietechnik GmbH
Duisburg, West Germany

Abstract

The interest of the cable industry in copper rod plants of small capacity still exists. Copper rod as a pre-product for cable factories is required in quantities of between 15,000 and 30,000 to/annum.

The demands made on the quality of the copper rod are valued more and more.

A plant design is being presented which comprises the technological advantages of large-scale plants but which is also economical for the low quantities produced.

Now that four of such small plants are operating successfully, four further cable manufacturers decided in favour of investing in such mini-production plants in 1988.

1. Introduction

Copper rod plays an important part in the industrialization process of every country. There is no electric motor without varnished copper wire for the motor windings, or power cables for the incoming energy.

Not to forget the telephone communication network, or the electrification of the railway systems where thousands of kilometers of contact wire are needed for the locomotives.

These examples give only a slight idea of the wide range of application for rolled copper rod. On the one hand, wire of hair-like fineness for the varnished-wire production is obtained, and, on the other hand, cables as contact wires for railway engines.

As far as the processing of copper is concerned, the treatment covers a wide field from copper cathodes to copper scrap. The same applies to the fuels where, depending on the country where this treatment takes place, unknown calorific values and pressure fluctuations have to be taken into account.

In view of the poor infrastructure in most of the developing countries, the production plants required are rather of the small to medium capacity range, i.e. plants between 20,000 and 60,000 t/y. We therefore found our earlier decision confirmed to develop a small-scale plant for the production of quality wire material.

The successful commissioning of the first of these small-scale plants at SAUDI CABLE in Jeddah/Saudi Arabia in 1984, proved that even in a small plant the production of rolled wire is of the same high standard as that of the large-scale KRUPP/HAZELETT copper wire plant. We had, in fact, reached our goal: quality copper wire produced in a KRUPP/HAZELETT small-scale facility.

About one year after our first contacts with our customers, we succeeded in securing three orders from several Chinese firms.

Within the first six months in 1986, these three plants went into operation and have ever since been working successfully (Fig. 1).

The three plants at

Changzhou Smeltery in Changzhou, Jiangsu Prov. (6-8 t/h),

Xiangtan Cable Works in Xiangtan, Hunan Prov. (8 t/h) and

Beijing Copper Plant in Beijing (14 t/h)

have a total installed capacity of 140,000 t/y which represents more than half of the total capacity installed in China.

Discussions held worldwide with potential customers encouraged us to go another step further in developing systems of even smaller capacities in the order of 4 - 7 t/h.

The success came soon: four reputed copper rod producers decided in favour of our novel tandem system with a capacity of 7 t/h.

2. Standard systems

The two small-scale facilities 8 C 10 and 12 C 10, and 12 C 10 L which are already operating in China as well as our latest developments, i.e. the types 5 C 8 and 7 C 10 fit harmoniously into the KRUPP/HAZELETT plant concept (Fig. 2). Four systems of the 7 C 10 type with a production capacity of 6 - 7 t/h will start operation early this year in four different countries, viz. Chile, Indonesia, Thailand and Greece where reputed copper rod and cable makers will be starting their production. All plant facilities are of the modular system. They start off with the furnace system where the alternatives are shaft melting, reverberatory or induction furnace types. Depending on the production capacities we are able to recommend five different casting machines: for very small capacities and medium capacities up to 14.5 t/h we offer the new Mini Caster type 29.88M, 29.88 and 29.112; for medium to maximum outputs we offer the 20.112 or 20.146 caster which have both proved very successful in operation. The rolling mill which has to reduce the cast bar to the final dimensions is selected from standardized stands with roll-ring diameters between 480 and 195 mm.

For wire cooling and descaling we have developed two systems as alternative solutions: a rod cooling and de-oxidation line and a lay pickling line.

There is also a choice between two coil forming systems: loop layer with coil forming chamber for spirally and orbitally arranged lays, or compact coiler for wound rod.

Mini plant design

The design of the mini plant shows all the features which are typical for our plant concepts (Fig. 3 and Fig. 8). In some details, however, we had to find new ways to ensure that after the production capacity has been reduced, profitability remains still the same.

Furnace

An ASARCO shaft melting furnace was selected. The furnace dimensions have practically not been changed. The size of the cathode to be charged has been increased in the last few years so that even in the mini plant the furnace inside diameter is still 1.8 m. The furnace height has been slightly reduced, although, relative to the melting capacity it is relatively high in comparison to the large-scale facilities. Our main target is to keep also the mini plants at the excellent thermal efficiency of 60 - 70 %.

The operation results proved that our decision to run the process without preheated air was right. The design of the burners - there are 10 of them in two rows - as well as the complete analysis and regulation system remained unchanged.

To simplify the cathode and scrap charging process, a skip hoist with charging basket has been provided. To determine the optimum method of discharging the bucket into the furnace opening, video recordings of an existing plant were made. The filling level of the shaft furnace is supervised by a TV camera and displayed on a monitor for information of the furnace operator or fork-lift truck driver.

The size of the holding furnace has been reduced to a capacity of 8 tons molten copper which represents a fully sufficient buffer of 40 - 60 minutes.

Generally, careful thought has been given to all design details to ensure that the scaled-down system was still cost-efficient in production. As far as this is concerned, our experience gained in furnace plants built earlier proved to be a great help.

The launders between the furnaces and the caster were adapted to low-flow volume and kept as short as possible to save energy.

Fig. 4 shows the furnace system with the two rows of burners of the ASARCO type. With our mini plants we have been able to approach a different category of customers who are confronting us with absolutely new problems with regard to the availability of fuels. Besides the calorific value which is in many cases too low, the main problem is created by the fluctuations of the calorific value in the fuels. We managed to get hold of this problem by developing a unit which keeps the calorific value constant.

3.2 Caster

The caster is based on the job-proven HAZELETT twin-belt principle. The major problem was the casting section. For reasons of quality we decided to have a section as large as possible despite the relatively small production rates.

For outputs up to 8 t/h we chose a casting section of 60×35 mm = 2100 mm^2, and a casting section of 55×50 mm = 2750 mm^2 for outputs up to 12 or 14.5 t/h.

Only for our smallest facilities 5 C 8 we have considered a casting section of 45×35 mm = 1575 mm^2 in view of the fact that otherwise the casting velocity would become too slow. However that casting section is still much bigger than in other systems of comparable output.

These casting sections permit a sufficiently high degree of reduction so that a finished rod of 8 mm dia. has undergone a hundred percent transformation of the casting structure. This is of particular importance because insufficient reduction leads to rod breakages during further processing in the drawing shop.

At the same time, these casting sections which are disproportionately large relative to the low casting rates, permit a reduction in casting speed to approx.75 % of the speed of large-scale plants. Therefore, the casting strands produced in the mini plant have the same fine-grained structure as those obtained from our plants built previously. The lower casting speed allowed, too, to reduce the length of the caster. The active mould length is now approx. 2.3 m.

As the largest castable size of the mini caster, a strand width of 75 mm and a thickness of 50 mm was selected. Since, in the large machines, the maximum strand width is 200 mm, we were able to markedly reduce the width of the casting belts to 368 mm which, in turn, leads to a reduction of the wear parts costs. The belt system, especially the return pulley diameter, is the same as in the large casters so that a maximum life of the casting belts is attained (Fig. 5).

To secure quality, we have adhered to the cooling nozzles which represent a special feature in the HAZELETT system.

Essential improvements - even in comparison with the large-scale plants - were achieved in the molten-metal feeding control. The casting spout protruding into the caster was slightly angled at the side and thus rendered it easier to watch the casting level. The size of casting spout and stopper was adapted to the casting rate. Precision of adjustment could be improved as result of the experience gained in the large-scale plants. Mini plants have a hydraulic adjustment system with adjustment steps of 0.05 mm.

The automatic casting level control profits also by these improvements. Besides the visual control system ALC used in large-scale plants, we tested another system for the mini plant, viz. a probe which measures the temperature of the upper casting belt. Both systems work excellently as a result of the sensitive control of the casting stopper.

3.3 Equipment between caster and rolling mill

Here the same machinery is required as for the large-scale plants. We took the opportunity to not only reduce the machine size, but also to redesign certain features on the basis of our experience gained from the 22 plants built earlier. Great emphasis was placed on saving costs by simplifying the manufacture. Out of the many design improvements, we mention only two:

+ Between the two double-cutters of the edge milling machine, a special scanning roller was arranged to ensure a uniform clearance between the cutters and the casting strand. Chamfering of the strand edges from the present 4 - 5 mm edge length could thus be reduced to approx. 2.5 mm, so that metal loss went down to 0.2 % due to edge milling.

+ All machines were of compact design and placed on a common base frame. This means not only manufacturing cost reduction, but easier assembly, too.

3.4 Rolling mill

After deciding about the casting sections in relation to the casting output and speed, the number of stands is determined, viz.:

Nine stands for the 8 C 10 system with an output of up to 8 t/h, and 10 stands for the 12 C 10 system with an output of up to 12 t/h (14.5 t/h).

When we decided about the rolling mill concept we wanted to include two special features which make our large-scale rolling mills outstanding:

1. The regrindability by 10 % for all rolls, because the operating cost factor must be particularly considered in mini plants, and the same applies to a great deal to the costs for the rolls.

2. Large roll-rings for the roughing stands to improve the forging effect of the strand at high reduction ratios.

We decided to build two in-line rolling block assemblies, i.e. one 3 or 4-stand roughing block with 360 mm roll-ring diameter, and one 6-stand finishing block with 220 mm roll-ring diameter. Both blocks are combined to form what is virtually one single block driven by a common DC motor (Fig. 6).

Compared with other large-scale plants, this configuration is cost-saving, because there is no longer any need for individually driven roughing or finishing stands, neither for the cooling line ahead of the two finishing stands, or the loop lowering facility.

In our large-scale plants we still have to use roughing stands, because the high rolling forces needed to produce the required strand dimensions of 130 x 60 mm, would render combined blocks uneconomical from the constructional point of view.

Considering the size of the casting cross section, rolling in one block may be given preference if the first stands have sufficiently large roll diameters and as long as they not only have a reduction effect but also a good "forging" effect. Therefore we use roll-rings of 360 mm diameter.

Depending on the rolling schedule, the lack of one or the other finishing stand is bound to lead to less flexibility in the production of various finished wire sizes. Nor is it possible to carry out a change of the rolls independently of the other stands in the block, for the oval and round passes which are very important for the surface finish of the rod. In a large-scale plant of 50 t/h capacity, the roll ring grooves are changed every 8 hours for reasons of quality; where not more than 8 t/h are put through, the necessity of roll change is less frequent. The longer intervals between roll changes in combination with the compactness of the unit make up for the slight disadvantage of reduced flexibility.

Looking at the roll pass design for the types 8 C 10 and 12 C 10, it becomes evident that also in those cases where there is a sole block configuration, it is still possible to obtain a large number of finished sizes (Fig. 7).

In connection with our smallest models 5 C 8 and 7 C 10 which lately have been designed with even smaller tonnages per hour and, consequently, smaller casting cross sections, we have decided in favour of an alternative solution for the rolling mill in order to cope with the abovementioned loss of flexibility in the production of a multiple range of finished wire sizes. The aim of this concept was to ensure the 10 % regrindability of the roll rings, as well as the use of large roll rings.

The concept of the 7 C 10 facility which is being built at present consists of 2 rolling stands each of the 10- stand and 8-stand rolling block arranged as tandem unit which are driven by a DC motor. Another independent change of the rolls after wear, and the production of a multiple range of finished wire is ensured in this tandem version with its drive by several DC motors. We may not miss to mention that this tandem version has also a favourable effect on the investment costs, as savings on the gear side make up for little additional costs on the electrical side.

Fig. 8 shows the layout of a KRUPP/HAZELETT plant in tandem arrangement.

The heat held by the small cross section is much lower so that the cooling system in the rolling mill has to act very sensitively to avoid too much cooling down of the rolled material. This is where the compact configuration of the rolling mill - block or tandem version - with a total length of not more than 7 - 8 m, as well as the cover hoods over the entire length turn out favourably. During maintenance work the hoods can be opened hydraulically (with safety locking in cases of pressure drop); they serve as a water-tight enclosure for the rolling mill.

3.5 Cooling lines, laying head, coil forming chamber

All the mini plants built by us are equipped with an alcohol deoxidation and cooling line, a laying head and a coil forming chamber with lifting table and loop layer for orbitally arranged rod lays on wood or steel pallets. All these facilities represent the most advanced technical standard for large-scale plants.

On repeated customer request we have lately been offering a system in which wire lays are collected on pallets placed on transfer cars and transformed into coils (Fig. 8). However, because of the lower wire speeds, special provisions were necessary in the coil forming system. The facility which applies a wax coating over the wire had completely been redesigned and is now integrated as a compact unit, including the tank with the valves and fittings, into the roller curve and pinch roll set in front of the laying head. This design improvement reduces manufacturing and assembling costs. (Fig. 9)

3.6 Electrical equipment - Electronic data logging system

The cost reduction is a side-effect of the rolling mill concept. While in large plants there are up to seven DC motors for the rolling mill, only one DC drive or a few small DC drives are necessary for the rolling block. Everything else meets the latest standards of large-scale plants, i.e. the freely programmable logic control, automatic speed control, auxiliary plant monitors, detectors searching for irregularities in the rolling mill area with automatic activation of the pendulum shear to cut the continuous strand entering the rolling mill.

To ease supervision of the process flow, and for logging important process data and at the same time continuously obtain reliable information on quality parameters of the rolled wire, we have developed especially for the needs of the KRUPP/HAZELETT system an electronic, computer- controlled recording and supervision system. The central control panel between the caster and the rolling mill is equipped with a VDU which, by pressing a pushbutton, displays mimic diagrams with the current data on electric power and gas consumption, or qualities and quantities of the rolled wire, etc.

Visual and audible fault signals are also emitted immediately after an irregularity has turned up. In addition, trend curves on important operation parameters can be displayed on request.

There is the same type of VDU in the laboratory (Fig. 10) where the results of the quality tests for each coil are entered. Together with the gross, tare and net weights automatically fed into the system through the weighing units, a record is made of each coil which can be printed out and stored on floppy disks.

In addition to this record, all the other operating parameters available at the time when this specific coil was produced can be recorded, and thus the production history of each coil can be traced.

On our customers' request, we will also equip our mini plants 7 C 10 and 5 C 8 with this data logger.

3.7 Auxiliary equipment

Hydraulic unit, oil lubrication system, emulsion unit, and circulating system for the cooling and deoxidation medium are, except that the sizes and recirculation quantities are smaller, very much the same as in the large-scale plants.

We have taken a few measures to reduce cellar space. The hydraulic system is now in a room on floor level, i.e. underneath the furnace platform. The recirculation systems for lube oil, emulsion and cooling media are installed in a pit behind the rolling mill.

Another outstanding feature is that, for the first time in mini plants, the emulsion unit operates in a closed circuit. From the rolling mill the emulsion flows through a pipe directly to the automatic vacuum filter which serves at the same time as storage tank.

4. Plant economy

Generally it has to be taken for granted that in mini plants the specific costs for rolling are higher due to the smaller production quantities. Unfortunately, however, producing smaller tonnages does not mean that the investment costs are reduced proportionately. In fact, most of the costs incurred for electricity, fuels, expendables, etc. needed for the production of one ton of copper rod are even higher than in large-scale plants.

Due to the concentration of the KRUPP/HAZELETT mini plants on a smaller area and by using central control units it was possible to reduce the personnel per shift to 5 or 6 operators.

The KRUPP melting furnace can be run with cheaper fuels, e.g. municipal gas, coke gas, etc. which results in considerable reduction of energy costs compared with other plant types.

Another cost benefit can be drawn from the use of less costly standard cathodes or copper scrap instead of high-grade cathodes. Such cost savings together with the high quality of the rod produced make mini plants a profitable investment.

5. Rod quality

Regarding the quality of rods, i.e. rupture strength, elongation, twist test, conductivity, spiral test, etc., mini-plants stand comparison with the large-scale KRUPP/HAZELETT plants.

The fact that the Chinese customer who runs the KRUPP/HAZELETT plants is meanwhile exporting considerable tonnages of copper rod, speaks for itself.

The surface oxide film is in the order between 300 and 350 A (30 - 35 micron) and is thus in the range of values obtained in our large-scale plants which use alcoholic descaling.

6. Flexibility of mini plants

It goes without saying that mini plants are capable of producing in addition to round material also other material such as flats, whenever this is required. In those cases, however, the rolling mill has to be adapted to these specific requirements. A current study has shown that besides rounds of 8 - 22 mm diameter, also small flats with a width of 18 - 40 mm can be produced.

7. Mini-plant layout

We have to revert again to the 3 plants built in China.

Although the plants consist basically of the same components, certain adaptations to the local conditions had to be made in the layout.

Changzhou Smeltery - plant type 8 C 10 (Fig. 11). A new building had to be erected for the plant so that we were able to use to a large extent our standard layout.

The furnace was equipped with an additional system for stabilizing the calorific value. The coke-oven gas available is stored in a large customer-furnished tank to give better mixing results and to serve as a buffer in bottleneck situations which occur in the daily supply.

Xiangtan Cable Works - plant type 8 C 10 (Fig. 12). This plant had to be fitted into the existing building area. Only the building for the furnace was new. The caster is of left-hand design. Behind the caster, the operation side changes from the left (rolling direction) to the right (rolling direction). This made it possible to build the rolling mill equipment of the same design as for Changzhou.

Beijing Copper Plant - plant type 12 C 10 L (Fig. 13). This plant with its output of 14.5 t/h is not only the largest built by KRUPP in China to date, but the largest all over the country. The charging equipment on the shaft melting furnace had to be turned by 90 deg. to meet the prevailing conditions at the site. On account of its higher production rate, the rolling mill has to reduce larger sections and has therefore ten stands instead of nine. New buildings had to be provided to accommodate the plant.

A characteristic feature for the mini plants type 7 C 10 currently under construction and partly going into operation at

    IKI Indah Cable/Indonesia

    Colada Continua Chilena S.A./Chile

    Thai Copper Rod Company Ltd./Thailand, and

    Halcor Metal Works S.A./Greece

is the layout according to Fig. 14. Here, too, the layout had to be modified and adapted to the prevailing conditions.

8. Summary

The KRUPP/HAZELETT mini plant has proved its reliability in industrial operation. The success we had in the past twelve years with the large-scale plants, which have a total installed capacity of 2.8 million tonnes of copper rod, are encouraging us to offer only the best to mini plant users. Thus these mini plant operating firms will be able to cope with the continually rising demands by their customers.

Like the large plants, mini plants are capable of processing the low-cost raw material such as standard cathodes and copper scrap into marketable rod. A variety of gases with varying analyses and fluctuation calorific values, and even oils of different qualities can be used in the melting furnace, provided they are properly prepared. Cost-efficiency and universal applicability of our mini plant are thus secured for the future.

Fig. 1 KRUPP/HAZELETT Mini Plant

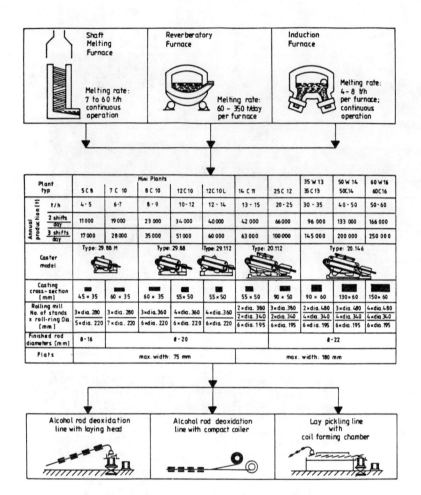

KRUPP / HAZELETT Standard Copper Rod Plants

The modular design and tailoring of each plant to individual specifications ensure that each user is provided with the most economic plant concept for his needs.
You can choose between:
three different types of furnace;
five twin belt casters with differing outputs;
two deoxidation systems, i.e. lay pickling or alcohol rod deoxidation;
two coil forming systems, i.e. elephant trunk with coil forming chamber for orbitally and spirally arranged lays or compact coiler for wound rod.

Fig. 2

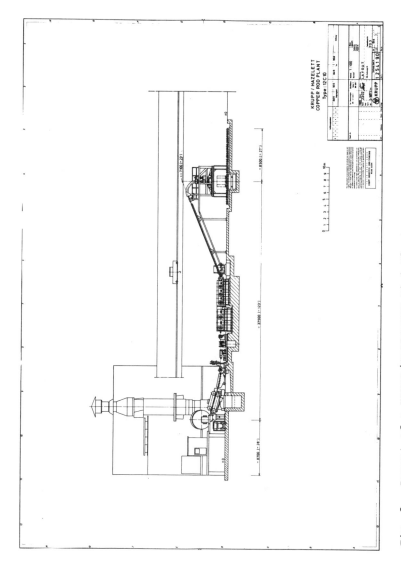

Fig. 3 Layout of a KRUPP/HAZZELETT Mini Plant

Fig. 4 Melting- and Holding Furnace

Fig. 5   Casting Machine - Metal Feeding

Fig. 6 KRUPP Rolling Block - 10 Stands

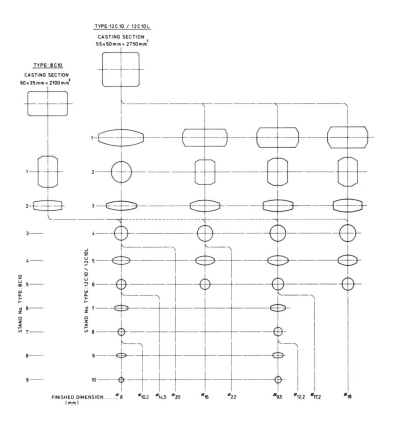

Fig. 7 Roll Pass Schedule

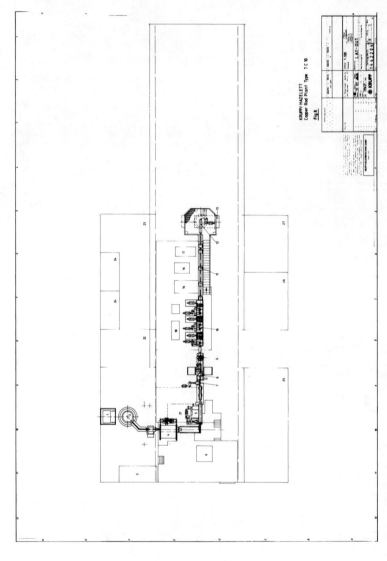

Fig. 8   KRUPP/HAZELETT Copper Rod Plant Type 7C10

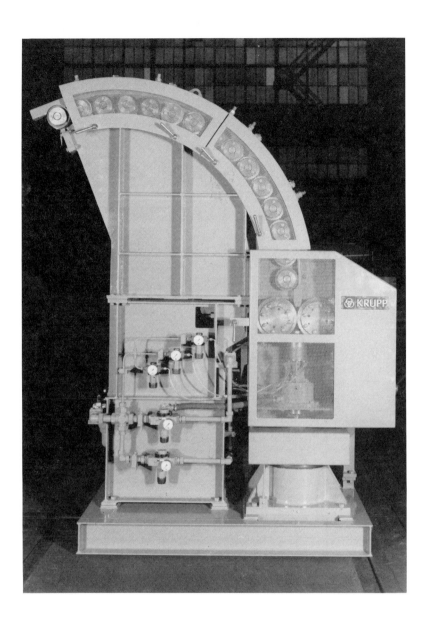

Fig. 9   Roller Guide, with Pinch Roll Set, Waxing System and Laying Head

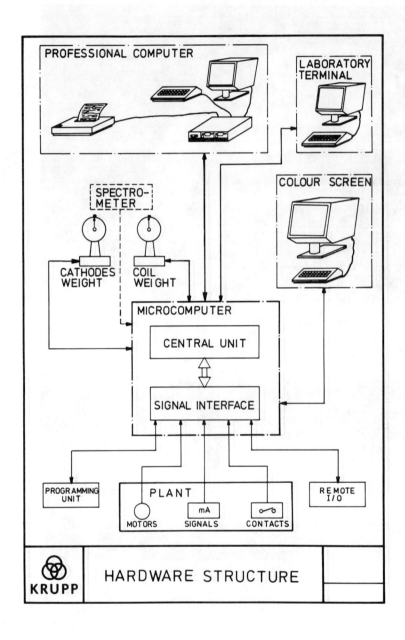

Fig. 10   Data Logger

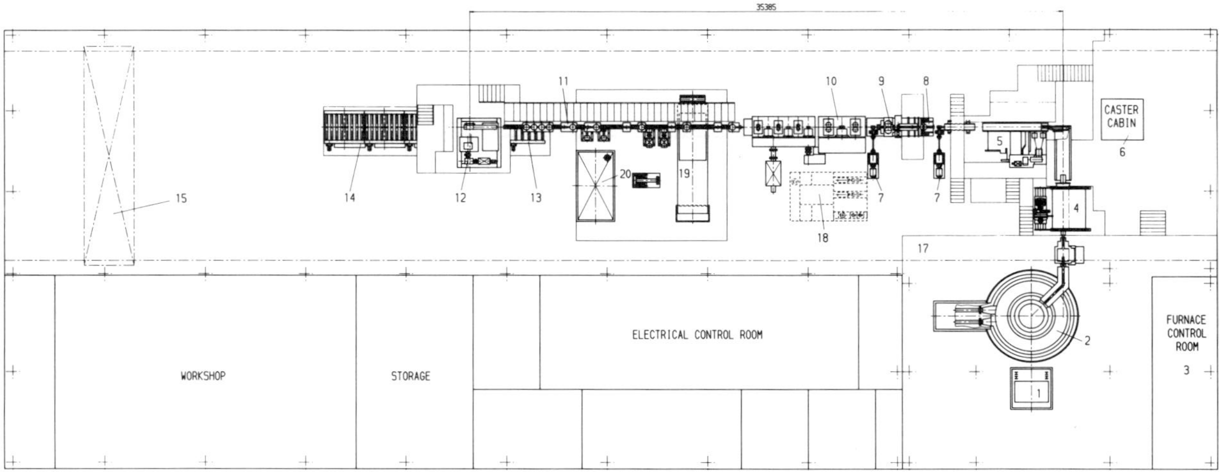

Fig. 11　CHANGZHOU SMELTERY

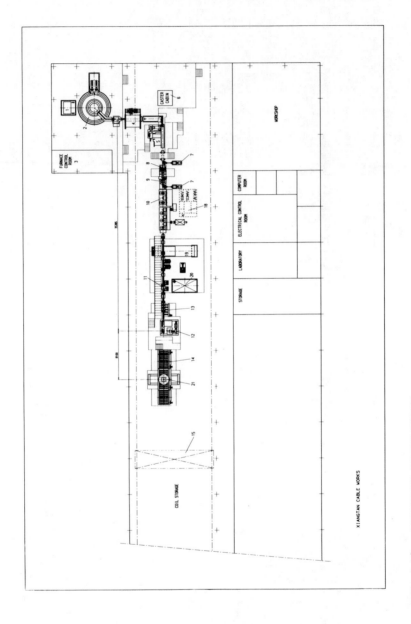

Fig. 12 XIANGTAN CABLE WORKS

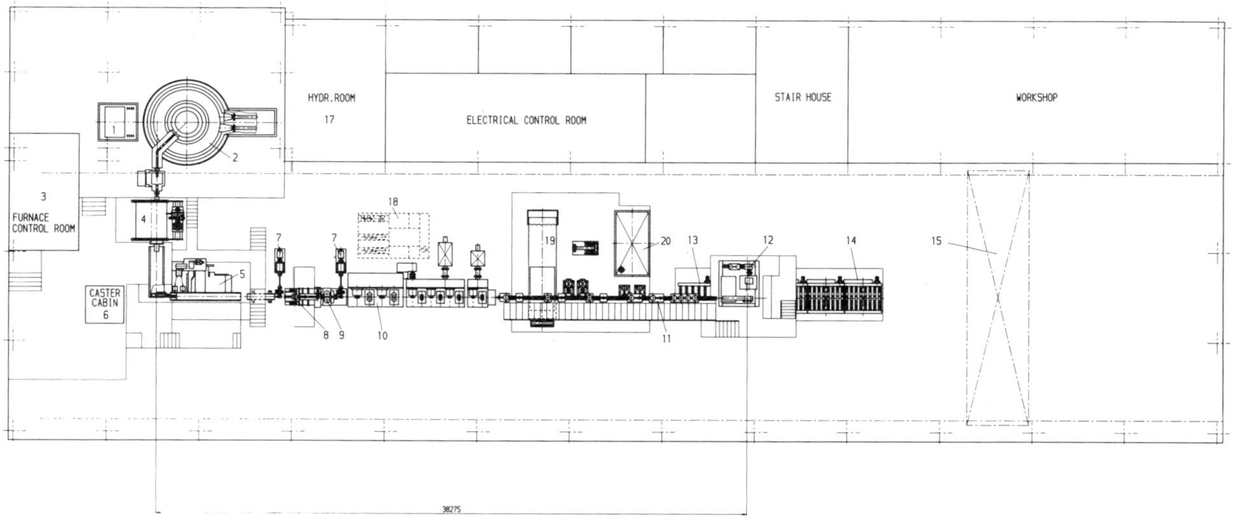

Fig. 13   BEIJING COPPER PLANT

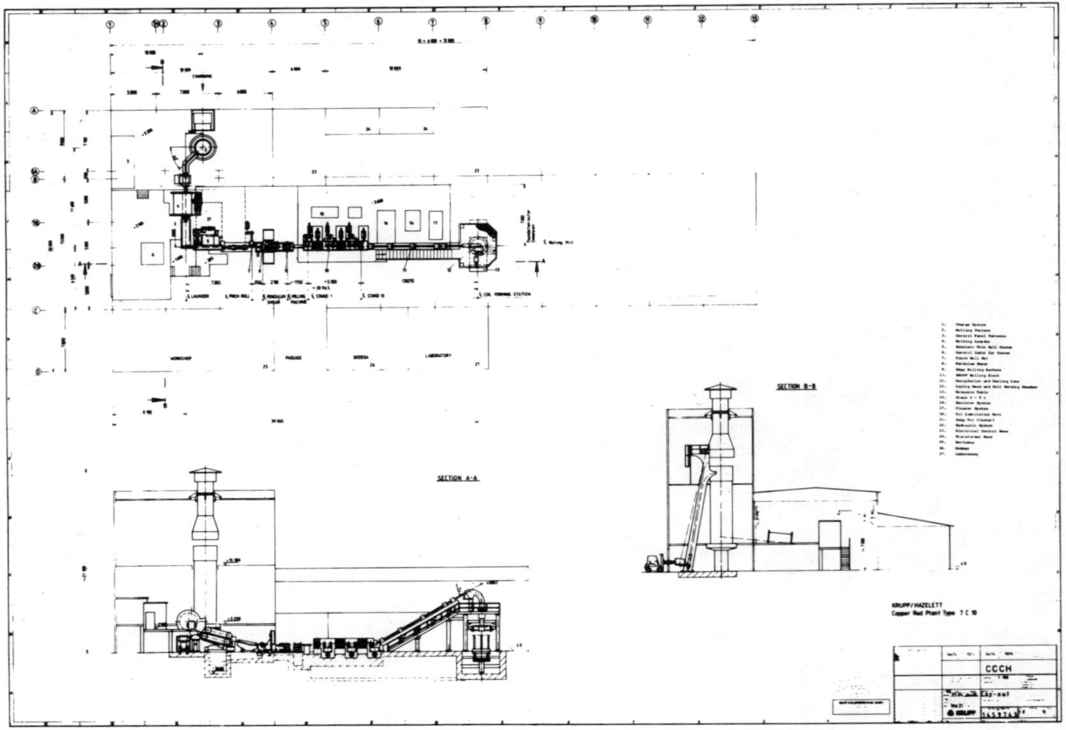

Fig. 14 KRUPP/HAZELETT Mini Copper Rod Plant Type 7C10

FLASH SMELTING OF PYRITE AND SULPHUR RECOVERY

J.A. Asteljoki[1] and T.P.T. Hanniala[2]

[1]Outokumpu Oy, Research Centre, P.O. Box 60, SF-28101 Pori, Finland
[2]Outokumpu Oy, Engineering Division, P.O.Box 86, SF-02201 Espoo, Finland

Abstract

Traditionally pyrite has widely been used for sulpuhric acid production by first roasting it and then sending the generated SO_2 gas for acid manufacture. The iron oxide calcine has, in some cases, been further used in iron production, but more and more it has been stockpiled forming an environmental problem.

During the last few years Outokumpu has developed the flash smelting of pyrite to produce environmentally safe slag and rich SO_2 gas. The gas can be efficiently used either for acid production or elemental sulphur production or even both. In this case the gas processing is more profitable than conventionally by virtue of the low gas volume and high SO_2 content. In addition to slag, also a small amount of matte can be co-produced.

This matte absorbs a considerable amount of precious metals, which frequently are found in complex pyrites, making their recovery possible. Also flash smelting makes it possible to treat very fine pyrites which normally are almost impossible to use in fluid bed roasters.

In this paper the new concept for pyrite processing and the corresponding pilot test results are described.

Introduction

Pyrite is one of the most common sulphide minerals. It can be met as pure mineral, matrix mineral in complex sulphide ores or it can be met as one of the main minerals in many non-ferrous sulphide ore deposits. In most cases it is utilized for sulphuric acid manufacture, but considerable quantities are also discarded. The known pyrite resources are enormous, but especially the exploitation of complex pyrites is limited because of the very fine grained mineralization together with relatively low non-ferrous metals contents. In many cases, however, the precious metals content is noteworthy making their recovery attractive.

Roasting and Sulphuric Acid. Today pyrite is mainly used for sulphuric acid production by roasting the pyrite and then ducting the gas to the sulphuric acid plant. The technology used is the well known and proven fluid bed roasting together with acid manufacture applying increasingly more the double contact technique. However, especially protection of environment has brought out some problems which have to be solved:

* In most cases the roasted calcine cannot be used in iron making because of the too high content of non-ferrous metals.
* The use of the calcine in cement industry is very limited owing to the high tonnages of calcine.
* The calcine has to be stockpiled causing remarkable dusting problems in the surroundings.
* The pyrite fines. When recovering non-ferrous metals from the complex pyrites the ore has to be ground very fine. As a result a big part of pyrite tailing is too fine to be used in fluid bed roasters, for which reason this has to be disposed of causing again severe dusting.

Complex Pyrites. As noted above the pyrites frequently contain sometimes even considerable amounts of precious metals and other valuable metals. By dead roasting the pyrite these are not recovered. It is, however, possible to develop flow sheets by which practically all valuable constituents can be recovered (Ref. 1). Leaching the precious metals from dead roasted calcine can be applied, but typically this is not profitable because of the high tonnages of calcine to be treated.

Production of Elemental Sulphur. As already noted above, pyrite is mostly used for acid manufacture. The high content of sulphur in pyrite makes its recovery as elemental sulphur attractive, too. However, this alternative has not been widely applied except in Outokumpu at the Kokkola Works in 1960's and 1970's (Ref. 2 and 3). The flow sheet is shown in Figure 1. At that time a pyrite smelter and sulphur plant was in operation where part of the sulphur was produced as elemental sulphur. In 1977 the plant was shut down because of the steeply decreased price of sulphur and strongly increased price of fuel. A similar type of sulphur plant was also built for four copper and nickel smelters in various parts of the world and still today two of those are in operation. Today the overproduction of sulphuric acid and the environmental protection have again raised the idea of producing elemental sulphur instead of sulphuric acid for consideration, but the cost and the lack of suitable technique has limited the application of this opportunity.

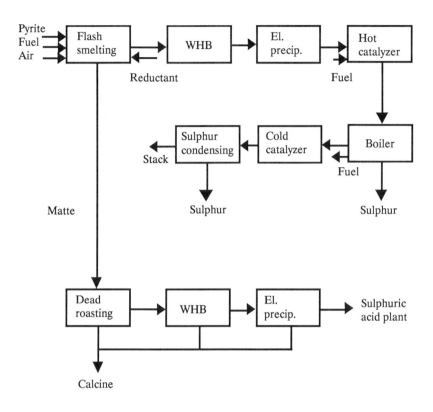

Figure 1 - Flow sheet of the pyrite smelter and sulphur plant at Kokkola.

In order to overcome the above mentioned problems and because Outokumpu has long traditions and a lot of experience in both roasting and smelting of pyrite on a commercial and on a pilot scale, it was decided to further develop the technology to meet today's conditions. This initiated a pilot test series where the target was to apply modern flash smelting technology for pyrite in order to produce high strength SO_2 gas and environmentally safe slag and in certain cases also to recover precious metals in pyrite.

New Concepts for Treating Pyrites and Producing Sulphur

As a solution to the above mentioned problems two new concepts for treating pyrites have been developed and pilot tested at the Outokumpu Research Centre. In the first case pyrite is totally oxidized in a flash smelting furnace to produce rich SO_2 gas and discardable slag. This is called the total oxidation smelting process. The second case differs from the first one in that a small amount of matte is co-produced absorbing considerable amounts of precious metals and some other metal values. This can be called the partial oxidation smelting process or the matte droplet smelting process.

Another important question is in which form should the sulphur content of pyrite be recovered. The most typical way is to produce sulphuric acid, but the use of flash smelting opens new opportunities to produce acid, liquid SO_2, elemental sulphur or any combination of them.

Total Oxidation Smelting

The flow sheet is presented in Figure 2. In the method pyrite is smelted in a flash smelting furnace with oxygen enriched air. The enrichment level is controlled so that the process is autogenous i.e. no additional fuel is used. Flux is also added to the feed. As a result of the exothermic reactions the charge smelts in the reaction shaft and liquid iron silicate slag is collected on the bottom of the flash furnace. Sulphur is totally converted to high strength SO_2 gas which is very suitable for acid production. The sulphur content of the slag is very low, less than 0.1 %, and all metals are in oxidic form dissolved in the slag. The slag is tapped and cooled into blocks or granulated depending on the further use of the slag.

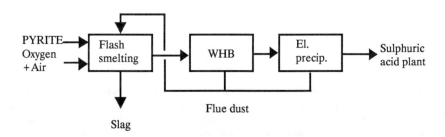

Figure 2 - Total oxidation smelting of pyrite.

Typical of flash smelting is that the feed material can be very fine. This makes it possible to utilize very fine grained pyrites which are not suitable for fluid bed roasting. The flash smelting method is proven (more than 30 smelters in the world) and thus is a safe way of treating pyrites.

Partial Oxidation Smelting

In the partial oxidation smelting process (Figure 3) the oxygen coefficient (the total amount of oxygen per tonne of pyrite) and flux feed are controlled so that a small and controlled amount of iron matte is formed under the slag layer. The matte absorbs most of the precious metals as well as copper which can be separately recovered. High strength SO_2 gas is again formed and the slag, which now contains 2-3 % of sulphur and is still safe for the environment, is cooled and discarded or sent for further use. Because of the small amount of matte produced this is called the matte droplet smelting process.

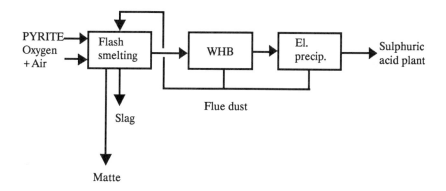

Figure 3 - Partial oxidation smelting of pyrite.

The most specific feature of this method compared to the previous one is that it makes it possible to recover most of the precious metals and copper in pyrite. The matte can be fed for instance to a copper smelter where the precious metals recovery is extremely high. Or the matte can be oxidized and the metal values recovered by leaching methods. Now the quantities to be treated are low compared to the quantities of roasted calcine. Or any other suitable method can be used.

Production of Sulphur

When sulphidic minerals, such as pyrite, are roasted, air or in some cases slightly oxygen enriched air is used. The amount of excess air required by the reactions to complete is high. The result is that the gas flow is high although very suitable by composition for sulphuric acid production. This fact causes the gas handling equipment and the acid plant to be large and hence also expensive.

The SO_2 gas produced in a flash smelting furnace has several characteristics which make it possible to consider other alternatives:

* low gas volume and
* rich in SO_2 by virtue of oxygen enrichment
* no free oxygen in the gas
* high temperature gas

Sulphuric Acid and Liquid SO_2. Under normal conditions sulphuric acid is produced. The low gas volume and high SO_2 content (more than 30 %) resulting from using the flash smelting options instead of fluid bed roasting decrease the size of the gas handling equipment. Typically the gas flow is 25 -30 % of that of a roaster gas with the consecutive decrease in costs. However, the high SO_2 strength of the gas makes it possible to produce also liquid sulphur dioxide by the cooling method. For instance part of the SO_2 can be recovered as liquid sulphur dioxide and the rest of the gas, containing still more than 20 % of SO_2, is used for acid production. After the O_2/SO_2 adjustment the gas still contains approximately 12 % of SO_2 which is close to the maximum content in a normal double contact plant (Figure 4).

Elemental Sulphur Production. The gas after a flash smelting furnace is very suitable for elemental sulphur production because of its high SO_2 strength, high temperature and the fact that there is no free oxygen in the gas. Compared to Outokumpu's previous operations (Ref. 2) at Kokkola there is today an important difference: the gas volumes are significantly lower. Under these circumstances the production of sulphur could be economically profitable, but the former process has a very harmful disadvantage - the tail gas contains too much sulphur containing components, for which reason a separate scrubber should be added ruining the profitability of the process. Moreover the process applied at Kokkola contained so many different gas cooling, heating and catalyzing stages that in any case the investment costs would be high. But some of the principles can be applied in a combined process where both elemental sulphur and sulphuric acid are produced.

Combined Elemental Sulphur and Sulphuric Acid Production. In this process alternative (Figure 4) the gas after the smelting section of the flash smelting furnace is reduced. Light naphtha, natural gas, pulverized coal or coke or any corresponding reductant can be used. The reduced gas is cooled in the waste heat boiler. The reactions taking place in the gas inside the boiler remain rather far away from the thermodynamic equilibrium. This means that the elemental sulphur yield is not so high as could be expected from theoretical calculations. Anyway a considerable amount of sulphur (about 50 - 60 % of the total sulphur content of the gas) is formed and this is recovered by condensation. The rest of the gas is burned - to have all sulphur in the gas as SO_2 - and ducted to the acid plant. By this kind of process it is possible to regulate the proportion of elemental sulphur and sulphuric acid production. At the same time the tail gas after the process is clean to be vented to the atmosphere.

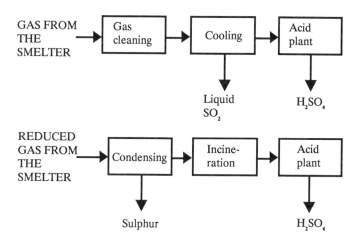

Figure 4 - Alternatives for sulphur production.

Pilot tests

In order to confirm the applicability of the presented concepts and to obtain sufficient engineering and design data large scale pyrite smelting pilot plant tests were carried out in 1987. The equipment used in the tests have been described in Outokumpu's previous papers (Ref. 4). The objectives of the tests were:

* to test the applicability of the process idea of producing only slag and high strength sulphur dioxide gas from pyrite in a flash smelting furnace
* to find out the optimum smelting conditions
* to find out the requirements for pyrite to reach high oxygen utilization in the furnace
* to find out the sulphur content of slag
* to test the applicability of the process idea of producing small amount of matte, slag and high strength sulphur dioxide gas
* to determine the partition coefficients of the valuable metals between matte and slag.

Three different pyrites were used in the tests. The trials were run in two campaigns of totally five weeks during which altogether 500 tonnes of pyrites were smelted. The compositions of the pyrites are shown in Table I.

Table I. Compositions of the pyrites smelted in the tests.

	Fe %	S %	Cu %	Ag g/t	Au g/t
Pyhäsalmi	46.5	49.8	0.39		
Almagrera	42.6	43.7	0.36	23	0.24
Tharsis	43.5	40.5	0.73	26-40	1.7-1.9

The Pyhäsalmi pyrite was Outokumpu's own domestic pyrite. The other two concentrates were from Spain from Minas de Almagrera S.A. and from Compania Española de Minas de Tharsis S.A. The grain sizes were as follows:

Pyhäsalmi	60 % minus 200 mesh
Almagrera	75-80 % minus 200 mesh
Tharsis	75 % minus 200 mesh

The Pyhäsalmi pyrite was only used to test the first - total oxidation smelting - process. The Almagrera pyrite was used for both processing tests while the Tharsis pyrite was used only for the second process - matte droplet smelting - tests.

The tests showed that both principles performed as calculated. They were easy to operate and control. The on-line availabilities ranged between 89 and 96 %, which are exceptionally high in pilot tests, and most of the downtime was due to equipment failures and checkings.

In the first campaign, where the total oxidation smelting was tested, three main parameters were followed

* oxygen utilization in the furnace
* sulphur content of slag
* magnetite content of slag

The main variable was the oxygen coefficient i.e. the total amount of oxygen per tonne of pyrite.

Figure 5 shows the sulphur content of slag vs. proportion of total iron as magnetite. One can see

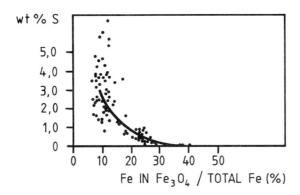

Figure 5 - Sulphur content of slag vs. proportion of total iron as magnetite.

that approximately one third of the total iron must be oxidized to magnetite in order to achieve a sulphur content lower than 0.1 % in slag. In this case the magnetite content is about 25 %. In liquid state this is totally dissolved and the slag can be easily tapped and does not form any build-ups in the furnace. This was also found in the tests.

From an economic point of view it is important that the oxygen utilization is high. This was found to depend on the grain size of concentrate. When smelting the Pyhäsalmi pyrite a few per cent of free oxygen was left in the gas while when smelting the finer Almagrera pyrite oxygen reacted completely. However, the fine concentrate results in a bit higher dust carry-over.

In the second campaign the distribution of precious metals was the main concern. This was tested by producing different amounts of matte. In Table II the results are presented.

The results show that - naturally - the more matte is produced the more precious metals is distributed to the matte (Figure 6). A more interesting point is that the lower the oxidation degree is i.e. the more matte is produced, the higher is the distribution coefficient. This phenomenon seems to be only apparent, because there were no

Table II. Results of matte droplet smelting tests.

Balance period		I	II	III	IV	V
Concentrate		Almagrera	Tharsis	Tharsis	Tharsis	Tharsis
Oxygen coefficient	Nm3/tonne	273.2	253.0	228.2	204.2	257.8
Ag in pyrite	g/t	23.0	40.0	26.0	27.0	30.0
Au in pyrite	g/t	0.24	1.8	1.7	1.8	1.9
Matte amount	% of pyrite	7.4	9.8	19.9	28.0	14.6
Ag in matte	g/t	93.0	209	111	98	108
Au in matte	g/t	1.26	7.8	7.5	6.7	6.2
Lm/s(Ag)	wt%/wt%	6.2	8.0	12.3	10.9	8.3
Lm/s(Au)	wt%/wt%	7.9	7.8	14.2	14.3	8.5
Recovery in matte (Ag)	%	35.4	49.5	76.8	84.3	60.8
Recovery in matte (Au)	%	41.0	48.8	79.1	87.5	61.3

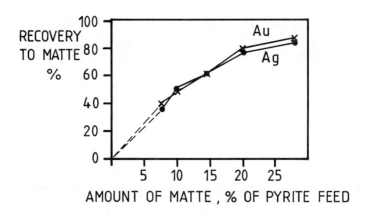

Figure 6 - Distribution of precious metals in matte.

significant differences between the compositions and oxygen or sulphur potentials in the slags and mattes. It is more evident that the settling conditions are the dominating factor. Firstly if the matte amount is high, the number of descending matte droplets is also high thus absorbing the small amounts of precious metals, which are evenly distributed in the slag. Secondly a high temperature and thus low viscosity favour the coalescing and settling velocity. This is supported by the test results as the bath temperature was 20 - 40°C higher in periods III and IV where the coefficients were the highest.

Environmetal Aspects

The protection of environment has risen among the most important aspects in the development work to be carried out within chemical and metallurgical industry. The new concepts presented in this paper are very important, because in addition to the economic aspects the environmental problems related to using pyrites in acid production can be overcome. Also the overproduction of sulphuric acid can be controlled by producing part of the sulphur as elemental sulphur, which is easy store and transport and can be utilized later on.

Dusting of Pyrite

In many cases pyrite is mined without producing any other mineral fractions. However, increasingly more complex pyrites are exploited and selective flotation is used to recover separate concentrates. Typically the mineralizations of complex pyrites are very fine grained for which reason the ore has to be ground extremely fine. After flotation the pyrite tailing has to be classified by hydrocyclones to remove the finest fractions which cannot be used in fluid bed roasters. Therefore a considerable amount of raw material is lost. The fine fraction has to be stockpiled, and today these piles of very fine pyrite have become a serious environmental problem.

As was discussed above, fine pyrite is very suitable raw material for a flash smelting furnace. By this means both valuable raw material is saved and the dusting problem is overcome.

Slag as Product

One of the key points in flash smelting of pyrite is to produce slag instead of roasted calcine. Typically the calcines contain too much impurities so that they cannot be used in iron and steel production. Also the tonnages of calcine are so high that cement industry can absorb only a vanishingly small part of the calcines produced. Therefore they have to be stockpiled. Now these piles have proved to create a serious problem by spreading dust everywhere in the surroundings.

In industrial areas where many of the roasters and acid plants are located there is no land available for storing increasing amounts of calcines. This means that the calcines have to be transported back to the mine sites causing additional costs and problems.

The production of slag is a safe solution for these problems. Firstly slag can be cast into blocks or it can be granulated so that the products do not make any dust. Secondly the slags are insoluble for which reason they can be and they are already today used as construction materials in different purposes. The slags even have positive economic value under certain circumstances.

Economic Aspects

In the following the profitabilities of the two new pyrite smelting concepts are studied. At first the total oxidation smelting process is compared to the normal fluid bed roasting and sulphuric acid production. In the latter part of this section the economics of the matte droplet smelting process is evaluated.

The assumptions used in the comparisons are as follows:

* Acid production 2000 t/d
* Oxygen enrichment at the autogenous level
* Composition of pyrite
    - Fe   40.8 %
    - S   44.9 %
    - Au   0.39 g/t
    - Ag   40 g/t
* Double contact acid plant
* Price of pyrite 40 USD/t

Total Oxidation Smelting vs Fluid Bed Roasting

As the first case the total oxidation smelting is compared to the normal fluid bed roaster and acid plant without taking into consideration the costs of calcine and possible waste pyrite stockpiling and any other environmental protection measures. In Figure 7 the internal rate is shown as a function of the price of sulphuric acid for both of the two cases.

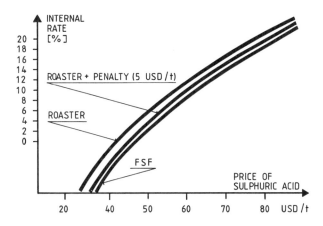

Figure 7 - Profitability of acid production. Single roaster line and single flash smelting line.

According to calculations the investment costs of the flash smelting process are higher (by 5 - 10 %) than those of a single roaster line at a low capacity level. This fact is reflected by the curves of Figure 7, which shows higher profitability for the normal roasting process. However, the difference is not very high and is actually compensated for by the transportation costs in the case where the roasted calcine has to be returned to the mine site. For instance if the calcine is subjected to a penalty of 5 USD/t, the profitabilities practically coincide.

Worth noticing is that the relative profitability of the two methods does not depend of the price of sulphuric acid.

At a higher capacity level where two or more roaster lines are required, the investment costs of a single flash smelting line (including the acid plant) are equal or possibly somewhat lower than those of a multiple line roasting plant (including the acid plant). One has to remember that a flash smelting furnace can be designed for a very high capacity so that always only one line is needed. The highest throughputs in flash smelters are at a level of 3000 t/d. Supposing that this is pyrite the corresponding acid production would be 4500 t/d.

As was noted the investment costs for the two high capacity cases are almost equal. However, the operating costs of the flash smelting case are approximately 10 % higher, but these may vary a lot according to the local conditions. The use of saturated steam is for instance important. The low value of saturated steam favors the flash smelting option, because steam can be effectively utilized in the production of oxygen and so the steam production and consumption are more in balance than in the roaster case.

Moreover if the advantages of the flash smelting alternative, listed below, are taken into account in the economic calculations, the total oxidation smelting process by flash smelting is economically more profitable than the conventional roaster:

- * The normally discarded fine pyrite can be utilized in flash smelting. This part of the feed is thus free of charge
- * Transportation cost of roasted calcine back to the mine or to a special waste area
- * Value of slag as saleable product
- * Utilization of the different sulphur production alternatives in the case of flash smelting.

Partial Oxidation Smelting i.e. Matte Droplet Smelting

As the second case the importance of recovering precious metals is discussed. The situation is now somewhat different from the first case, because now there are revenues from the precious metals, but at the same time the sulphuric acid production is slightly lower, because some sulphur is lost with matte.

When estimating the real value of the matte the following factors must be considered:

- * the total content of precious metals in matte
- * the effect of composition on the price
- * transportation costs for matte
- * treatment charge for matte
- * sulphur loss with matte

As was seen in the pilot results the recovery of the precious metals is the higher the more matte is produced. However, simultaneously the gold and silver concentrations in matte decrease and the total amount of matte increases resulting in both lower acid production and higher treatment costs for matte. So, it is evident that there is an optimum amount of matte which gives the highest profitability. This largely depends on the precious metals contents and copper content of the pyrite as well as several site specific factors.

In Figure 8 the presented factors are shown as an example. The calculations are based on slightly higher recoveries of gold and silver than those obtained in the pilot tests. It is expected that on a commercial scale this would be the case as a result of slightly higher temperature and longer retention time of slag on the furnace. One can see that with the above mentioned assumptions the optimum point is to produce 4 % matte (calculated from the pyrite feed). The effect of the transportation costs and treatment charge is noteworthy. Therefore it is important that the matte could be treated at the smelter and that a cheap and straightforward process to recover gold and silver could be found.

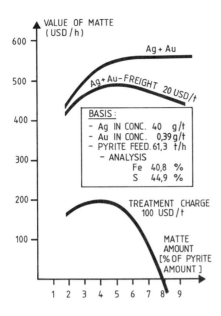

Figure 8 - The formation of the value of precious metals containing matte.

Conclusions

The applicabilities of the described two ways to smelt different pyrites were confirmed by the pilot tests. The both processes can be readily applied on a commercial scale, because they are totally based on proven technology. The profitability depends on several factors which have to be studied case by case.

Anyway the most important difference to the present roasting practice is in the environment protection aspect. The processes produce slag which is safe to store or it can be even utilized, and pyrite fines which normally cannot be utilized are suitable feed material to these processes. Still, even without the environmental aspects the processes are competitive especially if the capacity is high and the precious metals contents are high.

References

1. J.A. Asteljoki, S.P. Fugleberg and T.K. Tuominen; "The Outokumpu Flash Smelting Process for Complex Sulphide Ores", Complex Sulphides, ed. A.D. Zunkel, R.S. Boorman, A.E. Morris and R.J. Wesely, (Warrendale, PA: The Metallurgical Society, 1985), 595-607.

2. E.O. Nermes, "Flash Smelting of Pyrite Concentrate at Kokkola, Finland", Proceedings of the First International Flash Smelting Congress, ed. R. Seeste and T. Tuominen, (Helsinki, Finland, 1972), 181-215.

3. T. Tuominen, "SO_2-Reduction", ibid, 399-424.

4. T. Mäkinen and M. Kytö, "Pilot and Mini-pilot Tests of Flash Reactions", (Paper presented in the Bureau of Mines and Center fo Pyrometallurgy, Conference on Flash Reaction Processes, Salt Lake City, Utah, June 15-17, 1988), Flash Reaction Processes, ed. D.G.C. Robertson, H.Y. Sohn and N.J. Themelis, The Center for Pyrometallurgy, University of Missouri Rolla, Rolla, MO, 69-98.

PRESENT AND FUTURE TRENDS OF THE SINTERING-BLAST FURNACE PROCESS AT MHO

M. VAN CAMP

R & D
METALLURGIE HOBOKEN-OVERPELT
A. Greinerstraat 14
B - 2710   HOBOKEN
Belgium

D. CRAUWELS

Hoboken smelter
METALLURGIE HOBOKEN-OVERPELT
A. Greinerstraat 14
B - 2710   HOBOKEN
Belgium

Abstract

The traditionel lead smelter is progressively finding itself in competition with several industrially planned or installed new processes, such as : QSL, Lead Kaldo, Kivcet and Isasmelt. Stricter emission standards and reduced operating costs are the main incentives for introducing these smelting techniques in the primary lead industry. In the area of complex concentrate treatment as practised at Hoboken however, the traditional smelting sequence remains a viable and competitive processing route. Owing to the technology used and the peculiarities of this metallurgy, the emissions are under good control. Correct separation of the different phases is also a sophisticated operation which constitutes an important part of the development of new processes in complex metallurgy. Integration of new smelting techniques in our traditional lead-copper smelter creates new opportunities and remains as an objective requiring, however, considerable development effort.

In the foreseeable future, the sintering blast furnace circuit of Hoboken will remain an essential tool for treating complex concentrates and by-products. In order to cope with possible drawbacks of the process, MHO has already substantially invested in plant automation, data-logging, pollution abatement and integration of a new electric smelting process. Research and investment are in progress in order to meet any further demands placed on the operations.

Introduction

MHO has evolved from the integration of lead, copper and zinc operations located respectively at Hoboken, Olen and Overpelt. The synergy between the three plants and the underlying flowsheets have already been described (1).

The Hoboken smelter is devoted to a complex Pb-Cu metallurgy whereby lead and copper act as collectors for several valuable elements : precious metals (Ag, Au, Pt, Pd, Rh), antimony, bismuth, selenium, tellurium, tin and indium. In addition some of the zinc is collected in fume and sulphur is recovered as sulphuric acid. The processing methods are based on a combination of traditional processes used in lead and copper extraction and consist of the sequence sintering-blast furnace smelting - slag cleaning - converting.

Most primary lead is still produced worldwide through the sintering-blast furnace process. Reduced operating costs and stricter emission standards are the main incentives for the recent introduction of autogeneous smelting techniques in this industry. The following are already in commercial operation or at erection stage : QSL, Lead Kaldo, Kivcet, Isasmelt. By comparison the traditional processing sequence may suffer from various disadvantages.

On the sinter machine, high grade concentrates need to be downgraded to produce an acceptable sinter : this desulphurizing and agglomerating process mainly performed on updraught sinter strands, produces relatively weak SO2 gases with concentrations ranging from 2 to 6.5 %.

When the sinter is smelted, some of the residual sulphur is generally emitted as SO2 and released with the off-gases from the lead blast furnace.

The new smelting processes avoid the sintering step and claim to simplify charge preparation. Furthermore they use the heat of oxidation for smelting the charge and if fuel is used, it may be of cheaper quality.

Against this background some aspects of that development of complex smelting at the Hoboken smelter will be outlined.

Development of an alternative metallurgy at Hoboken

As a custom smelter, Hoboken treats a wide variety of materials : complex concentrates, drosses, Pb-Cu mattes, flue dusts, lead and copper bearing slags, slimes, cements, sweeps, electronic scrap, catalysts, speiss, etc. These materials vary widely in physical aspect, as well as in chemical composition and their heterogeneity requires the solution and control of metallurgical and physical problems associated with materials handling, charge preparation and separation of the different phases produced at the blast furnace (slag, matte, speiss, bullion and flue dust). To cope with these problems, the Hoboken smelter has continually expanded the capabilities of its processing circuits. This specialisation in complex feed products implies that changes in the direction of new processes have to take place progressively and require substantial development efforts to enable an equivalent complex metallurgy with these materials. This nevertheless implies that the sintering - blast furnace circuit remains the essential tool at Hoboken. MHO continues to invest in this circuit particularly in plant automation, data-logging, pollution abatement and recently in the addition of an electric slag cleaning furnace.

Besides these efforts a new process is being developed, according to which complex materials are smelted in two steps (2). The first step is characterized by a high oxygen potential and it is followed by a reduction of the slag in the second step. The first step is currently being investigated at Hoboken in a 7 MVA pilot electric furnace. Integration of this furnace in the smelter is expected to create advantages. One of these is a better separation of lead and copper which in a blast furnace requires repeated smelting stages to upgrade the copper matte. At higher pO2 and pSO2, lead and copper can more easily be separated. It is then possible to produce a rich copper matte, a relatively clean bullion and a practically copper free Pb-rich slag. This matte is treated in the Hoboken converters and the Pb slag is reduced in the blast furnaces.

First experiences with this new smelting process have shown that one of the major technical problems was the separation of the phases produced. It proved impossible to tap the four phases together and to separate them continuously in a forehearth, in contrast with current practice at the blast furnace. This will require substantial development efforts and it may be supposed that similar problems would be encountered with any of the commercially developed new lead smelting processes if applied to complex operation.

Present state of the sintering-blast furnace process

Particularly in the last decade Hoboken has continued to adapt and modernise its base smelting operations (3). A series of investments were made in order to meet stricter emission standards and to lower operating costs, increase productivity, improve working conditions and diversify feed materials.

In the sinter plant nearly all the equipment has been modernised except for the Dwight-Lloyd machine itself. The downdraught operation provides some advantages in pollution control outweighing for the particular conditions at Hoboken the higher productivity of an updraught operation. The sinter machine is regarded as a charge conditioner, where fine material will be dried, mixed, desulphurized, roasted and agglomerated to a quality product for blast furnace smelting. It provides approximately 40 % of the total blast furnace charge. Improvements in operation have required investments in charge preparation, sinter crushing, dry storage and proportioning together with an extended data logging and computer aided optimisation of the sinter process. Environmentally, the plant and its ventilation systems are able to keep the Pb concentration at a safe level in the working areas. Furthermore the processed sulphur is recovered and concentrated in a single gas stream with an average concentration of 5.5 % SO2, which is sent to the acid plant.

Through the years the operation of the blast furnace has been improved by the successive introduction of oxygen enrichment to 25 % ('69), blast preheat to 400°C ('72), improved continuous separation of slag, matte and bullion ('75), syphon tapping ('81) and increase of blowing rate ('85). All these contributed to improve working conditions and increase productivity. The smelting rate increased from 520 tonne of charge per day in the late sixties up to 750 tonne/day in 1986. Further investments were the installation of a new baghouse for ventilation gases ('83) and the start up of an electric slag cleaning furnace ('85).

In order further to increase productivity, to lower the overall emissions of the plant and to reduce labour and energy costs, the decision was taken to rationalize the smelting operations by production in two blast furnaces instead of three ('86). This implied enlarging furnace No 4 and required a new proportioning and feeding system, with a high degree of automation and computer control. An extensive data logging system was introduced continuously to monitor the most sensitive process parameters. Enlarging the furnace itself took place by increasing its length from 6 to 7.4 m with corresponding increase of the number of tuyeres and blast volume. A comparison of the main dimensions and data of the old and new is given in table 1.

Table 1 : Characteristics of blast furnace No 4

	Old	New
Number of tuyeres	38	48
Tuyere diameter (mm)	80	80
Width at tuyere level (cm)	131	131
Length at tuyere level (m)	6	7.4
Charge height (m)	4.3	4.3
Blast volume (Nm3/h)	12000	17000
Blast pressure (bar)	1.2	1.2
O2 in air blast (%)	25	25
Temperature of air blast (°C)	400	400
Smelting capacity (tonne/day)	750	1000

The layout of the new charging system is shown in fig. 1. It consists mainly of :
- a proportioning plant comprising 18 bins
- a main skip
- a surge bin at the top of the furnace
- an automatic charging car.

In the proportioning plant the bins are charged via pay-loaders and a skip elevator. Sinter is transported directly from the sinter plant to its respective bins by an enclosed belt conveyor, eliminating open handling and stockpiling and avoiding dust emissions. The main skip is hoisted to the top of the furnace and unloaded in a surge bin. From there a charging car travels automatically to the different charging positions along the charging hopper above the furnace. The system is able to proportion up to 17 different products with a weighing accuracy of 1 %.

The data logging system records the heat removed by each water jacket, the blast volume per tuyere, the pressure, temperature and volume of the air blast, temperature and pressure of process gases, etc. It enabled improvement of productivity by optimising the charging sequence of the furnace and ensuring a uniform blast distribution between tuyeres.

Compliance with SO2 emission standards

An important concern with the rebuilding of the blast furnace was whether it would be possible to meet future standards of SO2 emission. A campaign was undertaken to measure and evaluate existing emissions. Under normal operating conditions the SO2 content of the exhaust gases of our blast furnaces ranged between 20 and 50 ppm. This was measured after dilution at the blast furnace charge bay. Since the secondary air ingress at the charge bay is about equal to the blast furnace gases, the SO2 content of the undiluted process gases ranges between 40 and 100 ppm. These low values may be attributed to the sinter quality and the copper content of the blast furnace charge.

At Hoboken the charge of the sinter plant and the operating conditions of the Dwights are continuously controlled to obtain a high sinter quality. The method used is to control the mineralogy in order to obtain a refractory sinter of good reductibility. The sinter consists of melilite, lead-silicate, ferrite, lead-calcium-silicate, complex arsenic and antimony phases, metallic copper and as sulphur bearing components Cu_2S, $CaSO_4$ and a sulphate containing arsenate. Owing to the fact that PbS is not present this sinter does not give rise to roast-reduction reactions on heating. The sulphur emission is linked with the equilibrium at high temperature in the tuyere zone.

By the construction of a sulphur-oxygen potential diagram for Pb-Cu smelting the most important aspects of SO_2 evolution can be considered, as shown in fig. 2. The diagram is constructed for a temperature of $1100°C$. To locate the smelting processes on the oxygen potential scale, a previously established experimental correlation (4) has been used, relating the oxygen potential to the Pb-content of the slag produced. Lower lead contents, from 1 to 3 %, correspond to the blast furnace operation, whereas values above 10 % Pb correspond to the converting of our Pb-Cu mattes and to the sintering operation. The lead sulfide activity data were taken from Eric and Timucin (5) and the thermodynamic data from Barin, Knacke and Kubaschewski (6) (7).

For Pb rich smelting operations, the diagram gives equilibrium lines corresponding to Pb-Cu mattes with different Cu concentrations in equilibrium with slag at a PbO-activity of 0.1 : this is also approximately the activity of PbO in the Pb-rich slag produced in the pilot electric furnace. These lines represent the equilibrium between PbO and PbS according to the reaction

$$PbS + \tfrac{1}{2} O_2 = PbO + \tfrac{1}{2} S_2$$

In the same area, the pSO_2 line of 0.1 atm corresponds to the converting of Pb-Cu mattes and do also represent equilibrium in the final sintering conditions. For the Hoboken sinter, containing CuS_2 and metallic copper, these conditions are represented by the shaded area A. On the other hand, sinter which still contains PbS and metallic lead is situated in area B, thus at a lower pO_2. It can indeed be confirmed by microprobe analysis that sinter which contains metallic lead and PbS has appreciable Fe^{2+} concentrations, in contrast with normal Hoboken sinter where all the iron appears in the ferric state.

The blast furnace operating area, in the lower part of fig. 2, is drawn on the one hand from the previously mentioned correlation between oxygen potential and Pb-content of slag and on the other hand from sulphur potentials calculated as a function of possible matte compositions. The latter were calculated on the basis of a Pb-activity of 0.85, PbS activities in the $PbS-FeS-Cu_2S$ system as determined by Eric and Timucin and the equilibrium reaction :

$$PbS = Pb + \tfrac{1}{2} S_2$$

The resulting lines show that the pSO_2 of the smelting gases at tuyere level is a function of the degree of reduction and of the Cu content of the matte produced. Complex smelting where a strong reduction is required for a good matte and speiss separation and where relatively Cu rich mattes are produced (25-45 % Cu) will thus result in low SO_2 generation in the tuyere zone. As an illustration, an operation with a low grade matte and a 3 % Pb slag (point C) gives an equilibrium pSO_2 of 0.005 atm, compared with a more than tenfold lower pSO_2 for an operation with a 1.5 % Pb slag and a matte grade of 35 % Cu (point D). Measured SO_2 concentrations in the exhaust gases of our blast furnaces can thus be explained in terms of equilibrium conditions at the tuyere level, together with the absence of roast-reduction reactions in the shaft.

Some perspectives and future trends

For the sintering blast furnace process further developments will remain concentrated on the continual adaptation to varying feed materials and on further control and reduction of emissions in anticipation of even more stringent environmental regulations. Efforts will particularly be devoted to the following developments in the coming years.
- Optimisation of the treatment of intermediate products such as the Pb-rich slag from our first-stage electric furnace.
- Development of an improved exhaust gas cleaning system on the blast furnace for further emission abatement.
- Further automation and rationalisation of phase separations at the blast furnaces.

In parallel, the investigation and development of the new electric smelting process will continue.

Acknowledgement

The authors gratefully acknowledge the General Management of Metallurgie Hoboken-Overpelt for permission to present this paper.

References

1. R. Maes, F. Lauwers, L. Groothaert, "Processing of Complex Concentrates and By-Products at Metallurgie Hoboken-Overpelt", Proceedings of the Symposium on Complex Sulfides, The Metallurgical Society, November 1985, pp. 255-265.

2. L. Fontainas, R. Maes, "A two-step Process for Smelting Complex Pb-Cu-Zn Materials", Proceedings of the Symposium Lead-Zinc-Tin 80, The Metallurgical Society, February 1980, pp. 375-393.

3. A. Franckaerts, "Optimisation of the Lead-Sinterplant and Blast-Furnace Operations at Metallurgie Hoboken-Overpelt", Paper presented at the TMS-AIME, New Orleans, March 1986.

4. L. Fontainas, D. Verhulst, P. Bruwier, "Oxygen Potential Measurements in Lead Smelter Slags by means of disposable Solid-Electrolyte Probes", Canadian Metallurgical Quaterly, Vol 24, No 1, 1985, pp. 47-52.

5. H. Eric, M. Timucin, "Activities in Cu2S-FeS-PbS Melts at 1200°C", Metallurgical Transactions, Volume 12B, September 1981, pp. 493-500.

6. I. Barin, O. Knacke, "Thermochemical Properties of inorganic Substances", Springer Verlag, 1973.

7. I. Barin, O. Knacke, O. Kubaschewski, "Thermochemical Properties of inorganic Substances", supplement, Springer Verlag, 1977.

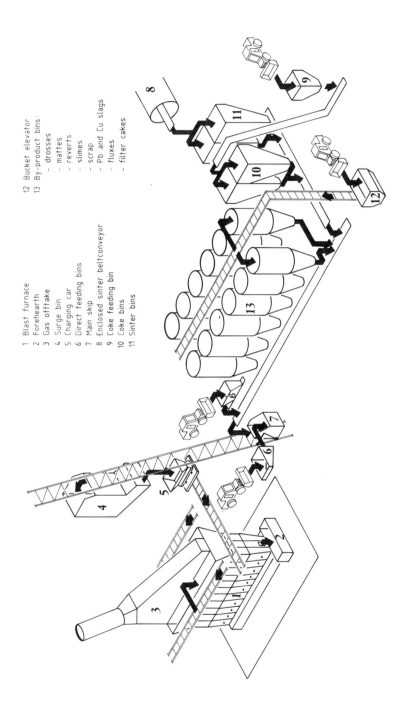

Fig. 1 : Charging system of the blast furnace

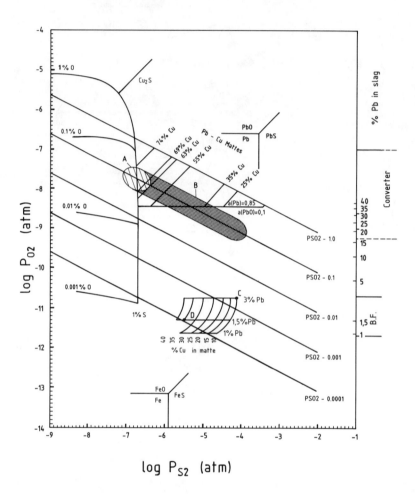

Fig. 2 : PbCu-diagram by 1100°C

HIGH GRADE MATTE OPERATION

IN THE MITSUBISHI CONTINUOUS PROCESS AT NAOSHIMA SMELTER

T. Shibasaki and K. Kanamori

Mitsubishi Metal Corporation
Naoshima Smelter and Refinery
4049-1 Naoshima-cho, Kagawa-gun,
Kagawa-ken, 761-31, Japan

Abstract

The copper content of concentrates to be smelted at Naoshima Smelter has recently become high. This tendency will be stronger in near future because of the high copper content concentrates from newly developed mines. For treatment of them there are many limitations from metallurgical and economical points of view. For example, fuel requirement at the smelting stage and excess heat generation at the converting stage increase substantially, when high grade concentrates are smelted at regular 68 % matte grade. High grade matte operation at 75 % matte grade solves these problems and besides it makes possible to increase the copper output from the existing facilities. This paper describes the concept and the results of feasibility study and test at the commercial plant of the high grade matte operation for high grade concentrates.

Introduction

Naoshima Smelter is one of custom smelters in Japan and its production capacity is 230,000 mtpy of anode. It has two smelting plants shown in Fig. 1. One is the continuous copper smelter which produces 90,000 mtpy of anode and the other is the conventional copper smelter which consists of a fluosolid roaster, a reberveratory furnace and three converters. Average copper content of concentrates treated is now 30 % Cu at Naoshima Smelter and it is predicted to become 33 % Cu in a few years because of increase of high grade copper concentrates. Even at present there are cases where copper content of the stock of concentrates becomes higher depending on the freight schedule. This situation makes our operations difficult from the technological and economical points of view. That is,

(1) When copper content of concentrates treated at the conventional reverberatory becomes high, matte grade goes up and consequently slag loss becomes high.
(2) If copper content of concentrates treated at the conventional line is kept as normal, one at the continuous line becomes higher relatively.
(3) when high grade concentrate is treated at the continuous line, fuel consumption at the smelting furnace increases and lance blowing rate also increases at the converting furnace.

To overcome such situation caused by treatment of high grade concentrates at the continuous line, operating flexibility was studied and test operation was carried out in the actual plant. The paper discusses the concept of the operation and the results of the test.

Features of the Mitsubishi Process

The Mitsubishi Process is developed by Mitsubishi Metal Corporation to realize higher productivity and lower operating cost of copper smelting (1). Conventional smelting process is batch process especially at converter operation which requires much cost for gas treatment and emission reduction. On the other hand the Mitsubishi Process is continuous process in which there are three furnaces, the smelting furnace, the slag cleaning furnace and the converting furnace, and these three furnaces are connected with launders. Fig.2 shows the conceptual flowsheet of the Mitsubishi Process. The major technical features are summarized below.

(1) Concentrates and fluxes are injected through top blowing lances to the furnaces. In this manner rapid smelting is attained and dust generation is minimized.
(2) High grade matte (65 % Cu) is produced in the smelting furnace. Slag and matte are separated in the slag cleaning furnace and copper loss is 0.5 - 0.6 % Cu in spite of higher grade of matte.
(3) In the converting furnace, matte is converted to blister continuously, forming the $Cu_2O-CaO-Fe_3O_4$ ternary slag.

Fifteen years operation proves these basic idea. Moreover, as following improvements have been done, productivity is increased.

(1) Coal has been mixed with concentrates to compensate heat balance of the smelting furnace.
(2) Oxygen enrichment has been increased up to 42 - 45 % from 32 % at the beginning of the commercial operation.
(3) Matte grade has been increased to 68 %.
(4) Furnace campaign has been extended longer by applying jackets to the furnaces.

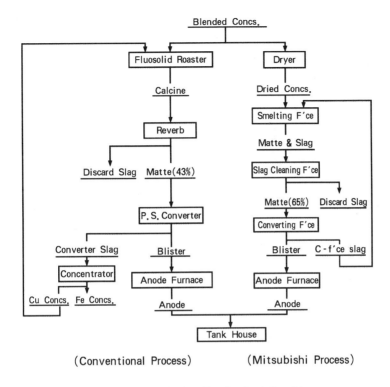

Fig. 1 Copper smelting flowsheet at Naoshima

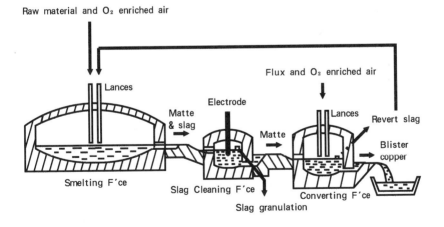

Fig. 2 Conceptual flowsheet of the Mitsubishi Process

In addition to these features, the Mitsubishi Process is considered a much more flexible smelting process than the conventional process for treating various kinds of concentrates and secondary materials, because on-line analysis of molten materials is applied, volume of molten materials in the system is small and many operation parameters are variable (2).

Change of Concentrate Grade

Naoshima Smelter which is a custom smelter has to treat many types of concentrates from abroad. But concentrates of chalcopyrite type which is 25 - 30 % Cu has usually been treated. The ratio of concentrate over 30 % Cu increases by the progress of flotation in these years. Moreover, concentrates of chalcocite and bornite types from newly developed mine, of which copper content exceeds 40 % Cu, has been increasing. Then yearly average copper content of concentrates is about 30 % at present at Naoshima. It is anticipated that it will become 33 % on 1994. Calcine roasted partially in a fluosolid roaster is fed to the reverberatory furnace of the conventional line at Naoshima. For the reverberatory furnace operation suitable matte grade is 42 - 44 % of copper. Removal ratio of sulfur in the fluosolid roaster is almost constant and if copper content of the concentrates becomes high, matte grade in the reverberatory furnace becomes high accordingly. This causes higher slag loss and build up of the reverberatory furnace bottom. If copper content at Naoshima Smelter becomes higher and copper content of concentrates treated at the conventional line is kept as same level as at present because of above reasons, one for the continuous line is estimated as below.

	1987	1994
av. cu % of conc.	30 %	33 %
for conventional line	29 %	29 %
for continuous line	31 %	37 %

Feasibility Study of High Grade Matte Operation

Heat Balance

High grade concentrate usually requires less oxygen to produce blister copper than regular concentrate. In particular, if high grade concentrate is smelted at regular matte grade, exothermic reaction heat decreases, and consequently much fuel is necessary at the smelting stage. On the other hand, in the converting stage oxygen requirement is increased and heat generation becomes bigger due to the increase of matte volume, being compared with treatment of regular concentrate at the same feed rate. These are indicated in Fig.3 and Fig.4. Fig.3 shows the fuel requirement for different matte grade at the smelting furnace. In case of 68 % matte grade operation fuel oil requirement is 800 l/h for regular concentrate (point A) and 1,300 l/h for high grade concentrate (point B). But it decreases to 700 l/h as same level as for treatment of regular concentrate by increasing matte grade to 75 % (point C). In Fig.4 lance blowing rate at the converting furnace is shown as a function of matte grade. In this case oxygen enrichment of the lance blowing is adjusted as excess heat is not generated in the heat balance. Blowing rate is increased from 12,000 Nm^3/h for regular concentrate (point A) to 18,000 Nm^3/h for high grade concentrate at 68 % matte grade (point B) but it is reduced to 12,000 Nm3/h at 75 % matte grade (point C). Since maximum capacity of the existing blowing facilities is 12,000 Nm^3/h, feed rate must be reduced to 30 tph unless matte grade is increased.

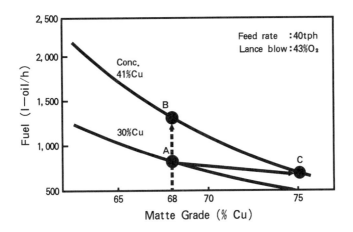

Fig. 3 Fuel requirement at the smelting furnace

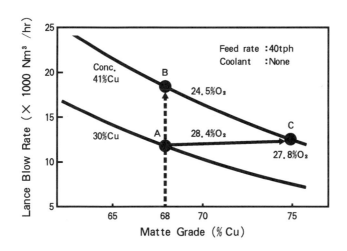

Fig. 4 Lance blow rate at the converting furnace

Magnetite Balance

Generally speaking the activity of magnetite at matte smelting stage is increased according to matte grade and it induces increase of slag loss. But slag loss increase is caused by inversion of magnetite balance too. Table I. shows magnetite balance estimated for different matte grade on treating different concentrate. In the case of 68 % matte grade operation of regular concentrate, magnetite is formed in the smelting furnace. On the other hand when high grade concentrate is treated at standard 68 % matte grade, magnetite balance is inverted. The equilibrium is expressed with following equations.

$$FeS + 3/2\ O_2 = FeO + SO_2 \qquad (1)$$

$$3Fe_3O_4 + FeS = 10FeO + SO_2 \qquad (2)$$

$$K = \frac{a_{FeO}^{10} \cdot P_{SO_2}}{a_{Fe_3O_4}^3 \cdot a_{FeS}}$$

Assuming a_{FeS} is 0.15 and a_{FeO} is 0.35, relation between $a_{Fe_3O_4}$ and P_{SO_2} is calculated and shown in Fig.5. In case that regular concentrates are treated at standard condition, generation of SO_2 is determined by the equation (1) and P_{SO_2} is considered to be in the range of 0.2 atm (point A). When magnetite is reduced in the molten bath, P_{SO_2} is considered to be 1 atm or more, because the reaction (2) proceeds towards right direction and foam of SO_2 is generated (point B). Thus, $a_{Fe_3O_4}$ increases even if matte grade is maintained at standard 68 % and the slag loss will be increased accordingly. Inversion of magnetite balance can be solved by increasing matte grade as shown in Table I. But it is inevitable to cause slag loss increase at high grade matte as 75 % Cu. However, high grade matte operation is effective for treating high grade concentrate for decrease of smelting cost and for increase of blister production in the existing plant.

Table I. Magnetite Balance at the Smelting Furnace

Concentrate	Regular, 30% Cu			High grade, 41% Cu					
Matte grade	Standard, 68% Cu			Standard, 68% Cu			High grade, 75% Cu		
	Mass kg	Fe_3O_4 %	Fe_3O_4 kg	Mass kg	Fe_3O_4 %	Fe_3O_4 kg	Mass kg	Fe_3O_4 %	Fe_3O_4 kg
<Input>									
Concentrate	100	1.0	1.0	100	1.0	1.0	100	1.0	1.0
C-Slag	9	60.0	5.4	12	60.0	7.2	4	60.0	2.4
Total			6.4			8.2			3.4
<Output>									
S-Slag	67	10.0	6.7	49	10.0	4.9	49	15.0	7.4
Matte	47	1.5	0.7	64	1.5	1.0	56	0.5	0.3
Total			7.4			5.9			7.7
Fe_3O_4 to be reduced			0			2.3			0

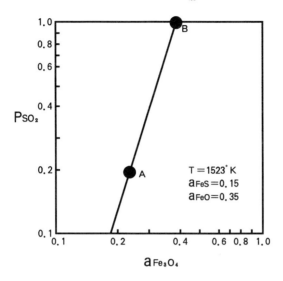

Fig.5 Relation between P_{SO_2} and $a_{Fe_3O_4}$

Test of High Grade Matte Operation

Test of high grade matte operation was carried out for one week. Operating data during test is shown in Table II together with data for regular matte operation. Feed rate of the test was decreased to 35 tph due to the capacity of the anode casting equipment. Blister production during the test was increased to more than 10,000 mtpm in spite of such limitation. Typical analysis data are shown in Table III.

The transition of slag loss and matte grade is shown in Fig.6 based on hourly on-line analysis data. Relation between matte grade and slag loss was plotted in Fig.7. Increase of slag loss was small up to 72 % matte grade, but it increased to 1.5 % at 74 - 75 % matte grade. Microscopic observation of slag obtained at 75 % matte grade showed that matte particles remaining in the slag were as few as during regular operation. Form analysis data of the slag as shown in Table IV showed that increase of slag loss was due to increase of dissolved copper.

As for the converting furnace operation no major problem was observed. Oxidation level in the converting furnace is controlled by means of analysing of copper content of the converting furnace slag as index. This control method seemed to be adequate at high grade matte operation too. Relation between copper content in the converting furnace slag and sulfur in blister copper is shown in Fig.8 together with data at regular operation. Also data of oxygen potential at the converting furnace were plotted on the previous data of regular operation in Fig.9 (3).

Table II. Test Results of High Grade Matte Operation

Operation		High grade matte operation	Regular operation
Concentrate grade,	% Cu	41	30
Matte grade,	% Cu	76	68
Smelting f'ce			
Feed rate,	mtph	34.6	40
Slag,	mtph	18.1	27
Matte,	mtph	19.1	19
Lance air,	Nm^3/h	10,300	14,500
Tonnage oxygen,	Nm^3/h	5,600	7,900
Oxygen concentration,	%	42	42
Pulverized coal,	kg/h	1,350	1,350
Converting f'ce			
Blister copper,	mtph	14.2	12.1
C-Slag,	mtph	0.8	3.5
Lance air,	Nm^3/h	7,800	10,200
Tonnage oxygen,	Nm^3/h	1,350	1,500
Oxygen concentration,	%	29.9	28.6
Monthly production (estimation)			
Concentrate feed,	mtpm	24,600	27,700
Blister copper,	mtpm	10,700	8,300

Table III. Analysis data

	Cu	Fe	S	SiO2	CaO	Al2O3
Concentrate	41.0	20.4	28.4	7.7	1.3	1.8
Matte	75.8	2.1	19.4	-	-	-
Discard slag	1.13	37.3	0.12	38.2	4.9	3.6
Blister	98.8	-	0.46	-	-	-
Converting f'ce slag	15.3	41.8	-	-	17.1	-

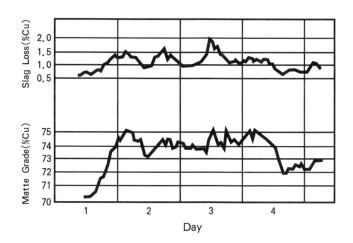

Fig.6 Transition of slag loss and matte grade during the test

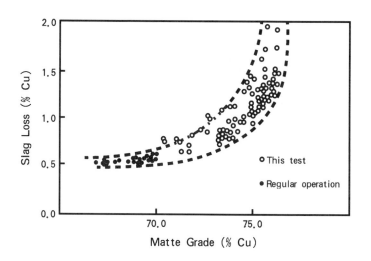

Fig.7 Relation between slag loss and matte grade

Table IV. Form Analysis of the Discard Slag

	Total Cu(%)	Chemical Form of Cu(%)		Matte grade when sample was taken
		Sulfide	Oxide	
Sample 1	0.90	0.05	0.85	75
Sample 2	1.36	0.06	1.30	76
Sample 3	0.55	0.06	0.49	68

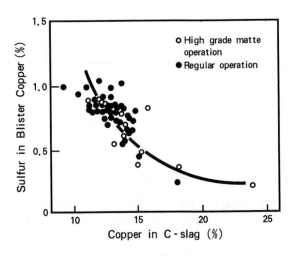

Fig. 8 Relation between Sulfur in blister and Copper in the Converting f'ce slag

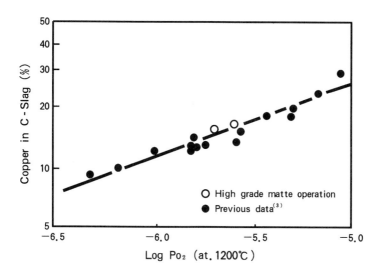

Fig. 9 Relation between copper in slag and oxygen potential at the converting furnace

Application to the Commercial Operation

High grade matte operation for high grade concentrate is effective as discussed above from the point of view of energy saving and productivity. Only increase of slag loss could be a problem. However, slag loss at 70 - 72 % matte grade is confirmed to be kept below 0.8 % Cu. Overall copper recovery was calculated at this condition and shown in Table V. It shows metal loss at 72 % semi-high grade matte operation of high grade concentrate was smaller than one at regular operation because of smaller slag fall. From this result semi-high grade matte operation is applied to the actual operation when the stock of high grade concentrate becomes large and it is required to keep the copper content of the furnace feed to the conventional line as normal.

Flotation test was carried out for the slag cleaning furnace slag taken at 75 % matte grade operation. Copper recovery was observed little on the granulated slag. Slow cooled slag sample was collected during the test and the flotation test at the laboratory was done. As a result of the test copper content of the tailing was lowered to 0.33 % Cu from 0.96 % of feed sample.

Table V. Analysis of Slag Loss in Semi-High Grade Matte Operation

Operation		Semi-High grade matte operation	Regular operation
Concentrate,	Cu %	41	30
Matte grade,	Cu %	72	68
Slag fall,	% of conc	49	67
Slag loss,	% Cu in slag	0.8	0.6
	Kg Cu/t conc.	3.9	4.0
Recovery of copper*, %		99.9	98.7

* Only slag loss is counted.

Increase of Productivity

Original furnace capacity of Naoshima continuous line was 4,000 mtpm at the smelting furnace of 8 m in diameter and at the converting furnace of 6.5 m in diameter. In 1983 feed rate was increased from 25 tph to 40 tph by adding oxygen and acid plant capacities, then anode production also increased to 7,500 mtpm without enlargement of the smelting and converting furnaces.

Higher throughput test was carried out at 50 mtph of concentrates for ten days in 1985 at the same furnace, consequently hearth efficiency of the smelting furnace is almost 23.3 T-conc/m2 D and this figure is much higher than that of the reverberatory furnace, 4.3 T-conc/m2 D (3). At this moment anode production from the process is estimated to be 9,500 mtpm.
Anode production is estimated to be 11,000 mtpm if high grade concentrate is smelted as high grade matte operation. Considering these test results, improvement of productivity at the Mitsubishi Process is shown in Fig.10 as annual production capacity with smelting furnace hearth efficiency.

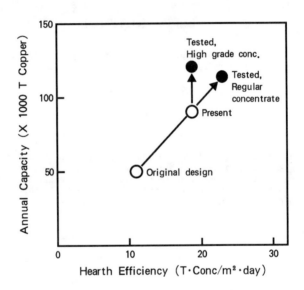

Fig.10 Improvement of productivity at the Mitsubishi Process

Summary

Possibility and advantages of high grade matte operation were studied for high grade concentrate treatment. The study showed that heat balances at the smelting and converting furnaces were improved and blister production at the existing plant was increased to more than 10,000 mtpm. Test of high grade matte operation was carried out and no major difficulty of operation was confirmed. Slag loss at 75 % matte grade exceeded 1.5 % Cu, but it is possible to recover copper from the discard slag by flotation. Semi-high grade matte operation at 70 - 72 % matte grade has been applied to the actual operation at Naoshima Smelter.

Acknowledgment

The authors wish to thank the management of Mitsubishi Metal Corporation for permission to publish this paper.

References

1. T. Nagano and T. Suzuki, "Commercial Operation of Mitsubishi Continuous Copper Smelting and Converting Process", "Extractive Metallurgy of Copper", ed. J.C. Yannopoulos and J.C. Agarwal (New York, NY : The Met. Soc. of AIME, 1976), PP.439 - 457.
2. T. Shibasaki, K. Kanamori and S. Kamio, "Mitsubishi Process - Prospects to the Future and Adaptability to Varying Conditions" (Paper presented at the 118th Annual Meeting of AIME, Las Vegas, February 1989).
3. M. Goto, S. Kawakita, N. Kikumoto and O. Iida, "High Intensity Operation at Naoshima Smelter", J. Metals, Sept.1986, pp.43 - 46.

FLASH SMELTING GETS A NEW DIMENSION:

CONTOP-CONTINUOUS SMELTING AND TOP BLOWING NOW IN INDUSTRIAL APPLICATION

G. Melcher

KHD Humboldt Wedag AG
Wierbergstrasse
5000 Köln 91, West Germany

Abstract

The application of the CONTOP Process has been successfully demonstrated in industrial scale in two copper-smelter plants.

The retro-fitting of the new technology in the existing smelting processes is described and the main design principles and the performance data are explained.

Based on the experience achieved future applications are discussed and design developments highlighted.

ASSESSMENT OF COPPER/ZINC SMELTER MODERNIZATION AT HUDSON BAY MINING

AND SMELTING (HBMS) - FLIN FLON, MANITOBA, CANADA

W.J.S. Craigen*, B. Barlin** and B. Krysa**

*CANMET, EMR, Ottawa, Ontario, Canada
**HBMS, Flin Flon, Manitoba, Canada

Abstract

A study of modernization alternatives aimed at improving productivity and reducing SO_2 emissions at the HBMS integrated smelter complex in Flin Flon, Manitoba, has been completed. Engineering design and economic feasibility studies on the most attractive alternatives (zinc pressure leach (ZPL) process for zinc and a Noranda Reactor for copper) were carried out, and mechanisms to implement the new technologies at HBMS are presently under discussion. This report compares the techno-economic viability of the proposed modernization with HBMS' present operations. The assessment indicates that the proposed smelter modernization will significantly reduce SO_2 emissions while ensuring the long term economic viability of the HBMS mine/mill/smelter operations in Manitoba.

Pages 382-398 pulled prior to publication
at the request of the authors due to legal ramifications.

Productivity and Technology
in the Metallurgical Industries
Edited by M. Koch and J.C. Taylor
The Minerals, Metals & Materials Society, 1989

RESTRUCTURING BOLIDEN'S RONNSKAR SMELTER

Bengt Lofkvist, P-O Lindgren, and Per G. Broman

Boliden Mineral AB
Smelting Division
S-93200 Skeleftehamn SWEDEN

Ronnskar has always been in a situation of expansion and reconstruction; therefore, I would like to brief you about our activities at the Ronnskar Works today and in the past.

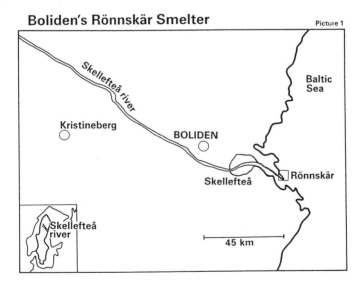

Picture 1

PICTURE 1   The Smelter is situated at the mouth of the Skellefteå River in the Northern part of the Gulf of Bothnia and was constructed in 1928 to enrich, melt and refine the gold-, copper and arsenic minerals discovered in the area covered by the Boliden Mines in the late 1920th.

PICTURE 2   The ores from the Boliden Mines contain deleterious elements such as Arsenic, Antimony, Bismuth, Selenium.

To enable recovery of the metals, Boliden has since the start been engaged in the developement of processes suited for the extraction of above mentioned metals. During the fifties and sixties the production of Lead, Arsenic and Selenium became viable, and during a twenty year period Bismuth was also produced in small quantities.

During the earlier days, the environmental impact of sulphur dioxide and metals was quite high. To protect the surrounding environment a stack, 145 meters high, was built in 1928 through which all the smelter off gases were discharged. Even before the smelter was built extensive research into methods for safe emission of gases and other waste materials was made. The eventual location of the smelter was only decided following similar research into climatic conditions. These historical efforts demonstrates Bolidens early commitment to the local environment, but today's critics perhaps consider this insignificant.

In 1955 the first sulphuric acid plant was built which transformed 25 % of the sulphur input into sulphuric acid.

Metals Production at Rönnskär Smelter

Picture 2

		1940	1950	1960	1970	1980	1984	1988
Copper	kton	10	26	38	52	56	103	94
Lead	kton		17	44	44	32	59	57
Arsenic	kton	3	8	12	16	9	7	9
Sulfur acid	kton			46	165	135	256	222
Sulphuric dioxide	kton					48	68	77
Zn-clinker	kton				31	18	29	26
As-metal	ton			16	257	1 004	719	1 041	135
Selenium	ton	68	23	80	66	5	68	
Bismuth	ton	209	757	1 231				
$NiSO_4$	ton		2	4	4	5	760	1 017
Gold	ton	7	2	4	4	5	5	7
Silver	ton	24	21	83	190	238	190	262

PICTURE 3   During the sixties the production of sulphuric acid increased and a new environmetal legislation in 1969 resulted in increased sulphur dioxide capacity during the seventies. These sulphur plants decreased the pollution drastically.

In 1964 a Zn-fuming plant was built, followed by a Lead-kaldo in 1975 and a Cu-kaldo in 1978. The Zn-plant reduced the content of Zinc in the outgoing slag, and the Lead-kaldo made it possible to treat the lead containing dusts from the copper smelter.

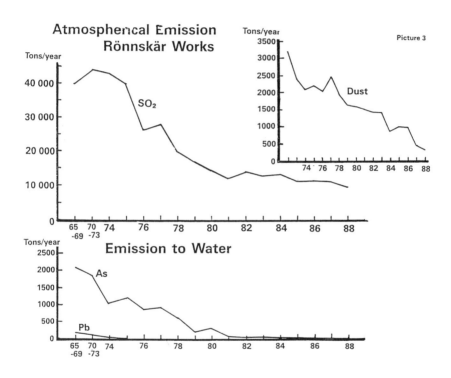

Picture 3

During the eighties nearly all remaining sources for discharge of fume have been equipped with security filters as a result of the concession by the Swedish Consession Board decision i 1979.

PICTURE 4   Here you can see what the smelter looked like in 1980. At this time the production was low, as was productivity, producing a disappointing financial result.

Flowsheet — Rönnskär works 1980

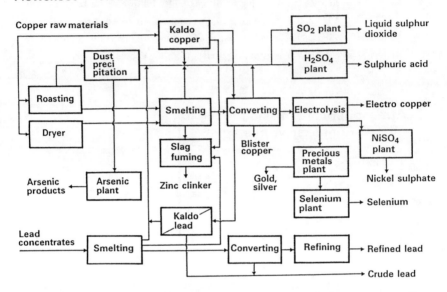

Picture 4

PICTURE 5   Efforts were made throughout the complex smelter to improve the situation. Nothing was left unquestioned, and drastic action was taken.

These measures resulted in nearly doubled production of copper in three years from 55 000 tonnes in 1980 to over 100 000 tonnes in 1983. Besides increased production the productivity improved tremendously. The higher level of production has remained unchanged until 1988.

The increased production was achieved by treating copper ashes and more electronic scrap in the Lead-kaldo and the purchase of concentrates from overseas. The mines owned by Boliden produced 60 000 tonnes of Copper annually, a level still kept today.

Due to increased productivity and a better copper-price in SEK (due to the devaluation of the Swedish Krona in 1982) the financial result became acceptable. During this particular time it was also decided to extend the copper refining capacity from 63 000 to 95 000 tonnes, in addition updating the Precious Metals facilities. These improvements were in use during 1986.

Rönnskär Copper Smelter
Parameters 1980—1984

Picture 5

		1980	1981	1982	1983	1984
Copper Production	tons/year	56 000	75 000	90 000	102 000	103 000
Productivity	ton/manhour	0,08	0,10	0,11	0,12	0,14
Consumption						
Slags and Residues	kton	25	19	29	53	51
Scrap	kton	13	12	20	30	24
Overseas Concentrates	kton	56	90	100	106	97
Boliden Concentrates	kton	177	217	232	252	220

PICTURE 6   The favourable conditions that were built up during the eighties came to a drastic halt in 1985-86, due to increased raw material and production (power, labour, environmental control) costs. As you are probably aware, many US smelters took action a few years earlier trying to restrain the rapid growth in expenses.

These unfavourable developments were not noticeable immediately at Rönnskär because of the earlier mentioned devaluation of the Swedish Krona as well as the increase in productivity.

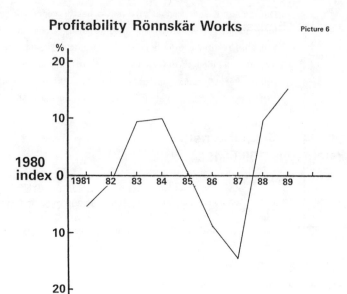

PICTURE 7   In a concession decision in 1986 conditions were drawn up for the smelter, one of the demands is to cut the sulphur dioxide discharge from 10 000 to 5 000 tonnes per year and cut the copper emissions by more than half. Another demand is to reduce the dust when handling materials. These demands must be achieved by 1991/1992.

In the early summer of 1986, Boliden was bought by the Trelleborg Group and a program was lined up to turn the immense negative results around.

The owners demand a 20 % profit on working capital for their subsidiaries.

As far as Rönnskär was concerned this seemed an impossible task when we feared to spend SEK 400 milj on environmental investments alone, according to the concession decision. A restructuring group was set up. The main task was to handle the stipulated environmental demands and to achieve an acceptable profit by using mines owned by Boliden as the main source for raw material.

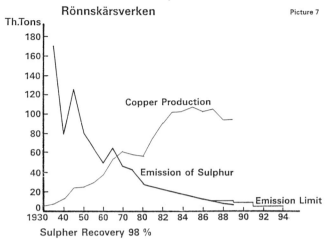

Picture 7

Sulpher Recovery 98 %

PICTURE 8   The investigation started in Januari 1987. Several people within Rönnskär were involved. In June 1987 (six months later) the result was submitted to the Board. The decision was taken to approve the new structure which meant the following changes:

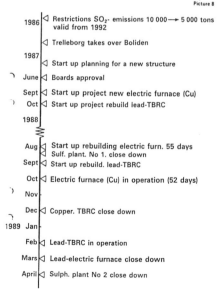

PICTURE 9   Out of the existing four smelters (Copper electric furnace, Copper kaldo, Lead electric furnace and Lead kaldo) the Copper kaldo and Lead electric furnace were to be shut down.

The 40 year old Copper smelter was replaced by a new one in the same spot. The furnace was raised one metre and equipped with a new off gas system which cuts the volume of gas by half. The new furnace is also equipped with a new efficient cooling system, which enables the production to increase by 20 %. The latter improvement together with a increased matte grade from 42 to 47 % is enough to compensate for the loss of copper production when closing the Cu-kaldo.

In the old system where the converters treated a 42 % matte, two converters had to work alongside 20 % of the time. The higher matte grade allows faster processing. Thus only one converter need to be in stack. This means that the volume of converter processgases is cut by half.

The Lead kaldo is equipped with a concentrate feeding system for melting and a new waste heat boiler. The Lead kaldo now produces raw-lead which is refined in the existing refinery. The Lead kaldo will produce the same annual production of Lead as before in 175 days. The remaining time is used for processing of electronic scrap which in melted form is transported to the copper converters.

By shutting the Lead furnace and the copper kaldo, and reducing the gas volumes by half from the copper smelter and the converters, it is possible to shut down the two oldest, most costly sulphur works.

The quantity of electro refined copper and lead produced will remain unchanged even with the new limited structure.

By shutting down two smelting units and two sulphur works and a lot less handling of material, significant expenses are cut trough less maintenance, transport and administration costs. During the past two years the workforce at Rönnskär has been reduced from 2 400 to 1 500 employed. All these factors result in an increased productivity of roughly 30 %.

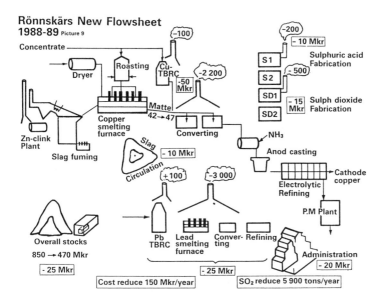

Rönnskärs New Flowsheet 1988-89 Picture 9

At the same time as the project for restructuring took shape, Rönnskär made efforts to lower stocks of material - rationalizing of capital. A new process for collecting slimes from the refinery and new routines in the Precious Metals plant have both reduced the metal stocks by a value of SEK 100 milj.

RESTRUCTURING OF THE COPPER SMELTER

PICTURE 10   In 1987 the project started by selecting a project group. Layout-work, planning and contact with suppliers took most of their time up until January 1988. More detailed planning was carried out as close as two months prior to shutdown of the copper smelter.

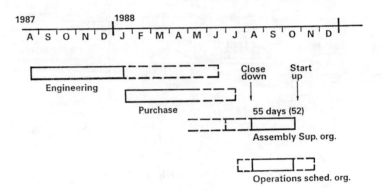

Project Schedule for Electric Furnace Rebuild Picture 10

PICTURE 11   Construction work entailed pulling the old copper smelter down and in the same spot reconstructing the prebuilt furnace on prebuilt foundations. Electrodes and charging systems were left unchanged. When the steel structure of the smelter had reached a certain level it was time for the refractories, followed by the installment of a new suspended roof. Other construction activities that had started months before were carried out even after the furnace was taken into use. A new computer system was introduced for the furnace and process control. During the installment the personel was trained on a simulator. The new smelter started up 3 days ahead of calculated 55 days. Approx 250 workers were instructed and prepared for the new smelter during the stoppage.

Schedule for Electric Furnace Rebuild Picture 11   Budget 48 Mkr
                                                        Final 50 Mkr

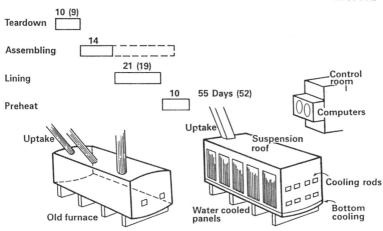

PICTURE 12   HOW WAS 8 WEEK SHUTDOWN OF THE MAIN COPPER SMELTER POSSIBLE WITHOUT LOSS OF PRODUCTION

The plan was to rebuild the copper plant without disturbing the production. A special group of people was selected from the production departement. After their training, experiments were made to test certain metallurgical capacities.

Each converter cycle should have it's rightful input of material. Each time there was a trouble or interference the presence of the planing group was highly appreciated to redirect the whole course of events.

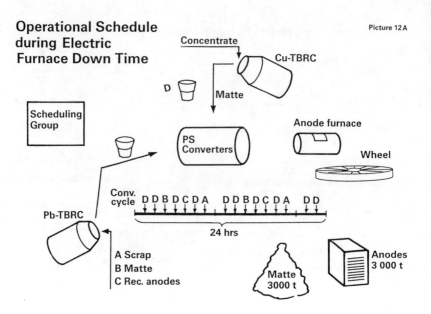

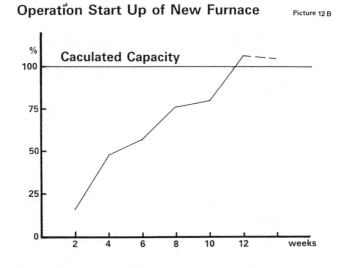

EPILOG

Rönnskär has always been subject to continuous reconstruction, either expansion or reduction. I hope that through this briefing you feel familiar with the background and history.

What will happen in the future? Most important though, is, that Rönnskär "feels" very strong and prosperous today. Competition worldwide is only a joyful challenge.

SO_2 ABATEMENT AND PRODUCTIVITY PROGRAMS AT

INCO'S COPPER CLIFF SMELTER

J. Blanco, H. Davies, P. Garritsen

INCO LIMITED
Ontario Division
Copper Cliff, Ontario
CANADA POM 1NO

ABSTRACT

Alternate methods of lowering SO_2 emissions while enhancing productivity have been evaluated. A program has been developed to retrofit the Copper Cliff Smelter to meet legislated SO_2 emission limit. The program entails process and equipment changes.

MINIMIZATION OF HEAT LOSSES

FROM FUEL FIRED HEAT TREATMENT EQUIPMENT

Dr. Hans-Otto Jochem

President

GAUTSCHI ELECTRO-FOURS LTD.
CH - 8274 Tägerwilen / Switzerland

The function of industrial heat treatment equipment consists in heating a product up to a required temperature, subject to a preset time cycle and certain temperature tolerances. The heating process may have the purpose to prepare the product for the hot-working process, such as rolling, pressing or forging or is used to improve the metallurgic characteristics of the product, e.g. by annealing or homogenizing. In any case, the temperature of the product has to be increased, controlled, and maintained during a specified time and, if necessary, decreased again.

In the following a representation of the minimization of the heat losses in this connection from fuel heated furnaces shall be established. Hereto the complete heat process will be divided in three sections:

a) Generation of a heat flow for the preparation of the required heat
b) Transfer of the caloric capacity of the heat flow onto the product
c) Cooling of the product to a temperature required in the further process

The solution process of these three problems will cause different types of heat losses, which will also require different measures for the minimization. These precautions will be described in the following

1. Generation of the heat flow

The generation of the heat flow will develop losses by poor combustion and flue gas.

1.1 Optimization of combustion

The energy contained in solid, liquid and gaseous fuel will be released by oxidation of the fuel. This release is optimal when as much oxidation agent, in the form of oxygen or oxygen mixtures, as theoretically necessary for a complete combustion is added to the fuel. This is called stoichiometric combustion. The theoretical combustion has to be achieved in practice as perfectly as possible. To ensure that there are no combustible particles in the flue gas and to avoid sooting, heating furnaces are operated with a small amount of excess air. Since the excess air also has to be heated, it results in significant heat losses (as shown in figure 1). The diagram represents the furnace efficiency for natural gas depending on excess air at different flue gas temperatures. At a waste gas temperature such as 1560 °F and an excess air of 25 % the furnace efficiency will decrease from 60 % to 40 %.

The adjustment of an optimal, theoretical combustion can be easily achieved by flow measurement and proportioning valves in the supply lines for air and fuel for each burner of the furnace. For better control of this adjustment - at full load as well as partial load range - a regulation of the adjustment by means of an electronic fuel/air ratio control or a continuous measurement of the oxygen content of the flue gas (solid electrolyte probe)

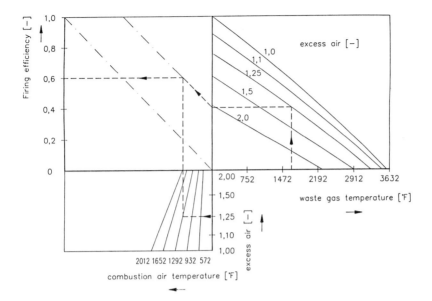

Fig. 1   Firing efficiency

will be preferable. Application of microprocessors will achieve a high degree of accuracy, which is important at fast variations of pressure, temperature and load where temporarily no excess or air shortage can be found.

Processors lead to energy savings by optimal combustion control, especially when the furnace will be operated under non-steady condition. This can be found when the product has different dimensions or weights. By means of a mathematic model the processor determines the optimal operating conditions and controls the furnace installation accordingly.

Basic requirement for a good theoretical combustion is, of course, a good burner design. The burner has to be designed in a way that in the flame area with the related ignition temperature and within the range of the ignition of fuel/air mixture each fuel particle gets in contact with the necessary oxygen. If this is not the case, unused fuel and oxygen will remain in the flue gas.

1.2 Heat recovery from flue gases

The generation of heat in fuel heated installations inevitably produce flue gas. This is the primary carrier of the heat flow which transfers the required heat to the product. The heat capacity for this purpose can only be partly used, because its temperature after heat transfer to the product or to a secondary heat flow in any case has to be above the end temperature of the product as otherwise no heat transfer from the flue gas to the product can be achieved.

The energy consumption of furnaces can be dramatically reduced by recycling the remaining heat in the flue gas back to the heat process.

For this purpose three methods are in use:

- Preheating of the combustion air and/or the fuel
- Preheating of the product
- Recirculation, i.e. partial feedback of the flue gas into the furnace chamber either directly by mixing with the combustion gas

1.2.1 Preheating of the combustion air and/or the fuel

It has to be said in advance that the preheating of fuel can take place within small range, because the hydrocarbons of the fuel will crack pyrolytically at elevated temperature. Above all the preheating of the combustion air should be used. Hereby the furnace efficiency could be increased considerably. The influence of the combustion air temperature to the furnace efficiency is shown in figure 1. Figure 2 illustrates the energy savings in relation to the preheating of the combustion air. The preheating is achieved by recuperators or regenerators. Recuperators are heat exchangers where the heat exchange takes place through a partition wall between the radiating flue gas and the heat absorbing air. At the regenerators the flue gases deliver their heat to the heat accumulators in cycles. If the heat accumulator is heated by the flue gas, a switchover takes place and then the combustion air is guided through the generator and releases the accumulated heat to the combustion air. The heat exchange can also be effected by design directly at the burner, i.e. by recuperator or regenerator.

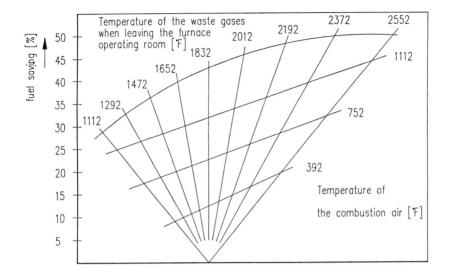

Fig. 2 Fuel saving by preheating of combustion air

This kind of burners is mainly applicable for modification of existing installations without combustion air preheating as modification can be achieved by simply changing the burners with the resulting benefit of important energy savings at low costs.

At indirect fuel heated furnaces operating with protection gas, radiant tube burners with integrated recuperator will be used. New installations will be preferably equipped with a central recuperator for the complete installation or for parts of the installation. In this case, the air preheating can be better controlled and operated more efficiently.

Recuperators are designed as both convection and radiation recuperators. At the heat absorbing side (combustion air) the heat transfer is always by convection, at the radiant side (flue gas) by radiation and convection.

While at low temperature of the flue gas the convection part predominates, the radiation part is higher at higher temperatures.

When flue gas temperature is below 1830 °F convection recuperators are normally employed; however, above 1830 °F radiation recuperators are used.

The heat transfer surfaces of a convection recuperator are made of heat resistant steel tubes (figure 3). Depending on the magnitude of the wall temperature, steel alloys of different quality are used. With radiant tubes as designing element for heat exchange surfaces, recuperators of this type can easily be adapted to given requirements. A convection type recuperator can be installed not only horizontally and vertically into the main flue but also at an incline. The temperature limits for the application are given by the material quality. If no corrosive components are contained in the flue gas, the limits for application of metallic recuperators are 1560 °F for the wall temperature, flue gas temperature 2100 °F and air preheat temperature 1200 °F.

Ceramic recuperators allow an air preheat temperature up to 1830 °F.

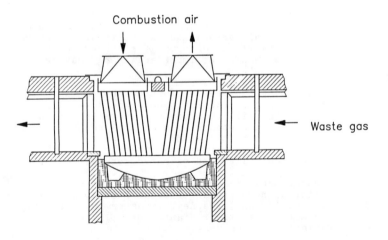

Fig. 3 Convection recuperator

Radiation type recuperators are built out of even cylindric inner and outer casings (figure 4). The radiant heat of the flue gas is transferred through the inner casing onto the combustion air in the gap between the inner and the outer casing. The application of radiation type recuperators is not economical if the furnace is operated and charged in batches. Since at times only slight air preselecting temperatures can be achieved, these recuperators are mostly used in connection with continuous heat-treating furnaces. In this case the radiation type recuperator can be used as part of the chimney whereby an exhaust blower is not necessary. The application of radiation type recuperators is limited by the temperature, at which the heat transfer by radiation is still effective.

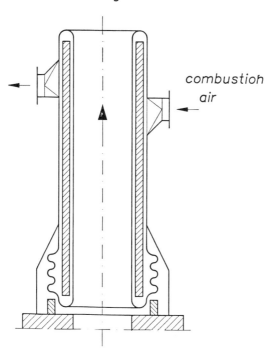

Fig. 4 Radiation recuperator

Beside the radiation and convection recuperators, mixed type recuperators have been found appropriate, as they are suitable for all industrial furnaces (figure 5). In this case the air to be preheated flows first from entrance manifold through a double casing into an upper collector and continues through rows of bended radiant tubes to the outlet collector ring. To improve the heat transfer in the tubes, these are equipped with twist baffles. The maximum flue gas temperature could reach up to 2460 °F, and the max. air preheating temperature up to 1200 °F. The lower useful limit for application is at 1290 °F of the flue gas. Compared to real radiation type recuperators, this compact style has a smaller overall height.

Apart from materials of the recuperators, there are still other limits for the air preheating. At temperatures below 355 °F to 395 °F, SO_2 and SO_3 sulphuric acids are produced which would destroy the flue gas systems of conventional style very fast. Lately, there are attempts to decrease flue gas temperatures further down to ambient temperature. This is possible by installing socalled drain boilers and heat pumps made of non-corrosive material. These installations are not yet working economically.

The upper limit for the air preheating is given by the related higher combustion temperature which in turn results in higher NO_x-formation. The legal regulations have to be met and, if necessary, installations for the removal of NO_x from the flue gas have to be provided.

It has to be determined for each case of application by calculations for optimization how much of air preheating and what kind of recuperator or regenerator achieve the highest economic advantage, as investment and installation costs are related to many different factors. For the profitability also the fuel costs are very important. For a payoff period of the recuperator we estimate approx. 1-3 years.

If the available energy quantity is limited, the payoff is probably less important than the desired production increase at an unchanged fuel quantity. A fuel saving of 20-25 % by installing recuperators can be easily achieved in most of the cases at economically justifiable costs. In case of a possible restriction of fuel supply, should be the

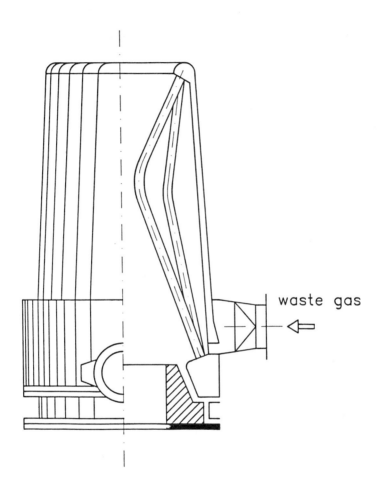

Fig. 5 Combined radiation and convection recuperator

investigation to install recuperators the first step in order to maintain or increase the production capacity.

As a rule of thumb, it can be assumed that each 210 °F of air preheating will result in approx. 5 % of fuel saving. Besides the fuel saving, the preheating of the combustion air will also increase the combustion temperature; each 210 °F of air preheating will achieve approx. 105 °F increase. As the intensity of heat transfer of the combustion gases to the product and the furnace walls will increase approx. in proportion to the fourth power of the abolute temperature, the capacity of the furnace will also improve. Furthermore, it is possible by this means to use fuel with low combustion temperature in processes which would require higher temperatures.

1.2.2 Preheating of the product

The preheating of the product by combustion gases leaving the furnace chamber is a good possibility to use the heating gas down to a temperature of 355 to 395 °F. The product is transferred through an unheated preheating zone through which the heating gases are guided after having left the heating chamber. If the heating gas exit temperature after heat transfer to the product can be successfully decreased to the same value as for the preheating of the combustion air, the energy saving will be the same. However, the preheating has the advantage that the combustion temperature will not be increased and the NO_x emission is consequently lower than at the preheating of the combustion air.

Figure 6 shows the trend of the product temperature and the fuel for continuous treating furnace using 50 % of the total furnace length for preheating in an unheated preheating zone.

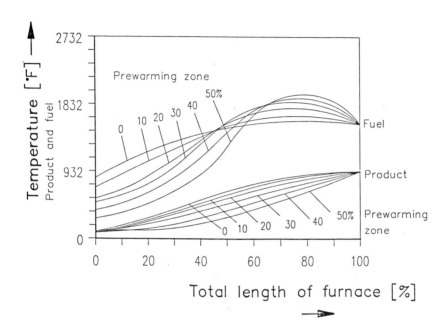

Fig. 6 Temperatures product and fuel at different portions of the prewarming zone of the total furnce length

Figure 7 demonstrates the increase of the fuel efficiency at extension of the preheating zone. The length of the preheating zone is limited by the temperature permanence characteristic of the refractory material in the heated furnace part. Furthermore, an excellent distribution of the heat flow in the preheating zone is absolutely necessary to avoid local melting of the product.

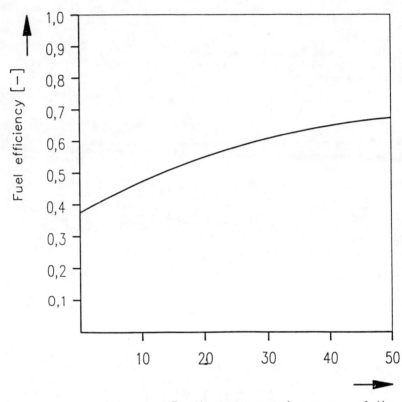

Fig. 7 Fuel efficiency when prewarming the product

1.2.3 <u>Recirculation</u>

During recirculation a partial flow of the flue gas will return directly to the furnace chamber or will be added to the heating gas and returned. A partial recuperation of the flue gas heat will be achieved by recycling of the flue gas, which requires a fan. The recycling will provide a constant temperature distribution in the heating chamber and a higher combustion speed in the furnace chamber. Consequently, heat transfer to the product will be improved with a better utilization of heat content of the heating gas. The feed back of the gases can also be achieved by an injector, which sucks the flue gases directly into the burners or into the furnace chamber, too.

2. Transfer of the heat content of the heat flow to the product

It is very important that the transfer of heat to the product is achieved within the shortest possible time. The shorter the time, the smaller is the furnace chamber in continuous heating furnaces with its heat losses and the shorter is the time during which heat losses can occur in furnaces with batch operated furnaces.

The heating operation is composed of two parts: the real period of heating and the period of temperature equilization during which the product is maintained at the achieved temperature to be equalized in the whole product within a permissible tolerance.

The heat transfer with aluminium is mainly achieved by convection as a result of the low temperature. The higher the speed of the heating flow to the product, the better the heat transfer. The high speed in the furnace chamber at direct heated furnaces is achieved by impulse burners. In addition, the speed can be increased by a recirculation fan (figure 8). The recirculation in furnaces indirectly heated by radiant tubes is achieved by recirculation fans only.

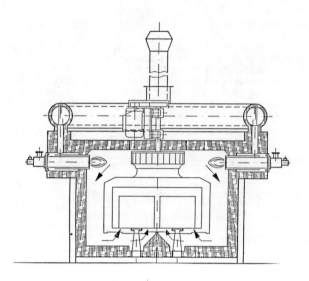

Fig. 8 Direct fired heat treatment furnace with radial fan

An important design consideration is the fact that the flow is guided through adequate channels with baffle plates to avoid pressure losses and to ensure that the flow really gets in contact with the product and that the flow does not break away and causes heat shadows. With good result the Coanda effect is used for this purpose. A faster heating can also be achieved by periodical change of flow direction of the heat flow by reversible operation, demonstrated for a furnace on figure 9. The heating in this case is more uniform.

The recirculation described in above chapter can also be used for the increase of the flow speed and therefore has a double efficiency.

It is important that during the heating stages and the holding time, the furnace losses are kept as low as possible. These are in essence:

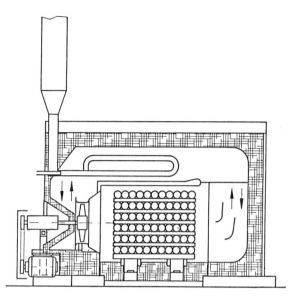

Fig. 9 Indirect fired heat treatment furnace with axial fan for reversing operation

2.1 Wall losses

The decrease of wall losses by better insulation materials has been a highly pertinent subject for a long time. Remarkable results have been achieved in this field. A better insulation could be achieved with decreasing wall thickness by using product made of ceramic fibre. In this connection the losses caused by accumulated heat can be kept lower.

However, heat losses caused by so called heat bridges, i.e. the uncontrolled heat flow of uninsulated furnace parts have been ignored. This especially applies to the furnace bottom, where even today structural sections of heavy wall thickness on high temperature are directly placed onto the foundation and transfer remarkable heat uncontrolled to earth where they disappear. For this case, different solutions have been investigated such as installing the complete furnace on steel balls in order to keep the heat transfer as low as possible. For the loss through the furnace bottom it is also essential that the soil under the furnace is dry. Through a wet soil four to five times more heat will get lost than through a dry soil.

2.2   Losses through openings

Heavy heat losses are created also by not tight furnace casings, this especially happens during opening and closing of the inlet and outlet doors. Labyrinth seals made of cloth or rope curtains can achieve considerable savings. Inspection holes and ignition holes have to be as small as possible and must be lockable.

To reduce losses caused by inherent design the furnace pressure control is very useful. The lower the furnace pressure is below the ambient pressure, the less
the heating gas can escape to the outside. However, furnace low pressure is as bad as excess pressure, as secondary air from outside the furnace will be introduced under suction into the furnace chamber and will cool down the heat flow and make the heat flow less efficient. Modern furnaces should not be designed without furnace pressure control.

2.3   Losses by coolants

In many cases particular components of industrial furnaces have to be cooled by air or liquid, to avoid overheating of construction elements. The heat of the drained coolant cannot be reused because the temperature level is too low. Modern furnaces are therefore operated without coolants as far as possible.

3. Cooling of the product to a specific temperature

During cooling of the product, valuable heat losses are created by the heat content of the product itself and by the cooling down of the product carrier.

3.1 Utilization of the heat content of the product

If the product in an annealing furnace is heated to a high temperature for subsequent milling, pressing, forging or hot working process, the heat content of the product will not be lost, but will be used in the subsequent process. However, if the product is cooled by air or water, the heat transferred to the coolant can be reused, e.g. to preheat the combustion air or the cold product. It has to be calculated from case to case, if such an utilization is economically useful. The efficiency is not so easy to achieve, as it is in case of the utilization of the flue gases, as the temperature level available is essentially lower.

3.2 Reduction of losses by product carriers

The mechanical product carrier such as furnace cars, glide shoes in pusher type furnaces, pallets etc. cool down after leaving the furnace and have to be heated again after charging.

The heat losses have been determined to be generally one third to half of outside wall losses on installed furnaces and should not be underestimated.

As a utilization by heat recovery cannot always be achieved or only at unproportionally high costs, these product carriers have to be designed in light construction and possibly the cooling has to be kept as low as possible by suitable arrangement as storage in insulated chambers.

4. **Thermal evaluation of heat treatment installations**

4.1 **Definition of efficiency**

For the evaluation of the quality of technical installations the efficiency is used, which is defined as the ratio of output to input. If different installations have to be evaluated according to the efficiency, all boundary conditions have to be equal. The different definition of boundary conditions has led to a large number of definitions for efficiency of heat treatment installations. It is correct that on fuel heated installations the quality of the heating and the furnace design have to be evaluated separately.

The thermal efficiency of an industrial furnace is basically the proportion of the available heat energy flow, which is brought into the heat treated product to the total heat energy transferred i.e. η_w = useful heat energy flow / total heat energy flow. The difference between these two energy flows consists of a multitude of single losses, which were discussed earlier. Beyond that, in fuel heated furnaces the energy of the fuel can only be transformed into heat by the furnace efficiency. The hot flue gases carry off part of the energy contained in the fuel.

The firing efficiency η_F is the proportion of the available heat to the furnace from the fuel to the utilized heat. The firing efficiency is consequently an evaluation standard for the quality of the heating systems. From the proportion of useful heat energy flow to the total heat energy flow available to the furnace, the furnace efficiency η_O can be calculated. This describes the quality of the furnace design as it puts into proportion the furnace losses to the useful heat.

The product of both efficiencies η_F and η_O is the total thermic efficiency of the industrial furnace per

$$w = \eta_F \times \eta_O$$

4.2 Heat balance

Better evaluation of installations is possible by drawing up a heat balance, whereby furnace components that are not operating well can be discovered and deviations from normal operation can be quantified.

Normally elementary heat balances for industrial furnaces are drawn up, which do not take into account at which temperature level the heat quantities are located, no statement is made to what extent an utilization of the respective heat quantity according to the second law of thermodynamics generally is possible. This would only be possible with energy balances which have been found to be too complicated to draw up in practice.

Figure 10 shows a summary energy balance of an annealing furnace with preheating for the charge and the combustion air. It clearly identifies what magnitude is the related heat losses, and to what extent they can be avoided. Ultimately it has to be decided if this will be done after a cost-profit analysis. However, in the interest of an economical output of our fuel resources, it would make sense to allow a longer time period for the amortization of energy saving investments, as it is the case for normal investments for rationalization.

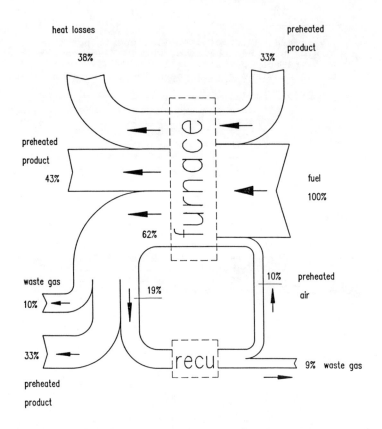

Fig. 10 Energy balance preheating furnace

Optimal Fuel Consumption Strategy for Heating Furnaces

Mohamed A. Youssef
Arab Maritime Transport Academy,

and

Ernest S. Geskin
New Jersey Institute of Technology, Newark, N.J. 07102

Abstract

A practical computational technique for analyzing energy use in heating furnaces is suggested. This technique determines the minimum amount of fuel, thermodynamically required, for heating and the operational conditions accomplishing this level of fuel consumption. Reasonable physical assumptions with respect to the heating process enable us to present the problem of fuel utilization in heating furnaces, batch and continuous types, as a lumped parameter dynamic optimization problem. The technique provides simplified equations for estimating the optimal firing rate, the optimal furnace temperature, and the minimum fuel consumption. The suggested technique enables us to determine the energy saving potential and to construct the practical algorithm of the temperature control. The application of the developed strategy on existing furnaces shows that a reduction of fuel consumption up to 12 % is feasible.

Introduction

Many industrial and power systems do not fully utilize their plant potentialities nor realize their performance indices. It is necessary, therefore, to develop methods of control and design that would enable us to utilize the full potential of a system; as well as, to find optimal systems in any particular sense. Heating furnaces, by themselves, are major consumers of oil and gas fuel. In some plants, furnaces consume up to 90% of the total energy used, [1]. The efficiency of fuel utilization in industrial furnaces provides a substantial opportunity for oil and gas economy. Most of the developments for improving energy utilization have been mainly in refractories, methods of construction, burner design, flame configuration, combustion process and heat transfer process. One of the most important means for the improvement in energy utilization is the optimization of furnace control. This could be achieved by the optimization of air-to-fuel ratio, furnace setpoint temperature and the schedule of furnace operation; as well as, by the implementation of the optimal control of furnace temperature.

The theory of optimal systems, based on the most recent advances in mathematics and engineering, has been rapidly developing during recent years. With advances in engineering and economics, increasingly more diverse and strict requirements are being demanded from automatic control systems. Control studies of large systems may often be useful when formulated as dynamic optimization problems. The association of such problems with the computational techniques that have been recently developed hold the promise of a great increase in the number of systems which may be useful in similar studies. One such system is the heating furnace. Previously, heating furnaces have been considered only as distributed-parameter systems [2, 3, 4, 5]. The development of the theory for optimal systems with distributed parameters presents a much more difficult problem than that for systems with lumped parameters. The basic difficulties are :

* The distributed-parameter systems are not characterized by a finite set of quantities, coordinates of the object, that vary only in time; but generally by a set of functions that show the dependence of the parameters on time and spatial variables or any combination. The distribution functions could be defined on sets of the most different types, these sets may have different dimension lines, surfaces or volumes. The same is also true for controlling actions.

* The dynamic behavior of systems with distributed parameters is described by complex functional equations for example, partial differential equations with complex boundary and initial conditions. Moreover, the character of the supplementary constraints accompanying the requirements of the actual problem statement becomes very involved.

As a result of the complexity of the process taking place in distributed parameter systems, it is in most cases difficult to obtain an analytic expression for the desired functions. Therefore, implementing optimal systems with lumped parameters becomes more practical.

A conventional definition of a lumped parameter dynamic optimization problem is :

Given a system described by the state variables $\underline{y} = (y_1, y_2,...,y_n)$ and control variables $\underline{u} = u_1, u_2,...,u_m)$, select the control $\underline{u}(\theta)$ defined on (θ_i, θ_f) such that :

$$I = \int_{\theta_i}^{\theta_f} \phi\ (\underline{y},\underline{u},\theta)\ d\theta \qquad (1)$$

is minimized, subject to the differential constraints

$$\frac{dy}{d\theta} = f(y,u,\theta) \tag{2}$$

and end conditions which might be of the form

$$y(\theta_i) = y_i \tag{3}$$

$$y(\theta_f) = y_f$$

In addition, there might be inequality constraints of the form

$$F(u_1, u_2,...,u_m) \leq 0 \tag{4}$$

In this paper, it will be shown that the problem of optimal control of a batch heating furnace may be meaningfully formulated as a problem of the same form.

The solution of this variational problem, of the minimization of fuel consumption in heating furnaces, determines the minimum amount of fuel necessary for the process performance and operating conditions. The obtained solution will be used to construct the furnace control algorithm. The bottleneck of any control system is the accuracy of information. By definition, the solution of the variational problem determines the state of a system for which the derivatives of the objective function with respect to the control variables equal zero. Consequently small deviations of process characteristics from their optimal levels have much less effect on the process results (in our case total fuel consumption) than similar variable changes far away from optimal levels. Furthermore, the goal of the variational computation could be defined as the estimation of an optimal interval for process characteristics rather than the determination of their exact values. The effect of the inaccurate information on the practical results of the optimal computation is less severe than its impact on other kinds of process analysis. In general, it could be anticipated that conventional information employed for process control is also sufficient for process optimization.

In previous work regarding optimal control of heating furnaces [6, 7, 8, 9, 10] the main assumption, following Thring and Reber [11], was that wall losses in the furnace are constant over the range of operating temperature. However, in practice there is no evidence from a carefully designed furnace model [12] that wall losses precisely are a constant; and in fact, it varies with the firing rate. Therefore, in present work this assumption will be dropped.

The Heating Furnace Model

Assumptions

In developing the mathematical model the following assumptions which lead to simplifying the physical system, in order to obtain a practical solution were made :

* The heated material is a "thin" body from the thermal point of view (characterized by the Biot number being less than 0.1)

* Thermal properties of the material are constant.

* Average value of the total heat transfer coefficient between the combustion gas and the load can be used.

* All fuel is burnt, and the heat loss from the furnace is primarily through stack and from the wall.

Heat Transfer Analysis

The material to be heated in the furnace is viewed as an infinite plate of thickness . Let the furnace temperature be denoted by the function :

$$T = T(\theta) \quad , \quad 0 \leq \theta \leq \theta_f \tag{5}$$

where

T = furnace temperature
θ = time
θ_f = heating time

Since the heated material would be considered as a "thin" body, its temperature distribution could be described as a function of time:

$$t = t(\theta) \quad , \quad 0 \leq \theta \leq \theta_f \tag{6}$$

where

t = material temperature

Applying the heat balance to the material for a time period $d\theta$, the heating process could be described by the relation

$$\frac{dt}{d\theta} = \frac{T-t}{\lambda_b} \tag{7}$$

with initial and final batch temperatures

$$t(0) = t_i \tag{8}$$

$$t(\theta_f) = t_f \tag{9}$$

where

$\lambda_b = \dfrac{mc_m}{Ah_t}$ = heating time constant

m = batch mass

c_m = material specific heat

A = batch heat transfer area

h_t = total heat transfer coefficient

t_i = initial batch temperature

t_f = final batch temperature

Equation (7) - (9) describe the process of heating of a lumped material in a batch furnace. These equations could be used to describe the heating of a massive batch, $Bi > 0.1$, as the first approximation. This could be achieved by the determination of λ_b as a function of the Biot number through the replacement of h_t by h_t' where h_t' is given by the equation [13,14]

$$h_t' = (1 + \sqrt{Bi}) h_t \qquad (10)$$

Hence,

$$\lambda_b' = \frac{mc_m}{Ah_t'} \qquad (11)$$

where

h_t' = modified heat transfer coefficient,

λ_b' = modified time constant,

Bi = Biot number

Heat Balance

The relation between firing rate, furnace temperature and material temperature could be constructed by the use of heat balance on a continuously operating furnace at steady state. For such a furnace assuming that all fuel is burnt, the thermal input should be balanced by the heat loss from the furnace in a total of three possible ways:

* Heat absorbed by the load.
* Heat carried away by flue gases.
* Heat losses from the walls and openings, etc.

The heat balance on the batch furnace could be expressed by the equation:

$$E + Q_g + Q_a = Q_m + Q_n + Q_w \qquad (12)$$

where

$E = H_g V_g$ = fuel firing rate

$Q_g = c_g t_g V_g$ = sensible heat rate of fuel

$Q_a = c_a t_a V_a$ = sensible heat rate of air

$Q_w = c_w t_w V_w$ = rate of heat lost with flue gases

Q_m = rate of heat absorbed by the load

Q_n = rate of heat losses from the walls

H_g = calorific value of fuel

c_g, c_a, c_w = specific heat of fuel, air, and flue gases

t_g, t_a, t_w = temperature of fuel, air, and flue gases

V_g, V_a, V_w = flow rate of fuel, air, and flue gases

The heat balance equation (12), could then be represented in the form

$$H_g V_g = Q_m + Q_l + c_w t_w V_w - c_g t_g V_g - c_a t_a V_a \qquad (13)$$

A coefficient of fuel utilization η_g may be defined by

$$\eta_g = \frac{H_g V_g + c_c t_g V_g + c_a t_a V_a - c_w t_w V_w}{H_g V_g} \qquad (14)$$

Then from Equations (13) and (14)

$$\eta_g E = Q_m + Q_n \qquad (15)$$

The adiabatic temperature of fuel combustion, T_{ad}, is determined under the assumption that all the heat evolved through combustion is spent only to raise the temperature of combustion products. From the definition, without preheating of air or fuel, the adiabatic temperature could be presented in the form :

$$T_{ad} = \frac{H_g V_g}{c_w V_w} \qquad (16)$$

With preheating of air and fuel, the fraction of heat recovered from the flue gases, r, is determined by the relation

$$r = \frac{c_a t_g V_g + c_a t_a V_a}{c_w t_w V_w} \qquad (17)$$

Equation (14) could be simplified by the use of the definitions of the adiabatic temperature, T_{ad}, and the coefficient of heat recovery, r. Dividing the numerator and denominator in equation (14) by $c_w V_w$, and making use of Equations (16) and (17), we obtain

$$\eta_g = \frac{T_{ad} + rt_w - t_w}{T_{ad}} \qquad (18)$$

Let us introduce the theoretical temperature of fuel combustion T_h, by the relation [7]

$$T_h = \frac{T_{ad}}{1-r} \qquad (19)$$

The theoretical temperature, T_h, is determined by the properties of fuel and the conditions of energy recovery in the furnace. At r=1, there is no heat loss with the flue gases and $T_h \to \infty$. If there is no heat recovery in the furnace, i.e., r=0, then $T_h = T_{ad}$.

By the use of Equation (19), Equation (18) is then expressed as

$$\eta_g = 1 - \frac{t_w}{T_h} \qquad (20)$$

Assuming that $t_w = T$, we can rewrite Equation (20) in the form

$$\eta_g = 1 - \frac{T}{T_h} \qquad (21)$$

Equation (21) represents the coefficient of fuel utilization in the furnace in the classical thermodynamic form.

Firing Rate

Now it is necessary to construct the relationship between fuel firing rate and process temperature. This relationship will eventually determine the total fuel consumption.

The firing rate E is expressed by Equation (15), which may be rewritten in the form

$$E = \frac{Q_m + Q_n}{\eta_g} \qquad (22)$$

Q_m and Q_l could be expressed by the relations

$$Q_m = h_t A(T - t) \qquad (23)$$

$$Q_n = UA_n(T - t_o) \qquad (24)$$

where

h_t = total heat transfer coefficient
U = overall heat transfer coefficient for the furnace wall
A = heat transfer area of the load
A_n = wall area
t_o = ambient temperature

Substituting η_g, Q_m and Q_n from Equations (21), (23) and (24) into Equation (22), we obtain

$$E = \frac{h_t A T_h (T-t) + UA_n T_h (T-t_o)}{T_h - T} \qquad (25)$$

Denoting

$$h_t A T_h = D_b$$

and

$$UA_n T_h = F_b$$

We can represent the firing rate as

$$E = \frac{D_b (T-t) + F_b (T-t_o)}{T_h - T} \qquad (26)$$

Equation (26) determines the firing rate in a batch furnace as a function of heating conditions.

The statement of the optimization Problem

As it follows from Equation (26), firing rate in batch furnaces and consequently the total fuel fired are functions of the variation of furnace temperature. As a result, the admissible change of T in the course of heating must be selected so that the required final metal temperature is achieved while total fuel consumption is minimal. The process is constrained by the furnace temperature limitation. The total fuel consumption could be described by the equation.

$$E = \int_0^{\theta_f} E \, d\theta \tag{27}$$

where E is the total fuel consumption. Substituting the value of E determined from Equation (26) in this equation, we obtain

$$E = \int_0^{\theta_f} \frac{D_b (T - t) + F_b (T - t_o)}{T_h - T} \, d\theta \tag{28}$$

In the process of heating metals in batch furnaces, the following two situations occur frequently:

(i) The time of heating process is fixed. In this case, it is required to control the heating optimally at minimum fuel consumption during a specified time.

(ii) The time heating process is unspecified, it is required to control the heating optimally at minimum fuel consumption during an optimal heating time.

The optimization problem with specified heating time

The problem of the minimization of the fuel consumption could be stated as follows:

Determine the function $T(\theta)$ which minimizes the functional

$$E = \int_0^{\theta_f} \frac{D_b (T - t) + F_b (T - t_o)}{T_h - T} \, d\theta \tag{28}$$

subject to

$$\frac{dt}{d\theta} = \frac{T - t}{\lambda_b} \tag{29}$$

$$T(\theta) \leq T_{max} \tag{30}$$

$$t(\theta) = t_i \tag{31}$$

$$t(\theta_f) = t_f \tag{32}$$

where

E = total fuel consumption

θ = time

θ_f = heating time

T = furnace temperature

T_{max} = maximum allowable furnace temperature

t = temperature of heated material

t_i, t_f = initial and final load temperature

Here, the system is described by the one-dimensional time dependent state variable $t(\theta)$ and the control variable $T(\theta)$. Equation (28) is the objective function of the optimization problem. Equation (29) and (30) determine constraints resulting from the conditions of heat transfer in the furnace. The initial and final process conditions are represented by Equations (31) and (32).

The Optimization Problem with Unspecified Heating Time

In the situation discussed above, the time of heating θ_f is given. The fuel consumption could be reduced still further if θ_f, as well as $T(\theta)$, is determined from the optimization conditions. In that case, the optimization problem can be stated as :

Find the function $T(\theta)$ and scalar θ_f which minimize the objective function given by Equation (28) subject to constraints given by Equations (29) - (32).

Optimal Heating Conditions

The optimal heating conditions could be determined from the minimization of the functional given by Equation (28) subject to the constraints (29) - (32). Assuming that the temperatures $T(\theta)$ and $t(\theta)$ are continuous and differentiable with respect to θ. Let us also assume that the function $E(T,t)$ is continuous and differentiable with respect to the variables T and t. Based on these assumptions, the optimal heating conditions could be determined by the use of maximum principle [15].

According to the maximum principle, for the solution of the stated optimization problem, it is necessary to construct the Hamiltonian function H given by the equation

$$H = \Psi \frac{T - t}{\lambda_b} - \frac{D_b(T - t) + F_b(T - t_o)}{T_h - T} \tag{33}$$

where the function $\Psi(\theta)$ satisfies the equation

$$\frac{d\Psi}{d\theta} = \frac{\partial H}{\partial t} \tag{34}$$

The maximum principle states that the optimal $T(\theta)$ maximize H for $0 \le \theta \le \theta_f$. Maximum of H is determined by the equation

$$\frac{\partial H}{\partial T} = 0 \quad , \quad \text{if } T < T_{max} \tag{35}$$

or by the equation

$$T = T_{max} \tag{36}$$

Optimization Problem Solution for Specified Heating Time

Determining Optimal Heating Condition ($T < T_{max}$) if the furnace temperature is less than the maximum m allowable value, the conditions of the optimization process are given by the system of equations

$$\frac{\partial H}{\partial T} = \frac{\Psi}{\lambda_b} - \frac{D_b(T_h - t) + F_b(T_h - t_o)}{(T_h - T)^2} = 0 \tag{37}$$

$$\frac{\partial \psi}{\partial \theta} = \frac{\Psi}{\lambda_b} - \frac{D_b}{T_h - T} \tag{38}$$

$$\frac{dt}{d\theta} = \frac{T - t}{\lambda_b} \tag{39}$$

where λ_b, D_b, F_b, T_h are assumed to be constants. The solution of this system results in the equations

$$t(\theta) = T_h - (T_h - t_i) \left(\frac{T_h - t_f}{T_h - t_i} \right)^{\theta/\theta_f} \quad (40)$$

$$T(\theta) = T_h - (T_h - t_i) \left(1 - \frac{\lambda_b}{\theta_f} \operatorname{Ln} \frac{T_h - t_i}{T_h - t_f} \right) \left(\frac{T_h - t_f}{T_h - t_i} \right)^{\theta/\theta_f} \quad (41)$$

Equations (40) and (941) determine the optimal process control. Substitution of t(θ) and T(θ) in equation (26) yields.

$$E_0(\theta) = \frac{D_b \dfrac{\lambda_b}{\theta_f} \operatorname{Ln} \dfrac{T_h - t_i}{T_h - t_f}}{1 - \dfrac{\lambda_b}{\theta_f} \operatorname{Ln} \dfrac{T_h - t_i}{T_h - t_f}}$$

$$+ F_b \left[\frac{(T_h - t_o)}{(T_h - t_i)\left(1 - \dfrac{\lambda_b}{\theta_f} \operatorname{Ln} \dfrac{T_h - t_i}{T_h - t_f}\right) \left[\dfrac{T_h - t_f}{T_h - t_i} \right]^{\theta/\theta_f}} - 1 \right] \quad (42)$$

From Equation (42), it follows that E_0, varies with time. Consequently, variable firing rate provides the optimal furnace conditions.

The minimum total fuel fired for the given conditions could be determined by substituting Equation (42) into Equation (28). After integrating and arranging, we obtain

$$E_0 = \frac{D_b \dfrac{\lambda_b}{\theta_f} \operatorname{Ln} \dfrac{T_h - t_i}{T_h - t_f}}{1 - \dfrac{\lambda_b}{\theta_f} \operatorname{Ln} \dfrac{T_h - t_i}{T_h - t_f}}$$

$$+ F_b \theta_f \left[\frac{(T_h - t_o)(T_f - t_i)}{(T_h - t_i)(T_h - t_f)\left(1 - \dfrac{\lambda_b}{\theta_f} \operatorname{Ln} \dfrac{T_h - t_i}{T_h - t_f}\right) \operatorname{Ln} \left[\dfrac{T_h - t_i}{T_h - t_f} \right]} - 1 \right] \quad (43)$$

E_0 is the minimum fuel consumption attainable under a given process characteristics.

The Determination of Optimal Heating Conditions ($T = T_{max}$). If in the course of heating the furnace temperature T, determined by Equation (41), reaches the maximum allowable value T_{max} before the load temperature reaches its final temperature, i.e. $t < t_f$. In this case, the minimum fuel consumption would still be determined by the maximization condition of the Hamiltonian function gives by Equation (33). However, Equation (35) does not describe the optimal behavior of the system; nevertheless, according to the maximum principle [15], Equation (36) determines the maximum value of the Hamiltonian function. Consequently, the optimal firing strategy is a two stage process. The first stage involves heating at a firing rate presented by Equation (42) as long as $T < T_{max}$; while at the second stage, heating is carried out at the maximum allowable temperature T_{max}.

Assuming that the final metal temperature at the first stage is t_l, then from Equation (41) it follows that,

$$\theta_1 = \lambda_b \frac{T_h - t_l}{T_{max} - t_l} Ln \frac{T_h - t_i}{T_h - t_l} \tag{44}$$

where

θ_1 = Heating time of the first stage

In the course of the second stage, the batch temperature changes from t_l to t_f and the furnace temperature is constant at T_{max}. The heating time of this period is

$$\theta_2 = \lambda_b Ln \frac{T_{max} - t_l}{T_{max} - t_f} \tag{45}$$

where

θ_2 = heating time of the second stage

According to the above Equations (44) and (45), since the heating time θ_f is specified, it follows that

$$\theta_f = \theta_1 + \theta_2 = \lambda_b \left[\frac{T_h - t_l}{T_{max} - t_l} Ln \frac{T_h - t_i}{T_h - t_l} Ln \frac{T_{max} - t_l}{T_{max} - t_f} \right] \tag{46}$$

by the use of Equation (46), t_l could be determined.

Hence, the optimal load temperature, the optimal furnace temperature and the optimal fuel consumption in the first stage could be determined by substituting t_l in Equations (40), (41) and (43).

Optimal load temperature, optimal firing rate and optimal fuel consumption in the second stage are given by the equations

$$t_2(\theta) = T_{max} - (T_{max} - t_l) e^{-(\theta - \theta_1)/\lambda_b} \tag{47}$$

$$E_{02}(\theta) = \frac{D_b (T_{max} - t_l)}{T_h - T_{max}} e^{-(\theta - \theta_1)/\lambda_b} + \frac{F_b (T_{max} - t_o)}{T_h - T_{max}} \tag{48}$$

$$E_{02} = \frac{D_b \lambda_b (T_{max} - t_1)}{T_h - T_{max}} (1 - e^{-\theta_2/\Lambda_b}) + \frac{F_b \theta_2 (T_{max} - t_0)}{T_h - T_{max}} \qquad (49)$$

where

t_1 = first stage final load temperature

t_2 = load temperature in the second stage

E_{02} = minimum firing rate in the second stage

E_{02} = minimum fuel consumption in the second stage

Finally, the total fuel consumption is the sum of the minimum consumption in both stages,

$$E_0 = E_{01} + E_{02} \qquad (50)$$

where

E_{01} = first stage minimum fuel consumption

Optimization Problem Solution for Unspecified Heating Time

Assuming that heating time is not prescribed, it could still be determined from fuel consumption minimization conditions; by varying E_0, it is possible to select a heating time at minimum total fuel fired. In other words, the fuel consumption could be reduced further if θ_f, as well as $T(\theta)$ and $E(\theta)$, is defined by the optimization conditions.

Optimal Heating Conditions ($T < T_{max}$)

Assuming that the furnace temperature does not reach the maximum allowable temperature in the course of the heating process, the optimization condition could still be described by the system of Equations (37) - (39). Therefore, the variation of t, T and E would be given by Equations (40) - (42), where the optimal heating time is determined from the condition,

$$\frac{\partial E_0}{\partial \theta_f} = 0 \qquad (51)$$

Solving equation (51), we obtain

$$\theta_0 = \lambda_b \delta (1 + \Lambda_b) \qquad (52)$$

Where

θ_0 = Optimal heating time

$$\delta = Ln \frac{T_h - t_i}{T_h - t_f} \qquad (53)$$

$$\Lambda_b = \sqrt{\frac{D_b\delta + F_b\gamma}{F_b(\gamma - \delta)}} \qquad (54)$$

$$\gamma = \frac{(T_h - t_o)(t_f - t_i)}{(T_h - t_i)(T_h - t_f)} \qquad (55)$$

By substituting θ_0 in Equations (40) and (41), the values of t and T could then be obtained. From Equation (52) and (42), it follows that the optimal firing rate is,

$$E_0(\theta) = \frac{D_b}{\Lambda_b} + F_b\left[\frac{(1+\Lambda_b)(T_h-t_o)}{\Lambda_b(T_h-t_i)\left(\frac{T_h-t_f}{T_h-t_i}\right)^{\theta/\lambda\delta(1+\Lambda_b)}} - 1\right] \qquad (56)$$

and from Equations (52) and (43), it follows that the minimum total fuel fired is,

$$E_0 = \frac{\lambda_b\delta(1+\Lambda_b)}{\Lambda_b}[D_b + F_b\lambda(1+\Lambda_b) - F_b\Lambda_b] \qquad (57)$$

where the constants δ, Λ_b and γ are given by Equations (53), (54) and (55) respectively.

Equation (57) determines the total fuel fired which is the minimum achievable for a given furnace construction. Equation (56) determines the firing rate which accomplishes that level of fuel consumption.

Optimal Heating Conditions ($T = T_{max}$). When the furnace temperature T reaches its maximum allowable value T_{max}, for $t < t_f$, the minimum fuel consumption is accomplished by a two-stage heating strategy: heating at firing rate presented by Equation (42) for $T < T_{max}$ and then at $T = T_{max}$. This actually follows from the condition of the maximization of the Hamiltonian function H.

Assuming that T reaches T_{max} when $t = t_l$. The corresponding time θ_1 could be determined by substituting t_l in Equations (44) and (52). Equating θ_l, we obtain

$$\frac{T_h - t_l}{T_{max} - t_l} = 1 + \Lambda_{bl} \qquad (58)$$

where

t_l = temperature of the load at the end of the first stage.

$$\Lambda_{bl} = \sqrt{\frac{D_b\delta_1 + F_b\gamma_l}{F_b(\gamma_{l1} - \delta_1)}} \qquad (59)$$

$$\delta_l = \operatorname{Ln} \frac{T_h - t_i}{T_h - t_l} \qquad (60)$$

$$\Lambda_l = \frac{(T_h - t_o)(t_l - t_i)}{(T_h - t_i)(T_h - t_l)} \qquad (61)$$

The solution of Equation (58) provides t_l.

Substituting t_l in Equations (56) and (57), we could then determine the firing rate and the minimum fuel consumption in the first heating stage. Equations (45), (48) and (49) determine the heating time, the firing rate and the minimum fuel consumption in the second heating stage. The total fuel consumption is determined from Equation (50).

Application of the Developed Technique

Having shown the analytical technique for determining the optimal conditions of fuel utilization in heating furnaces, it remains to demonstrate the applicability of such a technique towards an existing operating heating furnace. The required input information which must be acquired for the furnace analysis could be determined by simple measurements.

Input Information

The list of variables appearing in the constructed equations, to obtain optimal heating strategy for batch furnaces, determines the information which must be acquired for the furnace analysis. The required information includes characteristics of metals (m, c_m, k), furnaces (T_h, T_{max}, U, r) furnace and metal interaction (λ_b) and the process of heating (t_i, t_f, T, t, E, θ_f).

The values of t_i, t_f, T_{max}, c_m, k and m are usually known. Conventional instruments could be used for the monitoring of E, T and metal surface temperature t. If the load temperature gradient can be discounted, that load can be considered as a thermally thin (lumped) body and the value of t can be determined by the measurement of the load surface temperature. Otherwise, t should be determined by the measurement of the metal temperature at several different points of the load and subsequent averaging of this temperature.

The routine information about the conditions of combustion enables us to determine T_h. The acquisition of such information includes the measurement of air and fuel temperatures, flow rates, temperature and composition of the combustion products. The flow rate of the flue gases can be determined as a function of flow rate and composition of fuel.

The time constant λ_b can be computed directly if the total heat transfer coefficient h_t is known. Otherwise, λ_b should be determined by the simultaneous measurement of load temperature t and furnace temperature T, and substitution of the measured data into Equation (8). For the determination of λ_b it is convenient to represent Equations (8) in the form

$$\frac{t_k - t_j}{\theta_k} = \frac{T_j - t_j}{\lambda_b} \qquad (62)$$

where

t_j = Temperature of the metal at a fixed instant of time

T_j = Temperature of the furnace at a fixed instant of time

t_k = Temperature of metal in time θ_k after measuring t_j and T_j.

θ_k = Time interval.

The overall heat transfer coefficient of the furnace wall U is usually calculated if the wall construction, dimensions and thermal conductivity of each layer are known. However, in the calculation of the heat balance we assume that the heat loss from the furnace is primarily from the stack and walls, i.e., we have neglected some other losses such as losses through doors, cracks and heat accumulated in the lining. All these unaccounted losses and others will be referred to as "wall losses". Therefore, the value of the overall heat transfer coefficient will be calculated from the heat balance equation.

A Practical Example

A batch furnace at NORANDAL USA, Inc., Alabama, for heating aluminum ingots from room temperature up to 950° F was studied. The size of the ingot is 168"x60"x21". The ingot rests on the base of 168"x21". Natural gas is used as a fuel. The actual performance details of the furnace are summarized in Table I. The variation of the metal and combustion products temperature are shown in Fig. 1 as taken directly from recorders on the furnace.

Heating Process Characteristics

Applying equation (62) to the curves of the actual temperature variation (Fig. 1), we can determine the time constant. Thus, we obtain

$$\lambda_b = 3.43 \text{ hr}$$

From the definition of the time constant ($\lambda_b = mc_m/Ah_t$), we get

$$h_t = 6.4 \text{ Btu/hr.ft.}^{2o}F$$

The ingot can be considered as an infinite plate having a thickness s = 21" and heated from two sides; in this case the Biot number is

$$Bi = \frac{h_t \cdot s/2}{k} = 0.042$$

For such a value of the Biot number, the thin body hypothesis is accurate.

Since neither the air nor the fuel is preheated, the value of the fraction of heat recovered from flue gases is r = 0 and consequently, the theoretical temperature of fuel combustion is equal to the adiabatic temperature. Therefore, Equations (16) and (19) give

$$T_h = 4290 \text{ °F}$$

From the heat balance equation for the furnace, we can calculate the overall heat transfer coefficient, and consequently the value of the constants D_b and F_b are determined. Now we are in a position to determine the optimal heating conditions.

Table 1. Operating Conditions for NORANDAL USA Batch Furnace

Item	Characteristics	
Furnace	Dimensions (Length x width x height)	29.6' x 15.3' x 21.2'
	Hearth Loading	300,000 lbs
	Heating Time	7.0 hr
	Furnace Temperature	1100 °F
	Maximum Allowable Temperature	1200 °F
Load	Material	Aluminium
	Dimensions (Length x width x height)	168" x 60x21"
	Mass	20,000 lbs
	Number of Ingots per Charge	15 pcs
	Thermal Conductivity	132.0 Btu/hr. ft.°F
	Density	160.0 lbs/ft^3
	Specific Heat	0.20 Btu/lb.°F
	Inlet Temperature	70°F
	Outlet Temperature	950 °F
Fuel	Type of Fuel	Natural gas
	Calorific Value	1030 Btu/ft^3
	Flow Rate	19400 ft^3/hr
	Temperature	70 °F
	Specific Heat	0.029 Btu/ft^3.°F
Air	Flow Rate	194,000 ft^3./hr
	Temperature	70 °F
Flue Gases	Flow Rate	232,800 ft^3/hr
	Temperature	1100 °F
	Specific Heat	0.02 Btu/ft^3.°F

Optimal Heating Conditions

Since heating time is specified ($\theta_f = 7$ hr), the process could be described by Equations (40) - (43) as long as optimal furnace temperature does not exceed the maximum allowable temperature ($T_{max} = 1200°F$). According to Equation (41) the maximum value of T is obtained at the end of heating. The substitution of the different parameters into Equation (41) yields

$$T_f = 1333°F$$

The obtained furnace temperature exceeds 1200°F, therefore, heating is to be carried out in two stages. The metal temperature at the end of the first stage can be determined from Equation(46). For the given conditions

$$t_1 = 772.8°F$$

Now equations (44) and (45) give

$$\theta_1 = 5.15 \text{ hours}$$

and

$$\theta_2 = 1.85 \text{ hours}$$

Where θ_1 and θ_2 are the duration of heating of the first and second stage, respectively. Substituting the obtained results into Equations (40) - (43) and Equations (47) - (49) to obtain the optimal heating conditions for the first and second stage of heating, respectively.

The optimal furnace and metal temperatures are shown in Fig. 1. Fig. 2 shows optimal firing rate which accomplishes the minimum fuel consumption. Also, the optimal heat flow rate to the metal is shown in the same figure.

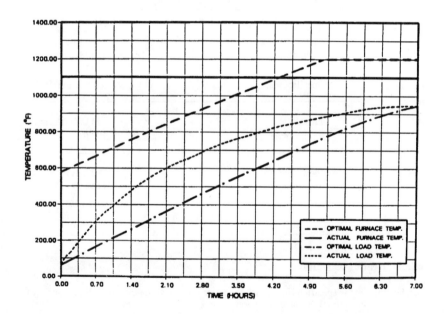

Figure 1 - *Actual and Optimal Temperature Profiles.*

Total energy consumption under the optimal conditions is

$E_0 = E_{01} + E_{02} = 126727568$ Btu

The actual energy consumption is

$E_u = H_g V_g \theta_f = 13987400$ Btu

Hence, the energy saving result from the process optimization is

$E_s = 9.4 \%$

Optimization of Heating Process Duration

Fuel consumption can be reduced by determining the optimal heating time θ_0 from Equation (52). Substituting the given process characteristics into this equation, we obtain

$\theta_0 = 6.5$ hours

From the obtained optimal heating time, it follows that the optimal heating is a two stage process, consequently the total fuel fired is

$E_0 = 126359600$ Btu

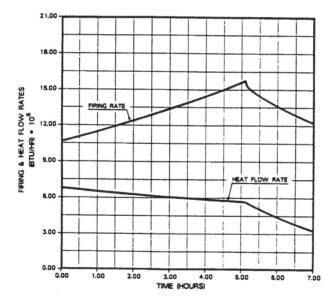

Figure 2 - *Optimal firing and heat flow rates*

which is the minimum fuel consumption achievable of the batch furnace with given design characteristics. The energy saving in this case is

$E_s = 9.65\%$

We can see that the effect of the optimization of process duration is insignificant. The effect of heating time θ_f on optimal fuel consumption E_0 is shown in fig. 3.

The implementation of the suggested optimal heating technique in that furnace caused a substantial reduction in energy consumption. That result is due to the fact that all assumptions, mentioned earlier, agreed with the actual working conditions. The Aluminum ingots are heated each as a thin body ((Bi < 0.1) ; the heat transfer coefficient and thermodynamical properties is rather low (1100 °F); the furnace temperature is equal to the combustion products temperature. Those conditions prove that all assumptions were satisfied in that case.

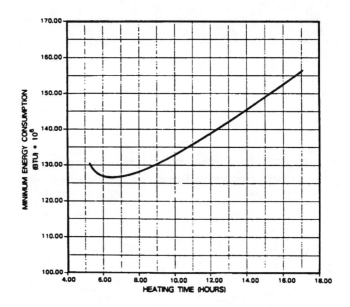

Figure 3 - *Minimum energy consumption at various heating times*

Discussion and Conclusion

The success of the mathematical approach of this study indicated the possibility of using it as a tool in studying similar systems. The approach has succeeded in deducing optimal conditions of fuel utilization in batch furnaces. The maximum principle application to the fuel consumption minimization problem, in heating furnaces, has resulted in a closed form solution. That solution determines the algorithm for optimal control and the criteria for estimating the control system operations. Although this study has demonstrated the maximum principle application to simplified heating conditions, the obtained solution could be applied to various types of batch and continuous heating furnaces.

The study has developed a technique that would estimate : an energy saving potential, optimal heating conditions and construction variations. Based on this technique, the obtained equations could be used to analyze various heating furnaces. For given process characteristics, Equations (40) - (50) have determined the optimal heating conditions. Firing strategies could be deduced by the determination of the firing rate or the furnace temperature which minimizes the fuel consumption.

Three firing strategies are possible for the assumed heating conditions. These strategies are deduced from Equations (40) - (58).

* If process duration θ_f is prescribed and $T < T_{max}$ throughout the entire heating process, then, the firing rate could be determined directly from Equation (42).

* If T_{max} is reached during the heating process, firing strategy would be a two-stage process. The first stage involves heating under an increasing firing rate until $T = T_{max}$. Then, in the second stage, T_{max} is stabilized by heating under a decreasing firing rate. First and second stage heating conditions are given by Equations (44) - (50).

* If process duration θ_f is selected from the optimization conditions. The optimal strategy is determined by Equations (52) - (58). The constant firing rate is the most common operation condition in batch furnaces. However, the result of this study has proven that a varying firing rate is more desirable in furnishing the optimal heating conditions.

The application of the suggested technique on existing batch furnace have shown that a reduction of 10% of fuel consumption is feasible. The implementation of the same technique on continuous furnaces (not shown in the paper), yield to fuel saving up to 12%. More work correlating the actual and optimal performances of existing heating furnaces would allow a greater confidence in the use of the technique for design purposes. That technique could be considered as a preliminary step in the computer aided furnace diagnostics.

Acknowledgment

The authors would like to acknowledge and thank NORANDAL USA, Inc., Alabama, for generously providing information on the batch furnace quoted in this paper.

Nomenclature

- A = Load heat transfer area
- A_n = Furnace wall area
- B_i = Biot number
- c = Specific heat
- D_b = $Ah_t T_h$
- E = Total fuel fired
- E_0 = Optimal total fuel fired
- E = Firing rate
- E_0 = Optimal firing rate
- F_b = $UA_n T_h$
- H = Hamiltonian function
- H_g = Calorific value of fuel
- h_t = Total heat transfer coefficient
- h_t' = Modified total heat transfer coefficient
- k = Load thermal conductivity
- m = Batch mass
- Q_a = Sensible heat rate of air
- Q_g = Sensible heat rate of fuel
- Q_n = Rate of heat losses from wall
- Q_m = Rate of heat absorption by the load
- Q_w = Rate of heat lost with flue gases
- r = Coefficient of heat recovery
- s = Load thickness
- T = Furnace temperature
- T_{ad} = Adiabatic temperature of fuel combustion
- T_h = Theoretical temperature of fuel combustion
- T_{max} = Maximum allowable furnace temperature
- t = Load temperature
- t_a = combustion air temperature
- t_g = Fuel temperature
- t_o = Ambient air temperature
- t_w = Flue gases temperature
- U = Overall heat transfer coefficient through wall
- V = Flow rate
- η_g = Coefficient of fuel utilization
- θ = Time
- θ_f = Heating time
- λ_b = Heating time constant
- ψ = Adjoint variable

Subscripts

- a = Air
- b = Batch furnace
- f = Final condition
- g = Fuel
- i = Initial or inlet condition
- k = Interval "k"
- n = Furnace walls
- m = Metal
- o = Ambiant
- 0 = Optimal
- s = Saving
- t = Total
- u = Actual
- w = Flue gas

References

1. T. Bailey and F. M. Wall, "Ethylene Furnace Design," Chemical Engineering Progress, 1978, 74, no. 7: 45-51.

2. A. G. Butkovskii and A. Y. Lerner, "The Optimal Control of Systems with Distributed Parameters," Automation and Remote Control, 1960, 21, no. 6: 472-477.

3. A. G. Butkovskii, "Optimum Processes in Systems with Distributed Parameters, "Automation and Remote Control, 1961, 22, no. 1: 13-21.

4. A. G. Butkovskii, "Some Approximate Methods for Solving Problems of Optimal Control of Distributed Parameter Systems, " Automation and Remote Control, 1961, 22, no. 12: 1429-1438.

5. Y. Sakawa, "Solution of Optimal Control Problem in a Distributed Parameter Systems, " IEEE Transactions on Automatic Control, 1964, AC9, no. 4: 420-426.

6. E. S. Geskin, "Optimization of Energy Consumption in Batch Furnaces, " The Metallurgical Society of AIME, TMS Paper Selection, A80-11, 1980.

7. E. S. Geskin, " The Second Law Analysis of Fuel Consumption in Furnaces," Energy, The International Journal, 1980, 5, no. 8-9: 949-954.

8. E. S. Geskin, " The Application of Automatic Control for Energy Saving in Heating Furnaces," (Proceedings of the 2nd IFCA Symposium on Automation in Mining, Mineral and Metal Processing, Montreal, Canada, 1980), 623-632.

9. E. S. Geskin, " The Minimization of Fuel Consumption in Heating Furnaces, " The Metallurgical Society of AIME, TMS Paper Selection, A 80-12, 1980.

10. E. S. Geskin et al., " Analysis of Energy Utilization in Heating and Melting Furnaces, " Energy Conservation Workshop VII, The Aluminium Association, 1983, 101-113.

11. R. H. Essenhigh et al., " Furnace Analysis: A comparative Study, " Combustion Technology - Some Modern Development (Academic Press, 1974).

12. D.E. Macllan, "Thermal Efficiency of Industrial Furnaces: A Study of the Effect of Firing Rate and Output, " (M.S. Thesis, The Pennsylvania State University, 1965).

13. E. M. Goldfarb, Thermal Processes in Metallurgy (Metalurgia, U.S.S.R., 1966).

14. S. I. Averin, Computation of Heating Furnaces (Gostechizdat U.S.S.R., Kiev, 1969), 294.

15. L. S. Pontryagin et al., The Mathematical Theory of Optimal Processes (Wiley Interscience Publishers, New York, 1962).

16. A. G. Butkovskii, Distributed Control Systems (American Elsevier Publishing Company, New York, 1969).

17. B. Golttfried and Y. Weisman, Introduction to Optimization Theory (Prentice-Hall, Englewood Cliffs, New Jersey 1973).

NON-FERROUS METALLURGY:

DECREASE OF ENERGY CONSUMPTION

P. Paschen

Department of Technology and
Metallurgy of Non-Ferrous Metals

Montanuniversität Leoben
Austria
Franz-Josef-Straße 18
A-8700 Leoben

Abstract

Metal production is energy intensive. About 6 % of the world's total energy consumption is for (ferrous and non-ferrous) metal production. In non-ferrous metallurgy, aluminium production is by far the biggest energy consumer and lead the smallest one.

Compared with the absolute theoretical minimum (derived from thermodynamics and the position of the elements in the Periodic System), modern lead metallurgy, in absolute figures, consumes the smallest amount of energy, followed by I.S.furnace zinc reduction, classical lead metallurgy and zinc electrolysis. Compared with the theoretical values, pyrometallurgy of zinc ranks first, followed by aluminium electrolysis.

Possibilities for energy saving are discussed for aluminium, zinc, copper, lead. Promising steps could be taken in the Bayer process, in aluminium and zinc electrolysis, in lower coke consumption in the I.S.furnace and in the lead blast furnace, in avoiding zinc distillation refining to a certain extent, in anode scrap and cathode melting after copper electrolytic refining, in copper winning electrolysis.

Some of these suggestions have been tested in laboratory scale experiments. In aluminium electrolysis, one good result could not be generalized because of inert anode material quality problems. Zinc electrowinning with synthetic pure electrolyte shows specific electricity consumption figures well below industrial practice. The concentration limits for the normal technical impurities were found out - with only one or two of them being present at the same time. Copper electrowinning results can only be presented at the Conference in Cologne in september. In tantalum fused salt electrolysis we could achieve excellent results by application of periodic current reversal - a drop of 33 % in specific energy consumption.

Introduction

Following the first law of thermodynamics, energy cannot be lost. We could be tempted to generously "consume" energy. Aside from its price, the problem is that the value of heat energy depends on its temperature. Technical energy transformation mostly ends with the release of energy at low temperature level. What we do, is not energy "consumption", but energy devaluation. The so-called consumption of energy is a transformation of a manifold convertible form of energy into one which is convertible only to limited extent.

The "big four" non-ferrous metals were produced in 1987 as follows (world-wide) (1):

$$\begin{array}{lrl} \text{Aluminium} & 16.3 & \text{Mt} \\ \text{Copper} & 10.2 & \text{Mt} \\ \text{Zinc} & 7.0 & \text{Mt} \\ \text{Lead} & 5.6 & \text{Mt} \end{array}$$

Taking into consideration the different recycling rates or the different percentage of primary metal production, one can assume that the energy need for the production of these four metals was about 5.10^9 GJ.

On the one hand, this is only 1.5 % of the world's primary energy consumption of 320.10^9 GJ (together with ferrous metallurgy it is 6 %), on the other hand, it represents costs in the order of magnitude of 25 billions of DM (25.10^9 deutschmark). Obviously it is worthwile for non-ferrous metallurgists to save some gigajoules.

Energy Need for Non-ferrous Metals' Production

We have recently revised literature data on energy consumption figures and have additionally done some own calculations. These calculations were made for the above mentioned "big four" non-ferrous metals and - as a specialty - for tantalum.

Fig. 1 shows the energy need for the production of 1 t of primary metal for iron/steel, aluminium, copper, zinc, lead and tantalum. The figures relate to the metallurgical process steps only, mining and ore beneficiation not being included (2), (3), (4), (5), (6), (7), (10).

The absolute theoretical minimum for the production of a metal is the enthalpy of formation of its compound, in our case its oxide. This amount of energy is necessary to destroy the oxide and reduce the metallic element (8).

Fig. 2 shows this energy need as a function of the ratio metal mass per mole of oxygen.

Table I shows the theoretical energy need values for 298 K, the values for industrial practice (reduction step only, no refining) with a conversion rate of 1 kWh = 3.6 MJ and the factor, demonstrating how much more energy is needed in practice than in theory. The next column in the table repeats the energy consumption in practice, but now with a conversion rate of 1 kWh = 10.0 MJ (36 % efficiency in converting primary energy sources into electricity) and - last column - the new factor based on this.

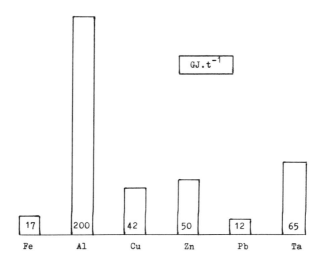

Figure 1 - Energy need for the metallurgical production of 1 t of primary metal

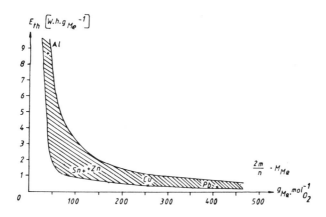

Figure 2 - Theoretical energy need for the oxide reduction as a function of the ratio metal mass per mole of oxygen

Table I. Theoretical and industrial energy need for metal reduction with two different conversion rates kWh-MJ

Metal	$GJ.t^{-1}$ theory	$GJ.t^{-1}$ industry x)	Factor x)	$GJ.t^{-1}$ industry xx)	Factor xx)
Fe	7.38	15	2.0	42	5.6
Al	31.20	54	1.7	150	4.8
Cu	1.35	18 - 29	13 - 21	50 - 80	37 - 59
Zn	5.33	18	3.4	50	9.4
Pb	1.06	11 - 18	10 - 17	30 - 50	28 - 47
Ta	1.57	18	11.5	50	32

x) 1 kWh = 3.6 MJ   xx) 1 kWh = 10.0 MJ

The table above can be discussed as follows: The first column shows the very high amount of energy needed to break up the Al_2O_3 molecule. All other oxides need considerably less, which is due to the decreasing amount of oxygen bound to 1 mole of metal. Lead is the optimum in that respect (atomic weight 207).

In the second column the best present industrial consumption figures are stated, of course for the reduction step only. For that reason there is no correspondence with fig. 1. In case of copper and lead different process routes have different values (the classical processes have the higher ones, the so-called direct reduction processes the lower ones). Energy conversion rate is physical theoretical with factor 3.6.

The "Factor" in column three compares practice and theory and indicates how much more energy our smelters consume than purely theoretically necessary. Aluminium, iron and zinc have the most favourable factors with classical copper metallurgy at the end. But this picture is not right, especially not for electrolytic processes, where primary energy sources have to be converted into electricity, which is done with bad conversion rates between 33 and 38 %. For table I we took 36 %, giving 1 kWh = 10 MJ. In case of hydroelectric power this is more favourable.

For that reason we have calculated (column 4) the industrial energy need with that conversion rate of 10 again and re-calculated the "Factor" too (column 5). Now these values are too high, because never all energy in a metallurgical flowsheet is only electricity, all other sources (fuel oil, natural gas, coal, coke) show better conversion rates. The truth will be in between column 2 and 4, respectively 3 and 5.

It can be stated that of all metallurgical reduction processes of "the big four" non-ferrous metals modern lead metallurgy is the less energy consuming process in absolute figures, followed by the I.S.furnace zinc reduction, classical lead metallurgy and zinc electrolysis. Compared with the theoretical values, pyrometallurgy of zinc ranks first (attention: zinc quality!), followed by aluminium electrolysis. The most energy wasting processes are the classical pyro-, hydro- and electro-winning copper productions.

Break-down per Metal and Unit Operation

Looking for possibilities to reduce the amount of energy needed per t of metal one has to distinguish the various unit operation steps for each metal.

Aluminium

Mining and bauxite transportation require between 10 and 15 GJ per t of Al_2O_3 or 20 to 30 GJ per t of metal: Electric machinery, crushing, diesel fuel for cargo ships etc. The Bayer process needs between 8.5 and 30 GJ per t of Al_2O_3 or 17 to 60 per t of metal. Bayer plants on the high end of that scale can bring their energy consumption down by wet grinding and milling, by using tube reactors with liquid salt heating instead of autoclaves with steam, by optimizing red mud handling and by utilizing fluid bed calcining. The lower limit seems to be 8.5 GJ per t Al_2O_3 = 17 GJ per t of aluminium to-day.

Fused salt electrolysis consumes at least 80 % of the total energy demand. 85 % of that is electricity, 15 % is contributed by anode burning. There are only two possibilities for reducing electricity consumption in the electrolysis: Decrease of cell voltage and increase of current efficiency.

A decrease of cell voltage (only 1.2 V of 4 V is decomposition voltage) can be achieved by lowering anode polarisation (anode surface, chemical composition of the electrolyte, current density), reduce ohmic losses of conductors and electrodes and the bath (conductivity of electrolyte at working temperature, electrode spacing).

An increase of current efficiency is very much linked to a decrease of the back-oxydation of aluminium at the anode. Involved in this very complex problem are many parameters. One has to decrease the electrolyte temperature, giving less solubility of aluminium metal, to increase the electrode spacing with the counter-action of higher voltage, higher ohmic bath resistance, higher temperature, increase current density, once more with the consequence of increasing temperature, increase the alumina concentration in the bath, optimize the cryolithe ratio $NaF:AlF_3$ to approximately 2.6 to 2.9. Besides back-oxydation of metal, there exists back-dissolution and carbide formation too. The consumption of anode carbon is also higher than theoretical. It can be reduced by lower current density and bath temperature and by better anode quality.

Zinc

The share of electrolytic zinc in total world zinc production is about 80 %, leaving the remainder predominantly to the lead-zinc-blast-furnace (I.S. furnace).

Of the total energy need of 50 GJ per t of metal for the hydrometallurgical route 15 % or 7.5 GJ is for roasting, sulphuric acid production, leaching and liquor purification and 80 % or nearly 40 GJ is for electrolysis. There are some minor possibilities for saving energy prior to electrolysis: Zinc dust, steam, dewatering resp. drying of the precipitation products of iron, lead-silver and the other impurities. In the electrolysis itself, the specific electricity consumption can - at a given current density - be lowered by a decrease of the electric resistance of the electrolyte (high sulphuric acid and low zinc concentration with the counter-action of increasing decomposition voltage, by applying higher temperature with the counter-action of decreasing overvoltage), by reduction of the electrode spacing and by bringing down all electrode polarisation effects (higher electrolyte circulation, better cathode surface).

Of course, lower current density reduces the energy consumption too. - An attractive possibility could be to use the 4.5 GJ of heat per t of zinc, which have to be sent to the atmosphere by cooling towers. The low temperature level and the small ΔT, however, obviously prevent this. - Good short-circuit control, optimum current transmittal construction and application of periodic current reversal PCR are further possibilities.

In pyrometallurgy of zinc 54 GJ are needed, of which 9 GJ may be deducted as a credit for lead production and low calorific off-gas. 10 % of the total is for sinter-roasting and sulphuric acid production, 74 % for the blast furnace (mainly coke) and 13 % for zinc refining.

The biggest single part is coke ($35\ GJ.t^{-1}$). The theoretical value lies well below (5.3 !). Considerable saving possibilities should be imaginable: Less reactive coke, high blast preheating temperature, input of other energy sources into the blast furnace, making better use of the furnace off-gas. Heat losses of the furnace through cooling water and radiation can certainly still be reduced too.

The zinc distillation refining process consumes 7 GJ. We have the impression that a better metal purchasing policy with blending of GOB with SHG zinc for various zinc applications could avoid part of this.

Copper

It makes a big difference whether chalcopyrite ores are processed via mining - flotation - classical or modern pyrometallurgy - electrolytic refining or whether it begins with poor oxydic ores and follows the leaching - electrowinning route. Open pit mining and ore dressing by flotation need 13 + 41 = 54 GJ per t of copper. Pyrometallurgy then starts with roasting and concentrate smelting included sulphuric acid plant (18 to 22 GJ) followed by electrolytic refining (7 to 14 GJ), giving a total of about 32 GJ per t of copper. In so-called direct smelting processes approximately 5 GJ less are consumed. The leaching plus electrowinning route needs 25 to 30 GJ for leaching, 20 to 27 for electrowinning and 2 to 3 for melting, a total between 47 and 60 GJ per t of copper. Remark: In fig. 1 an average of 42 GJ is given.

A possibility for saving energy in copper pyrometallurgy is at first to avoid the roasting plus reverberatory furnace smelting route. A reverb furnace is a poor heat exchanger and the heat losses to the environment are considerable (tremendous high off-gas volume). All flash smelting processes are much better and the use of pure oxygen in modern direct smelting processes can bring down the total energy demand to about 12 to 17 GJ per t. The heating value of the sulphur has to be used as far as possible.

In refining electrolysis the electrolytical process itself should need less than 3 GJ. Saving possibilities are in cathode and anode scrap melting with a tolerable maximum of 4 GJ, e.g. by use of a shft furnace.

Leaching plus solvent extraction are high energy consuming because of the big amount of liquor volume to be handled (pumps etc.) and because of the extraction media and sulphuric acid losses. Centrifuging could possibly improve this (counter-action: more electricity needed). The electrowinning electrolysis too can certainly still be improved with respect to minimize energy consumption. In principle, all fundamentals of electrolysis are valid as in case of zinc.

Lead

In classical pyrometallurgy of lead via sinter-roasting - blast furnace - refining the advantage of the high lead content of flotation concentrates cannot be used profitably because of the pecularities of the sinter-roasting process. The down-grading of concentrates with recycled sinter product is a process-theoretic nonsense.

Nevertheless, sinter-roasting together with sulphuric acid production needs only 1.5 to 2.0 GJ per t of lead - the high at. weight of lead being an inestimable benefit. Blast furnace smelting of sinter uses coke as the main energy source in the order of 6 to 7 GJ or between 9 and 12 GJ per t of lead bullion with all other energy consumers included.

Only the modern direct smelting processes make sufficiently use of the costless sulphur fuel in the concentrate. The energy need depends very strongly on the lead content of the feed to the smelting unit. GJ values between 2 and 10 are reported. The disadvantage of the highly productive modern processes with better space-time-yield is the high lead content in the primary slag, which has to be re-processed, often by use of electricity.

Lead refining is stepwise with numerous heating-up and cooling-down procedures, but the energy need seldom exceeds 2 GJ per t of refined lead.

Energy saving possibilities are either in coke consumption in the blast furnace (as in zinc blast furnace reduction the theoretical energy need is well below industrial practice values) or in making use of high lead containing concentrates in direct smelting processes with non-electric slag impoverishment.

Experimental Results

The experimental results in laboratory scale which we have achieved till now with the aim to reduce energy consumption refer to aluminium, zinc, copper and tantalum.

Aluminium

In aluminium electrolysis we tried to bring down electrode polarisation and electrode spacing by the use of inert anodes, made from CaO-stabilized zirconia ZrO_2, an oxygen ion conducting anode. Liquid silver served as the electrone conducting medium in contact with a graphite rod.

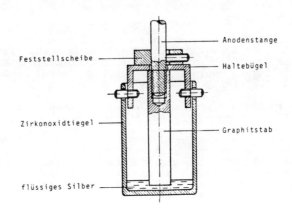

Figure 3 - Inert anode

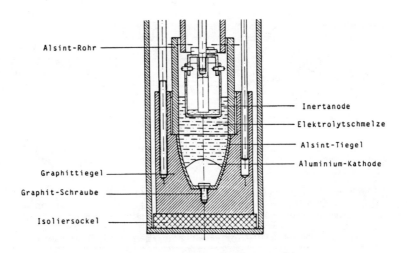

Figure 4 - Electrolytic cell with inert anode

With the low current density of 0.2 $A.cm^{-2}$, which was necessary for the stability of the anode, we reached an excellent current efficiency of 97 % - much better than the best value of 86 % with preliminary anode carbon tests. The chemical stability of the zirconia, however, was so bad (the stabilizing CaO dissolves in the electrolyte), that the tests had to be stopped after one hour. This is the unsolved problem.

Tests with periodic current reversal PCR showed a slightly lower cell voltage, which was desired, but poor current efficiencies between 27 and 51 %. This is due to a rapid re-dissolution of aluminium during the cathodic cycle. So, the PCR tests were a failure.

Zinc

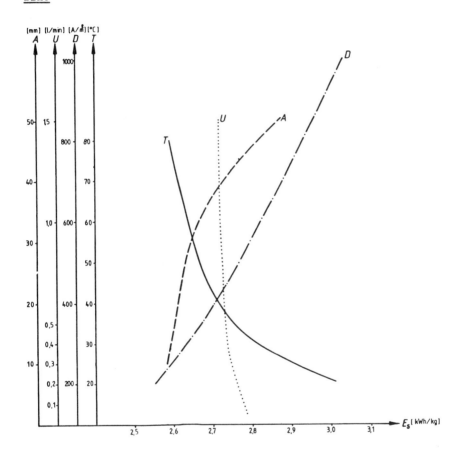

Figure 5 - Influence of temperature T, current density D, electrolyte circulation U and electrode spacing A on specific energy consumption in zinc electrolysis

In zinc electrolysis we began with a pure zinc sulphate electrolyte without any impurities (70 g/l zinc and 170 g/l free sulphuric acid) and tried to reduce specific electricity consumption by varying the parameters temperature (from 20 to 80 °C), current density (200 to 1000 $A.m^{-2}$), electrolyte circulation (0.1 to 1.5 $l.min^{-1}$) and electrode spacing (10 to 50 mm).

The results are shown in fig. 5. The specific electricity consumption ranges from about 3 to about 2.6 kWh per kg of zinc. Favourable are high temperature and high electrolyte circulation and low electrode spacing and current density. 3 and 2.6 kWh per kg correspond to 30 and 26 GJ per t - well below the industrial value of about 40.

We then added impurities to the synthetic electrolyte: Cadmium, copper, arsenic, cobalt, nickel, antimony, germanium, but always separately only one at the same time. The results are shown in fig. 6.

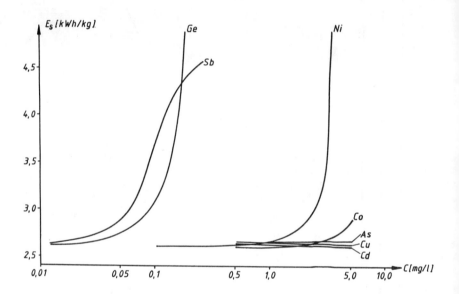

Figure 6 - Influence of content of impurities in zinc electrolyte on specific energy consumption

The bad results with cobalt and nickel and especially with antimony and germanium are due to local element formation and break-down of hydrogen overvoltage, which lead to re-dissolution of zinc already deposited before. In order not to surpass the favourable electricity consumption of 2.6 kWh per kg of zinc there are only allowed the following concentrations of impurities: 5 mg/l for As, Cu, Cd; 2 mg/l for Co; 1 mg/l of Ni; 0.02 mg/l for Sb and Ge.

It will take us a lot more time to test the various possibilities of adding at least two impurities at the same time. The first results we got with cobalt + nickel. With the conditions temperature = 35 °C, current density 400 A.m^{-2}, electrode spacing 20 mm, electrolyte circulation 0.6 l.min^{-1}, time 8 h we obtained fig. 7 and fig. 8.

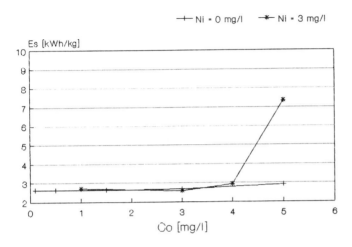

Figure 7 - Influence of nickel additions to cobalt containing zinc electrolyte on specific energy consumption

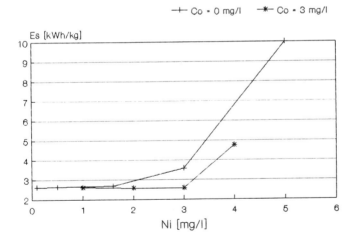

Figure 8 - Influence of cobalt additions to nickel containing zinc electrolyte on specific energy consumption

It seems that neither cobalt nor nickel accelerate each other's influence on increasing electricity consumption, but it is too early to elaborate on this.

Copper

We have just started the experiments on copper winning electrolysis. First results can only be given at the Conference in Cologne in september.

Tantalum

Tests on tantalum fused salt electrolysis were first conducted with direct current and showed the following results as a function of current density, fig. 9.

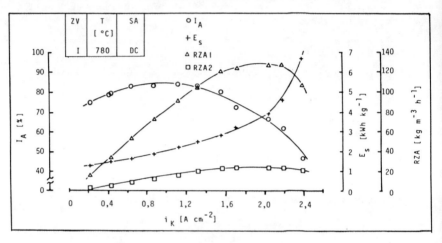

Figure 9 - Specific energy consumption +---+ in tantalum electrolysis as a function of current density

Electricity consumption is between 1.5 and 7.0 kWh per kg of tantalum. Periodic current reversal PCR decreased this value at 2.4 A.cm^{-2} to 4.7 - a 33 % reduction and a very good result.

References

1. Metallstatistik 1978-1987. Metallgesellschaft AG, Frankfurt, 1988

2. L. Holappa and P. Paschen, "Energy Consumption in Reduction-Melting Processes in Ferrous Metallurgy", Scandinavian Journal of Metallurgy, soon

3. G. Leuprecht and P. Paschen, "Verminderung des Energieverbrauchs in der Aluminiummetallurgie", BHM 132 (1987) 160

4. A. Fölzer and P. Paschen, "Verminderung des Energieverbrauchs in der Kupfermetallurgie", BHM, soon

5. A. Anzinger and P. Paschen, "Verminderung des Energieverbrauchs in der Zinkmetallurgie", BHM 133 (1988) 333 and BHM 134 (1989) 12

6. E. Schöll and P. Paschen, "Verminderung des Energieverbrauchs in der Bleimetallurgie", BHM, soon

7. W. Köck, "Optimierung der Raum-Zeit-Ausbeute und des spezifischen Energieverbrauchs bei schmelzflußelektrolytischen Gewinnungsprozessen von Metallen der IV. und V. Nebengruppe des Periodensystems", Dr. thesis, Montanuniversität Leoben, 1988

8. P. Paschen, "Verminderung des Energieverbrauchs im Metallhüttenwesen - Theoretische Grundlagen", BHM 133 (1988) 161

9. W. Köck and P. Paschen, "Anwendungsmöglichkeiten des Umkehrstromverfahrens in der Schmelzflußgewinnungselektrolyse", Metall, submitted

10. H. Hilbrans and P. Paschen, "Energieverbrauch bei der Kupfergewinnung aus sulfidischen und oxidischen Erzen", Erzmetall 34 (1981) 639

USE OF OXY-FUEL BURNERS IN SMELTING LEAD BATTERY SCRAP
IN SHORT ROTARY FURNACES

Dipl.-Ing.K.F.Lamm, Managing Director
Blei- und Silberhütte Braubach GmbH,
D 5423 Braubach, West Germany

Abstract

After discussing the theoretical principles of heat transfer, it is shown that the share of heat transition by convection is negligible against other mechanisms of energy input.

Operating short rotary furnaces with oxy-fuel burners requires substantial adaptation of existing equipment and metallurgical practice, or in case of new investment, different layout and design.

Major results are:

1. Energy savings
   60% fuel and electric energy
   Fuel savings may pay for oxygen

2. Effects on pollution control
   -decrease of SO2 and NOx rate at higher contents per m3
   -decrease of waste gas and flue dust production by 60%.

3. Metallurgical results
   Optimizing effect on batchwise operation, lower viscosity of slag and hence lower metal losses.

1. Theoretical principles of heat transfer

The continuous counter-current kiln or waelz-process has always been known as the optimum of heat recovery. The batchwise operating rotary furnaces or rotating reverbs and above all, the short rotary furnaces have always been glorified as the maximum of energy input by combining convective heat transfer plus direct radiation.
From this point of knowlegde, it would have been rather riskfull to replace the long and soft reverted flame of a normal burner by the well known short and sharp beam of an oxy-fuel burner.
But, going more to the details, the mechanisms of heat transfer revealed themselves to be of greater complexity than expected.

FIG.I shows already four different ways of energy input:

A) direct radiation of flame (gas phase) to surface of solid phase bed

B) reverb-effect - reflection + radiation of refractory lining itself. The reflecting surface is more than 60% of the total

C) convective heat transition from phase to surface of solid phase bed

D) heat transfer by "cooling" the high temperature reverb-area, direct contact between refractory lining and solid phase bed.(seems to be the most important effect)

FIG.II shows another mechanism

E) solid phase mixing effect, which might be named "solid state convection".

From this point of view , it seemed to be more successful to neglect or even to give up only one path of heat transfer out of five.

The importance of path E is demonstrated by FIG.III, i.e. rolling bed is better than a slumping one, or, as every rotary furnace operator ought to know, a sudden increase of temperature ("boosting") at the beginning of each batchwise operation results only in a liquid layer between charge and furnace bottom, which allows the formation of slumping bed.This additionally prevents the heat transfer following to path D.

FIG.IV is again showing the predominance of both mechanisms D plus E, and, of course, the advantage of a 100 % rotary furnace against an oscillating one.

2. Practical results

In spite of the fact, that most of the expected results have been evaluated or at least roughly estimated before, some of them surprised by their accuracy when occuring.

2.1. Saving of energy

Fuel savings of 53% have been proved over a long period of test runs. Including the non-producing periods of keeping warm during weekends, the savings reached 60% due to the fact, that oxy-fuel-burners can be much better controlled in the low-load range. In our case the figures have been 37 kg/h against formerly 110 kg/h. This is not the result of use of oxygen itself, but of the unsatisfactory control behaviour of a normal 360 kg-per-hour-burner in the range below 30% of maximum.

Another contribution to the saving of energy results from the electric power absorption of cooler and filter equipment due to 60% less waste gas volume.
At least both shares of energy saving will normally pay for the cost of oxygen.

2.2. Effects on pollution control

Among these effects, only the most interesting ones will be widely discussed.
The most impressive result was the function of the oxy-fuel-burner itself as a stable after-burner instead of an additional pilote-burner in the after burning chamber. It is obviously very important, to have always a good after-burning function and not having lots of unburned hydrocarbons out of the fuel itself or from organic material of the charge going through the filter and finally to the stack.
In this context it should be pointed out, that oxy-fuel-burners are the best guarantee against e.g. dioxine formation, because there are all requirements fullfilled:
a permanent flame of high temperature and enough time for a close contact and complete reaction due to small gas volumes or low gas particle speed.
The theoretical waste gas volume is less than 10% of the original one. Reducing the total gas flow rate to only 30-40% of the former quantity means in the same moment a considerable increase of hygienic ventilation air. Following to the decrease of waste gas volume and to corresponding gas speed, the flue dust production sloped down by the same amount (exact figures are from 193 kg per hour down to 57).

This means on the other hand, that particle size is now 100% below 1 micrometer, as coarse dust is no longer dragged along due to low gas speed.

NOx formation by higher flame temperatures was less than expected. About half of NOx is estimated to result from nitrogen content of fuel.

The total quantity of atomizing air at about 25 m3/h can be neglected. There are only some peaks (but always below 300 mg/m3) to be found during charging, when access of air to the burner is possible. This leads automatically to the idea of differently shaped furnaces or even to a continous process.

Other results may be found on the following table I

TABLE I

ENERGY SAVINGS

Fuel:	up to 60 %
Electricity:	up to 60 %

POLLUTION CONTROL

Waste gas:	70 % less
Flue dust:	60 % less
Hydro carbons:	100% less
Dust content:	< 5 mg/m3
Filter bags:	lifetime + 20 %
SO2 + NOx:	higher contents per m3, but decrease total output
HO2:	increase of dew point

METALLURGY:

Metal in Slag:	0,7 units less

2.3. Metallurgical results

The main influence on day-to-day operation of furnaces is a generally optimizing effect on the batchwise process, thus allowing to achieve 5,7 charges per day instead of 5,3. This is, in terms of increase of production, rather a poor effect. But it is the reduction process itself, which does not allow higher melting rates without increasing metal losses during slag formation.

On the other hand, the lower viscosity of slag shortens the tapping period and gives more reduction time.

In our case, it has never been planned to shorten the total batch cycle, but in other applications, e.g. melting of scrap or ingots, a considerable increase of melting rate can be achieved.

Also the decrease of average lead content in slag by only 0.7 units might be considered as a poor result, but this is a total amount of about 150 tons of lead per year, which is not only to be seen as more production but as less lead given to the slag deposit.

3. Adaptation of equipment and metallurgical practice

In case of already existing installations, at least the gas cooling and cleaning equipment need some reshaping. Air preheating units and waste heat boilers are no longer of any use.

Cooler and filter surfaces have to be reduced to avoid corrosion and operating temperatures below the increased dew point.

Installation of the oxy-fuel burner itself requires the most accurate control of fuel flow rate in order to maintain always the exact stoichiometrical ratio of $\lambda = 1.00$ (1 kg of heavy oil requires 2,2 m3 of oxygen).

Reflecting upon new investments, the choice of an oxy-fuel burner is one of the most elegant ways to come to small units with very low installation and operating costs. However, the idea of a waste heat boiler should be given up before.

After 7 years of experience with 2 furnaces on heavy oil and more than 2 years with 1 furnace on natural gas, we are prepared to switch over to 100% natural gas per March 1st 1991.

As already mentioned before, there is no use of "boosting" up temperatures (which in principle is possible by using an oxy-fuel-burner, you are able to destroy mostly everything).

On the contrary, the tendency to smaller units with reduced furnace volume (in order to avoid air access and NOx formation) gives way to the reflection, to treat at least the grids metal fraction and the PbO_2 /PbO/ $PbSO_4$ containing paste in a new type of continuously operating furnace.

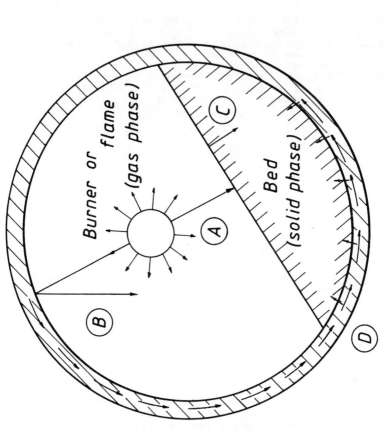

FIG. 1

heat transfer in a rotary furnace
[Pawliska, 1980 T.U. Berlin]

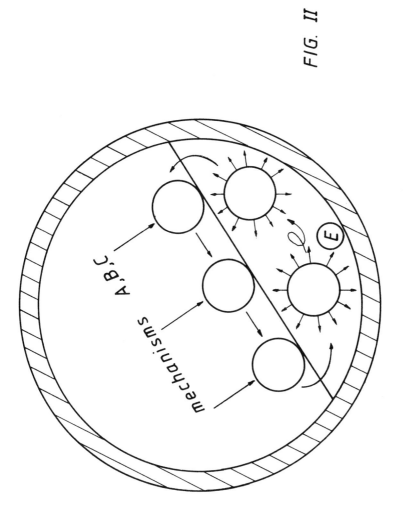

FIG. II

heat transfer in a rotary furnace
[Pawliska, 1980 T.U. Berlin]

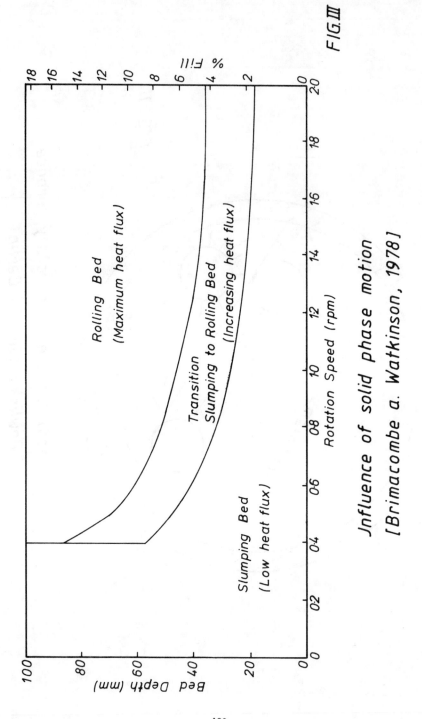

FIG. III

Influence of solid phase motion
[Brimacombe a. Watkinson, 1978]

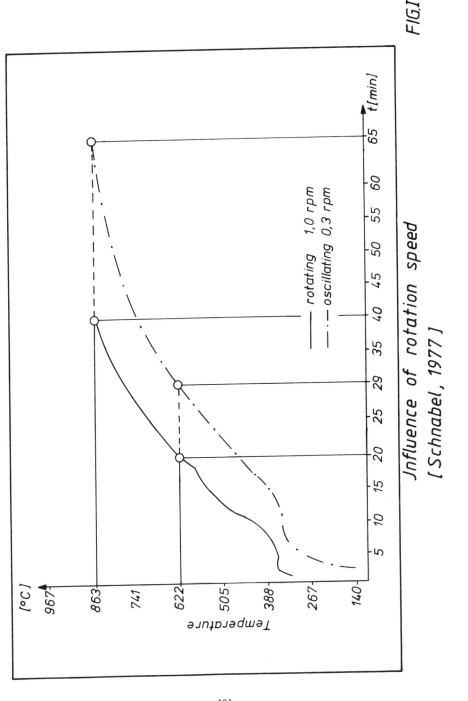

FIG.IV Influence of rotation speed [Schnabel, 1977]

LOW WASTE TECHNOLOGY FOR REPROCESSING BATTERY SCRAP

(a description of Blei- und Silberhütte Braubach - BSB Recycling GmbH)

K.F. Lamm, Dipl.-Ing.
Blei- und Silberhütte Braubach
BSB Recycling GmbH
D-5423 Braubach

A. E. Melin, Dr.-Ing.
Stolberg Ingenieurberatung GmbH
Consulting Engineers
D-5190 Stolberg/Rhld.

Abstract

Ways of processing battery scrap, advantages, disadvantages, constraints.
Targets of development of a new low waste technology:

- make use of most - if not all - constituents of old batteries and their conversion into highly valuable products,
- avoid non-usuable fractions which have to be dumped,
- drastic reduction of pollution emissions and long therm compliance to stringent pollution regulations considering acceptable operating and investment costs,
- improvement of the working place conditions.

Description of the battery breaking and separation system, of the desulphurization plant (where the sulphate in the paste is converted to sodium sulphate) and of the smelting/refining operations.

Introduction

Mining, smelting and refining of lead and silver at Braubach has an eventful history. First mining activities can be traced back to at least 1301 and first smelting operations to 1693. The actual "Blei- und Silberhütte Braubach" (BSB) was founded 90 years ago in 1899, and taken over by Metallgesellschaft AG in 1923.

In former times this smelter treated indigenous, and increasingly imported, concentrates to reach, shortly before World War II a yearly production of about 60.000 t of refined lead and 60.000 kg of silver.

Primary smelting (i.e. starting from lead concentrates) was discontinued in 1963, but the refinery still treated foreign lead bullion and some raw material diverted from secondary sources (i.e. from scrap). Scrap was the exclusive raw material source at BSB from 1977/78 on.

1977 also marked the beginning of a cooperation with the Italian Tonolli group, which had a strong experience in secondary smelting. Tonolli took shares in BSB which is now owned on a 50 % basis by Metallgesellschaft AG and IFIM International B.V., Amsterdam another Tonolli participation.

Generalities

Processing of battery scrap in secondary smelters mainly occurs

- without any pre-treatment (tel quel) by smelting whole batteries in a blast furnace, or
- after decasing, respectively breaking either with or without separation of the battery components in a blast furnace (rarely) or in rotary furnaces (more frequently).

Processing of whole batteries or decased battery scrap has led in the past to several problems, which - independently from the selected smelting process - concentrated in the gas-cleaning sections of the plant:

- the casings (in former times usually hard rubber, nowadays predominantly polypropylene), burn, and the gases of decomposition have to be transformed into compounds troublefree to the environment. This procedure (afterburning) requires high additional quantities of energy and multiple quantities of conditioning air.
  Furthermore, polypropylene is a valuable material worth recycling (not burning!).
- abt. 50 % of the separators (between negative and positive electrodes) are made out of PVC. By thermal decomposition of PVC chlorine and metal chlorides evolve and concentrate in the flue dust. As the latter has to be recycled, chlorine enriches and at a certain threshold a bleed-off for chlorine has to be found.
  The captation of some metal chlorides in bag filters may cause difficulties and additional environmental loads.

Other problems may occur

- in relation with the sulphur content of the feedstock
  with slag and matte fall
  with SO_2 emissions
- in relation with the antimony content of the feedstock
  with an overloading of the refining section (if and when the production of soft lead is intended).

From those drawbacks it appears that a foregoing separation of all constituents of scrapped batteries before melting resp. smelting is meaningfull.

Other existing concepts of separation plants, however, overlooked some relevant aspects like

- an increasing number of batteries (if not all) are delivered with their acid content
- elimination of PVC
- recuperation of clean polypropylene, etc.,

so that a new solution to an old problem had to be elaborated.

New technological developments at BSB

Targets of the developments which occurred at BSB since 1977 were:

- to make use of most - if not all - constituents of old batteries and their conversion into highly valuable products,

- to avoid non-usable fractions which have to be dumped,
- drastic reduction of polluting emissions and long-term compliance with stringent pollution regulations considering acceptable operating and investment costs,
- improvement of working place conditions.

With the processes actually in operation at BSB those objectives have already been fulfilled to a large extent:

- contrarily to current practice in other smelters the new plant of BSB is designed to treat complete old batteries including their acid fillings. This diluted acid (10 - 20 % H_2SO_4) can be used for metallurgical purposes or for pickling solution, avoiding the dissipation of tremendous quantities of acid (25 - 30 % of the weight of the complete battery!),

- the careful breaking technology of complete old batteries (developed by Tonolli) is a prerequisite for the clear separation of the various constituents in definite fractions: metallic lead, lead paste, PVC and ebonite (hard rubber) respectively polypropylene. This, in turn, is a prerequisite for the re-utilisation of clean polypropylene granules.

- The battery paste is a mixture of lead oxide and lead sulphates. In conventional practice, the metallurgical treatment of this paste is linked to a high slag fall and/or to high quantities of recycled products (lead losses!) as well as to high sulphur (SO_2) emissions (1).
  To avoid those drawbacks a desulphurization process was developed in which lead sulphates are transformed in lead carbonate and anhydrous sodium sulphate.
  Lead carbonate can be treated (reduced) without any problem in conventional metallurgical furnaces (e.g. short rotary furnaces) and anhydrous sodium sulphate (99.5 % purity, 10 ppm Pb) is another valuable product.

- Ebonite (hard rubber), polypropylene (PP) and PVC are also separated from each other. Clean polypropylene scrap - 99.5 % purity - is obtained as granules and ready for re-use.
  The hard rubber fraction, which is practically free from PVC (max. 1 %), can be used as additional fuel, for example in cement kilns.

  Research is currently in progress to find an outlet for the PVC scrap.

In summary, the processing of complete old batteries now yields:

- metallic lead (from grids, pole bridges, etc.), which is molten to obtain lead bullion which has to be refined,
- lead carbonate (from desulphurization of lead paste) with less than 1 % of residual sulphur, which is smelted to obtain a fairly clean lead bullion,
- anhydrous sodium sulphate (99.5 %, 10 ppm Pb) for sale
- clean polypropylene granules (98.5 % or 99.5 % after further processing) for sale
- ebonite (hard rubber) with max. 1 % PVC for sale
- PVC scrap and
- diluted sulphuric acid both for possible re-use

and the target of a nearly wastefree method of processing scrap batteries is reached at BSB.

The process at BSB

The basic steps of the process flow-sheet now in use at BSB are depicted in Fig. 1.

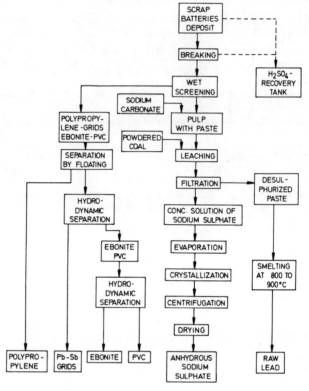

Fig. 1 Processing of old batteries at BSB (simplified flowsheet)

The plant can be divided in the following sections

    a) battery breaking and separation systems yielding metallic lead, lead paste, ebonite (hard rubber), polypropylene (PP), and PVC fractions,

    b) desulphurization plant where the lead paste is transformed into lead carbonate also yielding anhydrous sodium sulphate,

    c) polypropylene plant where PP fraction is upgraded to clean PP-chips,

    d) smelter where the metallic lead fraction is molten and the lead carbonate smelted to yield lead bullion which has to be further refined,

    e) refinery where the lead bullion obtained in the smelter is refined and, if required, alloyed to meet the various specifications of lead and lead alloy customers.

The equipment flowsheet of sections a) and c) is shown in Fig. 2.

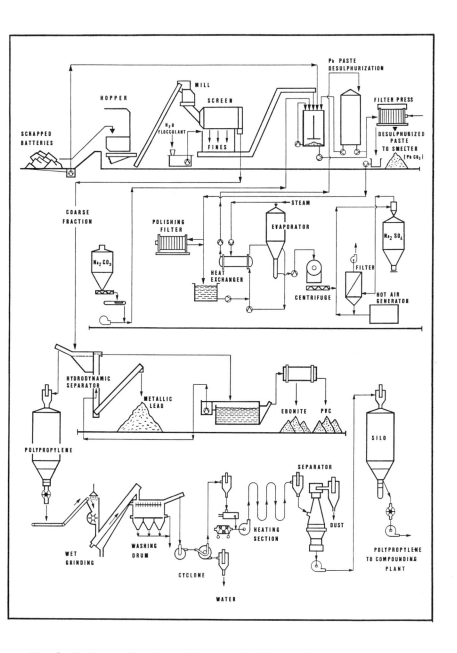

Fig. 2 Equipment flowsheet of battery breaking and separation system and polypropylene upgrading at BSB

Battery scrap is drawn from the stockyard (Fig. 3) by loaders and conveyed to a hopper (loader) which has side walls of stainless steel to resist to the corrosion of the acid still present in the batteries.

From there, the scrap is carried to the crusher via a plate conveyor of stainless steel.

Fig. 3 Stockyard for scrap batteries

All the a.m. sections (stockyard, loader, conveyor belt and crusher) are drained to collect the acid which is conveyed to the acid tank for further use.

Crushing is carried out by a hammer mill as to guarantee a careful breaking of the scrap.

The crushed material is discharged onto a wet screening drum where the separation of the paste from the remaining material occurs.

The latter is carried to the subsequent separation stage where polypropylene (PP), metallic lead (Pb), ebonite (hard rubber) and PVC are separated from each other.

PP is extracted from a separation tank and discharged into a bin, while Pb, hard rubber and PVC pass on to a further treatment into a hydrodynamic separator.

Lead grids and poles are extracted from the bottom of the hydrodynamic separator and conveyed to a bin for further treatment (melting).

Ebonite and PVC are now separated from each other and conveyed to separate bins.

All process waters are collected in a circuit and recycled in the plant without any discharge into the water treatment plant.

The PP fraction is ground, washed, cycloned, heated and fed to a compounding unit.

The lead paste is conveyed to section b) (desulphurization plant) see Fig. 4.

Fig. 4 Control room for lead desulphurization plant

Here, the paste mud is charged to reaction tanks where, in the presence of sodium carbonate, the following reaction takes place

$$PbSO_4 + Na_2CO_3 \longrightarrow PbCO_3 + Na_2SO_4$$

The desulphurized paste is separated from the concentrated solution of sodium sulphate by means of a filter-press.

After careful washing with water on the same filter the lead carbonate is discharged and stored for smelting.

The clear sodium sulphate solution is pumped to an evaporation unit where the anhydrous salt progressively separates. After centrifugation sodium sulphate is dried in a current of hot air and stored in a bin ready for dispatching.

Metallic lead and lead carbonate are molten respectively smelted in short rotary furnaces (section d smelter equipped with 4 short rotary furnaces) to produce lead bullion and a discardable slag.

The principle of short rotary furnace operations including process gas cooling and dedusting is shown in Fig. 5.

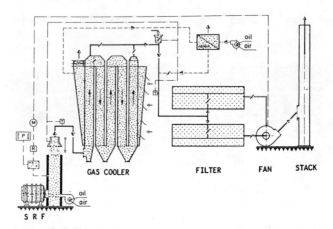

Fig. 5 Short rotary furnace with gas cooler, filter and stack

The short rotary furnace is charged by the front door with the material to be molten or smelted. The reaction(s) take(s) place under the action of the heat supplied by an oil burner (in the prevalent case oxyfuel burners are preferred (2)). Metal and slag are tapped from the charging side while the dust-loaden process gases leave the furnace through the rear end; they are first cooled in a forced draft cooler, after quenching with the air drawn around the furnace (sanitary air) then dedusted in a bag filter before being discharged (as clean gases) into the atmosphere.

The last section (e-refinery) is a classical but very simplified lead refinery. As lead bullion originating from secondary sources does not contain appreciable (and, consequently, valuable) amounts of copper, silver and bismuth, the steps of desilverizing, dezincing and debismuthizing (which are mandatory in primary lead refining!) can be by-passed and refining of secondary bullion is limited to

- drossing
- sulphur decoppering (to remove all but traces of copper, if any)
- softening (to remove tin antimony and/or arsenic, if required)
- alloying and final refining (if required, according to specifications)

The operations are carried out batchwise in kettles according to customers requirements. The metal is then cast in commercial shapes (45 - 50 kg pigs or jumbos, as required), bundled (if required) and ready for dispatching.

Summary

By introducing several new technologies BSB has achieved the goal to develop a nearly wastefree method for processing 60.000 t/a of battery scrap, Fig. 6.

Fig. 6 BSB smelter and a part of the city of Braubach

Most of the constituents of old batteries are recovered and upgraded to high value products (refined soft and hard lead, PP granules, hard rubber, anhydrous sodium sulphate and diluted sulphuric acid).

Compared to classical methods

- the amount of material to be dumped (e.g. slags) was reduced by 60 %
- the amount of lead contained in slags was reduced by 80 %,
- SO_2 discharges to the atmosphere were reduced by 90 %.

Moreover, the amount of fluxes required for smelting as well as the energy consumption were considerably lowered.

Fig. 7 shows a typical material balance for a Tonolli-battery breaking and separation system (CX) including smelting and refining of the products.

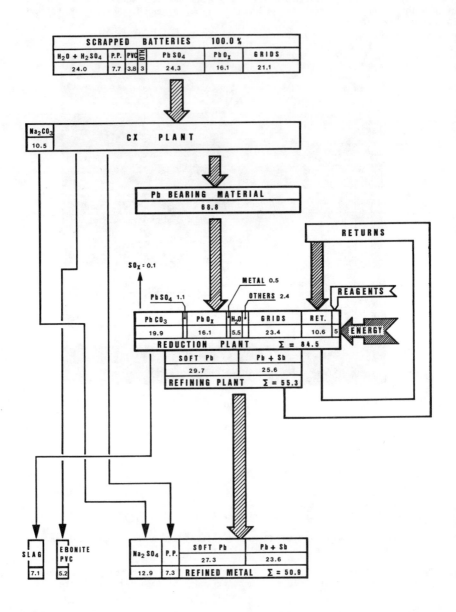

Fig. 7 Typical material balance for a Tonolly CX battery breaking and separation system including smelting and refining

Fig. 8 shows in an impressive way how the lead emissions could be lowered by the introduction of various new techniques at BSB since 1977.

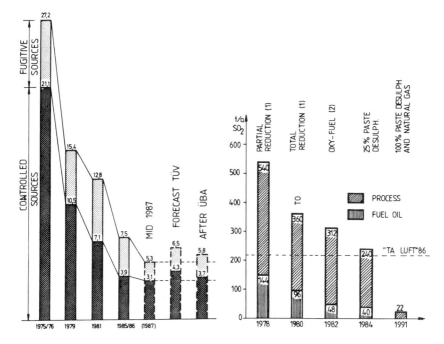

Fig. 8 Emission register of BSB 1975/76 to 1987 (in t Pb/year)

Fig. 9 SO_2 emission from the treatment of 60,000 t/a of scrapped batteries at BSB

SO_2-emissions have also been drastically reduced since the introduction of paste desulphurization, see Fig. 9

The development of these new technologies was sponsored partly by the German Environmental Protection Agency (Umweltbundesamt), the Ministry of Research and Technology (Bundesministerium für Forschung und Technologie), local authorities (Land Rheinland-Pfalz) and the European Community. Their kind interest and financial support is gratefully acknowledged.

The Tonoli CX breaking and separation system was developped by Dr. R. Capetti and Dr. M. Olper.

References

1. K.F. Lamm and A. Melin, "Beitrag zur Verhüttung von Akkuschrott" Erzmetall 33 (1980) 275 - 279
2. K.F. Lamm, "Use of oxy-fuel burners in smelting lead battery scrap in short rotary furnaces" TMS-AIME, Annual Meeting, New Orleans 1986

PROCESSING OF LEAD-ACID-BATTERY-SCRAP:

THE VARTA-PROCESS.

M.Koch and H.Niklas

KRUPP Industrietechnik GmbH, Duisburg and
VARTA Batterie AG, Hannover, Germany

Abstract

VARTA Batterie AG of Hannover, Germany, has developed a continuous battery scrap recycling process, based on direct smelting of whole spent lead acid batteries containing lead, lead compounds and organic material from casings and separators. The lead boullion from the furnace is being refined to battery grade lead alloys. By-products are minor amounts of slag and matte.

A reference plant of 25,000 tpy metal output has been in operation for many years, meeting the stringent regulations of environmental protection. The design and operation will be given in detail.

Recovering of lead from battery scrap is an old art in many developed countries. Some of you might remember the times when part of that job was done by hand, as our first slide brings back to memory.

Fig. 1

We are under the impression that at least some of the methods which presently are recommended for such recovery of lead are more or less an extrapolation of those historic methods of scrap breaking.

For 20 years VARTA has developed and operated a recycling process which has cut radically all ties and ropes to that history; mainly for two reasons: economy and reliability. We do think that both aspects can be demonstrated in our contribution to this conference.

Krupp Industrietechnik GmbH of Duisburg, Germany, holds a licence to VARTA's process since 1984 and is marketing the process world wide.

When the process was developed the battery cases were made of ebonite, i.e. hard-rubber, containing up to 70 % of inorganic filling material. After scrapping of the batteries the broken pieces, contaminated with lead-compounds, never had any market-value, rather the opposite: discarding created long lasting environmental problem areas. - Even now - with the battery cases generally being made of plastic material - we were quite unable to calculate any profit and economy for the complicated process-steps of breaking and separating various metallic and non-metallic fractions that would meet the expectations of our top management.

From such considerations the basic decision was derived to process the whole batteries, no breaking, no separation of anything. Through this rather simple basic concept we got rid of a whole string of equipment with all related consequences of investment-cost, labor, maintenance and repair.

Fig. 2 gives a schematic view of the few essential blocks of the process:

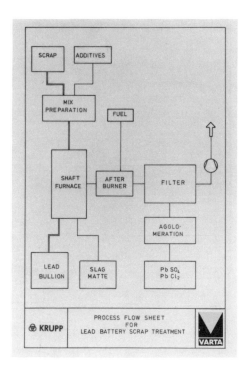

Fig.2

The first step is to prepare a coarse mixture of the scrap with some additives: coke, iron, limestone and a relatively large amount of slag, "return-slag" as we call it.

This mix is fed into a shaft-furnace which is a somewhat modernized form of the good old waterjacket furnace. Fig.3 gives a schematic view of the furnace. We think you all are familiar with the principle.

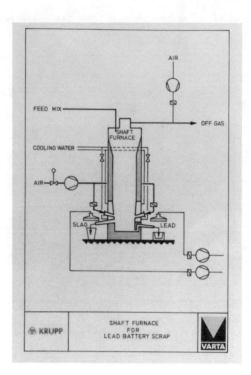

Fig.3

To give you an idea of the size: the furnace is about 7 m (23 ft) tall and has about 4 m² cross-section at the level of the tuyeres. The mixture of scrap and additives is slowly travelling down with increasing temperature. At first, in the upper part water is evaporated. Later, at 300 to 500°C the plastic material and the cellulose and resins of the battery separators are decomposed or "cracked". At this stage metallic lead parts melt and do react with the PbO_2 of the pos. electrodes.

Further down the PbO is reduced by the carbon-monoxide, later also the $PbSO_4$. Different from a rotary furnace, the shaft furnace is a strongly reducing piece of equipment. In any part of it we have surplus of CO. So, the oxides of sulphur are reduced to S^0, elementary sulphur which - of course - immediately reacts with the metallic lead to form PbS. So, the sulphur is eliminated from the gas phase. We'll come back to this point further on.

The slag we are using is melting only above 1000°C. So, it serves as inert carrier in the upper and middle part of the furnace and provides the porosity and homogenity required for uniform reaction across the furnace. This porosity-parameter is considered rather important. In consequence the air pressure at the tuyeres is unusual low, compared to other shaft furnace operations, i.e. 160..220 mm water column.

In the burning zone at the tuyeres-level we have a temperature of 1100..1150°C. This is also the temperature of the slag on tapping.

The lowest part of the furnace, the crucible, contains two liquid phases, the lead metal and - floating thereon - the liquid slag. The metal which is formed above this level has to pass through the liquid slag. But, since the temperatur on this level is much higher than for example in a rotary furnace, the viscosity of the slag is rather low and therefore the separation of lead and slag is rapid and exceptionally good. That means the lead content of the slag, after tapping, is low, about 1 % or less. Also, and equally important, the amount of slag that is newly formed is very low, about 50 kg per tonne of bullion. The loss of lead with the surplus-slag is only $\approx$ 0.05 % of the total lead input. - Under environmental aspects it is important that the slag is highly water resistant due to the fact that we are **not** using sodium-carbonat in the mix. The surplus-slag can be deposited without the penalty for hazardous material.

The tapping of the slag is discontinuous, every 10 to 15 minutes, whereas the lead is leaving the hearth through a siphon. It is cooled down to about 450° in a forehearth and then cast to ingots.

As indicated above, the return-slag is providing a uniform porosity within the furnace. But another important function is as caloric regulator. The heat generated in that process is relatively large. So, we use the recirculation of slag as buffer and carrier of heat.

An important aspect of metal smelting at least in Germany and some other European countries is the emission of SO_2. So, particular attention has to be paid to the sulphur. At present European battery scrap, after draining the liquid acid, contains about 39 kg S per tonne of lead. By legislation the emission of SO_2 in the off-gas is limited to max.800 mg/m³. So, either an expensive flue-gas scrubbing has to be installed or the sulphur has to be removed by chemical or metallurgical steps. In our "philosophy" we never were in favor of using wet processes or wet-chemistry in a smelter-plant. Therefore the traditional shaft-furnace practice has been modified to make the formation of <u>matte</u> the main outlet of the sulphur.

By far the largest portion of the sulphate and H_2SO_4 are converted to $S°$ and then to PbS. - Now, because of the amount of iron-scrap that is added with the mix, the equilibrium

$$PbS + Fe \longleftrightarrow Pb + FeS$$

is shifted far to the right hand side. The matte consists mainly of FeS and Fe, with 6..9 % Pb. The X-ray-analysis shows clearly that the Pb is contained as metal; lines of PbS are absent. The matte has 25..26 % S and is therefore an excellent carrier to eliminate the sulphur.

However, things are not quite as simple as they sound. Here is a part of the "proprietary know how" that is available for licencees only. Also, this appears to be an important contribution to the overall economy, compared to competing methods; provided they are subject to the same environmental protection laws.

Back to technical aspects: The phase [FeS + 0.35 Fe (+Pb)] is liquid at 1100°, but has very low solubility in the slag. The liquid mixture of slag + matte is tapped into crucibles where it separates easily into two layers. After solidification they can be separated mechanically.

The top of the furnace is carefully sealed to collect the flue-gas. The main components of this gas are CO_2, N_2, water-vapor and a few percent of carbon-monoxide, besides carbonblack, crack-products and lead compounds. Its temperature is about 200°C and it is further heated by direct firing of additional fuel. So, all combustible material is burned at 900 to 1000°. After cooling to < 200° it is passed to the bag-house. In the stack the dust concentration is below 2 mg/m³. -

The entire track of the gas is permanently kept at a slightly negative pressure level, so lead-bearing dust cannot escape. We think this is an important contribution to environment-protection .

To summarize and to revert to the initial claims of economy and reliability:

The overall economy is certainly a function of several components, mainly the restriction to few and simple process steps, inexpensive equipment; no expensive additives like sodium-compounds; avoiding hazardous deposits and their financial penalties.

The reliability is largely provided by restricting ourselves to rugged standard technology for smelter plants. Sophisticated equipment may be good-looking on the drawing board, but sometimes it looks rather poor after a couple of months of operation. So, that was not our choice.

Two sizes of plant are presently the standard-offer by Krupp: with 25,000 t metal-production per year and with 12,000 tpy respectively. Less than 12,000 t metal does not appear feasible with our system.

"Contibat" process for treatment of scrap lead batteries

Dr. Reinhard Fischer, Teichstraße 14, 5100 Aachen

Abstract

A number of processes for treatment of scrap lead batteries are known and in use.
The new "Contibat"-process has been in use for several years with good results. It differs from the other processes by only crushing and neutralizing the batteries after separation of acid. This material is then smelted in a rotary furnace by oxygen. During the continuous charging the plastics are burnt and lead is smelted simultaneously. Very little additional oil is necessary for smelting the slag. Halogens which are possibly produced from PVC are combined with the lead flue dust and eliminated by a fabric filter. The content of pollutants in the clean gas is lower then the German standards.

If there exists the possibility for recycling of Polypropylen the separation is possible.

The advantages of the process are:

1. Simple crushing plant
2. No plastic residue
3. Flue dust will be recycled into the furnace
4. Low quantity of waste gas
5. Low content of pollutants
6. Economical at capacities of 10.000 tpy.

In Germany there are ca. 30 Mio. cars and trucks. By calculation of an average working life of batteries of 3,5 years, a batterie weight of 15 kg, there is a yearly quantity of 130.000 t of scrap batteries. On a recoverable lead content of 55 % it is possible to produce 70.000 t of lead from this scrap.

By treatment of battery scrap this quantity of lead is recycled and does not contaminate the environment.

Old batteries contain ca. 10 % acid and 10 to 12 % plastics, the remaining are lead and lead compounds

Several processes exist for treatment of scrap lead batteries.
Varta smelt unbroken batteries, after separating of acid, in a blast furnace.
Batteries are mostly treated in a mechanical process, to separate acid and nonmetallics, and thereafter smelted in a rotary furnace.

As an example the process of Blei- und Silberhütte Braubach, which is based on a Tonolli-Process, is mentioned (4) Fig. 1.

Here batteries still containing acid are delivered. After a crusher the acid is collected and stored in a tank. Afterwards the oxide-fraction is separated by a wet rotating drum and dewatered by a filter press.
The screen overflow is divided by a cassifier into gridmetal, Ebonite with PVC and Polypropylene. Polypropylene is cut, classified and washed.

The heavy plastics from the classifier are united with the fraction of Ebonite and PVC and later on separated into PVC and Ebonite.

The Soda added transforms the leadsulfate of the active mass into lead-carbonate. Saleable sodium sulfate precipitated from the solution.

Oxide and gridmetal are smelted in a short rotary furnace.

These processes have common features:

Oxide and gridmetal are recovered separately.
Plastics are produced in 2 or 3 fractions.
Polypropylene is conveyed to a special plant, where it is treated to a saleable material.

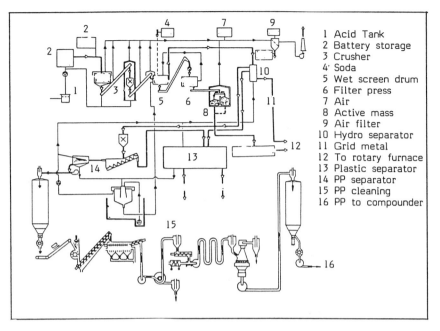

Figure 1 - Battery treatment plant Blei- und Silberhütte Braubach

The "Contibat"-process has been developed to have an economical process for treatment of batteries at smaller capacities, f.e. 10.000 to 30.000 t/a.
The flowsheet is shown in Fig. 2.

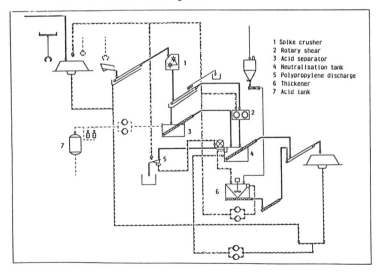

Figure 2 - Mechanical treatment plant

Slowly moving machines have been selected for crushing the batteries to avoid dust and aerosol formation.

Polypropylene, which in some cases has a certain market value, can be separated or conveyed into the furnace together with the other plastics, as PVC or Ebonite, and the metallic fraction.

The scrap batteries are fed into a vibrating feeder and conveyed by a chain conveyer into a spike crusher, where the batteries are opened to let flow out the acid.

The acid is separated by a vibrating screen and flows into a tank, where the slurry is settled. The slurry is transported by a screw into the neutralisation tank. The acid is stored in a tank for further use. The batteries, free of liquid acid, proceed to a rotary shear, are crushed and fall into the neutralisation tank. From here the solids are discharged by a screw and stored. If desired the floating Polypropylene can be discharged separately by a rake and can be brought to a further use.

The solution of the neutralisation tank is circulated, neutralized by lime and the content of solids is adjusted by settling part of the slime.
No waste water has to be treated.

The product of the mechanical treatment - a mixture of metal, active mass and plastics - is smelted simultaneously by oxygen in a rotary furnace.

The application of oxygen for the production of Nonferrous Metals has been published comprehensively by Paschen (1).

The operation of reverberatory furnaces with oxygen enriched air has been reported often during sessions of the German Society of Metallurgical Engineers, f.e.by Deininger (2).

Lamm (3) has given a report about the application of oxy fuel burners in rotary furnaces for smelting of battery scrap, that means, the metallic and metal containing fraction of old batteries after elimination of plastic parts.

The handling of liquid oxygen is of common usage, so that here only very shortly shall be dealt with.
Oxygen is delivered liquid by trucks and stored in tanks. It is evaporated by an atmospheric evaporator and used as gas.

Whereas the normal oil burners can be used for oxygen enriched combustion air, f.e. to 24-28 % O_2-content, the oxy fuel burners are of a special construction. The burner tube, which projects into the furnace, is water cooled. The fuel oil is atomized by compressed air and not by oil pressure.

That has the advantage that the atomization is independent from the throughput of oil, the atomization power can be kept constant by the compressed air so that changes of load don't influence the atomization.

Oil dust and oxygen are first mixed inside the furnace and not in a premixing chamber inside the burner. That diminishes the danger of explosion. By that construction the oxy fuel burner is decisively safer than a burner for oxygen enriched air, on which the oxygen normally is introduced before the ventilator. By that the possibility can not be avoided that oxygen comes into contact with oil or grease, which can cause fulminating.

For the combustion of 1 kg fuel oil (light) 11,05 m^3 air are necessary. Air contains 21 % oxygen which makes 2,32 m^3 O_2/kg oil or 2 m^3/l oil.

For the complete combustion in technical installations excess air of 1,1-1,2 is necessary. With oxy fuel burners no excess air is required.

The quantity of flue-gas (wet) for combustion with air is ca. 12 m^3/kg oil, for combustion with oxygen ca. 3,2 m^3/kg oil.
The flame temperature for combustion with air is at 1650-1700° C, with oxygen at 2200° C.

Whereas for combustion with air the heat transfer is mainly by convection for combustion with oxygen the heat is transferred by radiation. The retention time of flue gas inside the furnace increases considerably with oxy fuel burners, and is inversely proportional to the quantity of flue gas.

The advantage of combustion and the savings of oil do justify alone the use of an oil fuel burner only in case of a price relation of 1 m^3 O_2 to 1 kg of oil of 1:2.

Besides that a number of further advantages occur concerning metallurgy and operation which make profitable the use of oxy fuel burners for the production of NE-metals in a rotary furnace. For example should be mentioned:

Shortening of the time of charge and thereby increasing of performance.

Possibility of increasing performance without increasing the subsequent filter installation because of reducing of the flue gas.

Reduction of the flue dust and higher direct recovery of metal.

On the same capacity reduction of emissions by reduction of flue gas.

Thereby also reduction of power consumption of the ventilator.

Experiments have shown, that the combustion of plastic parts from batteries is completely and without formation of soot if oxygen will be used instead of air and if sufficient time for combustion inside the furnace is maintained.

Therefore the usual mode of running a rotary furnace has been changed.

Normally rotary furnaces are operated that the charge, consistant of lead material and admixtures, as coal, soda, iron etc., is fed into the empty furnace by means of special devices as f.e. vibrofeeder, charging mould.

When the furnace is full it is closed or the oil burner turned before it and the burner ignited. Then during cautious and intermittend rotating, the charge is smelted as far as to avoid fulminating of dusty material. Often then the first lead is already tapped and additional material fed into the furnace.

After a certain smelting time metal is tapped. Poor slag is smelted at increased temperature, possible after adding some further fluxes, after each or multiple charges.

At the process of smelting acid free batteries with plastics the following procedure is maintained.

The crushed batteries are fed into the furnace by a vibro- or screw-feeder more slowly than usual. During feeding the furnaces rotates slowly, continously or discontinously, and the burner is ignited. The oil injection is reduced and the oxygen addition increased.

By the retarded feeding and the instant ignition and combustion of the plastics it is avoided, that a nonregulated, sudden evaporation of the volatile parts of the plastcis and an uncomplete combustion occur. This is hardly avoidable if the furnace, filled with plastics containing batteries, will be ignited and turned.

The relation of fuel and oxygen and thereby the plastic burning potential of the furnace can be regulated so that a complete and soot free combustion of the plastics can be realized.

After terminating the feeding and the combustion of the plastics the oil injection will be increased and the charge will be smelted. The filling of the furnace which takes at the normal operation ca. 30 minutes is extended to ca. 90 minutes.

The charge consists of prepared and crushed batteries and
    8 % soda
    3 % iron borings
    4 % coal
of the battery weight.

The admixtures are fed together with the batteries.
The consumption figures per ton of batteries are:
    30 kg oil
    80 m^3 O$_2$

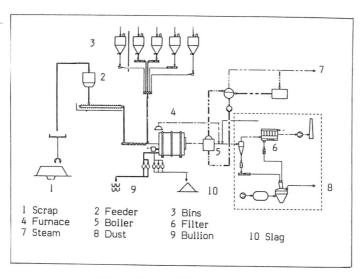

Figure 3 - Smelting plant

The smelting plant Fig. 3 consists of

- a feeding device

- a rotary furnace of 3 m ⌀

- an oxy fuel burner, built by Air Products, with a capacity of 20-150 kg/h oil and 50-300 m^3/h oxygen

- a waste heat boiler with a heating surface of 45 m^2 for flue gas cooling

- a subsequent fabric filter for dedusting flue gas and hygiene air.

The waste heat boiler has been installed instead of the usual aircooler.
The boiler has the following advantages against an aircooler:

The flue gas entrance temperature can be higher as with an aircooler which can not take entrance temperatures above 400° C because of material reasons.
The hot flue gas has to be cooled before the aircooler by adding hygiene or fresh air. That makes a great quantity of gases and because of a small temperature difference large heating surfaces are necessary.

With the waste heat boiler flue gas with 800-1000° C can enter directly from the rotary furnace into the boiler. The temperature of the steam can be determined so that the dew point of the flue gases will not be reached. Thereby the corrosion of the cooler can be avoided.

The steam produced can be used for many purposes better than the hot air of the air cooler, for which no use exists in the case of having only oxy fuel burners.

The "Contibat"-process has the following characteristics:

1. Acid containing batteries can be treated.

2. Polypropylene can be recovered.

3. No waste water treatment necessary.

4. Installations of slow running crushing equipment, which can operate without dedusting.

5. Neutralisation of adhering acid and therefore avoiding of evaporation of SO_3 inside the furnace.

6. Only one fraction metal + oxide + heavy plastics is produced, which makes a simple flow-sheet.

7. The heating value of the plastics is realized.

8. No problems with dumping of plastic residues containing small amounts of lead.

9. Existing chlorine from PVC will be precipitated together with lead flue dust and transformed into a rotary furnace slag on special flue dust campaigns.

References:

1. Peter Paschen: Sauerstoffanwendung in der Pyrometallurgie

   Erzmetall 33 (1980) S. 617, 34 (1981) S. 97

2. Lutz Deininger: Einsatz von Sauerstoffbrennern in Drehtrommel- und Raffinieröfen

   Paper presented at Meeting GDMB-Bleifachausschuß 1984

3. Karl-Friedrich Lamm: Einsatz von Sauerstoffbrennern am Kurztrommelofen

   Paper presented at Meeting GDMB-Bleifachausschuß 1985

4. Albert Melin : Verhüttung von Bleiakkuschrott

   (Recycling International, Berlin, EF-Verlag 1982) S. 749-758

Use of electric furnaces in non-ferrous metallurgy

Ch. König, G. Rath, T. Vlajcic

Mannesmann Demag Hüttentechnik, Duisburg, West Germany

Summary

Electric furnaces, till now chiefly employed for the production of ferro alloys and calcium carbide, are now used to an ever greater extent in non-ferrous metallurgy. The reason for the new situation is the increased value of the metals contained in the metallurgical residues as well as present demands for improved environmental protection. This paper explains electric furnace applications, the methods of energy input and present and future developments and tendencies.

1. Introductory and historical notes

Arc furnaces, which are characterized by the fact that the energy input is transmitted through electrodes, are becoming increasingly important. In the past, electric furnaces were chiefly used for the production of zinc, aluminium-silicon, lead, copper matte and nickel matte but now they play an ever more important role in non-ferrous metallurgy applications to an ever-increasing extent to minimize the metal content of slags received from upstream process stages or to process metallurgical residues. In former times, such residues were mostly taken to the dump or were smelted in shaft furnaces for the purpose of homogenization and the recovery of metals.

Non-ferrous metal smelting in electric furnaces started with copper and nickel ore smelting and served the purpose of concentrating the copper and the nickel content in the form of matte. Until then the classic smelters for copper and nickel had been the reverberatory furnace and in isolated cases the blast furnace and later the autogeneous or flash smelter.

As no indigenous fuel was available, the electric furnace to be used for copper and nickel matte was built in Sulitjelma, North Norway. The connected load of the furnace was 3,000 kVA. More furnaces of higher capacities were then built in Norway (6,000 kVA) and Finland (12,000 kVA) before the second world war started. In the meantime, copper matte and nickel matte furnaces of capacities as high as 33,000 kVA have been taken into service.

The reason for the increasing use of electric energy and the augmenting smelter capacities for copper and nickel is the smelting of platiniferous concentrates, low amounts of waste gas and dust and a higher metallurgical efficiency.

Pyrometallurgical smelting in electric furnaces to produce FeNi has found general and trend setting acceptance for oxidic nickel ores (and laterite, limonite, garnierite, serpentine).

It was exactly in connection with the production of FeNi from oxidic nickel ores that the development of electrometallurgical furnace designs in respect of shape, lining and shell cooling and smelting technology in general made fast progress. It was possible to raise FeNi furnace capacities up to 80,000 kVA. Fig. 1 shows a view to a typical NON-FERROUS-METAL furnace.

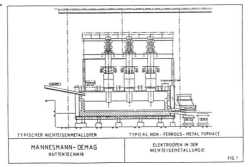

The above figure also illustrates a development which became trend setting in the following processes, namely work with a slag bath not covered with burden.

2. From electric furnace to non-ferrous metallurgy

As early as 1943, Swedish Boliden mining company started the smelting of sulphidic lead concentrates. Both process and equipment improvements continued until 1963. Today the capacity is 8,000 kVA and the annual production is 40,000 t of lead per year.

The Kivcet process is a combination of the flash smelting of
concentrates and electric slag cleaning. The first commercial
size furnace of the new type was commissioned at Sardinian
Samim in 1987. The connected load for the slag cleaning zone,
the electric furnace zone, is 9 MVA. Another furnace built in
Potosi, Bolivia, having a capacity of 4.2 MVA, has not come on
stream for politico-economic reasons.

A striking example of the development of the non-ferrous metal
industry towards electric furnace application is illustrated
by the following event. One of our Mannesmann Demag customers
is operating an oil fired furnace in which the slag of a
copper matte furnace is collected to allow residual copper to
settle down for recovery. Measurements revealed that only
about 10 to 15% of the energy input goes into the metallurgical process. The following disadvantages were recognized and
resulted in plans to reconstruct the furnace to enable energy
input through electrodes:

- Unsatisfactory thermal efficiency
- Excessive environment pollution by sulphurous combustion
  gases
- Limitation of the slag bath depth and therefore insufficient
  yield, because the energy acts on the slag surface only.
- Agitation of the slag by the burner gas stream and therefore
  low yield
- Problems in keeping the copper matte deposited on the hearth
  bottom at a temperature sufficiently high for tapping.

Some of our customers operating shaft furnace for recycle
materials and residues have developed interesting development
schemes. Shaft furnaces are to be replaced by electric
furnaces in two typical cases. It is intended to operate the
electric furnaces in a cyclic process, namely

- Continuous charging, smelting and reducing the material
  until a given amount of slag has accumulated,
- performing residual reduction,
- allowing metal inclusions in the slag to settle and
  segregate,

- tapping of the metal,
- draining the furnace of the slag with the exception of the hot heel.

The shaft furnace is to be retained in another typical case. Its inferior metallic yield is to be compensated by installing an electric furnace downstream of the shaft furnace. The electric furnace is to be used for residual slag reduction, minimizing the metal content in the slag and settling of the metal. Investment costs are expected to be much lower because the capacity to be installed will be lower.

DC furnace installation is also envisaged to achieve more efficient metal minimization in the slag by electrolysis and electrophoresis, depending on the amount of slag and the anticipated value of the metals to be extracted.

Electrically heated slag cleaning furnaces are rapidly gaining importance. The concentrates obtained in the flash or autogenous smelter contain copper and metal in the amount of 1 - 2 %. This percentage is even higher in the case of the Norandar converter.

3. **The electric furnace, a versatile smelting unit**

Electric furnaces have proved their worth in minimizing the metal content in slags. Experience has shown that copper and nickel contents can be reduced to levels lower than 0.6% in the ultimate slag if process control is adequate.

Charging liquid material taken from an upstream concentrating smelter may reduce the energy required for pure settling work to some 50 kWH per tonne of liquid charge. In general, solid recycle materials from other works departments and copper and nickel bearing dump materials are also charged through the furnace roof.

With electric heating it is possible to charge any combination between 100% of liquid material and 100% of solid material. Electric heating enables continuous smelting work to take place without much trouble. Magnetite separation from the slag can be counter-acted by slag temperature adjustment. Such temperature control is possible by varying loading and varying depth of electrode immersion in the slag. This method is specially important where converter slags and solid magnetite bearing materials are charged.

Electric furnaces have also found general acceptance for the cleaning of lead bearing slags obtained from lead shaft furnaces. They perform a considerable amount of reduction work, which is different from the metal content minimization in copper and nickel slags. Figure 2 refers.

	Pb-SCHLACKE AUS DEM SCHACHTOFEN Pb SLAG FROM SHAFT FURNACE	Pb-SCHLACKE NACH DER ABREINIGUNG Pb SLAG AFTER CLEANING
Pb	2-4%	0,5%
Zn	6-10%	4-5%
Sn	0,4-0,5%	0,2-0,3%
Cu	0,4-0,5%	0,2-0,3%
	SCHLACKENABREICHERUNG MINIMISATION OF METAL IN SLAG	
MANNESMANN DEMAG HÜTTENTECHNIK	ELEKTROÖFEN IN DER NICHTEISENMETALLURGIE	FIG. 2

The disadvantage of flame fired furnaces (Fig. 3) lies in the fact that the energy Q of burner flame B goes into the process only in the amount Q_1 through slag surface SL, the energy reflected by the roof included, whereas a major amount, Q_2 leaves the hearth interior H as lost energy. The slag assumes a high surface temperature but towards the hearth bottom it is even difficult to keep the material liquid or achieve the temperature which allows optimum viscosity and maximum segregation.

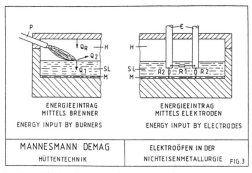

4. **Explanation of electric energy input**

Energy input (Fig. 3b) in electric furnaces is accomplished through electrodes E immersed in slag SL and by power flux through resistances R1 in the slag between the electrodces and resistances R2 in the slag under the electrodes.

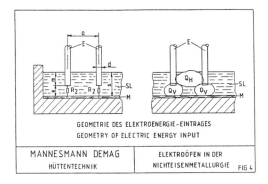

The important advantage offered by the electric furnace (Fig. 4) consists in the fact that by varying the geometric parameters of slag height s, electrode spacing a, electrode distance e from the metal bath M, and electrode immersion depth m it is possible to vary the amount of horizontal energy input Q_H between the electrodes and the amount of vertical energy input Q_V under the electrodes within a wide range.

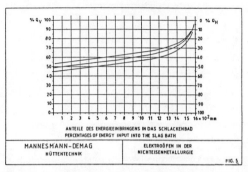

FIG. 5

The above diagram (Fig. 5) shows the percentages of vertical energy input % Q_V and the horizontal energy input %Q_H as a function of the depth of electrode immersion in the furnace with a slag bath 1,600 mm deep and 1,000 mm electrodes spaced 2,800 mm. Electrode spacing correction and its influence is also shown. It can be seen in our example case that a variation in electrode immersion depth permits a variation of the input under the electrodes in the amount of 50 to 90% of the total heat. Of course, influence is also possible by varying the slag depth or any other geometric dimension. Our diagram shows the influence exerted by varying the electrode spacing by +/- 400 mm.

Such variation would enable a variation of the horizontal and vertical energy input portions by +/- 5%. This clearly shows the nature of the mechanism of temperature control in the slag and the product bath.

The above control mechanism of energy input makes it possible to use an electric furnace downstream of any other concentrating smelter, as well, such as the QSL process or the Mitsubishi process.

The electric furnace ideally meets environmental protection requirements by its low amount of dust and waste gas generation and is additionally independent of the physical condition of the charge material (either liquid or solid) and allows easy slag temperature control.

5. Improvements in optimized energy input

Obtaining an appropriate temperature profile in the slag is important but equally essential is a uniform heat impingement on the entire slag volume to ensure that the full volume is uniformly metallurgically active. The conventional electric furnace (Fig. 6) does not satisfy this requirement optimally because the electrode arrangement, triangular in a round furnace, allows input heat propagation only in clover leaf shape, starting from the electrodes and associated bath resistances.

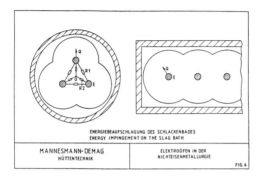

Mannesmann Demag engineers have therefore developed a unique furnace unit to solve this problem.

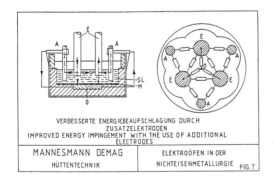

This new furnace (Fig. 7) is equipped with additional potential free electrodes A. These electrodes are included in the power route and heating system because they are connected with electrical system zero, which is the product bath. In addition to the usual resistance system among the electrodes and between the electrodes and the product bath there is now another resistance system, namely between the additional electrodes A and the main electrodes E. This transfers energy also to the outer areas of the slag bath and creates their equal metallurgical activity.

6. Conclusions

An overview of all present and intended uses of the electric furnace in the non-ferrous metal industry is presented in this paper. The high value of the metals contained in the wastes which have till now been dumped by the metal producing, electro plating, photo and data carrier industries as well as stringent anti-pollution legislation will quickly include the electric furnace into a variety of process roots. The electrode furnace, which is a very versatile, cost efficient and environmentally acceptable unit already now is being improved for a wider field of application.

Literature

1. H.I. Elvomder:
   The Boliden Lead Process
   AIME, Pittsburgh, November 28 - December 1966

2. R.P. Kuhn:
   Elektroofentechnik in der Metallurgie
   (Arc furnaces in metallurgy)
   GDMB, 19. Metallurgisches Seminar

3. G. Rath, G. Rottmann
   Moderne Schmelztechnologien zur Herstellung von Ferronickel
   (Modern smelting techniques for ferro-nickel production)
   Erzmetall 39 (1986) No. 2, pp 70 - 74

4. O. Emert, H. Winterhager et al
   Metallhüttenkunde für Eisenhüttenleute
   Verlag Stahleisen M.B.H. Düsseldorf

5. H. Hensgen, G. Wei, W. Wuth
   Reduktion von Kupferkonverterschlacke im
   Labor-Gleichstrom-Schlackenwiderstandsofen
   (Reduction of copper converter slag in the laboratory DC slag resistance furnace)
   Elektroofentechnik in der Metallurgie, pp 329-347
   VCH Verlagsgesellschaft, 1988

TREATMENT OF AUTOMOTIVE EXHAUST CATALYSTS

Volker Jung

Metal Research Department
Degussa AG, Hanau, West-Germany

Abstract

Catalysts accelerate chemical reaction without participating in it. A timely example of the use of a catalyst is the conversion of the pollutants generated by the combustion of fuel to non-toxic products. The problems posed by automotive exhaust emissions catalysts used in the automobile industry are described. The significance, occurrance, recovery processes, supply situation, and price movement of precious metals used as catalysts - platinum, palladium, and rhodium (hereafter referred to as PGM = platinum-group metals) - are discussed. The precious metal contents of automotive exhaust catalysts are recovered by hydrometallurgical or pyrometallurgical methods. These processes are explained and compared. Finally, the problems of logistics are mentioned and a brief glance into the future is attempted.

Introduction

Exhaust emissions from the combustion of vehicle fuels contain mainly carbon dioxide and water, but also carbon monoxide, uncombusted hydrocarbons, and nitrogen oxides. The latter cause considerable environmental pollution and must therefore be converted to the non-toxic subtances carbon dioxide, water, and nitrogen. These emissions represent the major portion of total pollutants except sulfur dioxide, the remainder is attributable to household waste, power generation, and industry, see Fig. 1. The first cars equipped with oxidation catalysts appeared as long ago as 1975 in the USA. The three-way catalyst (on a PtRh basis) made its appearance in 1980.

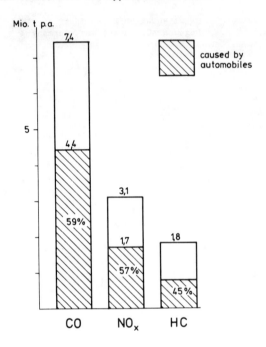

Figure 1 - Total air pollution in West-Germany in 1984 (1).

Automotive Exhausts Catalysts

Catalyst Types

A distinction is made between oxidation catalysts and three-way catalysts. In the oxidation catalysts, the hydrocarbons and carbon monoxide are converted to H_2O and CO_2 by the admixture of air. The nitrogen oxides do not take part in this process and are not converted. In the three-way catalyst, hydrocarbons, carbon monoxide, and nitrogen oxides are eliminated simultaneously by means of an optimally controlled fuel-air mixture.

Catalyst Structure and Composition

A basic distinction is made between metallic and ceramic substrates. Metal-substrate catalysts consist of corrugated stainless-steel foils of an FeCrAl alloy with added stabilisers such as yttrium and cerium, soldered into a steel can housing. The use of metal-substrate catalyst is currently limited to start-up catalysts and a small number of high-performance cars. Their share of the total catalyst population cannot be expected to exeed 10 % in the foreseeable future.

Ceramic-substrates used for automotive exhaust catalysts are available in forms of pellets and monoliths.

Pellets. Pellets are currently used primarily in large passenger cars and pick-up trucks in the USA. They basically consist of 85 % Al_2O_3, 3 % SiO_2, 2 % MgO, 3 % FeO, 1 % CaO, and 6 % other. The PGM content after coating is app. 0.05 %. The average catalyst weight per car is between 2 and 4 kg, depending on the model.

Monoliths. Over the last five years the ceramic monolith has emerged as the dominant form of substrate material for automotive exhaust catalysts with still rising tendency. The one-section (monolith) honeycomb element contains fine channels through which the exhaust gases are forced and in which the pollutants are converted. The monolith systems consist principally of cordierite with the following composition: 35 % Al_2O_3, 40 % SiO_2, 12 % MgO, 1 % FeO, 1 % CaO, and 11 % other. In this case, the PGM content after coating is 0.1 to 0.3 %. The average catalyst weight is app. 1 kg. In contrast to the surface-active alumina pellets, monoliths first must be covered with a γ-alumina slurry, the so-called "washcoat". On this washcoat the precious metal bearing solutions are applied in precisely-measured amounts and concentrations. About 1 to 2 g (0,03-0,06 oz) of precious metal per vehicle is used.

Significance of Precious Metals

Various combinations and proportions of the platinum-group metals platinum, palladium, and rhodium are used in automotive exhaust catalysts. The oxidation catalyst first used in the USA contained platinum and palladium in a ratio of 3:1 to 2:1, whereas three-way catalysts contained platinum and palladium in a range of 10:1 to 5:1 (2). The first catalysts introduced in Japan had only a palladium coating, then one consisting of palladium/rhodium. Nowadays, particularly with the export market in mind, platinum/rhodium catalysts are much more common. In Europe the catalyst with platinum/rhodium in a ratio of 5:1 has prevailed. There is a worldwide trend towards the monolithic catalyst with a platinum/rhodium ratio of 5:1 and a content of app. 1.5 g platinum per kg catalyst. The effects of individual platinum-group metals:

Platinum. Mainly for the conversion of hydrocarbons. Platinum is also relatively resistant to lead poisoning, exhibits considerable stability at high temperatures, is highly durable and acts rapidly in converting pollutants because of its fast warm-up.

Palladium. In addition to its catalytic action in the emissions oxidation process, it assists the conversion of hydrocarbons by platinum.

Rhodium. Rhodium has different valencies and therefore many catalytic functions. For that reason it is especially useful for the reduction of NO_x to N_2 by reaction with CO and H_2.

PGM Supply

Platinum metals are found only very rarely in pure form. They are normally combined with sulphidic or oxidic Cu/Ni or Cr ores. The main source of platinum is the South African Bushvelt Complex with its famous deposits at Merensky Reef and UG2 Reef. Ores containing high amounts of PGMs have been recovered from the following mines, see Table I. Table II shows the PGM supply 1987.

Table I. PGM content of Different Ores (g/t) (3,4)

Mineral Deposit	Pt	Pd	Rh	Pt:Rh
Merensky	3.2	1.4	0.16	20:1
UG2	2.5	20	0.54	5:1
Stillwater	6	2	0.21	30:1

Table II. PGM Supply 1987 (5)

Platinum : South Africa	2,520,000 oz	≙	78.1 t
Canada+USA	140,000 oz	≙	4.3 t
Other	40,000 oz	≙	1.2 t
USSR	400,000 oz	≙	12.4 t
total	3,100,000 oz	≙	96.0 t
Palladium : South Africa	1,090,000 oz	≙	33.8 t
North America	190,000 oz	≙	5.9 t
Other	90,000 oz	≙	2.8 t
USSR	1,790,000 oz	≙	55.5 t
total	3,160,000 oz	≙	98.0 t

Rhodium : no reliable data are available. In 1987 total rhodium production in the Western world was put at app. 310,000 oz ≙ 9.6 t. This figure ought to correspond roughly to consumption for the same period.

The proportions of each PGM vary between the individual sites, as shown in Table III.

Table III. PGM Percentage of Different Mineral Deposits (6)

	Pt	Pd	Ir	Rh	Ru	Os
Canada	43	45	2	4	4	2
South Africa	61	26	1	3	8	1
USSR	25	67	2	3	2	1

A considerable increase in production from the extension of existing mines and moreso from the opening of new mines, in South Africa in particular, is to be expected in the next few years. For example, recovery of the following additional PGM amounts can be expected (7):

Lebowa-Maandagshoek	84,000 oz	≙	2.7 t	(1994)
-Atok	58,500 oz	≙	1.9 t	(1989)
Impala-Karee	100,000 oz	≙	3.2 t	(1990)
Western Platinum	270,000 oz	≙	8.7 t	(1990)
Lefkochrysos	95,000 oz	≙	3.1 t	(1990)

PGM Price Movements

The ups and downs of the platinum price from 1978 to 1988 is shown in Fig. 2. In August 1987 the platinum quotation at the New York Stock Exchange touched 640 $/oz, before settling between 500 and 550 $/oz (5). Price movements for palladium were less spectacular. Over the past five years, prices fluctuated between 90 and 150 $/oz, hovering recently around 100 $/oz. Rhodium prices shot up by an astronomical rate. The sudden leap in demand for automotive exhaust catalysts was certainly responsible for this shift, with quotations rising from around 300 $/oz in 1983 to a current 1,200 $/oz.

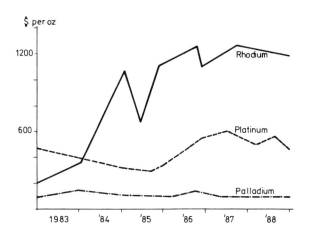

Figure 2 - Average US Dollar prices of PGM (5).

PGM Demand

Table III and Fig. 3 show the platinum, palladium, and rhodium demand in the Western world in the year 1987 (5). The tables reflect the fact that automotive exhaust catalysts contain only small amounts of palladium, but large propotions of rhodium.

Table III.   Demand of Platinum, Palladium and Rhodium in 1987 (in oz (t)).

	Platinum		Palladium		Rhodium	
	total	autocat.	total	autocat.	total	autocat.
North America	900,000 (27.9)	590,000 (18.3)	1,060,000 (32.9)	120,000 (3.7)		
Japan	1,650,000 (51.1)	295,000 (9.1)	1,420,000 (44.0)	65,000 (2.0)		
Europe	560,000 (17.4)	255,000 (7.9)	730,000 (22.6)	20,000 (0.6)		
Other	180,000 (5.6)					
total	3,290,000 (102)	1,140,000 (35)	3,210,000 (99.5)	205,000 (6.4)	310,000 (9.6)	226,000 (7)

These above mentioned figures correspond in general to those evaluated by Degussa AG, West-Germany.

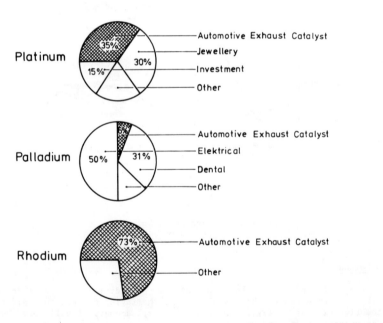

Figure 3 - PGM demand in percentage (8).

PGM in Automotive Exhaust Catalysts

Figure 4 shows the proportions of the precious metals platinum, palladium, and rhodium used in the manufacture of catalysts in 1987. The amounts of PGM used in the West since 1980 are shown in Fig. 5. In total, some 200 million vehicles, 4.5 million of which are in Europe, are fitted with a catalyst. They represent a "Mine on Wheels" (9). That corresponds to an amount of platinum of about 6,400,000 oz (200 t), palladium of about 2,000,000 oz (60 t), and rhodium of about 650,000 oz (20 t).

In 1988 around 40 million vehicles were produced, around 25 million or 60 % of which were fitted with three-way catalysts (10). That corresponds to catalyst-weights of app.

- 20,000 t in USA
- 15,000 t in Japan
- 5,000 t in Europe

Worldwide demand is currently running at around 40 million and is expected to rise to 60 million per annum by the year 2000.

The total number of catalysts on cars in the USA and Japan has remainded largely constant due to the fact that the legislation has not changed in the last years. The number of vehicles fitted with catalysts in Europe, on the contrary, is rising sharply. The forecasts for total newly-registered cars with catalyst are

1988 : 5 million
1991 : 8 million
1993 : 12 million.

The total number of cars fitted with a catalyst in West Germany progressed from only 15 % in 1986 to 40 % in 1987, 50 % in 1988, exceeding 60 % in 1989 (10).

In the last time many trials have been made to replace the expensive platinum by the much cheaper palladium, some of which were more or less successful.

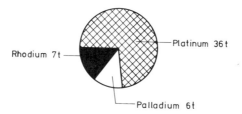

Figure 4 - Proportion of PGM in automotive exhaust catalysts (8).

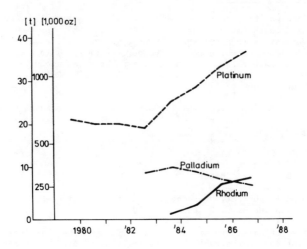

Figure 5 - PGM amounts fitted in automotive exhaust catalysts since 1980 (8).

The ore present at UG2 Reef is gaining increasing importance because it assures PGM for the automobile industry. This ore contains platinum and rhodium in a ratio of 5:1 which is preferred by automotive catalyst manufacturers (11). Platinum deposits worldwide are estimated to total 20,000 - 30,000 t (2), guaranteeing supplies for centuries to come. So the increased demand calculated for 1993 of + 31 % for cars, + 21 % for jewellery, and + 81 % for investment should be completely covered (12). As regards rhodium reserves, their extent in the USSR is largely unknown. Rhodium is mostly present together with platinum in the large South African deposits. The extent of rhodium reserves is believed to amount to 3,000 - 4,000 t. The amount of recycled spent automotive exhaust catalysts in 1988 is expected to be app. 20,000 t worldwide, about 5,000 t in Japan and about 5,000 t in Europe. The total amount of Pt recycled from automotive exhaust catalysts is expected to rise from app. 150,000 oz (5 t) in 1988 to 275,000 oz (8,6 t) in 1993 (8,13).

Treatment of Automotive Exhaust Catalysts

General Remarks

Because of the high value of the precious metals contained in automotive exhaust catalysts, their recycling is economically very interesting. Stable precious metal prices and large recycling quantities are necessary to justify the complex recycling processes viable. In addition, the limited natural resources of these metals, rhodium in particular, should be borne in mind. There are the following sources of automotive exhaust catalysts for recycling: production rejects, test cars, warranty claims parts and spent catalysts. They can be collected at workshops, muffler shops, and scrap dealers. First the cannings must be removed from the vehicles in order to get to the substrate (pellets or monoliths). Such procedures are normally not carried out by refineries. It takes considerable time to establish a suitable logistic structure.

In the case of pellets, a sample can be taken directly from the original batch following a careful mixing process, and then submitted to the analytic laboratory for examination.

The preparatory work for monolith system is more complicated. These materials are first crushed and ground. The resulting product is then mixed in order to homogenise the material and to take a representative sample. In most cases, so at Degussa AG, these precious metal bearing materials are not bought but toll refinded for the customer.

When the content has been analysed and the customer's approval received, recycling can commence.

Two basic methods of recycling are available:

Hydrometallurgical Methods

This type of method in turn consists of two process groups:

Dissolving of Substrates. Well-known is the dissolution by sulfuric acid or in an alkaline medium by means of sodium hydroxide under pressure. The first process results in an aluminate solution which is then further treated. In both cases, the precious metals are contained in the residue which is further leached with chlorine and hydrochloric acid for PGM dissolution. In the alkaline process, the presence of lead causes considerable interference as it is precipateted as lead hydroxide which is very difficult to filter off. Any silica content remains undissolved and also hinder the further processing of the precious metals. The γ - alumina converted into α - alumina in the course of a catalyst's lifetime makes the reprocessing much more difficult. It is not advisable to recycle monolithic substrates by these methods. Apart from the fact that only a small portion of this material is dissolved, some of the rhodium passes into the solution.

Dissolving of Precious Metals. Various methods with a wide range of precious metal yields are known, e.g. by means of hydrochloric acid and chlorine, hydrochloric acid and nitric acid or hydrochloric acid and hydrogen peroxide. A particular problem of these operations, in addition to the lower yield of PGMs, rhodium in particular, is the co-dissolution of aluminum, lead etc.. It is extremely difficult to separate the PGMs from the nonferrous metals in the dilute solution. The separation of the PGMs from the nonferrous metals by means of solvent extraction is one possibility (14). The disposal of chloridic waste water containing aluminum and lead is a further problem. The yields from this process, particularly of rhodium, are unsatisfactory. Several companies, especially in Japan, have evidently already acted accordingly and withdrawn from hydrometallurgical recycling of automotive exhaust catalysts.

Of the PGMs, palladium can be dissolved in a cyanidic solution but not platinum and rhodium. Cyanidic processing is nevertheless known to be used for recycling automotive exhaust catalysts.

Negative effects of hydrometallurgical recycling are:
- large quantities of waste water
- leached-out substrates are dumped
- precious metals losses
- little use of aluminate liquor/aluminum sulphate

Its advantages are that it works at low temperatures, the precious metal content is easily monitored (due to the homogeneous solution) and the precipitation processes etc. are easily to carry out.

Pyrometallurgical Methods

In general, pyrometallurgical methods of recycling automotive exhaust catalysts involve melting the ceramic substrate and simultaneously concentrating the precious metals in a collector metal. The precipitation of a separate precious metal phase is highly improbable in view of the very low PGM-content (< 0.3 %). Complete slagging of the substrate without losses of precious metals is of decisive importance in this process. Its main difficulties are the high melting point of the alumina pellets (appr. 2000 °C). These materials can therefore only be slagged by

- the addition of fluxes (e.g. lime)
- very high melting temperatures

Copper, nickel, lead, and iron are generally considered potential PGM collectors. None of them seems to be clearly superior. The collector metal is therefore selected more according to ease of processing and to subsequent wet-chemical stages. The collector metals iron and copper exhibit both advantages best.

The precious metals platinum, palladium, and rhodium are usually separated from the collector metal by sulfuric acid leaching; if copper is used as collector metal, electrolysis can also be used. Selective precipitation, e.g. by means of ammonium chloride and potassium chloride or solvent extraction, then produces a highly concentrated PGM solution which is then submitted for fine purification (15).

Compared with hydrometallurgical reycling of automotive exhaust catalysts, the pyrometallurgical method offers the following general advantages:

- high degree of concentration in a metal phase
- high precious metal yield
- can be carried out in a normal nonferrous furnace, e.g. blast furnace etc., with small amounts of by-products and residues

Pyrometallurgical Methods in Detail

Ceramic materials can only be melted by the addition of fluxes in a furnace capable of reaching only app. 1300 °C. More than 100 % lime or silica or a mixture of these must be added to pure alumina substrates. Iron oxide also lowers the melting point considerably. Only then processing in a blast furnace can be carried out. But even in case of melting down cordierite-monoliths with a lower melting point than alumina, addition of flux is necessary because of the forming of a sticky slag with bad separation between metal and slag. In most cases, copper or lead is used as collector for the precious metals. Virtually anything can be processed and recycled in these lead/copper blast furnaces, almost independantly of the shape and characteristics of the material. A decisive disadvantage, however, is the double amount of slag occurring.
In addition to high costs for melting energy, handling and storage etc., significant amounts of precious metals are lost in this slag. Rhodium in particular is dissipated into the lining or sump of the furnace. While the processing of PGM catalysts in large melting plants (e.g. copper blast furnaces and similar) has the advantage of rapid slagging and low melting costs, the subsequent concentration process from such dilute phases considerably complicates the process and therefore leads to higher costs. Obtaining the pure metal is a long and arduous process.

This conventional pyrometallurgical process has now been widely abandoned. Many refineries or nonferrous metallurgical plants use electrical high-temperature furnaces. These methods offer several advantages:
- compact dimensions and high capacity
- low amounts of slag, as max. 20 % flux added
- low precious metal losses
- temperature adapted to specific purpose
- pro-environmental, as very few emissions
- clear balance calculation

against which the following disadvantages are to be offset:
- high investment costs
- high electricity costs (depending on site)
- complex preparation of materials
- feed must be dry

A distinction is to be made between two types of high-temperature furnace for recycling automotive exhaust catalysts:
- Plasma furnaces
- Submerged-arc furnaces

Plasma furnaces

Plasma furnaces have been in use in the iron and steel industry, but of late also for the reduction of chromite ores and for processing dusts and automotive exhaust catalysts, etc. (16,17,18). Fundamental characteristics of the plasma method are high energy density, high temperatures, and short smelting times. The high radiation intensity and the extremely high plasma temperatures of more than 2000 °C at the gas/slag interface cause considerable problems for the lifetime of the lining. The introduction of plasma furnaces for recycling automotive exhaust catalysts was preceded by numerous tests, in particular to find suitable refractories. Only then was industrial use possible (Fig. 6) (19,20). The figures for the precious metal yield vary considerably. The relatively high lead content in certain US catalysts cause big problems both for the processing and for the environment.

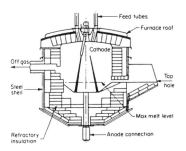

Figure 6 - Transferred arc plasma furnace (19).

Copper, nickel, and iron are used as collectors, with any iron inclusion in the slag removed magnetically. Lead cannot be used in view of the high melting temperature. Furthermore, these high temperatures cause reciprocal reactions between carbon and any oxides present. For instance, if the furnace is poorly controlled, silica may become reduced and form, if iron is present, an undesirable FeSi alloy, which is highly resistant to wet-chemical treatment, consequently hindering PGM enrichment (21). The whole process is governed by the power input and the material added, which then determine the temperature and viscosity of the slag.

The Iron and Steel Society (ISS) publication from 1987 provides a comprehensive survey of the general use of plasma furnaces and processes (22).

Submerged-Arc Furnaces

The treatment of oxidic materials with high melting points, such as chromite ores, nickel oxides etc. in a submerged-arc furnace represents a highly promising solution (Fig. 7). Here, the raw material itself serves as an electrical resistance and is liquified by the Joule effect (18,23). This means that the energy is not transmitted as radiation as in the plasma process, but is generated in the mixture of precious metal bearing feed and collector metal. If a liquid heel has been formed by small amounts of metal the feeding of the furnace begins. Therefore this process ensures extremely rapid, complete reaction with high precious metals yields. The electrical energy is transferred to the melt by a cooled conductive bottom of graphite or by metal plate elements. The opposing electrode is located in the furnace roof.

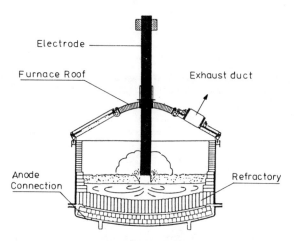

Figure 7 - Submerged-arc furnace, ASEA (23).

Direct current is normally applied, the polarity depending on the operating conditions and the raw materials. If the bottom electrode is the anode, more heat is generated in the bath. On the other hand, if the bottom electrode is the cathode, the electrophoretic effect, i.e. migration of metal ions, may have a positive effect on metal yield (24). The adaption of such a process, with iron or copper as precious-metal collector, is also conceivable for recycling automotive exhaust catalysts.

In both the plasma and submerged-arc furnace, the relatively thin metal layer on the furnace bottom is covered by a layer of slag ten times as thick. This slag is tapped off several times in the course of the smelting, depending on throughput speed; the metal remains in the furnace for a long time so that it becomes concentrated. Only when the precious metal content is sufficiently high for further operating, that means a couple of percents, it is poured out and delivered to a wet-chemical treatment.

PGM recovery rates

Data on precious metal yields should always be regarded with caution. The definition of "yield" alone lacks consistency. In some instances the entire amounts of precious metals, including residue which cannot be recovered economically, are taken as the percentage recovered; in other cases, only the precious metal content which can be directly reprocessed is considered. These economic aspects aside, fluctuation in the recovery rates for any one specific method does occur, and even moreso for the various process stages.

Certain published data can nevertheless provide a rough guide to the precious metal recovery rates from automotive exhaust catalysts which are technically possible. While a figure of 90 % to 95 % is usually stated for platinum and palladium (15,25,26), the comparable values for rhodium are much lower, about 75 % to 85 %. The main reason for this discrepancy are certainly losses into the furnace lining in pyrometallurgical processes. In the case of hydrometallurgical recycling processes the poor separation of the PGMs from other metals is probably the cause of a low yield. Nevertheless, some refineries state recovery rates of approximately 95 % for all PGMs (27).

Survey of Processes

Precious metal refineries and large producers of nonferrous metals are normally very cautious about publishing information on processes and internal data. It is nevertheless worthwhile attempting to collect the present state of information on automotive exhaust catalyst recycling, even if such information is incomplete and possibly already superseded by new developments. It is possible that certain process descriptions are based on suppositions.

The following use wet methods

Bureau of Mines	(cyanide leaching)
Canadian Platinum Refineries	(with aqua regia)
Degussa	(alkaline, acidic)
Heraeus	(alkaline)
Nippon Engelhard	
PGP	
Platinum Lake Technology	(solvent extraction)
Ryosho	(with aqua regia)
Takezaki	
Universals Oil Products	

The following prefer dry methods

Degussa	(blast furnace, electric furnace)
Hoboken (MHO)	(copper furnace)
Nippon Engelhard	
ORF	(plasma furnace)
Sumitomo	
Texas Gulf	(plasma furnace)

The following pursue wet-chemical or dry methods

    Catalytic Converter Refining
    IKEDA
    INCO
    Johnson Matthey
    US Platinum

These data are largely derived from press notes or trade periodicals, in particular "American Metal Market", "Metal Bulletin" and "Precious Metals (13,26,27, 28,29,30). The above lists do not take market shares into account. In many cases the incidence of used automotive exhaust catalysts does not reveal anything of the recycling locations. For instance, 65 % to 75 % of catalysts from the US are sent to other countries, Japan in particular (28). In Addition, neither method of wet-chemical nor pyrometallurgical processing has as yet prevailed over the other. Important factors governing decisions in favor of the one or the other are favorable site (transport question), energy costs, environmental protection regulations, methods of further-processing of the precious metal concentrates and, last but not least, precious metals recovery rates, which may vary considerably depending on the process and element (platinum, palladium, and rhodium) concerned.

It nevertheless appears that virtually all companies which have recycled automotive exhaust catalysts by large-scale wet-chemical methods have abandoned this venture sooner or later. They may well have done so for economic or environmental reasons. Degussa AG also prefers the pyrometallurgical treatment of automotive exhaust catalysts both the traditional way by blast furnace and the new way by high-temperature furnace.

Logistic Situation

A suitable logistic structure is necessary for the smooth supply of automotive exhaust catalysts for recycling according to the methods described above. The first collecting point is the repair shop, the scrap dealer or the muffler shop, where the exhaust systems are removed. Either the dismantling will be done here, or a sub-contractor can collect all complete cannings and then prepare them at a central site, depending on the local infrastructure. Only then can the pellets or monoliths containing the precious metals be transferred to a refinery for processing. Removing, collecting, dismantling, and transporting must constitute an interlocking network in order to ensure an economic catalyst recycling process. Even the purely mechanical process of dismantling is full of problems. Cutting, burning, sawing or cracking to open the metal can in order to reach the actual catalyst is labor-intensive. The amount of work involved in preparing pellet catalysts is by their very nature lower, as the pellets can simply be poured out once the canning is opened. On the other hand, some of the precious metal may be lost in the fines when a monolithic catalyst is sawn open, lowering the precious metal yield. The cost of these processes must remain in proportion to the value of the catalyst. For instance, the precious metal content of a medium-class European-built car with an engine capacity of app. 1.8 l is worth about 60 Deutschmark. The stainless steel canning will in any case be recycled as scrap.

The logistic structure appears to be relatively complete in Japan (31), whereas the geographical size of the USA presents big problems (32); collecting points and dismantling centers are now being established in Europe. In West-Germany precious metal refineries, car manufacturers, and scrap dealers etc. are already beginning to cooperate with each other. It would be very reasonable to fit the cars with canning-devices allowing a simple removing of the monolith-type catalyst by a flap or something like that.

Prospects

We all love the car. The modern world is inconceivable without it. But the natural environment in which we live is much more vital to us, and it must be our prime target to protect and preserve it.

We therefore all have to help accomplish this task. Unleaded fuel, the use of the catalyst to reduce emissions, and the recycling of valuable materials in order to protect resources are just some of the potential courses of action which have now taken on concrete shape. But the road to a rational symbiosis of nature and technology is still long and arduous.

References

1. Verband der chemischen Industrie e.V., Frankfurt/Main, West-Germany, Wald, Chemie und Umwelt, Febr. 1985, 2nd edition.

2. M.C.F. Steel, "Changing Patterns of Platinum Group Metals Use in Autocatalyst", Society of Automotive Engineers, 1988, no. 880127.

3. K.S. Liddell, L.B.McRae, and R.C. Dunne, "Process Routes for benification of noble metals from Merensky and UG-2 ores", published by the Institution of Mining and Metallurgy, London, 1985, 789-816.

4. George J. Hodges and Roger K. Clifford, "Recovering Platinum and Palladium at Stillwater", J. Metals, 40 (6) (1988), 32-35.

5. Platinum 1988, Interim Review (London, Johnson Matthey, 1988), 1-23.

6. Karl Kutzsche and Günter Häußler, "Aufkommen und Verwendung von Platinmetallen", Neue Hütte, 32(7)(1987), 266-271.

7. Metal Bulletin, 6 February 1989.

8. Platinum 1988 (London, Johnson Matthey, 1988), 1-60.

9. European Chemical News, 18 July 1988.

10. Verband der Automobilindustrie e.V., Frankfurt/Main, West-Germany, Tatsachen und Zahlen, 1988, 52. Folge.

11. B.H. Engler, E. Koberstein, and H. Völker, "Three-Way Catalyst Performance Using Minimized Rhodium Loadings", Society of Automotive Engineers, 1987, no. 872087.

12. Blick durch die Wirtschaft, West-Germany, 6 October 1988.

13. Metal Bulletin, 29 September 1988.

14. Private Communication, Platinum Lake Technology, 1989.

15. Wolfgang Hasenpusch, Symposium "Recycling im Verkehrswesen", 5/6 December 1989, Vienna, Austria.

16. W. Krieger et al., "Anwendungsmöglichkeiten von Plasma in der Prozeßtechnik" Berg- und hüttenmänn. Monatshefte, 132(11)(1987), 505-514.

17. C.A. Pickles, A. McLean, and C.B. Alcock, "Reduction of Iron-Bearing Materials in an Extended Arc Flash Reactor", Canad. Met. Quart., 24(4)(1985), 319-333.

18. K.U. Maske, Mintek, Report No. M178(1985).

19. J.R. Monk, "Application of Plasma to Metallurgical Processes", 5th Intern. Symp. on Plasma Chemistry, Edinburgh, August 1981.

20. James Saville, "Recovery of PGMs by Plasma Arc Smelting", 9. Intern. IPMI Conf., New York, 1985, 157-168.

21. James E. Hoffmann, "Recovering Platinum-Group Metals from Auto Catalysts", J. Metals, (40)(6)(1988), 44-44.

22. "Plasma Technology in Metallurgical Processing" (Grand Junction, CO, USA: Iron and Steel Society, 1987).

23. Sven-Einar Stenkvist, "12 years of dc arc furnace development leads to US order", Steel Times, October 1985, 480-483.

24. Joachim Höfler, "Reduktion NE-metallhaltiger Schlacken im Elektroofen unter besonderer Berücksichtigung der Verwendung gleichgerichteten Schmelzstroms" (Dr.-Ing.-Dissertation an d. RWTH Aachen, West-Germany, 1988).

25. Handelsblatt, West-Germany, 28 April 1988.

26. Chemical Engineering, 16 February 1987.

27. Chemical Week, 1st July 1987.

28. Am. Met. Market, 4 November 1987, 11 March 1988, and 25 October 1988.

29. Metal Bulletin, 1st May 1987.

30. Hanauer Anzeiger, West-Germany, 23 July 1988.

31. Andreas Brumby, "Der Kreislauf der Edelmetalle bei Autoabgaskatalysatoren", Metall, 42(7)(1988), 711-713.

32. Kai von Puttkamer, "Katalysator-Recycling-lohnendes Geschäft?", Automobil-Revue, Switzerland, 25(1988), 96-97.

FERROCHROME PRODUCTION FROM CHROMITE FINES VIA COAL-BASED

DIRECT REDUCTION IN ROTARY KILNS AND SUBSEQUENT MELTING IN ELECTRIC FURNACES

K.H. Ulrich and W. Janssen

Krupp Industrietechnik GmbH
Franz-Schubert Strasse 1-3
4100 Duisburg 14, West Germany

Abstract

This new technology uses a rotary kiln directly charged with fine grained chromite concentrate together with coal and flux material. The burden is heated to approx. 1450-1500°C at which temperature iron oxide, chromite, and partly silica are reduced to their metallic components. The material discharged from the kiln is - via a hot charging system - fed into an electric furnace with Söderberg electrodes. The furnace operates with an open bath and slag resistance heating. The kiln product is fed into the liquid slag where it is molten and the separation of gangue and the metal phase takes place. Necessary desulphurization is initiated by adjusting a basic slag and strongly reducing conditions through the presence of surplus char from the kiln operation. Power consumption in the electric furnace thus drops from 4000 kWh/t of FeCr for conventional processes to about 1100 kWh/t for this new technology.

Hydrometallurgy

THE DISPOSAL OF ARSENICAL SOLID RESIDUES

G.B. Harris and S. Monette

Noranda Technology Centre, 240 Hymus Boulevard,
Pointe Claire, Quebec, H9R 1G5, CANADA.

The environmentally safe disposal of arsenic-bearing residues from extractive metallurgical operations is a subject that has received much attention during the last ten years. The recent literature is reviewed, showing that ferric arsenates, and particularly those containing excess iron over that required to complex the arsenic, are the favoured form of arsenic disposal from hydrometallurgical operations. Data are presented which further reinforce these findings, that high iron (molar Fe/As >3) ferric arsenates are the most stable form of arsenical residue for disposal, particularly so if the residue also contains mixed base metals (Cu+Cd+Zn). Laboratory data show that samples of these compounds have continued to remain stable for periods up to three and a half years.

Data are also presented showing that real residues from plant and pilot plant hydrometallurgical operations conform in behaviour to expectations based on the laboratory study of high iron ferric arsenates. It is also shown that other arsenic compounds, such as iron and copper/iron speisses generated during the pyrometallurgical processing of arsenic-containing materials, initially tend to decompose, releasing soluble arsenic and iron, which, in a static environment, subsequently re-precipitate and remain stable, probably as a high iron ferric arsenate.

It is concluded that, given an understanding of the chemistry involved, arsenic resulting from extractive metallurgical operations can be disposed of in an environmentally acceptable manner.

INTRODUCTION

The elimination of arsenic from process streams, and its subsequent environmentally safe disposal has been a topic of great debate, great interest and great concern within the extractive metallurgical industry during the past ten years. This concern has been evidenced in two recent conferences,[1,2] each of which generated much discussion, but resulted in a certain polarisation of views.

Robins,[3-14] beginning in 1980, has undertaken a considerable study into the stability of metal arsenates, and in particular into the stability of ferric arsenates. Much of this work, however, has been theoretical, based on published thermodynamic data, and has indicated that simple (molar Fe/As ratio of 1:1) ferric arsenates are not sufficiently stable for safe environmental disposal. Such conclusions have been verified in practice.[15,16] At about the same time that Robins started to examine the theoretical aspects of arsenate compounds, work was inititiated at the Noranda Technology Centre to study the long term stability of arsenical solid wastes, with stability defined as <5 mg As/L being soluble into the liquid phase. A previous paper[15] has shown that:

(a) arsenic can be stabilised in residues over the pH range 4-7 providing that the Fe/As molar ratio is >3:1;

(b) the pH range of stability can be increased to 4-10 with the presence of small amounts of co-precipitated base metals, notably Cd+Zn+Cu;

(c) natural scorodite ($FeAsO_4 \cdot 2H_2O$) is much less soluble than published thermodynamic data would indicate.

During this same period, a number of other studies have been published supporting the conclusions that high iron ferric arsenates are significantly more stable than might otherwise have been thought based on thermodynamic data.[16-24] The work by Krause and Ettel,[16-19] in particular, show not only the effects of high iron content, but also the effect of crystallinity. Their data clearly show that (i) arsenic solubility passes through a minimum at pH 4, and (ii) crystallinity reduces the solubility of ferric arsenate (1:1 Fe/As molar ratio) by two orders of magnitude compared to the data published by Robins[3-14] and by the Chukhlantsev,[25] on whose data many of the conclusions regarding the solubility and stability of metal arsenates are based. Krause and Ettel studied the solubility of not only naturally occurring ferric arsenate (scorodite), but also synthetic scorodites produced in an autoclave by Dutrizac.[26] The results for both were similar, and were in agreement with data published by Dove and Rimstidt[20,21] and the present authors,[15] indicating that increased crystallinity obtained through formation at elevated temperature has a considerable beneficial effect on the stability of ferric arsenate. Such findings are important, since if the high iron ferric arsenates crystallise slowly with time into goethite and ferric arsenate (scorodite), then the arsenic contained therein should not be released into the environment. In fact, it is believed that ferric arsenate will crystallise to form scorodite more quickly than ferric hydroxide will to form goethite.[27]

Further evidence for the stability of high iron ferric arsenates has been generated by Stefanakis and Kontopoulos,[22-24] although this work was not as definitive as that of Krause and Ettel in that the solubility determinations were carried out for only 48 hours. Their data show a generally low solubility for high iron ferric arsenates, but also indicate that the solids were more stable when precipitated at lower pH (3 as opposed to 5 or 7) and lower temperature (33°C as opposed to 80°C). These findings regarding precipitation temperature and pH are at variance with previously published data,[15,16] and may be the result of incomplete washing of the solids and/or the formation of jarosites due to the presence of sodium. Figure 1, generated during the early stages of the study at the Noranda Technology Centre, shows that the precipitated iron arsenate solids required a great deal of washing to remove the last traces of mother liquor. This was particularly so when the solids were formed via precipitation with caustic soda, which was the case with Kontopolous et al.[24] Robins has also noted problems with the incomplete removal of mother liquor.[28] Nevertheless, these data are of importance, since they indicate what could happen in

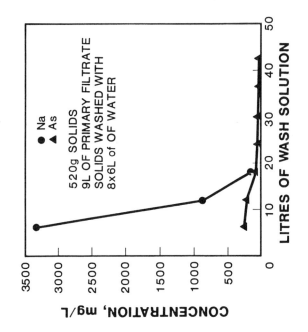

Figure 2. Effect of molar Fe/As ratio on solubility of arsenic from high iron ferric arsenates as a function of test pH

Figure 1. Washing of freshly precipitated iron arsenate solids

practice, and appropriate measures should be taken.

It can be concluded from this brief review of the literature that there is a growing body of evidence to support the view that high iron ferric arsenates are environmentally stable and an acceptable method for the safe disposal of arsenic resulting from extractive metallurgical operations. In this paper, data are presented to further reinforce this evidence, together with new data for other arsenical materials.

EXPERIMENTAL

The experimental procedure used throughout these studies has been described in some detail in a previous publication.[15]

Solids Preparation

Arsenate solids were prepared in a semi-continuous manner, at both pH 5 and pH 8, in order to more nearly reproduce the manner in which such solids would be precipitated in practice. In this procedure, metal ion solution, arsenic solution and precipitating reagent were fed continuously into a stirred tank reactor to prepare the required amount of solids. Sodium hydroxide solution and lime slurry were used as precipitants, and the solids were generated at 25°C, 80°C and 150°C, the latter in an autoclave, where it was not possible to control the pH accurately. The results for the solids precipitated at 150°C have been reported previously[15] and are not considered again in this paper. Solids were also prepared from both sulphate and chloride media. All solids were filtered, repulped in deionised water, and washed a sufficient number of times to give a constant arsenic and metal ion concentration in the wash solution, indicating the removal of all soluble ions originating from the entrainment of mother liquor. It was found that, in general, this required several stages of washing (Figure 1). In view of the very low arsenic solubility of a number of the precipitated solids, removal of all mother liquor was of prime importance.

Stability Testing

All tests were conducted at constant 25°C, with a solids to liquid ratio of 1:16. Wherever sufficient solids were available, 100 g (dry basis) were used, and where necessary, the particle size of the solids was reduced to minus 10 mm (0.375 in). The pH of the liquid phase was adjusted to 2-10 with either sodium hydroxide or sulphuric acid, and readjusted at each time of sampling, initially once every week, but subsequently every two or three months. Acetic acid, as specified in the United States Environmental Protection Agency EP Toxicity Procedure,[29] was not used, but rather sulphuric acid, which was deemed to be more appropriate for metallurgical wastes. The tests were continued until complete dissolution occurred, or, in the cases where stability was indicated, for a minimum of eighteen months. Most of these latter tests are still continuing. It should be noted that the tests were designed to determine stability rather than toxicity, but for comparative purposes, procedures similar to those outlined by the EPA[29] have been used.

Slurry samples were filtered on a 0.45 micron Millipore filter, and the arsenic determined by graphite furnace/atomic absorption spectroscopy using a Hitachi 180-80 spectrometer, with a detection limit of 0.01 mg As/L.

STABILITY OF METAL ARSENATES

Synthetic Laboratory Samples

A previous publication[15] has shown that simple (i.e. single metal, stoichiometric) metal arsenates, precipitated in an amorphous form are clearly unstable, and unsuitable as a vehicle for arsenic disposal. It was also shown that (i) atmospheric carbon dioxide had a considerable destabilising effect on a number of arsenates, notably those with highly insoluble carbonates, such as calcium, magnesium and cadmium, (ii) particularly in the case of ferric and copper arsenates, precipitating at higher temperatures (80°C and 150°C) resulted in solids with lowered arsenic solubility, and (iii) there was no significant difference in the stability/solubility of high iron ferric arsenates whether they were precipitated at pH 5 or pH 8, using lime (rather than caustic as used by Kontopoulos et al.[24]). It was further shown that high iron (Fe/As molar ratio >3) ferric arsenates were stable in the pH range 4-7, and that the presence of mixed base metals (Cu+Cd+Zn) extended the pH range of arsenic stability to 4-10. However, doubts have been expressed as to the longer term stability of these solids,[11,12] since, in some cases, test results were available for only five months.

The chemical analyses of the high iron solids referred to above are given in Table I.

TABLE I

CHEMICAL ANANLYSIS OF SYNTHETIC METAL ARSENATE COMPOUNDS

METAL ARSENATE[1]	CHEMICAL ANALYSIS, %						
	As	Fe	Cd	Cu	Zn	Ca	SO_4^{2-}
Fe/As 5.5:1, lime, 25°C, pH 8	4.85	20.0				14.9	34.8
(Fe+Cu+Zn+Cd)/As 1.1:1, lime, 25°C, pH 8	11.8	2.97	7.21	3.74	4.22	16.2	31.8
Fe/As 5.4:1, NaOH, 80°C, pH 5	10.3	41.9					3.72
Fe/As 10.3:1, NaOH, 80°C, pH 5	5.84	44.8					3.24
Fe/As 2.6:1, lime, 25°C, pH 5	7.55	14.6				11.8	27.8
Fe/As 3.1:1, lime, 25°C, pH 5	7.01	16.2				13.2	29.8
Fe/As 3.7:1, lime, 25°C, pH 5	6.48	18.2				14.5	32.7
Fe/As 6.7:1, lime, 25°C, pH 5	3.52	17.5				13.4	31.8
Fe/As 7.6:1, lime, 25°C, pH 5	3.08	17.9				13.6	33.0
Fe/As 6.7:1, lime, 25°C, pH 5 (chloride)[2]	10.4	52.1				0.41	0.06
Fe/As 3.3:1, lime, 90°C, pH 5	6.47	16.0				14.7	33.2
(Fe+Cu+Zn+Cd)/As 6.8:1, lime, 25°C, pH 5	3.58	15.6	1.15	1.62	0.80		11.2

1. Precipitated with lime or NaOH at temperature and pH indicated.
2. Precipitated from chloride medium, all others from sulphate.

N.B. Variation in chemical analysis for same metal arsenate precipitated under conditions due in part to presence of hydroxide, monohydrogen or dihydrogen arsenate.

In Table II, results of testing continued for a further period of eighteen months show that the conclusions reached previously[15] continue to remain valid. The data can be extrapolated (Figure 2) to show that a minimum Fe/As molar ratio of 3:1 is necessary to confer stability over the pH range 4-7. The data in Figure 2 also indicate that by increasing the Fe/As molar ratio, some degree of stability is also conferred in the alkaline region. There has been no indication that any of the high iron ferric arsenates are beginning to break down, except for the 3.1:1 Fe/As ratio at pH 10. This, however, might be expected, since the ratio is on the borderline for stability (Fe/As molar ratio of 3.1:1), and the tendency for ferric arsenates to break down is greatest at pH 10.

It can be seen from the results for the high iron plus mixed base metal arsenates that a considerable degree of stability is conferred on such solids with respect to arsenic. Recent, unpublished work conducted at the Australian Nuclear Science and Technology Organisation[30] has shown that base metals in general, and copper, cadmium and zinc in particular, can be readily incorporated into "deformed" lattice structures of goethite. Such deformations may be responsible for the increased stability observed with the presence of mixed base metals in the high iron ferric arsenate precipitates. It would seem appropriate that a study of the structures of these solids is warranted.

The data continue to demonstrate that increasing the iron to arsenic ratio increases residue stability (i.e. reduces arsenic solubility), that precipitation at higher temperature also confers greater stability, which is an observation opposite to that of Kontopoulos et al.[24] It is possible that these authors formed some jarosite through the use of caustic soda, which would have an adverse effect on solubility (see Plant and Pilot Plant Residues section). The two sets of data for Fe/As ratio of 6.7:1 suggest that the residue precipitated from the sulphate medium has a greater stability than its analogue precipitated from a chloride medium. This is particularly evident at pH 10, and suggests that the incorporation of gypsum (from lime, the precipitating agent) in the residue may possibly have a beneficial effect in enhancing stability.

TABLE II

STABILITY TESTING OF SYNTHETIC IRON ARSENATE COMPOUNDS

METAL ARSENATE[1]	pH 4		pH 7		pH 10		Natural pH[3]		
	Days	As mg/L	Days	As mg/L	Days	As mg/L	Days	pH	As mg/L
Fe/As 5.5:1, lime, 25°C, pH 8	942	0.14	942	<0.2	942	7.22	942	8.0	0.60
(Fe+Cu+Zn+Cd)/As 1.1:1, 25°C, pH 8	943	42.1	943	<0.2	943	2.63	943	8.4	<0.2
Fe/As 5.4:1, NaOH, 80°C, pH 5	911	0.37	911	0.85	911	394	911	4.2	0.09
Fe/As 10.3:1, NaOH, 80°C, pH 5	911	0.12	911	0.18	911	34.5	911	5.6	0.08
Fe/As 2.6:1, lime, 25°C, pH 5	730	0.30	722	0.60	722	30.6	722	4.6	0.32
Fe/As 3.1:1, lime, 25°C, pH 5	722	0.14	722	0.49	722	82.3	722	4.9	0.18
Fe/As 3.7:1, lime, 25°C, pH 5	723	0.09	723	0.34	723	11.1	723	4.4	0.12
Fe/As 6.7:1, lime, 25°C, pH 5	768	0.04	768	0.13	768	11.0	768	4.3	0.05
Fe/As 7.6:1, lime, 25°C, pH 5	767	0.03	735	0.49	735	1.28	735	4.0	0.11
Fe/As 6.7:1, lime, 25°C, pH 5 (chloride)[2]	735	0.12	735	0.48	735	238	769	4.9	0.03
Fe/As 3.3:1, lime, 90°C, pH 5	767	0.07	767	0.18	767	5.70	767	5.6	0.09
(Fe+Cu+Zn+Cd)/As 6.8:1, lime, 25°C, pH 5	649	0.06	649	<0.01	649	0.48	649	5.1	<0.01

All tests with 100 g solids (dry basis) in 1600 mL water at 25°C.
1. Precipitated with lime or NaOH at temperature and pH indicated.
2. Precipitated from chloride medium, all others from sulphate.
3. Solids slurried with distilled water only, no pH adjustment.

Plant and Pilot Plant Residues

The laboratory studies on the "synthetic" residues have provided, and continue to provide, a significant amount of valuable data as to the stability of arsenic residues. However, the efficacy of these data can only really be proven by the testing of residues generated under plant or pilot plant conditions. A number of such residues, the analyses of which are given in Table III, have been subjected to testing identical to that for the "synthetic" compounds, as shown in Table IV. It can be seen that these residues are extremely stable, in many cases even at pH 10, where the high iron ferric arsenates (without base metals) break down. One exception is the cobalt plant jarosite, with a particularly high arsenic content. Work by Dutrizac has shown that only a very small amount of arsenic can be incorporated into the jarosite structure, with the remainder being present as ferric arsenate.[31,32] Thus, in a jarosite with only a low arsenic content, such as the zinc plant jarosite, it can be postulated

TABLE III

CHEMICAL ANALYSIS OF ARSENIC-BEARING PLANT RESIDUES

MATERIAL	\multicolumn{10}{c}{SOLIDS ANALYSIS, % or (ppm)}

MATERIAL	As	Sb	Fe	Cu	Cd	Zn	Pb	Bi	Co	Ni	Ca
Gold plant effluent sludge[1]	(250)	(420)	1.87								37.8
Natural scorodite ($FeAsO_4 \cdot 2H_2O$)	29.2		22.1								
Lead plant slag[2]	0.59	(340)		0.27			4.02				
Cobalt plant jarosite[3]	6.13	(310)	18.2						0.32		7.98
Zinc plant jarosite	0.17		30.5								
Natural goethite	0.50	(190)	10.7	1.88	(100)	1.91	1.19	(40)	(400)	(300)	5.31
Smelter dust leach residue[4]	3.34	(140)	2.54		0.19	(580)					20.3
Copper arsenate/gypsum residue[5]	6.17	0.26	12.4			(71)	(200)			1.83	17.8

1. Effluent after cyanide destruction, treated with ferric iron and lime to produce sludge.
2. Iron silicate blast furnace slag.
3. From neutralisation with lime of pressure leach slurry at 95°C.[39]
4. From neutralisation with ferric iron and lime at pH 8 of liquor from leaching copper smelter precipitator dust.[40]
5. Neutralisation with lime of copper sulphate/arsenic liquor, generated by leaching purification residue from a copper tankhouse.

TABLE IV

RESULTS OF STABILITY TESTING OF ARSENIC-BEARING PLANT RESIDUES

MATERIAL	pH 4		pH 7		pH 10		Natural pH		
	Days	As mg/L	Days	As mg/L	Days	As mg/L	Days	pH	As mg/L
Gold plant effluent sludge	281	0.26	286	0.09	287	0.50	292	12.0	0.07
Natural scorodite ($FeAsO_4 \cdot 2H_2O$)	1405	0.50	1405	45.0	1405	516	1405	7.0	100
Lead plant slag	272	0.02	272	<0.02	272	0.3	272	6.4	<0.02
Cobalt plant jarosite	268	0.12	268	0.70	268	236	268	2.3	0.40
Zinc plant jarosite	917	0.06	917	0.02	917	0.31	917	6.6	0.02
Natural goethite	861	<0.01	892	0.05	892	1.10	892	8.3	0.06
Smelter dust leach residue	272	38.0	272	6.70	272	12.0	272	8.5	9.30
Dust leach residue calcined[1]	1129	259	1129	54.2	1129	23.5	1129	9.1	24.7
Copper arsenate/gypsum residue	582	68.0	582	7.10	582	8.50	582	9.3	8.90
Copper arsenate/gypsum calcined[1]	1129	70.1	1129	47.1	1129	2.53	1129	8.7	13.3

All tests with solid (dry basis) to liquid ratio of 1:16 at 25°C.
1. Calcined at 500°C for 4 hours.

that the arsenic is incorporated into the jarosite structure, and is, therefore, rendered stable. However, a jarosite with a high arsenic content (but formed under virtually identical conditions) will have a significant amount of free ferric arsenate, and it has previously been demonstrated[15] that the temperature of formation of such jarosites (95°C) is not sufficiently high to prevent the formation of unstable, amorphous ferric arsenate. The high iron content in this case is not sufficient to confer complete stability, since jarosite is formed rather than ferric hydroxide, and in the jarosite system there are two distinct iron phases, jarosite and ferric arsenate.[31,32] Ferric hydroxide, on the other hand, is well-known for its ability to strongly adsorb "impurities" such as arsenic.[13,33-38] In other respects, however, the cobalt plant jarosite, with its Fe/As molar ratio of 3.98:1 behaves as would be expected from the laboratory results on the "synthetic" high iron ferric arsenates, being stable in the low pH range 4-7.

The dust leach residue and the (iron-free) copper arsenate/gypsum residues showed significant arsenic solubility, which might be expected from their chemical analyses. However, these solids have not broken down beyond the levels indicated, and have remained "stable" at these levels, which can be regarded as being their solubilities, for several months. Calcining at 500°C did not decrease arsenic solubility, but rather had the opposite effect. Such residues cannot really be regarded as being suitable for direct disposal, however, due to their relatively high natural solubility.

Data for natural scorodite are included in Table IV, together with data for a sample of naturally occurring arsenical goethite. The results from these two tests are significant for two reasons. Firstly, the data for scorodite (crystalline ferric arsenate, molar Fe/As of 1:1) show that when in a crystalline state, stoichiometric ferric arsenate is indeed stable, and has remained so at pH 4 for over three years. This result is in accordance with data recently published by Krause and Ettel,[19] and would appear to confirm the claim that the stability of ferric arsenate is enhanced by two orders of magnitude when it is crystalline compared to when it is amorphous.[19]

TABLE V

CHEMICAL ANALYSIS OF SPEISSES

SPEISS			CHEMICAL ANALYSIS, %			
No	TYPE	Cooling Rate[1]	As	Fe	Cu	Sb
P1	Smelter speiss	-	44.5	54.2	-	2.65
P2	Smelter speiss	-	35.6	61.1	-	3.90
C1	Smelter speiss	-	40	40	10	1
C2	Smelter speiss	-	35	48	10	1
C3	Smelter speiss	-	30	55	10	1
C4	Smelter speiss	-	48	35	10	1
S1	Synthetic speiss	Slow	26.1	66.7	-	-
S2	Synthetic speiss	Fast	28.4	67.4	-	-
S3	Synthetic speiss	Slow	35.2	64.3	-	-
S4	Synthetic speiss	Fast	34.6	63.7	-	-
P3	Smelter speiss	Slow	35.8	60.7	-	2.43
P4	Smelter speiss	Fast	35.3	60.3	-	2.77
P5	Smelter speiss	Slow	28.9	67.0	-	2.95
P6	Smelter speiss	Fast	29.5	67.7	-	3.07
S5	Synthetic speiss	Slow	27.7	60.7	8.53	-
S6	Synthetic speiss	Fast	28.0	62.6	8.86	-
S7	Synthetic speiss	Slow	30.7	51.8	10.1	-
S8	Synthetic speiss	Fast	30.3	53.2	10.7	-

1. Speisses either quenched (fast cool) or allowed to cool naturally to ambient (slow cool).

Secondly, the results for the arsenical goethite are of significance, since it has been claimed[11,12] that the high iron ferric arsenates will ultimately crystallise, forming goethite, and in so doing release their arsenic. The data presented here suggest that this is not so, that arsenic incorporated into goethite, remains stable. The data for the high iron ferric arsenates shown in Table II suggest stability, and the data for the natural goethite reinforce this observation.

SPEISSES

A relatively common method of removing arsenic in pyrometallurgical operations is via the formation of iron or iron/copper arsenides (speisses). Similar compounds have been proposed as a vehicle for the safe disposal of arsenic.[41] At the Noranda Technology Centre, a number of speisses have been generated (Table V) as the result of concurrent pyrometallurgical investigations, and these compounds were subjected to stability testing. The pH values for the tests were different to those of the metal arsenates, since the testwork programs demanded that data be obtained at pH 2 for reasons other than stability testing. The effects of speiss cooling rate, and copper and antimony addition were investigated.

TABLE VI

RESULTS OF 24-H SPEISS TESTING

	SPEISS	pH 2		pH 5		pH 10		Natural pH		
		Days	As mg/L	Days	As mg/L	Days	As mg/L	Days	pH	As mg/L
P1	Smelter speiss	1	38.9	1	4.20	1	4.60			
P2	Smelter speiss	1	141	1	4.20	1	<0.2			
C1	Smelter speiss	1	<0.2	1	<0.2	1	<0.2			
C2	Smelter speiss	1	34.1	1	<0.2	1	<0.2			
C3	Smelter speiss	1	42.0	1	<0.2	1	<0.2			
C4	Smelter speiss	1	65.4	1	<0.2	1	<0.2			
S1	Synthetic speiss	1	1.20	1	0.06	1	0.28	1	10.0	0.33
S2	Synthetic speiss	1	0.68	1	0.58	1	0.58	1	10.5	0.50
S3	Synthetic speiss	1	1.05	1	0.10	1	0.43	1	9.5	0.28
S4	Synthetic speiss	1	0.80	1	0.40	1	0.48	1	10.4	0.50
P3	Smelter speiss	1	1.33	1	0.27	1	0.33	1	10.4	0.11
P4	Smelter speiss	1	1.73	1	0.73	1	0.43	1	11.4	0.58
P5	Smelter speiss	1	0.80	1	0.34	1	0.32	1	10.5	0.30
P6	Smelter speiss	1	0.93	1	0.33	1	0.25	1	10.4	0.33
S5	Synthetic speiss	1	24.0	1	2.43	1	0.93	1	10.0	2.13
S6	Synthetic speiss	1	19.5	1	2.13	1	1.63	1	10.2	1.85
S7	Synthetic speiss	1	39.5	1	2.48	1	4.95	1	10.9	1.80
S8	Synthetic speiss	1	24.0	1	1.93	1	1.48	1	10.0	1.93

The results of stability testing indicated (Tables VI, VII, VIII and Figure 3):

(a) that in general, speisses were sufficiently unstable after one day at pH 2 to exceed the EPA limit of 5 mg/L soluble arsenic (Table VI), but not at either pH 5 or pH 10;

(b) there was no obvious effect of cooling rate, Fe/As ratio, or whether the presence of copper was beneficial in promoting stability. The presence of antimony, however, appeared to promote increased arsenic solubility over the long term (Table VIII);

(c) all the speisses tested showed some degree of breakdown. However, as can be seen from Table VIII and Figure 3, the level of soluble arsenic increased to a maximum, which varied both in terms of concentration and in the time taken to achieve it, depending upon the speiss. Thereafter, the arsenic concentration decreased to a very low level, which would generally be regarded as being acceptable.

TABLE VII

MAXIMUM LEVELS OF ARSENIC IN SOLUTION DURING SPEISS TESTING

SPEISS		pH 2 Days	pH 2 As mg/L	pH 5 Days	pH 5 As mg/L	pH 10 Days	pH 10 As mg/L	Natural pH Days	Natural pH pH	Natural pH As mg/L
P1	Smelter speiss	130	2060	580	60.4	215	95.3			
P2	Smelter speiss	580	625	215	604	1	0.30			
C1	Smelter speiss	580	699	580	6.7	580	0.19			
C2	Smelter speiss	1	34.2	211	35.0	1144	*			
C3	Smelter speiss	1	42.0	439	10.2	1144	**			
C4	Smelter speiss	577	902	1144	*	577	2.30			
S1	Synthetic speiss	62	1.40	62	6.30	135	1.87	62	10.3	8.60
S2	Synthetic speiss	28	3.30	62	5.60	62	6.30	62	10.9	36.0
S3	Synthetic speiss	62	2.20	62	7.10	62	15.9	62	9.7	4.80
S4	Synthetic speiss	62	1.00	62	7.50	62	2.10	62	10.9	29.8
P3	Smelter speiss	137	310	61	29.0	61	44.5	137	10.6	51.2
P4	Smelter speiss	712	*	137	33.4	61	57.8	137	11.9	123
P5	Smelter speiss	62	4.00	27	6.50	62	19.7	62	10.8	37.6
P6	Smelter speiss	62	15.6	62	7.50	62	13.2	712	10.7	*
S5	Synthetic speiss	1	24.0	1	2.43	132	32.0	57	10.5	15.3
S6	Synthetic speiss	1	19.5	137	7.54	57	64.5	57	10.5	14.2
S7	Synthetic speiss	1	39.5	1	2.48	57	31.3	57	10.8	58.8
S8	Synthetic speiss	1	24.0	709	*	57	74.7	57	10.6	36.8

* indicates As level still rising, shown in Table VIII
** indicates As levelled out at a maximum, shown in Table VIII

This apparently unusual behaviour can be attributed to an initial breakdown of the speiss, generating both soluble iron and soluble arsenic. (Significant levels of iron were also noted in solution, generally equivalent to or greater than what would be expected based on arsenic concentration in solution and speiss chemical analysis). The iron subsequently oxidised under the influence of atmospheric oxygen, forming ferric hydroxide which re-precipitated. As ferric hydroxide precipitates, it is well known to act as an adsorbent for ions such as arsenic in solution.[13,33-38] This phenomenon was observed in virtually all of the speiss tests conducted, and suggests that, in a static environment (i.e. a closed system), the disposal of speiss may be acceptable. However, it is clear that many of the speisses are not suitable for direct disposal in a dynamic environment, where the arsenic can be washed away. The data demonstrate that, because soluble arsenic is released, speisses cannot in themselves be regarded as being suitable for direct disposal. But, in a situation where the ingress of air and/or water can be minimized, significant breakdown should be prevented, although any such dump will require constant monitoring.

TABLE VIII

RESULTS OF LONG TERM SPEISS TESTING

	SPEISS	pH 2 Days	pH 2 As mg/L	pH 5 Days	pH 5 As mg/L	pH 10 Days	pH 10 As mg/L	Natural pH Days	Natural pH	As mg/L
P1	Smelter speiss	1146	10.2	1146	1.68	1146	43.8			
P2	Smelter speiss	1146	91.8	1146	9.05	1146	0.01			
C1	Smelter speiss	1147	348	1147	6.2	1147	0.05			
C2	Smelter speiss	1144	0.04	1144	9.33	1144	7.34			
C3	Smelter speiss	1144	0.90	1144	1.12	1144	1.12			
C4	Smelter speiss	1144	726	1144	10.8	1144	0.03			
S1	Synthetic speiss	706	0.07	706	<0.01	706	<0.01	706	11.2	0.27
S2	Synthetic speiss	706	0.04	706	<0.01	706	0.03	706	11.5	0.01
S3	Synthetic speiss	706	0.05	706	0.02	706	0.52	706	10.9	1.50
S4	Synthetic speiss	706	0.26	706	0.18	706	0.02	706	11.4	0.04
P3	Smelter speiss	705	94.4	705	12.6	705	2.71	705	10.8	32.1
P4	Smelter speiss	712	330	712	16.9	712	11.1	712	11.6	84.3
P5	Smelter speiss	712	1.50	712	0.06	712	2.71	712	11.6	2.41
P6	Smelter speiss	712	0.38	712	0.12	712	1.80	712	11.0	34.0
S5	Synthetic speiss	707	0.04	707	0.21	707	0.40	707	11.1	0.28
S6	Synthetic speiss	707	0.26	707	0.03	707	1.50	707	11.1	0.31
S7	Synthetic speiss	708	0.31	708	0.01	708	0.26	708	11.6	4.81
S8	Synthetic speiss	709	0.20	709	7.82	709	0.36	709	11.5	2.71

The speiss data are generally consistent with the observations already presented, that high iron ferric arsenates are the most stable form of arsenic. The data in Table VIII indicate that over time, the soluble arsenic component of speisses will revert to a very low level, in many cases <1 mg/L. The resulting residue exhibits all the characteristics of the amorphous high iron ferric arsenate residues in terms of its solubility and stability.

DISCUSSION

The data presented in this paper, taken in conjunction with data recently published in the literature,[15-24] provide a very sound indication that arsenic emanating from extractive metallurgical operations, both hydrometallurgical and pyrometallurgical, can be disposed of in an environmentally acceptable manner. It has been demonstrated that crystalline ferric arsenate (the mineral scorodite) has a solubility of <1 mg As/L at pH 4, a finding in accordance with data recently published by Krause and Ettel.[19] It has further been shown that high iron ferric arsenates with a molar Fe/As ratio >3:1 have been stable for periods up to three years, in the pH range 4-7, and that this range can be extended to 4-10 if there are mixed base metals (Cu+Cd+Zn) present in the residue. These laboratory data have further been shown to be applicable to "real" residues generated in plant and pilot plant operation, and that they can be further extended to encompass compounds such as iron and copper/iron speisses.

It is of interest to note that verification of these findings can be found in actual practice. Both Giant Yellowknife Mines[42] and Inco's CRED plant[43] dispose of arsenical solids, with reported Fe/As ratios of 8-10:1 and 27:1 respectively, with no indication of soluble arsenic in the run-off waters. These disposal systems are dynamic, in that new material is constantly being added, and so represent a situation where there are both old and recent solids. The high Fe/As ratios, based on laboratory data, are ideal for stability, even at pH values >7.

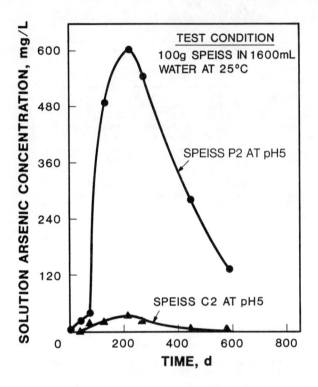

Figure 3. Stability of speisses as a function of time, showing a maximum in arsenic concentration

The true solubility of ferric arsenate is undoubtedly the dominating factor affecting the stability of the high iron ferric arsenate solids.[15] The data presented in this paper indicate quite strongly that given a knowledge of the system and an understanding of the chemistry involved in that system, it should be possible to generate an arsenical residue that will be environmentally stable. Structural analysis of the high iron ferric arsenates, both at the time of precipitation and after a period of some years under stability testing, could furnish further valuable information as to the behaviour of these residues.

CONCLUSIONS

It has been shown that precipitation of arsenic from hydrometallurgical process solutions, or the production iron or copper/iron speisses from pyrometallurgical operations, having Fe/As molar ratios >3 provide a potentially effective method for the disposal of arsenic. The incorporation of base metals into the hydrometallurgical residue enhances the stability of the residues. Nevertheless, until more longer term data are collected, and until a clearer understanding of the behaviour of these residues is obtained, it will remain necessary to continually monitor disposal sites, although such data that there are suggest that no leakage will occur.

ACKNOWLEDGEMENT

The authors wish to acknowledge Noranda Inc. for permission to publish this paper.

REFERENCES

1. Impurity Control and Disposal, Proceedings of the 15th Annual CIM Hydrometallurgy Meeting, Vancouver, August 1985.

2. Reddy, R.G. et al., Editors, Arsenic Metallurgy: Fundamentals and Applications, Proceedings of an International Symposium, Phoenix, Arizona, January 1988, TMS-AIME.

3. Robins, R.G., The solubility of metal arsenates, Paper D-1-3 in Fourth Joint Meeting, MMIJ-AIME, Tokyo, 1980, p. D25.

4. Robins, R.G., The solubility of metal arsenates, Met. Trans. B, 12B(1), March 1981, p. 103.

5. Robins, R.G., The solubility of metal arsenates II, Paper presented at 111th AIME Annual Meeting, Dallas, Texas, February 1982.

6. Robins, R.G. and Tozawa, K., Arsenic removal from gold processing wastewaters: the potential ineffectiveness of lime, CIM Bulletin, 75(840), April 1982, p. 171.

7. Robins, R.G., Stabilities of arsenic V and arsenic III compounds in aqueous metal extraction systems, in Hydrometallurgy: Research, Development and Plant Practice, Proceedings of the 3rd International Symposium on Hydrometallurgy (K. Osseo-Asare and J.D. Miller, Editors), Atlanta, Georgia, March 6-10, 1983, TMS-AIME, New York, p. 291.

8. Robins, R.G., The stability of arsenic in gold mine processing wastes, in Precious Metals: Mining, Extraction and Processing (V. Kudryk, D.A. Corrigan and W.W. Liang, Editors), Proceedings of a TMS-AIME/IPMI International Symposium, Los Angeles, February 1984, TMS-AIME, New York, p. 241.

9. Robins, R.G., The aqueous chemistry of arsenic in relation to hydrometallurgical processes, op. cit., reference 1, p. 1-1.

10. Robins, R.G., The solubility and stability of ferric arsenate: $FeAsO_4.2H_2O$ (scorodite), Paper presented at the 116th AIME Annual Meeting, Denver, Colorado, February, 1987.

11. Robins, R.G. and Glastras, M.V., The precipitation of arsenic from aqueous solution in relation to disposal from hydrometallurgical processes, in Research and Development in Extractive Metallurgy, AusIMM, Adelaide Branch, May 1987, p. 223.

12. Robins, R.G., Solubility and stability of scorodite, $FeASO_4.2H_2O$: Discussion, American Mineralogist 72, 1987, p. 842.

13. Robins, R.G and Haung, J.Y.C., The adsorption of arsenate ion by ferric hydroxide, op. cit., reference 2, p. 99.

14. Robins, R.G., Arsenic hydrometallurgy, op. cit., reference 2, p. 215.

15. Harris, G.B. and Monette, S., The stability of arsenic-bearing residues, op. cit.,

reference 2, p. 469.

16. Krause, E. and Ettel, V.A., Ferric arsenate compounds: are they environmentally safe? Solubilities of basic ferric arsenates, op. cit., reference 1, p. 5-1.

17. Krause, E. and Ettel, V.A., Solubilities and stabilities of ferric arsenates, in Crystallisation and Precipitation (G.L. Strathdee, M.O. Klein and L.A. Melis, Editors), Proceedings ISCAP '87, Saskatoon, Saskatchewan, Canada, October 5-7, 1987, p. 195.

18. Krause, E. and Ettel, V.A., Solubilities and stabilities of ferric arsenates, Hydrometallurgy, 1988 (in press).

19. Krause, E. and Ettel, V.A., Solubility and stability of scorodite, $FeAsO_4.2H_2O$: New data and further discussion, American Mineralogist 73, 1988, p. 850.

20. Dove, P.M. and Rimstidt, J.D., The solubility and stability of scorodite, $FeAsO_4.2H_2O$, American Mineralogist, 70, 1985, p. 838.

21. Dove, P.M., The solubility and stability of scorodite, $FeAsO_4.2H_2O$, M.S. Thesis, Virginia Polytechnic Institute and State University, Blacksburg, Virginia, March 1984.

22. Stefanakis, M. and Kontopoulos, A., Production of environmentally acceptable arsenites-arsenates from solid arsenic trioxide, op. cit., reference 2, p. 287.

23. Papassiopi, N., Stefanakis, M. and Kontopoulos, A., Removal of arsenic from solutions by precipitation as ferric arsenates, op. cit., reference 2, p. 321.

24. Kontopoulos, A., Stefanakis, M. and Papassiopi, N., Arsenic control in hydrometallurgy by precipitation as ferric arsenates,

25. Chukhlantsev, V.G., The solubility products of a number of arsenates, Russian Journal of Analytical Chemistry 11, p. 565.

26. Dutrizac, J.E. and Jambor, J.L., The synthesis of crystalline scorodite, $FeAsO_4.2H_2O$, Hydrometallurgy, 19(3), January 1988, p. 377.

27. Dutrizac, J.E., CANMET, Personal communication, June 1987.

28. Robins, R.G., University of New South Wales, Personal communication, September 1983.

29. EP Toxicity Test Procedure, United States Federal Register 45(98), May 19, !980, p. 33127.

30. Gerth, J., Australian Nuclear Science and Technology Organisation, Personal communication, February 1988.

31. Dutrizac, J.E. and Jambor, J.L., The behaviour of arsenic during jarosite precipitation: arsenic precipitation at 97°C from sulphate or chloride media, Canadian Metallurgical Quarterly 26(2), April-June 1987, p. 91.

32. Dutrizac, J.E., Jambor, J.L. and Chen, T.T., The behaviour of arsenic during jarosite precipitation: reactions at 150°C and the mechanism of arsenic precipitation, Canadian Metallurgical Quarterly 26(2), April-June 1987, p. 103.

33. Jones, C.J., Hudson, B.C. and McGugan, P.J., The removal of arsenic(V) from acidic solutions, J. Hazardous Materials 2 (1977/78), p. 333.

34. Leckie, J.O., Merrill, D.T. and Chow, W., Trace element removal from power plant wastestreams by adsorption/co-precipitation with amorphous iron oxyhydroxide, In Separation of Heavy Metals and Other Trace Contaminants, (R.W. Peters, B. Mo Kim,

Editors), AIChE, New York, 1985, p. 28.

35. Thorsen, G., Extractive metallurgy of copper, In Topics in Non-Ferrous Extractive Metallurgy (A.R. Burkin, Editor), Critical Reports on Applied Chemistry Volume I, John Wiley and Sons, New York, 1980, p. 1.

36. Pierce, M.L. and Moore, C.B., Adsorption of arsenite on amorphous iron hydroxide from dilute aqueous solution, Environmental Science and Technology 14(2), February 1980, p. 214.

37. Pierce, M.L. and Moore, C.B., Adsorption of arsenite and arsenate on amorphous iron hydroxide, Water Resources, 16, 1982, p. 1247.

38. Ghosh, M.M., Adsorption of inorganic arsenic and organo-arsenical on hydrous oxides, In Metals Speciation, Separation and Recovery (J.W. Patterson, R. Passino, Editors), Proceedings of an International Symposium, Chicago, July 1986, Lewis Publishers, 1987, p. 499.

39. Harris, G.B., Monette, S. and Stanley, R.W., Hydrometallurgical treatment of Blackbird cobalt concentrate, in Hydrometallurgy: Research, Development and Plant Practice, Proceedings of the 3rd International Symposium on Hydrometallurgy (K. Osseo-Asare and J.D. Miller, Editors), Atlanta, Georgia, March 6-10, 1983, TMS-AIME, New York, p. 139.

40. Harris, G.B. and Monette, S., A hydrometallurgical approach to treating copper smelter precipitator dusts, In Complex Sulfides: Processing of Ores, Concentrates and By-Products, Proceedings of TMS-AIME/CIM Symposium, San Diego, California, November 10-13, 1985, p. 361.

41. Tahija, D. and Haung, H., Hydrometallurgical formation of iron-arsenic compounds, op. cit., reference 1, p. 4-1.

42. Thomas, K.G., The complexity of gold extraction at Giant Yellowknife Mines, Paper presented at the Falconbridge Limited Gold Milling Seminar, Val d'Or, Quebec, Canada, October, 1985.

43. Opratko, V. and Steward, D.A., The removal of iron and arsenic from leach liquors at Inco's copper refinery, electrowinning department, Paper presented at 20th annual CIM Conference of Metallurgists, Hamilton, Ontario, August 25, 1981.

NICKEL, COBALT AND COPPER BEARING SYNTHETIC MODEL OXYHYDROXIDES

OF MANGANESE(IV)—RELEVANCE IN MANGANESE NODULE PROCESSING

Rakesh Kumar, R.K. Ray and A.K. Biswas

Department of Metallurgical Engineering
Indian Institute of Technology Kanpur
Kanpur-208016
INDIA

Abstract

Structure and leachability aspects of pure and doped synthetic oxyhydroxide of manganese viz. birnessite and δ-MnO_2, are investigated. Chemical analysis, surface area, XRD and TEM data indicated that (i) in terms of structural complexities the doped samples can be arranged as follows: sorption > ion exchange > coprecipitation and (ii) irrespective of method of doping, the doped elements Ni, Co and Cu occur in the structure of the host phases. The leaching data for the doped elements in sulfuric acid can be adequately explained in terms of structural complexities and crystal field stabilization energies (CFSE). The leaching of samples is accompanied by collapse of birnessite structure and formation of poorly leachable phases like nsutite and cryptomelane. All the samples are easily leachable in sulfurous acid. Chemical nature of elements and surface area can influence preferential dissolution of synthetic samples in sulfurous acid. The implications of the results are presented for nodules of different mineralogy.

Introduction

In spite of the fact that a large number of processes have been developed and patented for metal extraction from deep-sea manganese nodules (1-3), the variation in process parameters vis-a-vis nodules mineralogy and internal structure stands in the way of their universal application. There are as yet no published data available on the mechanism/s of incorporation (e.g. lattice substitution, sorption, coprecipitation etc.) of Ni, Co and Cu in host oxyhydroxides of manganese and iron in manganese nodules. Attempts to identify the exact mode of incorporation from leaching data have not been very successful due to multi-mineralic nature of nodules (1,4-6).

The phases in manganese nodules are fine grained and intimately intermixed giving rise to complex internal structure. It is difficult or rather impossible to extract a significant amount of a particular phase for meaningful and systematic characterization and leaching studies. Leaching studies on the well characterized synthetic analog of phases offer an opportunity to gain some insight into the complex nature of phases present in nodules.

The major interest in manganese nodules centres on the oxide constituents of manganese because Ni, Cu and often Co correlate positively with manganese in nodules (3,6-8). 10 A^o phase (todorokite/buserite), 7 A^o phase (birnessite) and δ-MnO_2 or vernadite are the predominant manganese oxyhydroxide minerals present in nodules (9,10). The structural relationships among different oxyhydroxides of manganese(IV) have been discussed in the literature (10-12). Birnessite is a representative member of sheet structure phyllomanganates and δ-MnO_2 is randomly stacked birnessite (12).

There have been several attempts to study the dissolution of pure oxides and hydroxides of Mn, Ni, Co and Cu in acids (13-17). The available literature on leaching of well characterized doped oxyhydroxides of manganese (relevant to nodules mineralogy) has been found to be scanty. Giovanoli (18) studied the interaction of buserites, doped with Ni, Co and Cu, with H^+ ions in dilute HNO_3. Although collapse of buserite structure and formation of mineral nsutite was noted during acidification reaction, its significance for leaching of doped elements was not highlighted. The chemical composition data for the leach liquors were also not reported.

The scope of this paper encompasses synthesis, characterization and leaching studies on pure and doped oxyhydroxides of tetravalent manganese, namely, birnessite and δ-MnO_2. The minerals were doped in ion exchange, coprecipitation and sorption mode. The leaching media employed were H_2SO_4 (non-reducing) and H_2SO_3 (reducing). The H_2SO_4-H_2SO_3 route is a promising hydrometallurgical route for the future processing of nodules and its technological importance has been pointed out in recent literature (19,20). The implications of characterization and leaching results are presented for manganese nodules genesis and processing.

Materials and Methods

Manganese Oxide Samples

The samples which were synthesized fall in two broad categories:
A. Birnessite samples: (i) Pure Na-birnessite, (ii) birnessite with Ni/Co/Cu-coprecipitation mode and (iii) birnessite with Ni/Co/Cu-ion exchange mode.
B. δ-MnO_2 samples: (i) Pure δ-MnO_2 and (ii) δ-MnO_2 with Ni/Co/Cu-sorption mode. Disordered birnessite, i.e. δ-MnO_2 was selected for doping in sorption mode because of its very high specific surface area which may impart unique surface properties.

The above samples will be referred to as B-O, V-O, B-R(C), B-R(I) and V-R(S). B- and V- mean birnessite and δ-MnO$_2$ (or vernadite). R is doped element (Ni, Co and Cu) and O, C, I and S refer to pure sample, coprecipitation, ion exchange and sorption mode (vide reference 3, 12 and 21 for the terminology used). Mode of doping refers to method of doping and not the mode of occurrence.

Preparation of Birnessites. The birnessite samples were prepared by dehydration of respective buserite (Bu) samples.

Na-birnessite was prepared by the method similar to that of Giovanoli et al. (22). 200 ml of 0.5M Mn(NO$_3$)$_2$.4H$_2$O solution is poured in a 1000 ml thermostatically controlled ($\sim$30°C) double walled cylinder fitted with a 30 mm, G-1 glass frit at the bottom. Oxygen is bubbled through the solution at a rate of 2.5 L/min and a solution of 55 g NaOH ($\sim$30°C) in 200 ml H$_2$O is added to the nitrate solution. The oxygenation is stopped after 4-5 hrs. The black precipitate (Na-buserite) is isolated in a centrifuge and thoroughly washed 6-7 times with triple distilled water. The above product is isolated and dried in an oven at 120°C for 24 hrs.

The method adopted for synthesis of B-R(C) samples was similar to B-O. In this case, same amount of NaOH solution was added to 200 ml solution containing both Mn(NO$_3$)$_2$.4H$_2$O and R(NO$_3$)$_2$.XH$_2$O. The Mn/R mole ratio in the solution was approximately 20.

B-R(I) samples were prepared from thoroughly washed Na-buserite (Bu) samples. Na-buserite equivalent to one synthesis (yields $\sim$10 g Na-birnessite on drying) was equilibrated with 750 ml of respective metal nitrate solution (0.67 M) for one week at room temperature. The pH was measured at the beginning and intermediate time intervals during the progress of ion exchange reaction. The initial and final pH's were as follows: 5.3 and 4.4 for Bu-Ni(I), 5.1 and 4.6 for Bu-Co(I) and 3.7 and 3.5 for Bu-Cu(I). The Bu-R(I) samples were isolated and washed NO$_3^-$ free (diphenylamine test). B-R(I) samples were obtained from Bu-R(I) samples by drying at 120°C as before.

Dried birnessite samples were ground and sieved to get -210 + 149 μm and -63 + 53 μm size fractions.

Preparation of δ-MnO$_2$ Samples. The procedure used for the synthesis of V-O was the one used by Buser et al. (23) to get the highly disordered variety of birnessite (referred to as δ-MnO$_2$ in this investigation and by Buser et al. (23) also). 18 ml of concentrated HCL (sp. gr. 1.181) was added to 250 ml of 0.4M boiling KMnO$_4$ solution. The product obtained was thoroughly washed and dried at 120°C.

V-R(S) samples were prepared from undried V-O. Undried and washed V-O (equivalent to one synthesis, 7.5 g after drying) was equilibrated with 4500 ml of 3.5 x 10^{-4}M R(NO$_3$)$_2$.XH$_2$O solution at pH 8.2 $\pm$ 0.2 and temperature 25 $\pm$ 1°C for 36 hrs. No base electrolyte was used. The solid samples were thoroughly washed and dried at 120°C. All the samples were ground to -53 μm size before storing.

Characterization Techniques

The concentration of doped elements in solutions were estimated by colorimetric method using sodium diethyldithiocarbamate (Na-DDTC) (24). Manganese was estimated by permanganate method using potassium periodate as an oxidizing agent (24). The average oxidation state of manganese (O/Mn ratio as MnO$_x$) in the solid samples was determined by oxalate method (25). The error limit of all the analytical methods used was within 2%.

Surface area was estimated by ethylene glycol monomethyl ether (EGMME) retention method. The details are presented elsewhere (3).

The techniques of X-ray diffraction (XRD) were employed to gather informations on the phase constituents and crystallinity of samples. Original samples (including buserites) and leach residues obtained in sulfuric acid leaching were examined by XRD. XRD traces were recorded on a Iso-Debyeflex 2002D diffractometer using Cr-K_α radiation and graphite monochromator. All the samples were scanned under identical conditions - 3.0° 2θ/min, tube voltage 40 kV and tube current 25 mA. Buserite samples were examined in moist condition; drying of sample results in collapse of structure to 7 A°-phase. XRD traces of samples were also recorded after intercalation treatment (26,27) with dodecylammonium chloride.

The characterization by XRD was complimented with transmission electron microscopy (TEM) in SAED and imaging mode. TEM studies were carried out on a Philips EM-301 electron microscope.

Leaching Experiments

Leaching was carried out in a multineck, cylindrical glass reaction vessel, immersed in a thermostatically controlled water bath. Sulfuric acid (1.0N, 50°C) and sulfurous acid (0.2 g SO_2 in 100 ml H_2O, 20°C) were employed as leaching media. -63 + 53 μm and -210 + 149 μm size fractions of birnessites were used in sulfuric and sulfurous acid leaching systems respectively. Stirring speed (900 r.p.m.) was constant in all the experiments.

Leaching data are presented in terms of fraction reacted (F_i) or 'leachability parameter' (NF_i) (i.e. F_i/SA, g m^{-2}; SA = initial specific surface area, m^2/g). The use of normalized fraction reacted (NF_i) was found to be essential in order to have a coherent basis for leaching data comparison among the different samples, eliminating the ever present factor of the magnitude of surface area.

Characterization of Samples

Chemical Composition and Surface Area

The chemical composition and specific surface area data in Table I are presented on weight basis of samples dried at 120°C and equilibrated over $CaCl_2$ in vacuum to get the constant weight. The samples in different subcategories, namely, coprecipitation, ion exchange and sorption, show similar and distinct compositional features viz. similar mole ratio 'y' and 'z' (where y = [Mn_T]/([Mn(II)] + [R]) and z = [Na] or [K]/[Mn_T], vide Table-I). Wide variation in water content of different samples needs to be noted. B-R(I) samples contain more water compared to B-R(C).

The sample B-Co(I) exhibits a (O/Mn) ratio of ~2, indicating presence of total manganese in +(IV) oxidation state. Leaching results for this sample (discussed later) have shown that at least 2.56% manganese is present in +(II) oxidation state. Taking this fact into account, the (O/Mn) ratio in this sample turns out to be 1.94 or less. The discrepancy between the observed and calculated oxidation state arises probably due to partial oxidation of Co(II) to Co(III) during synthesis. Co(III) present in the sample consumes excess amount of oxalate and in turn results in apparent increase in (O/Mn) ratio (25). Our contention regarding occurrence of Co(III) is supported by XPS data of Crowther et al. (28) for cobalt doped in Na-birnessite in ion exchange mode. Computed weight percent of oxidized cobalt in B-Co(I) sample turns out to be 4.4% or more.

Table I. Chemical Composition and Specific Surface Area (SA) of Samples

S. No.	Sample code	Chemical composition, wt %						x or (Mn/O)	Mole ratio			SA (m²/g)	
		Mn_T	Mn(II)	R	Na	K	H_2O		y	z		-210+149 μm	-63+53 μm
1	B-0	53.1	9.0	–	5.7	–	10.9	1.820	5.88	0.257		25.8	32.4
2	B-Ni(C)	51.0	6.6	2.71	8.3	–	6.5	1.864	5.62	0.384		35.0	35.1
3	B-Co(C)	52.0	6.1	2.74	8.2	–	5.0	1.877	6.00	0.378		25.8	38.6
4	B-Cu(C)	49.8	6.0	2.90	8.0	–	8.5	1.873	5.85	0.380		24.8	48.5
5	B-Ni(I)	50.5	9.0	10.6	0.24	–	8.9	1.813	2.69	0.011		45.1	59.0
6	B-Co(I)	42.6	-0.6	16.7	0.29	–	11.1	2.025	≤2.73	0.016		30.8	47.8
7	B-Cu(I)	44.7	2.8	15.4	0.28	–	10.4	1.935	2.77	0.015		30.6	55.2
8	V-0	48.5	<1.0	–	–	9.0	11.0	1.991	–	0.262		–	149
9	V-Ni(S)	48.5	<1.0	1.27	–	7.5	11.2	1.991	–	0.218		–	138
10	V-Co(S)	48.5	<1.0	1.27	–	7.8	10.7	1.991	–	0.226		–	138
11	V-Cu(S)	48.5	<1.0	1.32	–	7.5	11.1	1.991	–	0.218		–	138

Table II. Chemical Equivalent Balance in Birnessites

Sample code	$\sum \frac{z_i \text{ wt \%}}{M_i}$	CE_i, %					$\Delta_i = (CE_i)_{doped} - (CE_i)_{B-0}$			
		Mn(IV)	Mn(II)	R(II)	R(III)	Na^+	Mn(IV)	Mn(II)	R(II)	R(III)	Na^+
B-0	3.78	84.8	8.6	0	–	6.5	0	0	0	–	0
B-Ni(C)	3.92	82.3	6.1	2.34	–	9.2	-2.5	-2.5	2.34	–	2.7
B-Co(C)	4.01	83.3	5.5	2.10	–	8.9	-1.5	-3.1	2.10	–	2.4
B-Cu(C)	3.85	82.9	5.7	2.37	–	9.0	-1.9	-2.9	2.37	–	2.5
B-Ni(I)	3.72	81.7	8.6	9.71	–	0.03	-3.1	0.0	9.71	–	-6.47
B-Co(I)	3.66	79.5	2.5	11.4	6.1	0.04	-5.3	-6.1	11.4	6.1	-6.46
B-Cu(I)	3.65	83.5	2.8	13.3	–	0.03	-1.3	-5.8	13.3	–	-6.47

Mode of Incorporation of Doped Element. Comparison of doped birnessite compositions with pure birnessite (B-0) (Table I) reveals that incorporation of doped element in birnessite involves exchange with structural Mn(II), Na$^+$ and Mn(IV) or perhaps Mn(III) in the original buserite structure (the estimation of (O/Mn) ratio by oxalate method does not distinguish between Mn(III) and mixture of Mn(IV) and Mn(II)). The exchange with H$^+$ was additionally noted during synthesis of metal buserites in ion exchange mode as indicated by the decrease in pH.

The structure of birnessite (or buserite) consists of layers of manganese(IV) ions octahedrally coordinated to oxygen and or hydroxyl ions. The layers have a permanent negative charge caused by vacancies and/or diadochic substitution of Mn(II) or Mn(III) for Mn(IV). Successive interlayers are separated by an interlayer region containing Na$^+$, water molecules and Mn(II) ions (22). The doped divalent transition element may occupy the vacant lattice sites in (MnO$_6$) octahedra, substitute for Mn(II) in (MnO$_6$) octahedra and/or Mn(II) and Na$^+$ in the interlayer (9,26).

Chemical compositions of birnessite samples are presented in percent chemical equivalent (CE_i) (Table-II), to elucidate the nature of doped element substitution in birnessite structure. The chemical equivalent percent of an ion i with valency Z_i and atomic weight M_i is calculated using the following formula

$$(CE_i) = \left[(z_i \cdot \frac{wt\% \ i}{M_i})/\sum(z_i \cdot \frac{wt\% \ i}{M_i})\right] \times 100$$

In calculating (CE_i), it is assumed that (i) Mn(III) present in the birnessite structure is equivalent to mixture of Mn(II) and Mn(IV), (ii) exchange with H$^+$ is not significant and (iii) only Mn(IV), Mn(II), R(II), R(III) and Na$^+$ are involved in substitution.

The total number of chemical equivalent (i.e. denominator in CE_i calculation) in B-0 and doped samples reveal that the correspondence between B-0 and doped sample is not 1:1. The total number of equivalents shows upto 6% deviation compared with B-0. Assuming idealized situation in which (i) there is 1:1 correspondence between doped samples and pure birnessite and (ii) vacancies are not involved in exchange, the composition of doped samples are compared with B-0 in Table-II.

It appears (vide Table-II) that Mn(IV) in original birnessite phase may be substituted by divalent Ni, Co and Cu. Ionic radii and crystal field stabilization energy (CFSE) criteria are often invoked to explain the substitution reactions among transition metal ions (29,30).

Substitution of Mn(IV) by divalent ions is unlikely because of large difference in ionic radii of Mn(IV) (0.53 A°) and Ni(II) (0.69 A°), Co(II) (0.745 A°) and Cu(II) (0.73 A°) and loss in CFSE (the CFSE of Mn(IV), Ni(II), Co(II) and Cu(II) are 393 kJ/mole, 122.2 kJ/mole, 92.9 kJ/mole and 90.4 kJ/mole respectively). It is more likely that anomaly, regarding Mn(IV) substitution by divalent metal ions, results due to violation of assumption made e.g. 1:1 correspondence between doped and pure birnessite, role of vacancies etc. Variation in surface area and water content among birnessites (Table-I) indicates that incorporation of doped element is indeed accompanied by structural break-up and/or rearrangement. The substitution of Mn(IV) (r = 0.53 A°, CFSE = 393.3 kJ mol^{-1}) by low spin Co(III) (r = 0.545 A°, CFSE = 533.5 kJ mol^{-1}) is suggested by gain in CFSE and similar ionic radii. This also explains the very large difference in $CE_{Mn(IV)}$ value between B-0 and B-Co(I) (Table-II).

Other things being equal, there will be gain in CFSE due to substitution of Mn(II) (r = 0.83 A°, CFSE = 0) by Ni(II), Co(II) and Cu(II). Cheical composition data in Tables-I and II strongly suggest the substitution of Mn(II) by divalent doped elements in B-R(C). It is striking to note that doping in B-R(C) samples is associated with higher uptake of Na^+. It is proposed that incorporation of doped element in coprecipitation mode is associated with filling up of more sites in (MnO_6) octahedra, resulting in a more negative charge on (MnO_6) octahedra sheets and in turn higher uptake of Na^+.

Substitution of cobalt in B-Co(I) can be almost completely explained in terms of exchange with Na^+, Mn(II) and Mn(IV). Also there is a good match between equivalent of Cu(II) incorporated in B-Cu(I) and Mn(II) and Na^+ replaced (Table-II). Nickel in B-Ni(I) substitutes for Na^+. It is striking to note that against the background of CFSE, Mn(II) is not substituted by Ni(II) in B-Ni(I). It appears that in B-Ni(I) sample, where the (MnO_6) sheet structure is already built before ion exchange, the effect of gain in CFSE is offset by large discrepancy between ionic radii of Mn(II) (0.83 A°) and Ni(II) (0.69 A°). This is in contrast with B-Ni(C) where Ni(II) can occupy all those sites which are amenable to Mn(II) substitution in the growing buserite phase.

The sorption of Ni, Co, Cu on V-0 can be explained in terms of exchange with K^+. Sorption of Ni, Co and Cu by δ-MnO_2 results in almost complete removal of metal ions from solution phase. The change in surface area of δ-MnO_2 after sorption (which is an indication of structural rearrangement during sorption) needs to be noted (Table-I).

Phase Constituents, Structure and Crystallinity

X-ray Diffraction. XRD patterns of birnessite samples are presented in Figure 1.

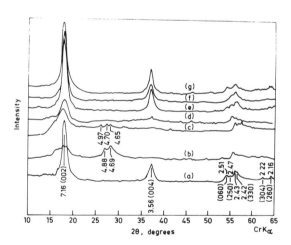

Figure 1 - XRD patterns of birnessite samples (a) B-0, (b) B-Ni(I), (c) B-Co(I), (d) B-Cu(I), (e) B-Ni(C), (f) B-Co(C) and (g) B-Cu(C).

XRD pattern of B-0 in Figure 1 corresponds to Na-birnessite ($Na_4Mn_{14}O_{27} \cdot 9H_2O$) (ASTM 23-1046). The doped samples contain the 7 A° line which is characteristic of birnessite (Figure 1). Additional lines appear between 4.5-5.0 A° for B-Ni(I) and B-Co(I) samples. These lines may be attributed to incomplete conversion during buserite to birnessite transformation (line at 4.8 (or (9.6/2) A° may result due to 002 reflection of disordered buserite)and/or formation of new phases. The line at 4.8 A° also corresponds to the 002 reflection in todorokite (ASTM 13-164). X-ray data are inadequate to elucidate this point any further.

Salient XRD features of birnessites and corresponding parent buserites (before and after intercalation) are summarized in Table-III. XRD features of buserites are included because structures of birnessites are expected to be governed by the parent buserites used for synthesis.

Table III. XRD Features of Birnessites and Corresponding Parent Buserites

Birnessite			Buserite		
Sample code	Intercalation response	$\Delta 2\theta_{002}$	Sample code	Position of 10 A° line	Intercalation response
B-0	–	1.0°	Bu-0	9.93	Yes
B-Ni(C)	Yes	1.1°	Bu-Ni(C)	10.04	Yes
B-Co(C)	Yes	1.1°	Bu-Co(C)	10.03	Yes
B-Cu(C)	Yes	1.1°	Bu-Cu(C)	10.03	Yes
B-Ni(I)	No	3.5°	Bu-Ni(I)	9.53	No
B-Co(I)	Poor	2.5°	Bu-Co(I)	9.64	Poor
B-Cu(I)	No	1.5°	Bu-Cu(I)	9.57	No

Bu-R(I) samples show marked decrease in 10 A° basal spacing, observed in Bu-0, to a lower value of 9.6 A°. These samples are also characterized by negative (for Bu-Ni(I) and Bu-Cu(I)) and poor (for Bu-Co(I)) response to intercalation by dodecylammonium chloride (Table-III). The position of the 10 A° line remains the same for the samples belonging to Bu-R(C) category which responds positively to intercalation treatment (Table-III). Thus XRD results for buserite samples indicate that doping of Ni, Co and Cu in ion exchange mode causes more severe structural changes as compared to coprecipitation mode. As compared to Bu-R(C) samples (simple layered structure), Bu-R(I) samples might have more complex structure (e.g. highly disordered layer, hybrid layer or tunnel structure) as indicated by intercalation-XRD results (Table-III).

Positive response of B-R(C) samples to intercalation treatment highlights the relationship between birnessites and parent buserites(Bu-R(C)) structure (Table-III). B-R(I) samples also have complex structures similar to Bu-R(I). Negative and poor response of parent Bu-Ni(I) and Bu-Co(I) samples respectively to intercalation, coupled with appearance of extra line between 4.5-5.0 A° for corresponding birnessites (Fig. 1), indicate the possibility of the formation of new phases during ion exchange.

Fewer number of lines observed in doped samples as compared to B-0 (Fig. 1), means that there is a decrease in the crystallinity of samples after doping. B-R(C) and B-R(I) samples show distinct behaviour with respect to line broadening for the 7 A° line (002 reflection). There is no line broadening for B-R(C) samples as compared with B-0 ($\Delta 2\theta \sim 1°$) (Fig. 1). This implies that there is no structural disorder and/or break up of crystals in c-direction of unit cell and line broadening results due to structural disorder and/or break up in a-, b-directions. Line broadening in B-R(I) samples is quite large and varies in the following order: B-Ni(I) (3.5°) > B-Co(I) (2.5°) > B-Cu(I) (1.5°). Surface area is inversely proportional to thickness for the sheet structure minerals. Break-up of crystals in <001> direction results in much higher specific surface area for B-R(I) samples as compared to B-R(C) (Table-I).

XRD pattern of pure δ-MnO_2 sample (V-0) showed three lines at 3.6, 2.4 and 1.4 A°. The sample synthesized in this investigation is more crystalline in nature (three XRD lines, SA ~ 150 m^2/g) as compared to the highly disordered variety (two lines, 2.4 and 1.42 A°, SA ~300 m^2/g) reported by Buser and co-workers (23,31). No new phase formation was detected in doped V-R(S) samples.

Transmission Electron Microscopy. To complement XRD data, TEM studies were carried out to determine the finer structural details and morphology of particles. A substantial number of electron diffraction patterns were recorded from the different samples. Some typical electron micrographs and corresponding SAED patterns are shown in Fig. 2. Indexing details for the spot and ring patterns are presented in Tables-IVA and IVB. The interplanar angles $\emptyset_1$ and $\emptyset_2$ in Table-IV refer to the angle between planes represented by interplanar spacings d_1 and d_2 and d_1 and d_3 respectively.

The electron micrograph and SAED pattern in Fig. 2(a),(b) is typical of large number of particles examined in B-0 and B-R(C) samples. The particles showed flake kind of morphology and preferred orientation perpendicular to c-direction (Fig. 2(a) and Table-IVA). Streaking in the SAED pattern was a common feature observed for all the samples (Fig. 2(b)). The structural similarities between the hexagonal Na-free birnessite cell (a_H = 2.84 A°, c_H = 7.27 A°) and orthorhombic Na-birnessite cell (a_o = 8.52 A°, b_o = 15.39 A°, c_o = 14.26 A°) have been illustrated by Giovanoli et al. (22). The streaking in the SAED patterns result due to topotactic intergrowth of the hexagonal cell with the orthorhombic cell (a_o = 3 a_H, $b_o \simeq \sqrt{3}.3\ a_H$, $c_o \simeq 2\ c_H$) (22). Unconfirmed rare occurrence of the mineral manganite was detected in sample B-Cu(C). The pattern indexed for manganite can also be indexed for Na-birnessite type structure with slightly large error (3.3%).

The electron micrograph and SAED pattern (Fig. 2(c),(d)) correspond to a typical particle observed in B-Ni(I). The SAED pattern (Fig. 2(d)) shows no streaking and can be indexed for both hexagonal cell of layered structure birnessite (or buserite) and orthorhombic cell of tunnel structure todorokite (variable 'a' parameter, multiple of 4.88 A° (32)). In both the cases the beam-direction is <001> (Table-IV). It may be recalled that XRD pattern of B-Ni(I) shows a diffuse line at 7 A° and additional strong line at 4.8 A°. Also the sample is characterized by negative response to intercalation indicating the absence of simple layer structure. Hence it is proposed that todorokite kind of tunnel structure phases with varying tunnel width may be present in B-Ni(I), if not as a separate phase at least as an artefact in the sheet structure phase. The conversion of sheet structure phase into tunnel structure phase might involve the movement of complete block of (MnO_6) octahedra sheet (33) and may be greatly facilitated by diffusion of exchangeable divalent cations in the structurally conditioned defects (34) and structural disorder like stacking fault and vacancies.

There are two sets of reflections which occur in the SAED pattern (Fig. 2(f)) of a typical particle (Fig. 2(e)) observed in B-Co(I). One set of reflections shows 'd' spacings and interplanar angles similar to B-Ni(I) (Table-IVA). The other set of reflections (d_1 = 2.62 A°, d_2 = 2.62 A°, d_3 = 1.52 A° and interplanar angles $\emptyset_1$ = 60°, $\emptyset_2$ = 30° (Table-IV)) could not be indexed for simple structure Mn(IV) oxide minerals. Chukhrov et al. (35)

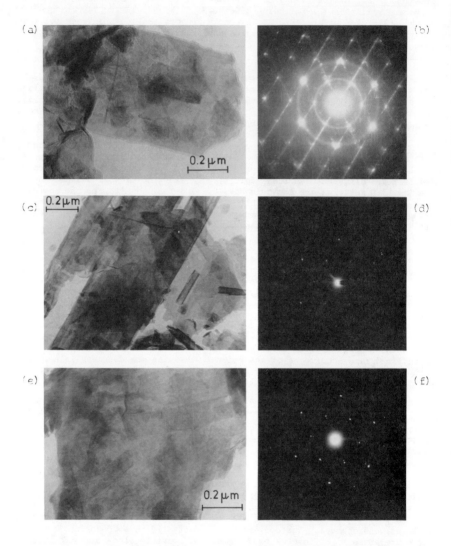

Figure 2 - Some typical electron micrographs and SAED patterns from birnessite samples (a), (b) B-Co(C) (similar features observed for B-Ni(C), B-Cu(C) and B-O), (c), (d) B-Ni(I) and (g), (h) B-Co(I).

Table IV. Indexing of Selected Area Electron Diffraction (SAED) Patterns

A. SPOT PATTERNS

Sample code	Experimental					Theoretical						(hkl)$_i$			Beam direction [HKL]	Possible mineral/Figure
	d_1 Å	d_2 Å	d_3 Å	$\emptyset_1^°$	$\emptyset_2^°$	d_1 Å	d_2 Å	d_3 Å	$\emptyset_1^°$	$\emptyset_2^°$		i=1	i=2	i=3		
B-O,B-R(C)	2.53	2.57	1.47	61.5	31.0	2.49	2.57	1.47	61.0	30.0		330	060	39̄0	001	BO/2(a),(b)
B-O,B-R(C)	2.42	2.42	1.45	60.0	30.0	2.46	2.46	1.42	60.0	30.0		100	010	110	001	BH/2(a),(b)
B-O,B-R(C)	2.46	2.51	1.44	58.0	29.0	2.49	2.49	1.42	58.1	29.0		330	330	600	001	BO
B-O,B-Ni(C)	4.17	4.11	2.40	60.0	30.0	4.27	4.20	2.45	60.5	30.5		200	13̄1	33̄1	02̄6	BO
B-Co(C)	2.16	2.19	1.27	60.5	30.5	2.14	2.20	1.25	59.0	29.0		400	260	660	001	BO
B-Cu(C)	4.53	4.53	2.57	61.0	30.0	4.40	4.40	2.57	62.0	31.0		130	130	060	001	BO
						4.49	4.55	2.60	59.6	29.6		200	110	310	001	MM
B-Ni(I)	2.44	2.44	1.41	60.0	30.0	2.46	2.46	1.43	60.6	30.3		21̄0	2̄10	02̄0	001	TO/2(c),(d)
						2.46	2.44	1.41	59.7	30.0		2̄10	400	6̄7̄0	001	TO
						2.46	2.46	1.42	60.0	30.0		100	010	110	001	BH
B-Co(I)	2.42	2.42	1.40	60.0	30.0	2.46	2.46	1.42	60.0	30.0		100	010	110	001	AH/2(e),(f)
	2.62	2.62	1.52	60.0	30.0	2.63	2.63	1.52	60.0	30.0		100	010	110	001	
B-Cu(I)	4.45	4.45	2.59	61.0	30.0	4.40	4.40	2.57	61.9	31.0		1̄30	130	060	001	BO
	2.10	2.10	1.21	59.5	30.0	2.06	2.09	1.19	58.3	28.9		420	350	730	001	BO

B. RING PATTERNS

Sample code	Observed 'd' spacings, Å	Possible mineral	Remark
B-O	3.54, 2.56$^+$, 2.46, 2.26, 1.51, 1.46, 1.42, 1.26	BO	BO – Na-birnessite (23-1046)*
B-O,B-R(C)	2.56, 2.26, 1.52	BO	BH – Na-free birnessite (23-1239)
B-Co(C)	4.90, 2.91, 2.52, 1.88, 1.66, 1.43, 1.07	–**	TO – Todorokite (13-164)
B-Ni(I)	2.46, 1.41	BH, TO	AH – Asbolane (40)
B-Co(I)	2.64, 2.46, 2.13, 1.52, 1.40, 1.21, 0.892, 0.781	AH	MM – Manganite (18-805)
B-Cu(I)	2.44, 2.25, 1.52, 1.41	BO, BH	*ASTM reference card

** Unidentified, observed only once.   + Underline means strong reflection.

have reported the occurrence of mixed hybrid layer asbolane minerals with two hexagonal sub-lattices (sub-lattice-I a = 2.834 A°, c = 9.34 A°; sub-lattice-II a = 3.04 A°, c = 9.34 A°) consisting of alternate arrangement of (MnO_6) and (CoO_6) octahedral sheets. It is interesting to note that the second set of reflections obtained for B-Co(I) can be indexed for sub-lattice-II, for the same beam direction <001> as for sub-lattice-I. Presence of sub-lattice-II may explain the presence of the line at 4.67 (9.34/2) A° in the XRD pattern of B-Co(I). Presence of reflections at d = 2.64 A° and d = 1.52 A° in the ring SAED pattern of particles (Table-IVB) can only be explained in terms of the second sub-lattice. Thus, SAED features coupled with poor response of B-Co(I) to intercalation treatment suggest that B-Co(I) is likely to have a mixed hybrid layer structure.

All the SAED patterns recorded from B-Cu(I) sample, can be indexed for a Na-birnessite kind of orthorhombic lattice (Table-IV). The particles show preferred orientation perpendicular to <001> direction. The B-Cu(I) sample is characterized by broadening of the (002) reflection, absence of lines between 4.5-5.0 A° and negative response to intercalation. The negative intercalation response of this sample may be assumed to be due mainly to stacking disorder in <001> direction, and not due to formation of a new phase as observed in B-Ni(I) and B-Co(I).

The foregoing discussion on doped birnessite samples shows that doping of Ni, Co and Cu may result in a variety of structural arrangement and/or disorder in the host phase. Thus, the deviations observed in the chemical equivalent balance (Table-II) are quite understandable.

V-O samples yielded two kinds of morphologies and SAED patterns (Fig. 3).

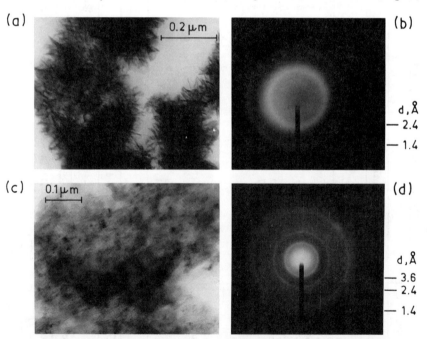

Figure 3 - Electron micrographs and SAED patterns from V-O.

The morphology shown in Fig. 3(a) is typical of δ-MnO$_2$ phase (32). The diffraction rings in SAED pattern of Fig. 3(b) and Fig. 3(d) correspond to d ~ 2.4, 1.4 A° and d ~ 3.6, 2.4, 1.4 A° i.e. the V-O sample may consist of birnessites having varying amount of random stacking of (MnO$_6$) octahedra sheets (23).

No separate phase of Ni, Co and Cu was detected in any of the doped samples.

Summary of Characterization Studies

The incorporation of Ni(II), Co(II) and Cu(II) in the host manganese oxide phases may take place via simple exchange reactions involving Mn^{2+}, K^+, Na^+ and H^+ or exchange reaction with electron transfer step (as observed in this work and by others for Co (28,29)). The exchange reactions can be partly understood in terms of CFSE and ionic radii of different ions.

The exchange reactions take place deep inside the bulk phases and are accompanied by structural breakdown, disorder or rearrangement, leading to formation of new phases. Thus, the mineral name birnessite and δ-MnO$_2$ can be used only in broad sense for the doped samples.

Since birnessite and δ-MnO$_2$ represent the same basic phase, the doped samples can be mutually compared to elicit the effect of the nature of doping. As revealed by surface area, XRD and TEM data, the doped samples can be arranged in terms of structural complexities* as follows: sorption > ion exchange > coprecipitation mode.

Although synthesis conditions employed in preparing the different samples are far from the conditions that may be actually encountered in natural systems, there are marked structural similarities between the synthetic and natural minerals observed in manganese nodules. The poor crystallinity, large surface area and morphological feature of synthetic V-O and V-R(S) samples are very similar to the δ-MnO$_2$ phase observed in natural manganese nodules. Synthetic B-O and B-R(C) samples represent those manganese oxyhydroxide phases in manganese nodules which have simple layered structure. The occurrence of simple layered structure phase in marine environment has been reported by Ostwald and Dubrawski (27). B-R(I) samples, which respond negatively or poorly to intercalation treatments by dodecylammonium chloride, are analogous to those phases in nodules which have complex structure (e.g. hybrid layer, large stacking and structural disorder etc.) in nodules. The 10 A° phases observed in the nodules investigated by the authors fall in this category (3). As in the case of natural manganese nodules, no separate phase of Ni, Co and Cu is observed in any of the synthetic phases. Thus the term 'synthetic structural analog of natural phases' is justified to describe the synthetic samples.

Leaching Behaviour of Synthetic Samples

Sulfuric Acid Leaching

The behaviour of pure and doped manganese oxyhydroxides with regard to the dissolution of manganese(II), and Ni, Co and Cu is illustrated in Figures 4 and 5 respectively. Fraction of metals reacted at some selected time intervals are given in Table-V. The dissolution curves (Figures 4 and 5) are characterized by an initial fast release of metals, in the first few minutes, followed by a very slow dissolution. The transition from initial fast release to slow dissolution is quite sharp for B-O and B-R(C) samples.

---

*Structural complexities are defined in terms of crystallinity, surface area, intercalation response and complexities observed in electron micrographs and corresponding SAED patterns.

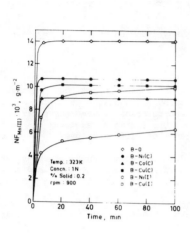
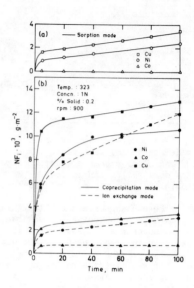

Figure 4 — Dissolution of manganese(II) from different birnessite samples in sulfuric acid.

Figure 5 — Dissolution of doped Ni, Co and Cu in sulfuric acid (a) sorption mode (V-R(S) samples) and (b) coprecipitation and ion exchange mode (B-R(C) and B-R(I) samples).

Table V. Fraction of Manganese and Minor Elements Reacted in Sulfuric and Sulfurous Acid Leaching at Different Time Intervals

Sample code	Sulfuric acid						Sulfurous acid			
	T = 5 min			T = 100 min			T = 0.5 min		T = 5 min	
	F_{Mn}	$F_{Mn(II)}$	F_R	F_{Mn}	$F_{Mn(II)}$	F_R	F_{Mn}	F_R	F_{Mn}	R_R
B-0	0.079	0.451	–	0.081	0.458	–	0.543	–	0.998	–
B-Ni(C)	0.051	0.377	0.208	0.051	0.377	0.373	0.258	0.308	0.879	0.909
B-Co(C)	0.046	0.349	0.080	0.046	0.349	0.137	0.465	0.529	0.954	0.966
B-Cu(C)	0.059	0.472	0.506	0.063	0.501	0.633	0.293	0.265	0.927	0.930
B-Ni(I)	0.046	0.251	0.089	0.069	0.377	0.183	0.406	0.379	0.860	0.966
B-Co(I)	0.057	–	0.029	0.060	–	0.035	0.352	0.378	0.899	0.945
B-Cu(I)	0.026	0.404	0.308	0.037	0.557	0.650	0.456	0.516	0.966	0.962
V-0	0.002	–	–	0.002	–	–	0.803	–	1.000	–
V-Ni(S)	0.001	–	0.129	0.001	–	0.325	0.803	0.789	1.000	1.000
V-Co(S)	0.001	–	0.000	0.001	–	0.000	0.803	0.693	0.995	1.000
V-Cu(S)	0.001	–	0.218	0.001	–	0.468	0.802	0.614	0.995	0.998

The XRD patterns of leach residues revealed that the leaching is accompanied by collapse of original structure (i.e. disappearance or reduction in the intensity of 7 A° and 4.8 A° lines) and formation of new phases like nsutite (ASTM 17-510) and cryptomelane (ASTM 20-908). X-ray diffraction evidence for the formation of cryptomelane was particularly strong in the residue obtained after leaching of B-Cu(C) sample. The formation of cryptomelane has not been reported in earlier study, by Giovanoli (18), on the interaction of doped layered structure oxyhydroxides with H^+ ions in HNO_3.

Nsutite and cryptomelane (T(1,1), T(1,2) structure, vide reference 11 for T(m,n) nomenclature of Mn(IV) oxyhydroxides) have a compact structure compared with more open original structure viz. birnessite (T(2,∞)). The leachability of elements, after the initial fast release, will be governed by the structure of the new phases formed. Phases with more compact structure are expected to have poor leachability. Thus it is likely that the poor leachability of phases at latter time intervals results due to structural collapse and formation of new more compact poorly leachable phases.

The nature of leaching curves (Figures 4 and 5) can also be interpreted in terms of different structural sites in the birnessite structure. Based on radiochemical studies, Buser (36) could distinguish three kinds of exchange rates in layered structure manganates(IV), of which half times were of the order of (i) minutes, (ii) half an hour and (iii) 40 hours, respectively. Buser (36) attributed the fastest reaction to the exchange in the intermediate layer and slowest to the exchange in the main layer edifice. The initial fast leaching of elements (Figures 4 and 5) is likely to be the result of leaching of ions from intermediate layers. The leaching of metal ions from intermediate layer result in the structural collapse and formation of new phases. The possibility of interlayer diffusion is significantly reduced after the structural change from birnessite to nsutite and cryptomelane. This means that both structural change and leaching of metal ions from main layer edifice can result in the slow nature of leaching after initial fast release of metal ions.

Six percent of total manganese is extracted from B-Co(I) sample after 100 minutes leaching (Table-V). Mn(IV) is practically insoluble in dilute sulfuric acid (17). It turns out that at least 2.56% manganese in B-Co(I) is present in +(II) oxidation state. Since the total amount of Mn(II) is not known precisely in B-Co(I), the $NF_{Mn(II)}$ values could not be calculated for this sample, and it is not included in Fig. 4. The variation of leachability parameter $NF_{Mn(II)}$ with time indicates that the variations in dissolution behaviour of Mn(II) from different samples cannot be explained on the basis of total initial Mn(II) content and surface area alone, and evidently the structure of phases has a role to play in the dissolution process. B-R(I) samples (complex structure) show more time-dependence for Mn(II) dissolution compared with B-R(C) or B-O (simple layered structure) (Fig. 4). The amount of Mn dissolved from $\delta-MnO_2$ samples was less than 1% of the total manganese content (Table-V).

Two significant observations were made for the dissolution of Ni, Co and Cu from the doped samples (Fig.5): (i) for the same doped element in different mode of doping, the degree of leachability is observed to vary in the following sequence: sorption < ion exchange < coprecipitation mode and (ii) for the same mode of doping, leachability of doped elements varies in following order: Cu > Ni > Co. As summarized before in the previous section on characterization, the complexities of structure increase among the different categories of samples as sorption > ion exchange > coprecipitation mode. It is striking to note that there is a direct relationship between the increasing structural complexities and poor leachability. Leaching sequence observed for the doped elements in any particular mode of doping (i.e. leachability of Cu > Ni > Co) indicates that the interaction of doped element with the host Mn(IV) phases is specific in nature. Significantly it may be noted that the leaching sequence observed for divalent Ni, Cu and Mn exhibit direct correlation with their CFSE (CFSE of Ni(II), Cu(II), Mn(II) are 122 kJ/mole, 90.4 kJ/mole and 0 kJ/mole) (Table-V, Fig. 5). The dissolution behaviour of cobalt is expected to be more complicated due to its oxidation in the structure of the host phase. Co(III) has a very large crystal field stabilization energy (CFSE for Co(III) = 533.5 kJ/mole).

It is amazing to note that in spite of very large surface area of δ-MnO_2, the doped samples (V-R(S)) show poorest leachability for Ni, Co and Cu (Fig. 5). The leaching behaviour of V-R(S) samples is well in contrast with iron bearing mineral goethite (goethite-R(S) samples) from which the doped elements are very easily leached (Rakesh Kumar, Biswas, A.K. and Ray, R.K., unpublished results). Buser (36) has proposed that the mineral δ-MnO_2 may be considered to consist of 'two dimensional crystals' (missing or very weak basal reflection in XRD). It appears quite likely that the exchange of doped element with K^+ (during synthesis) is followed by subsequent incorporation in the δ-MnO_2 structure. The leaching results for V-R(S) samples (Fig. 5) corroborate Burns's (37) contention that the elements are incorporated in the vacant lattice sites during sorption of Ni, Co and Cu on δ-MnO_2. The leaching data for Co in V-Co(S) indicate that Co(II) sorbed on δ-MnO_2 is completely oxidised (Fig. 5). The dissolution of tetravalent cobalt in sulfuric acid is not feasible from thermodynamic point of view (3). Our leaching data are in conformity with XPS data of Murray and Dilard (38), for Co sorbed on δ-MnO_2, which indicate the oxidation of Co(II) to Co(III).

<u>Possible Leaching Mechanisms in Sulfuric Acid</u>. Apart from the mass transfer steps in solution phase, the other kinetic steps which will be involved in the dissolution of metal ions from doped manganese oxyhydroxide phase are as follows:

1. Diffusion of hydronium ion (H_3O^+) from solid-solution interface to the doped metal ions (surrounded by O^{2-} or OH^- or H_2O ligands, octahedral coordination) in the crystal structure of host manganese oxyhydroxide.
2. Chemical reaction of hydronium ions with metal ion in the solid structure leading to the formation of hydrated metal ion $M(H_2O)^{z+}$.
3. Diffusion of hydrated metal ion from inside the solid to the solid solution interface.

Among the kinetic steps 1-3, steps 1 and 3 are mass transfer steps and step 2 is a chemical reaction step.

Diffusion in the solid phase may be facilitated by structural vacancies, interlayers etc. Since the size of hydrated metal ion $M(H_2O)^{z+}$ is bigger than the hydronium ion (H_3O^+), mass transfer step 3 is expected to affect the reaction kinetics and leachability of metal ions much more as compared with step 1. As we have noted from the leaching data that for a particular doped element the degree of leachability varies in the following order: sorption < ion exchange < coprecipitation; it is likely that the resistance for the diffusion of $M(H_2O)_6^{z+}$ will follow the reverse order.

The predominant mechanism of hydrated metal ion migration in V-R(S) (highly disordered structures) and B-R(C) (simple layered structure) samples are likely to be vacancy diffusion and interlayer diffusion (at least before collapse of B-R(C) structure) respectively. Vacancy diffusion is expected to be slower as compared with interlayer diffusion. Thus poor leachability of doped element from V-R(S) samples as compared to B-R(C) samples is understandable. B-R(I) samples which are characterized by complex structure, for example hybrid layer, layered structure with large stacking and structural disorder, will offer intermediate resistance (between coprecipitation and sorption mode) and in turn their leaching behaviour will fall between B-R(C) and

V-R(S) samples.

The structural collapse for a simple layered structure phase will be easier as compared to a phase with more complex structure. Easier structural collapse means more sharp transition from the initial fast leaching to the slow leaching. This possibly explains why the initial faster release of elements like Mn(II), Cu and Ni from simple layered structure phases is followed up by a more slower leaching as compared to complex structure phases (Figures 4 and 5).

The chemical reaction of metal ion, in solid oxyhydroxide structure, with hydronium ion can be represented by reaction mechanism shown in Fig. 6(a). The mechanism involves the formation of a seven coordinate transition state. The reaction mechanism (Fig. 6(a)) has been formulated by analogy with bimolecular substitution reaction in solution phase which bear resemblance with the leaching reaction (39,40).

The chemical reaction mechanism (Fig. 6) involves the following steps:

(a) a hydronium ion (H_3O^+) approaches the octahedrally coordinated central transition metal ion (M).
(b) an activated complex possessing a pentagonal bipyramidal symmetry is formed. The H_3O^+ ion forms the seventh group, and is bound through the lone pair electrons of oxygen atom.
(c) the activated complex disproportionates to give a water molecule and hydrated metal ion surrounded by five OH^- group. The process is repeated five time to yield a hexahydrated metal ion.

The reaction scheme proposed in Fig. 6(a) can be looked at in two different ways: (i) in terms of breaking of metal-oxygen or metal-hydroxide bond, the metal ion which form stronger chemical bond may be expected to show poor leachability and (ii) in terms of activation energy E_a for the formation of activated-transition-complex. Higher activation energy means poor leachability ($k = A \exp(-E_a/RT)$).

Since the ionic radii of O^{2-} or OH^- ligands are fixed, the radii of doped metal ions will be one of the most important factors to govern the bond strength between metal ion (M) and ligands. Element with lower ionic radius will have higher bond strength (bond energy $\propto 1/r$, metal ion bond with O^{2-} or OH^- is predominantly ionic in nature). The radii of d^n metals ion are governed by variations in the shielding of core electrons which results from crystal field effect. Figure 6(b) illustrate the relationship between electronic configuration, crystal field stabilization energy (CFSE) and ionic radius of elements of interest. It follows from Fig. 6(b) that elements can be arranged as follows in terms of bond strength: Co(III) > Ni(II) > Cu(II) $\sim$ Co(II) > Mn(II). Interestingly this sequence is consistent with the leachability order observed for the doped elements in a particular mode of doping (Figs. 4 and 5, Table-V).

The difference between the CFSE of metal ion in six coordinate initial state and seven coordinate transition state (i.e. crystal field activation energy (CFAE), ΔE_a) may be regarded as a contribution to the total activation energy (E_a). A large positive value of ΔE_a will slow down the reaction. The calculated values of CFAE for different ions are listed below (40):

Ion	Mn(II)	Co(II)	Ni(II)	Cu(II)	Co(III)*
Electronic configuration	d^5	d^7	d^8	d^9	d^6
CFAE (ΔE_a) (in 10 Dq)	0	0	4.26	1.07	8.52

* Low spin.

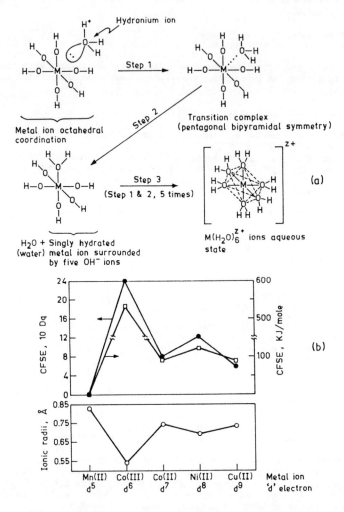

Figure 6 - (a) Reaction mechanism for octahedrally coordinated metal ion in solid oxyhydroxide structure (adopted from reference (39,40)), (b) relationship between electronic configuration, crystal field stabilization energy (CFSE) and ionic radii of some selected metal ions (30,39).

The above values of ΔE_a indicate that metal ions in the structure will be resistant to leaching in the following order: $Co(III) > Ni(II) > Cu(II) > Co(II) = Mn(II)$.

Thus both bond energy and activation energy concepts lead to the same end conclusion (supported by experimental results) regarding the leachability of metal ions in the structure of manganese oxyhydroxide. The similar conclusions are also expected to emerge for dissolution of pure oxide and hydroxide of doped elements. So the occurrence of doped element in the host manganese(IV) oxyhydroxides can be described as "occluded as part of structure in substitution mode". Occlusion follows from close correspondence between CFSE and leachability of doped element, and occurrence as part of structure in substitution mode follows from chemical equivalence balance calculations, role of structure in leaching and structural collapse accompanied by leaching of doped elements.

Sulfurous Acid Leaching

Leaching data (in terms of NF_i or F_i) for dissolution of different samples in sulfurous acid are presented in Figures 7-9 and Table-V.

Leaching experiments on the pure (B-O and V-O) and doped samples (B-R(C), B-R(I) and V-R(S)) indicated that the solubilization of manganese and doped elements was very rapid in sulfurous acid (Table-V). All the manganese and doped elements were extracted almost completely from the different samples (0.25 g) within 10 minutes with 250 ml of leach solution containing 0.0078 mole (0.2 wt % SO_2) of SO_2. Leaching results for doped elements are quite expected because the host oxide matrix containing Mn(IV) is solubilized in dissolved SO_2.

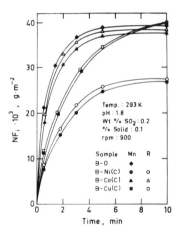

Figure 7 - Dissolution behaviour of birnessites doped with Ni, Co or Cu in coprecipitation mode compared with pure birnessite (leaching medium-sulfurous acid).

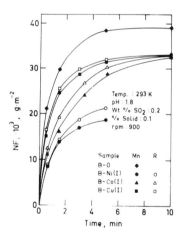

Figure 8 - Dissolution behaviour of birnessites doped with Ni, Co or Cu in ion exchange mode compared with pure birnessite (leaching medium-sulfurous acid).

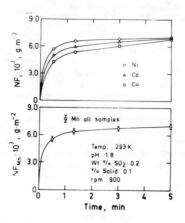

Figure 9 - Dissolution behaviour of V-O and V-R(S) samples in sulfurous acid.

All the doped birnessite samples (B-R(I or C)) show more resistance to leaching compared with pure birnessite (Figures 7 and 8). This indicate that the incorporation of doped elements in the birnessite structure leads to stabilization of doped phases with respect to pure birnessite. The doping of Ni leads to maximum stabilization.

The dissolution of doped elements correlates well with the dissolution of host manganese oxide phases (Figures 7-9). This corroborates our earlier contention that the doped elements occur as part of the host birnessite or δ-MnO_2 structure. Each synthetic sample shows its characteristic leaching curve. This means that each synthetic sample represent a distinct phase and that mineralogy of the host phase can influence the leaching behaviour of the doped element.

Non-preferential dissolution of manganese and doped element should result in 1:1 correspondence between NF_{Mn} and NF_R. The leaching curves in Figures 7-9 illustrate that the correspondence between NF_{Mn} and NF_R is not exactly 1:1. Small deviations from 1:1 correspondence, which are observed for B-R(C), B-Cu(I) and V-Ni(S), fall well within the analytical and experimental error limits (+2-3%). However deviations are particularly significant for B-Ni(I), B-Co(I), V-Co(S) and V-Cu(S). This may be attributed to structural inhomogeneity and/or preferential dissolution of doped or host element.

It may be noted that the concentration of SO_2 (the total amount of dissolved SO_2 was more than the stoichiometric requirement) in the leach liquor decreases from 0.2 wt % (t = 0 minute) to approximately zero at the end of leaching. Since the dissolution of manganese involves the reduction from +(IV) to +(II) oxidation state, it is likely that the dissolution of manganese will be more sensitive to decreasing concentration of SO_2. All the leaching curves for doped birnessites (Figures 7 and 8) reveal that dissolution of doped elements show some preference over manganese. Interestingly, this trend is reversed for V-R(S) sample i.e. manganese dissolves preferentially as compared to the doped elements. It is possible that manganese dissolves preferentially from V-R(S) samples because these samples have significantly large surface area (~140 m^2/g) as compared with B-R(I/C) samples (20-50 m^2/g). Large surface area of V-R(S) samples is associated with poor crystallinity and large structural disorder. Since crystal imperfections represent high energy regions, they can accelerate manganese (major) dissolution and lead to its preferential dissolution over the doped elements (minor). Thus the dissolution of manganese may be affected by the nature of the doped element and crystallinity and structural disorder in the host phase.

Unlike sulfuric acid leaching, no specific leachability order is observed for dissolution of doped element in sulfurous acid leaching. This indicates that dissolution mechanism/s operating in sulfurous acid are much more complex.

Nickel dissolution from B-Ni(I or C) samples show inferior leachability as compared with cobalt and copper dissolution from B-R(I or C) samples (Figs. 7 and 8). Leachability trends for Co and Cu dissolution differ in ion exchanged and coprecipitated birnessite samples. The possible different factors which could influence leachability sequence are (i) mineralogy of host phase i.e. structure, crystallinity etc., (ii) presence of oxidized cobalt in the structure, (iii) reduction of Cu^{2+} to Cu^{+}, etc. The influence of mineralogy on dissolution has been discussed before in this subsection. Preferential dissolution of Co(II) over Co(III) during dissolution of mixed valence cobalt oxide in sulfurous acid has been noted by Khalafalla and Pahlman (16). Cu^{+} ions which form due to reduction of Cu^{2+} in sulfurous acid are insoluble, and can catalyse the dissolution of tetravalent manganese by a reduction step (2).

Implications in Manganese Nodules Genesis and Processing

The synthetic phases were prepared under conditions which are far from the conditions encountered in the natural marine environment. However the synthetic phases had marked structural and chemical similarities with phases in manganese nodules. This makes the data on dissolution behaviour of synthetic phases directly relevant in nodules processing.

Mode of Occurrence and Leaching Behaviour of Nodules

As discussed in the previous sections on characterization and leaching of synthetic samples, we have observed that irrespective of the method of doping elements (Ni, Co and Cu) get incorporated in the structure of birnessite and δ-MnO_2. The theory that the doped elements have been structurally incorporated is supported by (i) equivalence balance calculations (Table-II), (ii) failure to detect any separate phase formation for Ni, Co and Cu by XRD and TEM, (iii) structural changes detected by XRD and TEM for doped samples, (iv) structural collapse accompanied with leaching of doped elements, (v) sulfuric acid leaching data (poor leachability of doped elements from V-R(S) samples could not be explained otherwise) and (vi) sulfurous acid leaching data (close correspondence between dissolution of doped element and host phase). The significance of these results is that the structure of manganese(IV) oxide phases in nodules has an important role to play in the uptake of Ni, Co and Cu in manganese nodules.

Fuerstenau and Han's (41) contention that for lattice substitution to be the mechanism of incorporation of minor element in manganese nodules, the minor element should show close correspondence with dissolution behaviour of major element, is not necessarily justified for oxyhydroxide phases of manganese in nodules; there are structural peculiarities associated with them. These structural features which were noted by Buser and coworkers (34,36) are highlighted below: (i) manganese oxide minerals in nodules are mixed valence e.g. Na-birnessite is sodium manganese(II,III) manganate(IV), and (ii) manganese oxide minerals, which are relevant to nodules, show structurally conditioned defects (such as lattice regions with different degree of order); this imparts to them pronounced reactivity even at room temperature. Higher reactivity will result in preferential dissolution of divalent Ni, Co and Cu in sulfuric acid without appreciably attacking the $Mn(IV)O_6$ framework structure.

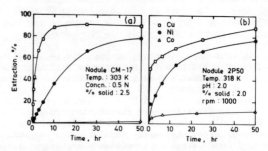

Figure 10 - Dissolution behaviour of Ni, Co and Cu from manganese nodules in sulfuric acid. Adopted from (a) Itoh et al. (19) and (b) Fuerstenau et al. (42).

Figure 10 illustrates some typical leaching curves of nodules in sulfuric acid under similar experimental conditions as employed by us for synthetic samples. Comparison of Fig. 10 with Fig. 5 (for synthetic samples) reveals some striking similarities: (i) leachability order (Cu > Ni > Co) for nodules is same as observed for synthetic manganese oxide samples (same mode of doping), (ii) for the same mode of doping the dissolution of Ni showed more time dependence compared with Cu, and (iii) leaching curves are characterized by an initial fast release of elements.

The leaching sequence observed for nodules is also consistent with the CFSE of Cu(II), Ni(II) and Co(III). In some investigations on sulfuric acid leaching of nodules (1,41) cobalt has shown higher leachability compared with Cu and Ni. Also fraction of Co reacted did not show any increase with time. Elemental correlations in manganese nodules have shown that quite often cobalt correlates positively with iron (43,44). Leaching results on G-Co(S), (G - means goethite) suggest that cobalt adsorbed on α-FeOOH remain predominantly in +II oxidation state (3). The higher leachability of cobalt may be attributed to positive association with iron and/or lower oxidation state. The CFSE of Co(II) (92.9 kJ/mole) is very similar to Cu(II) (90.4 kJ/mole).

Keeping in view the similarities in mineralogy and dissolution behaviour of synthetic phases with manganese nodules, it is proposed that (i) elements like Ni, Co and Cu are incorporated in manganese oxide phases of nodules in lattice substitution mode, (ii) multiple substitution sites are available in structure of host manganese oxide phases and (iii) the interaction of nickel, cobalt and copper with host manganese(IV) oxide phases in nodules is specific in nature i.e. chemical effects like CFSE are involved in the interaction.

Unlike the cobalt bearing manganese oxyhydroxide samples (e.g. V-Co(S)), there is no evidence to suggest oxidation of doped Co(II) in G-Co(S/C) (G - means goethite) to higher oxidation state (3). Halbach et al. (44) have suggested that cobalt can form robust complex with $\equiv FeH_2SiO_4^-$, and can be oxidized to higher oxidation state in strong electric field of Si^{4+}. There is no definite experimental evidence available to support this contention except the positive association of Co with Fe in radial and columnar type of morphology. The role of Mn^{4+} in oxidation of cobalt to +(III) state was suggested by Crowther et al. (28). Halbach et al. (44) have drawn analogy between the role of Mn^{4+} and Si^{4+} in oxidation of Co(II) to higher oxidation state. It may be pointed out that oxidation of Co(II) can be facilitated in strong electric field of Mn^{4+} because it can undergo reduction to lower valency state i.e. +II. This possibility is ruled out for Si^{4+}. The substitution of low spin Fe^{3+} (r = 0.55 A°) by low spin Co^{3+} (r = 0.53 A°), as suggested by some workers (9,43), is also open to question because low spin configuration for Fe^{3+} is unstable. This suggests that more definite evidence is required to support the occurrence of Co(III) in iron bearing phases in nodules.

The collapse of birnessite structure and formation of new phases like nsutite and cryptomelane have very important significance for the sulfuric acid leaching of nodules. Since nsutite and cryptomelane (T(1,1), T(1,2) type of structure) have more compact structure compared with birnessite (T(2,∞) structure), the structural changes during leaching of synthetic phases and similar phases in nodules can result in a drastic decrease in the dissolution rate.

The future attempts to minimize the degree of structural collapse, during leaching of elements, may have far-reaching implications for recovery of metals in sulfuric acid leaching. Hypothetically the problem lies in finding out element/s or their organic complexes which can enter into the structure of layered/tunnel structure phases and remain in structure during leaching without adversely affecting the dissolution of valuable minor elements present in the structure.

In contrast with sulfuric acid leaching, no specific leachability order for doped elements is observed in sulfurous acid leaching of nodules (1,5,16). Similar observations made in this investigation for sulfurous acid leaching of synthetic manganese oxide samples were attributed to complex dissolution mechanisms as discussed for synthetic samples. More complex mechanisms involved in sulfurous acid leaching make it unsuitable for prediction of mode of occurrence of minor elements in manganese nodules.

Significance of Internal Structure for Leachability of Nodules

The internal structure of manganese nodules and minerals present therein constitute important mineralogical factors in nodules processing. Some typical internal features of nodules are presented in Fig. 11. The internal

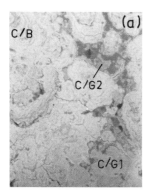
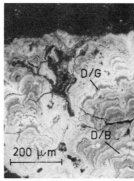
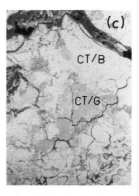

Figure 11 - Back scattered electron (BSE)-SEM micrographs of typical microstructural features observed in nodules from Central Indian Ocean Basin (a) columnar and radial pattern, (b) dendritic segregation and (c) compact type of morphology. Regions marked C/B, C/G1 and D/G consist of intimate association of $\delta-MnO_2$ + $FeOOH \cdot XH_2O$, CT/B and D/G represent 10 $A^°$ phase and CT/G and C/G2 are made up of Fe-rich aluminosilicates. Columnar and radial pattern - Ni and Cu are associated with Mn fraction and positively correlates with Mn/Fe ratio, Co correlates with Fe; dendritic segregation - 10 $A^°$ phase richest in Ni, Cu; compact type of morphology - Ni, Co and Cu are associated with Mn rich 10 $A^°$ phase, Ni and Cu show negative correlation with (Mn/Fe) ratio (3).

structure of nodules is governed by growth environment (3,8). Nodules which form in highly oxidizing environment or the one which grow in seamount regions have predominantly columnar and radial kind of morphology (Fig. 11(a)). Nodules with columnar and radial pattern kind of morphology (δ-MnO_2 + FeOOH. XH_2O assemblage) are expected to have inferior leachability of Ni, Co and Cu in sulfuric acid leaching (under identical conditions) compared with nodules having dendritic segregation as the prime morphology (forms in reducing environment, tunnel structure 10 A° phase (Fig. 11(b)) or compact type of morphology (forms in highly reducing environment, manganese mineral disordered buserite-todorokite assemblage (Fig. 11(c)). In brief, we want to emphasize that in addition to total (Ni + Co + Cu) content and abundance, predominant mineralogy is an equally important parameter in deciding the ore grade of nodules.

Published literature on extraction of metals from manganese nodules lacks detailed description of nodules used in terms of internal structure, mineralogy and elemental associations. We expect our data on synthetic phases to prove more useful if used in conjunction with leaching data on nodules having predominantly one kind of internal structure.

REFERENCES

1. D.W. Fuerstenau and K.N. Han, "Metallurgy and Processing of Marine Manganese Nodules", Mineral Processing and Technology Review, 1 (1983) 1-83.

2. B.W. Haynes, S.L. Law and Maeda, R., "Updated Process Flowsheets for Manganese Nodules Processing" (Information Circular 8924, Bureau of Mines, U.S. Department of the Interior, Washington, D.C., 1983).

3. Rakesh Kumar, "Characterization and Leaching Studies on Deep-Sea Manganese Nodules and the Synthetic Analogs" (Ph.D. thesis, Indian Institute of Technology, Kanpur, 1988).

4. H.L. Ehrlich, S.H. Yang, and J.D. Mainwaring, Jr., "Bacteriology of Manganese Nodules, Fate of Copper, Nickel, Cobalt and Iron during Bacterial and Chemical Reduction of Manganese(IV)", Z. Allg. Mikrobiol., 13 (1973) 39-48.

5. K.N. Han, and D.W. Fuerstenau, "Extraction Behaviour of Metal Elements from Deep-Sea Manganese Nodules in Reducing Media", Marine Mining, 2(3) (1980) 155-169.

6. A.C. Stiff, Q. Fernando, and H. Zeitlin, "Electron Microprobe Analysis of Pacific Ocean Ferromanganese Nodules", Marine Mining, 3(3-4) (1982) 271-284.

7. B.W. Haynes, S.L. Law, and D.C. Barron, "An Elemental Description of Pacific Manganese Nodules", Marine Mining, 5(3) (1986) 239-276.

8. P. Halbach et al., "Geochemical and Mineralogical Control of Different Genetic Type of Deep-Sea Manganese Nodules from the Pacific Ocean", Mineral. Deposita (Berl.), 16 (1981) 59-84.

9. B.W. Haynes et al., "Mineralogical and Elemental Description of Pacific Manganese Nodules" (Information Circular 8906, Bureau of Mines, U.S. Department of the Interior, Washington, D.C., 1983).

10. R.G. Burns, V.M. Burns, and H.W. Stockman, "A Review of Todorokite-

Buserite Problem: Implications to the Mineralogy of Marine Manganese Nodules", American Mineralogist, 68 (1983) 972-980.

11. S. Turner, and P.R. Buseck, "Todorokite: A New Family of Naturally Occurring Manganese Oxides", Science, 212 (1981) 1024-1027.

12. R. Giovanoli, "Manganese Oxide Minerals" (Paper presented at XIIIth Congress of International Society of Soil Science, Hamburgh, 13-20th August, 1986).

13. B. Terry, "Specific Chemical Rate Constants for the Acid Dissolution of Oxides and Silicates", Hydrometallurgy, 11 (1983) 315-344.

14. J.B. Hiskey, and M.E. Wadsworth, "Electrochemical Processes in the Leaching of Metal Sulfides and Oxides", Process and Fundamental Considerations of Selected Hydrometallurgical Systems, ed. M. Kuhn (SME-AIME, Denver, Colorado, 1981), 303-325.

15. J.D. Miller, and R.Y. Wan, "Reaction Kinetics for the Leaching of MnO_2 by Sulfur Dioxide", Hydrometallurgy, 10 (1983) 219-242.

16. S.E. Khalafalla, and J.E. Pahlman, "Selective Extraction of Metals from Pacific Sea Nodules with Dissolved Sulfur Dioxide" (RI 8518, Bureau of Mines, U.S. Department of Interior, Washington, D.C., 1981).

17. P. Ruetschi, and R. Giovanoli, "The Behaviour of MnO_2 in Strong Acidic Solutions", J. Appl. Electrochem., 12 (1982) 109-114.

18. R. Giovanoli, "On Natural and Synthetic Manganese Nodules", Geology and Geochemistry of Manganese, ed. I.M. Varentsov and G. Grasselly (Budapest: Akademiai Kiado, 1980), 159-202.

19. H. Itoh, A. Okuwaki, and T. Okabe, "Processing of Pacific Ocean Manganese Nodules" (Paper presented at 12th Offshore Technology Conference, Houston, Texas, 5-8 May, 1980), 359-364.

20. S.E. Khalafalla, and J.E. Pahlman, "Selective Extraction of Metals from Pacific Sea Nodules with Dissolved Sulfur Dioxide", J. Metals, 33(8) (1981) 37-42.

21. I.M. Varentsov et al., "Synthesis of Mn, Fe, Ni, Co Oxide-hydroxide Phases on Manganese Oxides: On a Model for Transition Ore Formation in Recent Basins", Acta Mineral. Perog. Szeged, 24(1) (1979) 63-90.

22. R. Giovanoli, E. Stahli, and W. Feitknecht, "Uber Oxidhydroxide des Vierwertigen Mangans mit Schichttengitter Natrium(II,III) Manganat(IV)", Helv. Chim. Acta, 53(2) (1970) 209-220.

23. W. Buser, P. Graf, and W. Feitknecht, "Beitrag zur Kenntnis des Mangan(II)-manganite und des $\delta-MnO_2$", Helv. Chim. Acta, 37 (1954) 2322-2333.

24. K. Kodama, Methods of Quantitative Inorganic Analysis, an Encyclopedia of Gravimetric, Titerimetric and Colorimetric Methods (New York, NY: Interscience, 1963).

25. J.W. Murray, L.S. Balistrieri, and B. Paul, "The Oxidation State of Manganese in Marine Sediments and Ferromanganese Nodules", Geochim. Cosmochim. Acta, 48 (1984) 1237-1247.

26. E. Paterson, J.L. Bunch, and D.R. Clark, "Cation Exchange in Synthetic

Manganates 1. Alkylammonium Exchange in a Synthetic Phyllomanganate", Clay Minerals, 21 (1986) 949-955.

27. J. Ostwald, and J.V. Dubrawski, "An X-ray Diffraction Investigation of a Marine 10 Å° Manganate", Mineralogical Magazine, 51 (1987) 463-466.

28. D.L. Crowther, J.G. Dilard, and J.W. Murray, "The Mechanism of Co(II) Oxidation on Synthetic Birnessite", Geochim. Cosmochim. Acta, 47 (1983) 1399-1403.

29. P. Loganathan, and R.G. Burau, "Sorption of Heavy Metal Ions by Hydrous Manganese Oxide", Geochim. Cosmochim. Acta, 37 (1973) 1277-1293.

30. P. Henderson, Inorganic Geochemistry (London: Pergamon Press, 1982) 353.

31. W. Buser, and P. Graf, "Differenzierung von Mangan(II) Manganit und δ-MnO_2 durch Oberflachenmessung nach Brunauer-Emmett-Teller", Helv. Chim. Acta, 38 (1955) 830-834.

32. F.V. Chukhrov et al., "Contribution to the Mineralogy of Authigenic Manganese Phases from Marine Manganese Deposits", Mineral. Deposita (Berl.), 14 (1979) 249-261.

33. L. Eyring, and L.T. Tai, "The Structural Chemistry of Some Complex Oxides: Ordered and Disordered Extended Defects", Treatise on Solid State Chemistry, Vol. 3, ed. N.B. Hannay (New York, NY: Plenum Press, 1975), 167-252.

34. W. Buser, "Radiochemische Untersuchungen und Festkorperverbindugen I. Isotopenaustauschversuche in System Festkorper-losung", Helv. Chim. Acta, 37(7) (1954) 2334-2344.

35. F.V. Chukhrov et al., "Mixed Layer Asbolan-buserite Minerals and Asbolans in Oceanic Iron-Manganese Concretions", International Geology Review, 25 (1983) 838-847.

36. W. Buser, and P. Graf, "Radiochemische Untersuchungen Festkorperverbindugen und Manganiten", Helv. Chim. Acta, 38(3) (1955) 810-829.

37. R.G. Burns, and V.M. Burns, "Mineralogy", Marine Manganese Deposits, ed. G.P. Glasby (New York, NY: Elsevier, 1977), 185-248.

38. J.W. Murray, and J.G. Dillard, "The Oxidation of Co(II) Adsorbed on Manganese Dioxide", Geochim. Cosmochim. Acta, 43 (1979) 781-787.

39. Roger G. Burns, Mineralogical Application of Crystal Field Theory (Cambridge: Cambridge University Press, 1970), 168.

40. F. Basolo, and R.G. Pearson, Mechanisms of Inorganic Reactions, A Study of Metal Complexes in Solutions (New York, NY: John Wiley and Sons, Inc., 1967).

41. D.W. Fuerstenau, and K.N. Han, "Extractive Metallurgy", Marine Manganese Deposits, ed. G.P. Glasby (New York, NY: Elsevier, 1977), 357-390.

42. D.W. Fuerstenau, A.P. Herring, and M. Hoover, "Characterization and Extraction of Metals from Sea Floor Nodules", Trans. AIME, 254 (1973) 205-211.

43. G.P. Glasby, and T. Thijssen, "Control of the Mineralogy and Composition of Marine Manganese Nodules by Supply of Divalent Transition Metal Ions", N. Jb. Miner. Abh., 145(3) (1982) 291-307.

44. P. Halbach, R. Giovanoli, and D. Borstel, "Geochemical Processes Controlling the Relationship between Co, Mn and Fe in Early Diagenetic Deep-Sea Nodules", Earth and Planet. Sci. Lett., 60 (1982) 226-236.

THE HYDROTHERMAL CONVERSION OF JAROSITE-TYPE COMPOUNDS TO HEMATITE

J.E. Dutrizac

CANMET
555 Booth Street
Ottawa, Ontario
Canada K1A 0G1

Abstract

The hydrothermal conversion of synthetic jarosite-type compounds to hematite has been investigated experimentally. The time-temperature relationships for the conversion reaction have been established, and it is shown that temperatures >200°C are required to produce Fe_2O_3 precipitates with low sulphate contents. Although high concentrations of $ZnSO_4$, $FeSO_4$ or Na_2SO_4 have little effect on the reaction, free H_2SO_4 concentrations >0.5 M are detrimental. High ferric sulphate concentrations also are objectionable, likely because of the hydrolysis acid produced during iron precipitation. The beneficial effect of hematite seeding on the conversion reaction is illustrated; greater than 100% seed recycle is desirable. Pulp densities as high as 140 g/L of jarosite can be employed provided that the seed additions are increased correspondingly. Finally, the behaviour of arsenic in the conversion process also has been elucidated; most of the arsenic remains with the hematite although ~25% of the total arsenic reports to the processing solution.

Introduction

Most zinc concentrates contain several percent iron, and most of this iron is solubilized in the hot acid leaching circuits required to achieve high zinc extractions in the roast-leach-electrolysis process which currently dominates world zinc production. The co-dissolved iron must eventually be precipitated in a readily filterable form to minimize processing costs and to ensure low losses of entrained zinc-bearing solution. The dissolved iron frequently is rejected as a jarosite-type compound ($MFe_3(SO)_2(OH)_6$ where M = Na, NH_4, K, H_3O, Ag, 1/2 Pb^{2+}, etc.) although other iron species such as goethite (α FeO.OH) and hematite (Fe_2O_3) sometimes are precipitated (1). Because of its simplicity and low cost, however, the jarosite process is favoured and is employed in at least fifteen zinc plants worldwide (2).

Despite its widespread commercial usage, the jarosite process is not without potential environmental problems (3). Jarosite residues typically contain only 25-30% Fe. The consequence is that a "typical" zinc plant producing 150,000 t/y of zinc from concentrates containing ~50% Zn and ~10% Fe also generates 100,000 - 120,000 t/y of jarosite residue. Because of the low bulk density of the residue, the corresponding annual volume of jarosite exceeds 60,000 m^3. Furthermore, the jarosite residue contains low concentrations of Pb and Cd that originate from the calcine used to neutralize the hydrolysis acid produced during jarosite precipitation. The residues also contain modest concentrations of water soluble Zn, Fe^{3+} and H_2SO_4 that arise from the washing limitations existing in most plants. The result is an environmentally suspect residue which occupies large land areas and requires continuous monitoring and control. These factors are a concern to all zinc producers using the jarosite process, but they are especially worrysome to those operations sited in populous industrialized regions.

One possible method of addressing the jarosite impoundment problem is to convert the jarosite to hematite. Such an action would result in a lower volume of residue and likely would produce a somewhat "cleaner" product; the possibility of making a marketable hematite from existing jarosite residues is remote. Several methods of converting jarosite to hematite are known. Simple thermal decomposition at temperatures above 700°C results in the formation of Fe_2O_3, H_2O and SO_2 gas which would have to be recovered (4,5). Although the ammonia of ammonium jarosite is volatilized, the alkali jarosites decompose with the formation of Na_2SO_4, K_2SO_4, etc. which would have to be leached from the hematite product. The process is also very energy intensive. Dissolution of the jarosite at low temperatures in concentrated H_2SO_4 or HCl media, often with simultaneous SO_2 reduction, effectively decomposes the jarosite, but simply regenerates an iron-bearing solution which requires further, and usually unspecified, treatment (6,7). More successful is the decomposition of the jarosite residue in ammoniacal media that results in the in situ precipitation of the iron. At low temperatures, a ferric hydroxide - limonite slime is formed (4,8) although at temperatures >100°C a filterable Fe_2O_3 product is generated (4). Sulphate contents of the hematite product made at >100°C range from 0.2 to 1.8% SO_4. An excess of ammonia is required to drive the reaction, and the resulting $(NH_4)_2SO_4$ solution would have to be treated by energy intensive evaporative crystallization.

A potentially more attractive procedure involves the direct conversion of the jarosite to hematite at elevated temperatures (>200°C) in weakly acidic H_2SO_4 media. The overall reaction sequence can be described by:

$$2\ MFe_3(SO_4)_2(OH)_6 + 6\ H_2SO_4 \rightarrow 3\ Fe_2(SO_4)_3 + M_2SO_4 + 12\ H_2O \quad (1)$$
$$3\ Fe_2(SO_4)_3 + 9\ H_2O \rightarrow 3\ Fe_2O_3 + 9\ H_2SO_4 \quad (2)$$

$$2\ MFe_3(SO_4)_2(OH)_6 \rightarrow 3\ Fe_2O_3 + M_2SO_4 + 3\ H_2O + 3\ H_2SO_4 \quad (3)$$

The overall reaction generates H_2SO_4, and the difficulty is that the accumulation of the acid eventually halts the precipitation of Fe_2O_3. One method of overcoming this difficulty is simply to control the initial acid concentration, temperature, etc. such that the overall reaction can proceed to an acceptable, if not complete, level. Another approach, and that adopted in the Outokumpu half-conversion process (9,10), is to add sufficient zinc ferrite to the jarosite such that the excess acid generated by Equation 3 is consumed by the comparatively rapid leaching of the ferrite:

$$3\ ZnFe_2O_4 + 12\ H_2SO_4 \rightarrow 3\ ZnSO_4 + 3\ Fe_2(SO_4)_3 + 12\ H_2O \quad (4)$$
$$3Fe_2(SO_4)_3 + 9H_2O \rightarrow 3Fe_2O_3 + 9H_2SO_4 \quad (5)$$

$$3ZnFe_2O_4 + 3H_2SO_4 \rightarrow 3ZnSO_4 + 3Fe_2O_3 + 3H_2O \quad (6)$$

The Noranda high temperature conversion process (i.e., Equation 6) achieves the same end, but avoids the intermediate formation of jarosite by carrying out all steps of the process at 220°C (11).

All of the above approaches necessitate the use of temperatures >200°C to ensure low residual concentrations of dissolved iron and a relatively clean hematite product. Furthermore, all of the approaches require a good understanding of the underlying chemical reactions which lead to the precipitation of hematite free of contaminating phases. In an effort to provide additional information on the parameters which affect the jarosite to hematite conversion reaction, a systematic study was undertaken, using synthetic sodium jarosite, to define the key conversion parameters and to quantify their effects. Special attention was given to the sulphate content of the product and to the role of hematite seeding. The behaviour of arsenic during the conversion reaction also was examined as arsenic is sometimes present in jarosite precipitates (12,13). The results of the various experiments and their impact on a possible jarosite to hematite conversion process are discussed in this report.

Experimental
Materials

Sodium jarosite was synthesized separately by reacting 0.5 M Fe^{3+} (as sulphate) - 0.2 M Na_2SO_4 solutions at pH=1.6 and 98°C for ~20 h in a 2L reaction kettle with good agitation (~400 rpm). Previous work (14) has shown that such conditions are near optimum for jarosite formation; the composition of the sodium jarosite was, in wt %: Na 4.02, Fe 33.22, SO_4 41.67. In some experiments an arsenate-bearing sodium jarosite also was used, and this was prepared by reacting 0.3 M Fe^{3+} (as sulphate) - 0.2 M Na_2SO_4 solutions containing 1 g/L As^{5+} as Na_2HAsO_4 at pH=1.0 and 150°C in a Parr 2L autoclave for 24 h. The product was single-phase and contained, in wt %: Na 4.05, Fe 33.8, SO_4 38.34, AsO_4 2.78. Reagent grade Fe_2O_3 containing 69.6% Fe and 0.76% SO_4 (!) was used to seed the reaction. All other chemicals were reagent grade, and only distilled water was used for the tests.

Procedures

The reactions were carried out using 1L of solution contained in the glass liner of a Parr 2L autoclave fitted with Ti internals and a Ti cooling coil. The solution and seed were added to the autoclave which was then sealed and inserted into a furnace preheated to the desired temperature. About 60 min was needed to heat the charge from 25°C to 240°C with this procedure. The experiment was timed from the moment the working temperature was obtained, and thus, some reaction inevitably occurred before the measurements began. At the end of the test, the charge was cooled to below 40°C in less than 5 min by the use of the water cooling coil. The autoclave then was opened and the contents were filtered, washed and dried at 110°C. When solution analyses were required, the filtrate and washings were collected, brought to volume, and then analyzed.

Product Characterization

The various products formed during the test program were analyzed chemically for Na, Fe, SO_4, etc.; the results are consistently reported as wt% SO_4 and, where applicable, as wt% AsO_4. All the products were examined by Guinier-deWolff X-ray diffraction analysis as this method is particularly effective for detecting trace levels of impurities and the degree of product crystallinity. Separate tests with pre-weighed mixtures of the various products confirmed that 0.25% of Na-jarosite, $Fe(SO_4)(OH)$ or Fe_2O_3 could be detected in any of the phases produced.

Results and Discussion
Effect of Temperature

Figure 1 illustrates the effect of temperature on the amount and composition of the precipitates formed when 40 g of Na-jarosite was heated for 5 h in 0.2 M H_2SO_4 media. No hematite seed was used for these experiments. The composition of the product remains relatively constant to 190°C, but at higher temperatures, the iron content increases significantly whereas the Na and SO_4 contents decrease markedly. The sodium content becomes negligible (<0.03%) at temperatures above 200°C but the SO_4 content is always several percent. The product yield curve (i.e., the net amount of product formed) approximately parallels that of sulphate or sodium and this suggests that the product yield curve is a good first approximation of the degree of jarosite decomposition, Equation 3.

X-ray diffraction study of the various products showed Na-jarosite to be the dominant phase, together with moderate amounts of Fe_2O_3 and traces of $Fe(SO_4)(OH)$, at temperatures <200°C. At temperatures >200°C, only Fe_2O_3 was detected although the SO_4 contents sometimes exceeded 5%. Despite the elevated SO_4 contents, however, the hematite powder diffraction pattern was sharp and coincided closely with the ASTM reference pattern.

A similar series of experiments was carried out in the absence of Fe_2O_3 seed but using an initial H_2SO_4 concentration of 0.3 M. The results are generally similar to those of Figure 1 except that the jarosite decomposition began at ~200°C rather than 190°C. X-ray diffraction study of the products formed in these experiments indicated Na-jarosite with moderate amounts of Fe_2O_3 and traces of $Fe(SO_4)(OH)$ at temperatures <200°C. Hematite is the dominant product formed above 210°C although the Fe_2O_3 sometimes is contaminated with traces of $Fe(SO_4)(OH)$. Interestingly, the intensities of the X-ray diffraction lines of the $Fe(SO_4)(OH)$ seem inadequate to account for the several percent SO_4 sometimes measured.

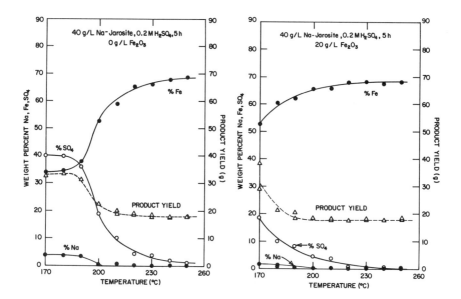

Figure 1 - Effect of temperature on the amount and composition of the products made by converting Na-jarosite to hematite in the absence of hematite seed.

Figure 2 - Effect of temperature on the amount and composition of the products made in the conversion reaction in the presence of 20 g/L Fe_2O_3 seed.

Figure 2 shows the effect of the reaction temperature on the amount and composition of the products made under conditions similar to those of Figure 1 except for the presence of 20 g/L Fe_2O_3 seed. As can be determined from the product yield curve of Figure 1, 20 g/L Fe_2O_3 corresponds to ~100% seed recycle. The presence of the seed has a significant and beneficial effect on the decomposition of the jarosite; the decomposition proceeds at an appreciable rate even at temperatures as low as 170°C; i.e., a significant amount of decomposition occurs during the heat-up period. Furthermore, the product yield curve suggests that the reaction is complete at 200°C in the 5 h retention time used. The Na content of the product is reduced to <0.14% Na at 210°C, and this indicates the near total destruction of the jarosite. Such a conclusion is supported by the X-ray diffraction results which revealed Na-jarosite and major amounts of Fe_2O_3 below 200°C, but only hematite at the higher temperatures. Despite the fact that the jarosite has been converted to Fe_2O_3, there is a further increase in the iron content of the product, and an inverse decline in SO_4, until ~220°C. Above this temperature, the Fe content levels off at ~67% and the SO_4 content remains nearly constant at <1% SO_4. This observation suggests that a conversion temperature of at least 220°C must be used to form hematite having a low SO_4 content.

Retention Time

The effect of retention time on the composition of the product made from 40 g/L Na-jarosite decomposed at 225°C in the absence of Fe_2O_3 seed is illustrated in Figure 3. Under the conditions used, little reaction occurs during the heat-up period, and the composition at "zero" time is essentially that of Na-jarosite. Retention of the charge at 225°C leads to the progressive conversion of the jarosite to hematite as is evidenced by the steady reduction in the Na and SO_4 contents and the increase in Fe. The reaction seems to be complete after two hours at 225°C although the product still contains ~4% SO_4. Furthermore, the data show considerable scatter suggesting a somewhat non-reproducible decomposition process. X-ray diffraction studies showed that Fe_2O_3 and Na-jarosite were the only phases present for times to 2 h, but that Fe_2O_3 and occasional traces of $Fe(SO_4)(OH)$ were the products made in the longer time tests.

Figure 4 presents the corresponding results realized in the experiments done at 225°C but in the presence of 20 g/L Fe_2O_3 seed. The presence of the seed has a marked influence on the conversion of the jarosite to hematite and on the reproducibility of the results. As suggested by the Na contents and the product yield curve, the reaction is nearly complete by the time the autoclave attains the 225°C reaction temperature (~60 min from 25°C). The reaction seems to be totally complete after 0.5 h retention time at 225°C, and this is reflected by the near constancy of the Na, Fe and SO_4 contents. The presence of the seed has shortened the time to complete reaction from 2 h to 0.5 h, and has also resulted in a lower "limiting" SO_4 content of ~1.5% SO_4 (cf. Figure 3).

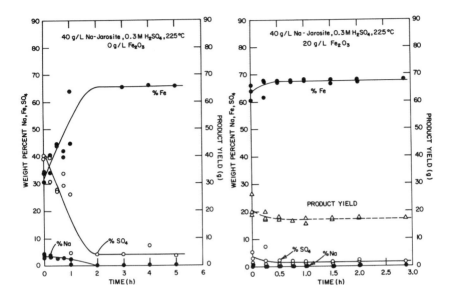

Figure 3 - The composition of the conversion products as a function of the retention time at 225°C in the absence of hematite seed.

Figure 4 - The net product yield and composition of the conversion products as a function of the retention time at 225°C in the presence of 20 g/L Fe_2O_3.

The X-ray diffraction studies indicated only Fe_2O_3 in the products made at 225°C for >0.5 h. The implication is that the presence of Fe_2O_3 not only accelerates the decomposition of the jarosite but also suppresses the undesirable formation of $Fe(SO_4)(OH)$ even for initial acid concentrations as high as 0.3 M H_2SO_4.

So effective is the presence of Fe_2O_3 seed that the conversion reaction proceeds at a satisfactory rate at temperatures as low as 210°C, and this aspect is well illustrated in Figure 5. The compositional data indicate that significant reaction occurred during the ~60 min heat-up period such that Fe_2O_3 is the dominant constituent of the product made at "zero" time. There is a further gradual reduction in the Na and SO_4 contents, with an inversely proportional increase in Fe, over the first hour at 210°C, but steady state compositions are realized at longer retention times. After 1.0 h reaction at 210°C in the presence of the seed, a product containing >67% Fe and ~1.5% SO_4 is realized that X-ray diffraction indicates to consist only of well crystallized hematite.

Effect of Fe_2O_3 Seed Additions

The above results indicate the beneficial effect of hematite seeding on the conversion of jarosite to hematite. The presence of the seed seems to aid both the decomposition of the jarosite as well as the production of a hematite product free of contaminants such as $Fe(SO_4)(OH)$. Although the influence of Fe_2O_3 seed on the jarosite decomposition reaction is a novel observation its effect on the precipitation of Fe_2O_3, even at temperatures as low as 135°C, has been noted by Randolph et al. (15). To determine the optimum amount of Fe_2O_3 seed needed to promote the conversion reaction, a series of experiments was done whereby 40 g/L of Na-jarosite was heated to 225°C for 1 h in the presence of various amounts of Fe_2O_3 seed. The data presented in Figure 6 indicate a considerable scatter in the results obtained in the absence of the seed. This finding is consistent with the data of Figure 3 and further serves to illustrate the stabilizing effect which the seed exerts on the overall conversion reaction. As evidenced by the product yield and composition curves, the extent of jarosite to hematite conversion increases with increasing seed additions to ~20 g/L Fe_2O_3, but levels off in the presence of greater amounts of seed. As the conversion reaction itself generates ~20 g/L of new Fe_2O_3, it can be concluded that ~100% seed recycle represents the minimum desirable amount to promote the conversion of jarosite to hematite.

Under the preferred conditions of ~100% seed addition, a "typical" product contains 0.01 - 0.02% Na, >68% Fe and ~1.5% SO_4. X-ray diffraction analysis of the products consistently indicated only well crystallized hematite.

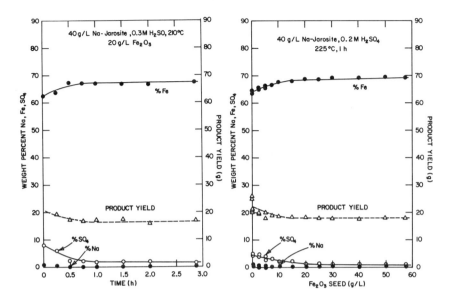

Figure 5 - The net product yield and composition of the conversion products as a function of the retention time at 210°C in the presence of 20 g/L Fe_2O_3.

Figure 6 - The influence of the amount of Fe_2O_3 seed addition on the net product yield and product composition for the conversion of 40 g/L Na-jarosite to hematite.

Because reagent grade hematite is employed as seed, the reaction product is "diluted" with pure Fe_2O_3. To investigate the effect of seed recycle on the composition of the product, a series of experiments was done whereby 40 g/L of Na-jarosite was reacted for 2 h at 225°C in 0.1 M H_2SO_4 media. Twenty grams of reagent grade Fe_2O_3 was initially used to seed the reaction, but the product of the first test was used as seed for the second experiment, and so forth. The cycle was repeated four times and the results are presented in Table 1. As expected, the net product yield remains nearly constant regardless of the source of the hematite seed. The product composition, however, undergoes a slight change as the Fe_2O_3 seed is recycled from one test to the next. In particular, the

Na content increases from 0.02 to 0.06% and the SO_4 content rises from 1.58 to 3.98%. There is a slight decrease in the iron content although all the Fe contents are >66%. There was a noticeable darkening and coarsening of the precipitates as the Fe_2O_3 seed was recycled although X-ray diffraction showed all the precipitates to be hematite.

Table 1 - Effect of Fe_2O_3 Seed Recycle on the Amount and Composition of the Products Made at 225°C

Cycle	Product Yield (g)	Na (%)	Fe (%)	SO_4 (%)	X-ray
0	-	0.03	69.6	0.76	Fe_2O_3 only
1	18.1	0.02	69.1	1.58	Fe_2O_3 only
2	18.8	0.03	67.9	2.75	Fe_2O_3 only
3	18.8	0.04	66.8	3.35	Fe_2O_3 only
4	18.9	0.06	66.8	3.98	Fe_2O_3 only

The results presented in Table 1 indicate that the SO_4 content of the hematite precipitate is not eliminated on recycle. This observation is supported by separate tests where the Fe_2O_3 conversion product was re-leached at 225°C, without any seed additions, in either water or 0.2 M H_2SO_4. There was no significant reduction of the SO_4 content in any of the re-leaching experiments. Because the hematite seed used in any commercial jarosite to hematite conversion process would almost certainly be obtained by recycling part of the product, it is likely that a commercial hematite product would contain ~66% Fe and ~4% SO_4. Surprising is the fact that such hematites are consistently shown by Guinier-deWolff X-ray diffraction analysis to consist only of well crystallized Fe_2O_3.

The Influence of Acid Concentration

The overall conversion of sodium jarosite to hematite generates sulphuric acid, and hence, the conversion reaction is expected to depend on the acid concentration.

$$2NaFe_3(SO_4)_2(OH)_6 \rightarrow 3Fe_2O_3 + Na_2SO_4 + 3H_2O + 3H_2SO_4 \qquad (7)$$

Figure 7 shows the effect of the initial H_2SO_4 concentration of the solution on the amount and composition of the product made by heating 40 g/L of Na-jarosite for 2 h at 225°C in the absence of Fe_2O_3 seed. Extensive decomposition of the jarosite occurs in the absence of H_2SO_4, and this is reflected by the low Na and SO_4 contents together with the elevated Fe content. The Fe and SO_4 contents of the products remain approximately constant to ~0.3 M H_2SO_4; at higher acid concentrations, however, the Fe content of the product drops sharply to ~30% Fe at 0.5 M H_2SO_4 and then levels off at higher acid concentrations. The SO_4 response is the inverse of that of Fe, and the limiting SO_4 content is ~55%. The product yield curve follows that of SO_4 although there is really little significant variation in the amount of product formed under these conditions. Significantly, the Na content of all the products is consistently <0.05% and this indicates that destruction of the sodium jarosite occurred in all the experiments.

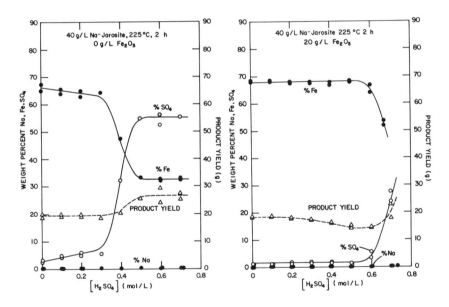

Figure 7 - Effect of the initial H_2SO_4 concentration on the amount and and composition of the conversion products made in the absence of hematite seed.

Figure 8 - Effect of the initial H_2SO_4 concentration on the conversion products made in the presence of of 20 g/L Fe_2O_3 seed.

X-ray diffraction study of the products made in 0 - 0.3 M H_2SO_4 indicated principally Fe_2O_3 with faint traces of Na-jarosite, estimated to be <0.5% jarosite. For acid concentrations >0.3 M H_2SO_4, $Fe(SO_4)(OH)$ became increasing important and was the dominant phase for acid concentrations >0.5 M H_2SO_4.

The results suggest that the jarosite decomposition reaction proceeds independently of the Fe_2O_3 precipitation reaction. The decomposition of the sodium jarosite is favoured by increasing acid concentrations, and the evidence suggests that this reaction proceeds to near-completion for all acid concentrations at the 225°C reaction temperature.

$$2NaFe_3(SO_4)_2(OH)_6 + 6H_2SO_4 \rightarrow 3Fe_2(SO_4)_3 + Na_2SO_4 + 12H_2O \qquad (8)$$

By contrast, the corresponding hematite precipitation reaction generates acid, and this reaction clearly is suppressed for acid concentrations >0.3 M H_2SO_4 in the absence of added hematite seed.

$$3Fe_2(SO_4)_3 + 9H_2O \rightarrow 3Fe_2O_3 + 9H_2SO_4 \qquad (9)$$

Acid concentrations >0.5 M H_2SO_4 entirely suppress Reaction 9, and the precipitation of $Fe(SO_4)(OH)$ is favoured.

$$Fe_2(SO_4)_3 + 2H_2O \rightarrow 2Fe(SO_4)(OH) + H_2SO_4 \qquad (10)$$

Figure 8 summarizes the results from the corresponding experiments done to evaluate the effect of acid concentration in the presence of 20 g/L Fe_2O_3 seed. Comparison with Figure 7 indicates that the presence of seed stabilizes the system to acid concentrations as high as 0.5 M H_2SO_4. The Na contents consistently are less than 0.02%, and the SO_4 contents are ~1.5% SO_4. The iron content of the precipitates typically is ~68% (cf. Fe_2O_3 = 69.6% Fe). X-ray diffraction analysis of the products made in 0 - 0.5 M H_2SO_4 indicated Fe_2O_3 with occasional faint traces of $Fe(SO_4)(OH)$ estimated to be <0.5% of the total mass of the precipitate. Above 0.5 M H_2SO_4, however, the SO_4 content rises sharply and the Fe content falls. X-ray diffraction study revealed increasing amounts of $Fe(SO_4)(OH)$ forming at the expense of the Fe_2O_3. The product yield curve is also different from that obtained in the absence of seed (Figure 7). The amount of product declines gradually with increasing acid concentrations >0.3 M, and this is likely a reflection of the increased solubility of Fe_2O_3 in the more acidic media. For initial acid concentrations <0.3 M H_2SO_4, the terminal iron concentration of the solution is typically only 1-2 g/L Fe^{3+}. The amount of product begins to increase only when $Fe(SO_4)(OH)$ is formed in large amounts at acid concentrations >0.5 M H_2SO_4.

The presence of Fe_2O_3 seed stabilizes the jarosite to hematite conversion process at 225°C; the principal role of the seed seems to be facilitating the Fe_2O_3 precipitation step as opposed to enhancing the rate of jarosite decomposition.

Effect of Ferric Ion Concentration

Jarosite precipitation is carried out in the presence of ferric sulphate solutions, and for this reason the effect of the Fe^{3+} concentration on the jarosite to hematite conversion reaction was ascertained. Figure 9 shows the effect of the initial Fe^{3+} concentration on the amount and composition of the products formed at 225°C in the absence of seed. The amount of product increases linearly for Fe^{3+} concentrations from 0 to 0.1 M. The composition of the product is relatively constant over this concentration range and is indicative of hematite. In fact, X-ray diffraction study of these precipitates indicated Fe_2O_3 together with moderate amounts of Na-jarosite. For Fe^{3+} concentrations >0.1 M, the amount of product increases significantly, and there is a marked increase in the SO_4 content and a corresponding decrease in Fe. X-ray diffraction study of these precipitates indicated progressively increasing amounts of $Fe(SO_4)(OH)$ although all the products contained minor amounts of both hematite and Na-jarosite.

Figure 10 shows the corresponding results obtained in the presence of 20 g/L Fe_2O_3 seed. The amount of product increases linearly to 0.3 M Fe^{3+}, but increases more significantly at the higher acid levels. The composition of the product remains constant to ~0.2 M Fe^{3+}, but thereafter, there are marked changes in the SO_4 and Fe contents. The samples prepared from solutions containing <0.2 M Fe^{3+} were shown to consist of Fe_2O_3 with occasional traces of jarosite or $Fe(SO_4)(OH)$. The products made in solutions having >0.3 M Fe^{3+} consisted mostly of $Fe(SO_4)(OH)$ with lesser amounts of hematite.

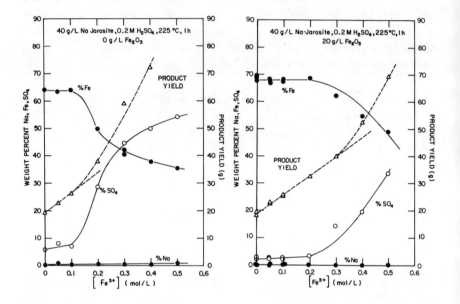

Figure 9 - Product yield and composition of the conversion products made in the presence of various initial Fe^{3+} concentrations but in the absence of hematite seed.

Figure 10 - Net product yield and composition of the conversion products made in the presence of various initial Fe^{3+} concentrations and in the presence of 20 g/L Fe_2O_3 seed.

That the amount of precipitate increases with increasing concentrations of Fe^{3+} indicates that the dissolved iron also is being hydrolyzed and precipitated. This reaction generates significant amounts of hydrolysis acid (Equation 9), and it is believed that it is the high concentration of hydrolysis acid rather than the ferric concentration per se which is responsible for the production of $Fe(SO_4)(OH)$ in preference to Fe_2O_3. That the Na contents are consistently <0.08% Na suggests that the elevated Fe^{3+} concentrations have only a minor stabilizing effect on the jarosite. The hematite seed encourages the precipitation of Fe_2O_3 rather than $Fe(SO_4)(OH)$, but even 100% seed addition is unable to overcome the deleterious effect of high Fe^{3+} concentrations and their associated hydrolysis acid. Any jarosite conversion process should be effected in solutions having as low a ferric iron concentration as possible.

Effect of $ZnSO_4$, $FeSO_4$ and Na_2SO_4 Additions

Any commercial jarosite to hematite conversion process likely would be conducted in solutions containing significant amounts of $ZnSO_4$ as well as modest concentrations of $FeSO_4$ and the alkali sulphate used in the jarosite circuit. Accordingly, a series of experiments was done to ascertain the effect of $ZnSO_4$, $FeSO_4$ and Na_2SO_4 on the conversion reaction carried out at 225°C in the presence of 20 g/L Fe_2O_3 seed.

Figure 11 illustrates the effect of $ZnSO_4$ concentrations in the range of 0-1.5 M on the amount and composition of the products made at 225°C. Under the conditions used, the product made in the absence of $ZnSO_4$ consists of Fe_2O_3 containing ~68% Fe and ~1.5% SO_4. Addition of $ZnSO_4$ to the solution has no significant effect on either the amount of product or its composition. An identical conclusion also was reached from a series of experiments carried out in the absence of Fe_2O_3 seed and for $ZnSO_4$ concentrations of 0-1.5 M. X-ray diffraction analysis showed all the products of Figure 11 to consist of hematite with occasional faint traces (likely <0.25%) of Na-jarosite. The hematites made in the presence of $ZnSO_4$ contained 0.19 to 0.67 wt% Zn, and the Zn could not be eliminated by washing the precipitates with either water or weakly acidic solutions. The belief is that the hematite incorporates~0.5 wt% Zn in its structure although such solid solution has not been unequivocally demonstrated. Aside from the minor degree of Zn incorporation in the Fe_2O_3, the presence of $ZnSO_4$ in the solution seems to have little effect on the jarosite to hematite conversion reaction.

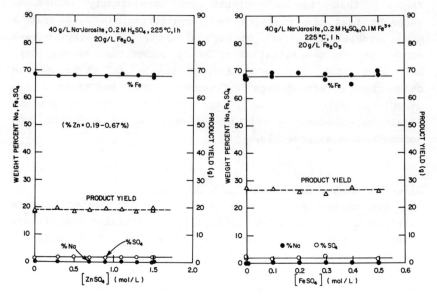

Figure 11 - Effect of $ZnSO_4$ concentration on the amount and composition of the conversion products made at 225°C in the presence of 20 g/L Fe_2O_3 seed.

Figure 12 - Effect of the $FeSO_4$ concentration on the amount and composition of the conversion products made from solutions also containing 0.1 M Fe^{3+} and in the presence of 20 g/L Fe_2O_3 seed.

Zinc processing solutions sometimes contain low concentrations of $FeSO_4$, and Figure 12 illustrates the effect of the $FeSO_4$ concentration on the amount and composition of the products formed at 225°C in the presence of 20 g/L Fe_2O_3 seed. All the solutions also contained 0.1 M Fe^{3+} (as sulphate) to simulate more closely industrial practice. Regardless of the $FeSO_4$ concentration, the amount of the product formed remains essentially constant and the composition of the hematite is consistently ~68% Fe and ~2% SO_4. Sodium contents are 0.02 to 0.06%. The experiments illustrated in Figure 12 suggest that the concentrations of $FeSO_4$ likely to be present in zinc processing solutions will have no significant effect on the jarosite to hematite conversion reaction, provided that the reaction is carried out in the presence of adequate amounts of hematite seed. The data also suggest that moderate variations in the Eh of the solution have little impact on the jarosite to hematite conversion reaction.

Addition of up to 0.5 M Na_2SO_4 in the presence of 20 g/L Fe_2O_3 seed also has no significant effect on either the amount or composition of the product made by reacting 40 g/L Na-jarosite in 0.2 M H_2SO_4 solution at 225°C for 1 h, Figure 13. In all instances, the product consists of hematite containing ~68% Fe and ~1.5% SO_4. The Na contents of the products ranged from 0.02 to 0.06%, and there was no apparent correlation between the Na_2SO_4 concentration of the solution and the Na content of the Fe_2O_3 product. The differences may simply reflect variations in the residue washing efficiency. It is concluded that, in the presence of hematite seed, modest concentrations of Na_2SO_4 in the processing solution will have a negligible effect on the jarosite to hematite conversion reaction.

In the absence of hematite seed, however, the presence of Na_2SO_4 in the processing solution has a significant, if somewhat erratic, effect on the conversion process. Figure 14 shows the product yield curve and the Fe and Na contents of the products as a function of the Na_2SO_4 concentration of the solution when no Fe_2O_3 seed was used. The SO_4 contents are not illustrated as these overlap the product yield curve in the range 0-0.3 M Na_2SO_4. As is commonly noted in the experiments done without seed, there is often considerable scatter in the results obtained. It is also evident that the system displays complex behaviour. In the absence of Na_2SO_4, the Na-jarosite undergoes significant decomposition to hematite as evidenced by both the compositions and the X-ray diffraction results. Addition of Na_2SO_4 enhances the stability of the jarosite phase such that Na-jarosite is the principal X-ray detected phase for the products made in solutions containing 0.1 to 0.2 M Na_2SO_4. Higher concentrations of Na_2SO_4 seem to destabilize the jarosite phase such that the products made in the presence of

0.4 or 0.5 M Na_2SO_4 consist predominantly of Fe_2O_3.

Although most zinc processing solutions likely would contain <0.1 M Na_2SO_4, it is clear that hematite seeding can play an important role in overcoming the stabilizing effect of such low Na_2SO_4 concentrations on the jarosite phase in a jarosite to hematite conversion process.

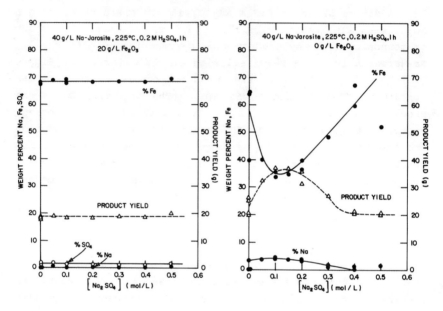

Figure 13 - Effect of the Na_2SO_4 concentration on the amount and composition of the conversion products made at 225°C in the presence of 20 g/L Fe_2O_3 seed.

Figure 14 - Effect of the Na_2SO_4 concentration on the amount and composition of products made at 225°C in the absence of hematite seed.

Effect of the Jarosite Pulp Density

The above experiments were done using a 40 g/L Na-jarosite pulp density that yield ~20 g/L of Fe_2O_3 product. From an industrial point of view, it would be desirable to use higher jarosite pulp densities to minimize the size of the autoclaves required in the residue treatment plant. Accordingly, the effect of the Na-jarosite pulp density on the jarosite to hematite conversion reaction was examined in the presence of hematite seed such that the mass ratio of Na-jarosite/Fe$_2$O$_3$ = 2/1. The experiments were carried out for 2 h at 225°C and in the absence of any initial free H$_2$SO$_4$; the results are presented in Figure 15 as a function of the amount of Na-jarosite used.

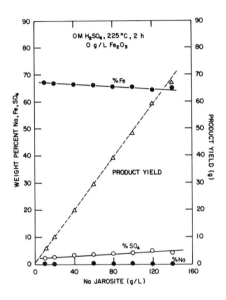

Figure 15 - The influence of the pulp density of Na-jarosite on the net product yield and product composition for conversion experiments carried out at a constant ratio of jarosite to Fe$_2$O$_3$ seed.

The net amount of product formed increases linearly with the quantity of jarosite charged into the reactor. Furthermore, the Na contents are <0.08% for all pulp densities, and this indicates the nearly total decomposition of the jarosite under all conditions studied. This conclusion is also supported by the X-ray diffraction data which showed only trace amounts of jarosite even in the products made from 140 g/L of Na-jarosite. The SO$_4$ contents increase from 1.16 to 2.18% as the jarosite pulp density increases from 10 to 140 g/L; the iron contents decrease slightly as SO_4 increases. Somewhat similar results were realized when the above experiments were repeated in the absence of Fe_2O_3 seed. The notable differences were Na contents as high as 0.2% and SO_4 contents to ~5%; X-ray diffraction analysis indicated modest amounts of Na-jarosite in the products made from the high pulp density experiments in the absence of added seed.

The above data suggest that jarosite pulp densities at least as high as 140 g/L could be used in a jarosite to hematite conversion circuit provided that the initial free acid concentration is low and that the amount of hematite seed is increased as the jarosite pulp density increases.

Behaviour of Arsenic

Arsenic, as AsO_4, is a common impurity in zinc processing circuits, and part of the arsenic reports to the jarosite precipitates made after hot acid leaching. Low concentrations of arsenate seem to be structurally incorporated in the jarosite although higher concentrations are present as an amorphous ferric arsenate or as crystalline scorodite (12, 13). The behaviour of arsenic during the conversion of jarosite to hematite is of environmental interest, and accordingly, a series of experiments was done to evaluate the deportment of arsenic. A Na-jarosite containing 2.78% AsO_4 was prepared separately at 150°C, and X-ray diffraction analysis showed only Na-jarosite. The implication is that all the AsO_4 is structurally incorporated in the jarosite. The arsenate-bearing sodium jarosite was converted to hematite by heating for 2 h at 225°C in the presence of various initial concentrations of H_2SO_4; 20 g/L of Fe_2O_3 seed was used for all the tests. The results presented in Figure 16 indicate that 70-80% of the arsenic remains with the hematite to give Fe_2O_3 precipitates containing up to 2.2% AsO_4.

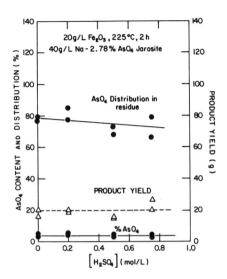

Figure 16 - The effect of H_2SO_4 concentration on the deportment of arsenic during the conversion of arsenate-bearing Na-jarosite to hematite at 225°C and in the presence of 20 g/L Fe_2O_3 seed.

X-ray diffraction of the arsenic-bearing hematite showed the presence of a few weak, and unidentifiable, extra lines which might be attributable to a distinct arsenate phase. It is also evident from Figure 16 that the percentage of the original arsenate remaining in the hematite decreases slightly with increasing H_2SO_4 concentrations. In all instances, however, a majority of the arsenic reports with the hematite. The solutions resulting from the conversion reaction contain 200-400 mg/L AsO_4. The observed deportment of arsenic during the conversion of jarosite to hematite is undesirable as it means firstly that the hematite product is contaminated with arsenic and secondly that the solution also must be treated for arsenic control.

Conclusions

Sodium jarosite is readily converted to hematite by hydrothermal reaction at temperatures >200°C. The reaction seems to involve the initial decomposition of the jarosite followed by the subsequent precipitation of hematite from the resulting acid ferric sulphate solution. The presence of hematite seed accelerates and stabilizes the conversion process, likely by promoting the Fe_2O_3 precipitation step that in turn drives the jarosite decomposition reaction by removing Fe^{3+} from solution and by generating H_2SO_4. A seed recycle of at least 100% is desirable. Under the preferred conditions, a reaction time of 0.5 h is adequate at 225°C whereas a 1.0 h retention time is needed at 210°C. Based on the totality of the results, a temperature of 220-230°C is required to provide rapid conversion coupled with low SO_4 contents in the final Fe_2O_3 product. Acid concentration also is a critical control parameter, and the acid concentration must be kept as low as possible. Even in the presence of Fe_2O_3 seed, initial acid concentrations >0.5 M H_2SO_4 lead to the formation of $Fe(SO_4)(OH)$ in preference to Fe_2O_3. High Fe^{3+} concentrations also are objectionable because of the H_2SO_4 generated during the hydrolysis of the ferric ion. In the presence of adequate amounts of Fe_2O_3 seed, the jarosite to hematite conversion reaction seems to be independent of the concentrations of $ZnSO_4$, $FeSO_4$ and Na_2SO_4. Although most of the experiments were done using a pulp density of 40 g/L of Na-jarosite, pulp densities as high as 140 g/L of Na-jarosite could be handled provided that the addition of Fe_2O_3 seed was increased proportionately. About 75% of the arsenic present in the original Na-jarosite stays in the hematite product, and the remainder passes into the processing solution. Also, the hematite product is consistently contaminated with 1-4% SO_4 despite the fact that sensitive Guinier-deWolff X-ray diffraction analysis indicates only well crystallized hematite. The mechanism of SO_4 incorporation is not currently known. Products made in the presence of $ZnSO_4$ contain ~0.25% Zn, and the belief is that the zinc is structurally incorporated in the hematite.

Acknowledgements

The author acknowledges that this work was made possible by the co-operation and encouragement of the Jarosite R & D Group which consists of Budelco B.V., Noranda Inc., Norzink A.S., Outokumpu Oy and Preussag-Weser-Zink GmbH. The author also is indebted to J.L. Jambor for the X-ray diffraction results and their interpretation and to O. Dinardo for assistance with the experimental program.

References

(1) J.E. Dutrizac, "The Physical Chemistry of Iron Precipitation in the Zinc Industry", Lead-Zinc-Tin'80, eds., J.M. Cigan, T.S. Mackey and T.J. O'Keefe, Warrendale, Pa: The Metallurgical Society, 1979, 532-564.

(2) V. Arregui, A.R. Gordon and G. Steintveit, "The Jarosite Process - Past, Present and Future", Lead-Zinc-Tin'80, eds., J.M. Cigan, T.S. Mackey and T.J. O'Keefe, Warrendale, Pa: The Metallurgical Society, 1979, 97-123.

(3) J.C. Taylor and A.D. Zunkel, "Environmental Challenges for the Lead-Zinc Industry", J. Metals 40(8) (1988) 27-30.

(4) W. Kunda and H. Veltman, "Decomposition of Jarosite", Metal. Trans. 10B (1979) 439-446.

(5) V.M. Piskunov, A.F. Matveev and A.S. Yaroslavtsev, "Utilizing Iron Residues from Zinc Production in the USSR", J. Metals 40(8) (1988) 36-39.

(6) J.K. Rastas and J.R. Nyberg, "A Hydrometallurgical Process for the Recovery of Lead, Silver and Gold, as well as Zinc, from Impure Jarosite Residues of an Electrolytic Zinc Process", U.K. Patent Application 2,084,555; September 30, 1981.

(7) A. Van Ceulen and C. Eusebe, "Traitement des Résidues de l'Hydrometallurgie du Zinc", Mem. Etud. Sci. Rev. Metall. 79(10) (1982) 569-576.

(8) A. Hernandez, A. Cuadra and J.L. Limpo, "Eliminacion del Hierro de las Soluciones Hidrometalurgicas como Limonita", Rev. Metal. Madrid 20(5) (1984) 331-336.

(9) J.K. Rastas, J.R. Nyberg, K.J. Karpale and L.G. Bjorkquist, "Hydrometallurgical Process for the Treatment of a Raw Material which Contains Oxide and Ferrite of Zinc, Copper and Cadmium", U.S. Patent 4,362,702, December 7, 1982.

(10) J.K. Rastas and J.R. Nyberg, "Hydrometallurgical Process for the Recovery of Lead, Silver and Gold, as well as Zinc, from Impure Jarosite Residues of an Electrolytic Zinc Process", U.S. Patent 4,366,127, December 28, 1982.

(11) A. Ismay, R.W. Stanley, D. Shink and D. Daoust, "The High Temperature Conversion Process", <u>Iron Control in Hydrometallurgy</u>, eds., J.E. Dutrizac and A.J. Monhemius, Chichester, England: Ellis Horwood Limited, 1986, 618-639.

(12) J.E. Dutrizac and J.L. Jambor, "The Behaviour of Arsenic During Jarosite Precipitation: Arsenic Precipitation at 97°C from Sulphate or Chloride Media", <u>Can. Metal. Quart.</u> 26(2) (1987) 91-101.

(13) J.E. Dutrizac, J.L. Jambor and T.T. Chen, "The Behaviour of Arsenic During Jarosite Precipitation: Reactions at 150°C and the Mechanism of Arsenic Precipitation", <u>Can. Metal. Quart.</u> 26(2) (1987) 103-115.

(14) J.E. Dutrizac, "Factors Affecting Alkali Jarosite Precipitation", <u>Metal.Trans.</u> 14B (1983) 531-539.

(15) A.D. Randolph, R.D. Williams and D.A. Milligan, "Hydrolysis of Iron on a Retained Crystal Bed", <u>AIChE Symp. Series</u> 74 (173) (1978) 84-96.

SOLVENT EXTRACTION OF PRECIOUS METALS BY S-ALKYL DERIVATIVES OF DEHYDRODITHIZONE AND DITHIZONE

M. Grote, G. Pickert and A. Kettrup

Department of Applied Chemistry
University of Paderborn,
D-4790 Paderborn, F.R.G.

Abstract

S-alkyl derivates of dehydrodithizone and dithizone were prepared as liquid analogues of certain gel-type and macroreticular ion-exchange resins. 5-Decylthio-2,3-diphenyl-2H-tetrazolium bromide (TD-2) and 1,5-diphenyl-3-decylthioformazan (D-2) were examined as solvent extractants from acid media of the platinum group metals, gold and some base metals, either individually or as a group.

The compounds were investigated basically with respect to their solubility in organic media with different polarity, their stability in contact with oxidizing agents and alkaline aqueous solutions and their affinity towards precious metals and base metals, and their ability to be regenerated. Extractant TD-2 dissolved in an appropriate solvent removes particularly iridium, osmium, platinum and gold effectively and reversibly by forming ion pairs, whereas D-2 forms extractable species with Pd(II) via selective chelation. The chelating agent can be regenerated completely. Treatment of different synthetic and real precious metal bearing process solutions by means of the extractants are subject of present investigations.

Introduction

This investigation supplements earlier and present investigations on different gel-type and macroreticular ion exchange resins (1-4). The polymers contain dehydrodithizone (P-TD) or dithizone (P-D) as functional groups which are attached via their sulfur atom to the polystyrene-matrix. Although the sorbents remove the primary platinum group metals and gold selectively and reversibly it was difficult for example to extract and to elute Ir(IV) quantitatively.

Furthermore, the chelating polymer P-D could not be utilized for cyclic separation procedures, as its capacity lowered drastically after one cycle of loading and elution. The question was to arise, if liquid analogues of these resins would reveal some certain advantages in the field of precious metal extraction. For this purpose S-alkyl derivatives of dehydrodithizone (TD-2) and of dithizone (D-2) have been synthesized, as shown in Fig.1. The synthesis of the extractants is based on a method described by HUTTON and IRVING who prepared some derivatives of dithizone (5,6). At first dehydrodithizone was reacted with decylbromide, yielding the tetrazolium bromide TD-2, which can be subsequently reduced to the formazan D-2 by action of ascorbic acid (Fig.1).

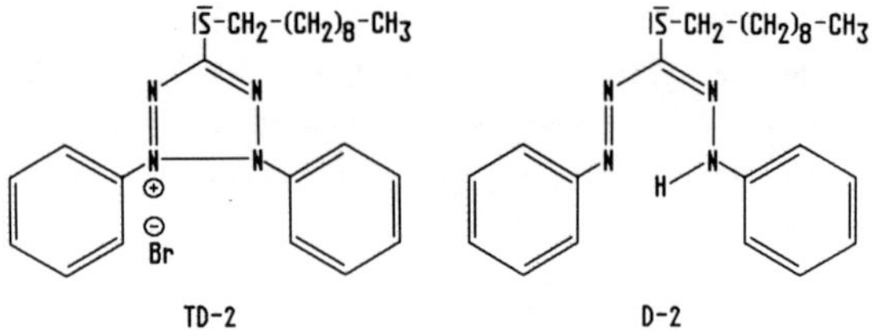

Figure 1 - Structures of the extractants TD-2 and D-2

Liquid anion exchanger TD-2

Basic properties

Because of the quaternary heterocycle TD-2 shows both with metal complexes and halides as counterions a fairly good solubility in a relatively polar organic solvent, e.g. dichloromethane, chloroform, various chloroethanes and chlorobenzene. However, in unpolar solvents like benzene, tetrachloromethane or n-hexane TD-2 is hardly soluble. The introduction of larger alkyl-chains (C_{14}, C_{16}) did not improve the solubility of the tetrazolium salts. The beige, crystalline needles of TD-2, isolated as bromide are stable against light, whereas the chloride form disproportionates gradually into the corresponding formazan and a condensed tetrazolium salt. Its structure is characterized by a linkage between the 2'-carbon atoms of the N^2- and N^3-phenyl rings. Therefore the extraction procedures in contact with hydrochloric acid media have to be carried out under exclusion of light. Under these conditions stability was confirmed in non oxidizing weak and strong acids. Nitric acid and alkalies tend to decompose the tetrazolium salt.

Extraction properties

Extraction studies were performed by shaking mechanically (145 min^{-1}) 10 ml of the aqueous phase together with 10 ml of the extractant solution at room temperature. TD-2 as bromide was dissolved in chloroform and then converted to the chloride form by three times treatment with 1 M HCl. The TD-2 concentration was 5 mmol/l for the extraction of individual metal chloro complexes, and 10 mmol/l were used for the simultaneous extraction procedures. The concentration of metal ions was 0.5 mmol/l for individual extractions and 0.2 mmol/l for each species in metal mixtures. All organic and aqueous solutions used were presaturated by each other prior to extraction.

The metal contents of the aqueous solutions were determined by plasma emission spectrometry (DCP).

Extraction of individual metals

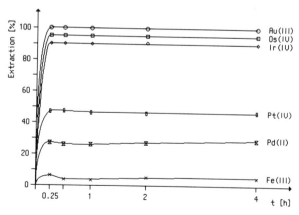

Figure 2 - Extraction of individual precious metals and Fe(III) from 1 M HCl by TD-2 as a function of time

Fig.2 shows the results of individual extraction procedures of some noble metals and Fe(III) from 1 M HCl. The maximum extraction rate is achieved after 15 min. The order of affinities found for the examined noble metal chloro complexes is Au(III) > Os(IV) > Ir(IV) >> Pt(IV) > Pd(II) >> Rh(III), Ru(III). Hence, the behaviour of TD-2 differs markedly from strong basic amine systems and weakly basic isothiouronium extractants as well (7,8).

Because TD-2 is predominantly acting as anion exchanger only small amounts of Fe(III) (about 5 %) were extracted under the conditions mentioned above. An increase of HCl-concentration to 4 mol/l raises the extraction yield for Fe(III) to 80 % due to the formation of $[FeCl_4]^-$ in the aqueous phase. By varying the HCl concentration the opposit effect could be observed in the case of noble metals. Particularly Pd(II) is markedly affected by strongly acidic media. The metal was extracted in a yield of 55 % from 0.5 M HCl and only 3 % from 5 M HCl.

Extraction of combined metals

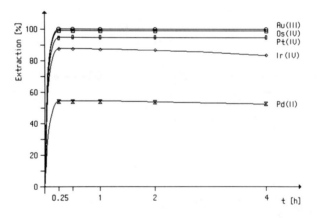

Figure 3 - Extraction of combined precious metals from 1 M HCl by TD-2 as a function of time

As typical for anion exchange mechanisms, the equilibrium of extraction of metal mixtures was reached after a few minutes. High yields were observed for Au(III), Os(IV) and Ir(IV). It is striking that the extraction (%) of the latter decreases slightly with increasing extraction time. This could be due to reduction of extracted Ir(IV) to poorly extractable Ir(III), followed by a release into the aqueous phase. This effect resembles results of sorption studies performed with these metals and the analogue of P-TD resins (2). Our supposition was confirmed by extraction data which resulted from metal mixtures containing Ir(III) and Pt(II) instead of oxidized species.

Synergistic effects were observed in the case of extraction of Pt(IV) and Pd(II). The extraction rates determined for the combined metals were two times higher as observed for the individual complexes. The extraction yield of Pt(IV) was always approximately doubled in comparison with Pd(II).

It must be added, that no remarkable extraction of Ru(III), Rh(III), Cu(II) and Ni(II) occured.

Stripping

Separation of individual metals by means of solvent extraction can be achieved principally by two different routes: - selective extraction of metal ions by an appropriate reagent or - coextraction of metals followed by selective stripping. Under the conditions already described it is possible to separate some precious metals as a group from base metals. Additional experiments using stripping agents such as perchloric acid, sodium thiocyanate and thiourea should ascertain further modes of separation.

Stripping by perchloric acid

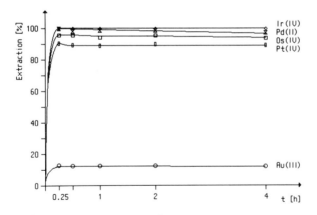

Figure 4 - Back extraction from simultaneously loaded TD-2 by 2 M $HClO_4$ as a function of time

The perchlorate anion is known to have a great affinity to quaternary ammonium cations, so that it should be possible to displace also metal complexes from loaded tetrazolium salts. Fig. 4 demonstrates the results of the back extraction procedure by means of perchloric acid.

With the exception of Au(III) the extracted metals could be removed from the extractant almost completely. Because of its single negative charge, size and polarisability the $[AuCl_4]^-$ complexe is strongly bonded to the tetrazolium cations, so its back extraction is difficult. By contrast Pd(II) is replaced rapidly by aqueous perchlorate solution which furthers the hydrolysis of $[PdCl_4]^{2-}$ to non extractable species (9).

Stripping by sodium thiocyanate

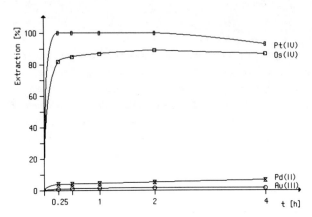

Figure 5 - Back extraction from simultaneously loaded TD-2 by 0.5 M NaSCN as a function of time

In addition to the simple displacing effect of perchloric acid, thiocyanate anions form complexes via several intermediate stages and equilibria by ligand substitution. In presence of coextracted Ir(IV) precipitates were observed at the interphase. In absence of Ir(IV) these effects could only be detected after stripping times of more than four hours. The square planar complexes of Au(III) and Pd(II) were retained more strongly in the organic phase than the octahedral species of Pt(IV) and Os(IV) complexes, which were released more efficiently from the extractant.

Stripping by thiourea

Due to the higher nucleophilic power of thiourea substitution reactions with tetrazolium bonded chloro complexes of precious metals ought to proceed more completely than with SCN^- anions.

As demonstrated in Fig.6 in the thiourea system cationic complexes are formed in relatively fast rates and stripped from TD-2. A problem is the formation of very poorly soluble mixed complexes in presence of combined Pd, Pt and Ir.

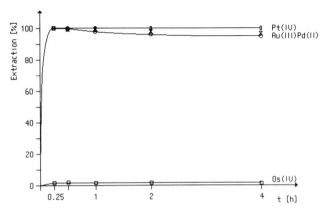

Figure 6 - Back extraction from simultaneously loaded TD-2 by Thiourea as a function of time

Without coextracted Ir(IV) the other precious metals can be back extracted in high yields. Only Os(IV) complexes, which are very inert to ligand substitution, remain nearly completely in the organic phase during the stripping step.

Separation of Pt(IV) and Pd(II)

As example for the applicability of differential stripping procedures the separation of Pt(IV) and Pd(II) has been investigated. The first step was the combined extraction of these platinum metals from base metals. The loaded extractant was then stripped by sodium thiocyanate. Pt(IV) could be removed completely together with a small amount of Pd(II) (10 % of the initial amount). The residual fraction of Pd(II) in the organic phase was finally recovered by an acidified thiourea solution.

Further investigations are in progress to test the utility of extractants based on tetrazolium salts. However, the restricted solubility of this class of compounds will recommend them mainly for analytical purposes.

Chelate forming extractant D-2

Basic properties

The solubility of the dark-red crystals of D-2 in organic solvents differs significantly from that of the corresponding tetrazolium salt. D-2 is highly soluble in polar solvents like chloroform, dichloromethane, tetrachloroethylene, chlorobenzene etc. and nonpolar solvents like kerosene, toluene or xylene.

The compound shows no interactions with light both for the crystalline state and in solution. In contact with hydrochloric acid solutions the extractants is stable even at 10 M HCl. Pretreatment with alkaline solutions (e.g. 0.01 M NaOH) results no changes of extraction characteristics. However, it is necessary to avoid contact to strong oxidizing agents, such as nitric acid or Au(III), otherwise D-2 will be converted into the tetrazolium salt TD-2.

Extraction properties

The extractions were carried out as already described with an aqueous to organic phase ratio of 1. The extractant was used at a threefold molar excess compared to the individual metal in the aqueous phase. Usually the concentration of D-2 was 3 mmol/l in different solvents.

In preliminary studies a multielement mixture of various base and precious metals was examined, revealing a high affinity of D-2 towards Pd(II), whereas Ir(IV) and Pt(IV) were slightly extracted and the base metals Fe(III), Cu(II) and Ni(II) only in traces (< 1.5 %). These results did not coincide to those reported by IRVING et al. (5,6) who investigated S-methyl-dithizone. This analogue to the S-decylderivate D-2 did not yield at all extractable metal complexes. It is somehow surprising, that the length of the alkyl chain attached to the sulphur atom should have such a strong influence on the extraction properties of dithizone derivatives.

Because of the superior affinity for Pd(II) compared to the other precious metals, the extractability of this metal by D-2 was examined in detail.

Effect of solvents on the extraction of Pd(II)

The effectivity of an extractant can be markedly influenced by the organic solvent. Fig. 7 shows the effect of various solvents on the extraction of Pd(II).

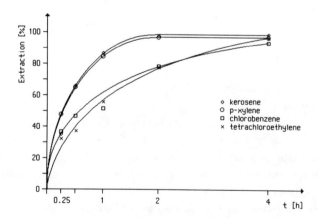

Figure 7 - Effect of solvents on the extraction of Pd(II) from 1M HCl by D-2

By every solvent used the precious metal can be separated completely from the aqueous phase. As can be seen the nonpolar solvents enhance the kinetic in comparison with the polar aliphatic and aromatic media. An extraction time of at least 4 hours was required in this cases to achieve equilibrium. Even then the 3-thiodecyl-formazan D-2 is much more reactive than formerly investigated phenylcarbamoyl-formazans (10). Because D-2 is easier soluble in xylene than in kerosene, the aromatic was chosen exclusively for additional extraction procedures.

Effect of chloride concentration on the extraction of Pd(II)

The Cl^--concentration in the aqueous phase was varied by adding different amounts of hydrochloric acid or sodium chloride. As demonstrated by Fig.8, increasing concentrations of acid deteriorates the extractability of Pd(II). Increasing portions of sodium chloride caused the same effects.

It does appear, that the retarding influence of Cl^- corresponds to the following equilibrium reaction (org = organic):

$$[PdCl_4]^{2-} + \overline{(D-2)org} = \overline{[PdCl_2] - (D-2)org} + 2Cl^-$$

The assumed 1:1 stoichiometry of the Pd(II)-complex of D-2 is based on elementary analysis and spectroscopic investigations of the isolated compound. Evidence is given, that the ligand coordinates without deprotonation.

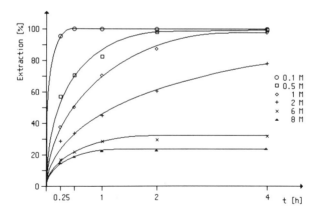

Figure 8 - Effect of HCl-concentration on the extraction of Pd(II) by D-2

Stripping of Pd(II)

Perchloric acid (2 mol/l), sodium thiocyanate (0.5 mol/l and 2 mol/l) and thiourea (5 % in 0.1 M HCl) were tested as stripping agents Due to strong chelation of Pd(II) perchloric acid is not able to displace the metal from the loaded organic phase. Higher concentrated solutions of NaSCN are more effective (2 M NaSCN: 50 % back extracted) within 15 min, but the nucleophilicity of the ligand is not sufficient to achieve a complete recovery of Pd(II). Quantitative stripping of Pd(II) after few minutes is possible by using an acidified solution of thiourea. The extractant D-2 can be regenerated without any loss of its loading capacity, wich was proved by more than ten repeated cycles of the extraction/back extraction procedure.

Extraction of combined metals

To investigate the selectivity of D-2 various solutions containing different mixtures of precious and base metals were treated with D-2. Synergistic and competing effects were observed in the case of the primary platinum group metals. Individually Pt(II) and Pt(IV) were extracted only poorly. The extraction yields obtained for Pt(II) varied between 0 % and 5 %, those of the Pt(IV) about 15 % depending on the acidity of the solution. In presence of Pd(II), however, the extraction yields determined at different HCl-concentrations revealed a strong interdependence of Pt(IV) and Pd(II) (see Tab. I).

Table I. Simultaneous extraction (%) of Pd(II) and Pt(IV) (each 0.5 mmol/l) by D-2 as function of acidity

HCl concentration [mol/l]	t = 15 min		t = 2 h	
	Pd(II) %	Pt(IV)	Pd(II) %	Pt(IV)
0.1	91.6	15.0	100	24.7
0.5	59.0	6.1	99.5	7.7
1	44.6	16.4	95.8	6.0
2	45.4	12.4	73.7	10.5
4	11.3	20.6	33.5	14.5
6	10.3	28.0	19.3	19.0

At low acidities (< 1 M HCl) Pd(II) is separated almost completely as can be expected from the basic properties of D-2. Significant amounts of Pt(IV) were coextracted from 0.1 M HCl (15 - 25 %) and from stronger acidic solutions (>2 mol/l HCl).

In a medium range of acidity an optimum Pd:Pt ratio is obtained, showing percentual yields lower than 10 % for Pt(IV). Anyway, this behaviour differs markedly from that observed for the single element solutions. Furthermore, with extending shaking time co-extracted Pt(IV) seems to be continuously displaced from the organic phase by an increasing concentration of chelated Pd(II). This assumption was verified by additional consecutive steps of loading, using different feed solutions which contained Pd/Pt-mixtures and Pd individually.

The kinetics observed are indicative of a more complicated behaviour of the S-decyldithizone, which may act as N,S-chelating ligand and at high acidities as ion-pair forming extractant as well.

The co-extraction of Pt(IV) does not impair the application of D-2 as selective extractant for Pd(II). It was found, that thiourea strips exclusively Pd(II), whereas minor amounts of Pt(IV) remain in the organic phase. They will be exchanged during the next loading step by extracted Pd(II) and released into the feed solution, so that a poisoning of the extractants is limited. Studies performed with synthetic and real refinery solutions show, that Pd(II) can be separated efficiently from large excess of base metals (Table II).

Table II. Simultaneous extraction of precious and base metals from 1 M HCl by D-2 as a function of time

Initial metal concentrations [mg/l]					
Ir(IV)	Pt(IV)	Pd(II)	Fe(III)	Cu(II)	Ni(II)
109.5	189.3	145.9	1130	334	440
time [min]		Extraction [%]			
Ir(IV)	Pt(IV)	Pd(II)	Fe(III)	Cu(II)	Ni(II)
15 1.9	5.5	17.5	1.8	~ 0	0.9
30 3.6	5.3	30.9	4.2	2.4	2.7
60 6.5	7.7	56.8	3.7	2.4	2.7
120 5.2	7.4	90.0	3.9	1.8	1.8
240 4.7	10.1	93.9	3.1	2.0	1.5
300 1.7	7.9	95.1	3.5	2.4	2.3

It is to conclude that, in contrast to its polymeric analogue, the extractant D-2 can be utilized in separation procedures. Furthermore, in comparison to commercial available thioethers and hydroxy-oximes (7,11) the compound D-2 can be considered as an efficient extractant for Pd(II).

The stability of the extractant allows the regeneration by thiourea and, hence, the development of cyclic procedures.

Problems which may arise from the treatment of solutions containing Pt(IV) seems not to be critical.

Because of the promising features of D-2 its applicability for the separation of Pd(II) from different process solutions is subject of additional investigations.

REFERENCES

1. M. Grote and A. Kettrup,
"The Separation of Noble Metals by Ion Exchangers Containing S-bonded Dithizone and Dehydrodithizone as Functional Groups", in: Ion Exchange Technology, D. Naden, D. McKee, M. Streat (Eds.) Ellis Horwood Ltd., London 1984, 618-625.

2. M. Grote and A. Kettrup,
"Ion Exchange Resins containing S-bonded Dithizone and Dehydrodithizone as Functional Groups Part 1. Preparation of the Resins and Investigation of the Sorption of Noble Metals and Base Metals", Analytica Chimica Acta, 172 (1985) 223-239.

3. M. Grote and A. Kettrup,
"Ion Exchange Resins containing S-bonded Dithizone and Dehydrodithizone as Functional Groups Part 2. Desorption Properties and Development of Separation Procedures for Gold and Platinum Group Metals", Analytica Chimica Acta, 175 (1985) 239 - 255.

4. M. Grote, M. Sandrock, and A. Kettrup,
"A Comparative Study of Various Dehydrodithizone Modified Resins for the Recovery of Precious Metals", in: Ion Exchange for Industry, M. Streat (Ed.), Ellis Horwood Ltd., Chichester 1988, 404-413.

5. T. Hutton and H.M.N.H. Irving,
"3-Carboxymethylthio-1,5-diphenylformazan: a Potential Terdentate Ligand with Unusual Properties", Journal of Chemical Society, London, Perkin Transactions, II, (1) (1980) 139-145.

6. H.M.N.H. Irving, A.H. Nabilsi, and S.C. Sahota,
"Part XXX. Complexes of Metals with S-Methyldithizone and the Methylation of Methyldithizonates", Analytica Chimica Acta, 67 (1973) 135-144.

7. M.J. Cleare, P. Charlesworth, and D.J. Bryoson,
"Solvent Extraction in Platinum Group Metal Processing", Journal of Chemical Technology and Biotechnology, 29 (1979) 210-214.

8. E.A. Jones, "DDTU as an analytical reagent", National Institute for Metallurgy, Report No. 1569 (1973) 1-16.

9. A. Warshawsky,
"Extraction of Platinum Group Metals with Ternary Amines and controlled Separation by Stripping with Nucleophilic Ligands", Separation And Purification Methods, 12 (1) (1983) 1-35.

10. M. Grote, U. Hüppe, and A. Kettrup,
"Solvent Extraction of Noble Metals by Formazans", Hydrometallurgy, 19 (1987) 51-68.

11. S. Daamach, G. Cote and D. Bauer,
    "Separation of platinum group metals in hydrochloric media: solvent extraction of Palladium (II) with dialkyl sulphide"
    in G.A. Davies (Ed.): Separation Processes in Hydrometallurgy
    Ellis Horwood Ltd., Chichester 1987, 221-228.

PRESSURE CYANIDATION: AN OPTION TO TREAT REFRACTORY GOLD

S. C. Girardi, O. M. Alruiz and J. P. Anfruns*

Department of Mining Engineering
University of Chile

* Las Palmas S.A.

Abstract

Pressure cyanidation at high temperature has been proposed as an alternative to treat precious metal resources. Different technical aspects such as process thermodynamics and kinetics, cyanide decomposition at high temperature and oxygen partial pressure effects have been studied in many research centres around the world. The main conclusion has been generally that this process is technically feasible and that leaching times can be significantly reduced.

Different Chilean gold resources were selected for an experimental study with a laboratory scale autoclave. Metallurgical results, and important variables such as oxygen pressure, temperature, cyanide concentration and particle size, were measured and analysed. The collected data were used to develop an empirical model of the process and to estimate the model parameters. The technical information obtained by these means was used to carry out a preliminary feasibility study of a commercial scale unit using pressure cyanidation. Evaluation of results indicates that this process has a significant potential as an economically attractive alternative for the treatment of certain refractory gold resources.

Introduction

Gold producers have been induced to treat more difficult and/or lower grade resources because of an increasing world demand for gold which in turn has pushed metal prices to higher values, see Figure 1. As a result, gold processing technology has been challenged to improve metallurgical results in those ores or concentrates where conventional processes do not respond efficiently.

Several new gold projects have started recently in many places around the world and another important number is near to start up. As a typical example, only a few years ago practically all the chilean gold production was closely associated with the copper industry where gold and silver were recovered as a dore metal by-product. Today the situation is rather different because approximately more that 75% of the chilean production of these metals are directly obtained as primary products.

The classical industrial practice used for the treatment of gold ores usually involves cyanidation as the main process. Cyanidation is one of the oldest and widely used industrial process to leach gold and silver ores. As in any other metallurgical alternative, there are some resources where acceptable precious metal recoveries and low reagent consumptions are very difficult or impossible to achieve. In such cases, the material (ore or concentrate) can be called a refractory material. The main reason for this behavior (refractoriness) can be understood by considering the ore mineralogical composition. Gold can be found finely included (or associated) in a complex matrix, commonly pyrite or arsenopyrite. Both minerals increase cyanide and oxygen consumption significantly during a leaching process.

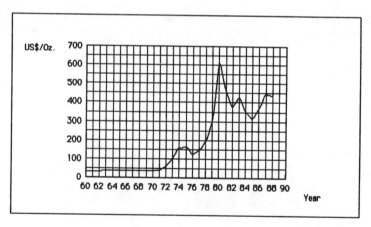

Figure 1 - Gold Price Evolution

Refractory Gold Process Options

The usual treatment method for refractory ores has been an oxidizing roast of the flotation concentrate (mainly sulfides) before cyanidation, to produce a porous calcine which is more amenable to a conventional leaching process (1). However this procedure is not considered attractive at present due not only to current standards of high product recoveries and low operating costs, but also because of pollution problems associated with it. The need of a more efficient treatment method for refractory gold ores has led to various developments and innovations in gold process technology. At least three alternatives have been recently proposed to solve this particular problem:

- Bio-leaching ------------------------------ conventional cyanidation
- Autoclave acid digestion ------------------ conventional cyanidation
- Direct pressure cyanidation

Bio-leaching has been used to accelerate the rate of oxidation of sulfides by the action of the thiobacillus ferro-oxidans bacterium. This method was initially applied to heap leaching operations with limited success. However more recent works have demostrated that the optimum reactor design is an agitated tank (2,3,4) The main constraints for the extensive use of this technique has been its high operating costs and long residence times required. A modified version of this reactor, which eliminates many of the previous problems, has been recently developed by Davy McKee. This process is claimed to offer lower capital and operating costs. According to Davy McKee's information, testwork carried out using this process has shown that 90% gold recoveries are possible to be achieved with leach residence times between 24 h to 72 h.

Another pretreatment method applicable to highly refractory gold resources, prior to conventional cyanidation, is autoclave acid digestion (pressure oxidation). Slurries with temperatures in the range of 90°C to 210°C and acidified by the action of sulphuric acid (pH: 1.5 - 2.0) are pressurized in autoclaves in order to oxidize the sulfide minerals. The usual process conditions are: process residence time within 1 to 4 hours, oxygen supply at an approximate rate of 40 kg per ton, and usual working pressures within the range of 10 to 20 atmospheres. After this pretreatment stage the slurry is neutralized adding lime to raise the pH over 10 and sent to cyanide leaching tanks. Results in gold recovery of approximately 90% have been reported by plants where this dual process is used (5).

Direct pressure cyanidation has been also tested in treating gold resources. This method has found succesfull applications in processing special gold ores at pilot plant and commercial scale (6,7). Gold extractions have been significantly improved compared with the usual low extraction obtained by conventional cyanidation (8) Batch (conventional autoclave) and continuous (pressurized pipe) reactors are the optional equipment used to carry out this process. Process kinetics is very fast and the total treatment time has been reduced as much as 10 to 50 times compared with conventional cyanidation for similar gold recoveries. Oxygen is injected into the reactor in a wide range of pressures, depending on operational and/or equipment design capabilities. Process is usually carried out above normal temperature but below 85°C to avoid areas of stability where the gold-water-cyanide system thermodynamics could generate unwanted compounds.

If a particular gold ore or concentrate is treated with any of the process alternatives mentioned above (and also combinations or extensions of them), different metallurgical results can be expected. The resource characteristics: gold grade, mineralogical composition of valuable and gangue species, interactions among gold and host minerals, size distribution, degree of liberation and so on, determine which of these gold beneficiation techniques is more efficient and amenable in technical, economical and also in environmental terms.

The Pressure Cyanidation Process

One of the most attractive hydrometallurgical processes recently proposed as a viable alternative to avoid the roasting-conventional cyanidation route has been the pressure cyanidation process. Good results in direct pressure cyanidation of auriferous sulfide ores were reported more than 10 years ago (9). More recently, Lurgi Chemie GmbH decided to investigate if gold leaching could be improved when a pipe type reactor (continuously operated autoclave) was used. Auspicious results were obtained because gold extraction could be accelerated by the effect of oxygen pressure and temperature.

The process alternative that the authors of this paper finally decided to study more closely was pressure cyanidation of ores and flotation concentrates. Both kinds of products are quite usual in Chile and it has been demostrated that a significant proportion of these resources does not respond efficiently to the traditional processing methods (10). The most common cause of this behavior could be explained, as it was mentioned before, because of the gold mineralogy and its usual association with species like pyrite, arsenopyrite and many other sulfides. In such cases gold can be found finely dispersed and/or located in the crystal lattice of these minerals, with the obvious difficulties associated with the traditional leaching process.

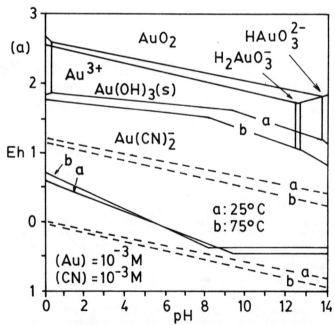

Figure 2 - Effect of temperature on the stability of dissolved gold species (11).

In any case, the accepted chemical reaction of gold dissolution can still be described by the equation 1:

$$2\ Au + 4CN^- + O_2 + 2H_2O \quad <=> \quad 2Au(CN)_2^- + H_2O_2 + 2OH^- \qquad (1)$$

However, taking into account the operational levels of temperature and O_2 partial pressure of a direct pressure cyanidation process it was found that there was a lack of basic information about gold-cyanide-water system thermodynamics, in conditions that exceeded the standard atmospheric pressure and temperature. A basic project was carried out to build stability diagrams, up to 125°C, for the Au-CN-H_2O and Ag-CN-H_2O systems. Valuable information was obtained for these simplified systems, with respect to the effect on the stability of cyanide soluble complexes and also on the cyanide oxidation. In Figure 2, the Eh-pH diagram describes the effect of temperature on the stability of dissolved gold. The results showed that cyanidation of gold and silver is not seriously influenced by moderate increases in temperature and that oxidized species are thermodynamically stable over a wide range of temperature (11).

The fundamental process kinetics was also studied, at least in a semiquantitative sense, using a simplified version of the rotating disc technique. Information about the effects of pressure and temperature on the rate of dissolution of metallic gold was obtained as a result of an autoclave experimental programme. Figure 3, summarizes the results obtained in these experiments. These results confirmed that the specific rate of gold dissolution can be substancially increased when temperature and oxygen partial pressure are above the standard values (12).

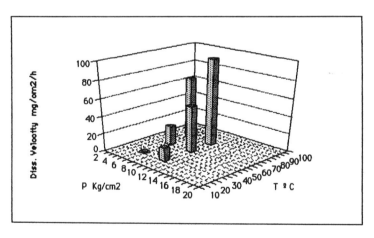

Figure 3 - Effect of temperature and pressure on the dissolution rate of metallic gold-rotating disc experiments (12).

Autoclave Leaching Experimental Programme

A 1000 ml Autoclave Engineering laboratory reactor was specially modified to carry out the pressure cyanidation experimental programme. This batch autoclave was equipped with an autonomous heating system, a pneumatic/ electric impeller agitation system, special devices to feed or take samples during testing and automatic temperature controllers.

Three different Chilean gold resources were selected to perform the autoclave experimental phase. Samples of a siliceous ore containing 7 gpt of Au (Gold Ore), a pyritic sulfide flotation concentrate containing 120 gpt of Au (Concentrate 1) and polymetallic flotation concentrate containing 122 gpt of Au (Concentrate 2) were used for the testing programme. The selected ore and flotation concentrates, were characterized using particle size, mineral identification and chemical analysis standard techniques: sieve testing, optical polarising microscopy, electron probe micro-analysis, fire assaying and atomic absorption spectrophotometry (13).

Examination of these materials revealed that in the case of the ore, gold was mainly in its native state, associated with a siliceous matrix rock and it was found in relatively fine sizes (20 micrometers). On the other hand, according to the chemical-mineralogical studies, both concentrates exhibited high grades of gold and silver, in the ranges of 120 to 160 gpt of Au and 400 to 500 gpt of Ag. In one of these concentrates, the most important mineral specie was pyrite (93%) and gold occured mainly as fine inclusions of electrum (15 micrometers) in pyrite. The other flotation concentrate was a mixture of different sulfide minerals, where sphallerite and chalcopyrite were the most important ones with gold and silver as relatively coarse size native metal particles (20-60 micrometers). In summary, mineral characterization showed that acceptable metallurgical results could be expected when conventional cyanidation is applied to the ore, however with both concentrates high reagent comsumptions and very low gold recoveries could be predicted if conventional cyanidation is used.

In order to simulate realistic processing conditions, no grinding was used in the case of both concentrates. On the other hand, the run-of-mine gold ore was ground to obtain an 80% under 200 mesh product. Sodium cyanide initial concentrations were selected taking into account both, theoretical reaction stoichiometry and empirical information. Enough leaching reagent was asured, in the case of the gold ore, using a 2 g/l dosis for the autoclave testing programme. For the concentrates, cyanide dosifications were estimated using empirically available information about relative dissolution rates on the different mineral species, including temperature effects (14). These doses were established so that the (dissolved O_2/CN^-) ratio at the end of each test was expected to be equal or greater than 6.

The experiments were carried out maintaining constant conditions of agitation, pulp solid percent by weight of 25% and a pH of 11 with NaOH. Oxygen pressure and temperature were selected as process variables. The experimental conditions were firstly defined by the autoclave system operating capabilities and also by the $Au-CN-H_2O$ thermodynamics. A range of temperature between 25°C and 75°C and a pressure range between 0.2 and 15.0 atm of O_2 were covered by the testing programme using experimental design techniques.

Results and Discussions

The experimental results indicated that temperature-pressure conditions can severely influence on gold cyanidation kinetics and recovery. Three-dimensional block diagrams, as the ones illustrated in Figures 4, 5 and 6,were useful to graphically compare process evolution at any time during the experiments. It was observed in all the tested conditions that the most significant effects were obtained when temperature was increased, however the highest kinetic values were always associated with both maxima: oxygen pressure and temperature conditions (15).

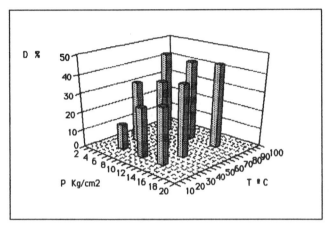

Figure 4 - Gold dissolution (%) after 45 minutes as a function of temperature and pressure (13).

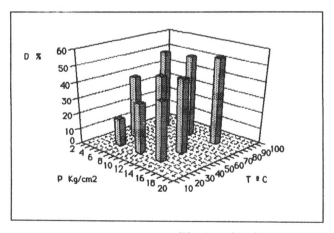

Figure 5 - Gold dissolution (%) after 90 minutes as a function of temperature and pressure (13).

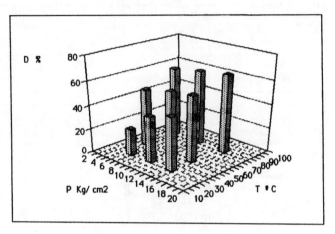

Figure 6 – Gold dissolution (%) after 180 minutes as a function of temperature and pressure (13).

Autoclave cyanidation process results obtained for the three selected gold materials were also systematically analysed using an empirical mathematical model. This model was built up considering the effects of the two main independent process variables (oxygen pressure and temperature) on cummulative recovery and dissolution velocity of gold. An average linear regression coefficient value of 0.99 was achieved when the mathematical model described by the equation 2 was tested using all the experimental data.

$$D = 1 - \exp\{-(K \cdot t)^\beta\} \qquad (2)$$

In equation 2, D is degree of dissolution, K is a lumped rate constant, t is time and β is a dimensionless factor. Highest values of K and β were obtained for the Ore, and the lowest ones were associated with the polymetallic concentrate. Both factors were calculated for each one of these experiments using two different polynomial functions (involving pressure and temperature). Curve fitting was evaluated and accepted for 70 experimental data points at a 5% of significance level, applying the Kolmogorov statistical test. The range of the calculated K and β values for each one of the tested products are listed in Table I.

Table I. Values of K and β for the Tested Gold Products

Product	K (min^{-1})	β
Gold Ore	0.038 - 0.049	1.1 - 2.3
Concentrate 1	0.004 - 0.008	0.9 - 1.7
Concentrate 2	0.001 - 0.006	0.3 - 0.4

The experimental results of this study were used to carry out a preliminary evaluation of a industrial scale pressure cyanidation unit. The design was projected using a six-stage autoclave system and accessory equipment to provide a residence time of 2 to 3 hours, with a processing capacity of 8600 tpy for a 120 gpt gold concentrate. A preliminary capital costs estimation was done using the procedure described by A. Mular (16). The operating costs were scalated from laboratory experimental information and plant practice data of similar tonnage operations. Considering an average gold price of US$ 400/oz this project showed an Internal Rate of Return of 330% and a Payback Period of only 3 months.

An independent experimental study involving another complex flotation product was also tested using a triple process option: pressure oxidation/conventional cyanidation and finally pressure cyanidation of the tailings. Results were highly attractive, because total gold recoveries up to near 100% were achieved (17). A more systematic study must be done to assess if this new alternative is commercially viable.

Conclusions

The main conclusions of this study can be summarized as follows:

1. The technical viability of the autoclave cyanidation process to improve metallurgical responses has been experimentally demonstrated with different Chilean gold products.

2. The most important effects over process kinetics were achieved when the maximum selected levels of temperature and oxygen pressure were used.

3. A preliminary feasibility study of an industrial scale autoclave cyanidation unit indicates that, when certain gold concentrates are considered, the profit of this project is highly attractive.

4. A fourth alternative to leach refractory gold resources could be added to the initial process option list: A two stage processing design at high pressure which includes an autoclave acid digestion followed by a pH adjustment and autoclave cyanidation.

Acknowledgments

The authors gratefully acknowledge the financial support provided by the Chilean Science and Technology Foundation (FONDECYT), Grant N°378-87. Appreciation is also expressed to Las Palmas S.A. for providing a valuable support to carry out this work.

References

1. N.C.Wall, J.C.Hornby and J.K.Sethi, "Gold Beneficiation", *Mining Magazine*, (1987), 393-401.

2. P. van Aswegen and A.K. Haines, "Bacteria enhance Gold Recovery" *International Mining*, (May 1988), 19-23.

3. Y.Attia, J.Litchfield, and L. Vealer "Applications of Biotechnology in the Recovery of Gold", in *Microbiological Effects on Metallurgical Processes* (Proceedings of AIME Annual Meeting, New York, 1985).

4. H.E.Gibbs, M.I. Errington and F.D. Pooley, "Economics of Bacterial Leaching", *Canadian Metallurgical Quarterly*, 24-2 (1985), 121-125.

5. G.D. Argal, "Perseverance and Winning Ways at McLaughling Gold", *Engineering and Mining Journal*, 187 (1986), 26-32.

6. H.B. Pietsh, W.M. Turke and G.H.Rathje, "Research of Pressure Leaching or Ores Containing Precious Metals", *Erzmetall*, 36 (1983), 261-265.

7. F.D. Wicks et al. "Mercure starts up its New Alkaline Pressure-Oxidation Autoclave Plant" *Engineering and Mining Journal*,189 (June 1988), 26-31.

8. D.R. Davis and D.B. Paterson, "Practical Implementation of low Alkalinity Pressure Cyanidation Leaching Techniques for the Recovery of Gold from Flotation Concentrates" (Gold 100 International Conference on Gold, vol 2, Johannesburg, 1986).

9. A.B. Abdullaev, A.M. Muminova, "Autoclave Leaching of Auriferous Sulfide Ores", *Soviet Journal of Non-Ferrous Metals*, 16-6 (1975), 91-92.

10. S.C. Girardi et al., "Pressure Cyanidation of Gold Ores", *Precious and Rare Metal Technologies*, ed. A.E. Torma and I.H. Gundiler (Elsevier Science Publishers B.V., Amsterdam, 1988), 183-191.

11. F.J. Arriagada et al., "A Thermodynamic Study of Au-CN-H_2O and Ag-CN-H_2O Systems up to 125°C" (unpublished Paper, AIME Metallurgical Transaction B).

12. J.P. Anfruns, "Gold and Silver Leaching in Tubular Reactors" (Final Report FONDECYT-CHILE, Project 115-85, 1986).

13. S.C. Girardi "Pressure Cyanidation: Feasibility Study and Pilot Scale Reactor Preliminary Design" (Final Report FONDECYT-Chile, Proyect 378-87, 1988).

14. F. Habashi, *Principles of Extractive Metallurgy, Volume II* (New York, NY: Gordon & Breach, 1970).

15. S.C. Girardi, O.Alruiz and J. Anfruns, "Autoclave leaching of Difficult Treatment Auriferous Resources" (Proceedings of ALAMET I Congress, Rio de Janeiro, 1988).

16. A.L. Mular, "The Estimation of Preliminary Costs" *Mineral Processing Plant Design*, ed A. Mular and R. Bhappu (Society of Mining Engineers, A.I.M.E., New York, 1980).

17. S.C. Girardi, "Acid Digestion-Cyanidation of Gold Concentrates" (Report MET-272/St. Joe Mining Co., Univ. of Chile, 1988).

ELECTROCHEMICAL STUDY OF SILVER

CEMENTATION ON LEAD IN CHLORIDE SOLUTION

K.H. Kim*, T. Kang* and H-J. Sohn**

* Department of Metallurgical Engineering,
** Department of Mineral and Petroleum Engineering,
College of Engineering, Seoul National University,
Seoul, Korea. 151-742

Abstract

The kinetic study of silver cementation on lead was carried out using rotating disk electrode technique.The cementation reaction was the first order with respect to silver ion concentration and the rate determining step was found to be mass transfer through the liquid film boundary layer with the activation energy of 3.27 kcal/mol. An increase in reaction rate was observed during the cementation reaction, which was probably due to the local turbulence caused by the deposits as well as the increase in total reacting area.

Introduction

Hydrometallurgical processing of galena developed by U.S. Bureau of Mines [1][2][3] consists of the leaching of galena with $FeCl_3$ in brine solution, precipitation of $PbCl_2$ from leach liquor followed by fused-salt electrolysis of $PbCl_2$. But it requires a purification step to remove silver and other base metals before precipitation of $PbCl_2$, and the recovery of silver as a by-product could affect the overall economics of this process significantly.

Haver and Wong [1] carried out the cementation of silver with lead powder to obtain over 90% recovery within 15 minutes and the reaction chemistry in chloride solution can be expressed as,

$$2 \, AgCl_n^{1-n} + Pb = 2 \, Ag + PbCl_n^{2-n} + nCl^- \qquad (1)$$

For most of the cementation reactions, the reaction rate is the first order with respect to the noble metal ion concentration and the reaction rate is controlled by diffusion in liquid film boundary layer except a few systems. Also, the morphology of the deposits affects the cementation reaction rate significantly, and these were reviewed on several occasions [4][5][6]. The kinetic studies of silver cementation with various metals in different solutions were performed extensively, e.g., Ag/Zn [7][8], Ag/Cu[9], Ag/Fe or Fe alloy [8][10], and Ag/Pb [11][12], mostly using rotating disk technique. Recently, Parga et al,[13] performed the engineering analysis of Merrill-Crowe plant data.

The purpose of this study was to investigate the kinetics of silver cementation reaction in chloride solution by electrochemical and direct chemical measurements using rotating disk system.

Experimental

Silver and lead metal (purity : 99.9%) disks of 1.5 cm diameter were mounted in epoxy resin for the rotating disk experiments. Immediately prior to each experiment, the disk surface was prepared by polishing through two grit

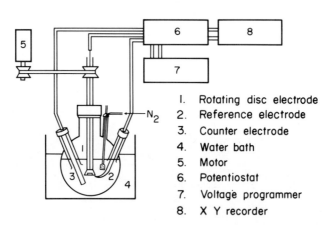

1. Rotating disc electrode
2. Reference electrode
3. Counter electrode
4. Water bath
5. Motor
6. Potentiostat
7. Voltage programmer
8. X Y recorder

Figure 1 - Schematic diagram of experimental set-up

sizes of silicon carbide emery paper with a final polish using a 3μm alumina powder. Distilled water and reagent grade chemicals were used to prepare the solutions and purified nitrogen was passed through the solutions to maintain the oxygen free atmosphere.

A typical three way electrode system (silver or lead working electrode, graphite counter electrode and Ag-AgCl reference electrode) was used for electrochemical measurements. Polarization data were obtained using a EG&G PAR Model 173 potentiostat and Model 175 potential sweep generator. Schematic diagram of experimental set-up is shown in Figure 1. For cementation experiments, 5cc of solution was withdrawn periodically and analysed for metal ion concentration with atomic absorption spectrophotometer.

Results and Discussion

The overall reaction chemistries described in equation (1) can be divided into two half-cell reactions due to the electrochemical nature, and can be expressed as,

$$\text{cathode}: \text{AgCl}_n^{1-n} + e^- = \text{Ag} + n\text{Cl}^- \tag{2}$$

$$\text{anode}: \text{Pb} + n\text{Cl}^- = \text{PbCl}_n^{2-n} + 2e^- \tag{3}$$

The potentiodynamic curves were obtained for the above half-cell reactions by electrochemical measurements to investigate the characteristics of each reaction.

Characteristics of polarization curves

The polarization curves of each half-cell reaction are shown in Figure 2. The experiments were performed at 30°C, and the initial concentration of silver for cathodic and lead for anodic reaction were 20 ppm and 80 ppm, respectively. Sodium chloride of 2.85 mol/l was added as a supporting electrolyte for the rest of experiments except otherwise specified.

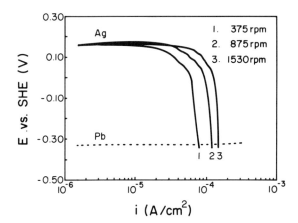

Figure 2-Polarization curves of silver and lead in chloride solution

As shown in Figure 2, the polarization curves of silver reached the limiting current region just below the rest potential, and the limiting current density increased with the increase of rotating speed. However, the polarization curves of lead showed little effect with the variation of rotating speed and initial lead concentration which indicated the reaction was reversible.

The anodic polarization curve intersects the limiting current region of cathodic polarization curves in this Evans diagram and the rest potential difference between anode and cathode reaction is 0.486V. These facts suggest that the cementation reaction would be controlled by mass transfer of silver ions in liquid phase as proposed by Power and Ritchie [14]. Furthermore, the general irreversibility of cementation reactions was established by Wadsworth [4], in which it was shown that if $z\Delta E^o$ is greater than 0.3V (0.97V for this study), the back reaction kinetics can be negligible. In here, z is the number of electrons involved in cementation reaction and ΔE^o is the equilibrium potential difference between anode and cathode reaction.

The limiting current density, i_ℓ can be expressed as follows with the assumption of negligible contribution from migration due to the presence of a large amount of supporting electrolyte,

$$i_\ell = \frac{DzFC}{\delta} \tag{4}$$

where, D is the diffusivity, F is Faraday constant, C is the concentration of silver ion and δ is the thickness of the diffusion boundary layer. The cementation reaction rate can be correlated with the current density as,

$$\frac{dn}{dt} = \frac{iA}{zF} \tag{5}$$

In here, n is the number of moles and A is the reacting surface area.

Cementation reactions

The cementation reaction rate was measured with the lead rotating anode in silver chloride-brine solution at 30°C and the solution was withdrawn at a specified time interval to analyze the silver ion concentration. The rotating speed was varied between 550-1755 rpm, which corresponded to the laminar flow conditon with the Reynolds number of $5 \times 10^3 - 5 \times 10^4$. As mentioned previously, the cementation reaction exhibits the first order kinetics in general, the integrated from of reaction rate is,

$$\log(C/C_o) = \frac{kA}{2.303 \, V} t \tag{6}$$

where C_o and C are the concentration of silver at t = 0 and t = t respectively. V is the volume of solution, k is the rate constant and t is time. The experimental results were found to fit the equation (6) as shown in Figure 3. The enhancement in reaction rate was observed after deposition of a certain amount of silver, which is typical for most of the cementation system in low level of noble metal concentration and it will be discussed in later section.

The rate constant k in equation (6) can be calculated from the slope in Figure 3, both for the initial and enhanced stages. If the reaction rate is controlled by diffusion in liquid film boundary layer, the rate constant can be expressed according to the Levich's theory [15] under the laminar flow con-

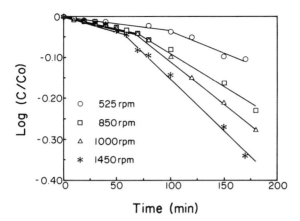

Figure 3 - Effect of rotating speed on the cementation of silver with lead.

dition as follows,

$$k = 0.62\, D^{\frac{2}{3}} \nu^{-\frac{1}{6}} \omega^{\frac{1}{2}} \qquad (7)$$

where, ν is the kinematic viscosity and ω is the angular velocity. A plot of k vs. $\omega^{\frac{1}{2}}$ is presented in Figure 4, and shows a good straight line passing through the origin for the initial stage.

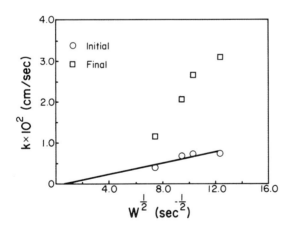

Figure 4 - Initial rate constant as a function of rotating speed.

Upon combining the equations (5), (6) and (7), the current densities can be calculated from the cementation data and were compared in Table 1 with the limiting current densities obtained from polarization curves.

Table 1. Comparisons between current densities calculated from the cementation data and limiting current densities. (30° C, 2.85 mol/l NaCl solution)

rpm	i (cementation, mA/cm²)	i_ℓ (mA/cm²)
525	0.0703	0.0868
850	0.120	0.111
1000	0.125	0.120
1450	0.137	0.144

The cementation rates calculated from the diffusion limited current densities are very close with those of directly measured values as shown in Table 1 in the initial stage.

Diffusion coefficient of silver ions

The diffusivity of silver can be obtained from the limiting current densities shown in polarization curves using equation (8).

Effect of temperature. The limiting current densities were obtained with the variation of rotation speed at 25°C and plotted in Figure 5. It shows a good linearity which obeys the Levich's equation. The diffusion coefficient of silver can be calculated from the slope in Figure 5 to obtain 9.16 x 10^{-6} cm²/sec using the kinematic viscosity of 1.13 x 10^{-2} cm²/sec for the system [16]. The same analyses were performed for different temperatures and the Arrhenius plot was made in Figure 6. The activation energy of 3.27 kcal/mol was obtained which supports the diffusion controlled reaction as mentioned previously.

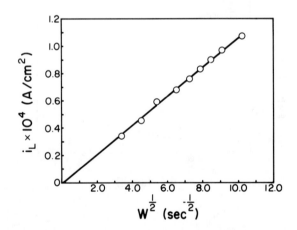

Figure 5 - Effect of rotating speed on the limiting current density.

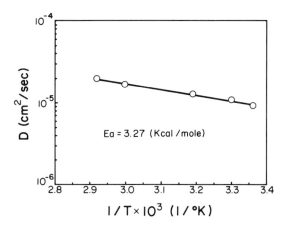

Figure 6 - An Arrhenius type plot for diffusivity vs. reciprocal of absolute temperature.

Effect of NaCl concentration.   The silver ions are present as silver chloro-complex in chloride solution, e.g. Ag^+, $AgCl^o$, $AgCl_2^-$, $AgCl_3^=$, $AgCl_4^{---}$ etc. The concentration of each species depends on the total chloride concentration, temperature and total silver ion concentration. Based on Seward's data [17], concentrations of these silver chloro-complexes were calculated with the variation of total [Cl$^-$] concentration at 25°C and presented in Table 2. With the increase of total [Cl$^-$] concentration, the concentration of the complex with a large ligand number increases as shown in the Table. Since $AgCl_3^=$ and $AgCl_4^{---}$ are the predominant species in brine solution, the diffusivity obtained previously would be an averaged value of these two complexes.

Table 2. The equilibrium distribution of silver-chloro complexes for different NaCl concentration at 25° C.

NaCl conc. (mol/l)	wt % of Ag-Cl complex				
	Ag^+	$AgCl^o$	$AgCl_2^-$	$AgCl_3^=$	$AgCl_4^{---}$
1.83	0.00	0.09	16.83	42.25	40.84
2.85	0.00	0.02	7.96	33.57	58.44
3.93	0.00	0.01	4.21	26.07	69.71
5.09	0.00	0.00	2.46	20.62	76.91

Diffusion coefficients were estimated for different NaCl concentrations and plotted in terms of ionic strength in Figure 7. Diffusivities decrease with the increase of NaCl concentration. This may indicate that the diffusivity of a complex with a large ligand number is smaller than that of a low ligand number, since the concentration of the former increased as the concentration of total [Cl$^-$] increases. In addition, ionic strength also increases simultaneously. Gosting and Harned [18] and also Stokes et al., [19] reported that the diffusivity decreases with an increase of ionic strength. Experiments were performed with the variation of $NaClO_4$ concentration to see the effect of the ionic strength at constant [Cl$^-$] concentration,   also shown in Figure 7. The decreasing trend of the diffusivity with the increase of ionic strength agrees qualitatively with the above studies. From the two curves in Figure 7, the diffusivity of silver complex tends to decrease due to the increase in ionic strength and ligand number as the concentration of NaCl increases.

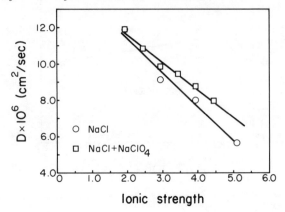

Figure 7 - Effect of NaCl and $NaClO_4$ concentration on the diffusivity of silver

Enhancement of cementation rate

As mentioned before, the enhancement of cementation rate was observed after the reaction proceeded to some extent. Strickland and Lawson [5] reported that these phenomena were observed when specific deposit mass was over 0.4 mg/cm², and it appeared at a deposit mass of 0.35-0.56 mg/cm² in this case. The enhancement of reaction rate is mainly associated with the product formed at the disk surface, which were shown in Figure 8. Miller [6] analyzed several cementation systems and reported that main cause of rate increase was due to the increase in surface area rather than micro-turbulence [20][21] or decrease of diffusion film thickness [22].

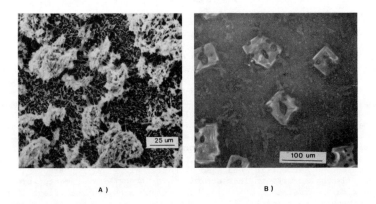

Figure 8 - Photographs of surface deposit (a) Ag/Pb and (b) Ag/Ag

The limiting current densities were measured in the absence and in the presence of a deposit of 1.37 mg/cm² to see the effect of increase in surface area, which are shown in Figure 9. The effective surface area was 2.27 times larger than of free surface, which is comparable with that of Beckstead and Miller [23]. However, the little enhancement in rate was observed at low rotating speed, which is in disagreement with Miller's analyses [6]. With the increases of rotating speed, the cementation rate was enhanced markedly as illustrated in Figure 9. This would indicate that the rate enhancement was mainly due to the increased mass transfer which was probably caused by the local turbulence in the presence of deposits, rather than the increase in effective surface area.

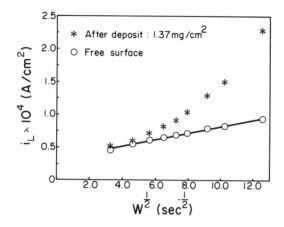

Figure 9 - Effect of surface deposit on the limiting current density

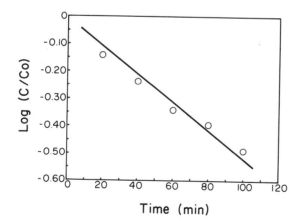

Figure 10 - First order plot for silver cementation with Pb powder

647

Figure 10 shows the cementation result with 0.25 grams of lead powders of 83 μm (mean diameter) at a stirring speed of 600 rpm. The cementation rate followed the first order kinetics as expected and the rate enhancement was not observed in this case. The reaction rate constant calculated from the slope in Figure 9 is approximately the same as that obtained from the limiting current density values at 1500 rpm in Figure 5. This would mean that actual cementation rate is closely related with the initial stage of rotating disk experiments rather than the rate enhancement region.

Conclusions

The cementation kinetics of silver on lead were investigated with the rotating disc technique. The reaction rate was the first order with respect to the silver ion and controlled by the diffusion of silver through the liquid film boundary layer. The diffusivity of silver complex decreased with the increase of [Cl^-] concentration. The enhancement of rate was observed after deposition of silver (0.35-0.56 mg/cm²) and this is probably due to the local turbulence near deposits to increase the mass transfer rate.

Acknowledgement

The authors wishes to express their gratitude to Korea Science and Engineering Foundation (KOSEF) for the financial assistance to this study.

References

1. F. P. Haver and M. M. Wong, "Ferric Chloride-Brine Leaching of Galena Concentrate", (U. S. Bureau of Mines Report of Investigation 8150, 1976)

2. F. P. Haver et al., "Recovery of Lead from Lead Chloride by Fused-Salt Electrolysis," (U. S. Bureau of Mines Report of Investigation 8166, 1976)

3. M.M. Wong, F. P. Haver and R. G. Sandberg, "Ferric Chloride Leach-Electrolysis Process for Production of Lead" Lead-Zin-Tin '80, ed. J. M. Cigan, T. S. Mackey and T. J. O'Keefe, (Warrendale, PA:The Metallurgical Society, 1980), 445-454

4. M. E. Wadsworth, "Reduction of Metals in Solution," Trans. AIME, 245 (1969), 1381-1394

5. P. H. Strickland and F. Lawson, "The Cementation of Metals from Dilute Aqueous Solution," Proc. Aust. Inst. Min. Met. No. 237 (1971), 71-74

6. J. D. Miller, "Cementation," Rate Process of Extractive Metallurgy, ed. H. Y. Sohn and M. E. Wadsworth, (New York and London: Plenum Press, 1979), 197-244

7. H. E. A. von Hahn and T. R. Ingraham, "Kinetics of Silver Cementation on Zinc in Alkaline Cyanide and Perchloric Acid Solutions, "Can. Met. Quart., 7(1) (1968), 15-26

8. J. A. P. Mathes, F. Lawson and D. R. Canterford, "The Cementation of Uncomplexed Silver Ions onto Zinc and onto Ferrous Alloys," Hydrometallurgy. 14(1985), 1-21

9. H. E. A. von Hahn and T. R. Ingraham, "Kinetics of Silver Cementation on Copper in Perchloric Acid and Alkaline Cyanide Solutions," Trans. AIME, 239 (1967), 1895-1900

10. M. L. Epockoposyan and I. A. Kakovskii, "An Examination of Copper and Silver Cementation Kinetics with Metallic Iron from Chloride Solutions," Tsvetn. Metal., 38(10) (1965), 15-19

11. K. H. Kim, H-J. Sohn and T. Kang, "Kinetics of Silver Cementation on Lead in Nitrate Solution," J. Korean Institute of Metals, 24(6) (1986), 663-668

12. R. M. Morrison, D. J. Mackinnon and J. M. Brannen, "Silver Cementation from Chloride Solution Using Rotating Disks of Copper and Lead," Hydrometallurgy, 18(1987), 207-223

13. J. R. Parga, R. Y. Wan and J. D. Miller, "Zinc-dust Cementation of Silver from Alkaline Cyanide Solutions - Analysis of Merrill-Crowe Plant Data," Minerals and Metallurgical Processing, 5(3) (1988), 170-176

14. G. P. Power and I. M. Ritchie, "A Contribution to the Theory of Cementation," Aust. J. Chem., 29(1976), 699-709

15. V. G. Levich, Physicochemical Hydrodynamics (2nd ed., Englewood Cliffs, N. J. : Prentice Hall Inc., 1962), 69

16. R. C. Weast ed., CRC Handbook of Chemistry and Physics, (64th ed., Boca Randon, Florida : CRC Press Inc., 1983), D-250

17. T. M. Seward, "The Stability of Chloride Complexes of Silver in Hydrothermal Solutions up to 350°C," Geochim. Cosmochim Acta, 40 (1976), 1329-1341

18. L. J. Gosting and H. S. Harned, "The Application of the Onsager Theory of Ionic Mobilities to Self-Diffusion," J. Am. Chem. Soc., 73(1951), 159-161

19. R. H. Stokes, L. A. Woolf and R. Mills, "Tracer Diffusion of Iodide Ion in Aqueous Alkali Chloride Solutions at 25 C," J. Phys. Chem., 61(1957), 1634-1636

20. P. H. Strickland and F. Lawson, "The Measurement and Interpretation of Cementation Rate Data," International Symposium on Hydrometallurgy, ed. D. J. Evans and R. S. Shoemaker, (New York : AIME, 1973), 293-330

21. E. C. Lee, F. Lawson and K. N. Han, "Precipitation of Cadmium with Zinc from Dilute Aqueous Solutions : 1- under and Inert Atmosphere," Trans. IMM 84 (1975), C81-C87

22. H. Majima, M. Mamiya, H. Tanaka and K. Matsuda, "Physical Chemistry of Copper Cementation," World Mining and Metals Technology V.2, ed. A. Weiss, (New York : AIME, 1976), 567-584

23. J. D. Miller and L. W. Beckstead, "Surface Deposit Effects in the Kinetics of Copper Cementation by Iron," Met. Trans. 4(1973), 1967-1973.

RECOVERY OF ZINC FROM BLAST FURNACE DUST

OF METALLURGICAL COMPANY IN YUGOSLAVIA

M.Stamatović, S.Milošević, N.Canić, L.Mihovilović and T.Živanović

Institute for Technology of Nuclear and
Other Mineral Raw Materials, Beograd, Yugoslavia

Abstract

The treatment and disposal of blast furnace (BF) and electric arc furnace (EAF) steelmaking dust have been given increasing attention in recent years for two principal reasons. Firstly, the dust is now classed as a hazardous waste material under U.S. EPA regulations which prohibit its disposal in dumps other then controlled landfill sites (Federal register, Vol 45. No 98, Section 261.32, page 33124) because of the leachability of its toxic constituents (Zn, Pb, Cd, and Cr). Classification of dust as a hazardaus waste material in Yugoslavia, if and when guidelines are formulated, might mean that if the dust is to be disposed of, it would need to be encapsulated or transported to a controlled landfill site. Secondly, considering the depleting reserves of primary mineral resources for the recovery of metal values, which may be economically recoverable, depending on the cost associated with desposal resulting from enforcement of the EPA regulations.

A long term research program is in progress in Institute to establish a technological process for the recovery of zinc from blast furnace dust of metallurgical company in Yugoslavia. This paper shows the technical feasibility of caustic leach-electrowining process for the recovery of zinc from blast furnace dust. Zinc recovery as zinc dust is more than 90% and zinc dust so obtained may be used successfully for cementation of impurites in the hydrometallurgical process. The process has been studied both on the small and pilot scales and process engineering studies indicate that it is economical.

Introduction

The treatment and disposal of blast furnace and electric arc furnace steelmaking dust have been given increasing attention as a hazardous waste material under U.S. EPA regulations which prohibit its disposal in dumps other then controlled landfill sities (Federal Register, Vol 45. No. 98, Section 261.32, page 33124) because of the leachability of its toxic constituents (Zn, Pb, Cd, and Cr). Its EPA - assigned hazardous waste number is K061 (1). Classification of dust as a hazardous waste material in Yugoslavia, if and when guidelines are formulated, might mean that if the dust is to be disposed of it would need to be encapsulated or transported to a controlled landfill site.

Secondly, considering the depleting reserves of primary mineral resources for the recovery of metal values, which may be economically recoverable, depending on the cost associated with disposal resulting from enforcement of the EPA regulations.

Several studies and surveys have been carried out to examine various aspects of dust treatment but few commercial plants are in operation. Technology employed uses pyrometallurgical techniques (2-4). Commercial scale hydrometallurgical plants are not known to exist although small-scale plants are frequently more economically attractive than their pyrometallurgical counter parts which generally require economies of scale to be viable.

This paper shows the technical feasibility of a caustic leach - electrowinning process for the recovery of zinc from blast furnace steelmaking dust. Zinc recovery as zinc dust is more than 90% and zinc dust so obtained may be used successfully for cementation of impurities in the hydrometallurgical process. The process has been studied both on the small and pilot scales and process engineering studies indicate that it is economical.

Recent Options in Dust Treatment Technologies

The problems associated with dust disposal in the U.S, Japan, Canada, England etc. have become more acute since dust has been listed as a hazardous waste under the Resource Conservation and Recovery Act (RCRA) of 1976. The high cost of disposal has generated new interest in methods of recovery of metal values which may include recycling of pelletized dust and residues. Furthermore, disposal of dust is complicated by the fact that the dust is extremely fine and difficult to handle unless it is first pelletized.

Several hundred thousand tons of dust are generated each year and represent a potential source of several metals, depending on the composition of the feed to the process. Recovery or recycling of these metals may be a possible alternative to the disposal options. Incomplete characterization of dust in the past has made the selection of recovery or disposal options difficult. Several processes, both pyrometallurgical and hydrometallurgical, have been proposed for dust treating and these are well documented in literature.

Pyrometallurgical plants in the Japan, W.Germany and U.S. treat EAF dust in a Waelz kiln for recovery of lead and zinc as an impure Waelz oxide intermediate product (typically containing 50-55% Zn and 6-7% Pb) which requires further pyrometallurgical processing for reduction to metals. An electrothermic shaft furnace process called PLASMADIUS using plasma heat, piloted by SKF Steel Engineering in Sweden, is currently being commercialized to recover Zn, Pb, Ni, and Fe from EAF steelmaking dust (5-7).

The hydrometallurgical processes are primarily directed at recovery of zinc and lead, leaving a residue that might be recycled for recovery of iron with a bleed stream discharged to a controlled landfill site. Experiments have been conducted with most leaching media including the strong mineral acids and bases (8-11). The main problems to overcome relate to efficient selective removal of zinc, lead, and cadmium from iron. Recently Thorsen et al, have described a process of leaching such wastes with a liquid - organic phase, Versatic 911, containing a cation exchanger. After extraction step, chlorides are washed out with water, followed by stripping of zinc and other metal values by sulphuric acid. From the zinc sulphate solution, zinc is recovered through the conventional process.

Basic Characterization of Dust

The formation, physical and chemical characterisation of blast - furnace dust has been studied to determine if this dust is a suitable material for recovery of zinc. Blast - furnace dusts are considered to be composed primarily of a self - agglomerated collection of microfine physically and chemically complex particles consisting principally of oxides. The composition of the dust used in the present study is shown in Table I., and average values of zinc and iron for the period of ten years are given in Table II, (Metallurgical Company Smederevo - Yugoslavia). All chemical analyses were performed in an atomic absorption spectrometer after dissolution of the samples in concentrated hydrochloric acid and subsequent dilution. It can be seen that values of iron and zinc are dominant. No correlation between chemistry and furnace size, dust collecting system or other factors has been established. The major conclusion which can be derived from Table II. is that the composition of dust (zinc and iron content) is influenced directly by the type of feed being melted. The content of lead in dust is very low therefore no problems associated with electrowining of zinc exist.

Table I. Chemical Composition of Dust from Blast Furnace Used in This Study

Element +	wt. %
Fe	21.970
Zn	17.280
Si	8.250
Al	0.310
Ca	4.220
Mg	5.775
Na	0.185
K	0.400
Pb	0.200
Mn	0.190
Cd	0.003
Cr	0.011
Ti	0.006
Cu	0.014
S	0.920
Cfix	39.950
P	nil

+Arithmetic averages of dust from Metallurgical Company Smederevo - Yugoslavia

The granulometric analyzes carried out according to Tyler classification, and the contents of zinc and iron in samples of fractions less than 63 um are shown in Tables III and IV, respectively. Individual particles are generally spherical and range in size with the majority from 33 um to 0.0 um (63.17 wt.%) see Table III. Recycling or disposal of dust is complicated by the fact that the dust is extremely fine and difficult to handle. It can be seen from table IV. that with decreasing of fractions size the content of zinc increases while the content of iron decreases.

Table II. Average Values of Iron and Zinc for the Period of Ten Years

Year	Zn (wt.%)	Fe (wt.%)
1977	6.26	19.24
1978	49.57	8.35
1979	53.59	1.33
1980	3.05	43.24
1981	7.03	36.28
1982	12.21	31.92
1983	17.21	28.18
1984	2.80	21.28
1985	8.15	21.24
1986+	12.54	13.68
1987+	17.28	21.79

+ Arithmetic averages determined in ITNMS

Table III. Granulometric Analysis

Sample (um)	wt.%	wt.%	wt.%
150	5.09	5.09	100.00
150 - 100	5.80	10.89	94.91
100 - 74	7.38	18.27	89.11
74 - 63	6.86	25.13	81.73
63 - 44	7.80	36.83	70.97
33 - 23	13.03	49.86	63.17
23 - 15	11.28	61.14	50.14
15 - 11	8.82	69.96	38.86
11 - 0.0	30.04	100.00	30.04

Table IV. Zinc and Iron Contents of 63 um Fractions

Sample (um)	M%	Zn%	Fe%	%DZn	%DFe
- 63	25.13	9.74	8.38	14.66	14.19
63 - 33	11.70	6.69	36.30	4.68	28.62
33 - 11	33.13	17.00	16.76	33.66	37.41
11 - 0.0	30.04	26.19	9.77	47.03	19.78
	100.00	16.69	12.73	100.00	100.00

Dust samples are examined by X-ray powder diffractometry to indentify most abundant phases. Philips generator and diffractometer with CoK radiation was used. X-ray diffractometry showed that the largest components of blast furnace dust as received were hematite and magnetite in combination with zinc oxide, a background of smaller peaks being attributable to the various small quantities of other oxides that were present.

The predominant mechanism of dust formation is by deposition and oxidation of material from the vapour phase on condensed nuclei such as fugative dust. The high amount of iron in the dust means that certain fractions display magnetic properties. Dust phases have been classified according to their magnetic material variations over a wide-range (5-50%). Zinc oxide is dominantly present in the non magnetic fractions. The properties of zinc present as oxide and ferrite have a significant effect on treatment and recovery options because the former is soluble to a varying degree in most leachants whereas ferrites tend to be insoluble. It has been considered that the various conditions under which the dust is formed influence the zinc solubility significantly.

Table V. Typical Dust Analyses of Magnetic Separation Treatment

Sample	M%	Zn%	Fe%	M%Zn%	M%Fe%	%DZn	%DFe
MF	16.13	11.46	34.21	184.85	551.81	14.32	40.28
NMF	83.87	13.19	9.37	1106.25	816.19	85.68	49.72
	100.00	12.91	13.68	1291.10	1368.00	100.00	100.00

MF - Fraction of magnetic material
NMF - Fraction of non-magnetic material

Lead, like zinc, often occurs in small particles within an agglomerate, or as small regions within a large particle. Lead rich regions have been shown to exist on the surface of iron-rich particles but lead is not expected to be incorporated into the ferrite structure. Blast - furnace dust used in this investigation contained zinc in the form of oxide with very low content of lead.

Experimental Procedure

Early literature refers to the solubility of zinc and lead oxides in caustic soda and to the possibility of producing zinc dust from the electrolysis of sodium zincate solution. When zinc is present mainly as oxide, the material can be treated directly with sodium hydroxide. When a substantial amount is combined as ferrite, a reduction step may be desirable to maximize zinc recovery. The solubility of iron in sodium hydroxide is limited which fadevours the treatment of high-iron content materials.

Electrowining of zinc powder from wastes can be carried out successfully in alkaline medium. Several reviews have appeared in the literature on the deposition of zinc powder using an alkaline bath (12-14).

During this investigation it was found that magnetic separation and ionic flotation method for segregation of the zinc from iron-rich components had no effect, therefore these experiments were not the scope of this paper.

CAUSTIC LEACHING EXPERIMENTS. Small-scale experiments of caustic leaching was carried out in 1000-cm^3 beaker with the final goal to determine the basic parameters of process. The influence of leaching time, concentration of sodium hydroxide, temperature, and solid/liquid ratio on the leachability of zinc from blast furnace dust were studied. The contents of zinc and iron were measured in dust residue after leaching step and in the leach liquid. The insoluble leach residue was washed with water and the filtrate was kept for recycling until the content of zinc in it become sufficiently high for electrolysis.

For conducting pilot-scale experiments a mixer capacity of $30m^3$ and settling tank of $45m^3$ were used. The optimal conditions of leaching established in small-scale leaching tests were used in pilot-scale.

ELECTROLYSIS OF SODIUM ZINCATE SOLUTION. For small-scale experiments a PVC cell of dimension 300x180x120 mm was used for electrolysis. Two stainless steel anodes of 150x100 mm and one cathode of the same size were used. distance between them was 30 mm. The sodium zincate solution earlier obtained was electrolysed at 3.4-3.5 V and a current density of 250-350 A/m^2 to give a fine

zinc dust deposit was used. The deposit was stripped every four hours, washed with water, dried in an atmosphere of nitrogen and the content of zinc in deposit was determined.

A DC rectifier giving 400 A was used in pilot-scale experiments. Three PVC cells of 700x500x400 mm were connected in series for electrodeposition. Mild steel anodes and stainless steel cathodes of 480x310 mm were used. The process as given in small-scale experiments was followed.

Results and Discussion

Caustic leaching experiments were performed to determine the influence of leaching time, concentration of sodium hydroxide in leaching medium, temperature, and solid/liquid ratio on the leachability of zinc from blast furnace dust. Some of major experimental results that were obtained when the dust was exposed to caustic leaching tests are shown in Table VI. According to experimental results the optimal parameters for caustic leaching tests are, leaching time of 3h, 30 wt% sodium hydroxide solution as leaching medium, temperature of 40°C, and solid/liquid ratio of 1:5. The results confirmed that blast-furnace dust contained zinc in the form of oxide and the segregation of zinc from iron rich components is possible without envolving additives and high temperatures. Pilot-plant experiments were performed based on process parameters derived from small-scale experiments. Small-scale experiment results are certified by pilot-scale experiments, recvering zinc from iron-rich components in more than 90%.

Table VI. Effect of Leaching Time, Leaching Medium, Temperature, and S/L Ratio on Zinc Removal from Blast Furnace Dust

Time (h)	Leaching medium wt. %	Temperature (°C)	S/L ratio	Zinc removed wt. %
0.5	10	35	1:10	5.01
1.0	10	35	1:10	5.03
3.0	10	35	1:10	5.68
6.0	10	35	1:10	5.68
9.0	10	35	1:10	5.69
24.0	10	35	1:10	5.59
3.0	20	35	1:10	11.48
3.0	30	35	1:10	12.22
3.0	30	20	1:10	5.20
3.0	30	40	1:10	15.76
3.0	30	70	1:10	15.80
3.0	30	40	1:3.5	10.80
3.0	30	40	1:4	10.27
3.0	30	40	1:5	15.70

In the small-scale electrolysis experiments, keeping the spacing between anode and catode fixed at 30 mm and a voltage of 2.8-3.0 V, a black sticky deposit was obtained. An optimum voltage for obtaining a fine zinc dust deposit was 3.4-3.6 V. The effect of applied current density on the shape of the deposit was also studied. At current densities ranging from 100 A/m² to 900 A/m² the shape of the deposit was examined at 40 times magnification under a microscope. At low current densities (100-200 A/m²) a grey coloured, slimy deposit was obtained while at higher current density (400-900 A/m²), the deposit consisted of large flakes or needles. A current density of 250-350 A/m² was maintained in all following experiments, as the deposit obtained under this condition was a fine zinc dust powder. Table VII gives the chemical analysis of an average zinc dust deposit.

Table VII. Chemical Analysis of Zinc Dust Deposit

Constituent	Assay (percentage)
Total zinc	99.5700
Metallic zinc	97.3000
Lead	0.0250
Copper	0.0010
Iron	0.0080
Cadmium	0.0001

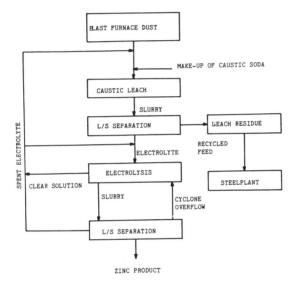

Figure 1. Simplified General Process Flowsheet

Conclusion

The caustic leach-electrowining process appears to be a technically feasible method of recovering zinc from blast furnace dust. Blast furnace dust can be leached very successfully in 30 wt.% sodium hydroxide solution for 3h at 40°C with liquid/solid ratio of 1:5 for segregation of zinc from iron-rich blast furnace dust. The sodium zincate solution when electrolysed at 3.4-3.6 V and 250-350 A/m² gives a fine zinc dust deposition. Recycling of pelletized dust and residues is possible.

The proposed process is flexible in scale and can be applied to small, medium or large-scale operations. It is relatively clean in terms of work-place and ambient air emission. Liquid-solid separation is however, somewhat difficult with concentrated sodium hydroxide solution and very fine solids but can be achieved with conventional equipment.

Magnetically-induced decantation may eventually offer some improvements in this respect. Solid wastes, if any, are less toxic than the feed material but bleed-off liquid may have to be treated before discharge.

References

1. Krishann,E.R., "Recovery of Heavy Metals from Electric Arc Furnace Steelmaking Dust" (paper presented at the Summer National AIChE Meeting, Cleveland, Ohio, Aug. 29-Sept. 1, 1982).
2. Hissel, J., Frenay,J., and Herman,J., "Pilot Study of a Caustic Soda Treatment Process for Reducing the Zinc and Lead Content of Waste Products from Iron-and Steel-Making Processes" (paper presented at the First Process Technology Conference at Washington, D.C.,1980).
3. Freany,J.M., and Hissel,J., "Zinc and Lead Recovery from Iron - and steel Making Dusts; Pilot Study of a Caustic Soda Process" (paper presented at TMS-AIME Annual Meeting, Los Angeles, Feb. 1984).
4. Ojima,Y., Ishikawa,Y., and Yasukawa,M., "Production of Zinc Oxide for Zinc Smelting Process from Steelmaking Dusts at Shisaka Works" (paper presented at TMS-AIME Annual Meeting, Los Angeles, Feb. 1984).
5. Adams,C., "Recycling of Steel Plant Waste Oxides-A Review", (CANMET Report 79-73, March 1979).
6. Maczek,H., and Kola,R., "Recovery of Zinc and Lead from Electric-Furnace Steelmaking Dust at Berzelius" Jour. of Metals, Jan. (1980), 62-68.
7. Higley,L.W.Jr., and Fine,M.M., "Electric Furnace Steelmaking Dust - zinc as raw material", (Rep.Invest.U.S. Bur.Mines 8209, 15, 1977).
8. U.S.Patent 4 071 357, 1978.
9-11. U.K.Patents, 1 600 287; 1 600 022; 1 568 362.
12. Thomas, B.K., and Fray,D.J., Leaching of Oxidic Zinc materials with chlorine hydrate, Metall, Trans., 12B (1981), 281-286.
13. Calusary,A., Electrodeposition of Metal Powders,(Amsterdam, Elseiver, 1979), 215-234.
14. Mantell,C., Electrochemical Engineering, (New York, NY: Mc Graw Hill Publishing Companx, 1960), 192-210.

EXTRACTION OF ZINC FROM COMPLEX POLYMETALLIC SULPHIDE CONCENTRATES

V. N. Misra

Kalgoorlie Metallurgical Laboratory
Chemistry Centre, Department of Mines
PO Box 881, Kalgoorlie, 6430, Western Australia

Abstract

The complex polymetallic sulphide ores, which are in great abundance in various parts of the world, seem to be less responsive to conventional processing techniques because of their peculiar mineralogy. In view of the scanty information available, the present study was initiated to investigate the kinetics and mechanism of selective pressure leaching of a complex polymetallic sulphide concentrate at 120-150°C and oxygen pressure 800-2000kpa in dilute sulphuric acid solutions for 15 to 150 minutes to extract 95% of the zinc in the pregnant solution while most of the lead and silver remained in the residue along with most of the pyrite.

This selective pressure leaching process appeared to obey the diffusion controlled mechanism with an apparent activation energy of 55kJ/mole. The effects of concentration of the leachant, temperature and time of leaching, particle size, oxygen pressure and stirring on the dissolution rates were investigated. The experimental results of the present study suggest the viability of a process in which $H_2SO_4+O_2$ is used as leachant.

Introduction

An extensive computer literature search indicates that complex deposits, which are in finely disseminated form, seem to be less responsive to conventional processing techniques (1, 2, 3). The dilemma posed by these ores is that the production of separate concentrates of acceptable grade involves a heavy loss of valuable minerals to the tailings, while the production of bulk concentrate with reasonable mill recovery involves a similar loss of values to smelter slags and residues. Also cost intensive fine grinding of these ores is essential to liberate the individual minerals; however under the conditions of very fine particle size, flotation kinetics and overall industrial flotation performance become extremely poor.

There is no technology available today to obtain significantly higher concentrate grade and recoveries from complex sulphide ores. A new metallurgical process, suitable for these finely disseminated (with considerable intergrowth), ores, and yielding high metal recoveries at moderate operating cost, must therefore be developed (4, 9, 10). The metallurgical problems associated with complex sulphide ores may be tackled by adopting pyrometallurgical and/or hydrometallurgical methods. The former processes can be considered if the sale of sulphuric acid is economically feasible, but no single process will successfully treat these polymetallic concentrates. Hydro-metallurgical processing appears to be the solution particularly in areas where the sale of the by-product acid is not possible and where it is becoming increasingly difficult to meet environmental standards (11, 12). In this latter process, it would be possible to selectively dissolve sphalerite, galena and chalcopyrite leaving pyrite virtually unattacked. Further, operating parameters could be selected such that, during the treatment, sulphide sulphur attached to the base metal sulphides could be converted into elemental sulphur. This would reduce waste disposal problems due to sulphate ions in aqueous effluents.

That enormous interest has been aroused in the treatment of complex sulphide ores is obvious from the number of conferences held on this topic and the numerous other research publications appearing to date (4-21). A variety of leachants have been suggested, e.g., $HCl+O_2$, $H_2SO_4+O_2$, $FeCl_3$ solutions, $Fe_2(SO_4)_3$ solutions and $CuCl_2$ solutions. Out of the various processes proposed, only two processes viz Sherritt Gordon in Canada (14, 15, 18, 21) and Minemet Recherche in France (13), have received considerable attention. At Sherritt Gordon, the concentrate is leached in sulphuric acid and air under pressure at a high temperature for a fixed period. Zinc and copper are selectively dissolved whereas lead is converted into lead sulphate and lead jarosite [$Pb_{0.5} Fe_3(SO_4)_2 (OH)_6$] and joins the residue along with the pyrite; any iron minerals that go into solution are reprecipitated. After filtration, the pregnant solution is purified and then electrolyzed to extract zinc metal; and the acid produced during electrolysis is recycled to the leaching step. Pyrite and elemental sulphur can be separated from the residue by flotation; however, the extraction of lead from the jarosite sets an acute problem since lead in this form is insoluble in the commonly used reagents.

In the Minemet process, the concentrate is leached in a solution of $CuCl_2$ (40 g/l Cu^{2+}) near the boiling point and at atmospheric pressure. The non-ferrous metal sulphides (ZnS, PbS, $CuFeS_2$) go into solution as chlorides, pyrite is unattacked, and elemental sulphur is generated.

In order to keep CuCl, PbCl$_2$, and AgCl in solution, NaCl of the order of 250 g/l is added to the leachant. The solution is purified by precipitation FeOOH at pH 2.6 with air injection. Lead and silver are recovered by cementation with zinc and copper metals respectively. Thus the final solution will comprise of ZnCl$_2$, CuCl, CuCl$_2$, and NaCl. Zinc is then extracted by diethyl hexyl phosphoric acid and is stripped from the organic phase by H$_2$SO$_4$; the strip solution is electrolyzed in the conventional way. The leachant is regenerated by oxidation of acidified CuCl. In this process, no lead jarosite is generated, but there are a few other drawbacks such as:

a) The precipitation of iron as an extra step.

b) The necessity of separating the complexed cuprous chloride from zinc chloride.

c) Indispensability of adding a large quantity of sodium chloride to keep cuprous chloride in solution.

d) Requirement of oxidizing cuprous chloride to cupric chloride for recycle.

The process proposed by Mizoguchi and Habashi (12) is similar to the Sheritt Gordon Process (21) in which sulphuric acid is replaced by hydrochlric acid to minimize the presence of sulphate ions in solution. This restricts the formation of lead jarosite. The main drawback of this process is that the direct recovery of zinc from chloride medium is not as yet technically advanced.

In the present study, a series of tests were conducted to investigate the leaching behaviour of a natural complex polymetallic sulphide ore in sulphuric acid/oxygen leachant with a view to understand the basic kinetics of zinc extraction reaction.

Experimental Details

Materials

Complex sulphide ores of Australian origin was used in the study. After the ore had been ground to -75µm bulk concentrate was prepared in the laboratory by flotation. The chemical analyses of the ore and bulk concentrate are given in Table 1. Mineralogical examination showed that the principal minerals, sphalerite, chalcopyrite and galena, were finely laminated, extremely fine grained and intimately intergrown with non-economic minerals such as dolomite and pyrite.

Table I. Chemical Analysis of Complex Sulphide Ore and Bulk Concentrate, Wt%

Elements	Complex Sulphide Ore	Bulk Concentrate Produced by Flotation
Zn	8.6	18.0
Pb	3.5	14.0
Cu	0.18	0.4
Ag	32.0 g/t	96.0 g/t

The chemicals used in the experiments were all of reagent grade, and distilled water was used to prepare the solutions. Sodium lignin sulphonate was used as a surface active agent to ensure rapid and complete leaching. Dilute sulphuric acid was used as the leachant.

Apparatus

Leaching experiments were conducted in a 600ml titanium autoclave with a stirring speed of 1000 rpm supplied by Parr Instrument Company, Moline, Illinois. The reactor head possessed three valves for gas inlet, gas release and liquid sampling. The gas inlet valve was connected to oxygen gas tanks. The liquid sampling valve was attached to the same fitting with gas inlet valve and connected to the dip tube immersed in the solution. With this arrangement, a sample representative of the leach solution could be taken. X-Ray diffraction and atomic absorption spectrometry were used for identification of phases in the residue, and for analysis of lead, zinc, silver, copper and iron in residues and the pregnant solution. The pH of the solution was measured with a Fisher pH meter.

Experimental Procedure

Known weights of the solid sample (30g) were charged to 200ml of lixiviant maintained at the desired temperature and oxygen pressure inside the pressure reactor. The oxygen pressure reported here is the difference between the total pressure before and after the oxygen introduction. At predetermined intervals 5-cm^3 aliquots were withdrawn for analysis. At the end of each experiment, heating of the autoclave was shut off, the oxygen supply was stopped, and the reaction mixture was quenched to room temperature by introducing cold water in the cooling coils. The autoclave was emptied and its contents filtered using a vacuum filter. The solution and washings were adjusted to one litre. Leach residues were dried for two hours at 80°C and then analysed for their contents. The clear filtrate was further purified by precipitating iron and recovering metal such as copper and cadmium by cementation. The purified pregnant liquor was concentrated and electrolysed to obtain zinc.

Chemical Analysis

Zinc was determined by chelatometry with EDTA, using xylenol orange as indicator. Sulphate ion was determined gravimetrically as $BaSO_4$. The elemental sulphur of the leached residues was determined by the Soxhlet method using carbon disulphide as a solvent. Acetate-soluble lead in the leach residue (non-sulphide lead) was determined by heating 3g of residue with 50ml of 20% ammonium acetate solution followed by chelatometry with EDTA, using xylenol orange as indicator. In some cases, usually at high acidity or in the early stages of leaching, crystals of $PbCl_2$ formed in the leach solution on standing. The determination of lead in solution was therefore carried out in the following fashion. The pregnant solution was cooled to about 10°C, and the crystals formed were filtered off, dissolved in 20ml of 20% ammonium acetate and the lead was determined in filtrate and in the ammonium acetate solution, the sum represents lead in solution. Residues were examined by the optical microscope and by X-ray diffraction. In some cases the residue was extracted by CS_2 to remove elemental sulphur, then by ammonium acetate solution to remove $PbSO_4$, then subjected to decantation and gravity separation in water to separate the iron oxide fraction from the unreacted sulphide; such a method was found to be simple and effective in isolating the products of the reaction.

The influence of leachant concentration (0.5-2.0N), temperature (120°-150°C), time of leaching (15-150 min.), particle size (10-100μm) and oxygen pressure (800-2000kPa) on the leaching efficiency of the zinc were investigated.

Results and Discussion

The rate of zinc extraction was studied over the temperature range of 120-150°C. The results, shown in Figure 1, confirm that increasing temperature increases the leaching rate.

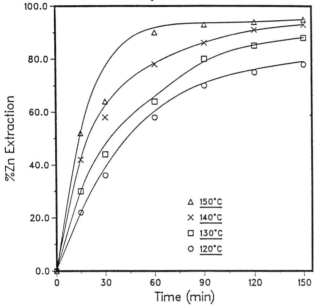

Figure 1. Effect of temperature on the recovery of zinc

Figure 2 shows the results of experiments conducted with dil.H_2SO_4 concentration varying from 0.5 to 2.0N. The increase in sulphuric acid concentration from 0.5 to 1.00N increases the leaching rate, but a further increase tends to level the rate. This observation has been reported by previous investigators who found that the extraction exhibited a maximum point as the acid concentration increased. However, the decreasing effect was not observed in this study, probably because the acid concentration was not high enough to pass the critical point.

The effect of particle size on the rate of zinc extraction was examined by measuring the reaction kinetics of four size fraction.

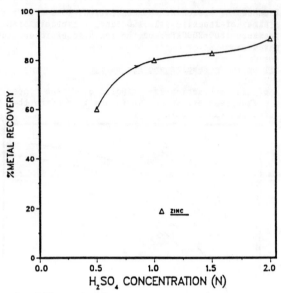

Figure 2: Effect of sulphuric acid concentration on the rate of recovery

Figure 3 demonstrates that the faster reaction rate is observed with the smaller particle size.

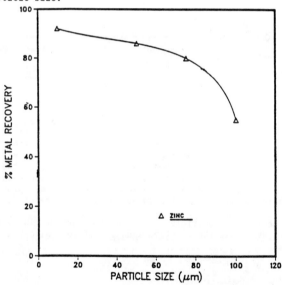

Figure 3: Effect of particle size on the rate of recovery

Figure 4 shows the effect of oxygen pressure on the rate of leaching of sulphide minerals (-75μm size) carried out at 120°C for 60 minutes in 1N H_2SO_4 solution. In a separate experiment it has been observed that leaching in the absence of oxygen is slower than when oxygen is present. X-ray diffraction analysis of the residue showed that at 150°C practically all of the zinc sulphide dissolved while the PbS did not. On the other hand, when leaching was conducted in the presence of oxygen, the recovery of zinc increases rapidly with increased oxygen pressure up to 1000kPa then slower at higher oxygen pressure (1500kPa). As indicated by X-ray diffraction, at 1500kPa oxygen pressure mainly lead precipitation as jarosite was achieved, but at 1000kPa oxygen pressure mainly $PbSO_4$ was precipitated.

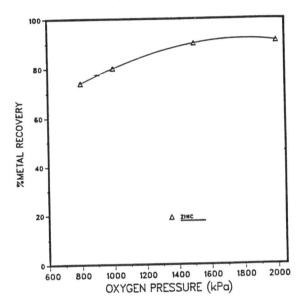

Figure 4: Effect of oxygen pressure on the rate of recovery

The main objective of the leaching experiments was to evaluate the kinetics of leaching complex sulphide concentrates by dil.H_2SO_4+O_2 and to recover zinc metal. The kinetics of dissolution as mentioned earlier, are strongly dependent on factors such as temperature, particle size, oxygen pressure and leachant concentration.

The chemistry of leaching complex sulphide concentrates by sulphuric acid plus oxygen can be represented by a scheme of equations. One of the most important feature of these reactions is the production of elemental sulphur. The reactions occurring are as follows:

$$ZnS + H_2SO_4 + 0.5O_2 \rightarrow ZnSO_4 + H_2O + S° \quad (1)$$

$$PbS + H_2SO_4 + 0.5O_2 \rightarrow PbSO_4 + H_2O + S° \quad (2)$$

$$CuFeS_2 + H_2SO_4 + 1.5O_2 \rightarrow CuSO_4 + FeOOH + 0.5H_2O + 2S° \quad (3)$$

The zinc and copper go into solutions while the lead is transformed into sulphates and lead jarosite and is retained in the residue together with pyrite and elemental sulphur.

In the leaching process, the rate determining step is either chemical reaction at the surface of the solid or diffusion of reactant or product through a layer of residue or insoluble reactions product. A review of these mechanisms is available in the literature (22) rate-control by diffusion through product layer best describes the leaching rates observed in this investigation. If one approximates the ore particles as spheres, then using the 'shrinking core' model, when chemical processes alone are rate limiting, the relationship between the fraction conversion, α, and time t, is:

$$[1-(1-\alpha)^{1/3}] = k_c\, t \quad (4)$$

where k_c is the specific rate constant for chemical control. When the rate of diffusion (k_d) of the reactants through the product layer is rate limiting, then:

$$[3/2 - \alpha - 3/2\,(1-\alpha)^{2/3}] = k_d\, t \quad (5)$$

By plotting these expressions, the values by k_c and k_d can be obtained from the slope of the resulting line.

Figures 5 and 6 show the chemical and diffusion controlled models. The experimental data treated according to the kinetic model correlated well with diffusion control shown in Figure 6.

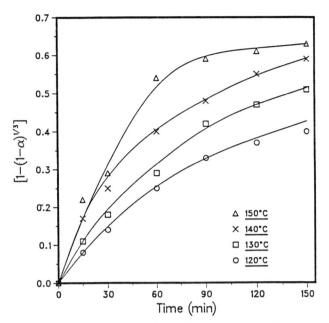

Figure 5: Plot of $[1-(1-\alpha)^{1/3}]$ against time for the extraction of zinc

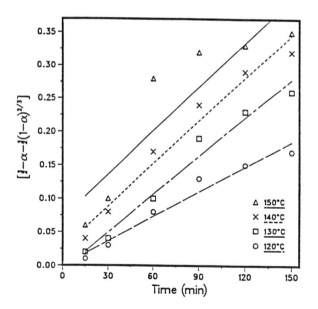

Figure 6: Plot of $(3/2-\alpha-3/2(1-\alpha)^{2/3}]$ vs time for the extraction of zinc

The Arrhenius plot of link vs 1/T is shown in Figure 7.

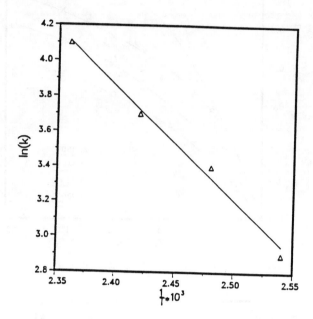

Figure 7: Plot of ln(k) against $\frac{1}{T} * 10^3$

The activation energy obtained from the slope is 55kJ/mole which appears rather high from the usual figures of 16-20kJ/mole for a diffusion controlled process. However, Munoz, Miller and Wadsworth (19) have also reported high activation energies of 70-85kJ/mole for diffusion controlled processes in the leaching of chalcopyrite by $Fe_2(SO_4)_3$. Therefore the relatively high activation energy, obtained in this work for diffusion control mechanism, is consistent with the findings of earlier researchers.

The recovery of zinc with increasing speed of agitation which also suggests the process is diffusion controlled. Acid leaching is a heterogeneous dissolution process and hence agitation helps to keep the particles in suspension for efficient reaction, due to good liquid-solid contract throughout the leaching. Since higher pulp density- necessitates more agitation, a pulp density of 0.15 has been maintained. Figure 1 also shows that zinc recovery increases gradually with increasing time. After about 120 minutes at 150°C and in initial 1N H_2SO_4 solution, more than 95% of the zinc has been dissolved.

The experimental results of the present investigation suggest the viability of a process similar to the Sherritt Gordon Process in which $H_2SO_4+O_2$ is used as leachant. The benefits would be increased by reaction rates and improved lead recovery since the information of lead jarosite is minimised due to judicious selection of parameters. A flowsheet for the proposed treatment method is given in Figure 8. The detailed investigation on the processing of leached residue to recover lead and silver are in progress. Although the proposed method for the extraction of zinc from complex sulphide concentrates seems promising, large scale trials are necessary to evaluate process optimisation and economics.

Conclusions

The present study on the extraction of zinc from complex sulphide concentrates in sulphuric acid clearly demonstrates the following:

a) Complex sulphide concentrates can be dissociated at 150°C and oxygen pressure of 1000kPa in 1N H_2SO_4 solution for 60 minutes to yield 95% of the zinc in the pregnant solution.

b) Leaching of zinc is markedly faster in the presence of oxygen.

c) Leaching of complex suplhides in $H_2SO_4+O_2$ leachant is diffusion controlled through a boundary layer; the process is strongly dependent on temperature and agitations and the activation energy is 55kJ/mole.

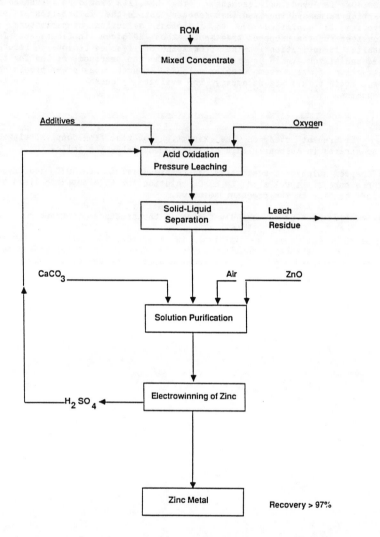

FIGURE 8 - FLOWSHEET FOR ZINC, COPPER & CADMIUM EXTRACTIONS

References

1. D. R. Weir, "Some Recent Developments in Hydrometallurgy" (Can. Metall. Quarterly, 23, (1984), 353-364.

2. G. Barbery et al., "Exploitation of Complex Sulphide Deposits: A Review of Processing Options From Ore to Metals" (Proceedings Complex Sulphides Ore, I.M.M. London, Ed. M J Jones, (1980), 135-150.

3. G Barbery et al., "Complex Sulphides Ores: Processing Options" Proc. Mineral Processing at a Cross Roads, Problems and Prospects (Ed. B.A. Wills and R. W. Barley), (1986), 157-194.

4. V.N. Misra "Oxidative-Pressure Leaching Behaviour of Complex Sulphide Ores" (Paper presented at Western Australian School of Mines Conference 1988, R and D for the Mineral Industry, Kalgoorlie, 5-7 November 1988) 249-255.

5. S. Guy et al., "Solubility of Lead and Zinc Compound in Ammonical Ammonium Sulphate Solutions" Hydrometallurgy, 8, (1982), 251-260.

6. S. Guy et al., "Cupric Chloride Leaching of Complex Cu/Zu/Pb Ore" Hydrometallurgy, 10, (1983), 243-255.

7. "Complex Metallurgy 1978", M. J. Jones (Ed.), Inst. Metall. London, 1978.

8. "Complex Sulphide Ores" Joint Conf. of Inst. Min. Metall. London, and Consiglio Nazionale delle Richerhe, Rome, 1980, M. J. Jones (Ed.) Inst. Min. Metall., London.

9. R. S. Salter, and R. S. Boorman, "New Developments in R.P.C. Sulphation Roast Process - Technology and Application", (Zinc 1983 C.I.M. Metallurgical Society's 13th Annual Hydrometallurgical Meeting August 1983).

10. D. M. Muir et al., "Leaching of the McArthur River Zinc - Lead Sulphide Concentrate in Aqueowschloride", Proc. Australas, Inst. Min. Metall. 259, (1976), 23-35.

11. J. Dutrizac, "Ferric Sulphate Percolation Leaching of Pyritic Zn-Pb-Cu Ore" C.I.M. Bulletin, 72, (1979), 109-118.

12. T. Mizoguchi, and F. Habashi, "The Aqueous Oxidation of Complex Sulphide Concentrate in HCI Acid" Inst. J. of Miner. Processes, 8, (1981), 177-193.

13. J. M. Demarthe, and A. Georgeaux, "Hydrometallurgical Treatment of Complex Sulphides", Complex Metallurgy 1978, M. J. Jones (Ed.), Inst. Min. Metall. London, 113-120.

14. G. L. Bolton, N. Zubrycki, and H. Veltman, "Pressure Leaching Process for Complex Zinc-Lead Concentrate", J. Laskowski (Ed.), Intern. Min. Proc. Congres, Warsaw, (1979), 581-607.

15. V. N. Mackiw, and H. Veltman, "Recovery of Zinc and Lead From Complex Low Grade Sulphide Concentrate by Acid Pressure Leaching" C.I.M. Bulletin, 70, (1967), 80-85.

16. F. S. Wong, et al., "Recovery of Zinc From Sulphide Flotation Tailings" Congress, Metall. E. L. Fielding, and A.R. Gordon, Ed. 4, (1986), 13th C.M.M.I., 187-195.

17. M. C Shaji, et al., "Hydro-electro Metallurgical Recovery of Cu, Pb and Zinc From Complex Sulphide Ores" Trans. Ind. Inst. of Metals, 36, (1983), 400-405.

18. E. G. Parker, "Oxidation Pressure Leaching of Zinc Concentrates" C.I.M. Bulletin, 74, (1981), 145-150.

19. P. B. Munoz, et al., "Reaction Mechanism for Acidic Ferric Sulphate Leaching of Chalcopyrite" Metall. Trans. B, 10B, (1979), 149-158.

20. P.C. Ruth, et al., "Kinetics of Dissolution of Sulphide Minerals in Ferric Chloride Solution: Application to Complex Sulphide Concentrate" Trans. Inst. Min. Metall. 97, (1988), 159-162

21. H. Veltman, and G. L. Bolton, "Direct Pressure Leaching of Zinc Blends with Simultaneous Production of Elemental Sulphur. A State of-the Art Review", Erzmetall, 33, (1980), 76-83.

22. H. Y. Sohn, and M. E. Wadsworth, Rate Processes of Extractive Metallurgy 1979 Plenum Press, New York.

RELATION BETWEEN THE LEACHABILITY OF ZINC-CONCENTRATES

IN Fe(III)-H_2SO_4 MEDIA AND THEIR ELECTRIC CONDUCTIVITY

X.Y. Xiong and M. Jacob-Dulière

Service de Métallurgie des Métaux Non-Ferreux
Faculté Polytechnique de Mons,
Mons, Belgium

and

J. Sterckx
Metallurgie Hoboken-Overpelt
Overpelt, Belgium

Abstract

In the goethite and hematite processes, prior to iron precipitation, Fe(III) is reduced to Fe(II) by using zinc concentrates. It has been established that the reactivity of the ZnS in the concentrate is not only grain-size dependent but also influenced by other factors such as the iron-content and mineralogical composition of the concentrates. Comparing the reactivity in "reduction conditions" of 15 concentrates of quite different origins, it was possible to classify them into two distinct groups. The groups' different behaviour can be correlated to their semi-conductor properties. The development of a rather simple conductivity measuring equipment permits us to predict the leachability of concentrates in H_2SO_4-$Fe_2(SO_4)_3$-media.

1. Introduction.

Three known processes are applied in the hydrometallurgical zinc refining operation for separating Fe from the zinc-bearing medium: i.e. the jarosite, goethite and hematite processes.
Whereas for jarosite precipitation, trivalent Fe, present in the solution, is the start material, trivalent iron shall first be converted into its bivalent form in the two other processes on account of technical and/or economical reasons.

These processes have already repeatedly been described in literature and we shall therefore not detail them further here.

We wish to tackle one specific problem, related to the required conversion of iron into its bivalent form, which happens on industrial scale in the so-called REDUCTION STEP.

In order to situate this operation in the flowsheet, we give on figure 1 a simplified survey of the goethite flowsheet, as applied in the Overpelt plant. A more detailed review of MHO's Overpelt zinc operation has already been subject of a prior publication (1).

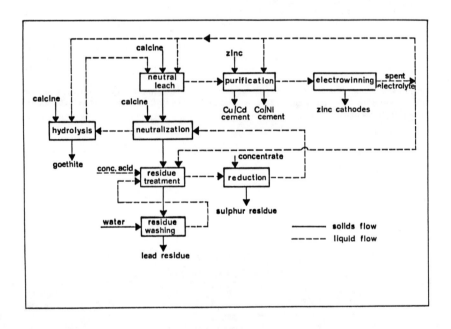

Figure 1 - Flowsheet of the Overpelt hydrometallurgical zinc plant

2. The reduction step and its effect on the economy of the zinc plant

The most appropriate way to convert Fe(III) into Fe(II) in a zinc flowsheet is to make use of Zn sulphide, according to following reaction:

$$Fe_2(SO_4)_3 + ZnS \longrightarrow 2\ FeSO_4 + ZnSO_4 + S$$

and the form in which this ZnS is added in a Zn-plant is obviously the unroasted Zn-concentrate.

The use of concentrate implies, however, for "custom-smelters" such as MHO an additional difficulty in the economic running of the Zn production. It appeared clearly from practical experience that all concentrates are not equally appropriate to carry out this reduction-reaction which might give rise to problems for a smelter such as Overpelt that cannot obtain supplies with a constant composition, since the origin of its input concentrates diverge strongly.

Reactivity of Zn-concentrates in the reduction step - one among the steps of a continuous system - is important indeed on account of the following features:

- Lower reactiviy implies for a given Fe-input and a constant retention time of the Fe-bearing solution in the reactors that a larger quantiy of concentrate has to go through the reduction-step.
  This can lead to an excess charge of both the thickeners and filters used to separate the sulphur-bearing residue, resulting unavoidably in a forced lowering of the flow through the reactors, the thickeners and the filters of said operations and hence into a decreased overall zinc production.

- Moreover, the sulphur-residue obtained, has to be recycled to the roasting unit.
  For a smelter such as Overpelt, this might have serious additional economic consequences, since the goethite process requires an operation with two different types of calcine :
  . The first type is intended for normal processing via the different leaching steps and consists preferably of concentrate mixtures with high content of recoverable secondary metals.
  . The other type acts as a neutraliser in the hydrolysis step and since the final residue shall be considered as lost, it contains as low a Fe- and secondary metals content as possible.
  During the roasting step the iron generates indeed Zn-ferrites, which are not soluble in the "soft" leaching operation of the hydrolysis.
  It appears from experience, however, that concentrates with higher Fe-content are most suitable for the reduction step. Moreover, these concentrates contain some valuable recoverable secondary metals too, which makes it worthwhile to recycle only the sulphur residue obtained with the mixture intended for normal leaching.

Concentrate n°	Zn	Fe	Cu	Pb	Cd	S	SiO_2	CaO	MgO	Al_2O_3	Total	Fe/Zn in weight
1	53.20	9.20	1.00	0.10	0.28	32.20	1.50	0.12	0.14	0.25	97.99	0.17
2	50.50	10.22	0.32	2.35	0.09	32.09	0.90	0.28	0.09		96.87	0.20
3	49.10	8.60	0.14	2.30	0.16	30.80	3.70	0.66	0.38	0.84	96.68	0.18
4	38.50	11.80	0.17	3.10	0.15	37.60	5.60	0.16	0.09	1.39	98.56	0.31
5	52.90	6.40	0.33	1.10	0.26	31.40	4.70	0.60	0.37	0.30	98.36	0.12
6	56.07	7.20	0.04	0.40	0.18	33.90	0.21	0.37	0.14		98.51	0.13
7	52.80	2.60	0.17	1.90	0.20	31.40	1.40	1.97	0.47	0.38	94.79	0.05
8	48.40	8.27	1.90	4.12	0.46	29.60	2.88	1.26	0.15		97.04	0.17
9	61.80	2.00	0.21	1.20	0.15	32.60	0.40	0.50	0.28		99.14	0.03
10	53.80	6.70	0.04	1.60	0.06	34.60	2.10	0.22	0.06	0.29	99.47	0.12
11	53.70	4.22	0.47	1.60	0.13	31.10	3.20	0.80	0.38	0.45	96.75	0.08
12	57.00	5.00	0.08	1.40	0.28	33.00	0.40	0.80	0.10		98.06	0.09
13	51.60	6.43	0.35	3.10	0.19	33.20	2.79	0.22	0.30	0.56	99.24	0.12
14	40.00	14.30	0.55	4.20	0.05	34.40	0.60	0.07	0.06	0.14	94.37	0.36
15	60.80	3.01	0.18	0.82	0.33	31.90	2.20	0.12	0.02		99.38	0.05

Table I : Analytical composition of concentrates (% weight)

Concentrate n°	>150	>125	>88	>62	>44	>31	>22	>16	>11	>7.8	>5.5	>3.9	>2.8	>1.9	S.S.* m^2/cm^3
1	2	4.8	8.8	24.1	39.3	55.4	70.1	79.5	84.3	89.3	95.2	96.9	99.6	100	0.309
2	2.5	2.5	8.6	20.4	35.3	42.0	50.6	65.1	74.2	79.3	86.1	93.1	96.6	100	0.492
3	21.3	21.9	27.3	36.6	52.0	61.8	66.2	80.8	88.0	89.1	93.1	96.1	98.9	100	0.371
4	37.9	40.6	49.0	57.1	63.5	74.6	80.9	88	91.9	91.9	95.0	98.6	99.8	100	0.310
5	12.8	19.3	31.7	32.2	48.9	61.3	71.0	81.2	87.5	92.8	92.8	95.6	99.9	100	0.322
6	16	18.4	33.1	54.0	64.2	70.0	76.6	83.0	90.8	90.8	90.8	97.2	99.6	100	0.300
7	5.2	7.4	15.0	15.6	25.5	42.2	55.3	66.1	74.5	82.9	90.0	93.8	96.9	100	0.472
8	20.4	24.9	37.8	58.4	67.1	71.5	77.9	86.5	92.5	95.2	96.3	96.5	99.6	100	0.259
9	6.2	12.2	30.5	45.7	52.3	66.0	71.6	80.7	87.2	87.2	87.2	92.1	100	100	0.351
10	2.0	8.8	21.4	26.9	40.0	55.0	66.3	74.7	82.0	86.1	89.0	94.3	100	100	0.368
11	6.3	12.4	33.3	38.6	50.2	62.7	68.2	84.0	89.7	89.7	90.6	95.5	98.6	100	0.323
12	7.8	12.6	29.9	51.9	67.4	76.0	76.2	89.2	95.8	95.8	95.8	97.2	100	100	0.218
13	2	8.5	19.0	35.4	49.5	55.0	69.5	73.3	76.3	85.5	89.1	92.8	98	100	0.395
14	0	0	0	1.5	5.7	16.2	28.7	41.7	54.7	68.4	79.6	89.0	94.7	100	0.699
15	6.8	15.9	40.9	49.5	59.8	79.3	83.8	87.8	92.7	96.3	96.3	98.4	100	100	0.204

* Specific surface computed by Microtrac

Table II : Grain size distribution (microns) - % oversize

- Since during this recycling operation a wet product ($\pm$ 50 % humidity) has to be introduced in a thermic process, an increased quantity of sulphur residue to recycle shall have a negative effect on the thermic balance of the furnaces and on the water balance of gas treatment.

On account of these facts, it is therefore obvious that the reactivity of the concentrate, applied in the reduction step, can play an important part in the overall economy of a zinc plant.

It is therefore obvious that we, as custom smelters, have been paying for the last few years full attention to those parameters which influence the reactivity of the concentrates in the reduction-step and are proper to the very concentrates.

3. The concentrates and their behaviour at the reduction step.

The study has a double aim.
- On the one hand, we aimed at developing a simple reactivity test, that should allow us to determine rapidly reactivity of concentrate versus reduction step.
- On the other hand, we also tried to give a more fundamental explanation for the difference stated in the behaviour of the concentrates.

We restricted ourselves to some 15 concentrates of different origins, which give, however, a good idea of the distribution in types of concentrates processed at Overpelt. Their composition is given in table I, their grain-size distribution in table II.

Reactivity test

Research aimed first at finding a chemical comparative test between the various concentrates. Finally we came to a standard-test, starting from an industrial solution leaving the leaching section.

An amount of 0,5 g of concentrate per g of $Fe(III)$ present in the solution is added to 3 litres of this filtered industrial solution; it is then made to react for 4 hours at 68 °C controlling thereby the evolution of the $Fe(III)$ content in the solution.

After 4 hours the residue is filtered and it can be assayed for different elements if the conversion rate of the different sulphides needs to be established.

The test allows to determine the weight of $Fe(III)$ reduced to $Fe(II)$ per kg of dry concentrate.

Reactivity index I_{TV}

The results obtained are compared with a standard concentrate, being a concentrate that was mainly introduced with good result into the industrial reduction step :

$$I_{TV} = \frac{\text{g Fe(III) reduced per g dry concentrate}}{\text{g Fe(III) reduced per g dry reference concentrate}}$$

Table III gives the results of the reactivity test of the 15 concentrates, their reactivity Index I_{TV} and - by way of comparison - their total Fe-content.

Table III. Results of the reactivity rest.

Concentrate n°	Fe total Wt %	I_{TV}	kg Fe(III) reduced per kg of dry concentrate	t residue S per t reduced Fe(III)
1	9.20	103	0.680	0.68
2	10.22	100	0.662	0.63
3	8.60	97	0.645	0.81
4	11.80	86	0.571	1.00
5	6.40	75	0.495	1.21
6	7.20	68	0.450	1.3
7	2.60	65	0.429	1.49
8	8.27	60	0.398	1.51
9	2.00	52	0.346	1.96
10	6.70	50	0.331	2.05
11	4.22	49	0.323	2.09
12	5.00	47	0.308	2.24
13	6.43	41	0.271	2.56
14	14.30	31	0.202	3.44
15	3.01	28	0.188	4.4

Specific surface is supposed to have an important influence on the heterogeneous reduction-reaction rate. Therefore this parameter was further examined in the first place.

The concentrates ground up to 100 % minus 44 µm were subjected to the reactivity test : an example of the result obtained is given in table IV.

It appears from this data that complementary grinding has a strong influence on the reactivity of some concentrates; nevertheless, other parameters also act, due to lasting differences, on this reactivity.

Merely on the basis of the chemical compostion, it is not possible either to give a clear explanation on the difference in behaviour under reduction-circumstances; it is generally stated, however, that concentrates with a rather high Fe-content score better than purer Zn-suphides, but that within the range of concentrates with a high Fe-content, there are still significant differences (see table III).

A possible explanation for this phenomenon was searched in the difference in mineralogic occurence of iron in the concentrates.

Table IV. Evolution of reactivity of concentrates after grinding.

Concentrate n°	as cut		crushed	
	< 44 µm % Wt	kg Fe(III) reduced per kg dry concentrate	< 44 µm % Wt	kg Fe(III) reduced per kg dry concentrate
2	60,7	0,662	100	0,909
7	74,6	0,429	100	0,495
15	40,2	0,188	100	0,767

4. Mineralogic analysis and leaching tests on 10-45 µm fractions

To evidence the role played by iron in the concentrates on their reactivity at leaching, it was first necessary to know the mineralogic composition of the concentrates and particularly the nature, composition and quantity of each iron containing mineral.

The development of a mode of operation (2) enabled to achieve this analysis. The mineralogic analysis of the 15 concentrates under survey are given in table V.

The ponderal ratio of the marmatites, (Zn,Fe)S, varies from 62 to 89%, their iron content ranging from 0.5 to 11.1 % in weight (see table V).

Iron in the concentrates is mainly devided between marmatite and pyrite, as shown in table VI.

To evidence the specific part played by iron in the concentrates, the effect of other parameters had to be restricted; among others, the grain-size of the concentrate and the complexity of the leaching solution.

The study was conducted systematically on the grain-size fraction 10-45 µm of the concentrates, which have first been checked to be sure that neither their chemical nor their mineralogical analysis show a marked difference with regard to those of the concentrates as cut (3).

The leaching tests on the 10-45 µm fractions were achieved according to a mode of operation similar to that of the industrial laboratory; however, the initial solution - including but Fe(III), Zn(II) and H_2SO_4, respectively at the rate of 20, 80 and 90 g per liter - was synthesized.

Results are expressed according to index I(10-45 µm):

$$I(10\text{-}45\ \mu m) = \frac{\text{g of Fe(III) reduced per g dry 10-45 um fraction}}{\text{g of Fe(III) reduced per g dry reference concentrate}}$$

Concentrate	Marmatite	Pyrite	CuFeS2	PbSO4	ZnSO4	Fe2(SO4)3	S	gangue	Total	Fe/(Zn+Fe+S) in marmatite
1	83,33	3,82	2,89		5,80	0,14	0,64	2,01	98,63	7,78
2	80,66	7,82	0,92	3,44	4,12	0,82	0,62	1,27	99,67	7,53
3	82,89	3,29		3,37	2,05	0,57	0,22	5,58	97,97	8,34
4	64,74	9,67		4,54	2,57	0,46	8,96	7,24	98,18	11,08
5	84,34	2,94		1,61	2,20	0,07	0,22	5,97	97,35	5,94
6	88,40	4,77		0,60	3,04	1,36	0,17	0,72	99,06	5,20
7	82,24	2,88		2,78	3,93	0,72	0,04	4,22	98,31	1,29
8	75,38	3,61		5,48	6,03	3,23	0,17	0,15	98,34	6,46
9	91,38	2,17		1,39*	2,02	1,72	0,08	1,18	99,94	0,56
10	79,74	11,86		1,55	2,00	0,93	0,09	2,67	98,84	1,15
11	80,87	4,85		2,34	3,48	0,11	0,20	4,83	97,38	2,39
12	87,33	4,17		2,05	3,04	1,50	0,27	1,30	99,66	3,02
13	74,48	10,44		4,54	2,77	0,32	0,26	3,87	97,18	1,99
14	62,20	17,34	1,58	6,15	7,85	2,26	0,21	0,87	98,46	8,05
15	93,07	0,00		1,30	2,15	0,11	0,02	2,34	98,99	3,23

* PbS

Table V. Mineralogic analysis of concentrates (weight %)

Table VI. % Weight of concentrates in the form of Fe, of Fe included in marmatite and of Fe included in pyrite.

Concentrate n°	Total Fe % weight	Fe marmatite % weight	Fe pyrite % weight
1	9.20	6.50	1.78
2	10.22	7.09	2.62
3	8.60	6.91	1.53
4	11.80	7.17	4.50
5	6.40	5.01	1.37
6	7.20	4.60	2.22
7	2.60	1.06	1.31
8	8.27	4.87	1.68
9	2.00	0.51	1.01
10	6.70	0.92	5.52
11	4.22	1.93	2.26
12	5.00	2.64	1.94
13	6.43	1.48	4.86
14	14.30	5.12	8.07
15	3.01	2.98	0.00

Figure 2 compares these indexes I(10-45 µm) to the reactivity indexes I_{TV} found in the industrial laboratory on the corresponding concentrates.

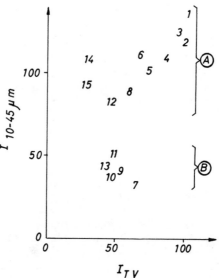

Figure 2 - Reactivity indexes of concentrates and their fraction 10-45 µm

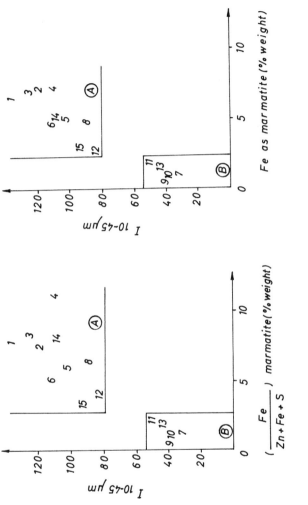

Figure 3 - Effect of the iron content of marmatites on the 10-15 μm fractions reactivity

Figure 4 - Effect of the iron include in marmatite on the 10-45 μm fractions reactivity

Besides occasional considerable differences, a collection of the concentrates into two groups is noted : group A with $I(10-45~\mu m)$ systematically higher than their corresponding I_{TV} and group B (concentrates n° 7, 9, 10, 11 and 13) with an opposite behaviour.

When comparing the reactivities $I(10-45~\mu m)$ with the iron content in marmatites, (figure 3) and the percentage of iron in the shape of marmatite (figure 4), a trend to improved reactivity with an increased iron content is noted in both cases; the most reactive fractions (group A) have marmatites with more than 3 % in weight of iron.

Complementary surveys (3) made on the leaching tests of the 0–45 μm fractions show that if the dissolved zinc quantity varies strongly with the composition of their marmatite, the pyrite of the concentrates remains systematically but little attacked. The composition of the marmatites of the leaching residues remains comparable with that of the starting concentrates : zinc and iron in the marmatites are proportionally dissolved to the Zn/Fe ratio in the marmatites of the treated concentrate. The formed orthorhombic sulphur crystallizes at the surface of the marmatite grains.

5. Measurement of electric conductivity of the 10–45 μm fractions

As the iron of the marmatites influences the leaching reactivity of the 10–45 μm fraction, it can be hoped that the electric conductivity of these materials develops in parallel with this reactivity.

Indeed, marmatite is a semiconductor of the p-type and the association of a growing proportion of iron in marmatite leads to an increased electric conductivity (4) (5).

We developed a measurement test of electric conductivity of the concentrates in order to examine whether the evolution of the measurement on these mineralogically heterogeneous materials remains associated with the high iron content of their marmatite.

A. Description of the electric conductivity test

To determine the electric conductivity of these powdery materials, measurement takes place under compression on a concentrate-pellet, that had first been dried for 10 hours at 80° C.

Measurement is made at controlled temperature and under a 0.55 to 0.65 MPa compression, ensuring a good measurement reproducibility and an insignificant deformation of the crystals.

The mould used (figure 5), made of EPOXY resin, includes a metallic resistance to heat the sample. The temperature is controlled by means of a thermometer placed in the upper part of the piston. After passing a continuous high-tension (1000 V) current, drops of potential V1 and V2 (in V) are noted, enabling to compute the electric conductivity σ (in $ohm^{-1}.cm^{-1}$) of the pellets with a thickness L (in cm) and a section S (in cm^2) :

$$\sigma = \frac{L}{(\frac{V1}{V2}+1).10^{-6}.S}$$

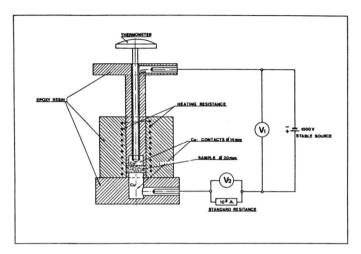

Figure 5 - Apparatus for the measurement for electric conductivity of the concentrates

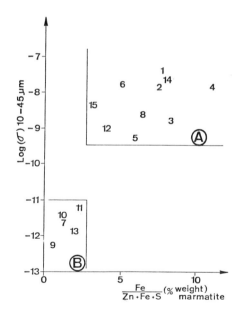

Figure 6 - Effect of the iron content of marmatites on the 10-45 μm fractions electric conductivity (in $ohm^{-1} \cdot cm^{-1}$ at T = 25° C and P = 0,62 MPa)

B. Results of the electric conductivity tests.

Shown on a diagram (figure 6) with in abscissa the iron content of the marmatites and in ordinate the electric conductivity (logarithmic scale), the 10-45 um fractions of the 15 concentrates are clearly classified in two groups, identified previously.

Both groups A and B differ from each other not only by their chemical reactivity (figure 2), and by their iron-marmatite content (figures 3 and 4), but also - and even more distinctly - by the difference in magnitude of their electric conductivity. The dividing line between both groups can approximatively be found again for marmatites at about 3 % in weight of iron.

On the other hand, both groups of concentrates differ neither by their pyrite proportion (figures 7a and 7b), nor by their total iron content (figures 7c and 7d).

The change in electric conductivity of the 10-45 μm fractions seems thus to depend mainly on the quality (and the quantity) of their marmatite.

Iron should be mainly present in the marmatite in the form of Fe(II) by substituting Zn(II) in the face-centred cubic of the blende and in a much smaller proportion, in the form of Fe(III) in interstitial position in the blende (6) (7).

Spectra of electronic spin resonance (E.S.R.) recorded on concentrates of both groups (figure 8) show the much more marked presence of Fe(III) in the concentrates of the A group than in those of the B group.

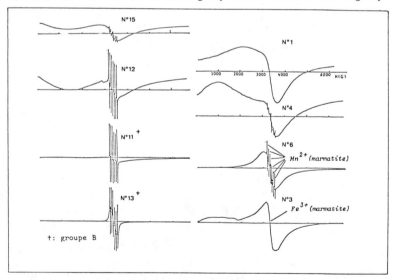

Figure 8 - R.S.E. spectra on 10-15 um fractions (T=300°K, v=9,50 GHz)

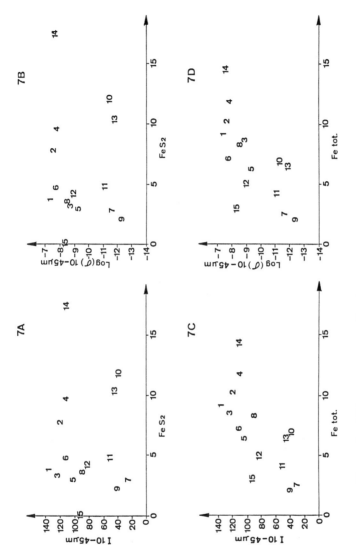

Figure 7 – Effect of pyrite (% Wt) and of Fe total (% Wt) on the 10-45 μm fractions reactivity and electric conductivity (in ohm^{-1}.cm^{-1} at T=25°C and P=0,62 MPa).

By applying the hypothesis of "hopping conduction" (8), we explain the existence of both groups of concentrates as follows : higher Fe(III) contents of the A concentrates increase the probability of creating positive holes due to the hopping of electrons from Fe(II) to Fe(III); the higher the concentration of holes, the higher the electric conductivity.

Further to the electrochemistry of semiconductors, an increased concentration of positively loaded carriers (free holes) of a p-type semiconductor favours the anodic reaction and unfavours the cathodic reaction (9).

anodic reaction : $ZnS = Zn^{++} + S + 2e-$

cathodic reaction: $2 Fe^{+++} + 2e- = 2 Fe^{++}$

Increase in reactivity with the electric conductivity of the concentrates justifies the direct S-production mechanism during leaching of the marmatite.

6. Effect on the specific surface

Figure 9 shows that for 10-45 μm fractions with adjoining electric conductivity, the reactivity increases with the specific surface.

Here is a combined influence of both criteria : electric conductivity and specific surface of the concentrates on their reactivity. Electric conductivity is, however, of primary importance : at low conductivity, the specific surface has but little influence on the reaction rate; at high conductivity, on the contrary, the specific surface has a marked influence.

It can be hoped that the electric conductivity measurement method described under 5.A might serve as a simple and fast test to characterize the more or less good reactivity when leaching the concentrates with adjoining specific surfaces.

It should be further examined whether this conductivity test remains significant when applied to concentrates as cut, with a less homogeneous grain size distribution and a more variable specific surface.

Results from the electric conductivity tests made on concentrates as cut are shown in figure 10 : a clear distinction is found between conductivities of the two groups of concentrates.

Figures 11.a and 11.b show the evolution of the reactivity index of the concentrates as cut and their fractions 10-45 um in function of their electric conductivity.

Only the concentrates as cut 12, 14 and 15 seem, with their low I_{TV} index, not to meet the expected influence of their strong electric conductivity.

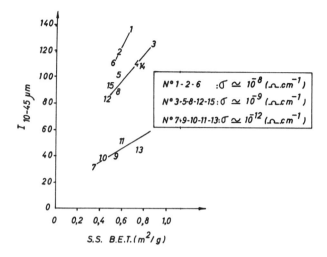

Figure 9 — Effect of specific surface on the leachability of 10-45 μm fractions with adjoining electric conductivity

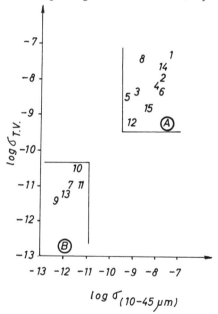

Figure 10 — Electric conductivities (in $ohm^{-1}.cm^{-1}$ at 25°C and 0,62 MPa) of concentrates and their 10-45 μm fraction

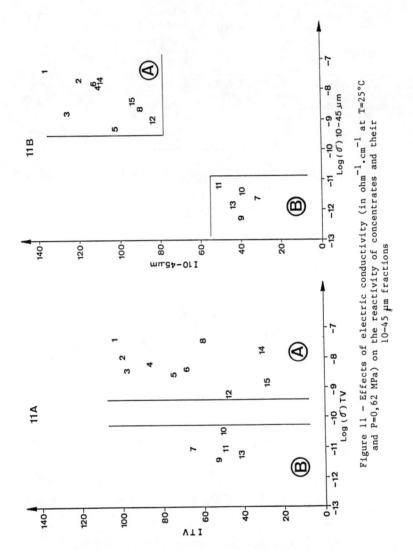

Figure 11 - Effects of electric conductivity (in $ohm^{-1}.cm^{-1}$ at T=25°C and P=0,62 MPa) on the reactivity of concentrates and their 10-45 μm fractions

The low reactivity of the as cut concentrate 14 can be explained by its trend to float during the leaching tests : stirring secures no good mixing of the pulp, this behaviour can be justified:
- either by the importance of the ultra-fine grain-size fraction (see table II)
- or by the retention of tensio-active agents, originating from flotation, on the very fine particles of this concentrate.

The concentrates as cut 12 and 15 are characterized by a small specific surface compared with the other concentrates (table II). This small specific surface justifies their low index I_{TV}.

When examining (figure 12) the evolution of these indexes I_{TV} in function of the specific surface (computed in m^2/cm^3), the favourable effect of the increased specific surface on the reactivities of the concentrates of adjoining electric conductivity is confirmed (ex : concentrates 15, 12, 8, 6, 5, 3); the more marked effect of the specific surface on concentrates of the A group is confirmed too.

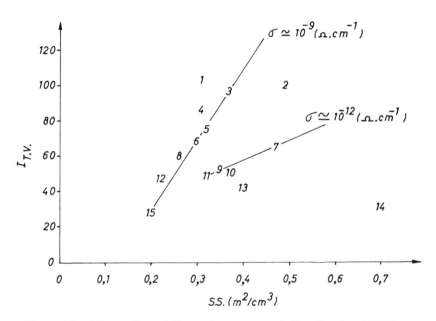

Figure 12 - Effect of specific surface on reactivity of concentrates with adjoining electric conductivity.

This finding is in conformity with the evolution of the reactivities of the concentrates after grinding, shown on table IV: n° 2 and n° 15 concentrates belong to the group with a high conductivity (marmatite with a high iron content), the n° 7 to the other group

7. Conclusions

The proven effect of the iron content of the marmatites (intrinsic effect) and the specific surface (extrinsic effect) on the reactivity of dissolution of the sulphured zinc concentrates can be used to select reactive concentrates and their possible preconditionning by additional grinding.

The proposed simple and fast test leading to the determination of electric conductivity in the concentrates, together with the classic test on chemical reactivity, enables to identify the concentrates with a "high electric conductivity" which are all potentially reactive and to foresee, among all concentrates, those of which the leaching reactivity could be markedly improved by an additional grinding.

This test on electric conductivity of the concentrates constitutes thus an important contribution for the zinc industry, not only for the hematite and goethite ways, but also for the direct pressure leaching of blendes.

Acknowledgement

The authors gratefully acknowledge the General Management of Metallurgie Hoboken-Overpelt for permission to publish this paper.

References

1. E. Van den Neste, "Metallurgie Hoboken-Overpelt's ZincElectrowinning Plant", 6th Annual C.I.M. Hydrometallurgical Meeting, C.I.M. Bulletin, August 1977, 173-185.

2. X.Y. Xiong and M. Jacob-Dulière, "Détermination de la Teneur en Fer des Marmatites des Concentrés Sulfurés de Zinc", ATB Métallurgie, XXVIII, (3-4)(1988).

3. X.Y. Xiong et M. Jacob-Dulière, Influence du Fer des Concentrés Sulfurés de Zinc sur leur Semi-conductibilité Electrique et leur Réactivité Chimique", ATB Métallurgie, XXVIII, (3-4)(1988).

4. M. Telkes, "Thermoelectric Power and Electrical Resistivity of Minerals, Amer. Mineral., 35, (1950), 536-555.

5. J.D. Keys, J. L. Horwood, T.M. Baleshta, L.J. Cabri and D.C. Harris "Iron-iron Interaction in Iron-containing Zinc Sulphide", The Canadian Mineralogist, 9, (1967) p. 453-467.

6. P.G. Manning, "Absorption Spectra of Fe(III) in Octahedral Sites in Sphalerite", The Canadian Mineralogist, 9 (1967) 57-64.

7. L.J. Cabri, "Density Determinations : Accuracy and Application to Sphalerite Stoechiometry", The American Mineralogist, 54 (1969), 539-548.

8. R.T. Shuey, "Semiconducting Ore Minerals", (Amsterdam : Elsevier Scientific Publishing Company, 1974)

9. X.Y. Xiong, "Comportement du Fer de Concentrés Sulfurés de Zinc en Hydrométallurgie", (Thèse de Doctorat en Sciences Appliquées, Faculté Polytechnique de Mons, Belgium, 1988).

ZINC RECOVERY BY SOLVENT EXTRACTION

A. Selke and D. de Juan Garcia

Non-Ferrous Department Lurgi GmbH, Frankfurt, FRG
Research and Process Development Department
Espanola del Zinc S.A., Cartagena, Spain

Abstract

The worldwide growing interest in treating low grade primary raw materials and the need to recycle, for environmental reasons, secondary raw materials has led to the development of several new processes for the economical recovery of valuable metals.

In the conventional hydrometallurgical route for the production of zinc i.e. the roast-leaching-electrowinning process, waste water, drosses and residues are generated with appreciable amounts of valuable metals such as zinc, copper and cadmium.

In April 1988, a new solvent extraction plant for the recovery of zinc but also copper and cadmium from waste materials yieled by the existing electrolytic zinc plant went into operation at Espanola del Zinc S.A., Cartagena, Spain.

Brief History

In May 1984 the first paper introducing the results obtained on laboratory scale for the recovery of zinc from leaching residues of the electrolytic zinc plant Espanola del Zinc S.A., Cartagena, Spain was presented to the International Mining and Metallurgy Convention held in Barcelona, Spain (1).

Between May 1984 and October 1985 further tests on a continuous basis were carried out in a mobile pilot plant with a throughput of 400 liters/hour.

In 1986 the engineering work started and finally in April 1988 the new solvent extraction plant with a design capacity of 6,000 tpy cathodic zinc started its operation.

The papers 2, 3, 4 and 7 presented in various occasions, described the stepwise development of the EXCINRES process.

Process

The process EXCINRES which means the Extraction of Zinc from Residues is protected by the Spanish patent No. 526.618 (5) a further patent is under application (6).

This process has been developed for the recovery of zinc from diluted sulphate solutions or waste waters as they are normally discarded in electrolytic zinc plants or other industries using sulphate metals salts. Copper and cadmium are also recovered as cementates which can be easily recycled to the existing Cd-plant.

At Espanola del Zinc S.A. in Cartagena, Spain this process, which is fully integrated in the existing electrolytic zinc plant, is mainly used for the recovery of the water soluble zinc ($ZnSO_4$) from stockpiled and fresh leaching residues.

Due to the selectivity of the zinc extraction offered by this process, it is also possible to treat secondary raw materials with higher amounts of harmful impurities than in the conventional roast-leaching-electrowinning process.

The extracted zinc which is stripped from the solvent by return cell acid is plated on conventional aluminum cathodes in the existing zinc cellhouse. Copper and cadmium, if present, are precipitated from the return raffinate by cementation with metallic zinc dross from the cathode zinc melting plant.

The EXCINRES process consists mainly of the following steps:

- the residues leaching with return raffinate from the zinc extraction step and others waste effluents,
- the solid-liquid separation of the diluted zinc bearing solution from the insoluble residues,
- the solvent extraction of zinc with Di-(2-ethylhexyl) phosphoric acid (D2EPHA),
- the solvent scrubbing with zinc sulphate solution,
- the stripping of the extracted zinc from the solvent with return cell acid from the cellhouse,
- the solvent regeneration with hydrochloric acid (HCl)

- the extraction of iron with Tri-butyl-phosphate (TBP) and the recovery of hydrochloric acid and
- precipitation of copper and cadmium (if present) from the return raffinate.

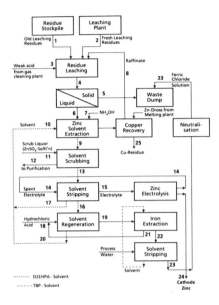

Figure 1 - Block Diagram of Espanola del Zinc

Based on the operating results from April 1988 until March 1989 the main flows of the zinc circuit as indicated in Figure 1 were as follows:

No.	Description	tpd	m3pd	Zn (%)	Zn (gpl)
1	Stockpile Residue	110.0	--	3.5 - 6.0	--
2	Leaching Residue	240.0	--	2.5 - 5.0	--
4	Zn Solution	--	1,970.0	--	5.0 - 10.0
5	Residue to Dump	330.0	--	max. 0.1	--
7	NH4OH Solution	--	35.0	--	--
8	Raffinate	--	2,000.0	--	max. 0.1
9	Loaded Solvent I	--	3,000.0	--	3.0 - 6.0
10	Unloaded Solvent I	--	3,000.0	--	max. 0.040
11	Scrub Liquor I	--	100.0	--	max. 160.0
12	Scrub Liquor II	--	100.0	--	max. 130.0
13	Loaded Solvent II	--	3,000.0	--	3.0 - 6.0
14	Spent Electrolyte	--	850.0	--	50.0 - 60.0
15	Electrolyte	--	850.0	--	60.0 - 80.0
16	Unloaded Solvent II	--	1,500.0	--	max. 0.040
24	Cathodic Zinc SHG	18.0	--	+ 99.99	--

The EXCINRES process is a classical solvent extraction process adapted to the recovery of zinc from leaching residues yielded by a conventional electrolytic zinc plant.

The stockpiled residues are reclaimed from the storage dump by pay loaders and unloaded to a spiral classifier before being fed to a wet ball mill, where they are disintegrated before being pumped to the leaching section. The fresh leaching residues yielded from the vacuum drum filters are collected in a repulping tank.

Both residues are leached at pH 2.0 with return raffinate from the solvent extraction section in agitator tanks where the zinc sulphates are quickly removed. Contact times of less than 10 minutes are enough to ensure a zinc extraction of more than 98 %.

The solid-liquid separation is carried in two counter-current steps using one 18 m diameter thickener in each step.

Figure 2 shows a view of the whole solvent extraction plant with the leaching and liquid-solid separation section in the foreground and the zinc extraction section at the background.

Figure 2 - View of the EXCINRES Plant at Espanola del Zinc S.A., Spain

Depending on the residues characteristic flocculants at a rate of 60 to 100 g/t residues are added to improve the settling rate. The thickener underflow of the second solid-liquid separation step is pumped to the dump, whilst the thickener overflow of the first solid-liquid separation step is filtered into filter presses in order to ensure, that the solid content of the Zn-bearing solution entering the solvent extraction section does not exceed 10 mg/l.

The addition of flocculants in the range as mentioned above did not influence the phase separation in the extraction circuit.

The extraction of zinc from the sulphate solution is done by an organic solvent consisting of 12 % D2EPHA and 88 % kerosene according to the simplified equation:

$$2\;[(D2EPHA)H] \;+\; ZnSO_4 \;<\!\!-\!\!-\!\!-\!\!-\!\!>\; (D2EPHA)_2\;Zn \;+\; H_2SO_4 \qquad (1)$$

As it can be seen from Figure 3, this reaction is pH-dependent and therefore it is used for the extraction of zinc at pH values of 1.7 to 2.5 and for its reextraction at higher acidities.

In order to achieve an effective solvent loading it is mandatory to maintain a constant pH level during the whole zinc extraction. Therefore the sulphuric acid generated by the reaction (1) is neutralized by the addition of ammonium hydroxide (NH4OH).

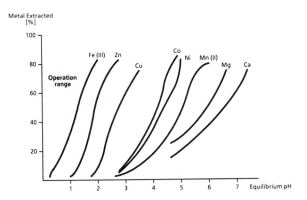

Figure 3 - Extraction of Metals by D2EPHA from Sulphate Solutions

To ensure a zinc extraction rate higher than 98 %, two extraction steps are provided. Each extraction step has a cylindrical mixer with a volume of abt. 15.0 cu.m. and a settler with an effective settling surface of 50 sq.m. Both are made of FRP.

Figure 4 shows the schematic and Figure 5 the real plant arrangement of a mixer-settler system in the extraction section.

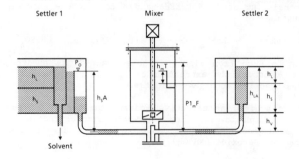

Figure 4 - Schematic Arrangement Mixer-Settler

Figure 5 - Plant Arrangement of a Mixer-Settler System

In order to minimize the losses of organic solvent, the raffinate after the zinc extraction but before being returned to the residues leaching section flows through a coke bed coalescer, where the organic solvent is separated from the solution and recovered. This solvent is returned to the main solvent circuit.

Figure 6 shows schematically how the coke bed coalescer works.

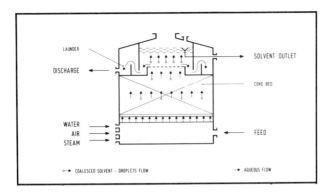

Figure 6 - Typical Diagram of a Coke Bed Coalescer

The coke bed is from time to time regenerated with water and compressed air.

Since the EXCINRES process is fully integrated in the existing roast-leaching-electrowinning circuit, the stripped electrolyte that goes after the zinc reextraction from the solvent extraction section to the zinc cellhouse, has to meet the same purity specifications as the electrolyte obtained in the main zinc winning circuit.

In order to grant that the impurities levels as far as copper, cadmium, nickel and cobalt are concerned are maintained, a solvent scrubbing section has been considered.

The scrubbing or purification of the Zn-bearing organic solvent takes place according to the following simplified equation:

$$(D2EPHA)_2 Me + ZnSO_4 \longrightarrow (D2EPHA)_2 Zn + MeSO_4 \qquad (2)$$

In the scrubbing section which consists of one mixer and one settler with the same characteristics as those in the solvent extraction, practically all the two-valences metals are removed from the Zn-bearing organic solvent.

As scrub liquor purified neutral solution with 140 to 160 gpl of Zn as $ZnSO4$ from the existing purification section is used. This solution, which after passing the scrubbing section contains practically all the two-valances metals removed from the organic solvent, is returned to the existing leaching and purification plant.

The purified Zn-bearing organic solvent is stripped in two countercurrent steps using return cell acid from the zinc cellhouse. The mixer-settler arrangement is identical to that in the zinc extraction section.

In order to prevent any contamination with organic solvent of the loaded electrolyte that goes to the zinc cellhouse after the reextraction steps, two pressure absorptions towers filled with inorganic materials have been provided. With this absorption system it has been possible to maintain in continuous operation a level of organic solvent in the electrolyte of 1 to 3 ppm.

The copper and cadmium not extracted by the organic solvent remains in the raffinate and therefore a further step for its recovery is required.

Depending on the copper and cadmium content in the raffinate, one third up to one half of the total raffinate flow goes to the copper and cadmium removal section before being sent to the residues leaching section.

Copper and cadmium are co-precipitated from the raffinate with metallic zinc dross from the existing zinc melting plant in three agitator tanks. The obtained Cu/Cd-residues are filtered in one filter press and sent for their further use to the existing Cd-plant.

As shown in Figure 3 all the Fe(III) contained in the Zn-bearing solution entering the zinc extraction section is co-extracted with the zinc by the organic (D2EPHA)-solvent.

The co-extracted iron is not removed from the organic solvent by the stripping return cell acid. Therefore, in order to avoid a Fe(III) build-up in the organic solvent, that not only would affect the extraction of zinc but also the operation of the whole solvent extraction section would become very difficult as a consequence of the higher viscosity of the organic solvent, a regeneration step is provided.

This regeneration step which in principle is a separate small solvent extraction section for iron, consists of the following steps:

- iron removal of the (D2EPHA)-organic solvent with hydrochloric acid (HCl),
- washing of the (D2EPHA)-organic solvent with water,
- iron extraction of the $FeCl_3$-solution with Tri-(Butyl)-Phosphate (TBP) and
- stripping of $FeCl_3$ of the (TBP)-organic solvent with water.

The regeneration of the (D2EPHA)-organic solvent, i.e. the removal of Fe(III), is carried out with 5 to 6 molar hydrochloric acid in one step that consists of one mixer and one settler. The iron is removed as a chlorine complex.

The regenerated organic, before being recycled to the Zn-extraction section, undergoes a water washing in a washing step consisting of mixer-settler arrangement, where the remaining chlorides in the organic solvent are totally removed.

The mixers have a volume of 7.0 cu.m. and the settlers a settling surface of 30.0 sq.m. In both steps the mixers and settlers are made of FRP lined with PVC.

The aqueous $FeCl_3$ solution obtained from the regeneration step, not only contains the removed iron but also unused free hydrochloric acid, which for environmental and economical reasons can not be discarded.

Therefore, a second solvent extraction circuit has been provided in order to extract the ferric chloride ($FeCl_3$) from the aqueous $FeCl_3$-solution and to recover the free hydrochloric acid.

This solvent extraction circuit uses a mixture of 12 % TBP, 12 % Isodecanol and 76 % of kerosene. Isodecanol is used in the same concentrations as the extractant (TBP) as a stabilizer to prevent the formation of a third liquid phase.

The ferric chloride complex is stripped from the organic solvent with water in a single step. The organic solvent is recycled to the extraction step, whilst the aqueous FeCl3-containing solution is neutralized before being discarded to the storage dump.

In both steps, the mixers have a volume of 7.0 cu.m and a settling surface of 18.0 sq.m. and they are made of FRP lined with PVC.

Conclusions

The EXCINRES plant with an investment of approx. DM 12.0 Million is able to produce 18 to 20 tpd of cathode zinc. As the operating results from April 1988 until March 1989 have proved, the process could be integrated within the existing electrolytic zinc plant without any substantial difficulties.

The average main consumption figures related to the production of one ton of zinc in solution to the existing zinc cellhouse are as follow:

- Ammonia (100 % NH3) ....: 430.0 kg
- (D2EPHA)-organic solvent ....: 2.5 kg
- (TPB)-organic solvent ....: 0.1 kg
- Isodecanol ....: 0.1 kg
- Kerosene ....: 17.0 kg
- Flocculant ....: 1.8 kg

The process is highly automated and practically the whole EXCINRES plant is supervised by only two operators per shift.

References

1. Jose Luis del Valle, Diego de Juan Garcia, Recuperacion de Metales en Forma soluble. Aplicacion a la Recuperacion de Zinc en Forma de Sulfato de los Residuos en Espanola del Zinc S.A. (Congreso Internacional de Mineria y Metalurgia, Barcelona, Spain, 1984).

2. D. de Juan Garcia, R. Lehmann, Recuperacion de Zinc soluble de los Residuos de la Hidrometalurgia del Zinc mediante la Extraccion por Disolventes Organicos (IV Asamblea General del CENIM, Madrid, Spain, 1985).

3. R. Heng, R. Lehmann, D. de Juan Garcia, Zinc Recovery from Hydrometallurgical Zinc processing Residues by Solvent Extraction (München, Germany, 1986).

4. Jose Luis del Valle, D. de Juan Garcia, El Proyecto EXCINRES (Congreso Internacional de Mineria y Metalurgia, Oviedo, Spain, 1988).

5. Spanish Patent No. 526.618.

6. Spanish Patent Application No. 8800209.

7. D. de Juan Garcia, Comparacion de la Extraccion del Fe(III) en Medio Cloruro con diversos Agentes de Extraccion (Congreso Internacional de Mineria y Metalurgia, Oviedo, Spain, 1988).

SIMULATION OF CONTINUOUS LEACHING

OF GALENA WITH $FeCl_3$ IN HYDROCHLORIC ACID SOLUTION

C-K Lee,* J-B Ryu,* H-J Sohn* and T. Kang**

* Department of Mineral and Petroleum Engineering
** Department of Metallurgical Engineering, College of Engineering, Seoul National University, Seoul, KOREA, 151-742

Abstract

The oxidative dissolution of galena was carried out with ferric chloride as an oxidant in NaCl-HCl solution and the reaction rate was found to be controlled by pore diffusion through the product layer. Predictions were made both for the continuous as well as the batch leaching of galena powder based on the above information, particle size distribution, and the fluid behavior within the reactor. The results showed in excellent agreement between the predicted and the experimental values. The design curves for this system were obtained in terms of Pb recovery and the number of stages required at each specified lead recovery.

Introduction

Galena has been processed conventionally by roast-sintering, smelting and fire refining or electrolytic refining to produce metallic lead. Recently U.S. Bureau of Mines [1][2][3] initiated a series of tests to produce Pb metal by hydrometallurgical route to avoid SO_2 gas and fine lead particle emission. Galena was leached in hydrochloric acid-brine solutions with $FeCl_3$ as an oxidant, followed by crystallization of $PbCl_2$ from leach liquor and fused salt electrolysis. Demarthe and Georgeaux [4] analyzed the economics of Minemet chloride process which consisted of two stage leaching of galena and electrolysis. Murphy et al., [5] conducted continuous leaching of galena and obtained 99 % recovery within 15 minutes at 95°C. The overall leaching reaction can be expressed as,

$$PbS + 2FeCl_3 = PbCl_2 + 2FeCl_2 + S° \qquad (1)$$

and the leaching kinetics were studied by several authors.

Dutrizac [6] carried out the leaching of galena crystals in a wide range of experimental conditions and reported parabolic kinetics in which the dissolution rate was controlled by the outward diffusion of $PbCl_2$ through the pores in elemental sulfur layer. Kim et al., [7] suggested that the reaction rate was controlled by chemical reaction at galena/sulfur interface below 57°C, which were contradicting with the above results. Rath et al., [8] also confirmed Kim's results for a variety of experimental conditions. Morin et al., [9] found that the leaching rate can be expressed in terms of mixed kinetic model [surface reaction plus pore diffusion] and took into account the self blockage of pores in sulfur layer due to the precipitation of $PbCl_2$. Fuerstenau et al., [10] also indicated that the pore diffusion of ferric chloride was rate determining step.

Non-oxidative leaching of galena also can occur as follows in hydrochloric acid [11] [12],

$$PbS + 2H^+ = Pb^{++} + H_2S \qquad (2)$$

but the reaction rate is relatively very slow compared with that of the oxidative leaching although this can be compromised to some extent by using the Clause reaction as mentioned by Wadsworth [13].

In order to clarify the discrepancy, the leaching of galena crystals and ground powders were carried out in this study with a variation of some important parameters such as the concentration of $[H^+]$, total $[Cl^-]$, $[Fe^{+++}]$, $[Pb^{++}]$, and $[Fe^{++}]$ as well as the temperature.

Experimental

Basically three types of experiments were performed in this study, i.e. electrochemical measurements, batch type leaching and continuous leaching. The samples used were obtained from Yeon Hwa Mine in Korea.

Electrochemical measurements were carried out using three way electrode system with potentiostat/galvanostat [EG&G PAR model, 173]. Galena crystals were cut into approximately 1 cm³ of volume and mounted in epoxy resin to be used as working electrode. Also these crystals were ground into powder for leaching experiments. Solutions were analyzed for various metal ion concentrations with atomic absorption spectrophotometer.

In continuous leaching tests, solution was heated up to a specified temperature in reactor and lixiviant reservoir, then galena powders were fed into the reactor to be leached up to a mean residence time as a batch mode. Continuous leaching was started immediately by adding the lixiviant solution and galena powder into the reactor simultaneously. Schematic diagram of experimental set-up is shown in Figure 1.

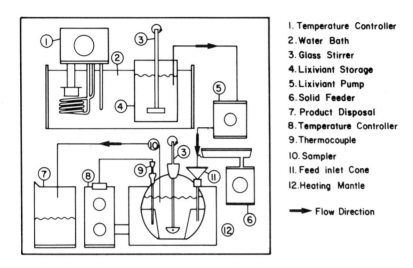

1. Temperature Controller
2. Water Bath
3. Glass Stirrer
4. Lixiviant Storage
5. Lixiviant Pump
6. Solid Feeder
7. Product Disposal
8. Temperature Controller
9. Thermocouple
10. Sampler
11. Feed inlet Cone
12. Heating Mantle

→ Flow Direction

Figure 1- Schematic diagram of continuous leaching experiments

Results and Discussion

Reaction chemistry of leaching of galena can be devided into two half-cell reaction as,

$$\text{cathodic}; \quad Fe^{+++} + e^- = Fe^{++} \tag{3}$$

$$\text{anodic}: \quad PbS = Pb^{++} + S^° + 2e^- \tag{4}$$

Paul et al.,[14] suggested that the anodic reaction can also be put into following scheme through the electrochemical measurements,

$$PbS = Pb^{++} + S^- + e^- \tag{5}$$

$$S^- = S^° + e^- \tag{6}$$

and at the same time, non-oxidative leaching can also takes place.

Leaching of galena crystal

Galena crystal mounted in epoxy resin was immersed in $FeCl_3$-HCl-brine solution and the dissolution rate was measured as a function of time.

Effect of temperature. The temperature dependency on leaching rate is shown in Figure 2. The reaction rate was increased linearly with the increase of temperature up to 60°C, which agrees well with the results of Kim et al., [7] and indicated the reaction rate was controlled by surface reaction. Above 60°C, dissolution rate became parabolic which was the main interested region in this study.

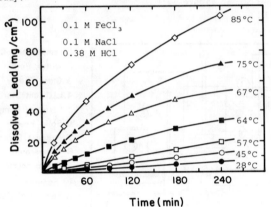

Figure 2 - The effect of temperature on the leaching rate of galena

Effect of $[Fe^{+++}]$, $[H^+]$ and total $[Cl^-]$ concentration. Figure 3 illustrates the effect of $[Fe^{+++}]$, from which the reaction rate was increased with the increase of $[Fe^{+++}]$ concentration. Also the $[H^+]$ ion increased the dissolution rate markedly as shown in Figure 4 and these effects were analyzed in kinetic section.

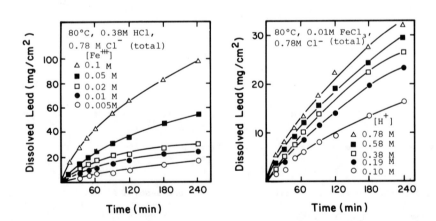

Figure 3 - The effect of ferric ion concentration on the leaching rate of galena

Figure 4 - The effect of hydrogen ion concentration on the leaching rate of galena

The dissolved lead ions make lead chloro-complexes in solution such as $[PbCl^+]$, $[PbCl_2]$, $[PbCl_3^-]$ and $[PbCl_4^=]$. The concentrations of these complexes depend on the thermodynamic conditions. As shown in Figure 5, the total $[Cl^-]$ concentration up to 1.0 mol/l had little effect on leaching.

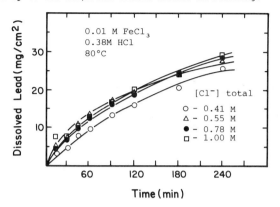

Figure 5 - The effect of chloride ion concentration on the leaching rate of galena

Effect of $[Pb^{++}]$ and $[Fe^{++}]$ concentration. In order to study the effect of $[Pb^{++}]$ and $[Fe^{++}]$ present initially, electrochemical measurements were performed since the amount of dissolved lead complexes during leaching was relatively small to differentiate between the dissolved $[Pb^{++}]$ and $[Fe^{++}]$ and the ones initially present by A.A. technique. Figure 6 shows the effect of $[Fe^{++}]$ concentration initially present on leaching. It retarded the reaction rate (corresponding to the current density) slightly. The general trend of polarization curves are similar to those reported by Skews [15] and Paul et al.,[14] and the peak would be attributed to the formation of elemental sulfur which passivated the galena surface. The same was the case for the effect of $[Pb^{++}]$ present as shown in Figure 7 from which the back reaction of reaction (1) could be negligible.

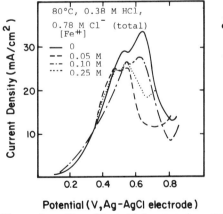
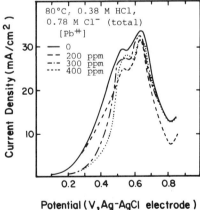

Figure 6 - The effect of ferrous ion concentration on the anodic polarization of galena

Figure 7 - The effect of lead ion concentration on the anodic polarization of galena

Reaction Kinetics. As mentioned previously, there are some discrepancies on the rate controlling step in the leaching of galena with $FeCl_3$ in brine solution. In Figure 2, the leaching response was parabolic at high temperature which indicated the pore diffusion is rate controlling step rather than the surface reaction. Also the topochemical nature of leaching reaction is clearly illustrated in Figure 8-a

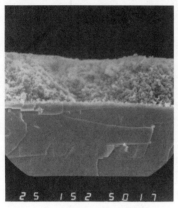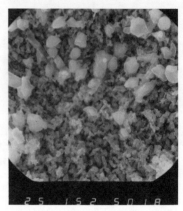

Figure 8 - Photographs of leached galena crystal a) side view b) top view

If the concentration of diffusing species in aqueous phase remains constant during the course of reaction, the resulting expression for the product layer diffusion process in semi-infinite crystal can be expressed as [16],

$$m^2 = k_p t \qquad (7)$$

where m is the amount of lead leached per unit area of galena, t is the time and k_p is the parabolic rate constant which is,

$$k_p = \rho_g D_e C_{Fe^{+++}} \qquad (8)$$

In here, ρ_g is the molar density of galena, D_e is the effective diffusivity of diffusing spieces and $C_{Fe^{+++}}$ is the concentration of [Fe^{+++}] in solution. Figure 9 shows the plot of m^2 against the time which gives good straight line

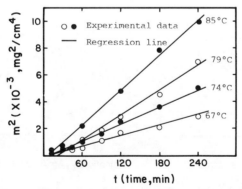

Figure 9 - A plot of m^2 vs. time for various temperature

as indicated in equation (7). From the rate constant, k_p obtained from the slope in Figure 9, the apparent activation energy was found to be 17.1 kcal/mol and the diffusivity at 75°C was 2.95 x 10^{-6} cm²/sec.

The same analysis was performed for the data shown in Figure 3 and also good straight line was observed for a wide variation of [Fe^{+++}] concentration. From this, the rate constant was plotted against the $C_{Fe^{+++}}$, which shows the first order dependency on [Fe^{+++}] concentration.

Leaching of galena powders

To test the validity of the above pore diffusion model, galena crystals were ground into powders to obtain various size ranges, and batch-type as well as continuous leaching experiments were carried out. Preliminary tests showed the effect of stirring speed had little effect on the leaching rate above 600 rpm and subsequent tests were performed at 600 rpm.

Leaching in batch type reactor. The product layer diffusion model for the spherical particles can be expressed as [16],

$$1 - (2/3)\alpha - (1-\alpha)^{2/3} = k_p' t \quad (9)$$

and

$$k_p' = \frac{8 \gamma D_e C_{Fe^{+++}}}{\rho_g} \frac{1}{d_o^2} \quad (10)$$

where α is the fraction reacted, γ is stoichiometry and d_o is the initial diameter of particles. The left hand side of equation (9) vs. time is plotted in Figure 10, which shows the good linearity. The rate constants obtained from Figure 10 vs. $[1/d_o^2]$ should be linear as indicated in equation (10), and shown in Figure 11. The diffusivity calculated from the slope is 2.35 x 10^{-6} cm²/sec, which is smaller than that of single crystal leaching.

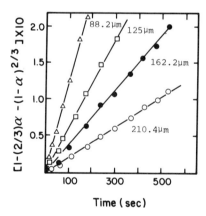
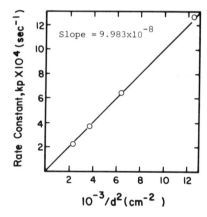

Figure 10 - A plot of 1-2/3α-(1-α)$^{2/3}$ vs. time for monosize galena particles

Figure 11 - A plot of parabolic rate constant vs. the square of the inverse initial particle diameter

However galena particles shown in Figure 12 are cubic rather than sphere, which was also confirmed by Fuerstenau et al., [10]. The sphericity factor, ϕ, was introduced into the rate constant k_b' to correct this effect, and $k_p'' = \phi k_p'$. Substituting ϕ (= 0.806 for the cubic particle) into the equation (9) gives the diffusivity of 2.92×10^{-6} cm^2/sec, which agrees well with the previous one.

Figure 12 - Photograph of galena powder showing cubic nature

The prediction on the leaching response was made with the diffusivity obtained from the above analysis for the Gamma size distribution, which can be expressed as,

$$f_3(d) = \frac{\lambda^p d^{p-1} \exp(\lambda d)}{\Gamma(p)} \qquad (11)$$

where $f_3(d)$ is a mass density function, λ and p are constants and d is the particle diameter. As shown in Figure 13, the predicted and the experimental values agree well to confirm the validity of the model suggested previously.

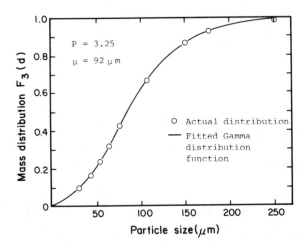

Figure 13 - A plot of fraction reacted vs. time for the Gamma size distribution

Continuous leaching of galena. In continuous leaching, the fluid behavior within the reactor should be known in order to predict the conversion of galena. KCl solutions were used as a tracer to find the degree of mixing within the reactor. The concentration of KCl was monitored as a function of time for a unit step function input of KCl as illustrated in Figure 14. It fits the perfect mixing case as expected, where the concentration of tracer in the outlet can be expressed [16],

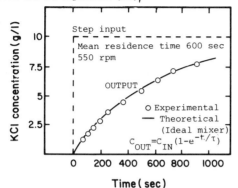

Figure 14 - A plot of tracer response for the step function input

$$C_{out} = C_{in} [1 - \exp(-t/\tau)] \qquad (12)$$

where C_{out} and C_{in} are the concentration of KCl in outlet and inlet, respectively and τ is the mean residence time. The age distribution function, $E(t)$ [17] can be obtained from the equation (12),

$$E(t) = [1/C_{in}][dC_{out}/dt] = [1/\tau]\exp(-t/\tau) \qquad (13)$$

Figure 15 shows the leaching response in continuous reactor for various mean residence time. Since the leaching was carried out as batch type before

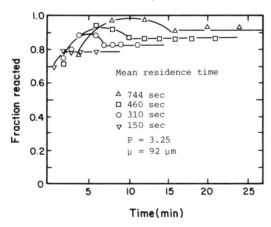

Figure 15 - A plot of fraction reacted vs. time for the continuous leaching

it reached the mean residence time, initially the fraction reacted was always higher than that of the continuous leaching. The steady-state conversion value was obtained approximately two minutes later as shown in Figure. The overall fraction reacted of galena particles can be predicted by combining the informations on the individual particle leaching kinetics, fluid behavior and particle size distribution.

Hence, the predicted value for the overall fraction reacted at steady-state operation is as follows,

$$\alpha(d) = \int_0^{t(d)} \alpha(d,t) \, E(t) \, dt \quad (14)$$

where $\alpha(d,t)$ is the conversion of particle size d at time t and the upper limit of integration $t(d)$ can be expressed as follows,

$$t(d) = \left(\frac{\rho_g}{24 \, YD_e C_{Fe^{+++}}}\right) d^2 \quad (15)$$

The conversion calculated based on equation (13) was compared with the experimental value in Figure 16. The agreement is excellent from which the validity of previous model can be confirmed.

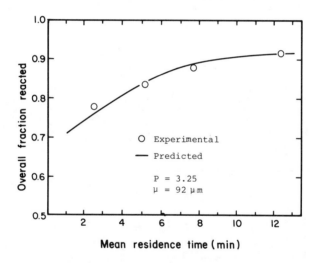

Figure 16 - Comparisons of overall conversion between experimental and predicted at 75°C

Based on the above information, the number of stages required were calculated in terms of conversion and mean residence time for the leaching of galena and shown in Figue 17. For the 98% recovery of lead, three stages of leaching are necessary with a total mean residence time of 14 minutes and the other data can be obtained from the design curves presented in the Figure 17.

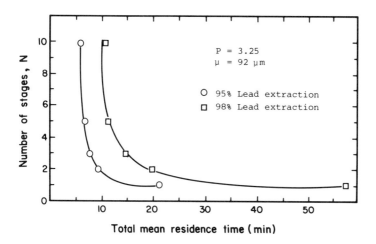

Figure 17 - A design curves for the continuous leaching of galena powders

Conclusions

The oxidative leaching of galena was performed with ferric chloride as oxidant in NaCl-HCl solution and the reaction rate was found to be controlled by pore diffusion through the product layer at high temperature above 60°C. Batch leaching data of galena power were correlated well with the above model which was modified to accomodate the particle shape factor. The predictions were made for continuous leaching of galena and the results showed an excellent agreement between the predicted and the experimental values. The simulated results indicated that three stages of leaching were necessary with a total mean residence time of 14 minutes for the 98% recovery of lead and the other design curves were obtained in terms of Pb recovery and the number of stages required at each specified lead recovery.

References

1. F. P. Haver and M. M. Wong, "Ferric Chloride-Brine Leaching of Galena Concentrate," [U.S. Bureau of Mines Report of Investigation 8105, 1976]

2. F. P. Haver et al., "Recovery of Lead from Lead Chloride by Fused-Salt Electrolysis," [U.S. Bureau of Mines Report of Investigation 8166, 1976]

3. M.M. Wong, F. P. Haver and R. G. Sandberg, "Ferric Chloride Leach-Electrolysis Process for Production of Lead," Lead-Zinc-Tin '80, ed. J. M. Cigan, T. S. Mackey and T. J. O'Keefe, [Warrendale,PA:The Metallurgical Society, 1980], 445-454

4. J. M. Demarthe and A. Georgeaus, "Hydrometallurgical Treatment of Lead Concentrate," Lead-Zinc-Tin '80, ed. J. M. Cigan, T. S. Mackey and T. J. O'Keefe, [Warrendale, PA:The Metallurgical Society, 1980], 426-444

5. J. E. Murphy, B. Eichbaum and J. A. Eisele, "Continuous Ferric Chloride Leaching of Galena," Minerals and Metallurgical Processing, 3[1] [1985], 38-42

6. J. E. Dutrizac, "The Dissolution of Galena in Ferric Chloride Media," Met. Trans. 17B [1][1986], 5-17
7. S. H. Kim, H. Henein and G. Warren, "An Investigation of the Thermodynamics and Kinetics of the Ferric Chloride Brine Leaching of Galena Concentrate," Met. Trans. 17B [1][1986], 29-39
8. P. C. Rath, R. K. Paramguru and P. K. Jena, "Kinetics of Sulfide Minerals in Ferric Chloride Solution, 1: Dissolution of Galena, Sphalerite and Chalcopyrite," Trans. Instn. Min. Metall. 97[1988] C150-158
9. D. Morin, A. Gaunand and H. Renon, "Representation of the Kinetics of Leaching of Galena by Ferric Chloride in Concentrated Sodium Chloride Solutions by a Modified Mixed Kinetics Model," Met. Trans. 16B [1][1985], 31-39
10. M. C. Fuerstenau et al., "Kinetics of Galena Dissolution in Ferric Chloride Solutions," Met. Trans. 17B [3][1986], 415-423
11. Y. Awakura, S. Kamei and H. Majima "A Kinetic Study of Nonoxidative Dissolution of Galena in Aqueous Acid Solution," Met. Trans. 11B [3][1980], 377-381
12. C. Nunez, F. Espiell and J. Garcia-Zayas, "Kinetics of Nonoxidative Leaching of Galena in Perchloric, Hydrobromic, and Hydrochloric Acid Solutions, " Met. Trans. 19B [4][1988], 541-546
13. M. E. Wadsworth, "Sulfide and Metal Leaching Reactions," Min. Eng. 37 [6] [1985], 557-562
14. R. L. Paul et al., "The Electrochemical Behavior of Galena [Lead Sulfide]-I. Anodic Dissolution," Electrochimica Acta, 23[1978], 625-633
15. H. R. Skews, "Electrowinning of Lead Directly from Galena," Proc. Aust. Inst. Min. Met. No. 244 [1972], 35-41
16. O. Levenspiel, Chemical Reaction Engineering [2nd. ed., New York, NY : John Wiley & Sons, Inc., 1972], 372
17. D. M. Himmelblau and K. B. Bischoff, Process Analysis and Simulation ; Deterministic System, [New York, NY : John Wiley & Sons, Inc., 1968], 64

Acknowledgements

The authors wish to express their gratitudes to the Korea Science and Engineering Foundation [KOSEF] for the financial assistance to this study.

Effectiveness of Glue in Actual Tankhouse Electrolyte

Controlled by the CollaMat[R]-Method

Bernd E. Langner, Peter Stantke
Norddeutsche Affinerie Aktiengesellschaft, Hamburg, FRG

Abstract

With a fully automized measurement which analyses the kinetics of electroadsorption of glue on a copper cathode (CollaMat[R]), concentration of active glue down to less then 30 ppb can be determined with high accuracy and reproducibility. The degradation of glue in an industrial tankhouse electrolyte has been studied. It could be shown that the degradation kinetics is 1st order and the half lifetime for active glue is between 60 minutes and more than two hours. From the results consequences for the tankhouse practice can be seen.

Introduction

The electrolytic refining of copper seems to be one of the most simple chemical processes:

$$Cu^0 - 2\ e^- = Cu^{2+}\ /\ Anode$$
$$Cu^{2+} + 2\ e^- = Cu^0\ /\ Cathode$$

However this apparently simple reaction equation says nothing about the quality of the deposited copper and about the refining conditions in the tankhouse, because this depends on many factors:

I) impurities in the anodes which report to the electrolyte or to the anode slime. The soluble impurities can be electrochemically reduced on the cathode and components of the slime can migrate to the cathode after being suspended.

II) the structure of the deposited copper. A bad copper structure leads to a high number of short circuits and to the incorporation of impurities.

While the amount of impurities in anodes cannot be influenced in a wide range, as it is determined by the assay of the raw material for the production of the anodes, the structure of the copper deposit can to a certain extent (dependent on the current density) be influenced by addition agents. The main objective using addition agents is to achieve a smooth copper deposit, which prevents the deposition of impurities of the anode occurring at irregular locations on the cathode.

Table 1 - Additives used in copper electrolyte

	glue	thiourea	chloride
Addition in g/t copper	56	40	50
Concentration in the electrolyte in mg/l	1 - 5	5 - 10	60

Beside chloride ions common addition agents in a copper tankhouse are thiourea and glue (s.Table 1) in very low concentrations of less than 10 mg/l. Although there is a great deal of literature about theories as to how addition agents produce a smooth cathode ((1)(3)(8)), there is no theory which can predict the best addition agent in a specific copper tankhouse. Basically, one can assume that thiourea catalyses the germ formation and glue stops the growth of germs to a certain extent.

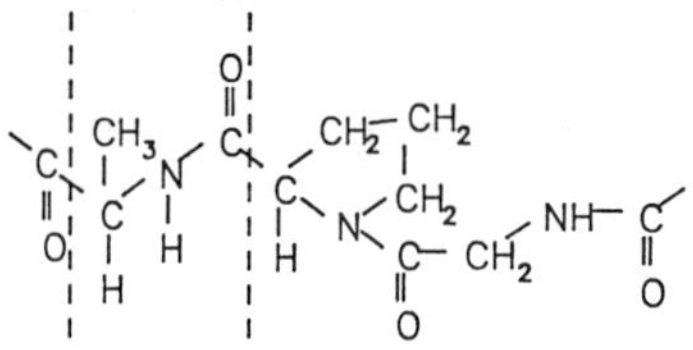

Fig. 1 - Molecular structure of glue

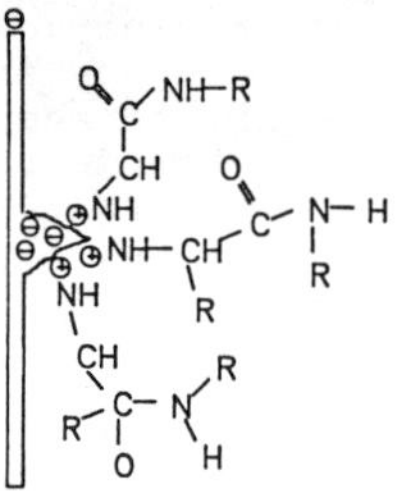

Fig. 2 - Glue adsorption on a cathode

Action of glue on the copper deposition

Glue is a natural product which consists of a complicated mixture of polypeptides with molecular weights between a few thousand and some millions (s.Fig.1). The action of glue on the cathode can be explained in a simplified way as follows (s.Fig.2):
The adsorption of glue on the cathode occurs mainly at locations with a high field strength (4). Places with a high field strength on a cathode surface are edges, nodules and needles. At these preferred places an electroadsorption of the polar glue molecules takes place, so that the field strength is decreased and the glue forms an isolating layer on for example a needle. As a result, the growth of the needle which could lead to a short circuit is stopped. After the normal copper layer has grown up to the peak of the needle the glue is desorbed and can be adsorbed again on another place with a high field strength. As seen in Fig.2 glue partly covers the cathode surface and increases the effective current density leading to a higher cathode polarization. Fig.3 a-c show cathodes which are produced with different glue concentrations in the electrolyte. While an optimized concentration leads to a smooth

cathode (Fig.3b) too low concentrations result in a high number of nodules on the surface. However an overdosing of glue also produces nodules, probably because most of the surface is blocked, leading to a high effective current density and a high overpotential voltage.

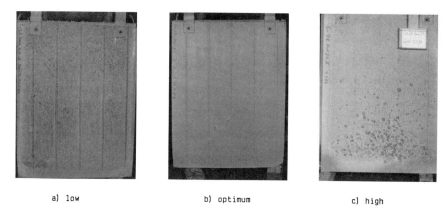

a) low   b) optimum   c) high

Fig. 3 a-c -Copper cathode with different glue concentrations in the electrolyte

Determination of glue in the tankhouse electrolyte

In contrast to the defined chemical compounds chloride and thiourea, which can be determined by chemical analysis (e.g. thiourea by polarography (2)), glue is a complex mixture of polypeptides with a broad molecular weight distribution (5), which cannot be determined simply by analysis. Furthermore, apparently only glue molecules within a certain molecular weight range benefit the deposition of copper in a tankhouse. According to the assumed theory (Fig.2) glue should reduce the effective surface of the cathode and consequently an increase of cathode polarization should be observed. Therefore, some investigations ((6)(7)(8)) on the determination of active glue concentration are based on measurements of cathode polarization. Nevertheless, there are some major difficulties with these measurements using a conventional approach:

I) cathode polarization is not only dependent on glue concentration but also on other tankhouse parameters (6) like acid content, state of oxidation of the electrolyte (As(III)/As(V), Cu(I)/Cu(II)).

II) measurements often are only poorly reproducible because of an irreversible change of the surface of the sensor cathode by adsorption of glue, which is not fully desorbed after each measurement.

III) measurements need a reference electrode, which is not stable over a longer period in the environment of a tankhouse electrolyte.

NA concept

Therefore, for the measurement of a static cathode polarization for the glue determination a new calibration is required after each change of electrolyte characteristics otherwise wrong glue concentrations result. Because of the disadvantages of the conventional procedures a new method was developed by Norddeutsche Affinerie (CollaMat^R, patent pending), which exhibits following advantages:

   I) no influence of electrolyte composition and state of oxidation
  II) no reference cell necessary
 III) high sensitivity down to about 20 ppb active glue
  IV) a simple robust sensor combined with adequate software.
      Fig. 4 shows a picture of the CollaMat^R apparatus.
      A measuring cycle functions as follows:

1.) Feed and return hose are suspended in the cell.
2.) Input of data on location and start of automatic programme. Phases 3 to 7 follow on automatically.
3.) The measuring cell is filled and circulated.
4.) The measuring system is regenerated.
5.) Measuring for 10 minutes at a constant solution throughflow.
6.) Evaluation of the potential time curve. After about 30 seconds the measuring value is given in mg glue/l.
7.) The measuring cell is emptied.

Total duration of a measuring cycle is about 13 minutes.

Fig. 4 - CollaMat apparatus

The calibration curve in Fig.5 shows the good linearity between the glue concentration and the measured value obtained with glue which has just bee added.

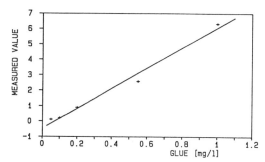

Fig. 5 - Calibration curve for glue concentration in the tankhouse electrolyte

From tankhouse practice it is known that glue is consumed during the electrolytic process:

- by adsorption on anode slime
- by incorporation into the cathodes
- and, most important, by decomposition.

The decomposition is caused mainly by hydrolysis, that means that molecules with high molecular weights are split in smaller molecules, which have only poor activity towards the copper deposition. Fig.6 shows the kinetics of the glue decay in actual tankhouse electrolyte. Reaching a maximum after about 15 - 60 minutes, the concentration decreases in an expotential way, presumably because hydrolysis takes place. The peak of the curve indicates that not only low but also very high molecular weights - present in the electrolyte after the addition of fresh glue - have an impact on the copper deposition. Taking only the time after the maximum into account, by a logarithmic plot (Fig.7) a kinetics pseudo-first-order reaction can be derived, which is typical behaviour for hydrolysis reactions. Half lifetimes of

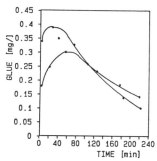

Fig. 6 - Decay of glue in a tankhouse electrolyte

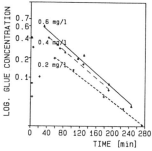

Fig. 7 - Logarithmic plot of glue concentration against time at different starting concentrations

active glue are in the order of 90 to 200 minutes, which means that after this time only half of the added glue is still active. Under normal tankhouse conditions with residence times in a single bath of about 2 hours there is a decomposition of about 50 % glue from the first cathode to the last one. It is imperative to know the decay rate of the glue for the cell design and to adjust the flow rate. The decay rate glue is dependent on different parameters like acid concentration and the origin of the used glue.

Results of measurements in the tankhouse

By experimenting in the plant the best glue concentration range has been determined both by visual inspection of the cathodes and by a statistical evaluation of shorts over a period of 6 months. Although in a tankhouse there are a lot of parameters affecting the number of short circuits other than glue concentration, Fig.8 demonstrates that the number of shorts increases if the glue concentration is too low. As a result of these investigations a range of an optimum glue addition procedures has been established, which improve copper deposition conditions. In Table 2 the optimal data for NA tankhouse conditions are shown. The difference in glue concentrations between inlet and outlet can be explained by the decomposition reaction as illustrated by Fig. 6.

Table 2 - Glue concentration in the inlet and outlet of the cells vs. quality of copper deposition

glue (inlet) (mg/l)	glue (outlet) (mg/l)	copper deposition
$\geq 0,95$	$\geq 0,50$	bad deposition
0,65 - 0,95	0,30 - 0,50	partially bad deposition, upper limiting range
0,35 - 0,65	0,15 - 0,30	good deposition optimum dosing of glue
0,20 - 0,35	0,12 - 0,15	partially bad deposition, lower limiting range
0,15 - 0,20	0,06 - 0,12	significant deterioation of the deposition
$\leq 0,15$	$\leq 0,06$	very bad deposition

Since February 1988 a portable CollaMat[R] device has been used in day-to-day tankhouse practice by the operators. Fig.9 points out that even small changes in glue addition -like an increase from 14 g glue/cell/day to 14.5 g glue/cell/day- lead to measurable higher concentrations of glue in the electrolyte. The detection limit is less than 30 ppb. This high sensitivity and high accuracy allows, if necessary, irregularities in the parameters which influence the hydrolysis of glue to be detected and corrected as well.

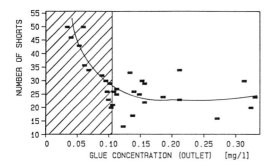

Fig. 8 - Number of shorts vs. glue concentration in the outlet of the cell

In Fig. 10 the number of short circuits is plotted against current density both for the time before the CollaMat^R has been used and after February 1988, starting with a monitoring of glue by the CollaMat^R. It can be seen that the number of short circuits has decreased by about 10 to 15%. The average current efficiency increased by about 1% at a high level of 95% - 96%.

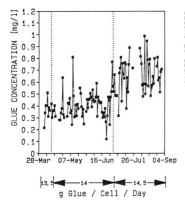

Fig. 9 - Continuous monitoring of glue in the plant

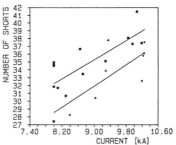

■ Period before introduction of CollaMat
+ Period since March '88, after introduction of CollaMat

This graph is based on data from the copper tankhouse of Norddeutsche Affinerie from September '86 to September '88.

Fig. 10 - Current density vs. number of shorts in the tankhouse

Conclusions

The CollaMat method allows glue to be monitored in a copper tankhouse down to concentrations of about 20 ppb. It is possible to optimize glue addition, to test different sorts of glue and to adjust flow rates and flow characteristics according to the decay of glue in the cells. It could be shown that the method can be adapted to different compositions of the electrolyte. Within the normal range of concentrations there is no interaction with other additives like chloride, thiourea, and avitone.
The compact robust CollaMat^R unit can be designed both for stationary operation and as a portable device with an accumulator for power supply, if it is necessary to monitor glue in several electrolytes.

References

(1) Z. Görlich, "Influence de la Thiouree sur les Proprietes des Depots Cathodiques de Cuivre", Ann.Chim.Fr, 4, 215 (1979)
(2) M. Goffmann and T.L. Jordan, US Patent No.4,474,649 to ASARCO Incorporated, Oct. 2, 1984
(3) D.F. Suarez and F.A. Olsen, "Nodulation of Copper Cathodes by Electrorefining Addition", The Electrorefining and Winning of Copper, ed.J.E.Hoffmann (Metallurgical Society,Inc., Warrandale, 1987),145.
(4) R. Winand, "Industrial Electrodeposition of Copper.Problems Connected to the Behaviour of Organic Addition", Application of Polarisation Measurement in the Control of Metal Deposition, ed. F.H. Warren (Elsevier Science Publisher, Amsterdam, 1984), 133.
(5) E.Sauer, "Tierische Leime und Gelatine", (Springer-Verlag,1958)
(6) T.N.Anderson, R.D. Budd and R.W. Strachan, "A Rapid Electrochemical Method for Measuring the Concentration of Active Glue in Copper Refinery Electrolyte which Contains Thiourea", Metallurgical Transactions, 7B, 333 (1976)
(7) S. Krzewska, L. Pajdowski, H. Podsidaly and J. Podsidaly, "Electrochemical Determination of Thiourea and Glue in the Industrial Copper Electrolyte", Metallurgical Transactions, 15B, 451 (1984).
(8) K. Knuutila, O. Forsen and A. Pehkonen, "The Effect of Organic Additives on the Electrocrystallization of copper", The Electrorefining and Winning of Copper,ed.J.E.Hoffmann (Metallurgical Society,Inc., Warrandale, 1987),145.

COLUMN LEACHING OF SCRAPPED PETROCHEMICAL CATALYSTS

S. Kelebek* and P.A. Distin

Department of Mining and Metallurgical Engineering, McGill University,
3450 University Street, Montreal, Quebec, H3A 2A7, Canada

*Present address: Energy, Mines and Resources Canada, CANMET/Coal Research
Laboratories, P.O. Box 1280, Devon, Alberta, T0C 1E0, Canada.

Abstract

Scrapped petrochemical catalysts are becoming an important secondary source of several metals. Many spent catalyst pellets and extrudates have high porosity, and their metals content may be leached using a static bed of uncrushed material, a procedure which produces more concentrated solutions than obtainable with crushed catalyst in a stirred reactor. However, it is difficult to create a selective leaching system that avoids codissolution of both the contained metals and alumina/alumino-silicate substrate.

Scrapped sulphided nickel/tungsten hydrocracking catalyst pellets have been leached in water or weak sulphuric acid solutions using a tubular reactor. The purpose was selective extraction of nickel over tungsten and aluminum, an objective promoted by catalyst preoxidation. Results are presented for the leaching of pellets without deliberate preoxidation, and for those preoxidized in air under various conditions. Also described is the effect of small nitric acid additions that provide an oxidizing leachant. Undesirable side-reactions occuring in the pellets at high preoxidation or leach temperatures are discussed.

Introduction

The petroleum refining and chemical process industries discharge large amounts of poisoned, spent catalysts. These contain various combinations of nickel, cobalt, molybdenum, tungsten, chromium, vanadium, zinc or platinum group metals supported on substrates such as alumina, silica or alumino-silicates of various types, and in the shape of pellets or extrudates. The metals may be present in different forms including, for example, oxides, or compounds such as nickel molybdate or tungstate. Sometimes the catalyst is presulphided for enhanced surface activity (e.g. nickel/molybdenum or nickel/tungsten catalysts).

In many cases, these scrapped materials are considered hazardous waste, the handling and disposal or which is an important environmental issue. The processing of these residues for recovery of their metallic values is an attractive disposal option, if this can be carried out economically. However, many recycling processes proposed in recent years have suffered from poor recoveries and incomplete metal separations, while a depressed metals market has recently prevailed. Several processing schemes have been described for various metals/substrate combinations. These methods include high temperature chlorination (1,2), sodium carbonate roasting followed by dissolution in water (1,3), leaching in solutions of sodium hydroxide (1,4,5), ammonium hydroxide with ammonium carbonate or ammonium sulphate (1,4,6), or sulphuric acid (1,7). In general, catalyst leaching has been carried out under atmospheric or high pressure conditions using a stirred reactor.

Recently, our laboratory has been studying 'column leaching' of scrapped automotive (8-11) and petrochemical catalysts (7). In this method, the catalyst is leached as a static bed in a tubular reactor using an upward flow of leachant. Generally, this percolation-type leaching system is suitable for uncrushed catalysts due to their high porosity. Additionally, metal concentrations in a major fraction of the product solution are much higher than obtainable using a stirred reactor at relatively low solids loading, and a solid/liquid separation step is avoided.

In the present work, a spent nickel/tungsten hydrocracking catalyst was leached in a laboratory scale tubular reactor. Based on results obtained using a stirred reactor to leach a scrapped nickel/tungsten hydroprocessing catalyst, previous workers (1) have proposed a flowsheet in which tungsten is leached into sodium hydroxide solution with essentially no coextraction of nickel. Subsequently nickel is extracted into sulphuric acid solution. In both these leaches, some aluminum from the substrate also dissolves. This sequence of leaching steps, using 'column leaching', has been investigated in our laboratory. Initial results showed that, while at least 90% tungsten recovery was possible, sodium hydroxide leach kinetics were surprisingly variable in duplicate runs under nominally the same conditions, an effect that was traced to the storage history of the spent catalyst. Oxidation during storage produces water soluble nickel/sulphur/oxygen compounds which initially dissolve during sodium hydroxide leaching, with subsequent reprecipitation of nickel, presumably as hydroxide. If highly oxidized catalyst is leached as a static bed, with more quiescent solution flow conditions than in a stirred reactor, this hydroxide precipitate tends to block catalyst pores and impede tungsten leaching. Also fine particulates, considered to be nickel hydroxide, were sometimes observed in the product solution leaving the column.

This paper describes water or sulphuric acid 'column leaching' of spent nickel/tungsten catalyst that has either been deliberately preoxidized under controlled conditions in air, or, is leached 'as received', but under oxidizing conditions provided by nitric acid addition. Using the data obtained, the potential for selectively extracting nickel before sodium hydroxide leaching of tungsten is evaluated (i.e. reverse order of leaching sequence described above).

Experimental

Leaching tests were carried out using a spent nickel/tungsten hydrocracking catalyst with chemical analysis shown in Table I.

Table I. Spent Catalyst Analysis

Element	Weight %
Nickel	3.84
Tungsten	3.90
Aluminum	12.33
Iron	2.05
Vanadium	-
Sulphur	1.95
Carbon	0.10
Surface Area	136 - 140 m^2/g

Vanadium and iron contents were determined because these metals are present in petroleum residue and tar sands, and are common catalyst poisons (12). The spent catalyst contained about 2% iron, with no vanadium content. In addition to aluminum, silicon was present, although a silica analysis was not performed. Total sulphur and carbon levels were about 2% and 0.1% respectively. The catalyst consisted of 1.5 mm diameter 5-10 mm long cylindrical pellets, with surface area, determined by the B.E.T. method, of about 138 m^2/g.

The leaching system, described in detail previously (7), consisted of a tubular glass reactor, with height and diameter of 1.2 m and 5.1 cm respectively, and equipped with heating tapes. In each test, between 1.0 and 1.2 kg of uncrushed pellets were leached as a static bed, preliminary tests, with stirred slurries in a beaker, having shown that nickel could be extracted as readily from uncrushed catalyst as from crushed material. A thermometer was immersed in the pregnant solution near its outlet at the top of the reactor. An upward flow of leachant was maintained at about 1.5 l/h, the solution volume filling the charged column being 1 l. The product solution was continuously collected in a series of flasks, then analysed by atomic absorption spectrophotometry.

Spent catalyst was received in a sealed drum, and was leached within one day after removal from the container, or following a preoxidation step. This latter procedure consisted either of spreading the pellets on a tray and exposing them to air at room temperature for several months, or heating the catalyst in air at up to 600°C for a few hours. In the former case, the tray of pellets was disburbed once a week to expose fresh surfaces to air.

Results and Discussion

Data for the leaching of catalyst which was not deliberately preoxidized are given in Figure 1.

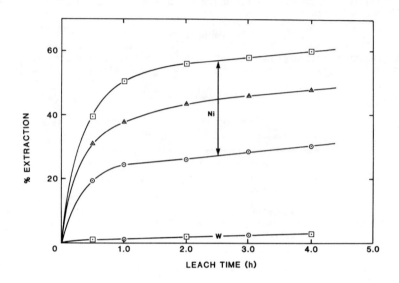

Figure 1 - Leaching of non-preoxidized catalyst at 80°C in 50 g/l sulphuric acid solutions with nitric acid contents of 0 g/l (o), 2.0 g/l (▲), 6.0 g/l (□).

At 80°C, about 30% of the nickel, but 16% of the aluminum, were extracted into an inlet leach solution containing only sulphuric acid at 50 g/l. An increase in sulphuric acid concentration above 50 g/l. gives mainly increased aluminum dissolution. However, addition of nitric acid provides oxidizing conditions that promote leaching of sulphided nickel, overall nickel recovery reaching 60% at 6.0 g/l. nitric acid, with only a marginal increase in aluminum extraction above 16%. With more than 6.0 g/l. nitric acid, significant evolution of nitrogen oxides from nitric acid reduction was noted, although the amounts involved were not measured. Less than 5% of the tungsten was extracted under any of the conditions of Figure 1.

An alternative to catalyst leaching under oxidizing conditions is to preoxidize the pellets in air. Figure 2 shows how nickel extraction, without nitric acid addition, depends on preoxidation time at room temperature, the leachant being water or a sulphuric acid solution. In each case, the catalyst was leached for 4 hours, at which point the nickel content of the outlet solution had dropped from between 23 and 8 g/l. to less than 0.3 g/l. About 27% of the nickel was water soluble even without intentional preoxidation. Only a marginal increase in this extraction resulted from adding up to 50 g/l. sulphuric acid. However, the effect of sulphuric acid becomes pronounced following lengthy exposure of the catalyst to air at 25°C. After a 6 month preoxidation period, about 55% of the nickel was water soluble, an extraction that increased to nearly 80% with 50 g/l. added acid. Tungsten dissolution increased slightly with preoxidation time, and reached a maximum of 11% extraction into water after 6 months of preoxidation. As expected, tungsten extraction decreased with increasing acid level.

Figure 3 shows the effect of oxidation temperature on subsequent nickel extractions, which were not improved following oxidation for more than the 6

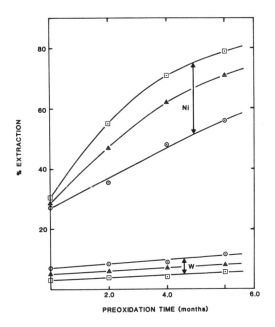

Figure 2 - Effect of preoxidation time at 25°C on leaching of catalyst at 80°C in water (o), 25 g/l. sulphuric acid (△), 50 g/l. sulphuric acid (□).

hour periods used. The nickel contents of the final outlet solutions were less than 0.3 g/l. Nickel recovery increased with increasing preoxidation temperature up to 300°C, beyond which a marked drop in extraction was seen. Also nitric acid addition improved nickel recoveries only if the preoxidation temperature did not exceed 300°C. Although use of excessively high oxidation temperatures could result in pellet sintering with a consequent suppression of nickel extraction, it is unlikely that the temperature reached in the present work would be sufficiently high to cause this effect. In reality, both pore volume and surface area increased with increasing oxidation temperature up to 600°C, at which the latter was 27% higher than for non-preoxidized catalyst.

It is probable that chemical rather than physical changes occuring in the pellets during oxidation at over 300°C are mainly responsible for suppressing subsequent nickel extraction. This is supported by the observation that the usual green colour of the outlet solution became yellow when catalyst preoxidized at over 300°C was leached. These effects may be due to the high temperature conversion of sulphur/oxygen compounds of nickel and tungsten to relatively acid resistant forms. The yellow solution colouration accompanying suppressed nickel dissolution may be associated with the behaviour of tungsten, especially if in compounds with nickel. For example, the nickel component of any nickel tungstate present may dissolve (13), leav-

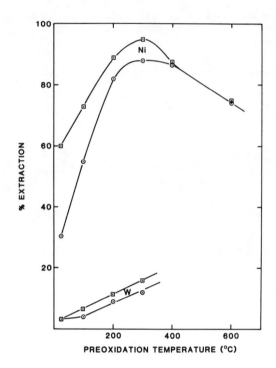

Figure 3 - Effect of 6 hour preoxidation periods at various temperatures on leaching of catalyst at 80°C in 50 g/l. sulphuric acid solutions with nitric acid contents of 0 g/l. (o), 6 g/l. (□).

ing a solid residue of tungstic acid:

$$NiWO_4 + H_2SO_4 \longrightarrow NiSO_4 + H_2WO_4 \tag{1}$$

Under the experimental conditions, equation 1 is probably a simplification in that, intially, nickel tungstate crystals may totally and momentarily dissolve within pores where local acidity is lower than that of the solution bulk. Upon transfer away from dissolution sites, dissolved tungsten would react giving hydrated oxides, $WO_3 \cdot xH_2O$, partially in yellow colloidal form. Tungstic acid may be considered to be a mixture of hydrated oxides (14), while acidification of a tungstate solution commonly gives yellow colloids (15). As shown in Figure 3, tungsten extraction rises with increasing preoxidation temperature. Substantial increases in apparent tungsten dissolution following oxidation at above 300°C were observed, but include tungsten in colloidal form. It should also be noted that the yellow colouration was not due to dissolved iron, which was present at less than 5 ppm. in all cases.

As described above, nickel extraction is lowered following preoxidation at above 300°C. Unexpectedly, nickel recovery from preoxidized catalyst is also suppressed if the leach temperature is too high. While rates of nickel dissolution from 'as received' pellets always increased with increasing temperature under otherwise the same conditions, preoxidized catalyst was leached most rapidly at about 80°C, above which nickel extraction is retarded. Figure 4 compares leaching behaviour at 80°C and 95°C for catalyst preoxidized at 300°C.

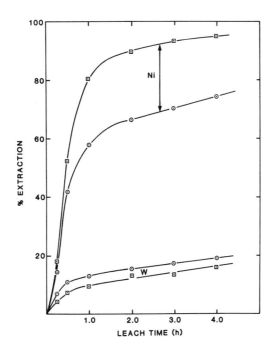

Figure 4 - Leaching of catalyst, preoxidized for 6 hours at 300°C, in 50 g/l. sulphuric acid with 6 g/l. nitric acid at 80°C (□), 95°C (o).

After 4 hour leaches, nickel extraction reached 95% at 80°C, but only 75% at 95°C. This effect was always observed for preoxidized catalyst, including material oxidized for several months at room temperature.

Despite decreased nickel extraction at 95°C (Figure 4), acid consumption was higher than at 80°C. Values of pH (measured at 25°C) for the outlet solution progressed from 3.1 to 1.0, and from 2.2 to 1.0 for the leaches at 95°C and 80°C respectively. The combination of reduced nickel extraction, but with increased acid consumption may be due to solid tungstic acid formation through oxidation of tungsten oxides by nitric acid. Nickel dissolution is markedly suppressed at 95°C by rapid production of tungstic acid which consumes nitric acid and coats the pores of the pellets. During catalyst preoxidation, it is assumed that tungsten sulphides are converted to oxides, the former being relatively unreactive when 'as received' catalyst is leached.

The presence of new tungsten-containing compounds, when formed during catalyst leaching, was seen as pale greenish-yellow deposits both on the black pellets and on internal reactor surfaces, especially at heating tape locations. Also, use of scanning electron microscopy showed that crystalline tungsten-bearing surface products were present on leached pellets.

Figure 5 shows how the nickel concentration in successive increments of outlet solution typically varies with leach time for conditions giving 95% extraction. The dissolved nickel level quickly reaches its maximum value, with most of the remaining reaction period representing the washing of leached nickel from the pellet bed.

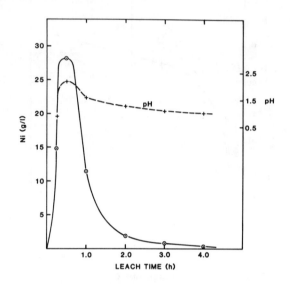

Figure 5 - Nickel concentration and pH in outlet solution during leaching of catalyst, preoxidized for 6 hours at 300°C, in 50 g/l. sulphuric acid solution with 6 g/l. nitric acid at 80°C.

Likewise, the pH of the outlet solution rises quickly to a maximum, before slowly reverting to the value for inlet leachant. In the present work, the conditions of Figure 5 gave the best combination of high nickel extraction (95%) and degree of selectivity for dissolution of nickel over that of tungsten and aluminum (16% and 38% extracted respectively). Previous workers (1) have proposed flowsheets for treating solutions generated by the leaching of scrapped petrochemical catalysts.

Conclusions

The sulphuric acid leaching response of scrapped sulphur-containing nickel/tungsten hydrocracking catalyst depends strongly on its oxidation state. Deliberate preoxidation of spent catalyst in air gives more readily soluble nickel-containing compounds than those present in unoxidized material. Provision of oxidizing conditions during catalyst leaching, with a small nitric acid addition, also promotes nickel extraction.

With preoxidation at above 300°C, the formation of relatively insoluble nickel-containing compounds gives decreased nickel extraction. If preoxidized catalyst is leached at over 80°C, nickel dissolution is also retarded, but by rapid formation of solid tungstic acid that blocks catalyst pores.

Assuming the goal of sulphuric acid leaching is high, but selective, recovery of nickel over tungsten and aluminum, the best extractions obtained were 95%, 16% and 38% respectively. Although this leaching behaviour is much better than achievable treating non-oxidized catalyst, either in a column or stirred reactor, subsequent dissolved metals separation and solution purification are nevertheless necessary. Various alternatives for solution treatment are currently being evaluated.

References

1. B.W. Jong and R.E. Siemens, "Proposed Methods for Recovering Critical Metals from Spent Catalysts", Recycle and Secondary Recovery of Metals, (P.R. Taylor, H.Y. Sohn, and N. Jarrett, eds.), TMS-AIME Symposium, Fort Lauderdale, December 1-4, 1985, 477-488.

2. G. Gravey, J. LeGroff and C. Gonin, "Preparation of Anhydrous Metallic Chlorides from Waste Catalysts", U.S. Patent 4,182,747, January 8, 1980.

3. H. Castagna, G. Gravey, and A. Roth, "Process for Recovering Molybdenum Values from Spent Catalysts", U.S. Patent 4,075,277, February 21, 1978.

4. W.A. Millsap and N. Reisler, "Cotter's New Plant Diets on Spent Catalysts - and Recovers Mo, Ni, W, and V Products", Eng. and Min. J., 179 (5) (1978), 105-107.

5. S. Toida, A. Ohno, and K. Higuchi, "Process for Recovering Molybdenum, Vanadium, Cobalt, and Nickel from Roasted Products of Used Catalysts from Hydrotreatment Desulfurization of Petroleum", U.S. Patent 4,145,397, March 20, 1979.

6. G. Gutnikov, "Method of Recovering Metals from Spent Catalyst Hydrorefining Catalysts", U.S. Patent 3,567,433, March 2, 1971.

7. S. Kelebek and P.A. Distin, "Nickel Extraction from a Nickel/Tungsten Spent Catalyst Using Column Leaching", J. Chem. Tech. Biotechnol., 44(4) (1989), 309-326.

8. F.K. Letowski and P.A. Distin, "Platinum and Palladium Recovery from Spent Catalysts by Aluminum Chloride Leaching", Recycle and Secondary Recovery of Metals, (P.R. Taylor, H.Y. Sohn, and N. Jarrett, eds.), TMS-AIME Symposium, Fort Lauderdale, December 1-4, 1985, 735-745.

9. F.K. Letowski and P.A. Distin, "Platinum and Palladium Recovery from Scrapped Catalytic Converters by a Chloride Leach Route - Laboratory Results and Pilot Plant Design", Separation Processes in Hydrometallurgy, (G.A. Davies, ed.), Soc. Chem. Ind. (UK) Symposium, Manchester, U.K., July 5-8, 1987, 68-76.

10. P.A. Distin and F.K. Letowski, "Recovery of Precious Metals from Materials Containing Same", Canadian Patent 1,228,989, November 10, 1987.

11. F.K. Letowski and P.A. Distin, "Development of Aluminum Chloride Leach Process for Platinum and Palladium Recovery from Scrapped Automotive Catalysts", New Precious Metal Technologies, IPMI Symposium, Montreal, Canada, June 11-15, 1989, In press.

12. D.S. Thakur and M.G. Thomas, "Catalyst Deactivation in Heavy Petroleum and Synthetic Crude Processing; A Review", _Appl. Catalysis_, 15(2) (1985) 197-225.

13. K. Osseo-Asare, "Solution Chemistry of Tungsten Leaching Systems", _Metall. Trans._, 13B (1982), 555-564.

14. M.L. Freedman, "Tungstic Acid", _J. Amer. Chem. Soc._, 81 (1959), 3834-3839.

15. G.D. Rieck, _Tungsten and Its Compounds_ (Pergamon Press, 1967), 38, 100.

AN ENGINEERING APPROACH TO THE ARSENIC PROBLEM IN THE EXTRACTION OF NON-FERROUS METALS

Norbert L. Piret, Dipl.-Ing.

Stolberg Ingenieurberatung GmbH
Consulting Engineers
D-5190 Stolberg/Rhld.

Albert E. Melin, Dr.-Ing.

Stolberg Ingenieurberatung GmbH
Consulting Engineers
D-5190 Stolberg/Rhld.

Abstract

Arsenic in non-ferrous metal smelter feed constitutes a major source of concern, since it increases production costs, interferes with Me-extraction, deteriorates product purity, presents an environmental hazard and faces problems of disposal. Proper engineering is a prerequisite to minimize costs due to internal As-recirculation, insufficient As-upgrading in intermediates and emission control. The optimization of the treatment of As-bearing materials relates to the Me/As ratios in each of the operating steps and the respective enrichment factor thereby. Considering these, it is possible for a particular feed to establish the most direct way to achieve the required Me/As ratio in the final product at the highest possible Me-recovery, minimize the number of intermediate products and optimize the As/Me ratios therein and to generate one final As-containing product. It will be demonstrated under which conditions hydrometallurgy is preferable to pyrometallurgy and whether a disposable As-bearing or a potential commercial by-product should be produced. The approach is illustrated by a number of practical examples.

Arsenic, its ocurrence, production and consumption patterns and its impact on non-ferrous metal production

With an overall abundance of 1.5 to 2.0 ppm, arsenic is relatively scarce, but it is found in high concentrations in sulphide deposits (60 ppm or more). It occurs mainly as sulphides, arsenides sulfosalts or oxidation products thereof in association with iron, copper, nickel, cobalt, lead, antimony, silver and gold, as arsenic bearing minerals (e.g. $FeAsS$, Cu_3AsS_4, $Cu_{12}As_4S_{13}$, $Cu_{12}Sb_4S_{13}$, $NiAs_2$ $(Co,Fe)AsS$, $CoAs_2$, Ag_3AsS_3) or contained within sulphides in solid solutions (the As-content in pyrite commonly ranges 0.02 to 0.5 %).

According to their importance, the major arsenic-bearing metaliferous deposits are the enargite-bearing Cu-Zn-Pb-deposits, arsenopyrite bearing pyritic Cu-deposits, native Ag-deposits, Ni-Co-arsenide deposits and arsenical refractory Au-deposits. The deposits being mined today represent a quantity of arsenic in excess of consumption, therefore, no deposit is currently mined solely for its arsenic content.

The bulk of the arsenic is produced worldwide as a byproduct from copper and lead smelting. Assuming an average of 6.5 kg As per t of copper and 0.8 kg As per t of lead reserve, the arsenic contained in the world copper and lead reserves is estimated at 2.1 million tonnes. Of this figure about 50 % can be recovered, so that recoverable reserves can be estimated at 1.0 million tonnes. Half of these deposits are located in Chile (260 kt), USA (50 kt), Canada (50 kt), Mexico (40 kt), Peru (40 kt) and Philippines (40 kt), with the remainder principally in Europe (France, Sweden), Africa (Namibia) and Oceania.

Commercial-grade arsenic trioxide with a minimum 95 % As_2O_3 content represents more than 90 % of the arsenic produced. Arsenic metal, with a minimum purity of 99 % represents less than 5 % of the total production.

The world refined production expressed as arsenic trioxide amounted in 1987 to 55,000 t. This production is distributed between the production countries as follows:

As_2O_3-world production 1987 (t)

Belgium	3,000
Canada	3,000
Chile	6,000
France	10,000
Mexico	6,000
Namibia	2,000
Peru	1,000
Philippines	5,000
Sweden	10,000
USA	-
USSR	8,000
Others	1,000

The production of metal is in the range 2,000 - 3,000 t/year.

The arsenic consumption in the USA in 1987 amounted to 23,000 t and the uses were distributed as follows:

Woood preservatives	68 %
Herbicides, fungicides	23 %
Glass, ceramic industry	3 %
Non-ferrous metal alloys	3 %
Others	2 %

Arsenic metal is primarily used for non-ferrous metal alloying.

The above statistical data, which are partly based on data from Reference 1, show that:

1. Arsenic interferes very often with the production of non-ferrous base-metals and precious metals as well as with the treatment of pyrites as a source of sulphuric acid. However, in the majority of cases the quantities of arsenic involved are too small to justify its conversion to a commercial product.

2. The overall quantity of arsenic set-free during the treatment of these non-ferrous metals exceeds the consumption of arsenic-based end products. As a matter of fact, it is well known that considerable amounts of arsenic exist in stockpiled arsenical flue dusts, in residues and dumps that have accumulated - not always under environmentally acceptable circumstances - from past non-ferrous metal processing activities.

On the other hand, considering that the products from non-ferrous metal processing must be virtually free of arsenic, in the production flowsheet there will be provided, in most of the cases, at least one separation step involving the metal and arsenic, namely during the metal refining. Consequently, there will be at least one outgoing arsenic-bearing product either as a waste product or an intermediate or a commercial product. Factors, which are important with regard to the selection of the type of arsenical byproduct are product quantity, costs of impoundment of arsenical wastes, costs of conversion to a commercial product.

With regard to arsenic emissions, it is important that the containment of arsenic is according to the environmental regulations not only in respect to the external limits of the processing operation but also as to any of the processing units located inside of the operation. Since the environmental legislation, in most industrialized countries, is stringent, as is shown in the table below, the necessity to comply to them unavoidably increases investment and operating costs. The number of process units involving arsenic must, therefore, be kept as low as possible.

Environmental legislations as to Arsenic Emissions	USA	FRG	Sources
Emission to air (mg/Nm^3)	$0.5^{1),3)}$	$1.0^{2)}$	1) WHO
Industrial water (mg/l)	$1.0^{3)}$	$0.1^{4)}$	2) TA-Luft 86 ($\geqslant 5$ g/h)
Domestic water (mg/l)	$0.05^{3)}$	$0.04^{5)}$	3) EPA, provisional
			4) 39 AbwVwV
			5) Trinkwasser VO 1980

For this reason, the unit operations involved in non-ferrous metal processing have been examined in respect to their ability and aptitude to effect good metal-arsenic separation. At the same time a method of evaluation has been outlined.

Classification of process steps as to metal/arsenic ratio

Pretreatment of primary materials

It is common to metallurgical operations that they possess the capability to deal with a particular impurity as long as its concentration does not exceed a certain limit, corresponding to a maximum tolerable ratio of impurity (Y) to metal (X). Hence, for the primary material input the ratio $Y_I/X_I \leqslant k$.

Primary materials such as concentrates for which $Y_I/X_I \leqslant k$ are usually denominated "conventional", whereas in the case $Y_I/X_I > k$, they are called "dirty" concentrates.

If that particular ratio is exceeded, a pretreatment of the primary material becomes necessary.

The purpose of the pretreatment is to lower the ratio Y_I/X_I to a normal level by eliminating part of Y or to convert Y to a species so that its effect is nullified.

For the pretreatment step, the enrichment factor e for element X to impurity Y in the main output phase should be as high as possible.

In the case of the impurity Y being arsenic, there are mainly three types of primary materials for which pretreatment is required:

- high arsenic copper concentrate in which $X_I/Y_I = Cu/As \geqslant 10$,
- refractory gold ores, in which gold is associated with arsenopyrite mineralisation,
- complex tin concentrates and preconcentrates.

There are available pyro- and hydrometallurgical operations for performing the pretreatment:

- pyrometallurgical operations:
    * partial roasting of copper concentrates,
    * dead roasting of refractory gold ores or concentrates,
    * dead roasting of tin concentrates,
    * fuming of tin preconcentrates.

- hydrometallurgical operations:
    * preleaching of copper concentrates in a sodium sulphide solution,
    * oxidative pressure leaching or other oxidative leaching processes for refractory gold ores or concentrates.

Treatment of the primary materials

Treatment of the primary materials involves separation of the main metal(s) from the bulk of the other elements present. Thereby the ratio of any of the elements present to the metal does not exceed a critical value. With regard to a particular impurity Y, there may be elimination of it or not. The enrichment factor of X to Y can vary widely, but usually operating parameters can be optimized in this respect.

Refining of the metal

For the material entering the refining step again the ratio Y/X is not allowed to exceed a predetermined value.

It is obvious that for a given combination of metals and impurities, particular processing steps may not be required in one or more of these 3 stages. With regard to arsenic, attention is to be given to its elimination as well in the treatment as in the refining stage, occasionally necessitating also a pretreatment stage.

In the final metal only traces of arsenic are permitted according to specifications on purity requirements:

Metal	Cu	Zn	Pb	Ni	Co	Sn	Iron ore
As (ppm)	1	1	1	1	0.5	<100	<500
X_F/As_F *	10^6	10^6	10^6	10^6	2.10^6	$>10^4$	$>10^3$

* order of magnitude

In the case of iron ore ex pyrite concentrate roasting, the As content may not exceed 500 ppm in relation to the iron content of 65 % in the iron ore.

Relationship between enrichment factor and distribution factors

The effectiveness of a process step with regard to the separation of element X from element Y may be evaluated by means of the enrichment factor which is based on distributions.

Definition of Distribution

The following diagram is the basis for the present analysis:

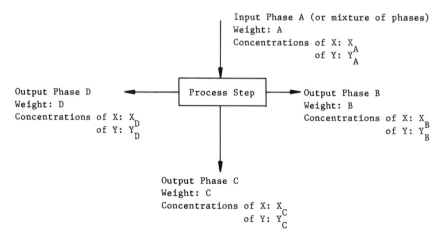

Mass balances

Relating to element X:
$$X_A \cdot A = X_B \cdot B + X_C \cdot C + X_D \cdot D$$

Relating to element Y:
$$Y_A \cdot A = Y_B \cdot B + Y_C \cdot C + Y_D \cdot D$$

Distribution Factors:

Relating to Element X : d_X

Relating to element Y : d_Y

in Phase B:
$$d_{X,B} = \frac{X_B \cdot B}{X_A \cdot A} \qquad d_{Y,B} = \frac{Y_B \cdot B}{Y_A \cdot A}$$

in phase C:
$$d_{X,C} = \frac{X_C \cdot C}{X_A \cdot A} \qquad d_{Y,C} = \frac{Y_C \cdot C}{Y_A \cdot A}$$

in phase D:
$$d_{X,D} = \frac{X_D \cdot D}{X_A \cdot A} \qquad d_{Y,D} = \frac{Y_D \cdot D}{Y_A \cdot A}$$

whereby:
$$d_{X,B} + d_{X,C} + d_{X,D} = 1 \quad (1) \qquad d_{Y,B} + d_{Y,C} + d_{Y,D} = 1 \quad (2)$$

Definition of enrichment factors

The enrichment factor for element X to element Y in one given output phase of a process step is defined as the ratio of the concentration of element X to element Y in that particular phase divided by the equivalent ratio in the input phase(s).

Hence:
in the phase B:
$$e_B = \frac{X_B/Y_B}{X_A/Y_A} = \frac{X_B \cdot B}{X_A \cdot A} : \frac{Y_B \cdot B}{Y_A \cdot A} = d_{X,B}/d_{Y,B} \quad (3)$$

in the phase C:
$$e_C = \frac{X_C/Y_C}{X_A/Y_A} = \frac{X_C \cdot C}{X_A \cdot A} : \frac{Y_C \cdot C}{Y_A \cdot A} = d_{X,C}/d_{Y,C} \quad (4)$$

in phase D:
$$e_D = \frac{X_D/Y_D}{X_A/Y_A} = \frac{X_D \cdot D}{X_A \cdot A} : \frac{Y_D \cdot D}{Y_A \cdot A} = d_{X,D}/d_{Y,D} \quad (5)$$

The above equations show that the enrichment factor of X to Y in a given phase is equivalent to the ratio of the distribution factor of X in that phase to the distribution factor of Y in that phase.

Incorporation of the relationships (3), (4) and (5) in the equation (2) results in the equation:

$$\frac{d_{X,B}}{e_B} + \frac{d_{X,C}}{e_C} + \frac{d_{X,D}}{e_D} = 1 \qquad (6)$$

Hence, if one of the distributions $d_{X,B}$, $d_{X,C}$ or $d_{X,D}$ is known as well as the compositions of the incoming phase(s) and of the outgoing phases, then the system is completely defined.

Practical implementation in flowsheet development

For good separation of X and Y from each other in a particular output phase of the process step under consideration, the value of e must be very large or small. In a phase for which the value of e equals 1, there is no enrichment of X to Y as compared to the input phase(s). In the flowsheet sequence the phases must be processed successively until the required final product purity as well for element X as for element Y is attained.

The flowsheet development is based on the compositions of the input materials and on the desired endproducts. Therefore, the value of the overall enrichment factor, e_o, is known:

$$e_o = \frac{X_F/Y_F}{X_I/Y_I}, \text{ (whereby I = Initial and F = Final)}$$

Thus, if, for one particular mass stream (main - or sidestream), the number of steps is n, then:

$$(X_I/Y_I) \times e_1 \times \ldots \times e_n = X_F/Y_F$$
and $e_1 \times \ldots \times e_n = e_o$

This value e_o must be attained utilizing the least number of streams and the least number of steps in each of the streams. This is particularly true if the element Y represents arsenic, since each process step presents a potential source of environmental contamination which requires special consideration.

In pyrometallurgical process steps the number of output phases usually is three (e.g. metal-, slag-, gas phases), occasionally this number may be four (e.g. metal-, matte-, slag-, gas phases) or five (metal-, matte-, speiss-, slag-, gas phases). Also in pyrometallurgical unit operations separation is mostly incomplete, which means that the distribution factors differ from zero or from unity. In addition, distribution factors may fluctuate widely depending on the operating conditions in the type of process under consideration. This fact, of course, does not facilitate the process flowsheet development.

In hydrometallurgical processing the number of output phases frequently does not exceed two (e.g. aqueous phase, solid phase). Furthermore, unlike in pyrometallurgical unit operations, the degree of separation is, usually high, meaning that the distribution factors are in the vicinity of unity or zero, resulting in values for the enrichment factor which are high or low (i.e. in the vicinity of zero). Because of this characteristic, in pyrometallurgical process flowsheets in multiple occasions the incorporation of hydrometallurgical separation steps has proven to be highly beneficial in order to achieve rapid elimination of some of the byproducts.

Separation of arsenic by means of evaporation

The process of separation consists of two parts:

- Separation of the arsenic from the condensed phase(s)
- Separation of the arsenic from the arsenic bearing gases.

As to the separation of the arsenic from the condensed phase(s), the latter may be a solid phase or a mixture of solid phases, in which case the removal occurs by means of roasting or can be a liquid phase or phases, in which case the arsenic elimination occurs during smelting.

The arsenic can be eliminated from a condensed phase under different forms. The compounds As_4O_6, As_4, As_4S_4 and $AsCl_3$ all have sufficiently high vapor pressures above a given temperature to enable their evaporation. Figure 1 shows that, besides $AsCl_3$, As_4O_6 possesses by far the highest vapor pressure. However, which compound will be predominant will depend on the actual partial pressures of oxygen and sulphur. Figure 2 shows the stability areas of the gaseous compounds in the system As-S-O at 2 temperatures. Also indicated is the area of stability of non-volatile arsenic pentoxide.

The elimination of arsenic through the gasphase is normally conducted under the conditions of stability of As_4O_6, which are indicated in Figure 2. This stability area is shifted towards higher O_2 potential at increasing temperature and the increase of the partial pressure of SO_2 expands it. The presence of sulphur is beneficial for the elimination of arsenic as As_4O_6 although it does not necessarily mean that evaporation is in the form of sulphide (7). Depending on conditions As can, however, be removed as the sulphide (e.g. as in the case of sulphidizing vaporisation) or as metal (As_4), as in the case of vacuum refining of copper matte and vacuum refining of arsenic metal.

As_4O_6: A – B: Modification ortorombic (arsenolite), A'–B': Modification monoclinic (claudetite)
As_4: C – D, $(As_2S_3)_2$: E – F, $AsCl_3$: G – H, As_4S_4: E' – F'

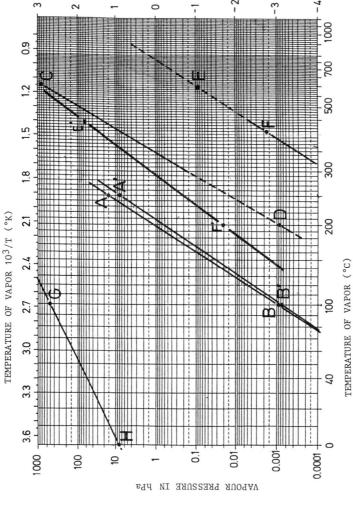

Fig. 1: Vapor pressure of As and As-compounds as a function of temperature
Data from: Metallurgical Thermochemistry, 5th Edition, 1979, Kubachewski and Alcock

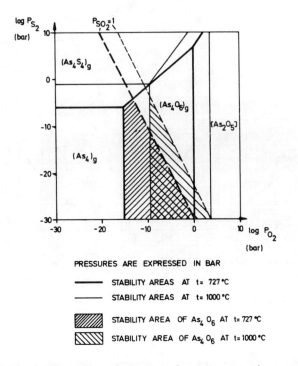

Fig. 2: Areas of predominance of gaseous arsenic compounds

Separation of arsenic from gases

Separation of dust and fumes from metallurgical gases is commonly performed in two stages, the first stage removing the bulk of the dust load and the second stage carrying out the final cleaning.

Final gas cleaning step. Arsenic in flue gases is in the form of the highly volatile arsenic trioxide. The concentration of gaseous As_2O_3 in the form of arsenolite (As_4O_6) in the saturated gas is given in figure 3 as a function of temperature. According to the diagram the concentrations of volatile arsenic at sea level are as follows:

°C	70	100	110	130
As[1] (mg/Nm³)	0.9	12	29	148

[1] as As_4O_6

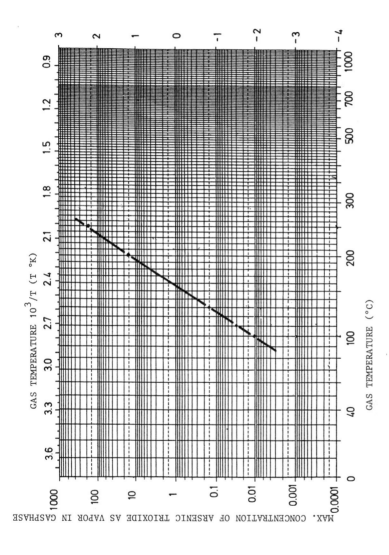

Fig. 3: Concentration of gaseous As_2O_3 as a function of temperature

- Dry gas cleaning system. In order to achieve 1 mg/Nm³ in the clean gas as required by the environmental regulations or to have sufficiently low levels of arsenic prior to entering the acid plant, gases must, firstly, be cooled to below 70 °C, if a dry gas cleaning system is being used, to ensure that the residual concentration of arsenic in the vapor phase is below that limit and, secondly, the collection efficiency of the dry gas cleaning system must be so that less than 0.5 mg/Nm³ are allowed to pass the collection device as solid particles. In the case of a baghouse in order to fulfill this requirement, the linear gas velocity must not exceed 0.5 m/min, corresponding to a gas volume of 30 m³/m².h.

Despite the fact that the size of the baghouse could be increased as required, the operation of a baghouse at low temperature is most difficult on account of:

- exceeding the dew point of sulphuric acid in the case of SO_2-bearing gases.
- the expenditure for cooling the gases to the low temperature which is required.

Furthermore, there is one other difficulty which is encountered during cooling of arsenic trioxide vapor bearing gases: As_2O_3 crystallizes out on cold surfaces as the cooling of the gases proceeds. Therefore, quenching or shock-cooling of the gases to the temperature of dry filtration should be provided. Shock-cooling is conducted by means of dilution of the gases with cold air e.g. using a device allowing the air to enter in the center of the gas stream (8). This system was installed at the arsenic roasting plant of Campbell Red Lake Mines (9) and the one of El Indio (10). The devices appear to be effective in preventing As_2O_3-crust building. However, the size of the baghouses increases in the proportion to the volume of air added and the dilution makes the gases unsuitable for H_2SO_4-production.

Due to the fact that, by means of dry gas cleaning, neither the environmental requirements nor the technical requirements for an acid plant can be fulfilled, a wet gas cleaning system will be necessary in most of the cases.

- Wet gas cleaning system. There are various methods for performing wet gas cleaning of arsenic-bearing waste gases, which are characterized by differing collection efficiencies. The differences in performance are related to the specific efficiencies of the transfer of the particles from the gas phase to the aqueous phase.

The following 3 systems have in simplified form the following characteristics:

	Collection efficiency	Power consumption	Investment costs
Wet scrubber	lowest	highest	lowest
Ionising washer	medium	medium	medium
Wet electrostatic precipitation	highest	lowest	highest

- Products obtained in the final gas cleaning system. In the dry gas cleaning system, the product purity will depend on the collection efficiency of the primary gas cleaning system as well as on the composition of the gas. For arsenic trioxide, collected in the final gas cleaning system As_2O_3-qualities in excess of 95 % have been reported (10).

In the wet gas cleaning system, arsenic will be washed out and eventually dissolve in the acidic solution.

Primary gas cleaning sytem. The primary gas cleaning system is normally a dry gas cleaning system, operating at higher temperature (400 - 300 °C) and designed for collection of the bulk of the dust load of the gases. With regard to arsenic in the gases, if it is in trivalent form, it will be present in the vapor phase (as As_4O_6) and, consequently, pass the collection device. Hence, as long as the arsenic is in the vapor phase, there will be excellent opportunity for separation of arsenic from the bulk of the dust, which could then be recycled without recirculation of arsenic.

It is known that the tendency for the formation of arsenic pentoxide (As_2O_5), which has a low vapor pressure, is small, even in the presence of excess O_2, unless metal oxides are present in the gas stream which can build stable metal arsenates.

In the figure 4 the equilibrium for the formation of ferric arsenate at 900 °C is shown. Similar diagrams can be constructed for the other metal oxides such as PbO, ZnO, etc. which are commonly present in metallurgical gas streams.

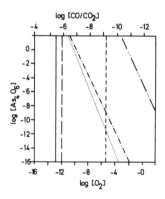

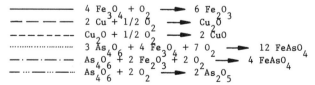

$$4\, Fe_3O_4 + O_2 \rightarrow 6\, Fe_2O_3$$
$$2\, Cu + 1/2\, O_2 \rightarrow Cu_2O$$
$$Cu_2O + 1/2\, O_2 \rightarrow 2\, CuO$$
$$3\, As_4O_6 + 4\, Fe_3O_4 + 7\, O_2 \rightarrow 12\, FeAsO_4$$
$$As_4O_6 + 2\, Fe_2O_3 + 2\, O_2 \rightarrow 4\, FeAsO_4$$
$$As_4O_6 + 2\, O_2 \rightarrow 2\, As_2O_5$$

Fig. 4: Equilibrium diagram for reactions of some oxides of arsenic, iron and copper (T = 900 K)

To prevent formation of metal arsenates the oxygen potential should be as low as possible and the temperature of the primary gas cleaning as high as possible. From the present survey conducted with regard to the subject of producing a primary dust as low as possible in arsenic, it can be concluded that there is room for improvement in existing metallurgical operations.

Separation of arsenic from gases in the form of sulphide. It has been proposed to separate As from gases in the form of As_2S_3, which is less volatile and which could be condensed. Arsenic sulphide could be generated directly during the roasting step (10, 11) or could be formed by injection of sulphur-bearing materials into the gasstream (12). An evaluation of this proposal is difficult at this stage, since new technology (condensor for As_4S_4 vapor) needs be developed. Furthermore, separation of primary dusts from fumes may be rendered difficult.

Roasting of arsenic bearing materials

To obtain efficient removal of As by roasting, it has to be ensured that the operation takes place within the stability area of one of the volatile arsenic compounds. Purpose of the roasting is the elimination of sulphur. The behavior of As during roasting can best be analysed with the aid of the S_2-O_2-As-Fe-Cu diagram, represented in figure 5.

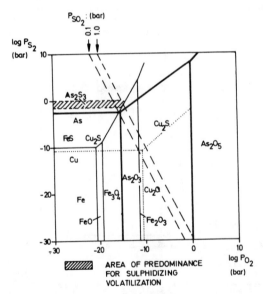

Fig. 5: Stability diagram of the system S_2, O_2, As, Fe, Cu at 727 °C

It is necessary to prevent the the formation of non-volatile pentoxide, which would combine to the stable metal arsenate compound. For this purpose the oxidation potential must be sufficiently low. The stability diagram at 1000 K indicates that the iron in the calcine must be at least in the magnetite form (Boliden process) or in the sulphide form (partial roasting).

The simplified diagram of figure 6 shows the areas of stability of FeS, Fe_3O_4, Fe_2O_3 and $FeAsO_4$ as a function of temperature for a given P_{SO2}. Increased temperature will enlarge the stability area of Fe_3O_4 and FeS, therefore, As elimination will be enhanced.

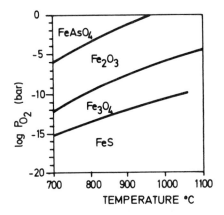

Fig. 6: Stability diagram of solid phases in the system Fe-As-O-S as a function of temperature

Reaction rates for arsenic elimination in roasting slow down towards the end of roasting, due to the fact that solid state transfer is involved. Retention times, however, are determined by the desired degree of sulphur elimination. The longer the retention time for sulphur elimination, the more complete the elimination of arsenic. This is illustrated in figure 7 by means of arsenic profiles during multiple hearth roasting. Thereby, the conditions must be such that there is no opportunity for the formation of non-volatile compounds.

- Partial roasting: Hereby the calcine will still contain FeS, there will be no tendency for formation of stable metal arsenates. As elimination will depend on retention time. Multiple Hearth (MH)- and one-stage Fluidized Bed (FB)-roasting are applicable.

- Magnetite roasting: In this case the calcine will be in the Fe_3O_4 form As elimination will be satisfactory. The Boliden process is based on one stage FB-roasting to magnetite (13).

- Dead roasting: The calcine will be principally in the Fe_2O_3 form with residual sulphur. As elimination can be achieved by means of
  * MH-roasting
  * Two stage FB-roasting as in the BASF-process (14).

- Reducing roasting: To prevent formation of ferric arsenate, in some occasions the roasting is carried out under reducing conditions. This is especially the case, if the required degree of sulphur elimination is high such as in the dearsenizing roasting of complex tin concentrates (7), but has also been applied in partial roasting of copper concentrates (15), to obtain low residue As levels (0.3 %).

- Sulphidizing roasting: In the sulphidizing evaporation pretreatment process of Outokumpu (11), it was proposed to operate in the P_{S2} range 0.05 - 0.1 bar at 600 - 800 °C, which would result in 0.1 % As in calcine, independently of the degree of desulphurisation.

Roasting operations for arsenic removal have specific fields of application, which are described hereafter.

Pretreatment of copper concentrates by means of roasting. Pretreatment of copper concentrates by partial roasting for arsenic removal has been applied commercially for a long period of time (16). This operation was performed in MH furnaces and was characterized by low throughputs, high maintenance and energy costs as well as difficulties with ambient control.

Despite of this, MH-roasting has found renewed interest, as is illustrated by the Nesa/Nichols MH-roaster of El Indio (15, 17). The latter has a throughput of 200 t/day and a production capacity of 20 t/day of As_2O_3. Roasting decreases the sulphur content from 37 to 22 % and the arsenic from 7 to 0.3 % (i.e. 95 - 97 % As elimination), corresponding to the enrichment factor of 35. The good elimination of As, despite the high throughput rate is due to the fact that roasting is conducted under very reducing conditions and high temperatures according to the findings of (18).

Pretreatment of copper concentrates by means of FB-roasting is in operation at Pasar on Lepanto concentrates. The 180 t/day one-stage Lurgi FB roaster (19) operates under less stringent reducing conditions and achieves an enrichment factor of 10 (1 % As in calcine). By this means an acceptable quality of copper concentrates is produced from the commercial point of view.

Improvements to the FB-roasting technology for As (and Sb) removal by means of increasing the retention time were proposed in (20).

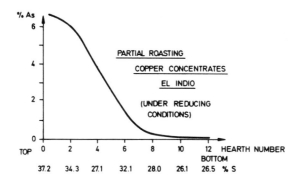

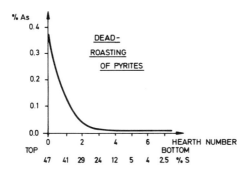

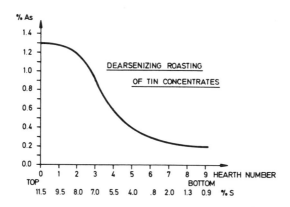

Fig. 7: Arsenic concentration on the hearths of commercially operated multiple hearth roasters. Profiles for various types of materials

Pretreatment of gold concentrate by roasting. Pretreatment of gold concentrates by roasting is conducted with the objective to render the gold amenable to leaching. Thereby, most of the sulphur in the concentrate is oxidized to SO_2 on account of which the concentrate particles become porous and gold accessible to leach attack. For this reason, it is not a prerequisite that arsenic is removed. It may also remain in the calcine e.g. as ferric arsenate.

Commercial application of pretreatment of refractory Au concentrates by roasting include:

- one stage FB-roasting (Dickenson, Cochenour-Willans (21)),
- two stage FB-roasting according to BASF-process, (Giant Yellowknife (22, 23, 24)), Campbell Red Lake (9, 21), New Consort Mine (25)).

Operating results are as follows:

	Cochenour Willans			Campbell Red Lake			Giant Yellowknife			New Consort Mine (S.A.)		
Stages	1			2			2			2		
Feed (t/d)	15			60			130			33		
	Feed	Cal-cine	Extrac-tion	Feed	Cal-cine	Extrac-tion	Feed	Cal-cine	Extrac-tion	Feed	Cal-cine	Extrac-tion
% S	18.4	1.6	93	20.2	1.6	96	17.5	2.5	88	6.6	0.7	92
% As	6.3	1.3	84	9.5	1.0	81	7.6	1.0	89	6.8	1.0	88
As_2O_3-Prod. (t/d)	n.a.			6.3			n.a.			n.a.		
Overall As_2O_3-recovery from gases (%)	n.a.			99.8			n.a.			n.a.		

In these cases relatively high arsenic elimination was obtained, likely due to the fact that sulphur in calcine was relatively high and conditions still reducing.

For the purpose of Au recovery from Greek arsenopyrite concentrates, as a pretreatment method, a one-stage FB-roasting under oxidizing conditions was proposed (METBA) (26). The elimination of S was 99 % and of As only 55 %. The arseniferous cinders (0.5 % S, 2.8 % As) were found to have a slight As solubility in water, which necessitated the incorporation of a dearsenification leaching step (150 g/l H_2SO_4-solution at 60 °C) prior to subsequent thiourea leaching. The latter resulted in more than 94 % PM extraction.

The same concentrates were also submitted to a dearsenizing roasting with air (equivalent to dead roasting at 2 % residual S) and under inert atmosphere (equivalent to partial roasting at 30 - 35 % residual S), resulting in a final As content of 0.1 - 0.3 %). Subsequent cyanide leaching gave 75 % respectively 88 % Au-extraction (27).

These examples demonstrate that the determination of the optimum roasting conditions for pretreatment prior to Au leaching is most important.

Dearsenizing roasting of complex tin concentrates. Complex tin concentrates grading 35 to 45 % Sn are associated with pyrites and also arsenopyrites. Prior to reducing smelting, it is essential to eliminate sulphur because of the high volatility of SnS, which would form during reducing smelting, causing excessive fume generation. Also the presence of arsenic is undesirable because of the need to eliminate it during refining. Dearsenizing roasting of complex tin concentrates aims at simultaneously achieving low S and As in the calcine ($S \leqslant 1$ %, $As \leqslant 0.1$ % - 0.2 %). This requirement necessitates narrow control of operating parameters. Commercially this process is applied in MH-furnaces at the EMV tin Smelter at Vinto (28, 29) and in a rotary furnace at the Albert Funk Tin Smelter at Freiberg (30, 31).

Intermediate arsenic-bearing products, such as "dry" Fe-As-dross, are advantageously returned to the primary roasting step. The enrichment factor of about 10 is achievable. The addition of carbon appears beneficial for As elimination.

Roasting of tin-bearing speiss. Roasting of speiss generated during blast furnace smelting of Pb-Cu primary material was performed at MHO (98) and is believed to have been replaced by an oxidative pressure leaching process in sulphuric acid medium.

Roasting of speiss formed during reduction smelting of tin raw materials was conducted at Capper Pass in MH-furnaces, which have now been replaced by a FB-roaster (32).

The composition of the speiss at Capper Pass is reported as follows:

	Sn	As	S	Fe	Ag
%	17 - 20	10 - 20	5 - 10	50 - 70	0.03 - 0.06

For adequate elimination of As, the roasting is performed as a partial roast at about 750 °C resulting in a calcine composition:

	S	As
%	5 - 8	0.6 - 1.0

For this purpose, pyrite is added to have a S/As-ratio of about 2 in the feed. The quantity of pyrite to be added is not negligible, depending on the As content of the speiss, and introduces an equivalent amount of iron into the circuit. As elimination is reported to be in excess of 90 %. Hence, an enrichment factor of about 10 is obtained.

The capacity of the operation is small and estimated at 5 to 10 t speiss/day depending on the As-content. Arsenic is removed from the gases by wet scrubbing using river water, which is discarded after dilution to below the permissible As-limit.

Roasting of cobalt-arsenides. Concentrate from Bou Azzer (Morocco), containing 50 % As and 10 - 12 % Co, were submitted to partial roasting removing 70 % of the arsenic as As_2O_3 (43). The calcine, analysing Co 18 %, As 17 %, was treated hydrometallurgically by Métaux Spéciaux S.A. (P.U.K.). The process included HCl-Cl$_2$-leaching, Fe-As-removal, Au-recovery from leach solution, solution purification and cobalt electrowinning with the simultaneous regeneration of chlorine (chlorine leach process).

Production capacity was 1,500 t Co/year. The operation was halted in 1982 due to exhaustion of the morrocan deposit.

The simplified flowsheet is shown in Figure 8.

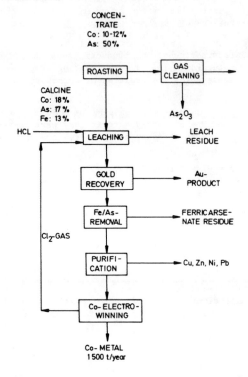

Fig. 8: P.U.K. process for Co-arsenide treatment (43)

<u>Arsenic behavior during pyrite roasting and cinders treatment</u>. Traditionally, pyrites have been a source of sulphur for the production of sulphuric acid. Thereby, pyrite cinders are generated, which contain almost all of the impurities which were initially present in the pyrites.

The composition of the pyrites can vary within a wide range, as can be seen from the following analysis:

Fe %	S %	As %	Sb %	Zn %	Pb %	Cu %	Ag ppm	Au ppm
42-49	45-52	0.05-0.4	0.01-0.05	0.2-4	0.1-1.5	0.04-1.5	10-60	0.5-2

The production of H_2SO_4 from pyrites more or less implies the subsequent treatment of pyrite cinders, since their impoundment becomes more and more difficult. To convert pyrite cinders to iron ore the non-ferrous metals must be removed to a large extent.

The maximum level of impurities tolerated in iron ore, based on the quality requirement of the iron, is as follows:

	Fe	S	As	Cu	Zn	Pb
%	65	$\leqslant 0.5$	$\leqslant 0.05$	$\leqslant 0.1$	$\leqslant 0.1$	$\leqslant 0.02$

The above analysis does not imply that iron/steel producers will readily accept this type of iron ore due to the possible implications to the operation because of the presence of the impurities.

With regard to arsenic, it normally needs be separated from the iron during the primary roasting of pyrites, since during the subsequent cinders treatment process for the removal of non-ferrous metals, there are no commercial facilities which are able to eliminate it.

- Primary roasting One stage FB-roasting of pyrites does not eliminate arsenic to a sufficiently low level, so that the cinders will be arsenical from the point of view of iron ore.

There are three processes which can produce pyrite cinders, which have an arsenic content low enough to be acceptable as an iron ore ($\leqslant 0.05$ %). The enrichment factor in these processes is about 10.

- Multiple hearth roasting:
  * applied in Spain and Portugal
  * hematite calcine
  * high residual sulphur (3 - 4 %)
  * coarse grained (0 - 6 mm)

- Boliden 1-stage FB-roasting (13, 33)
  * applied in Sweden (Hälsingborg)
  * magnetite calcine
  * low residual sulphur (ca. 1.5 %)
  * fine grained (90 % < 0.5 mm)

- BASF 2-stage FB-roasting (14)
  * applied in Barreiro (Quimigal) (34)
  * hematite calcine
  * low residual sulphur (ca. 1.0 - 1.5 %)
  * fine grained (90 % $\leqslant$ 0.4 mm)

- Cinders treatment There are two commercial processes available for the removal of non-ferrous metals from pyrite cinders. None of them, however, removes arsenic to an appreciable extent.

  - chloridizing roasting: as was applied at DKH (till 1983) and still in application at Quimigal (34)
    * suitable for hematite calcine, less for magnetite calcine
    * suitable for coarse calcine less for fine grained material (due to subsequent leaching)
    * poor removal of lead

- chloridizing volatilization: by means of the Kowa Seiko process as operating in Japan (Tobata) (35, 36, 37) and installed but inoperative at Barreiro (Quimigal) (34).
  * grain size of pyrite not important since pelletized
  * good removal of non-ferrous metals, including lead and precious metals
  * iron ore in the form of burnt pellets.

A third process, the Montedison chloridizing volatilization process (38), was developped especially for the removal of both non-ferrous metals and arsenic, but no commercial plant has been set-up. This process consists of two steps. The chlorination step under oxidizing conditions which is preceded by a FB magnetite roasting because of the necessity that the calcine is in the magnetite form. Thereby, the arsenic is volatilized.

The solutions, originating from leaching of the calcine ex the chloridizing roasting or from scrubbing of the metal chloride-bearing gases ex the chloridizing volatilization, are treated for arsenic removal and metal recovery.

Hydrometallurgical removal of impurities from one-stage FB-roasted arsenical pyrite cinders (0.5 % As, 3 % Zn, 4 % S) was investigated (39).

Using Cl_2 alone as leaching agent, resulted in poor extraction of Zn (40 %) and As (60 %) and 3 % iron codissolution, due to the presence of zinc ferrites, whereas other metals (Au, Ag, Cu, Pb) were satisfactorily extracted. Using Cl_2 and HCl resulted in 80 % Zn and As extraction but also in a prohibitive iron solubilization of 8 %.

In either case, the quality of the leach residue did still not correspond to the requirements of an iron ore, so that subsequent treatment by means of the Kowa Seiko process would still be necessary in case of iron production.

It was, therefore, concluded that the first alternative - simple Cl_2-leaching - would be adequate and economically more attractive as a process for the recovery of non-ferrous metals from pyrite cinders alone rather than for the production of iron ore (40).

Conclusions. Roasting operations present an excellent opportunity for separation of arsenic from the metal to be recovered. The selection of operating conditions is, thereby, most critical. The design of the gas cleaning must provide the possiblity of collecting the dust carry-over and the volatilized arsenic separately.

Table 1 presents a summary of selected roasting operations with regard to As elimination.

Table 1: Summary of selected roasting operations and their relationship to arsenic elimination

Pos.	Type of roasting	Residual sulphur	Form of the calcine (major)	Furnace type	Application to:	Purpose of process	Arsenic removal	Enrichment factor	Operation	References
1a	partial	20 - 25	FeS	FB	Cu conc.	pretreatment for As removal	good	10	Pasar	(19)
1b	partial	25 - 27	FeS	MH	Cu conc.	pretreatment for As removal	good	10	Boliden El Indio (*)	(16) (15)
1c	partial	20 - 25	FeS	MH	Cu conc. Cu/Ni conc.	S-removal for matte grade cont.	n.a.	-	-	-
1d	partial	20 - 25	FeS	FB	Cu conc. Cu/Ni conc.	S-removal for matte grade cont.	n.a.	-	Inspirat., Falconbr.	-
1e	partial	5 - 7	FeS/ Fe$_3$O$_4$	FB	Sn-speiss	As-removal	good	10	Capper Pass	(32)
1f	partial	n.a.	n.a.	MH	Co-As conc.	As-removal	60 - 70 %	ca. 3	P.U.K (Bou Azzer)	(43)
2a	dead	≤ 1.0	Fe$_2$O$_3$	MH	complex Sn conc.	pretreatment for S/As-remov.	good	10	EMV	(28, 29)
2b	dead	1 - 1.5	Fe$_2$O$_3$	FB	Cu conc.	S-removal	none	1	Brixlegg	(42)
2c	dead	< 1.0	Fe$_2$O$_3$	FB	Zn conc.	S-removal	none	1	-	-
2d	dead	< 2.0	Fe$_2$O$_3$	FB	pyrite conc.	S-removal	none	1	-	-
2e	dead	3 - 5	Fe$_2$O$_3$	MH	pyrite conc.	S-removal	good	10	Barreiro Tharsis	(34) (41)
2f	dead	< 0.1	Fe$_2$O$_3$	FB	pyrite conc.	S-removal	none	1	Inco/ Falconbridge	-
3	magnetite	1.5	Fe$_3$O$_4$	FB	pyrite conc.	S/As-removal	good	10	Boliden (Hälsingb.)	(13, 33)
4a	dead BASF-2 stage	1.0 - 1.5	Fe$_2$O$_3$	FB	pyrite conc.	S/As-removal	good	10	Barreiro	(14, 34)
4b	dead BASF-2 stage	1.5 - 2.0	Fe$_2$O$_3$	FB	gold conc.	S/As-removal	good	10	Campbell, Giant Yellowknife	(9, 21) (22, 24)
5	sulphating	< 1.0	Fe$_2$O$_3$	FB	Cu-Co conc. pyrrhotite conc.	sulphation of non-ferrous metals	none	1	-	-
6	chloridizing	0.1 - 0.4	Fe$_2$O$_3$	MH	pyrite cinders	sulfation of non-ferrous metals	none to little	ca. 1	DKH, Barreiro	(34)

(*) strongly reducing conditions

Deportment of arsenic during smelting

During smelting, elimination of As through the gaseous phase will depend on the activity of the arsenic in the condensed phases. The presence of more than one phase renders the evaluation more complex. Examples of condensed phases in various systems are given below:

Number of phases	Type of condensed phase	Application
1.	Slag	- Tin fuming - Zinc fuming - Waelz process
2.	Slag, matte Slag, metal	- Copper smelting - Cu-slag cleaning - Tin smelting - Secondary copper smelting - Ferro-nickel smelting
3.	Slag, matte, metal Slag, speiss, metal	- Pb blast furnace smelting - Pb/Zn blast furnace (ISP)
4.	Slag, matte, metal, speiss	- Pb/Cu blast furnace smelting (MHO) - Pb blast furnace smelting (Oroya)

The tendency of arsenic to distribute to any of these liquid phases rather than to the gas phase depends on the affinity of arsenic:

- for slag components, which is low because of the requirement of low metal losses to slag in smelting.
- for matte components, which increases with decreasing S/Me ratio.
- for the non-ferrous metals, which is appreciable but which depends on the particular metal.
- for metallic elements to form stable compounds (speiss).

The elimination of arsenic from the smelting circuits occurs:

- through the final slags.
- through the gas phase and the products recovered therefrom (dusts, fumes, sludges).
- other intermediate products, which require separate treatment (speiss, As-rich mattes, etc.).

Volatilization from one sole condensed phase

Fuming of low grade tin concentrates/preconcentrates and slags. In fuming of low grade tin concentrates, the 2 phases involved are a molten slag and a gas phase. The tin content of the concentrates is absorbed in the slags and converted to volatile tin sulphide upon addition of pyrite. Volatised SnS is postcombusted to SnO_2. As present in the concentrates follows Sn to a large extent.

The simplified reactions are:

Sn-dissolution in slag: $SnO_2 + CO + slag \longrightarrow SnO.slag + CO_2 \uparrow$
Pyrite dissociation: $FeS_2 \longrightarrow FeS + 1/4\ S_4^\circ \uparrow$
Volatilization of Sn: $FeS + SnO.slag \longrightarrow (SnS)_g \uparrow + FeO.slag$
Postcombustion: $- (SnS) + 2\ O_2 \longrightarrow SnO_2 + SO_2 \uparrow$
$- 1/4\ S_4^\circ + O_2 \longrightarrow SO_2 \uparrow$

The same reactions apply to the fuming of tin slags in which Sn is already dissolved as SnO. Usually, the Sn content of slags is about 10 % but As is very low (28).

A tin fuming operation presents excellent opportunity for separation of Sn and As in the gasphase, by performing the gas cleaning in 2 stages:

- First stage: at 350 to 400 °C, removing the tin fumes.
- Second step, at low temperature, removing the As_2O_3, after desublimation.

This system was adapted at La Palca (113) and at Freiberg (31). At the low grade tin smelter at Vinto (EMV), which discontinued its operation in 1983, treatment was conducted in two stages using a cyclone furnace in the first stage and a fuming furnace for slag detinning in the second stage. Gas cleaning, however, was performed in one stage and at relatively elevated temperature (150 - 200 °C) so that, unfortunately, the As-content was split between tin fumes and final gas (44).

Fuming slags contain 0.3 - 0.4 % Sn and 0.08 - 0.1 % As. The enrichment factor in the fumes depends mainly on the selectivity of the separation during first stage gas cleaning and may vary between 5 and 15.

Fuming of zinc from lead- and/or copper-smelting slags. The fuming of zinc slags produces a discard slag, a Zn/Pb oxide fume and eventually a matte phase. About 90 % of the arsenic present reports to the oxide. Hence, there is no separation between Zn and As (45).

Volatilisation of zinc from solid residues: Waelz process. Arsenic in the feed to a Waelz kiln originates mainly from 2 sources:

- zinc leach residues,
- dusts, fumes, sludges from non-ferrous metal operations.

According to (46, 47), only 10 % to 23 % of the As is volatilized, the remaining part reporting to the slag phase. The enrichment factor of Zn to As in fume is about 4 in the latter case.

Volatilisation of lead during zinc oxide clincker process. In the clincker process for fumed ZnO, Pb is volatilised in the form of PbO as well as As (in the form of As_2O_3). According to (48), the distributions are as follows:

%	Zn	As	Pb
Clincker	87	15	22
Pb-dust	13	85	78

The enrichment factor of Zn to As in clincker is, thereby, 87/15 = 5.8.

Reducing smelting processes

Reducing smelting of tin. Primary smelting of tin, traditionally, is performed in a bath type furnace:

- reverberatory (stationary or rotary),
- electric furnace.

Arsenic reports mainly to the metal (80 - 85 %) with the remaining split over dust and slags.

Depending on the level of arsenic present, either a separate phase due to the combination of iron and arsenic to the compounds FeAs of Fe_2As will be formed, known as speiss, or, at low As level, this phase will crystallize from the metal upon cooling during the so-called "drossing" operation (49).

Thus, for arsenic removal from complex tin concentrates, in principle, two process alternatives are possible:

- Selective removal of most of the arsenic by means of roasting prior to reducing smelting. The residual As is removed by "drossing". This is the conventional practice.
- Separation of the arsenic as a speiss during reducing smelting.

In either case, a refining stage is necessary.

Reducing smelting of lead and zinc/lead
- Blast furnace smelting of lead. In the blast furnace smelting process of lead, some elimination of As through the gas phase occurs during sinter roasting and in the blast furnace itself. However, a considerable amount of the arsenic is associated with the bottom products of the furnace.

If the As-content of the charge is low, it will be dissolved in the bullion and, to an appreciable extent, be separated from the lead together with copper during drossing, according to its solubility as a function of temperature. The dross can be converted to Cu/Pb-matte or be treated as such.

If the As-content is high and at high activity of the metals such as Cu, Ni, Co, besides the Cu/Pb-matte phase, there will be generated a separate speiss phase in which the As is concentrated, such as at MHO and La Oroya. Conditions for formation of speiss are outlined in (50, 51).

Figure 9 shows the distribution of As in reduction smelting of lead.

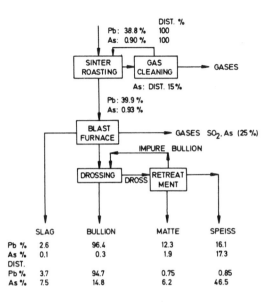

Fig. 9: Pb and As distribution in lead blast furnace smelting according to (52, 53)

<u>Direct smelting processes for lead</u>. Not much data are available on the deportment of arsenic during direct Pb-smelting. It is to be expected, however, that considerable volatilisation of As will take place in the first stage.

In the case of the QSL-process (54), it is stated that As will follow the dust, according to figure 10.

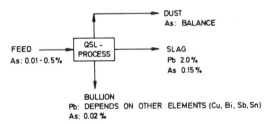

Fig. 10: Deportment of As in QSL-process (54)

<u>Zinc-lead blast furnace smelting (ISP)</u>. In the ISP, operating under strongly reducing conditions, the behavior of arsenic is governed by the reducing conditions itself and by the activity of the copper, present in the charge.

During sintering, normally 10 to 15 % of the arsenic in the feed is volatilised. Usually, all of it is being recycled to the sinter feed, together with the lead fume.

Volatilisation of arsenic in the ISF is usually small due to the strongly reducing conditions prevalent and can be suppressed to very low levels in the presence of small amounts of copper (0.5 %). The arsenic, which is volatilised, is largely taken up by the lead in the condenser and separated from it, as zinc arsenide, upon cooling. Residual As in the zinc is easily eliminated by means of addition of sodium.

The bulk of the arsenic, and all of it if 0.5 % Cu is present in the sinter, will report to the bottom products of the ISF: Bullion, slag and speiss. Matte will normally not be formed, except at abnormally high sulphur inputs (>1.5 % S in sinter, > 1.0 % S in coke).

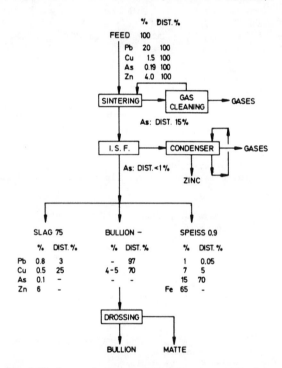

Fig. 11: Simplified mass balance of Pb, Cu, As in ISP (55, 56)

The simplified flowsheet is presented in figure 11. It is shown that most of the As is brought out as speiss, the enrichment factor being 6.10^{-4}.

Reducing smelting of copper. Reducing smelting of copper is performed in a blast furnace, in an electric furnace or, directly, in a converter and is applied on secondary copper, which normally does not contain arsenic.

In one particular case, the Brixlegg Process, dead roasted copper concentrates were submitted to reducing smelting in an electric furnace (42). As expected, the black copper obtained was high in As (Cu 92.5 %, As 1.01 %). Assuming negligible As volatilisation during FB-dead roasting of the concentrates, the enrichment factor from concentrate to black copper was 1.6.

Matte smelting

In matte smelting the As deportment is different from reducing smelting due to the high volatility of the As compounds in this phase. Unlike in reducing smelting, As will be separated from the metal principally through the gas phase. It is, therefore, important to pay particular attention to the gasphase for adequate Me-As separation in the same way as is being done in roasting of As-bearing materials and to ensure that As elimination occurs in the most direct way.

Matte smelting of copper concentrates. As elimination required during matte smelting of copper concentrates depends on two factors:

- the net As input with concentrates (and eventually other materials). This is best expressed as the ratio of Cu to As in the feed: $X_I/Y_I = k_I$.

- the permissible As-input into the refining, which may vary considerably. The As content of anodes fluctuates from 0.01 to 0.3 %. This is also best expressed as the ratio of Cu to As in the anodes: $X_A/Y_A = k_A$.

In figure 12, there is presented a nomogram from which the required fractional As elimination, E, ($0 \leqslant E \leqslant 1$), can be read as a function of Cu/As in the feed to obtain the desired Cu/As in the anodes. For instance, for a $(Cu/As)_I = k_I = 40$ and $(Cu/As)_A = k_A = 1000$, the required As elimination is 96 %. To obtain this high elimination of As, the 3 process steps:

- smelting
- converting
- slag cleaning

are available, with their respective gas cleaning operations.

The deportment of arsenic in copper smelter practice, particularly in smelting and converting, has been surveyed by Weissenberg et al (57).

The main conclusions of this review with regard to the As destributions to the gas phase, $(As)_G$, and to the slag phase, $(As)_S$ were as follows:

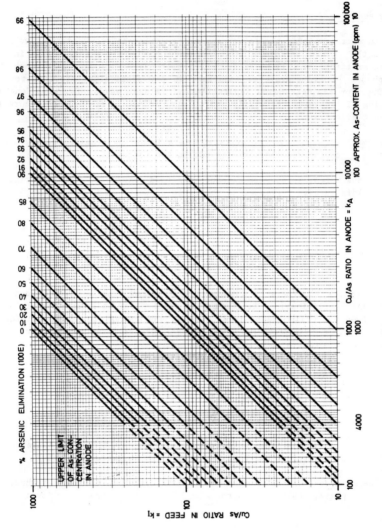

Fig. 12: Relationship between Cu/As ratio in feed (k_I) and in anode (k_A) as a function of % As-elimination (100 E) from circuit: $k_A^I = k_I \cdot 0.98 \cdot / \cdot (1 - E)$

- Smelting
  - Reverberatory smelting: $(As)_G$ $(As)_S$
    - High feed   50 – 70   10 – 20
    - Low feed (As < 0.2 %)   5 – 35   15 – 55
  - Electric furnace smelting: 10 – 25   50 – 70
  - Flash smelting: 75 – 85   5 – 20

  From the figures of the above table it is clear that flash smelting gives better elimination through the gas phase than bath smelting.

- Converting

  In converting, the distribution of As depends largely on the matte grade. At low matte grade, the formation of metallic copper is delayed and the activity of As remains high during a longer period of time. Hence, a longer proportion of the As is volatilised when converting a low grade matte. The following illustrates this influence for reverberatory and flash smelting:

Matte type	% Distribution	
	$(As)_G$	$(As)_S$
- Reverberatory smelting	75 – 90	5 – 25
- Flash smelting	40 – 50	20 – 30

  The difference in behavior between reverberatory and flash smelting is likely due to the fact that the As content in flash smelting matte is normally lower than the one in reverberatory matte.

- Slag cleaning

  This process step which was not discussed in (57), will be discussed separately later.

According to the survey of (57), the amount of As remaining in the blister may vary between less than 1.0 % up to 5 % of the As in the charge. Thereby, there is a tendency for:

- high As charges to result in a lower % of As reporting to blister than for low As charges.
- reverberatory smelting to give a higher % As reporting to blister than for flash smelting.

The wide range of As-distribution obtained in the survey (57) for one particular type of process is due not only to variations in charge composition but also to differences in operating conditions (temperature, gas volume, gas compositions).

Furthermore, it is obvious that the actual As elimination from the smelter circuit will depend on the fraction of it being recirculated, which must be taken into consideration.

Sources for As recirculation might be:

- dusts from primary and secondary gas cleaning
- matte or concentrate from slag cleaning.

The general smelter flowsheet, shown in figure 13, presents normal and potential outlets for As:

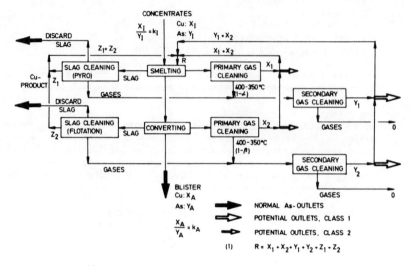

Fig. 13: General smelter flowsheet considering potential recirculation of smelter intermediate products

The effect of As concentration in the feed on the amount of recirculation of smelter dusts which can be tolerated to achieve 0.1 % As in the anodes was studied in detail as a function of matte grade for a particular flash smelting system (58). However, thereby incomplete consideration was given to the dependency of As elimination during smelting and converting on the % As in the respective feeds.

In the smelter survey of reference (57), it had already been shown that As elimination as % of input is more important at high As levels than at low levels.

According to the comprehensive work by several japanese investigators (59 to 63), which was based on thermodynamical data, the fractions of As reporting to the gasphase increases dramatically with the As in charge as well in smelting (figure 14) as in converting (figure 15A, 15B):

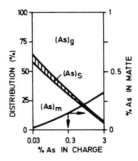

Fig. 14: As distribution (%) and concentration in matte at final smelting stage as a function of % As in charge (60)

T = 1,300 °C,   S = 0.3 (= flash smelting)
P_{SO2} = 0.1 bar,   % Cu in matte: 56

The graphical relationship indicates that, under the above conditions, the factor of As volatilized could be represented by an equation such as:
$(As)_G$ = 0.3 log (% As) + 0.8.

A. P_{SO2} = 0.1 bar   B. P_{SO2} = 1 bar

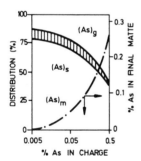 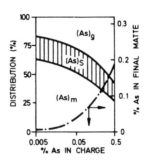

Fig. 15: Distribution of As (%) and concentration in matte at final converting stage (60)

T = 1,250 °C   S = 1.0 (= PS-converting)
P_{SO2} = 0.1 bar (A)   % Cu in matte: 80
       1.0 bar (B)

Furthermore, according to the same investigators, the ability of the process step to saturate the gas phase with As vapor, i.e. the intensity of contact between the gas phase and the condensed phases and the rate of As transfer, determines the elimination through the gas phase (figure 16A, 16B).

It was determined that in flash smelting the degree of saturation is relatively high, S corresponding to about 0.3, whereas in reverberatory smelting, the value of S is only about 0.08 (figure 16C). This explains the lower volatilisation of arsenic in reverberatory smelting assuming the fact that the volume of gas is up to 4 times as high as in flash smelting.

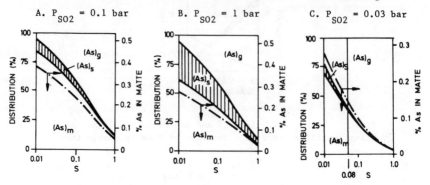

Fig. 16: Distribution of As (%) and concentration in matte at final smelting stage as a function of degree of saturation (60)

Flash smelting: S = 0.3
T = 1,300 °
P_{SO2} = 0.1 bar (A)
P_{SO2} = 1.0 bar (B)
% As in charge: 0.3
% Cu in matte: 56

Reverberatory: S = 0.08
T = 1,300 °C
P_{SO2} = 0.03 bar (C)
% As in charge: 0.3
% Cu in matte: 42

The effects of temperature and of oxygen enrichment are presented in figure 17A and figure 17B respectively.

A. Temperature (60)

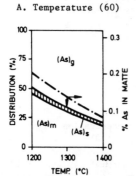

B. O_2-enrichment (60)

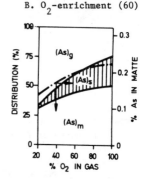

Fig. 17: Distribution of As (%) and concentration in matte at final smelting stage as a function of temperature and O_2-enrichment

T = 1200 - 1400 °C, S = 0.3
P_{SO2} = 0.1 bar,
% Cu in matte: 56
% As in charge: 0.3

T = 1300 °C, S = 0.3
P_{SO2} = 0.1 - 1.0 bar
% Cu in matte: 56,
% As in charge: 0.3

It can be seem from figure 17A that a high temperature is favorable for As volatilisation, in accordance with the findings of the cyclone processes (Flame Cyclone Process, Contop Process).

Figure 17B indicates that O_2-enrichment increases the As fraction reporting to slags and diminishes the As fraction volatilised, due to the lower gas quantities.

Moreover, during smelting initial As volatilisation at low matte grade is high and decreases rapidly with lowering As content. In the same way as with partial roasting of copper concentrates, the As elimination rises with increasing initial S/Cu ratio (figure 18).

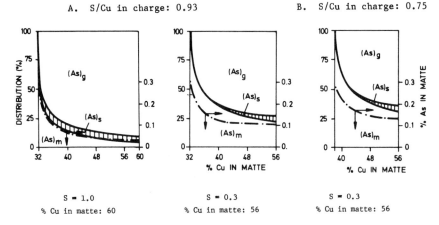

Fig. 18: Variation of As-distribution (%) and As-concentration in matte during smelting (60, 62)

$T = 1300$ °C, $S = 0.3 - 1.0$
$P_{SO2} = 0.1$ atm
% Cu in matte: 56 - 60
% As in charge: 0.3

Similarly, during converting to white metal (Cu_2S), initial As volatilisation decreases rapidly with increasing matte grade (figure 19A) and the As-content of the white metal is the lowest for the lowest matte grade, i.e. for the highest initial S/Cu ratio in the matte.

Fig. 19A 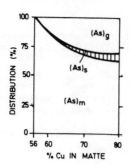 Fig. 19B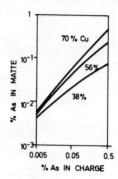

Fig. 19: Behavior of As during blowing to white metal (60)

T = 1250 °C, S = 1	T : 1250 °C, S = 1
P_{SO2} = 0.1 bar	P_{SO2} = 0.1 bar
% As in charge: 0.1	% As in charge: 0.005 - 0.5
% Cu in charge: 56	% Cu in charge: 38 - 70
% Cu in final matte: 60 - 80	% Cu in final matte: 80

The multitude of dependencies of the As-distribution on the operating conditions, shown above according to the investigations referred to in (60, 61, 62), is an explanation for the wide range of data obtained in the smelter survey by Weissenberg (57).

The most important conclusions are:

- the high elimination of As achievable through the gas phase for processes with high S-values (flash smelting, etc.).
- the increase of the distribution of As to the gasphase with increasing % As in the charge for processes with high S-values, as well in converting as in smelting (e.g. in flash smelting the % As in matte increases from 0.03 to only 0.3 % for an increase in the feed from 0.03 to 3 %. This is a very important consideration when circulation of As-bearing intermediates is involved).
- the increased elimination of As through the gas phase with increasing S/Cu ratio of the feed.

The incorporation of slag cleaning into the smelter flowsheet results in lowering the effect of arsenic elimination from smelter circuits in slags. There are, basically, two processes to clean copper smelter slags (64):

- pyrometallurgical slag cleaning, usually conducted in electric furnaces.
- slag cleaning by means of flotation of slowly-cooled slag.

- Pyrometallurgical slag cleaning

Pyrometallurgical slag cleaning is, in principle, performed under reducing conditions at an O_2-potential less than 10^{-8} bar, in the presence of sulphides.

In the case of settling of the slag, without addition of a reducing agent, reducing conditions are already brought about due to the presence of the electrodes. A matte phase is formed according to the sulphides present in the slag.

Addition of reducing agents and sulphide-bearing materials results in an O_2-potential of 10^{-10} - 10^{-11} bar. According to (60), a substantial fraction of arsenic is concentrated in the matte phase: Very little volatilization of As takes place ($\leqslant 10$ - 15 %) and the distribution coefficient slag/matte at 1300 °C is reported to be 10^{-3}.

In a recent investigation on pyrometallurgical reduction of converter slag, using pulverized coal injection (65), the following results with regard to Cu and As were obtained:

	Wt-%	Cu		As	
		%	Dist. %	%	Dist.%
Converter slag	100	4.85	100	0.08	100
Metal	4.8	85.3	85	1.4	83
Discard slag	127	0.53	14	<0.01	8
Dust	-	-	-	-	9

In conclusion, with pyrometallurgical slag cleaning, there is little opportunity to separate arsenic from copper, the enrichment factor in the recovered product being in the vicinity of one.

- Slag flotation

The effectiveness of slag flotation will depend, in first instance, on the crystallisation behavior during the process of slow cooling of the slag, and, secondly, on the conditions prevalent at the end of the metallurgical process, in which the slag was generated.

Data on deportment of As in slag flotation are not plentiful. The ones reported deal primarily with converter slag flotation. The following table gives an overview of the recoveries of As obtained during industrial Cu-converter slag flotation:

	% As	Dist. %	% As	Dist. %	% As	Dist. %	% As	Dist. %
Converter Slag	0.01	100	-	100	-	100	-	100
Concentrate	0.046	64	-	83	-	56	-	32
Tailings	0.004	36	-	17	-	44	-	78
Reference		(70)		(66)		(67)		(68)

The large fluctuations in recovery figures would indicate that parameters during the process of Cu recovery from converter slags are subjected to large variations.
It is likely that slag flotation presents better opportunity to separate As from Cu than pyrometallurgical slag cleaning (69).

Matte smelting of arsenical gold concentrate (Salsigne). In one particular case, matte smelting of arsenical Au concentrate is performed using a blast furnace. The Au-Fe-matte is treated for Au recovery. The flowsheet is presented in figure 20.

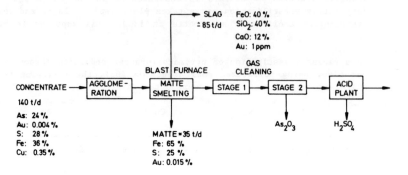

Fig. 20: Matte smelting of gold concentrate (Salsigne) (71)

It has been considered to convert the above flowsheet to a hydrometallurgical process consisting of roasting and Au leaching (72, 73).

Elimination of arsenic in pyrorefining

Copper fire refining

A recent survey of copper electrorefining practice (74) indicates that, though, in the majority of cases, the arsenic content of anodes does not exceed 0.1 %, however, levels as high as 0.3 % are encountered. The question is whether pyrorefining of copper with the purpose of lowering the As-content from 0.3 down would present an attractive proposition.

There are two potential processes for removal of arsenic by means of pyrorefining. These are:

1. Slagging of arsenic by means of alkaline fluxing of blister copper.
2. Vacuum distillation of arsenic from molten matte.

Due to the affinity of arsenic for copper in the molten state, its removal from blister by means of vacuum distillation is not effective (85, 86, 87, 88).

A considerable amount of work has been done in the field of alkaline fluxing using Na_2CO_3 (75, 76, 77, 78, 79, 80, 81, 82). Addition of Na_2CO_3 flux in an amount of 5 % of the weight of molten copper in the temperature range 1,150 - 1,250 °C results in a rapid elimination of the arsenic from the copper, e.g. from 0.3 to less than 0.03 %. The assumed reaction is:

$$2 As + 5 O + Na_2CO_3 \longrightarrow 2 NaAsO_3 + CO_2\uparrow$$

whereby As and O represent the species in solution in the molten copper.

Some copper reports to the soda slag, which requires further treatment. The enrichment factor of Cu to As is 10.

The investigations available on vacuum refining of matte (83, 84) indicate that at 1,150 - 1,200 °C and at 1 mm Hg (133 Pa), independently of the initial concentration in the range 0.15 to 0.5 %, a final concentration of 0.03 % As in molten copper could be obtained similarly as to the alkaline fluxing. The rate of elimination depends, of course, on other factors such as the ratio surface area to mass of the melt. An enrichment factor of the same order of magnitude as for alkaline fluxing could, therefore, be achieved by this means.

Besides occasional applications of alkaline fluxing of molten copper prior to anode casting in some operations, there are no commercial installations existing at this time, despite the good metallurgical results which could apparently be achieved. This may be due to the fact that available technology can handle the actual situation and, hence, does not justify the investment for further development and implementation of a new process.

Lead pyrorefining

The arsenic content of lead bullion may vary widely in the range 0.1 to 1.5 %.

High levels of arsenic in the feed to Pb-lead blast furnaces result in high arsenic in the copper drosses recovered from the lead bullion, as discussed ealier.

Arsenic removal by means of the subsequent Harris process depends on the tendency for As-, Sn and Sb-oxides to form sodium arsenates, -stannates and antimonates, when reacted with NaOH. The oxidation agents are sodiumnitrate ($NaNO_3$) or air. Thereby the oxidized species are transferred to the bath surface to form the soda slags. Reactions relative to arsenic are:

with $NaNO_3$: $\quad 2 As + 2 NaNO_3 + 4 NaOH \longrightarrow 2 Na_3AsO_4 + N_2\uparrow + H_2O$
with O_2: $\quad\quad 4 As + 5 O_2 + 12 NaOH \longrightarrow 4 Na_3AsO_4 + 6 H_2O$

The rate of reaction with O_2 is low. Therefore, the use of air will be restricted to pyrometallurgical refining of low arsenic bullion (e.g. as is the case in secondary lead refining).

The removal of arsenic is very complete, especially because of the exchange reactions occuring between the oxidized species of antimony and the residual metallic arsenic in liquid solution:

$$As + NaSbO_3 + 2 NaOH \longrightarrow Na_3AsO_4 + Sb + H_2O$$

Tin pyrorefining

Refining by chemical reaction above the melting point. Removal of arsenic from crude tin is not possible by means of oxidation, instead it is performed by means of formation of intermetallic As-compounds, which are insoluble in molten tin at a given temperature:

- By means of liquation with decreasing temperature, part of the dissolved arsenic together with dissolved iron forms an Fe-As insoluble intermetallic compound that segregates from the melt in the form of a dross. This can, advantageously, be separated from the melt in a "dry" state using a centrifuge. The crude tin may, thereafter, still contain as much as 0.3 % arsenic. The Fe-As dross can be submitted to an oxidizing treatment (e.g. roasting) to remove the arsenic as As_2O_3.

- By means of addition of strong electropositive metals which build intermetallic compounds with arsenic and which are insoluble in molten tin (89):

    * Sodium: Selective removal of arsenic alone is possible. Residual As: 0.025 - 0.030 %.
    * Aluminium: Both arsenic and antimony are separated from the melt together. Residual As: 0.008 %.

Both drosses separated from the tin have the tendency to remain "wet" even after centrifuging. Separation of As from Sn or of As from Sn/Sb can be performed by means of an oxidizing treatment or an oxidizing treatment followed by a reduction to recover metallic Sn or Sn/Sb-alloy.

Refining of tin by vacuum distillation. Application of vacuum distillation for refining of molten tin is commercially being applied in 2 types of furnace designs:

- Russian system: Temperature: 1400 °C
  Condensate in liquid state.
- Redlac system: Temperature: 1200 - 1300 °C
  Condensate in solid state.

Removal of As is impeded by formation of intermetallic compounds (SnAs, Sn_3As_2), which lowers its vapor pressure, but is improved at increasing temperature (1400 °C). Final As-concentrations below 100 ppm can, nevertheless, be obtained, if the bulk of the As has been removed by means of a preceding chemical treatment step.

The condensation of the distillate in the liquid form is reported to create difficulties in the presence of high As-levels due to the formation of solid intermetallic compounds in the condensate (90). Therefore, in the Russian system there is a limitation with regard to the As-concentration of the feed to the vacuum unit.

It may, therefore, be concluded that refining of tin by means of vacuum distillation is:

- not appropriate for As-removal.
- the Russian system results in better removal than the Redlac-system.
- at high As-levels, the Redlac-system should be utilized.

Vacuum distillation of arsenic

Metallic intermediates high in As can be treated at 900 to 1000 °C in a vacuum distillation furnace and be converted to a crude metallic distillate. Arsenic distillation occurs, thereby, in the form of As_4 (91).

Arsenic separation from solution

Precipitation of arsenic from solutions

With regard to the precipitation of arsenic from solutions, it is necessary to distinguish between the purpose of the precipitation step and the requirements associated with it:

1. Arsenic removal as a process step to separate arsenic from solution or from other dissolved metals: Selectivity of the precipitation is required.

2. Arsenic removal from waste effluents as a polishing step: Usually selectivity of the precipitation is not required.

3. Arsenic precipitation for rejection as a stable compound: This has been dealt with separately in a subsequent section because of its importance.

It is, of course, possible that two of the above mentioned purposes can be fulfilled in one and the same precipitation step.

For the purpose of separation of the bulk of the arsenic from other metals in solution or from the solution by means of precipitation, there are several methods which can be utilized depending on the conditions:

Arsenic is in the pentavalent state.
- Precipitation as ferric arsenate in the pH range 2 - 3: Examples are the purification of leach solutions which contain iron and arsenic. Thereby the precipitate can be separated from the leach solution together with the residue, as in the case of neutral leaching of zinc calcine or as in other particular examples (70, 92). In that case the arsenic will have to be eliminated during the further treatment of the residue (e.g. as in the treatment of neutral leach residue at Akita zinc (93) or as speiss in the electric smelting of arsenic-bearing lead residue from converter slag leaching at Naoshima (79). The iron/arsenic precipitate can also be separated directly as in the dust treatment scheme at the Kosaka Smelter of Dowa Mining (94) or as in the treatment of copper-arsenic cementation products (95) or as proposed in a scheme for the treatment of arsenic-containing lead smelter flue dusts (96).

- Precipitation as calcium arsenate at pH 2: Examples are:
  * Treatment of soda slags from lead pyrorefining (97, 98).
  * Precipitation of calcium arsenate at Equity Silver (99).

- Precipitation as $MgNH_4AsO_4 \cdot 6H_2O$: Precipitation of arsenic from ammoniacal solutions in the pH range 9.5 to 10.5. Residual arsenic solubility is 50 mg/l.

Arsenic is in the trivalent state
- Precipitation as As_2S_3 below pH 4:
  * Purification of nickel electrolyte for Cu and As at Thompson Refinery (100).
  * Removal of As and Cu from acidic zinc sulphate leach solutions in the zinc leach residue treatment.

For the purpose of recovery of As from waste effluents the treatment with sulphide is quite effective though not selective. It has been applied frequently, especially for effluent treatment in Japanese Smelters (100, 101).

Aqueous reduction of arsenic species

Metallurgical applications of aqueous reduction of arsenic are frequent especially in connection with the separation of arsenic from solution as a solid.

In solution arsenic may be present in the pentavalent state as arsenic acid (H_3AsO_4), which has a high solubility. Using SO_2 it is easily reduced to the trivalent arsenious acid (H_3AsO_3) (102), which has a much lower solubility than arsenic acid. This property is used for arsenic removal from solutions by means of crystallisation.

Arsenious acid can be reduced to metallic arsenic (As°) by a relatively mild reduction using Cd, Fe or Sn metal (103) but arsenic acid cannot.

Both arsenic acid and arsenious acid are reduced by Zn and Al metal to the minus trivalent state, yielding poisonous arsine gas (H_3As). In the presence of sufficient copper, or other metals such as cobalt, nickel, very insoluble metalarsenides are formed in neutral solution, avoiding the evolution of arsine gas. On this principle, the purification of neutral zinc leach solution, using As_2O_3 as an activator is based.

Electrolytically, in copper electrodeposition, dissolved arsenic is equally reduced from solution at high overpotential (low copper concentration), giving rise to the production of arsine in the absence of copper or of Cu_3As if sufficient copper is still present. With regard to arsenic formation during cathodic reduction of dissolved arsenic species, it is referred to (104, 105, 106).

Arsenic recovery from solution by means of crystallisation as arsenic trioxide

There are several processes in operation to recover commercial arsenic trioxide from solutions.

Crystallisation of As_2O_3 by SO_2-reduction of arsenate solutions. This As_2O_3 recovery system was introduced at the Toyo Smelter of Sumitomo in 1983 (101). The operation has a capacity to produce 60 t/month of pure arsenic trioxide.

In this process an arsenate solution, prepared from arsenic sulphide precipitate by means of selective oxidation of the arsenic, is the starting material. The dissolution is controlled so as to obtain the following composition of solution:

	As (t)	As^{3+}	As^{5+}
g/l	60 - 70	15	45 - 55

After passing through an activated carbon tower, the arsenate solution is reduced at a temperature of 15 - 20 °C by means of SO_2 gas containing 7 - 8 % SO_2 according to the reaction:

$$2 H_3AsO_4 + 2 SO_2 \longrightarrow As_2O_3\downarrow + 2 H_2SO_4 + H_2O$$

The solubility of As_2O_3 at that temperature corresponding to 15 g/l As^{3+}, about 75 % of the arsenic present crystallises as As_2O_3 (Fig. 21).

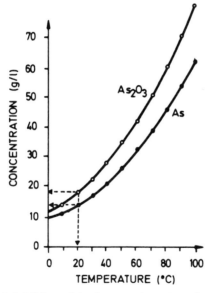

Fig. 21: Solubility of As_2O_3 as a function of temperature

The utilization of SO_2-gas is about 50 % and the waste gas is returned to the acid plant. Air is injected after the reduction stage to eliminate dissolved SO_2.

The arsenic trioxide slurry is filtered, washed and passed through a centrifugal separator for dehydration to 2 - 3 % H_2O and through a sealed rotary dryer to 0.1 % H_2O. Product purity is 99.9 % As_2O_3.

Crystallisation process from acidic scrubbing solutions (13). Crystallisation for the recovery of arsenic trioxide from SO_2-gas scrubbing solutions is widely applied. It is based on the fact that at any temperature the solubility of arsenic trioxide in aqueous solutions of sulphuric acid diminishes with increasing acid concentration up to a concentration of sulphuric acid of about 60 % (50 Bé, 957 g/l H_2SO_4), as shown in the diagram of Fig. 22.

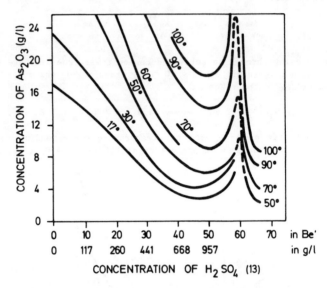

Fig. 22: Solubility of As_2O_3 in aqueous sulphuric acid solutions at various temperatures (Temperature in °C)

Hence, in the wet gas cleaning process as developed by Boliden, after dry removal of dust, SO_2-gases are washed in a scrubbing tower with a circulating dilute sulphuric acid solution, in which the arsenic trioxide vapor dissolves and crystallises out upon solution saturation.

A small stream is bled for removal of the As_2O_3-crystals by means of solid-liquid separation and the solution is returned to the circuit. The purity of the As_2O_3 crystals is about 90 %, due to the presence of other impurities which have been washed out. Further refining is required.

Refining of As_2O_3 can be performed:

> either hydrometallurgically (as at Boliden) by means of dissolution in water in an autoclave directly heated with steam followed by crystallisation by means of vacuum evaporative cooling

> or pyrometallurgically by sublimation and subsequent cooling according to a well-defined temperature profile.

Separation of As_2O_3 by vacuum crystallisation. The crystallisation of As_2O_3 by vacuum cooling can be applied on clear solutions containing arsenic as arsenic trioxide. Normally solution temperature is 60 °C or more and is cooled stagewise in a series of crystallisers (3 to 4) to 15 - 20 °C by means of vacuum evaporation. Since the solubility of arsenic is very dependent on temperature (see Fig. 21), crystallisation from the initial As-concentration ($\geqslant$ 30 g/l) to 13 - 15 g/l occurs. The arsenic trioxide crystals are separated by means of solid-liquid separation and the mother liquor is recirculated to the step where the arsenic solution is generated. Crystallisation behavior is inhibited by the presence of organics, which should be removed prior to crystallisation. The purity of the product obtained is better than 99.5 %.

The origin of the As_2O_3-bearing solution can be of primary nature, that means the solution is directly generated in the production plant, as is the case at the Onahama Smelter (107), where it is generated in the gas cooling section. It can be of secondary nature, it means that the solution is generated by leaching of flue dust from smelters (108, 109) or sludges from gas washing plants (110, 111).

Crystallisation of As_2O_3 by means of simultaneous solution cooling and reduction of arsenate ion. This method was proposed as a means to recover the arsenic contained in lead refining residues as As_2O_3 crystals (112). This method is, of course, a combination of the processes described under Section 7.3.1 (Reduction of As^{5+} by means of SO_2) and Section 7.2.3 (Solution cooling).

Conclusions. With regard to the production of As_2O_3 crystals from arsenic bearing solution there are three processes which have found commercial application. They are:

- Reduction of a highly concentrated As^{5+}-containing solution, using SO_2,
- Dissolution of arsenic trioxide in an already arsenious acid saturated, dilute sulphuric acid solution,
- Cooling by means of vacuum evaporation of an arsenious acid solution.

Disposal of arsenic-bearing materials

Disposal of waste arsenic products and arsenic-bearing residues

If for metallurgical reasons or because of economics, no commercial arsenic product can be produced, the arsenic input into a metallurgical operation needs to be converted into a compound which can be disposed in an environmentally safe manner.

There are several reasons on account of which the production of an arsenic product is prohibitive, amongst others:

- The arsenic is eliminated from the process in a residue or in tailings in which its concentration is too low. This is e.g. the case in the tailings from cyanide leaching of refractory gold ores, which

had been submitted to oxidative pressure leach pretreatment.
- The quantity of arsenic involved is too small.
- Due to the characteristics of the process, the arsenic is eliminated from process in a concentrated form as a compound which can be considered as environmentally stable. This is the case for hydrometallurgical operations in which the purification of the leach solution includes an iron-arsenic precipitation step.

In all these cases, the arsenic has to be converted to a stable compound, which has virtually no solubility in natural waters and which does not alter its characteristics as a function of time e.g. due to weathering (atmospheric oxidation, effect of CO_2 in the air) or due to the presence of reducing species such as sulphides. The matter of stability of arsenic compounds in dumps has been investigated intensively over last past years.

<u>Disposal of arsenic as calcium arsenate</u>. Until some years ago, it was common practice to dispose arsenic in the form of calcium arsenate (and/or calcium arsenite) (113, 100, 97, 98, 114, 99), because of the precipitation of arsenate from solution using lime being simple and inexpensive (Fig. 23).

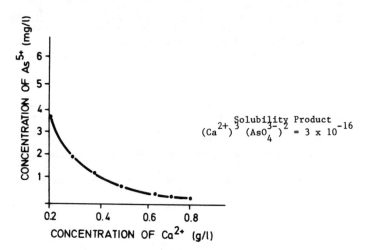

Fig. 23: Concentration of residual arsenic (5+) by means of the calcium arsenate precipitation method (114)

Because of the fact that due to the effect of atmospheric CO_2 the calcium arsenate precipitation does not produce a residue stable enough for dumping (115, 116), this practice is largely being discontinued. It is either replaced by the precipitation as ferric arsenate or by an arsenic recovery process (100, 97, 98).

<u>Disposal of arsenic as ferric arsenate</u>. There has been a great deal of discussion on the stability of ferric arsenates in dumps, especially in respect to the effect of atmospheric CO_2 which, based on thermodynamic calculations, increases the arsenic solubility (115, 118, 119). This problem is rendered very complicated because of the existence of a large number of compounds in

the system Fe-As-H_2O having different solubilities and because of the compound formed is dependent on the conditions during the precipitation.

Thorough investigations of this problem have demonstrated that the precipitation of basic ferric arsenates with a Fe/As molar ratio of 4 (i.e. weight ratio of 3) or more results in 100 to 1000 times lower arsenic solubility over the pH range 3 to 7 than the simple ferric arsenate or scorodite ($FeAsO_4 \cdot 2H_2O$) (120, 121). These findings are in agreement with earlier results of experiments on pilot dumps (95).

Furthermore, there is also evidence that the conditions under which the ferric arsenate has been precipitated also affect the stability of the precipitate. In one process for hydrometallurgical treatment of smelter dusts, flue dusts were autoclave leached at 140 °C and at an oxygen partial pressure of 345 kPa in a solution containing 18 g/l ferrous iron and about 50 g/l H_2SO_4 which resulted in good copper and molybdenum extractions and the simultaneous precipitation of a stable ferric arsenate residue. It is claimed that the stability of the ferric arsenate is due to the crystalline nature obtained under these conditions (122, 123). This may also apply to the ferric arsenate present in the tailings of cyanide leaching of refractory gold ores submitted to an oxidative pressure leach pretreatment.

Disposal of arsenic as metalarsenate using other precipitating agents. Other metals have been proposed for fixation of arsenate amongst others barium (124) and lead (125). It is claimed that these metalarsenates have lower solubility of arsenic and are less prone to arsenic redissolution due to the influence of atmospheric CO_2 than basic ferric arsenates. At least in the case of barium arsenate, some doubt appears to have risen with regard to its claimed stability (126).

Disposal of arsenic as arsenic trisulphide. This compound has reportedly a very low solubility at a pH below 4 but the arsenic solubility increases at higher pH due to the formation of the complex $As_3S_6^{3-}$ (117). Because of this characteristic and because not enough is known about its stability against atmospheric oxidation, arsenic trisulphide should not be considered as a suitable compound for disposal of arsenic.

Conclusions. From this analysis it may be concluded that the only environmentally acceptable manner for disposal of arsenic bearing residues for the time being is in the form of basic ferric arsenates having a molar ratio Fe/As equal or in excess of 3 to 4. Other precipitating agents which generate arsenate compounds having a lower arsenic solubility most likely will be economically prohibitive. Even if environmentally-safe arsenic residues can be generated, the costs of dumping these wastes will constitute an important factor against that practice. In 1985, in one particular case the cost of disposal of arsenic bearing residues at a licensed site was reported to be US$ 220 - 230 per wet tonne (99). Considerations should, therefore, be given, whenever possible, to the possiblity of generating a product rather than a waste.

Evaluation of feasibility of fixation of arsenic in metallurgical slags

The feasibility of fixation of arsenic in metallurgical slags as a means of disposal of arsenic in an environmentally acceptable manner has been examined (130). The technical and economical requirements for fixation of arsenic in slags are the following:

1. The fixation of arsenic in slags must be performed at elevated temperature. For the fixation of the arsenic in the specific form under which it was generated, the availability of molten slags is required. Hence, the fixation of the arsenic could only be carried out using molten slags produced from the same operation, for economical reasons.

2. The cooled slag must be chemically inert and not be altered chemically due to weathering or other natural influences.

3. The slag must have a high solubility for arsenic.

With regard to the capacity of slags to take up arsenic, it is well known that alkaline slags have a high solubility for trivalent arsenic and even more for pentavalent arsenic, due to its more pronounced acidic character (82). However, these slags are not stable, their main arsenic compounds being water-leachable (calciumarsenate, sodiumarsenate, etc.). Furthermore, the fact that slags are very often submitted to a reducing treatment to lower slag losses is likely to diminish the slag capacity for taking-up arsenic, because of the high volatility of arsenic trioxide which prevails under these conditions (127).

Trials to fix arsenic in the form of calciumarsenate, originating from hydrometallurgical treatment in pyrometallurgical slags failed, due to the decomposition of this compound (128). As a matter of fact, there is evidence that arsenic in silicate slag is present in the metallic form (129).

As a consequence of the above findings, it may be concluded that:

1. Pyrometallurgical slags from non-ferrous metal smelting are not very well suited as a potential for fixation of arsenic, due to their low capacity for taking it up. The conditions under which fixation of arsenic in slags are reported to function well do not appear to be very well applicable to these slags.

2. Smelters using slag flotation as slag cleaning process have better opportunity to eliminate arsenic in slags than smelters using a pyrometallurgical slag cleaning process.

3. As a result, the elimination of arsenic from pyrometallurgical operations will to a large extent have to occur through the gas phase, from which arsenic has to be collected and converted to a suitable form for its elimination.

Treatment of process intermediates and effluents

Treatment of copper refinery electrolyte bleed

In copper anodes the major impurities, which dissolve to a certain extent into the electrolyte, are nickel and arsenic. The levels of Ni and As in anodes may attain 0.7 % and 0.3 % respectively.

The fraction of As reporting to the electrolyte reaches about 90 % at 0.3 %, whereas only about 50 % of the Ni reports to the electrolyte at high Ni level in the anode. The volume of bleed of electrolyte can be read from the diagram of Fig. 24, taking into consideration the respective impurity levels in the anode and electrolyte concentrations.

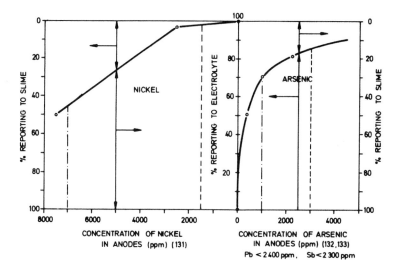

Fig. 24: Distributions of arsenic and nickel between anode slimes and electrolyte

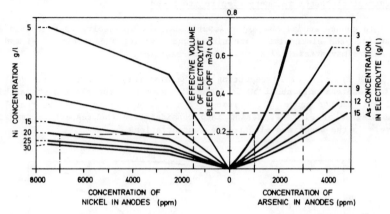

Fig. 25: Volume of electrolyte bleed and Ni and As contents of elekctrolyte for: - Ni/As ratio in anode = 7: — — —, Ni/As ratio in anode = 0.5: — · — · —

It can be seen from the diagrams of figure 24 and figure 25 that the amount of electrolyte to be bled for purification will be dictated primarily by the arsenic level of the anodes, in anodes having high As contents, since the maximum tolerated concentration of arsenic in electrolyte is limited.

Vice versa, if the Ni/As ratio in the anodes is high, the concentration of arsenic in the electrolyte due to the bleed for nickel will be low.

For the removal of arsenic from electrolyte bleed there are two processes commercially applied today:

1. Electrolytic removal of As (and the other impurities Sb, Bi) jointly with copper (as Cu_3As).
2. Liquid-liquid extraction H_3AsO_4 from strong or moderately strong acid electrolyte using TBP.

The flowsheets are presented in Figures 26 and 27. Whereas the second method is only being employed by two refineries (MHO and Copper Refineries), the first and original method is still in use at the majority of copper electrorefineries taking into consideration certain improvements, introduced to prevent formation of poisonous arsine (AsH_3) gas or to eliminate the effects of it having been formed (104, 105, 106).

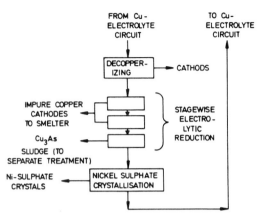

Fig. 26: Electrolytic purification of electrolyte bleed

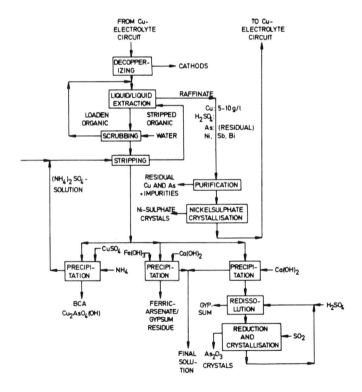

Fig. 27: Solvent extraction (for arsenic only) of electrolytic bleed with 3 alternatives for arsenic elimination

Two methods are being used to prevent arsine formation during electrolytic removal of arsenic and other impurities:

- Noranda Process, implemented at the CCR-Refinery: Electrowinning using current reversal with minimization of AsH_3 evolution (134, 135).
- Sumitomo Process, applied at the Niihama copper refinery (136): Stagewise depletion of As at well-defined Cu to As ratio results in high current efficiency for As removal (+ 95 %). The ratio of Cu to As is maintained by addition of copper electrolyte.

Alternative processes for treating Cu_3As sludge have been discussed elsewhere.

With regard to the removal of arsenic by means of solvent extraction, it has to be observed that the raffinate contains still 5 to 10 g/l copper as well as the other impurities, moreover about 5 to 10 % of the arsenic originally present in the electrolyte. These elements need be removed prior to the nickel sulphate crystallisation which is an additional step and generates also Cu-As sludge - though a smaller quantity of it.

However, if the Ni/As ratio in the anode is high, then the arsenic concentration in the electrolyte will be low, with the result that the solvent extraction process for arsenic removal will not be justified.

With regard to the final elimination of arsenic, either arsenic trioxide crystals can be produced or it can be converted to a ferric arsenate/gypsum residue for disposal. At Copper Refineries the arsenate is being used for the production of BCA (bicupric arsenate) according the reaction:

$$H_3AsO_4 + 2\ CuSO_4 + 4\ NH_4OH \longrightarrow Cu_2AsO_4.OH\downarrow + 2\ (NH_4)_2SO_4 + 3\ H_2O$$

which is used in the production of wood preservatives (CCA).

The different possibilities for removal of arsenic from copper electrolyte have been described in detail in (137). The data on liquid/liquid extraction of pentavalent arsenic from strong acid solutions are given in (138, 139, 140, 141).

Recently two other extraction methods have been disclosed. The first extractant, denominated XI-104 of Henkel AG, will not only remove As, but also Sb and Bi (142), whereas the second one, using an aliphatic long chain alcohol is more suited for the extraction of trivalent arsenic than for pentavalent (143).

Preparation of arsenic bearing solutions from arsenical subproducts
(flue dusts, sludges, precipitates)

Arsenical subproducts are generated in non-ferrous metal production. In the past, some of these subproducts were either recirculated to an appropriate process step or, when as a result, the recirculating load became too large, stockpiled. The reason for it was that there was available neither a suitable method of treatment nor favorable economics to produce a commercial arsenic product nor legislation requiring conversion to an environmentally stable residue.

Today, because of the environmental implication involved with it, stockpiling of arsenical subproducts is difficult. Hence, arsenical subproducts generated in the process will have to be converted directly to commercial products or to residues which can be impounded in an environmentally acceptable manner. Old stockpiles of arsenic bearing materials may have to be retreated within a limited period of time due to new environmental legislation (see e.g. Nerco Con Mine, 111).

Arsenic subproducts or intermediates from metallurgical processing are:

- Arsenic trisulphide (As_2S_3): This is a precipitate originating from sulphide treatment of mine and mill waters or plant waste effluents prior to discarding of the effluent.

- Arsenical flue dusts: These dusts are collected in the primary gas cleaning section, which normally should operate at a temperature between 400 and 350 °C, and not lower than 300 °C. At this temperature trivalent arsenic will still be volatile (as As_4O_6) and pass the gas cleaning section (e.g. the dry electrostatic precipitator). Even in the presence of oxygen, there is little tendency for the formation of pentavalent arsenic which is not volatile, unless the gas contains metaloxides (MeO) which build stable metal arsenates according to the reactions:

$$3\ MeO + (As_2O_3)_{gas} + O_2 \longrightarrow Me_3(AsO_4)_2\downarrow$$

Such metaloxides are, for instance, PbO and ZnO. For this reason, it would be expected that the majority of the arsenic collected in the primary gas cleaning stage is in the pentavalent state. The collection of arsenic in the primary gas cleaning stage could possibly be diminished by conducting the first gas cleaning step at an early stage and by maintaining a low oxygen potential in the gas. This procedure would allow the recycling of the primary dusts, without increasing the recycling load of arsenic.

- Arsenical sludges: In these the arsenic is present in the trivalent state. They are commonly generated in the wet gas cleaning system from the scrubbing of the condensed As_2O_3 fumes.

- Slags from lead pyrorefining: These slags usually contain sodiumarsenate (Na_3AsO_4) but sodiumarsenite (Na_3AsO_3) may also be present.

- Soda slags from copper pyrorefining: In these arsenic is present as water-soluble sodiumarsenate. These slags contain also appreciable amounts of copper.

- Copper-arsenic cementation products: These can be produced during the purification of zinc neutral leach solution. They contain besides Cu and As, also zinc.

- Cu-As sludges: They originate from copper electrolyte bleed-off purification. These contain besides Cu_3As appreciable amounts of copper.

Hydrometallurgical conversion of As_2S_3 to an arsenic solution. With regard to the treatment of impure As_2S_3 precipitates, three processes, one described in the literature (144) and two commercially applied (100, 101), are briefly discussed here. Their flowsheets are represented in Figures 28A and 28B respectively.

In the first process of Figure 28A, ammonia leaching is used to dissolve As_2S_3 as thioarsenite and oxygen is supplied under pressure at 120 °C to oxidize arsenite to arsenate and sulphide to sulphate. Arsenic is separated from solution to a large extent as magnesium-ammonium-arsenate. The precipitate is dissolved in sulphuric acid and the arsenic acid is reduced with SO_2-gas or sodium sulphite to crystallise As_2O_3.

In the second process of Sumitomo (Fig. 28B), the oxidation step is circumvented by a displacement leach at 70 °C using $CuSO_4$ according to the reaction:

$$3\ CuSO_4 + As_2S_3 + 3\ H_2O \longrightarrow As_2O_3\downarrow + 3\ CuS\downarrow + 3\ H_2SO_4$$

and subsequent slurry cooling. Separation of CuS and As_2O_3 is conducted by selective oxidation to arsenic acid, using air. The arsenic acid is then reduced using SO_2 under formation of As_2O_3 crystals. By this relatively simple method, as applied at Toyo Smelter, 60 t/month of As_2O_3 having 99.9 % purity is produced.

The third process was used to selectively dissolve arsenic from an As_2S_3-CuS precipitate (100) obtained in the purification of nickel electrorefining electrolyte at Inco's Thompson refinery. Oxidation with air was conducted at 80 °C and pH 6.5. After filtration, the CuS residue was sent to the smelter, the arsenic acid solution was treated with lime to precipitate calciumarsenate.

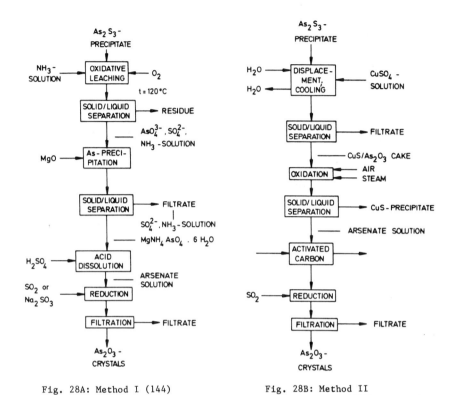

Fig. 28A: Method I (144)

Fig. 28B: Method II
Sumitomo Process (101)

Fig. 28: Hydrometallurgical treatment schemes for conversion of As_2S_3 precipitate to As_2O_3

<u>Pyrometallurgical conversion of As_2S_3 to As_2O_3</u> (Ashio, (145) and Fig. 29).
One japanese smelter has selected and installed a pyrometallurgical process to convert impure As_2S_3 filtercake to produce refined arsenic trioxide in powder form.

The process is performed in 2 stages: In the first stage sulfide is combusted and arsenic volatilised as As_2O_3. Gas is cooled slowly and the condensed As_2O_3-vapor collected in the condenser. Final gas cleaning occurs in a bag filter. In the second stage As_2O_3 is revolatilized and condensed as refined As_2O_3.

Production amounts to 120 t/month of 99 % As_2O_3.

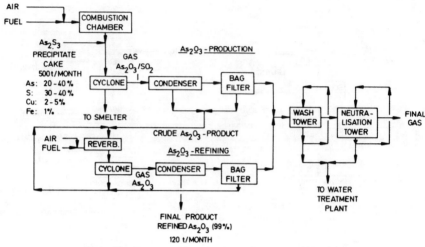

Fig. 29: Pyrometallurgical production of As_2O_3 from As_2S_3 - Ashio Smelter (145)

Conversion of As in flue dusts to an arsenic bearing solution. Flue dusts should, normally, contain most of the arsenic in the pentavalent state, since at the temperature at which the flue dusts are collected (400 - 300 °C), the arsenic trioxide will be still in the vapor state and pass the filter. This, however, is not always the case.

In (108, 109) a method is described for hydrometallurgical treatment of arsenic flue dusts originating from copper and lead smelting. The compositions of the flue dusts were:

%	As	Cu	Pb	Zn	Fe	Cd
Cu-Smelter	32.0	4.0	11.8	15.8	1.3	1.0
Pb-Smelter	15.6	0.2	55.0	4.1	0.1	2.3

It was demonstrated that a H_2SO_4 leach at pH 1.0 and 90 °C resulted in a good As, Zn and Cd-extraction. Arsenic trioxide is crystallised by reduction of the arsenate using SO_2 and solution cooling.

In (146), a process is proposed for hydrometallurgical treatment of smelter dust, in which more than 50 % of the arsenic is present in the trivalent state. For the purpose of separation, it is suitable for it to be either completely in the trivalent or the pentavalent state. The trivalent state can easily be obtained by means of SO_2 reduction. Complete oxidation of As^{3+} to As^{5+} in acidic solutions is difficult to achieve, using oxygen, even in the presence of iron and copper (as a catalyst for the oxidation of iron), the rate of reaction being low. Stronger oxidation reagents such as peroxide, chlorine should be used in acid medium (Fig. 30).

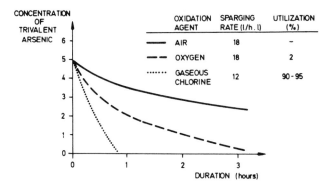

Fig. 30: Rate of oxidation of As^{3+} in aqueous leach
solutions containing iron, copper and 0.7 M H_2SO_4 (146)

Although at a pH above 3.0 higher reaction rates can be obtained even with O_2, autoclave oxidation at an oxygen pressure of 4 bar and at 90 to 95 °C was implemented by INCO at the Copper Cliff Refinery to ensure complete precipitation (Fe 10 mg/l, As 0.2 mg/l) at a pH of 3.9 to 4.1 (188).

In a third case (147), high temperature acid leaching (at 120 °C) without oxidation was reported to result in an excellent separation of Cu and As, the former remaining in the residue.

In these three examples for treating flue dusts the majority of the arsenic was extracted but other metal values such as copper only partly, whereas lead remained in the leach residue, therefore, the residue needs be recycled to the process.

In another example relating to copper smelter dust (122, 123), complete dissolution of the metal values Cu, Zn, Mo is achieved by performing the acid leaching under oxygen pressure at 140 °C. Thereby, iron and arsenic as well as other impurities (including lead if present) are being rejected in the residue.

From the above examples, it may be concluded that the selection of the leach process for flue dusts will depend on the type of the flue dusts, on the suitability of recovering arsenic or rejecting it, on the process operating costs.

In (148) a method is described to selectively dissolve arsenic from an arsenic/antimony containing oxidic material such as a flue dust. Oxidative leaching of the arsenic using an oxidizing agent such as peroxide at an elevated temperature of 80 °C results in selective extraction of arsenic, leaving the antimony in the residue. Arsenic is recovered from solution by means of crystallisation. It would be interesting to know whether this process could be applied to obtain an arsenic-free antimony residue, which would be a valid contribution to the antimony extraction from sulphidic raw material by means of fuming (149). The antimony oxides, produced in the fuming operations, will contain arsenic, depending on the As content of the input material.

Conversion of As present in sludges from wet gas cleaning/cooling. The sludge commonly contains arsenic in the trivalent state since it consists of condensed fume. Leaching in a weak sulphuric acid solution at elevated temperature (90 - 95 °C) and countercurrently results in more than 90 %. As extraction, as is the case in the retreatment plant of Nerco Con's mine (110, 111).

At Onahama (107), conditions in the gas scrubbing/cooling system section are such that arsenic remains in solution.

After filtration of solids, arsenic trioxide is crystallised by means of vacuum evaporative cooling. The production is:

	As_2O_3 (t/mo)	Quality % As_2O_3
Nerco Con	450	99.8
Onahama	40	99.2 - 99.6

In the case of Nerco Con's mine, the arsenic trioxide reclamation scheme presents the additional advantage of precious metals recovery from the arsenic sludge residue, which originated from refractory gold ore roasting.

Treatment of arsenical lead-copper dross. Arsenical lead-copper dross may contain 6 to 11 % As. Several methods have been proposed for the treatment of lead-copper dross.

Sulphuric acid oxidative leaching of the dross results in the arsenic reporting to the copper sulphate leach solution from which it has to be precipitated e.g. as ferric arsenate.

Oxidative ammonia leaching generates a NH_3-$Cu(NH_3)_4^{2-}$-SO_4^{2-} solution containing the arsenic as AsO_4^{3-}. Its removal could be performed in a bypass stream:

- by means of precipitation as $MgNH_4AsO_4 \cdot 6 H_2O$ according to (144) or
- by means of distillation of NH_3 and crystallisation of Na_3AsO_4 (150) which may be economically unattractive.

Hydrometallurgical treatment of metallurgical speiss. Although roasting can be applied and was, in fact, commercially applied some years ago, (98, p. 837), it is most likely that a hydrometallurgical oxidation process is better suited, since the valuable metals usually present (cobalt, nickel, precious metals), give better opportunity for their recovery.

Hydrometallurgical treatment of speiss has been investigated intensively on a laboratory scale by means of:

- pressure oxidation in H_2SO_4 solutions (153 to 158),
- oxidative ammoniacal leaching of speiss (159),
- sodiumhydroxide pressure leaching (160),
- oxidation in chloride medium (161).

Pressure oxidation of speiss in the sulphate system has found commercial application amongst others at MHO (151) and at Nickelhütte Aue (S10). At the former arsenic is being recovered as arsenic trioxide crystals whereas the latter applies the process of simultaneous electrodeposition of Cu and As. The flowsheet of MHO with regard to nickelspeiss is shown in figure 31.

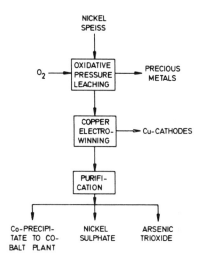

Fig. 31: MHO-process for treatment of speiss (151)

The ammoniacal system allows milder conditions with regard to the oxidation step, but arsenic removal and the build-up of sulphates render the system more complicated, whereas NaOH pressure leaching is unsatisfactory.

The chloride medium also presents excellent opportunity for the hydrometallurgical treatment of speiss (161) not only from the point of view of oxidative dissolution but also in respect of cobalt/nickel separation by means of liquid/liquid extraction and solution purification. The arsenic, however, will have to be removed by means of precipitation.

Treatment of arsenical soda slags from lead pyrorefining. Treatment of arsenical soda slags has found wide commercial application in lead pyrorefining for the treatment of the salts produced in the Harris process (97, 98). The objective of the treatment is the recovery of the sodium hydroxide and the metals contained in it. Normally the soda slag from the arsenic removal step contains also tin and some antimony which necessitates a four stage leaching/selective precipitation/solid-liquid separation to separate the following products:

- dilute NaOH-solution,
- antimonate residue,
- calcium stannate,
- calcium arsenate.

The process scheme as applied by MHO (98) is shown in Figure 32A.

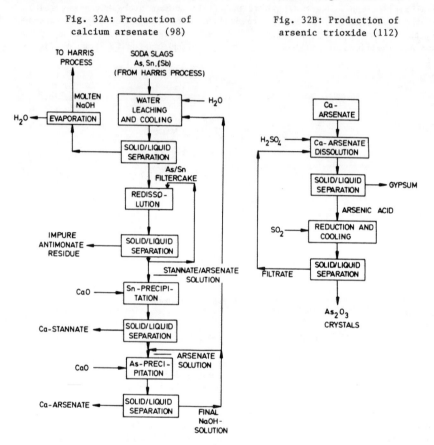

Fig. 32A: Production of calcium arsenate (98)

Fig. 32B: Production of arsenic trioxide (112)

Fig. 32: Treatment of soda slags

Calcium arsenate is a waste product, but because of its residual solubility, impoundment of Ca-arsenate has become environmentally unacceptable. The conversion of calcium arsenate to arsenic trioxide according to the flowsheet of Fig. 32B has been proposed and tested at BHAS (112).

The reactions are as follows:

Dissolution: $Ca_3(AsO_4)_2 + 3H_2SO_4 + 6H_2O \longrightarrow 2H_3AsO_4 + 3CaSO_4 \cdot 2H_2O\downarrow$
Reduction: $2H_3AsO_4 + 2SO_2 + 2H_2O \longrightarrow 2H_3AsO_3 + 2H_2SO_4$
Crystallisation: $2H_3AsO_3 \longrightarrow As_2O_3\downarrow + 3H_2O$

Total: $Ca_3(AsO_4)_2 + H_2SO_4 + 2SO_2 + 5H_2O \longrightarrow As_2O_3\downarrow + 3CaSO_4 \cdot 2H_2O\downarrow$

Treatment of soda slags from copper pyrorefining. The treatment of soda slags generated during the pyrometallurgical removal of arsenic (and antimony) by means of alkaline fluxing could be, in principle, carried out in a similar way, as the soda slags originating from lead pyrorefining (162). It has been reported, however, that a water-leaching would only result in incomplete dissolution of the arsenic from the copper residue, so that there would be an undesirable recirculation of arsenic with the Cu-bearing residue.

Another method which has been tested but not commercially applied is the acid dissolution of soda slags according to:

Dissolution: $2\,Na_3AsO_4 + 3\,H_2SO_4 \longrightarrow 3\,Na_2SO_4 + 2\,H_3AsO_4$
Reduction: $2\,H_3AsO_4 + 2\,SO_2 + 2\,H_2O \longrightarrow 2\,H_3AsO_3 + 2\,H_2SO_4$
Crystallisation: $2\,H_3AsO_3 \longrightarrow As_2O_3\downarrow + 3\,H_2O$

Total: $2\,Na_3AsO_4 + H_2SO_4 + 2\,SO_2 \longrightarrow As_2O_3\downarrow + 3\,Na_2SO_4 + H_2O$

This method results in good separation of copper and arsenic, however, the alkaline fluxing agent is lost and a solution of sodiumsulphate is generated, which has to be disposed of.

Treatment of Cu-As cementation products and electrorefinery decopperizing sludges. In such products arsenic is present in the As^{3-} state as Cu_3As. Concentration of Cu (40 - 60 %) and As (5 - 25 %) can vary widely and free metallic copper is usually present. In the case of a cementation product, Zn is the main accompanying element and Sb in the case of electrorefining sludge (5 %).

The copper cementation product can be dissolved easily using oxygen and sulphuric acid. After oxidation to AsO_4^{3-} and separation of the arsenic as ferric arsenate, a copper sulphate solution is obtained.

The above process was applied at DKH (Fig. 33A) (95, 163). Of course, it would be feasible to precipitate the arsenic as calcium arsenate, redissolve the precipitate with sulphuric acid and crystallise As_2O_3 after reduction using SO_2 (Fig. 33B)

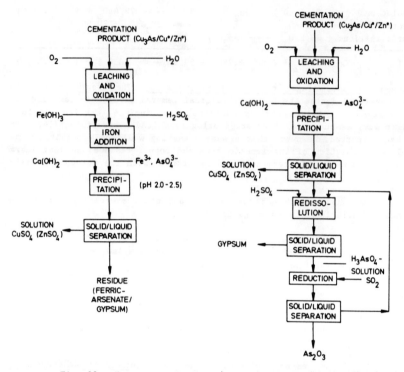

Fig. 33: Treatment of copper/arsenic cementation product

The Cu-As decopperizing sludges, containing virtually only Cu and As, could advantageously be used for the production of arsenic-based wood preservatives (CCA) (164, 165).

In (166), for oxidation of Cu_3As concentrated sulphuric acid baking at 250 °C was suggested, followed by dilute acid leaching. Extractions of both Cu and As were reported to be 97 % or more.

In (167), complete dissolution of Cu_3As, prepared artificially, was reported by means of leaching in 1000 g/l H_2SO_4 at 100 °C in 4 hours.

Finally, processes used for the treatment of speiss could also be applied for the treatment of Cu_3As sludges.

<u>Hydrometallurgical pretreatment processes</u>

<u>Hydrometallurgical pretreatment of copper concentrates</u>

Two processes, both based on sodium sulphide leaching according to the reactions:

$$2 \text{Cu}_3\text{AsS}_4 + 3 \text{Na}_2\text{S} \longrightarrow 3 \text{Cu}_2\text{S}\downarrow + 2 \text{Na}_3\text{AsS}_4$$

have been described.

In the first process, which was operated commercially at Equity Silver Mines (99), the purpose was selective leaching of Sb and As from the minerals tetrahydrite and enargite, leaving Cu in the residue. In the operation, after selective Na_2S-leaching, the solution was oxidized in a first step in an autoclave to insoluble Sb-antimonate, soluble As-arsenate and sodium sulfate. After filtration of the Sb-antimonate (to market), Ca-arsenate (for disposal) was precipitated in a second autoclave and, finally, the sodium sulphate was crystallised as salt cake (to market). The simplified flowsheet is as follows (figure 34):

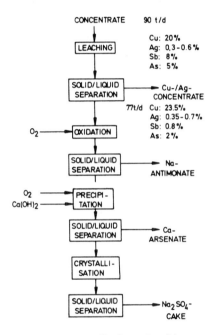

Fig. 34: Equity Silver mines Na_2S-preleaching process (99)

The enrichment for Cu to As was about 3, for Cu to Sb about 12. With regard to As, the separation was incomplete. Moreover, the operation was reported as "difficult" both from technical as from economical point of view. The grades of Sb and As in the orebody declining sooner than anticipated, preleaching has been discontinued some time ago.

In the second process, reported in (168), the leaching operation is similar. The recovery system is different, however. Arsenic is eliminated by non-oxidative crystallisation as thiosalt, which is converted to As-sulphide and H_2S-gas upon acidification. The proposed sulphide regeneration system is complex and a simplified one is shown in the flowsheet of the figure 35.

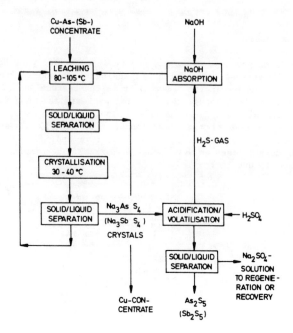

Fig. 35: Modified Na_2S-preleaching process (168)

Hydrometallurgical pretreatment of refractory gold concentrates/ores

Hydrometallurgical pretreatment of refractory gold concentrates or ores consists in an oxidation treatment in order to oxidize the sulphides, principally pyrites and arsenopyrites, to sulphates. Thereby, the gold which was finely disseminated within the sulphides, is made amenable to subsequent cyanide leaching.

Three types of processes have been developed for pretreatment of refractory gold ores/concentrates:

- Oxidative pressure leaching.
- Biological leaching.
- Wet chemical oxidation.

Oxidative pressure leaching. The oxidative pressure leaching pretreatment as operated at McLaughlin (169, 170, 171) results in an extraction of 92 % Au in the subsequent cyanide leaching of the refractory gold ore. Without pretreatment only 50 to 60 % Au extraction is obtained.

The pressure oxidation operates at 160 - 180 °C and 40 - 45 % solids concentration and a pressure of 1,300 kPa. Retentiom time required is 1.5 to 2 hrs, and oxygen consumption 0.1 t/t ore and final acidity is 15 - 20 g/l H_2SO_4. More than 95 % of the sulphur is oxidized to sulphate and iron and arsenic are precipitated according to the simplified reactions:

$$FeS_2 + 15/4\ O_2 + 2\ H_2O \longrightarrow 1/2\ Fe_2O_3 + 2\ H_2SO_4$$
$$FeAsS + 11/4\ O_2 + 1/2\ H_2O \longrightarrow FeAsO_4 + 1/2\ H_2SO_4$$

Unfortunately, silver reports to a large extent (90 %) to the residue due to precipitation as argentojarosite. The flowsheet is as follows (figure 36):

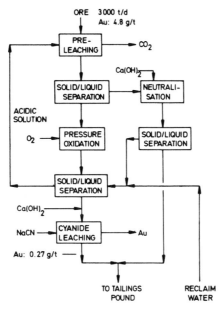

Fig. 36: Pretreatment of Au-ore by means of pressure oxidation (171)

In Brasil, a similar plant (Sao Bento), started operation in 1987 with a production of 2 t Au/year starting from an ore containing on average 12 g/t Au, 18 % S and 7 % As (172, 173).

Biological leaching. Intensive investigations on the biological oxidation of refractory gold ores has now resulted in the commercialisation of this process. At the Fairview Mine of Gencor (174) 12 t/d of flotation concentrates analysing Au 145 g/t, S 29 %, As 6 % are pretreated according to the BIOX-process. Subsequent cyanide leaching yields 95 % Au-recovery, exceeding by 5 % the recovery in the existing roasting facilities. The ferric arsenate bearing residue is reported to be environmentally acceptable.

Giant Bay has propagated a bioleach process specific for concentrates (Biotankleach) and for ores (Bioheapleach) (175). A commercial-size plant is reportedly under construction.

Wet chemical processes. A direct oxidation process using nitric acid for the oxidative pretreatment at atmospheric pressure and moderate temperature (85 °C) was developed by Hydrochem Development under the name Nitrox-process (176, 177, 178, 179). The simplified flowsheet is as follows (figure 37):

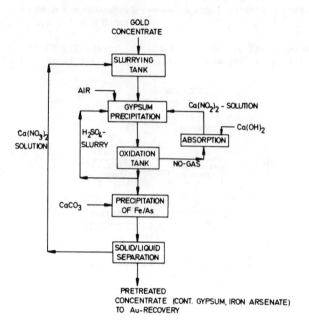

Fig. 37: Nitrox-process for refractory Au-concentrates

The reactions in simplified form are given below:

- Precip. of Gypsum/
  HNO$_3$ Regener. $Ca(NO_3)_2 + H_2SO_4 + 2 H_2O \longrightarrow CaSO_4 \cdot 2 H_2O\downarrow + 2 HNO_3$
- Oxidation   - $3 FeS_2 + 18 HNO_3 \longrightarrow Fe_2(SO_4)_3 + Fe(NO_3)_3 + 3 H_2SO_4 + 15 NO\uparrow + 6 H_2O$
  - $3 FeAsS + 14 HNO_3 \longrightarrow 3 FeAsO_4\downarrow + 14 NO\uparrow + 3 H_2SO_4 + 4 H_2O$
- NO-Absorption/   - $2 NO + Ca(OH)_2 \longrightarrow Ca(NO_2)_2$
  Oxidation   - $Ca(NO_2)_2 + O_2 \longrightarrow Ca(NO_3)_2$
- Neutralisation/ $H_2SO_4 + CaCO_3 + H_2O \longrightarrow CaSO_4 \cdot 2 H_2O\downarrow + CO_2\uparrow$
  Precipitation/ $Fe_2(SO_4)_3 + 3 CaCO_3 + 9 H_2O \longrightarrow 2 Fe(OH)_3\downarrow + 3 CaSO_4 \cdot 2 H_2O\downarrow + 3 CO_2\uparrow$
  Ca-Nitrate Gen. $Fe(NO_3)_3 + 3/2 CaCO_3 + 3/2 H_2O \longrightarrow Fe(OH)_3\downarrow + 3/2 Ca(NO_3)_2 + 3/2 CO_2\uparrow$

The reactions show that, in the overall scheme, the sulphide is oxidised to sulphate, which, by addition of lime/limestone, is converted to gypsum. The oxidation is performed by nitric acid which is regenerated with air. A drawback is that the oxidation to sulphate is not complete, some of the sulphide being converted to elemental sulphur, which reports to the Au-concentrate and is detrimental to subsequent cyanide leaching. Basically, 3 % of the FeS_2-sulphur and roughly 50 % of the FeAsS-sulphur is converted to S°.

The Nitrox process relies on recycling low solid reactor exit slurry to provide H_2SO_4 to combine with Ca-nitrate and Ca-nitrite. Obviously, there is a balance to be maintained on sulphuric acid and calcium recycles. Particular attention will have to be given in this connection to the composition of the material treated, with regard to the FeS_2 to FeAsS ratio.

To make-up for the HNO_3-losses, which are reported to amount to 2 % of the HNO_3 per cycle, it is proposed to pass a fraction at the oxidation air through a plasma g to produce NO_x.

The Fe/As ratio in the precipitate at 3 to 1 is reported to be satisfactory with regard to As-solubility. Long term precipitate stability testing is being conducted as a part of a pilot plant program.

A particular advantage of the Nitrox process is that silver recovery is of the same magnitude as the Au-recovery (90 to 95 %), unlike with oxidative pressure leaching.

It is considered that the Nitrox process may be particularly attractive for the direct treatment of complex Co-Ni-As concentrates.

Direct hydrometallurgical treatment of arsenical concentrates

There exist a number of arsenide-sulphide deposits, from which concentrates could be produced which are not amenable to conventional processing. Various processes have been developed for the treatment of such materials, but in the majority of cases the flowsheet consists of a sequence of known process steps. Table 2 summarizes the main process steps, proposed for the treatment of several types of arsenic-bearing concentrates.

Table 2: Proposed process flowsheets for direct hydrometallurgical treatment of complex arseniferous concentrates

	Type of Concentrate	Leaching	Main metal recovery process	Arsenic elimination	Others	Elimination of sulphur	Ref.
1.	Ag/As/S	oxidation using HNO_3 at 125 °C	precipitation as AgCl	separate ferric arsenate precipitation	sulphide precipitation for recovery of other metals	elemental sulphur	180
2.	Ni/As	oxidation using HNO_3 at 25 °C	undefined	separate ferric arsenate precipitation	---	not applicable	181
3.	Cu/As/S	oxidative pressure leaching (using Ag catalyst at 100 °C)	crystallisation of $CuSO_4$	solvent extraction of arsenic and crystallisation of As_2O_3	---	elemental sulphur and sulphate	182
4.	Co/As	Pressure oxidation using HNO_3/air at 80 °C	Co as metal-powder (H_2-reductant in NH_3-sol.)	separate precipitation as ferric-arsenate	---	not defined	183
5.	Cu/Co/As/S[1]	oxidation using $CaCl_2$-O_2 at 115 °C	solvent extraction + electrowinning of copper	ferric arsenate precipitation during leaching	Co precipitation as sulphide for further processing	gypsum	184
6.	Cu/Co/As/S[1]	pressure oxidation at 195 °C	copper solvent extraction and electrowinning	jarosite precipitation and separate ferric arsenate precipitation	Co-solvent extraction and electrowinning	gypsum and jarosite	185
7.	Cu/Co/As/S[1]	pressure oxidation at 150 °C	cobalt hydroxide precipitation and redissolution in spent electrolyte; Co-electrowinning	jarosite precipitation and Cu/As precipitation using H_2S	solution purification using iron exchange	gypsum and jarosite	186
8.	Cu/Co/As/S	two stage oxidation/reduction roasting followed by NH_3-leaching	solvent extraction and electrowinning	not specified	---	SO_2	187

[1]) Blackbird Cu-Co-As concentrate

The variety of process schemes shown in Table 2 would stress the need for a thorough evaluation of existing flowsheets to define the optimum route for such difficult-to-treat concentrates.

Conclusions

The present survey on the practical systems for As elimination from metallurgical circuits demonstrates the necessity for the incorporation of, at least, one process step which presents adequate opportunity of separation of arsenic from the metal to be recovered.

In pyrometallurgical processing, since metallurgical slags, which are generated in the process under reducing conditions, do not present sufficient capacity for As outlet, elimination through the gas phase or generation of an As-rich intermediate phase (such as a speiss) are the only two adequate methods.

The gas phase presents an excellent opportunity for the separation of As from the metal due to the possibility of application of the two stage gas cleaning process. This system has been applied with good result in the de-arsenizing roasting schemes, but less so in smelting of arseniferous primary materials.

Intermediate As-bearing products can be recycled or submitted to a separate treatment. If recycled, they must only be recycled to process steps which present good As separation from the metal and the number of process steps, to which As-bearing intermediates are recycled, must be minimized.

If separate treatment is applied, hydrometallurgical treatment is the best option, not only due to the high degree of separation achievable but also because of the availability of the two alternatives to recover As as a by-product or to convert it to a disposable solid residue.

In hydrometallurgical processing, metal/arsenic separation is readily achievable to a large extent by means of precipitation, liquid-liquid extraction, crystallisation or by other means.

Processes presenting the availability to eliminate As separately (either as a product or as a waste residue), i.e. in a more concentrated form rather than in combination with a leach residue should be given the preference, due to the potential environmental liability of large amounts of dilute-arsenic-bearing wastes.

For As removal from effluents satisfactory processes are available (e.g. sulphide precipitaton), which generate an As-intermediate product. If clean air standards according to regulations issued in industrialized countries have to be applied, final stage dry gas cleaning becomes extremely difficult to implement and a wet gas cleaning system is unavoidable.

In the case of the need to dispose of arsenic, the production of an environmentally stable As-bearing waste is not without problems, and attention should be given to generate a basic ferric arsenate having molar Fe/As-ratio in excess of 3 to 4.

The knowledge of the enrichment factor of the metal to be recovered to the arsenic enables evaluation of the most useful process steps allowing the determination of the most direct route for As elimination from the metallurgical circuit under consideration.

References

1. "Arsenic", US Bureau of Mines, Preprint from Bulletin 675, 1985.

2. "Arsenic Metallurgy - Fundamentals and Applications", Proceedings of Conference, TMS-AIME Annual Meeting, Jan. 1988.

3. "Arsenic Emissions from Primary Copper Smelters" - Background Information for Proposed Standards, EPA, Feb. 1981.

4. Copper Studies, CRU Ltd., May 1988, 9 - 12.

5. "Impurity control and disposal", Proceedings of the 15th Annual Hydrometallurgy Meeting of the CIM, Vancouver 1985, Papers No. 1 - 12, Arsenic control and disposal.

6. "The Electrorefining and Winning of Copper", Proceedings of Conference, TMS-AIME, Annual Meeting, Feb. 1987.

7. P.A. Wright, "Extractive Metallurgy of Tin", 2nd Edition, 1982, Section 6.2.1, 83 - 87.

8. US Patent 4,126,425, "Gas mixer for sublimation purposes", 1978.

9. J.O. Burckle, "Arsenic Emissions and Control: Gold roasting operating", Environmental International, Vol. 6, 1981, 443 - 451.

10. C.L. Kusik and R.M. Nadkarni, "Pyrometallurgical removal of arsenic from copper concentrates", see reference 2, 337 - 349.

11. H. Tuovinen and P. Setala, "Removal of Harmful Impurities from iron, copper, nickel and cobalt concentrates and ores", TMS-AIME, A82-4, 1982.

12. German Patent 3,003,635, "Verfahren und Vorrichtung zur Entarsenierung arsenhaltiger Materialen".

13. E. Wicklund, "Die Nutzung von Pyrit als Grundlage für die Produktion von Schwefelsäure und Eisenoxyd", Aufbereitungstechnik, 6, 1977, 285 - 289.

14. H. Wolf and W. Gösele, "BASF Double Stage Process for turbulent layer roasting of arsenical pyrites", Sulphur, No. 78, Sept./Oct. 1968, 20 - 25.

15. E.H. Smith et al, "Selective Roasting to dearsenify enargite/pyrite concentrate from St. Joe's El Indio Mine", Complex Sulphides, TMS-AIME, 1985, 421 - 440.

16. "Copper Smelting in Boliden's Rönnskär Works described", Journal of Metals, Mar. 1954, 331 - 337.

17. US Patent 4,226,617, "Method for treating a mineral sulphide", 1980.

18. A. Landsberg et al, "Behavior of Arsenic in a static bed during roasting of copper smelting feed", US Bureau of Mines, R.I. 8493, 1980.

19. See reference 4, 12.

20. G. Lindquist and A. Holmström, "Roasting of complex concentrates with high arsenic content", Sulphide Smelting, TMS-AIME, 1963, 451 - 472.

21. K.P. Wright, "Fluid Bed Roasting Practice in the Red Lake Camp", CIM-Bulletin, Aug. 1961, 595 - 600.

22. E.O. Foster, "The collection and recovery of gold from Roaster Exit Gases at Giant Yellowknife Mines Ltd.", CIM-Bulletin, Jun. 1963, 469 - 475.

23. L. White, "Giant and Con underpin Yellowknife Economy", Engineering and Mining Journal, Febr. 1982, 79 - 83.

24. L. Connel et al, "Roasting process at the Giant Yellowknife Mine", Paper presented at 20th Annual Conference of Metallurgists, CIM, Aug. 1981.

25. D.W. Penman, "Metallurgical aspects of the treatment of refractory ores from Barberton", see reference 5, paper Nr. 8.

26. L. Moussoulos et al, "Recovery of gold and silver from arseniferous pyrite cinders by acidic thiourea leaching", Precious Metals: Mining, extraction, processing, TMS-AIME, Febr. 1984, 323 - 335.

27. M. Stefanakis et al, "Dearsenifying roasting and gold extraction from a refractory pyrite concentrate", see reference 2, 173 - 197.

28. J. Lema Patino, "The fuming of tin slags in the extractive metallurgy of tin", in: Proc. 4th World Conference on Tin, Kuala Lumpur, Vol. 3, 1974, 113 - 148.

29. G. Murillo H., "Balance global de fundición de estaño Alta Ley de 1977 y 1983", Revista Metalúrgica, No. 7, FNI-UTO, Jul. 1986, 34 - 39.

30. "Albert Funk Tin Smelter", see reference 7, 200 - 201.

31. L. Müller et al, "Entwicklung der Zinnproduktion im VEB Bergbau- und Hüttenkombinat Albert Funk, Freiberg", Neue Hütte, 24, H. 10, Oct. 1979, 374 - 379.

32. J.A. Litten et al, "Removal of arsenic from speisses formed during smelting of tin concentrates and residues by use of a fluidized bed roaster", Extractive Metallurgy '87, IMM, London, 1987, 743 - 766.

33. K.G. Görling, "Die Arsenabscheidung bei der magnetitbildenden Röstung von arsenhaltigen Schwefelkies-Flotationskonzentraten, Erzmetall, Bd. 12, 1959, 553 - 557.

34. R.M. Motta Guedes et al, "Integrated treatment of pyrite cinders at Quimigal-Barreiro", Complex Sulphides, TMS-AIME, 1985, 457 - 470.

35. Y. Okubo, "Kowa Seiko Pelletizing chlorination process - Integral utilization of iron residues", Journal of Metals, March 1968, 63 - 67.

36. Anon, "Kowa Seiko Process for overall utilization of pyrites", Sulphur, No. 69, Mar./Apr. 1967, 29 - 31.

37. M. Yoshinaga et al, "Tec-Kowa Pelletizing Chlorination Process, its establishment and development", Complex sulphides, TMS-AIME, 1985, 457 - 470.

38. Anon, "Upgrading pyrite cinders for iron and steel production", Sulphur, No. 106, May/Jun. 1973, 52 - 57.

39. C. Nuñez Alvarez and J. Vinals Oliá, "Study of spanish pyrite cinders: Cl_2-HCl leaching of fluidized-bed arsenical cinders"
    Part 1, Trans. IMM, Sect. C, Dec. 1984, 162 - 172.
    Part 2, Trans. IMM, Sect. C, Dec. 1984, 173 - 179.

40. C. Nuñez Alvarez et al, "A non-integral process for the recovery of gold, silver, copper and zinc values contained in spanish pyrite cinders", 15th Int. Min. Proc. Cong., IMPC, Cannes, 1985, 328 - 337.

41. G.K. Strauss and K.G. Gray, "Complex pyritic ores of the Iberian Peninsula and their beneficiation, with special reference to Tharsis Company Mines, Spain", Complex Sulphide Ores, IMM, Rome, Oct. 1980, 79 - 87.

42. P. Kettner et al, "The Brixlegg electrosmelting process applied to copper concentrates", TMS-AIME, A72-48, 1972.

43. G. Gravey and F. Peyron, "A french producer of cobalt: Métaux Spéciaux S.A.", International Conference on Cobalt Metallurgy and Uses, Brussels, Nov. 1981.

44. E.A. Müller, "Verflüchtigung von Zinn aus armen schwefelhaltigen Konzentraten im Zyklonofen", Erzmetall, 30, 1977, 2, 54 - 60.

45. O. Sundström, "Die Schlackenverbaseanlage der Hüttenwerke Rönnskär der Boliden Gesellschaft", Erzmetall, 22, 1969, 3, 123 - 131.

46. "Oroya Metallurgical Operations", 5th Issue, 1972, 63 - 64.

47. R. Estel, "Erfahrungen bei der Verarbeitung von Hüttenwerksrohstoffen nach dem Waelzverfahren", Abfallstoffe in der NE-Metallurgie, 17 Metal. Seminar, GDMB, H. 47, 1986, 145 - 158.

48. O. Knacke et al, "Die Verflüchtigung von Blei aus technischem Zinkoxyd", Erzmetall, 9, 1956, H. 6, 261 - 270.

49. See reference 7, 158 - 164.

50. L. Fontainas et al, "Some metallurgical principles in the smelting of complex materials", in: Complex Metallurgy '78, IMM, Bad Harzburg, Sept. 1978, 13 - 23.

51. R. Maes, Résumé du travail, "Quelques principes métallurgiques de la fusion de matières complexes", ATB-Metallurgie, XXII, 3, 1982, 119 - 130.

52. See reference 46, 38 - 42.

53. L. Harris, "Lead smelting improvements at La Oroya", Pyrometallurgical processes in nonferrous metallurgy, Pittsburgh 1965, 197 - 215.

54. P. Fischer, "The QSL-process", paper presented at CIM Conference of Metallurgists, Aug. 1983.

55. J.L. Bryson and P.M.J. Gray, "Recovery of copper in the Imperial Smelting Furnace", Trans IMM, Section C, Vol. 77, Jun. 1968, C72-84.

56. C.J.G. Evans and P.M.J. Gray, "Influence of raw material composition in the zinc-lead blast-furnace", Symp. Advances in Extractive Metallurgy, IMM, London, Oct. 1971, 565 - 589.

57. I.J. Weissenberg et al, "Arsenic distribution and control in copper smelters", Journal of Metals, Oct. 1979, 38 - 44.

58. M.A. Cocquerel and M.F. Shaw, "Arsenic Control in modern Smelting Processes", Sulphide Smelting, Vol. 2, TMS-AIME, Nov. 1983, 961 - 975.

59. A. Yazawa, "Thermodynamic Considerations of Copper Smelting", Canadian Metallurgical Quarterly, Vol. 13, 1974, 443 - 453.

60. K. Itagaki and A. Yazawa, "Thermodynamic Evaluation of Distribution of Arsenic, Antimony and Bismuth in Copper Smelting", Sulphide Smelting, Vol. 1, TMS-AIME, 1983, 119 - 142.

61. K. Itagaki and A. Yazawa, "Thermodynamic properties of arsenic and antimony in copper smelting systems", Symposium on Complex Sulphides, TMS-AIME, San Diego, 1985, 705 - 722.

62. K. Itagaki, "Thermodynamic evaluation of distribution behaviour of VA elements and effect of the use of oxygen in Copper Smelting", Metallurgical Review of MMIJ, Vol. 3, 1986, 3, 87 - 100.

63. T. Nagano, "Progress of copper sulphide continuous melting", in: Physical Chemistry of Extractive Metallurgy, Edit. Kudryk & Rao, TMS-AIME, 1985.

64. W.R.N. Snelgrove and J.C. Taylor, "The recovery of values from non-ferrous smelter slags", Canadian Metallurgical Quarterly, Vol. 20, 1981, 2, 231 - 240.

65. Y. Mori and T. Kimuza, "Cleaning of copper converter slag by coal injection", Metallurgical Review of MMIJ, Vol. 3, 1986, 3, 141 - 154.

66. J. Minoura and Y. Maeda, "Current Operation at Kosaka Smelter and Refinery", Metallurgical Review of MMIJ, Vol. 1, 1984, 2, 138 - 156.

67. D.B. George et al, "Minor Element behavior in copper smelting and converting", World Mining and Metals Technology, Joint MMIJ-AIME Meeting, Denver, Sept. 1976, 534 - 550.

68. P.J. Mackey et al, "Minor Elements in the Noranda process", Paper at 104th Annual Meeting AIME, Febr. 1975.

69. S.C.C. Barnett, "The methods and economics of slag cleaning", Mining Magazine, May 1979, 408 - 417.

70. T. Suzuki, H. Uchida and H. Mochida, "Converter dust treatment at Naoshima Smelter", Preprint at 106th TMS-AIME Meeting, Mar. 1977.

71. "Goldaufbereitungs- und Verhüttungsanlage in Salsigne", Erzmetall, 39, 1986, 2, 88.

72. M. Lesoille et al, "Treatement de pyrite et arsenopyrite aurifère par grillage et cyanuration", Industrie Minérale, - Minéralurgie, Jun. 1977, 3, 194 - 202.

73. A. van Lierde et al, "Développement du nouveau procédé de traitement pour le minerai de Salsigne", Proc. SIM Congress, Besançon, May 1982, 847 - 874.

74. J.H. Schloen, "Electrolytic Copper Refining - Tank Room Data", see reference 6, 3 - 18.

75. S. Monden et al, "Impurity elimination from smelter copper by alkaline flux injection", Advances in Sulphide Smelting, TMS-AIME, Nov. 1983, 901 - 918.

76. K.G. Lombeck et al, "Überlegungen zur Entfernung von Arsen und Antimon aus Kupfer durch selektive Oxidation und Salzschlackenbehandlung", Metall, 37, 2, Febr. 1983, 144 - 147.

77. M. Devia et al, "Kinetics of Arsenic and Antimony elimination by slagging in copper refining", Proc. of 2nd Int. Symp. on Metallurgical Slags and Fluxes, TMS-AIME, 1984, 643 - 647.

78. T. Nakamara et al, "The Removal of Group Vb Elements (As, Sb, Bi) from molten copper using a Na_2CO_3 flux", Canadian Metallurgical Quarterly, Vol. 23, 1984, 413 - 419.

79. L.V. Kojo et al, "The Thermodynamics of Copper Fire Refining by Sodium Carbonate", in: Proc. of 2nd. Int. Symp. on Metallurgical Slags and Fluxes, TMS-AIME, 1984, 723 - 737.

80. L.V. Kojo et al, "Thermodynamics of antimony, arsenic and copper in Na_2CO_3-slags at 1473 K", Erzmetall, 37, 1984, 21 - 26.

81. T. Nakamura, "A Study on the Fire Refining of Crude Copper by Alkaline Carbonate Fluxes", see reference 6, 33 - 46.

82. A. Yazawa and Y. Takeda, "Dissolution of arsenic and antimony in slags", Metallurgical Review of MMIJ, Vol. 3, No. 2, Nov. 1986, 117 - 130.

83. H. Kametani et al, "A fundamental study on the vacuum treatment of molten matte and white metal", Trans. JIM, Vol. 14, 1973, 218 - 222.

84. R.L. Player, "Removal of impurities from copper matte by vacuum treatment", Proc. of 6th Int. Vacuum Metallurgy Conference, 1979, 214 - 224.

85. K.G. Lombeck et al, "Untersuchungen zur Arsen-, Antimon- und Wismutabtrennung aus Kupferschmelzen durch Vakuumdestillation", Metall, 36, 11, Nov. 1982, 1192 - 1196.

86. H. Kametani et al, "A fundamental study on vacuum lift Refining", Trans JIM, 1972, 13, 13 - 20.

87. R. Harris, "Vacuum Refining copper melts to remove Bismuth, Arsenic and Antimony", Metallurgical Transactions B, Vol. 15B, June 1984, 251 - 257.

88. E. Ozberk and R.I.L. Guthrie, "Evaluation of vacuum induction melting for copper refining", Trans. IMM, Section C, 94, Sept. 1985, 146 - 157.

89. See reference 7, Section 11.1.2, 227 - 230.

90. S.C. Pearce, "Developments in the smelting and refining of tin", in Lead, Zinc, Tin '80, TMS-AIME 1980, 754 - 770.

91. R. Zapata, "Destilación del arsenico", Revista Metalurgica, No. 7, FNI-UTO, Jul. 1986, 45 - 48.

92. H. Kudelka et al, "Copper electrowinning at DKH", CIM Bulletin, Aub. 1977, 186 - 197.

93. T. Ohtsuka et al, "Progress of zinc residue treatment in the Iijima Refinery", TMS-AIME A78-7, 1978.

94. E. Mohri and M. Yamada, "Recovery of metals from the dusts of flash smelting furnace", in World Mining and Metals Technology, Joint MMIJ-AIME Meeting Sept. 1976, 520 - 533.

95. G. Zintl, "Auslaugeverhalten metallhaltiger Fällschlämme", Erzmetall, Bd. 26 (1973), 2, 60 - 65.

96. P.A. Bloom, J.H. Maysilles and H. Dolezal, "Hydrometallurgical treatment of arsenic-containig lead smelter flue dust", US Bureau of Mines, RI 8679, 1982.

97. K. Emicke, G. Holzapfel and E. Kniprath, "Bleiraffination bei der Norddeutschen Affinerie", Erzmetall, 24, 1971, 5, 205 - 256.

or K. Emicke, G. Holzapfel and E. Kniprath, "Lead Refining and Auxiliary By-Product Recoveries at Norddeutsche Affinerie (NA)", TMS-AIME Symposium Lead and Zinc, 1970, 867 - 890.

98. J.L. Leroy, P.S. Lenoir and L.E. Escoyez, "Lead Smelter Operation at N.V. Metallurgie Hoboken S.A.", TMS-AIME Symposium Lead and Zinc, 1970, 824 - 852.

99. C.R. Edwards, "Engineering the Equity Concentrate Leach Process", Complex Sulphides: Processing of ores, concentrates and by-products, TMS-AIME/CIM Symposium, Nov. 1985, 197 - 219.

100. J.R. Boldt and P. Queneau, "The Winning of Nickel", Section Electrowinning with Sulphide anodes, 367 - 368

101. M. Fujimori et al., "New Sumitomo process and operations for recovery of arsenic trioxide from arsenic residue", Mineral Processing and Extractive Metallurgy, IMM, Kunming, 1984, 55 - 62
or: TMS-AIME, A84-27, 1984,

102. B.R. Palmer et al, "Reduction of Arsenic acid with aqueous Sulphur dioxide", Metallurgical Transactions B, Vol. 7B, Sept. 1976, 385 - 390.

103. B.R. Palmer et al, "Electrochemical Reduction of Arsenic (3+) with Cadmium Metal", Metallurgical Transactions B, Vol. 6B, Dec. 1975, 557 - 563.

104. F. Habashi and M.I. Ismail, "Health Hazards and Pollution in the metallurgical Industry due to phosphine and arsine", CIM Bulletin, Aug. 1975, 99 - 103.

105. M.L. Cotton et al, "Removal of Arsine from process Emissions", The Metallurgical Society of CIM, Annual Volume 1977.

106. J.H. Davis, "The Thermodynamics of Arsenic Evolution", J. Electrochem. Soc., May 1977, 722 - 723.

107. H. Kohno, Y. Sugawara and M. Hashimoto, "Operation of arsenic trioxide manufacturing plant", Onahama Smelting & Refining Co. Ltd., Preprint, TMS-AIME, 112th Annual Meeting, 1983.

108. B.W. Madson, H. Dolezal and P.A. Bloom, "Processing arsenical flue dusts", TMS-AIME, A81-63, 1981.

109. US Patent 4,401,632, "Recovery of arsenic from flue dust", 1983

110. "Cominco's new arsenic trioxide plant soon to go on stream in Yellowknife", Engineering and Mining Journal, Nov. 1982, 39, 43

111. D.H. Anthony, "Nerco Con Mine Arsenic plant - Environmental Management through Resource Recovery", see reference 2, 135 - 143.

112. B.R. Kretschmer and A.M.H. Wauchope, "Recovery of arsenic from lead refinery residues", Proc. Symposium on Extractive Metallurgy, Nov. 1984, Aus. IMM, Melbourne, 193 - 197.

113. O. Hinojosa and A. Nogales, "Instalación y marcha del proceso de volatilización de la Palca", Revista Metalurgica, No. 7, FNI-UTO, Jul. 1986, 15 - 21.

114. D. Laguitton, "Arsenic removal from gold-mine waste waters: Basic Chemistry of the Lime Addition Method", CIM Bulletin, Sept. 1976, 105 - 109.

115. R.G. Robins and K. Tozawa, "Arsenic removal from gold processing waste waters: the potential Ineffectiveness of Lime", CIM Bulletin, April 1982, 171 - 174

116. R.G. Robins, "The stability of arsenic in gold mine processing wastes", Precious Metals: Mining, extraction, processing, TMS-AIME, 1984, 241 - 249.

117. R.G. Robins, "The Stability of Arsenic (V) and Argenic (III) Compounds in aqueous metal extraction systems", Hydrometallurgy '83, TMS-AIME, 1983, 291 - 309.

118. R.G. Robins, "The aqueous chemistry of arsenic in relation to hydrometallurgical processes", see reference 5, paper No. 1.

119. R.G. Robins and M.V. Glastras, "The precipitation of arsenic from aqueous solution in relation to disposal from hydrometallurgical processes ", Proc. of Conference: Research and Development in Extractive Metallurgy, May 1987, The Aus IMM, Adelaide Branch, 223 - 229.

120. E. Krause and V.A. Ettel, "Ferric Arsenate Compounds: are they environmentally safe? Solubilities of basic ferric arsenates", see reference 5, Paper Nr. 5.

121. G.B. Harris and S. Monette, "The stability of arsenic-bearing residues", see reference 2, 469 - 488.

122. J.D. Prater, W.J. Schlitt and K.J. Richards, "Kennecott process for treating copper smelter flue dusts", in: Process Fundamentals Consid. selected hydromet. Systems, Chapter 13, TMS-AIME, 1981, 143 - 152.

123. US Patent 4,149,880 (1979), "Recovery of copper from arsenic containing metallurgical waste materials".

124. Canadian Patent 1,139,466 (1983), "Removal of arsenic from aqueous solutions".

125. P. Comba, P.R. Dahnke and L.G. Twidell, "Removal of arsenic from process waste water solutions", see reference 2, 305 - 319.

126. R.G. Robins, "The solubility of Bariumarsenate: Sherrit's Bariumarsenate Process", Metallurgical Transactions B, Vol. 16B, June 1985, 404 - 406.

127. M. Kawahara, T. Doi and K. Morinaga, "The Behavior of Arsenic in Slag Melts", Nippon Kogyo Kaishi, 95, 1979, 877 - 882.

128. R.H. Lau, et al., "Vaporization Studies of Calcium arsenate under Neutral Conditions and in Contact with Copper Pyrometallurgical Slags", Metallurgical Transactions B, Vol. 14B, June 1983, 171 - 174.

129. D.M. Dabbs and D.C. Lynch, "Analysis of the molecular Form of Arsenic in Silicate Slag", Sulphide Smelting, Vol. 1, TMS-AIME 1983, 143 - 169.

130. A.K. Mehta, "Investigation of new techniques for control of smelter arsenic bearing wastes", Vol. I, EPA-600/2-81-049a, July 1981

131. De Decker et al, "Leaching of Copper Refinery slimes", see reference H7, page 593, figure 2.

132. J.P. Demaerel, "The behaviour of arsenic in the copper electrorefining process", see reference 6, 195 - 209.

133. P.L. Claessens et al, "Behaviour of minor elements in copper electrorefining", in: Raffinationsverfahren in der Metallurgie, Paper F 2, GDMB, Hamburg, Oct. 1983, 253 - 265

134. US-Patent 4,083,761, "Arsenic removal from electrolytes with application of periodic reversal", 1978.

135. J. Thiriar et al, "Electrolyte Purification with minimum arsine gas evolution using periodic reversal of current", Hydrometallurgy '83, TMS-AIME, 1983, 705 - 720.

136. T. Toyabe et al, "Impurity control of electrolyte at Sumitomo Niihama Refinery", see reference 6, 117 - 128.

137. G. Zarate C., "Alternative processes for copper refining electrolyte purification", Recycle and Secondary Recovery of Metals, TMS-AIME, 1985, 589 - 607.

138. A. de Schepper, "MHO solvent extraction process for removal of arsenic from Olen copper tankhouse feed", Extractive Metallurgy '85, IMM, 1985, 167 - 188.

139. J. O'Kane et al, "Solvent extraction applied to removal of arsenic from tankhouse electrolyte at Copper Refineries Pty. Ltd.", Extractive Metallurgy '85, IMM, 1985, 951 - 965.

140. Canadian Patent 1,070,504, "Method for removing arsenic from copper electrolytic solutions or the like", 1980.

141. US-Patent 4,115,512, "Method for removing arsenic from copper and/or nickel bearing aqueous acidic solutions by solvent extraction", 1978.

142. W. Schwab and H. Kroke, "Purification of copper tankhouse electrolyte by solvent extraction", see reference 2.

143. A. Baradel et al, "Extraction of arsenic from copper refining electrolyte", Journal of Metals, Febr. 1986, 32 - 37.

144. K. Tozawa, Y. Umetsu and T. Nishimura, "Hydrometallurgical Recovery or Removal of Arsenic from Copper Smelter Byproducts", TMS-AIME, A78-65, 1978.

145. K. Murao and S. Nako, "Recovery of heavy Metals from the Waste Water of sulphuric Process in Ashio Smelter", World Mining and Metals Technology, Proceedings Joint MMIJ-AIME Meeting, Sept. 1976, 808 - 816.

146. G.B., Harris and S. Monette, "A hydrometallurgical Approach to Treating Copper Smelter Precipitator Dusts", Complex Sulphides: Processing of Ores, Concentrates and By-Products, TMS-AIME/CIM Symposium, Nov. 1985, 361 - 375.

147. Jia-Jun Ke, Rui-Yun Qiu and Chia-Yung Chen, "Recovery of metal values from copper smelter flue dusts", Hydrometallurgy, 12, 1984, 217 - 224.

148. US Patent 4,218,425: "Process for extracting arsenic from oxidic materials", 1980.

149. E.P. Müller, "Die Verflüchtigung von Antimon aus sulfidischen Vorstoffen im Zyklonofen", Erzmetall, 31, 1978, 6, 267 - 275.

150. UK Patent Appl. 2,089,775 A, "Removal of copper from arsenical drosses", 1982.

151. R. Maes et al, "Processing of complex concentrates and byproducts at MHO", in: Complex Sulphides, TMS-AIME, 1985, 255 - 265.

152. H. Götze and D. Löwe, "Drucklaugung von Nickelrohstoffen - ein Beitrag zur Intensivierung der Nickelmetallurgie", Freiberger Forschungshefte, Bd. 210, 1979, 117 - 128.

153. F.E. Pawlek and H. Pietsch, "Anwendung des Druckaufschlusses auf Erze und Hüttenzwischenprodukten", Erzmetall, Bd. 10, 1957, 8, 373 - 383.

154. J. Gerlach and F.E. Pawlek, "Pressure leaching of speiss", Unit Processes in Hydrometallurgy, TMS-AIME, Feb. 1963, Vol. 24, 308 - 325.

155. FRG-DAS 1,161,432, "Verfahren zur Aufarbeitung arsenidischer und/oder antimonidischer Hüttenzwischenprodukte", 1964.

156. J. Blanderer, "Die metallurgische Speissen und ihre Verarbeitung", Erzmetall, Bd. 17, 1964, 5, 247 - 263.

157. H. Fries, J. Gerlach and F.E. Pawlek, "Beitrag zur Drucklaugung von Speissen", Erzmetall, Bd. 18, 1965, 10, 509 - 514.

158. J. Gerlach et al, "Der Einfluß des Gitteraufbaus von Metallverbindungen auf ihre Laugbarkeit", Erzmetall, Bd. 25, 1972, 9, 448 - 453.

159. J. Gerlach, F.E. Pawlek and H. Traulsen, "Beitrag zur Drucklaugung von Speissen in ammoniakalischen Lösungen", Erzmetall, Bd. 18, 1965, 12, 605 - 660.

160. J. Gerlach, F.E. Pawlek and G. Schäde, "Beitrag zur Drucklaugung von Speissen in verdünnter Natriumhydroxid-Lösungen", Erzmetall, Bd. 24, 1971, 4, 173 - 177.

161. A. van Peteghem et al, "Chemische Verfahren zur Verarbeitung von nickel- und kobalthaltigen Vorstoffen (chloridischer Weg)", in GDMB-Symposium Nickel, Sept. 1970, 147 - 154.

162. B. Strake, "Entfernung von Arsen und Antimon aus Kupfer durch eine Soda-Schlackenbehandlung in der Rührpfanne", VDI, Reihe 3, Nr. 115, 1986, 19 - 22.

163. Ger. Pat. 2,342,729 (1975), "Precipitation and separation of arsenic from copper containing solutions".

164. O. Hyvärinen et al, "Electrorefining practice at Outokumpu Oy", see reference 6, 377 - 383.

165. D.W. Hoey et al, "Modern Tankhouse design and practices at Copper Refineries Pty. Ltd.", see reference 6, 271 - 293.

166. A.H. Mirza, "Formation and stability studies of iron-arsenic and copper-arsenic compounds from copper electrorefining sludge", see reference 2, 37 - 58.

167. J. Dewalens et al, "The leaching of ß-Cu_3As by oxygen in concentrated sulphuric acid solutions", Bull. Soc. Chim. Belg., 82, 11 - 12, 1973, 711 - 714.

168. R.M. Nadkarni and C.L. Kusik, "Hydrometallurgical removal of arsenic from copper concentrates", see reference 2, 263 - 286.

169. D.N. Skillings Jr., "Homestake Proceeding with its McLaughlin Gold Project", Skillings' Mining Review, Jan. 22, 1983, 3 - 6.

170. R. Guinevere, "McLaughlin Project: Process, project and construction development", Paper XIV, First Int. Symposium on Precious Metals Recovery, Jun. 1984.

171. G.O. Argall, "Perseverance and winning ways at McLaughlin Gold", Engineering and Mining Journal, Oct. 1986, 26 - 32.

172. Anon, "Sao Bento produzindo a plena capazidade", Minerios, May 1987, 113 - 114.

173. A.C. de Araujo et al, "Gold mining in Brazil: Past, present and future", in: Gold Mining '87, SME-AIME, Chap. 6, 74 - 91.

174. P.C. van Aswegen and A.K. Haines, "Bacteria enhance gold recovery", International Mining, May 1988, 19 - 23.

175. A. Bruynesteyn et al, "The Biotankleach process", Gold 100, Proc. of Inst. Conf., Vol.2, SAIMM, 1986.

176. G. van Weert et al, "Prochem's Nitrox process", CIM Bulletin, Vol. 79, Nov. 1986, 84 - 85.

177. G. van Weert et al, "The Nitrox process for treating gold bearing arseno-pyrites", TMS-AIME, A87-40, 1987.

178. G. van Weert, "Design and Operating results of the Nitrox process", Paper at Second Int. Gold Conf., Nov. 1988, Vancouver.

179. G. van Weert, "An Update of the Nitrox process", Randol Gold Forum '88, 209 - 210.

180. W. Kunda, "Treatment of complex silver arsenide concentrate", Proc. of 4th IPMI Conf., 1980, 39 - 57.

181. A.S. Block-Bolten, "Separation of arsenic from nickel", Process Mineralogy II, Applications in Metallurgy, TMS-AIME, 1982, 187 - 195.

182. K.P.V. Lei and T.G. Carnahan, "Effect of silver on oxidative leaching of an arsenical copper sulphide concentrate", TMS-AIME, A88-4, 1988.

183. US Patent 2,805,936, "Leaching of arsenide ores", 1957.

184. G.A. Smyres et al, "Calcium chloride-oxygen leaching and metals recovery from an arsenical copper-cobalt concentrate", US Bureau of Mines, RI 9020, 1986.

185. R.O. Dannenberg et al, "Recovery of cobalt and copper from complex sulphide concentrates", US Bureau of Mines, RI 9138, 1987.

186. G.B. Harris, S. Monette and R.W. Stanley, "Hydrometallurgical treatment of Blackbird cobalt concentrate", in: Hydrometallurgy, TMS-AIME, Mar. 1982, 139 - 150.

187. A.R. Rule et al, "Recovery of copper, cobalt and nickel from waste mill tailings", Proc. of 5th Mineral waste utilization symposium, Apr. 1976, 62 - 66.

188. P.M. Tyroler et al, "Hydrometallurgical processing of INCO's Pressure Carbonyl Residue", in Extractive Metallurgy of Nickel & Cobalt, TMS-AIME, Jan. 1988, 391 - 401.

IMPROVEMENTS OF HYDROMETALLURGICAL PROCESSES BY APPLICATION OF

THE ACID RETARDATION PROCEDURE - ACTUAL SITUATION AND DEVELOPMENTS -

M. Gülbas*, R. Kammel**, and H.-W. Lieber**

*GOEMA Dr. Götzelmann KG, Vaihingen/ Enz
and
**Institute for Metallurgy, Technical University of Berlin

Abstract

Metal cations can be separated from free acids by a diffusion process which is carried out by feeding the spent solution into a particle bed of a special ion exchange resin. The metal cations are passing the resin bed immediately while the free acid to a high degree is eluated from the resin afterwards just by use of water. Therefore, this procedure has been termed acid retardation.

Due to the fact, that valuable acids can be recovered in this way, equipments for acid retardation have been applied successfully during the past few years in the metal finishing industry, mainly for the regeneration of sulutions for scale removal, pickling, acid dipping, bright dipping, anodizing, chemical machining, etc.

For the application of this procedure in hydrometallurgy, investigations have been started for the separation of Mg and Mn from electrolytes in zinc winning plants as well as for the removal of As from spent electrolytes in copper refineries. Results of these research projects will be reported under technical and economical aspects.

Introduction

In the European Community, metals like cadmium, chromium, copper, lead, mercury, nickel, zinc etc. are regarded as "hazardous materials", and effluents resulting from their production or processing have to be treated in the near future by use of the best available technology. As well, process solutions have to be regenerated and, as a rule, can no longer be discarded. The volumes of solid waste have to be decreased by preventing losses of materials (e.g. recycling, closed-loop operation, recovery of water and materials contained in effluents), or by reuse of materials, and only if there is no chance for a further utilization of water, chemicals or metals, waste water can be drained and solid waste can be disposed of.

Under these aspects, in the metal finishing industry new procedures for environmental protection have been developed to meet these new regulations and to comply with considerably reduced residual concentrations of the constituents in the effluents to be drained into the sewer.

Spent acid solutions in most cases contain free acids to a high extent, which in the conventional route have been neutralized for precipitation of heavy metals. But recovery of free acids can save part of the acid consumption, alkali for neutralization, and smaller amounts of neutral salts will pollute the sewer.

In hydrometallurgy, similar problems have to be solved in connection with bleeding solutions for the limitation of impurity build-up. Therefore, investigations have been started with this promising procedure and the results have indicated that acid retardation can bring about many advantages compared with the conventional mode of operation.

Experimental procedure in laboratory-scale

Hatch and Dillon (1) have alternately passed an acidified metal salt solution and water through a column filled with an anion exchange resin and found out that the eluation of the free acid is clearly retarded, whereas the metal salt is passing the resin bed immediately.

The concentrations of the liquids resulting from the elution process have been presented in Fig. 1.

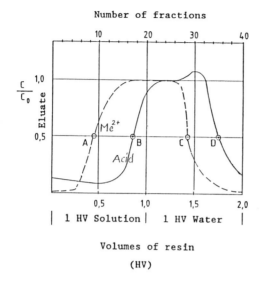

Fig. 1: Acid retardation effect

There is a marked separating effect, and after collecting the eluate solutions alternately in two different tanks, these storage bins contain a weak acid solution with a high metal concentration and a strong acid with a decreased metal content.

It has been confirmed that this effect greatly depends on the type of resin applied and that the reason for the process of acid retardation is the different rate of diffusion of metal ions, dissociated acids and acid molecules, respectively.

Strongly basic anion exchange resins are most suitable for this application. The ratio of the height of the column and the diameter, as Hartinger (2) has found out, is very important for the retardation effect, as can be seen from Fig. 2.

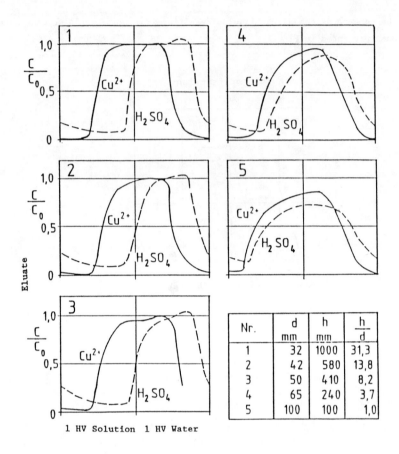

Fig. 2: Influence of the geometry of the column on the separation of copper and sulfuric acid by acid retardation (example: $CuSO_4 + H_2SO_4$)

The degree of separation also depends on the initial concentration of the free acid. In very dilute solutions, retardation becomes less distinct (Fig. 3), whereas acid retardation can be applied for acid solutions with a low metal content as well as for higher metal concentrations (Fig. 4).

Acid retardation has a wide range of application, because strong acids like hydrochloric acid, sulfuric acid, nitric acid and acid mixtures

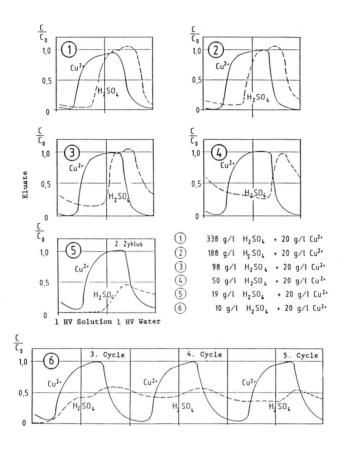

Fig. 3: $CuSO_4$-H_2SO_4-solutions with 20 g/l Cu^{2+} and different concentrations of free acid

have been tested quite successfully. For phosphoric acid and hydrofluoric acid, the applicability has to be checked individually. The formation of metal complexes can have a detrimental effect on acid retardation, but in some cases the separation can even be improved due to the very steep shape of the elution curves.

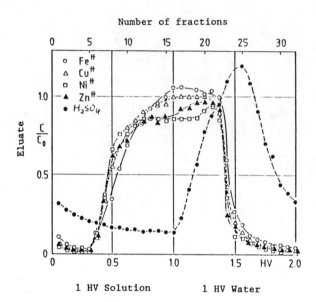

Fig. 4: Separation of a solution containing 42 g/l Ni^{2+}, 3 g/l Cu^{2+}, 15 g/l Zn^{2+}, 7 g/l Fe^{2+} and 95 g/l free sulfuric acid

Outline of technical equipments

The main part of an industrially used equipment for acid retardation is the column containing the special ion exchange resin. For alternating feeding the column with the metal containing acid and water as the eluant, dosing pumps as well as apportioning vessels can be used. The volumes of the liquids leaving the resin bed can be separately adjusted by independently operated valves to meet the requirements resulting from the eluation curves. In this way, different types of solutions can be treated individually to attain optimum results with regard to the concentration ratios of the materials to be separated.

An equipment with a capacity of about 300 litres per hour of acid metal salt solution is presented in Fig. 5.

Fig. 5: Equipment for regeneration of electrolytes for Al anodizing (height of equipment: 2 m) photo by courtesy of GOEMA

This appliance is operated automatically, and this type of column is a proven system, used in many metal finishing plants for different processes.

Industrial applications

In the metal finising industry, one main application of the acid retardation process is the regeneration of electrolytes for anodizing of aluminium. These solutions mostly consist of 150 to 275 g of sulfuric acid per litre but should not contain more than 10 g/l of aluminium, to maintain a high specific conductivity and to guarantee oxide layers of optimum quality. As an aluminium concentration of more than 2 g/l markedly decreases the conductivity, for an optimum current density a higher cell voltage and a growing energy demand for cooling are necessary. Therefore, a continuous regeneration of electrolytes for anodizing of aluminium is an important process step for environmental protection. Moreover, savings for high purity sulfuric acid, alkali for neutralization, sludge disposal and energy have been attained. That's why acid retardation is a very profitable procedure for every anodizing shop (3).

Acid retardation has also been applied under technical conditions for the regeneration of etchants, bright dips and pickling solutions. In all these cases, large amounts of free acids have been saved, and the conditions for metal recovery, by which metal containing sludge can be avoided, have been improved (4).

In order to make use of acid retardation in hydrometallurgy, investigations have been started to find out optimum conditions for the removal of impurities and for acid recovery in zinc electrowinning as well as in copper refining plants. In every zinc tank house, large volumes of electrolytes have to be discarded for a constant concentration level of Mg and Mn. Many tests have been run to confirm that Mg and Mn can be easily removed from the electrolytes, thus returning the major part of the sulfuric acid directly into the electrolyte conduit system (Fig. 6). Zinc can be simply recovered from the Mg and Mn eluate by precipitation with a controlled pH (Fig. 7).

For the removal of As und Ni from electrolytes in copper refineries, laboratory-scale tests have indicated that several different portions of eluates can be collected separately to concentrate the impurities under such conditions that they can be eliminated at the lowest possible expense (Fig. 8).

Economy

In the near future, companies processing "hazardous materials" such as As, Cd, Cr, Cu, Ni, Pb, Zn etc., have to comply with new legislative rules for environmental protection, demanding the use of a new low-waste technology. Under these aspects, acid retardation can effectively contribute to the application of improved procedures, combined with low expenditure. Modest costs for investments, energy and labour, small floor space requirements and big savings for fresh acids, neutralization chemicals, sludge disposal and lower metal losses guarantee that such an investment pays for itself within a short period of time. Therefore, in most anodizing shops the time of amortization of the investment costs for an acid retardation equipment has been less than one year. In etching and pickling plants, expenditure for acid retardation pays for itself even faster.

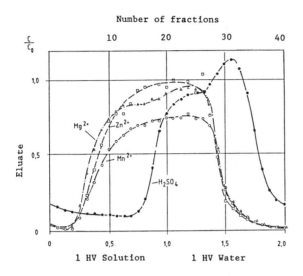

Fig. 6: Separation of H_2SO_4 and the sulfates of Mg, Mn and Zn by acid retardation. 201 g/l H_2SO_4; 46,9 g/l Zn; 9,4 g/l Mn; 7,9 g/l Mg.

In zinc electrowinning plants, bleeding off impurities in a conventional way requires the treatment of several ten m^3 of electrolyte per day, leading to large volumes of sludge and everlasting environmental problems. Therefore, a rapid repay of the investment is obvious.

The removal of As and Ni in copper refining can be performed in quite different ways. But a pre-separation of As and Ni, returning part of the liquid directly into the tank house, and the treatment of smaller volumes of bleeding electrolytes will certainly contribute to cost savings and will make it easier to meet the environmental requirements.

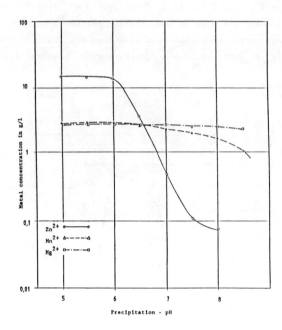

Fig. 7: Precipitation of Zn^{2+} as a hydroxide in the presence of Mg^{2+} and Mn^{2+} from fractions 6-15

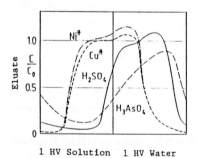

Fig. 8: Separation of Cu^{2+}, Ni^{2+} and As^{5+} from sulfuric acid by acid retardation
Feeding solution: 9,5 g/l Cu^{2+}, 30 g/l Ni^{2+}, 21 g/l As^{5+} and 430 g/l free H_2SO_4

Summary

Acid retardation is a proven procedure for the recovery of free acids from spent metal bearing solutions. The process is based on different diffusion rates of metal ions and acids into a special ion exchange resin. For acid recovery, no chemicals are necessary. Therefore, this recovery technique can contribute to meet the environmental requirements. The procedure has been made use of in metal finishing already, but there are good chances for other applications, e.g. in hydrometallurgy.

References:

1. Hatch, M. J. and J. A. Dillon: Acid Retardation. Industrial and Engineering Chemistry 2 (1963) p. 253-263.

2. Lieber, H.-W.: Entwicklung emissionsarmer Technologien zur Gewinnung von NE-Metallen unter besonderer Berücksichtigung der Hydrometallurgie. Forschungsbericht BMFT, März 1987, S. 154.

3. Hartinger, L.: Lehr- und Handbuch der Abwassertechnik. S. 322, 370. Verlag Ernst & Sohn, Berlin, 1985.

4. Lieber, H.-W., W. Götzelmann, M. Gülbas, L. Hartinger and R. Kammel: Retardation and its use in hydrometallurgy. Paper, annual convention of GDMB, Saarbrücken, June 10, 1988, to be published in Erzmetall.

Light and Other Metals

ADVANCES IN APPLICATION OF

ZIRCONIUM, NIOBIUM & TITANIUM ALLOYS THROUGH POWDER METALLURGY

A. F. Condliff

Teledyne Wah Chang Albany
Albany, Oregon

Abstract

Alloys of zirconium, niobium and titanium have enjoyed expanded current and proposed uses in new areas. This expansion has been due in large part to the utilization of powder metallurgy (P/M) technology and the variety of properties these alloys exhibit. Applications, actual and proposed, of these alloys, utilizing P/M technology, have extended into the chemical, medical, automotive, aerospace and defense systems industries. The properties that make these applications possible are high room and elevated strengths, outstanding corrosion and erosion fabricability. One zirconium alloy (zirconium + 2.5% niobium + 0.4% oxygen) is of interest in orthopedics as a surgical implant alloy. An alloy of niobium (niobium + 30% hafnium + 9% tungsten) is of interest for aerospace applications due to outstanding high temperature strength. Titanium powder parts produced with very low residual chlorine have produced fatigue properties better than wrought titanium alloys, when enhanced by heat treatment.

Introduction

Although zirconium, titanium and niobium alloys have had commercial applications for many years, each alloy has been relatively restricted in diversification. One of the reasons for this is the fact that they are more costly than the more common engineering alloys, thus, they are only used if economically justifiable. For this reason, near net shape (NNS) powder metallurgy (PM) technology is attractive as one route to lower material and manufacturing costs.

The major application for zirconium alloys is in nuclear power electrical generators, as it has been for over 30 years. It is used as a structural metal and also to contain the uranium oxide fuel. The three main properties exhibited by zirconium in this application are low thermal neutron absorption, high mechanical strength and excellent resistance to corrosion by high temperature, high pressure water and steam.

Titanium and titanium alloys have had their major applications in aircraft and aerospace. This is due to the high strength-to-weight ratios exhibited by these alloys. The most widely used titanium alloy, Ti-6Al-4V, has a yield strength in excess of 650 MPa (100,000 psi) and a density of 4.43 gm/cc (.160 pounds/cu. in.).

The use of niobium and niobium alloys (excluding ferroniobium or alloy additions) increased about thirty years ago as high temperature alloys used in space nuclear power generators. Although this field is currently re-vitalized in the United States, the largest application for niobium today is as the major component of niobium-titanium superconducting alloys for construction of the powerful magnets needed for nuclear particle physics development and devices such as magnetic resonance imagers (MRI) for medical diagnosis.

Powder Metallurgy Technology

Powder Metallurgy (PM) technology can be very useful in at least three ways to expand the potential applications of zirconium, titanium and niobium. The first is that of economics. These metals and their alloys are relatively expensive, thus, near-net-shape processing can reduce both material and labor costs. Secondly, PM techniques can expand the use of difficult-to-fabricate alloys that exhibit otherwise attractive properties. Very often, the most difficult process step in traditional ingot metallurgy (IM) is in working a cast ingot structure to a wrought structure, especially with difficult-to-work alloys. Figure 1 compares routes from ingot to mill products using both PM and IM processing.

The third area considered is that of improving the microstructure of alloys regarding both chemical homogeniety and intermetallic compound formation. Both of these characteristics of alloys are normally improved if rapid cooling techniques are utilized to produce the powders. Chemical segregation and large intermetallic particles can preclude successful IM processing and detract from final mechanical properties. Conversely, very fine grained, rapidly cooled powder particles, with extremely small, uniformly dispersed intermetallic particles will normally enhance both processing and final mechanical properties.

The recent advances in hot isostatic pressing (HIP) equipment and technology have greatly advanced the potential to produce 100% dense P/M parts, either for end use or for further metallurgical processing.

Zirconium, titanium and niobium and their alloys are available in three basic types of powders as shown in Table I. Also shown in Table I are the characteristics of each type. The application will determine which type of powder should be used. Sponge fines can be used in non-critical, low performance applications due to the lower cost, however, a small amount of porosity will be present in the final product. Hydride-dehydride (HDH) powder is used for high performance parts that require 100% density and properties such as high fatigue limits. The rapidly cooled spherical powders also produce 100% dense parts and exhibit outstanding properties. Additionally, they are normally lower in oxygen and nitrogen content and are more fabricable in the as-HIP'ed condition.

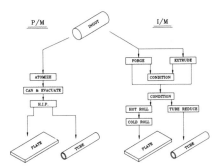

Figure 1 - PM/IM Process Steps

Table I - Types of Zirconium, Titanium, and Niobium Powders

TYPE	CHARACTERISTICS
Sponge fines *	Irregular shape, generally - 80 mesh (<177 microns) Cold compactible Lowest purity Lowest cost
Hydride/Dehydride	Irregular shape, generally - 80 mesh (<177 microns) Cold compactible Higher purity Moderate cost
Spherical	Spherical shape, generally - 20 mesh (<850 microns) Rapidly cooled properties Not cold compactible Highest purity Highest cost

Except Niobium

Figures 2 and 3 show a sample of -80 mesh (177 microns) HDH zirconium powder and -20 mesh (850 microns) electron beam atomized (EBA) niobium alloy WC-3009 spherical powder respectively.

Figure 2 - -80 Mesh Hydride-Dehydride Zirconium Powder

Figure 3 - -20 Mesh Electron Beam Atomized Niobium Alloy Spherical Powder

Figures 4 and 5 show microstructures of Zircadyne[R] 705 alloy (zirconium + 2.5% niobium) in the cold isostatic pressed and sintered condition (CIPS) and in the CIPS plus hot isostatic pressed condition (CHIP) respectively. Residual porosity is apparent in Figure 4 and is eliminated in Figure 5.

[R] Registered trademark of Teledyne Wah Chang Albany

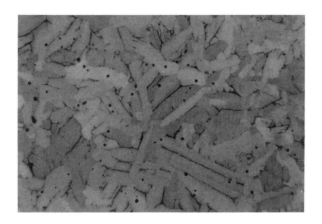

Figure 4 - Zircadyne 705 in the CIPS Condition.   50X

Figure 5 - Zircadyne 705 in the CHIP Condition   50X

Potential Applications Through PM Technology

Zirconium and Zirconium Alloys

In the recent past, there has been a significant increase in the amount of zirconium used as a material of construction in the chemical process industry (CPI). This is due to the outstanding corrosion resistance of zirconium to most of major mineral acids, organic acids, caustics, salts and organic solutions used throughout the CPI. The corrosion resistance is due to the inert and impervious nature of the zirconium oxide that forms naturally and immediately on the surface of zirconium when exposed to oxygen. If the oxide is breached, it reforms rapidly, as long as even a small source of oxygen is available.

Mill products of zirconium, such as sheet, plate, pipe, tube, etc., are readily and economically available. Ancilliary items such as valves, fasteners, stub ends, etc., are comparatively costly due to the fact that considerable labor costs and material loss is incurred in manufacturing such items. PM technology can help to reduce costs, thus promoting the economic benefits of using zirconium. Figure 6 is an example of 3/8-16 nut pre-forms made by the CHIP process using HDH powder. Typically, nuts are made by machining hexagonal bar, drilling, tapping and tumbling. The PM pre-forms are ready to be tapped and tumbled. Figure 7 shows a 1 inch -300 psi valve that typically is machined from solid forged billet stock. Shown in 7a. is a valve PM pre-form that has also been produced by the CHIP process using HDH powder. 7 b. shows the finished machine valve. The material savings is 40% over solid billet and the machining costs are reduced by 30%.

Figure 6 - 3/8-16 Zirconium Nut Pre-forms

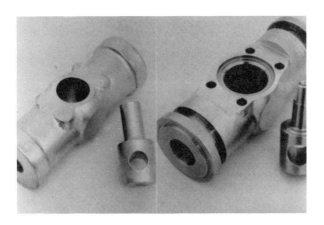

Figure 7 - 1" - 300 psi Zirconium Valve Pre-form

As shown in Table II, the corrosion resistance of of PM zirconium alloys is essentially unchanged from that of the IM structure corrosion resistance.

Table II. Corrosion Rates of PM Zircadyne 702

PROCESS	65% H_2SO_4 (Boiling)	70% HNO_3 (Boiling)	20% HCl (Boiling)	20% HCl + 500 ppm $FeCl_3$ * (Boiling)
CIP + Sinter	1.8, 1.9	.2, .2	.2, .2	1.5, 1.1
CIP + Sinter + HIP	1.3, 1.1	.1, .1	.1, .1	1.0
I/M	2.3	<.1	<.1	2.2

* Some slight pitting was observed

Another example of the versatility of zirconium that has been enhanced by PM technology is in the area of pyrophoric devices. Zirconium is a reactive metal, as are titanium, aluminum, magnesium and others, and can be ignited in the presence of oxygen if in finely divided form. However, zirconium emits an intense white light when ignited, compared to all others. Because of this brilliant luminescence, several applications have developed. Figure 8 shows an assembly of porous zirconium pellets tightly mounted on a graphite rod. When an electrical current is passed through the rod, while oxygen flows over the pellets, the pellets ignite and the light emitted is used to pump an optical laser which is put to several applications. Figure 9 is a microstructure of the interconnecting porosity in each pellet. This can be controlled to produce the desired combusting time duration. The porous structure fills with oxygen prior to ignition, thus, producing a brilliant light source.

Figure 8 - Zirconium Porous Pellets on Graphite Rods

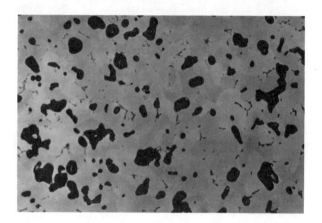

Figure 9 - Porous Microstructure of Zirconium Pellets

A third new potential application for zirconium through PM is in higher strength alloys, with no sacrifice in corrosion resistance. It is well known that zirconium can be strengthened by alloying with oxygen, with some accompanying loss in ductility. It is rather difficult to accurately control the oxygen content through IM, but is easily accomplished through PM by controlling the particle size (thus, the particle exposed surface area).

Figure 10 is a curve showing the relationship of oxygen content of Zircadyne 705 to the mechanical strength properties. It is seen that yield strengths over (680 MPa) 100 ksi can be attained, with significant residual ductility. Since zirconium has long been known to be a very biocompatible metal (1), it is of interest as a potential alloy for medical implants.

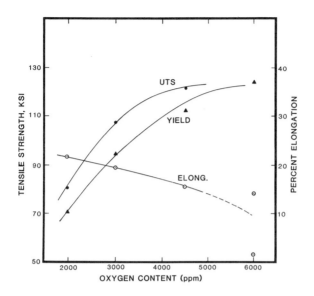

Figure 10 - Effect of Oxygen on the Strength of Zircadyne 705

Titanium and Titanium Alloys

For several years, titanium alloys made by CHIP of sponge fines powder have had a few noncritical applications. However, applications of any consequence for the more costly spherical powder have not materialized, even though this type of powder is capable of producing reliable critical parts for aircraft and aerospace applications. The reasons for this lack of use are unclear. More recently, HDH powder has been shown to also produce critical properties, such as fatigue performance, that can actually exceed wrought properties which are accepted as the standard. Accurate costs for this type of powder are yet to be determined since it has not been produced in production size quantities due to lack of demand. The cost is estimated to fall between sponge fines and spherical powder. Figure 11 shows the fatigue limits for HDH titanium -6% Al - 4% V test samples made by blending the elemental powders, CHIP'ing, and heat treating as follows:

HDH - Heat to 1025°C, water quench, hydrogenation to 0.8 wt%H at 600°C, dehydrogenation at 760°C.

BUS - Heat to 1025°C, water quench, 815°C age.

The base line properties are produced by the CHIP process. This work was sponsored by the U.S. Air Force Materials Laboratory, Wright-Patterson AFB, Ohio (2).

The development of high fatigue performance titanium alloy PM parts has also had some interest in other industries, such as parts for high performance automobiles, where load-reversal and fatigue are a limiting factor.

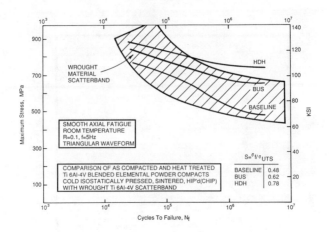

Figure 11 - Blended Elemental PM Ti-6 Al-4V Fatigue Properties (2)

Niobium and Niobium Alloys

During the decade 1960-1970, there was much interest in the United States in developing refractory metal alloys for use in space nuclear power programs. Most of the interest centered on niobium and tantalum alloys. Literally, hundreds of niobium alloys were proposed and produced in small quantities to study. Many of the alloys had attractive properties, but were difficult or impossible to fabricate using traditional IM. The niobium alloys that exhibited the best high temperature strengths contained tungsten, which provided the strength while also rendering the alloy difficult to fabricate. Thus, alloys of lower performance were further developed because they were fabricable. Recent revitalized interest in space nuclear power, and other fields in the United States, caused Teledyne Wah Chang Albany to develop processing of a niobium-tungsten-hafnium alloy using PM technology. The alloy was attractive, not only for strength, but also for improved resistance to oxidation due to the formation of a complex Nb-Hf oxide that retarded further oxidation.

Previous attempts to produce alloy WC-3005 (Nb-9%W-30%Hf) through vacuum arc IM had proven futile. This was mainly due to the difficulty in processing a vacuum arc melted ingot through to a forged structure due to the very high strength of this alloy at elevated temperatures.

It was found that by HIP'ing EBA spherical powder into the shape of a plate, bar or pipe, there was relatively little difficulty experienced in further hot and cold working of the alloy. The EBA powder is made by electron beam melting drops of prepared alloy onto a spinning disc, atomizing the drops into tiny spheres. Figure 12 shows the microstructure of the spheres to be very fine dendrites, compared to IM ingot structure.

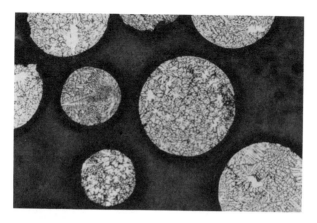

Figure 12 - WC-3009 EBA Spheres   50X

Tensile test bars made by HIP'ing EBA spherical powder had yield strengths at room temperature exceeding (689 MPa) 100 ksi and surprisingly high ductility. This is shown in Figure 13 where WC-3009 is compared to a similar alloy, WC-3015 (Nb-15%W-30% Hf), and several other niobium base alloys. Nickel based Alloy 718 is also shown for comparision. Ironically, the first interest to be shown in WC-3009 came not from the aerospace field, but from ordnance, as a potential liner alloy for gun barrels to be utilized in advanced gun systems being developed. Figure 13 shows two of the test liners which were produced by HIP'ing EBA spherical powder and subsequently successfully test fired. This application is currently in development.

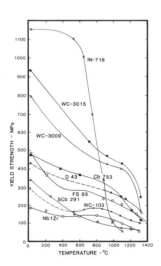
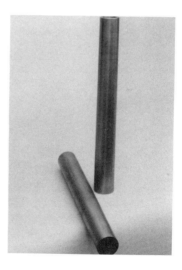

Figure 13 - Yield Strength-Temperature Curves for Several Niobium Alloys & Alloy 718

Figure 14 - WC-3009 Gun Barrel Liners, 63 mm O.D. x 38 mm I.D. x 510 mm L.

Further development of this WC-3009 is underway in the area of producing mill products from HIP'ed spherical and HDA powders. The EBA spherical powder compacts exhibit good cold ductility directly after HIP'ing. It is believed that this is due to the refinement of the size and distribution of intermetallic compounds that form during solidification. Figure 15 shows a typical microstructure of IM wrought WC-3009 and Figure 16 shows that of HIP'ed EBA spherical powder. It is apparent that the more rapidly cooled EBA spherical particles contain smaller and more uniformly distributed intermetallics, resulting in higher fabricability. This PM structure can be readily hot and cold rolled to sheet, plate, bar, etc. Subsequently, it has also been found that HDH powder can be HIP'ed to flat or round form and hot worked to plate or bar (3). Direct cold working prior to hot working is unsuccessful in this case.

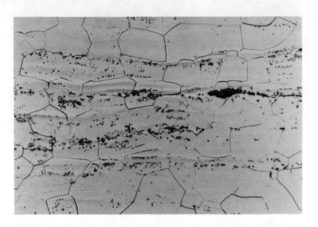

Figure 15 - IM Wrought WC-3009 Microstructure   200 X

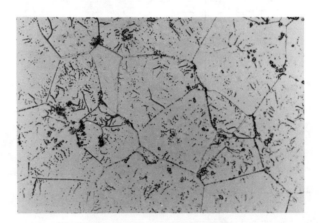

Figure 16 - PM EBA Spherical Powder
WC-3009 Microstructure 200X

PM has also made possible new approaches to produce special devices that have application in space nuclear power generators and elsewhere. Space nuclear power generators are designed to utilize liquid sodium or lithium as a heat transfer medium. This design requires the use of heat pipes to condense the hot gaseous coolant to liquid and return it to the hot zone. Figure 17 shows the details of a heat pipe that is constructed of a 18.75 mm niobium-1% zirconium tube lined with a porous layer of spherical niobium powder. The design also includes three porous arteries running the length of the pipe. As the hot sodium or lithium gas travels through the tube, which is exposed to cold space vacuum on the outer surface, it condenses and returns by capillary movement through the pores in the liner to a collector. The arteries aid in faster return of the condensed liquid.

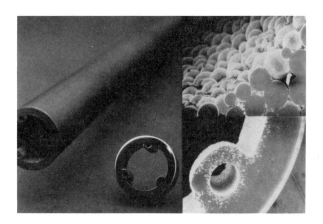

Figure 17 - Niobium-1% Zirconium Heat Pipe Lined with Porous Niobium Spherical Powder

Summary

Zirconium PM technology has created potential increases of applications in chemical processing, through cost reduction of ancilliary items of equipment construction, such as valves and fasteners. PM has also made possible the controlled use of a costless alloy addition, oxygen, to create higher strength zirconium alloys with no sacrifice of corrosion resistance. It has also created new uses in pyrotechnics through the utilization of controlled porosity, and thus, combustion characteristics.

Titanium alloys produced through PM, have been shown to equal or exceed the critical mechanical properties required for aircraft or aerospace applications, thus, should see wider use in these areas as well as several others where fatigue limits and momentum reversal are important. The near-net-shape concept would reduce the costs considerably compared to traditional IM processing. This is predicated on the use of the spherical or HDH powders.

PM technology has been the key to development of higher strength niobium alloys that heretofore have been limited in application due to extreme fabrication difficulties. These have been primarily associated with conversion of a cast ingot structure to wrought form. PM bypasses this problem area, and at the same time, can improve fabricability and mechanical properties of mill products and final near-net-shape parts.

References

(1) P. G. Liang, A. B. Ferguson, E. S. Hodge, "Tissue Reaction in Rabbit Muscle Exposed to Metallic Implants", Journal of Biomedical Research, Vol. 1 (1967), 135-149.

(2) D. Eylon, R. G. Vogt, F. H. Froes, "Property Improvement of Low Chlorine Titanium Alloy Blended Elemental Powder Compacts by Microstructure Modification," MPIF Annual Powder Metallurgy Conference & Exhibition, May 18-21, 1986, Progress in Powder Metallurgy, Vol. 42, 625-634.

(3) C. C. Wojcik, "Evaluation of Powder Metallurgy Processed Nb-30Hf-9W (WC-3009)," International PM Conference Proceedings, June 5-10, 1988 APMI-MPIF, Modern Developments in Powder Metallurgy, Vol. 19, 187-200.

VAW-Technology for Alumina-Production

According to the Bayer-Process

Dr. Wolfgang Arnswald

Aluminium Oxid Stade GmbH

Abstract

The production of Al_2O_3 according to the Bayer-process includes a number of rather varying basic operations of industrial processing techniques. In Germany - a country without own bauxite resources, with high costs for energy and labour and a high standard of environmental protection - the production of Al_2O_3 is only economically possible in plants with a very highly developed technology. VAW - Germany's biggest aluminum producer - had a leading part in almost all stages of developing the Bayer-process. This refers mainly to the development in the areas of the tube digester technology incl. wet oxidation, the red mud filtration, mechanical precipitation and the fluidized bed calcination. This technology will be presented. Operational results of the alumina plant in Stade, erected 1971 - 73 in line with the at that time latest VAW-know-how, and meantime modified further, are to demonstrate what can be achieved nowadays in the areas of raw material and energy utilization, repair, personnel and environmental protection.

Introduction

The production of currently about 28 million mt Al_2O_3 (85 % of which for the smelters) in the Western world is extracted from bauxite according to the 1887 patented Bayer-process.

Basis of the Bayer-Process

The Bayer-process (1) is a circuit process, consisting of the extraction of aluminum hydrate contained in the bauxite with caustic soda liquor at elevated temperature, the separation of the solid residue (red mud) and the following crystallization ("precipitation") of gibbsite from the pregnant liquor setting free simultaneously caustic soda from the supersaturated aluminate liquor obtained after cooling. The following simplified reaction pattern describes the process for a gibbsite bauxite:

$$Al(OH)_{3(bauxite)} + NaOH \xrightarrow{T_1} Na\,Al(OH)_4 + \text{red mud to deposit}$$

$$Al(OH)_{3(refined)} + NaOH \xleftarrow{T_2} Na\,Al(OH)_4$$

Bases for the process are the temperature-dependent solubility of the aluminum hydrate and the characteristic stability of the normally metastable supersaturated aluminate liquors. This phenomenon in particular permits their handling during the separation of the red mud generated at digestion without precipitation of gibbsite. Only through addition of big quantities of seed the supersaturation is gradually decreased. The characteristics of the Bayer-process have remained up to now nearly unchanged, since the results of this "one-stage refining process by recrystallization" are remarkable indeed. For instance the Fe_2O_3-content between bauxite and alumina can be reduced by approx. 3 decimal powers.

VAW-Technology for Alumina Production

Vereinigte Aluminium-Werke AG (VAW) has produced beside aluminum also alumina since its foundation in 1917. At the end of the war in 1945 the company had three alumina plants with a total capacity of 330,000 mt.

After 1948, when the production of aluminum and alumina was allowed again in West Germany, reconstruction of the two remaining plants started, which were partly destroyed, respectively damaged.

Modernization and capacity expansion enabled VAW to adapt their plants, working according to the Bayer process, to the progress of engineering and apparatus technique and to act innovatively itself.

In the course of this an important development was started and brought to complete operational maturity, especially in the areas of technology of the bauxite digestion, the red mud filtration, and the calcination.

The development of the VAW-technology was always carried out under the aspect that in Germany, a country without bauxite mines, with high costs for energy and labor, and a high level of environmental protection, at longterm, the Al_2O_3-production will only be economically possible, in plants with highly developed technology.

Therefore, especially great care was taken that

- in the new units bauxite types of all known deposits could be processed with high recovery
- the liquor productivity is optimized and the energy requirement minimized (2)
- the significance of environmental protection is considered adequately

The Stade alumina plant of AOS (Aluminium Oxid Stade GmbH) was constructed from 1971 - 1973 according to the, at that time, most up-to-date VAW-know-how and has been modified continuously in the meantime (Fig. 1). Its current annual capacity is close to 0.7 million mt Al_2O_3.

Fig. 1: Alumina Plant Stade

A simplified flow sheet whereupon this plant is based is shown in Fig. 2.

It is showing the numerous quite different unit operations which belong to the Bayer process. The steps marked with an asterix are featuring the application of modern VAW-technology, which is explained more detailed in the following.

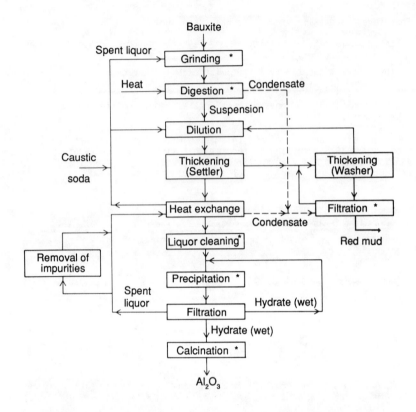

Fig. 2: Simplified flow sheet of the Bayer-process according to VAW-Technology

Grinding Facility

Through the introduction of DSM-screen classification the bauxite wet-grinding in the rod mills can be done in a closed circuit, whereby only the oversize particles are recycled to the mill (3). Fig. 3

The demand of electrical energy for the total mill unit could be reduced to 8 kWh/mt bauxite by avoidance of overgrinding. A ground material is received, which makes good digestion recovery possible, and also causes very little wear of the tubes by erosion over a longer time. Moreover this material has an advantageous effect on the red mud separation. Through evaluation of cinematographic investigations a form of grinding-plate has been developed, which makes it possible to lengthen the endurance of the mill lining considerably.

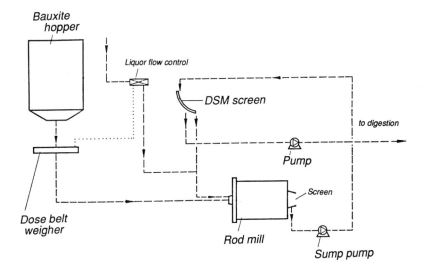

Fig. 3: Bauxite grinding in closed system

Tube Digester Units

The solubility of the aluminum hydrate types contained in the bauxite increases considerably with the alkali concentration and the temperature (4).

Also the dissolving rate is influenced accordingly. This means that the reaction velocity is determined by chemical processes below 150° C and by diffusion at higher temperatures.

The aluminum minerals in the bauxite refer to gibbsite (γ-AlOH)$_3$), boehmite (γ-AlOOH) or diaspore (α-AlOOH), or mixtures of these minerals, of which the degree of solubility respectively dissolving velocity decreases in this sequence.

Therefore, gibbsite bauxite can be digested for industrial purpose already at 150° C and medium Na_2O-concentrations whereas boehmite or even diaspore bauxites require higher Na_2O-concentrations and temperatures as high as possible.

When VAW was confronted with the question how its digestion facilities should look in the future, the decision was based on units with low heat energy consumption, wherein digestion of all bauxite mixes at a medium Na_2O-concentration and higher temperature was possible, i.e. tube digester units. (5)

The diagram of such a unit, which is distinctly different from the normally used autoclaves, is shown on Fig. 4.

The high digestion temperature of 270° C (also 280° C were tested successfully) made it possible to reduce the Na_2O-concentration in the digestion liquor to the level used for precipitation. There is no need for heat intensive evaporator units to keep up two different caustic

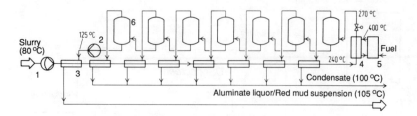

Fig. 4: Scheme of a Tube Digester
1 Diaphragm piston pump, 2 Centrifugal pump, 3 Tube reactor - heat exchanger - , 4 Salt heat exchanger, 5 Salt heater, 6 Flash Tanks

concentrations for digestion and precipitation. The required quantity of wash water for the process is obtained by flash evaporation right in the heat exchanger part of the tube digester unit.

The extraction takes place in the tube reactor at turbulent flow. Heat and mass transfer are at their best, retention time behavior and material conversions are ideal.

The heat supply to the digester can take place via a steam heat exchanger. However, more appropriate is the use of a heat carrier salt. It is easy to adjust the temperature difference between digester slurry and heat carrier salt to 100 - 150°, thus only requiring comparatively small heat exchange surface. No material problems exist, normal steel can be used.

As the bauxite/caustic liquor suspension is already mixed completely when it is preheated in the tube digester units with the help of flash-vapor, the usual attack of corrosion of the caustic liquor is considerably reduced so that also here normal steel is used.

Generally oil or gas serve as sources of primary energy. VAW's latest development in cooperation with Lurgi has lead to a coal based alumina plant energy concept in which also low quality fuels (waste coal) can be used to heat the heat carrier salt, respectively to produce steam for the general use of the alumina plant (6).

Also the wet oxidation of organic C-compounds in the liquor circuit has been developed by VAW, whereby pure oxygen is fed into the tube digester units, a procedure which has been performed at the Stade Plant for years.

The tube digester units, which have been operating since 1973 with running times of over 90 % without remarkable disturbances, show that this technology has ripened and is safe to operate with regard to pumping (diaphragm piston pumps 340 m³/h, 100 bar) and to a specific cleaning procedure of the reactor.

The capacity of each tube digester unit in Stade is around 175,000 mt per year. The main cost factors for alumina production are bauxite and heat energy, which are both favorably influenced by the tube digester technology (recovery over 96 % of theory from boehmite-containing bauxites).

Red Mud Filtration

The digested bauxite slurry contains only the so-called red mud as solid matter, which is separated, after thickening it in the settler to approx. 400 g/l solids by drafting the underflow. This is done in either a 4-7 stage washer cascade, or in only 2 washer stages, followed by red mud filtration. The technology of the red mud filtration on vacuum drum filters with a filter area of up to 100 m² and roller take-off was developed by VAW (7) (8). It means savings of alkali and heat energy, as well as remarkably little space requirement. The washed, and up to 40 - 50 % dewatered red mud with only 0.5 % water soluable Na_2O-content offers a good basis both for dumping at a deposit appropriate to environmental requirements and for further utilization.

Liquor Cleaning

The liquor which leaves the red mud settler still contains up to 50 mg/l Fe_2O_3 in the form of suspended, partly also colloidal red mud. This solid matter has to be removed, as otherwise it would contaminate the produced alumina. This filtration is done with Kelly-filters, whereby as a filter aid either lime slurry or, according to VAW-technology, $Al(OH)_3$ is used (9). To the liquor, which has been already cooled down to about 75° C, some grams of $Al(OH)_3$ are added as seed and as filter aid. With this a precipitation occurs, promoting also the separation of the finest Fe_2O_3-particles. After filtration of the aluminum hydrate/red mud mixture, a liquor with $<$ 10 mg Fe_2O_3/l is obtained, also when processing bauxites, with which Fe_2O_3-separation is a problem. The aluminum hydrate of the filter cake is redissolved, returned to the clarifying settler, so that in this stage of the process no additional filter residue is produced. The liquor cleaning stage is a very important part of the process, although, measured by the small quantities separated, rather costly.

Precipitation

Crystallization of the aluminum hydroxide from the supersatured caustic liquor cooled down from 75° to 55° C is performed within 40 to 50 hours. In order to recover as much hydroxide as possible from the pregnant liquor, which means to keep the recycled liquor volume low, the decomposed suspension subjected to filtration partly contains up to 600 g/l aluminum hydroxide as seed or more.

Whereas previously precipitation was carried out batchwise in mechanically agitated tanks with volumes from 300 - 600 m³, later on systems of continuously working precipitator cascades with cone-type tanks and air lifts prevailed.

In cooperation with Ekato, VAW developed the continuous precipitator cascade with mechanical agitation (10). These tanks with up to 3500 m³ volume are equipped with socalled "Multistage-Impulse-Counter Current" (MiG) agitators. Fig. 5 is showing such a decomposer tank optimized for this type of agitator. The demand of power could be decreased to 0,01 kW/m³, agitation is going easy on crystals and is so reliable in operation, that e.g. after a 90 minutes power breakdown the settled seed could be resuspended completely. Hence the results of former model tests were confirmed (Fig. 6).

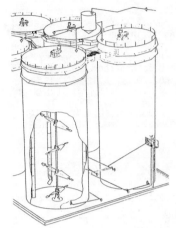

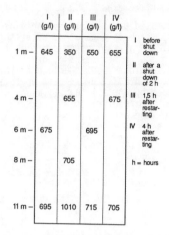

Fig. 5: MIG-agitated 3000 m³-precipitator cascade

Fig. 6: Solid distribution in the suspension after a shut-down

Calcination

The dehydration of the wet hydrate took formerly place in rotary kilns, like those used in the cement industry. In order to reduce heat consumption heat exchanger rotary kilns were introduced.

Studies with the aim to produce a material of very high purity, since the abrasion of the brickwork impaired the SiO_2-content in the alumina, finally led to the development of the circulating fluid-bed calcination. Lurgi and VAW developed jointly this process over a period of 10 years, until it was ready for production, so that in Stade the circulating fluid bed calcination could already be exclusively applied (11).

The scheme of such a unit for 700 mt/day is shown on Fig. 7.

Meantime this system has been adopted worldwide in units up to 1,700 mt/day.

Through best possible heat recovery from product and waste gas, as well as insulating the unit accordingly, the heat energy consumption could be reduced to 3.1 GJ/mt Al_2O_3. This means a saving of up to 35 % vs. the former rotary kilns.

The fluid-bed calciners are very flexible as far as throughput and temperature are concerned.
At temperatures between 1000 and 1150° C an alumina with a specific surface between 30 and 80 m²/g (BET) can be produced, as it is required for the HF-adsorption in dry srubbing dust removal of modern reduction plants.

The units in Stade proved easy to be serviced and very reliable. Running times of about 95 % are usual. Space requirement is low.

Meantime Lurgi/VAW have modified one of the existing units, normally producing alumina for reduction, by adding one stage to the system, so that also alumina specialities can be produced.
For this a so-called high-temperature reactor is intercalated. In a second combustion stage, at temperatures around 1,500° C, material with a BET-surface of <3 m²/g is produced.

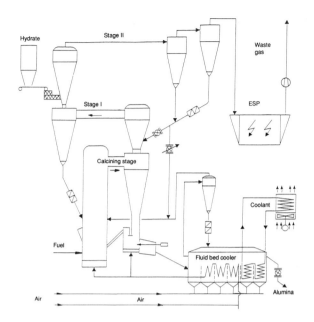

Fig. 7: Principle flow sheet of a circulating fluid bed calciner

Other Plant Units

VAW has also developed very special technology and know-how in other sectors of the processing stages, not mentioned in detail previously, for instance:

- red mud thickening by using exclusively synthetic flocculants,

- operation of a flat bottom thickener as settler or washer,

- fast running 3-disc-filters with a filter area of 112 m² for hydrate filtration which can also be equipped with a washing facility,

- red mud pumping over a distance of more than 6 km with the help of diaphragm piston pumps.

In the area of red mud utilization the construction of deposit dikes with the help of lime stabilized red mud has been developed and tried out successfully (12). Furthermore, research projects were developed, ready for putting into practice, for utilizing red mud as

- Filling material for road construction with bitumen (red filler) (13)
- Addition to the silt of brackish areas with the purpose of obtaining a soil suitable for farming
- Aid to reduce the phosphate elution from marshy soil agriculturally utilized.

Environmental Protection

The alumina plants operating with the Bayer-process are in general compatible to ecological requirements. At the Stade alumina plant the following relevant environmental data are attained, respectively kept, applying VAW-technology:

Noise: <45 dB(A) near the plants' fence.

Waste Water: The cooling water leaves the plant with the rain water practically unpolluted. At very rare occasions, when the pH-value rises to 9, the water flow to the Elbe river is stopped, and the water is taken into the liquor circuit.
The rain water of the red mud deposit area is treated in a two-stage purification plant; similarly the plant's sewage water is treated.

Waste Gases: The special method of combustion in the fluid bed guarantees NOx-values of only a few mg/Nm^3. Also the dust content is far below the future limit of 50 mg/Nm^3 dry.

The residual gases from the bauxite digestion, which are not condensable and very odorous, are controlled and burnt.

Red Mud Deposit Area: The sealing of the deposit against the ground water is warranted, as constant measurements over 15 years have proved. A one year measurement, according to "TA-Luft" requirements, of the deposit's dust emission came to an average value of 0.07 $g/m^3/d$.

Summary

The consistent application of the latest technology - mostly based on VAW-know-how - resulted at the Stade alumina plant, AOS, in a cost situation which at least partly compensates for the disadvantage of location in the Federal Republic of Germany, due to optimum bauxite utilization, saving in energy, personnel, and repairs. Investment costs up to now haved amounted to 600 million DM.

Last year's results were as follows:

Production:	0.690	million mt Al_2O_3 layout 0.6 mill.
Bauxite recovery based on the theory:	96	%
Heating Energy:	8.15	GJ/mt Al_2O_3
Electrical Energy:	265	kWh/mt Al_2O_3
Repair costs:	26	DM/t Al_2O_3
Personnel:	490	employees

References

1. Bielfeldt, K. - Winkhaus, G. Aluminiumverbindungen, Winnacker/Küchler, Chemische Technologie Bd. 3: Anorganische Technologie II, 4. Auflage (1983)

2. Bielfeldt, K. - Winkhaus, G. Herausforderung an die Verfahrenstechnik der Aluminiumoxiderzeugung. ALUMINIUM, 57 (1981) pp. 653-656.

3. Wargalla, G. Die Naßvermahlung von Bauxit. ERZMETALL, 28 (1975) pp. 336-337

4. Kämpf, F. Der Bauxitaufschluß im Bayer-Verfahren. ALUMINIUM-Verlag Düsseldorf, special print of the documentation of 8th Int.LMT (1987) Leoben-Vienna pp. 30-37

5. Bielfeldt, K. - Arnswald, W. Probleme des Bauxit-Aufschlusses. ALUMINIUM, 43 (1967) pp. 355-360

6. Beisswänger, H. - et al. New Coal Based Alumina Plant Energy Concept, Light Metals (1984) pp. 357-372

7. Kämpf, F. - Tusche, J. Erfahrungen bei der Filtration von Rotschlamm auf Drehfiltern. ERZMETALL XX (1967) pp. 402-409

8. Tusche, J. - Heber, W. Red Mud Separation By Filtration. Light Metals (1979) pp. 83-105

9. DP Nr. 960 811

10. Arnswald, W. Agitation of Decomposers. 100th AIME-Ann.Meet. New Yord (1971)

11. Schmidt, H. - et al. Practical Experience with the Operation of LURGI/ VAW Fluid Bed Calciners. 105th AIME-Ann.Meet. Las Vegas (1976) Light Metals (1976)

12. Lotze, J. Experiences in Red Mud Disposals. ERZMETALL, 35 (1982) pp. 526-533

13. Thome, R. - Wargalla, G. Rotfüller aus Rotschlamm für den bituminösen Straßenbau. ALUMINIUM, 54 (1978) pp. 655-656

A NEW IN-LINE ALUMINUM TREATMENT SYSTEM USING NONTOXIC GASES AND A GAS-PERMEABLE VESSEL BOTTOM

Debabrata Saha and J. Scott Becker, and L. Gluns

Air Products and Chemicals, Inc., 7201 Hamilton Boulevard, Allentown, Pennsylvania 18195-1501 and Air Products, GmbH, Klosterstr., D 4000 Dusseldorf 1, Germany

Abstract

A new in-line aluminum treatment system has been developed to meet the industrywide need for a low-capital-cost, easy-to-operate and -maintain, and environmentally safe system. This system can be used to remove dissolved hydrogen and promote the flotation of inclusions. The simultaneous injection of sparging gases into the melt with a blanketing atmosphere over the melt surface produces a clean, high-quality product. A small amount of nontoxic reactive gas, preferably sulfur hexafluoride, can be added to the inert gas (nitrogen or argon) when especially high-quality aluminum is required. The nontoxic SF_6 gas enhances hydrogen removal and provides a protective layer over the melt surface which inhibits back diffusion of hydrogen. Some of the notable features of the unit are its stationary gas dispersion medium (porous refractory bottom), heated gas plenum, baffles, submerged metal exit port, and drain plug. The unit is based on a unique gas-permeable refractory bottom in the treatment vessel which provides effective gas-metal mixing without the need for a complex, submerged, rotating gas-dispersion mechanism. Elimination of the rotating mechanism substantially reduces both equipment cost and maintenance requirements, while gas bubble distribution is not adversely affected.

Introduction

Molten aluminum and its alloys are routinely treated for hydrogen removal in order to eliminate gas porosity in the final products. Several in-line degassing systems based on submerged rotary impellers have been developed over the years to meet the needs of the aluminum industry for producing gas-free, consistent-quality metal at a rapid rate. Although the systems are generally efficient and reliable, they are considered too costly and complex for the majority of aluminum producers.

In order to meet the needs of the industry for a low-capital-cost, easy-to-operate and -maintain, and environmentally safe system, Air Products and Chemicals, Inc. has developed the SIGMA™ (Submerged Injection of Gases into Molten Aluminum) In-line Aluminum Degassing

System[1] which uses a stationary, low-maintenance gas-injection device to remove dissolved hydrogen and promote the floatation of inclusions. Elimination of the rotating gas-injection mechanism substantially reduces both equipment cost and maintenance requirements as well as making the system simple to use.

This paper discusses several criteria for effective degassing, and how the design features of the SIGMA™ system meet these criteria. Also discussed is the importance of using high-purity inert gases. Plant data on the use of the SIGMA™ unit are included.

Why Degassing is Needed

The solubility of hydrogen in pure aluminum is dramatically different between the liquid and solid phases of aluminum at its melting point of 660.4°C (1220.7°F). At this temperature and at one atmosphere pressure of hydrogen, molten aluminum can dissolve 0.46 cc/100 gms of hydrogen, whereas the solid holds only 0.05 cc/100 grams under equilibrium conditions. Hence, as liquid aluminum cools and begins to solidify, most of the hydrogen present comes out of solution and forms gas bubbles (porosity) which are entrapped in the solidified metal unless the metal is degassed before casting. In addition, the soundness of a casting is further deteriorated if the molten metal temperature at casting exceeds the melting point by a substantial margin. This results because the hydrogen solubility in liquid aluminum doubles for each 200°F increase in metal temperature. For example, the solubility is 0.63 and 1.24 cc/100 grams of aluminum at 700°C (1292°F) and 800°C (1472°F), respectively. Therefore, tight control of gas content prior to casting is a very important step in aluminum processing.

The SIGMA™ Reactor

The balance of this paper discusses the theory behind and application of a new in-line reactor in which the process gas or gas blend is injected into molten aluminum from the bottom of the treatment vessel. This bottom is made of a special gas-permeable refractory. This refractory is nonwetting by molten aluminum and has a mean pore diameter three times smaller than that of standard alumina porous materials. As a result, the molten aluminum is well saturated with extremely small gas bubbles. Use of baffles in the vessel ensures intimate gas-metal mixing before the treated metal leaves the vessel. The static capacity of the vessel is sized according to the casting rate so that adequate residence time is available for efficient degassing. A skim door can be provided, as an optional feature, to facilitate dross removal. The outlet port of the vessel is designed so that the treated metal leaves the vessel below the dross layer, thus resulting in a cleaner metal. Heat for holding metal temperature between drops is provided by electric resistance radiant heaters. Figure 1 is a schematic of a typical vessel.

The Theory of Degassing

Aluminum degassing is influenced by chemical thermodynamic and kinetic factors. These factors have been reviewed by Engh, Sigworth, Bortor, Pehlke et al. in several publications.[2,3,4,5] When the sparge inert gases are introduced near (via rotary nozzle) or at the bottom (via porous media as in the SIGMA™ unit) of a melt, hydrogen diffuses to the ascending gas bubbles and leaves the melt. During hydrogen removal, as the partial pressure of hydrogen in the bubbles increases and approaches equilibrium with the melt, the hydrogen removal stops. The volume of sparge gases required to remove a certain volume of hydrogen can be calculated from the following formula[2,3]:

$$\frac{1}{[\%H_t]} - \frac{1}{[\%H_i]} = \frac{200 \cdot f_h^2 \cdot G}{M \cdot K^2} t \qquad (1)$$

where $\%H_i$ = Initial wt % hydrogen (before degassing)
$\%H_t$ = Wt. % hydrogen at time, t
f_h = Henrian activity coefficient of hydrogen
G = Flow rate of sparge gas, kg-mole/sec
t = Degassing time, sec
M = Wt. of metal treated, kg
K = Equilibrium constant for hydrogen dissolution

f_h is 1.0 for pure Al, 1.48 for alloy 356, 1.78 for alloy 319. K is 1.09×10^{-4} at 1350°F.

Thus, in order to reduce hydrogen content from 0.2 cc/100 gms (0.179 wppm) to 0.1 cc/100 gms (0.0896 wppm) in 1000 lbs. of molten pure aluminum within 240 seconds, the theoretical flow rate of sparge gas required is 6.27×10^{-6} kg-mole/sec or 17.9 SCF per hour. In this sample calculation, 100% efficiency has been assumed which means that pH2 in the bubble reaches equilibrium with the hydrogen content in the metal before the bubbles leave the melt.

Thermodynamic equilibrium has been assumed in the above calculation. In practice, kinetic factors, in most cases, determine the effective hydrogen removal rate. The rate at which hydrogen moves from the bulk metal into the ascending gas bubbles is generally controlled by the diffusive transport through a thin, stagnant layer of fluid (boundary layer) surrounding the bubble. Under such conditions, the gas removal rate can be expressed as[3]:

$$[\%H_t] / [\%H_i] = \exp - (k \cdot \rho \cdot A \cdot t)/M \qquad (2)$$

where k = Empirical mass transfer coefficient, m/sec
ρ = Density of liquid aluminum, kg/m³
A = Surface area of bubbles, m²

The mass transfer coefficient can be predicted from the following correlation[6]:

$$k = 0.0128 \left(\frac{D_H \cdot u}{d_B}\right)^{0.5} \qquad (3)$$

where, D_H = Diffusivity of hydrogen, cm²/sec
u = Bubble ascending velocity, cm/sec
d_B = Bubble diameter, cm.

Thus equation (3) indicates that as bubble diameter decreases, the mass transfer coefficient increases. Although the bubble rising velocity decreases with smaller bubbles (explained later on), the adverse effects of lower velocity on k is less compared to the positive effect of a smaller d_B.

Bortor[3,5] found in a series of experiments that argon and nitrogen give virtually the same values of k. However, chlorine gives consistently higher values of mass transfer coefficient, k. Graphical presentation of his experimental data also indicates that, as the bubble radius decreases, k increases, which in turn improves degassing efficiency (equation 2 above and 4 below). Other researchers also indicate the similar inverse relationship between bubble size and k.

Equation (1) represents the thermodynamic situation whereas equation (2) represents a diffusion-controlled situation. In order to better evaluate a degassing system, a dimensionless hydrogen concentration expression, $\psi/[\%H]$, has been proposed[3]:

$$\psi/[\%H] = \frac{(k.\rho.A.K^2)}{(800.\ fH^2.G)} \cdot \frac{1}{[\%H]} \qquad (4)$$

Here ψ represents the ratio of the ability of hydrogen to diffuse into the bubbles during their ascent to the capacity of the sparge gas to remove hydrogen. The efficiency of sparge gas utilization, as shown in Figure 2[3], increases as the dimensionless hydrogen concentration increases, and thermodynamic equilibrium (equation 1) is approached when $\psi/[\%H]$ exceeds 3. The other end of the curve indicates less efficient gas utilization, and hydrogen diffusion controls the hydrogen removal rate (equation 2) when $\psi/[\%H]$ is less than 0.3.

The efficiency of sparge gas utilization is increased as the total surface area (A) increases, as revealed by equation (4). Also, according to Fick's Law, the diffusion rate is enhanced by increasing the metal-gas interfacial area. As the interfacial area increases, the $\psi/[\%H]$ also increases. For a given volume of gas, the interfacial area doubles if the bubble diameter is halved (Figure 3). Hence, as the bubble diameter decreases, $\psi/[\%H]$ increases rapidly and the process can approach 100% efficiency (thermodynamic equilibrium).

Criteria for Degassing

A countercurrent flow of molten metal and sparge gases (either inert alone or a blend of inert and reactive gases) is the basic principle behind most degassing systems. Through a mass transfer mechanism, the hydrogen dissolved in the molten metal diffuses into the ascending gas bubbles. The gas bubbles are most often generated by a gas-injecting device (rotating gas impellers, nozzles, etc.), and initially they have virtually zero hydrogen partial pressure. Atomic hydrogen diffuses into the bubble, changes to molecular form inside the bubble, and leaves the melt along with the gas bubbles. The hydrogen partial pressure of the gas bubbles increases as the bubbles ascend in the melt continuously picking up hydrogen from the melt.

The following criteria for effective degassing have been considered in designing the SIGMA™ unit.

- Minimum bubble diameter of the sparging gas
- Maximum number of bubbles and total gas-liquid interfacial area
- Maximum residence time of sparging gas bubbles in the melt
- Sufficient residence time of metal in the degassing reactor
- Sufficient metal flow rate
- Efficient sparge gas utilization
- Preheating of the sparge gas
- Maximum mass transfer of hydrogen
- Minimum boundary layer oxide film formation
- Maximum mixing power and time

In addition to the above design considerations, the SIGMA™ unit was engineered to meet the following additional criteria:

- Minimum floor space requirement
- Ease of dross skimming
- Easy to operate and maintain
- Nonpolluting
- Simplicity of design

The SIGMA™ in-line aluminum degassing system incorporates the above criteria, among others, to meet the needs of the Aluminum Industry.

Depending on cast house layout, metal flow rate, and frequency of alloy changes, the reactor is custom designed to meet specific user's needs. For example, the inlet and outlet ports of the reactor can be oriented to match an existing metal flow pattern from the furnace to the casting station. A tilting device can also be included to provide complete drainage so as to reduce metal loss where alloy changes are quite frequent and alloy mixing cannot be tolerated. The reactor static capacity is designed to allow adequate metal residence time for a specific metal flow rate.

Evaluation of Critical Design Parameters

When a gas is introduced into a liquid, the gas bubbles ascend at the following rate[7]:

$$u = 1.02 \, (g \cdot d_B/2)^{0.5} \qquad (5)$$

where u = Bubble ascending velocity, cm/sec
  g = Acceleration due to gravity, cm/sec^2
  d_B = Diameter of gas bubble, cm

Equation (5) shows that as the initial bubble diameter decreases, the rising velocity of the bubbles also decreases (Figure 4), resulting in longer residence time in the melt for the smaller bubbles.

The total number of bubbles, N_B, created in the melt, during one second can be estimated by:

$$N_B = \frac{G}{4/3 \, \pi \, (d_B/2)^3} \, 22.4 \times 10^6 \qquad (6)$$

where G = Flow rate of sparge gas, kg-mole/sec

The total surface area of bubbles in the melt (gas-metal interfacial area) is then given by:

$$A = \pi \, d_B^2 \cdot N_B \cdot t_B \cdot 10^{-4} = \frac{6 \, G \, h}{u \, d_B} \, 22.4 \times 10^2 \qquad (7)$$

where h = Depth of the melt where gas is injected, cm
t_B = Average residence time of bubbles in the melt, sec
= h/u

As can be seen from the above expression (7), as the bubble size decreases for a given sparge gas flow rate, the metal-gas interfacial area increases (Figure 3) which promotes hydrogen diffusion from the melt to the ascending gas bubbles according to equations (2) to (4) and Fick's Law.

The above analysis clearly indicates the critical importance of understanding the bubble size. The following empirical relationship developed by Davidson and Amick[8] can be used to do this:

$$d_B = 0.54 \, (22.4 \times 10^6 \, G \, d_n^{0.5})^{0.289} \qquad (8)$$

where d_n = Outside diameter of the gas injecting nozzle, cm

Many of the above equations relating to bubble formation were derived by investigators for the situation where gas was injected through a nozzle. In order to apply the above equations to the porous refractory media concept, used in the SIGMA™ system, it is essential to estimate the flow of sparge gas through individual pore openings (similar to a nozzle system). In effect, the porous media can be assumed as a collection of multiple nozzles. The following steps are therefore used to estimate the gas flow rate through an individual pore of the media:

1. Determine the number of pores in the hot face of the porous media which are in contact with the molten aluminum.

   Assuming all pores at the media surface are circular in cross-section and the total opening area in the hot face reflects the porosity of the material, then the number of pore openings at the hot face, N_p, becomes:

$$N_p = \frac{a \cdot (0.01) X}{\pi \cdot (d_p/2)^2 \, (10^{-8})} \qquad (9)$$

where a = Surface area at the hot face of the porous media, cm²
X = Apparent porosity of the media, %
d_p = Mean pore diameter, microns

2. If the gas flow rate through the entire section of the porous media is G kg-mole/sec, then the sparge gas flow rate through an individual pore, G_p, becomes:

$$G_p = \frac{G \times 22.4 \times 10^6}{N_p}, \quad cc/sec \quad (10)$$

3. The mean pore diameter is essentially the nozzle inside diameter. The outside pore diameter (similar to nozzle outside diameter, d_n) must be assumed to be about three times the mean pore inside diameter based upon visual observation of the refractory. A factor of three leaves considerable space at the hot face of the porous media not made up of equivalent nozzles. In addition, this assumption will make various estimates in this paper conservative from the process efficiency viewpoint. Hence,

$$d_n = 3(d_p)(10^{-4}) \text{ cm} \quad (11)$$

Having derived equivalent parameters for porous media, we can now apply the published expressions for estimating various parameters assuming nozzle injection to predict degassing efficiency in the SIGMA™ reactor.

To illustrate a porous media scenario, sample calculations are made for the SIGMA™ in-line treatment unit with the following data:

Porous Media Surface Area (a)	= 648 in² or 4180 cm²
Mean Pore Diameter (d_p)	= 40 microns (0.004 cm)
Apparent Porosity of Media (x)	= 25%
Number of Pores (N_p)	= 83.2 million (calculated, equation 9)
Equivalent Outside Diameter (d_n)	= 0.012 cm (calculated, eq. 11)
Sparge Gas Flow Rate	= 400 SCF/hr or 140.4x10⁻⁶ kg-mole/sec
Gas Flow Rate Through Each Pore (G_p)	= 37.8x10⁻⁶ cc/sec (Calculated, equation 10)
Metal Treatment Rate	= 275 lbs/min or 2.08 kg/sec
Height of Metal Bath (h)	= 18 inches or 45.72 cm

Bubble diameter, d_B = $0.54 (q \times d_n^{0.5})^{0.289}$ cm
= $0.54 (0.0000378)^{0.289} (0.012^{0.5})^{0.289}$ cm
= 0.015 cm

A bubble diameter of 0.015 cm produced with a porous media-based reactor compares very favorably with typical bubble diameters of 0.2 cm reportedly generated with rotary nozzles[2].

Ascending Velocity, u = $1.02 (g \times d_B/2)^{0.5}$ cm/sec
= $1.02 (980.7 \times 0.015/2)^{0.5}$ cm/sec
= 2.77 cm/sec

Obviously the injection of sparge gas through the bottom of the reactor maximizes the bubble residence time in the melt as opposed to gas injection through any device inserted from the melt surface.

Number of bubbles created in the melt, $n = \dfrac{G_p N_p}{4/3 \, \pi \, (d_B/2)^3}$ per sec

$= \dfrac{0.0000378 \, (83.2 \cdot 10^6)}{4/3 \, (3.14)(0.015/2)^3}$ per sec

$= 1780 \times 10^6$ per sec

The very large number of bubbles insures minimization of transport distance of hydrogen from the bulk metal to individual gas bubbles as well as saturation of the melt with bubbles.

Liquid metal-gas interfacial area, $A_T = \dfrac{6 \, G_p \, h}{u x d_B}(n_p) \, (10^{-4}) \, m^2$

$= \dfrac{6 \, (0.0000378) \, (45.72)}{2.77 \, (0.015)} \, 83.2 \, (10^6)(10^{-4}) m^2$

$= 2,076 \, m^2$

We can also estimate the "gas hold-up" fraction (ϕ) of the bath volume using the following expression[2]:

$A_T = 6 \times 10^2 \, V_m \cdot \phi/d_B$

where V_m = Volume of aluminum melt, m^3

In a typical SIGMA™ reactor, $V_m = 0.2 \, m^3$, therefore, $\phi = 0.29$ which indicates a substantial volume of bubbles within the melt. This is particularly valid in this reactor where the bubbles are injected into the melt through the entire melt cross-section from the bottom.

When the values of the various parameters calculated above are plugged into the dimensionless expression (4), $\psi/[\%H]$ exceeds the value of 3. Consequently, the SIGMA™ system should approach thermodynamic equilibrium (equation 1). A mass transfer coefficient of 5×10^{-4} m/sec and equilibrium constant of 1.09×10^{-4} have been used in calculating the $\psi/[\%H]$ value. Since the reactor approaches thermodynamic equilibrium, the efficiency of sparge gas utilization also approaches 100%.

Sparge Gas Composition

Use of pure nitrogen as a sparge gas provides the undesirable possibility of forming aluminum nitride (Figure 5). AlN not only increases the inclusion content of the metal but also can hinder hydrogen diffusion through the gas-metal interface where an AlN film can exist. (Pilling-Bedworth ratio of AlN is greater than unity, indicating a passivation (blockage) of the bubble surface which hinders further hydrogen diffusion.) The addition of a small amount of chlorine has traditionally been used to etch out the passivating AlN film. This is the major mechanism by which chlorine improves the mass transfer coefficient as previously discussed.

Use of argon also eliminates the formation of a passivating nitride layer and subsequent inclusions. Consequently, the need for chlorine can often be avoided when argon is used.

An alternative to chlorine is sulfur hexafluoride (SF_6) which is safer to handle and generates greater heat of reaction compared to the chlorine reaction. The high in situ temperature produced can enhance hydrogen diffusion across the gas bubble boundary layer. In addition, the reaction products of SF_6 treatment are very resistant to hydrolysis, resulting in an environmentally safer process (halide emission is minimal). Also, SF_6 provides halogen atoms (similar to chlorine), which some researchers believe may inherently increase the mass transfer coefficient, resulting in a further improvement in degassing efficiency.

In addition to sparging of the melt in the reactor, the capability of blanketing the melt surface provided by the reactor lid also offers additional benefits. Since the presence of moisture above the melt surface would encourage back-diffusion of hydrogen into the degassed melt, a blanket of dry inert gas can be beneficial.

Effects of Sparge Gas Purity

The question of sparge gas purity has recently been raised as lower-purity gases sources have become available.

In any aluminum degassing operation, sparge gases are introduced at or near the bottom of the melt via porous media, nozzles, or rotors. As the bubbles ascend from the injection point, they remove hydrogen from the melt by a five-step process[2]:

a. Dissolved hydrogen (atomic) in the melt diffuses from the bulk metal to the vicinity of bubbles.
b. As the hydrogen atoms reach the boundary layer, surrounding the bubble, they diffuse through the thin boundary layer.
c. Hydrogen is then adsorbed onto the bubble surface and desorbed from it as molecular hydrogen.
d. Hydrogen as a gaseous species diffuses into the interior of the sparge gas bubble.
e. Finally, the bubbles leave the melt through the upper surface, removing hydrogen along with it.

Diffusion of hydrogen through the boundary layer is rate-limiting. According to Fick's Law, the rate of diffusion is inversely proportional to boundary layer thickness. Any presence of oxygen in the gas bubble effectively contributes to the boundary layer thickness. The reason is that an Al_2O_3 film forms and grows instantaneously if the sparge gas contains oxygen-sourcing impurities like O_2, H_2O, or CO_2. This film inhibits the transport of hydrogen through the boundary layer into the bubble (see Figure 5). Moisture not only forms an oxide film but also impedes the degassing process by reducing the driving force for hydrogen removal. In the extreme case, moisture introduced through sparge gases can even reverse the degassing process, i.e., add hydrogen to the molten aluminum.

The diffusion rate of hydrogen through an oxide film can realistically be retarded to the extent that degassing becomes unacceptable in a production environment. If the concentrations of the oxygen-sourcing impurities are very high, the oxide film can be continuous and up to several monolayers thick, resulting in a complete stoppage of hydrogen diffusion into the bubbles. Hence, even with optimum gas-metal stirring and a very large number of extremely small bubbles, the presence of oxygen-sourcing impurities in the sparge gas can adversely affect the degassing process in a significant manner.

A model[9] has been developed to estimate the thickness of Al_2O_3 film on the inside of the sparge gas bubble. This thickness (δ) is given by:

$$\delta = r \, [1 - [1 - 0.00076533 * C * [1 + 0.0002699 * h]]^{0.33}] \quad (12)$$

where r = Radius of the sparge gas bubble at the melt temperature, cm
C = Volume % of oxygen from all sources in the sparge gas
h = Depth of the melt where the bubble is injected, cm

The above expression emphasizes the strong impact of bubble size and oxygen-sourcing impurities in the sparge gas and the minimal influence of melt depth on the alumina layer thickness. Figure 6 shows the effects of oxygen contamination on oxide layer thickness at various bubble diameters.

Obviously, the degassing process becomes more efficient as the bubble diameter decreases and/or purity of the sparge gas increases. For example, when the sparge gas contains 0.1% (1000 ppm) total oxygen, the theoretical oxide layer thickness can be calculated to be 36 angstroms if the bubble diameter is 2.54 cm (open end wands) or 3.6 angstroms if the bubble diameter is 0.25 cm (porous media or nozzles). The same oxide layer thickness (3.6 angstroms) can be achieved with an open-ended wand (2.54 cm bubble diameter) if the total oxygen content in the sparge gas is reduced to 100 ppm. A melt depth of 45 cm or 18 inches has been used in this estimation. In summary, a tenfold decrease in oxide layer thickness requires the same tenfold reduction in either bubble diameter or oxygen content in the sparge gas if the same degassing results are to be achieved.

Assuming that at the aluminum melt temperature, the oxide layer can grow only in multiples of crystalline layer thickness, equation (12) can be applied to estimate the critical total oxygen content in the sparge gas. Above some critical content, the oxide layer will cover the entire bubble surface with at least one monolayer of crystalline alumina. This can result in blockage of the diffusion of hydrogen across the gas-metal interface (Figure 5). If the oxygen content is less than the critical value, the oxide layer will still form but some areas at the gas-metal interface will remain uncovered, thus allowing hydrogen to diffuse through the interface into the interior of the bubble. Researchers studying the rate of hydrogen solution in aluminum have confirmed that the adsorption and desorption of hydrogen occurs rapidly at an oxide-free surface. The critical oxygen concentration can therefore be calculated from the above assumptions using the value of the lattice parameter along which direction an alumina crystal will grow. Assuming the c-axis growth of a hexagonal alpha alumina cell, the relevant value is 13 angstroms.

For a melt depth of 45 cm, if the bubble diameter is 2.54 cm, the critical oxygen (O_2, H_2O, and CO_2) concentration in the sparge gas is 357 ppm. If the oxygen impurity is higher, the 2.54 cm gas bubble will be completely surrounded by at least one monolayer of crystalline alumina. However, by changing the gas injection method, if the bubble diameter is reduced from 2.54 to ~0.25 cm, the oxygen-sourcing impurity in the sparge gas can be relaxed to 3628 ppm or 0.36 vol% without having a continuous monolayer. As shown earlier, each tenfold decrease in bubble diameter allows a tenfold increase in oxygen impurity level.

The above discussion indicates the importance of purge gas purity. Two sources of nitrogen, pressure or vacuum swing adsorption (i.e., PSA/VSA) and cryogenically produced liquid nitrogen (LIN), (or liquid argon (LAR)) are available. LIN contains less than 13 ppm O_2 from all sources (O_2, H_2O, and CO_2) whereas noncryogenically produced nitrogen typically contains between 5,067 and 20,067 ppm O_2 for PSA-nitrogen or between 564 and 20,064 ppm O_2 in the case of VSA-nitrogen. If LIN is used, bubble diameter does not become a critical factor because there is simply insufficient oxygen to cover the entire bubble surface at any practical bubble size (up to 29 cm). However, the range of maximum diameter becomes 0.16 to 0.04 cm for PSA and .52 to 0.04 cm for VSA. Consequently, the economic advantage of using PSA or VSA nitrogen instead of LIN will be lost when the requirement of product quality is considered.

Finally, while it is imperative that the bubbles not be surrounded completely by oxide film, obviously the less area covered, the better is the degassing efficiency. Hence, the maximum allowable bubble diameter or O_2 impurity in the sparge gas should be substantially less than the critical diameter or concentration calculated above in order to achieve optimum degassing efficiency and allow for normal process variations. In summary, cryogenically produced nitrogen (argon) offers substantial advantages over noncryogenically produced gases in aluminum degassing.

Preheating of Sparge Gases

When sparge gas at ambient temperature (20°C) is injected into molten aluminum at 710°C (1320°F), the thermal expansion of the gas is about 3.4 times, since the thermal conductivity of gas is high compared to its specific heat, and it can be assumed that the bubbles reach the melt temperature shortly after they leave the injection device. As the gas volume increases due to thermal expansion between the initial and final gas temperatures, the bubble diameter increases (by 50%) which adversely affects degassing efficiency, as discussed earlier. Another side effect of an increase in bubble diameter is an increased bubble ascending velocity (22% increase), resulting in a significant reduction in bubble residence time in the melt. This also affects degassing negatively.

In the SIGMA™ system, the gas is preheated to about 700°F before entering the melt, because the volume of the gas plenum located below the porous bottom in the reactor allows adequate time to preheat the gas before it passes through the porous media. This results in a reduction in volume expansion to 1.5 times instead of 3.4 times with cold gas or an increase of only 15% in bubble diameter as compared to 50%. The preheating of the sparge gas thus improves degassing efficiency.

Apart from controlling bubble expansion, preheating of sparge gas offers a side benefit. Preheating of gas to 700°F is equivalent to about 2444 Kcal/kg-mole (N_2) or 1739 Kcal/kg-mole (Ar). If 400 SCFH (11.3 NM^3H) of nitrogen is injected into an in-line treatment reactor about 1242 Kcal/hr, or 1.44 KWH, will not be extracted from the melt if the gas is preheated to 700°F. In the SIGMA™ system, preheating of gas is accomplished utilizing the heat which has already been lost from the reactor through the porous media by conduction. It is a simple case of recuperation of heat.

Mixing Power and Time

The work done by a gas bubble injected into a melt consists of two components: one relates to the displacement of a liquid "column" as the bubble rises, and the second to the expansion of the bubble as the hydrostatic pressure around it diminishes. Nakanishi and Szekely[10] have defined the concept of mixing energy per unit time or the power dissipated in mixing the bath as the product of work done by a rising bubble times the number of bubbles injected per unit of time. The relationship between mixing power and the gas flow rate, as calculated by Themelis and Goyal[11], is as follows:

$$P_m = 2 \left(\frac{Q'}{22.4}\right) RT \ln(1 + \rho m * g * h/P_a) \qquad (13)$$

A useful operating factor is ϵ, the ratio of mixing power per ton of melt.

$$\epsilon = \frac{P_m}{M} = 2(10^6) G \frac{RT}{M} \ln(1 + \rho m * g * h/P_a) \qquad (14)$$

where
- Q' = Gas flow rate, litres/sec
- P_m = Mixing power input (Joules/sec)
- G = Gas flow rate (kg-mole/sec)
- R = Universal gas constant, 8.314 J/mole °K
- T = Temperature, 1000°K
- ρm = Density of molten aluminum, 2.64 gms/cc
- g = Acceleration due to gravity, 980.6 cm/sec²
- h = Metal depth, cm
- P_a = Atmospheric pressure, 1.016x10⁶ dynes/cm²
- M = Metal weight, Kgs

Mixing time, t_m, is defined as the time required to attain complete mixing in a stirred bath. Nakanishi and Szekely related the concept of mixing time to the mixing power by the following expression[10],

$$t_m = 800 \, \epsilon^{-0.4} \qquad (15)$$

where t_m = Mixing time, sec
ϵ = Mixing power input, watts/ton

For the SIGMA™ system,

Q' = 3.15 litres/sec, h = 45.72 cm, M = 453 Kg,

Mixing power input per ton of melt, ε = 2,173 Joules/sec

The mixing time, tm, calculated using equation (15), is therefore 37 secs. Since equation (15) was derived using 50- to 200-ton argon-stirred steel ladles, the constant term (800) will be much lower for the much smaller aluminum degassing systems. Hence, the mixing time may be well within the bubble residence time of (metal depth/bubble ascending velocity or 45.72/2.77) 16.5 seconds. In any case, the metal residence time in the SIGMA™ reactor is about 240 seconds, hence the melt has adequate time to be throughly mixed before exiting the reactor.

Plant Operating Data

A 275 lbs/minute capacity SIGMA™ unit has been in use for more than six months, treating more than 12 million lbs of 6000-series alloys. Using about 400 SCF per hour of argon, hydrogen has been consistently reduced by at least 50% when the incoming metal contains 0.2-0.3 cc/100 grams of hydrogen. Brief experiments using small additions of SF_6 to the sparge gas have shown SF_6 to be beneficial, however, more work is needed to quantify the results. A blanketing gas of nitrogen was also utilized effectively.

Vacuum gas samples have shown (Figure 7) the elimination of all porosity in most cases, or a great reduction in pore size, if any porosity is found, compared to furnace degassing with argon and hexachloroethane. The working environment has significantly improved because of the absence of chlorine-containing emissions.

Installing the reactor upstream of the filter bowl (Figure 8) has allowed a comparison of vacuum-solidified samples before and after the filter bowl (20 ppi filter). An examination of the samples indicates comparable quality, leading to the conclusion that the floatation of inclusions in the degassing reactor is as effective as a positive filtration device using a 20 ppi filter.

The design of the reactor assembly allows a quick change-out of the porous bottom, when required. Using proper scheduling, the change-out can be completed in an eight-hour period. Preliminary indications are that bottom refractory life should average one year.

Conclusions

The SIGMA™ system is a newly available in-line aluminum degassing system which can be installed in any cast house with minimal modification to the existing plant layout. It provides an attractive alternative to primary and secondary aluminum producers to switch either from furnace (batch) degassing to in-line degassing or from an expensive, more complex, rotary impeller-based in-line system to a lower-cost alternative.

Since the reactor provides extremely small bubbles, which become intimately mixed with the melt, the degassing process approaches thermodynamic equilibrium, resulting in very efficient degassing. Figure 9 shows the bubble pattern in a water model wherein porous refractory bottom is the gas-injecting medium.

The system also offers an alternative to the toxic chlorine or chlorine-containing gaseous/solid reagents. The use of sulfur hexafluoride as a reactive gas is very promising and provides an environmentally safer work place.

In any degassing operation the quality of inert gas is very critical, especially with respect to oxygen-sourcing impurities. Pure argon, followed by cryogenically produced nitrogen, is preferred as the inert gas.

The SIGMA™ system may provide an answer to the industry's long-standing need for an effective, low-cost, simple in-line system.

References

[1] Saha, D., Heffron, J.F., and Becker, J.S., "A New In-line Aluminum Treatment System Using Nontoxic Gases and A Gas Permeable Vessel Bottom", Proceedings of the 4th Int'l Aluminum Extrusion Technology Seminar, Vol. 1, pp. 39-44.

[2] Sigworth, G.K., and Engh, T.A., "Chemical and Kinetic Factors Related to Hydrogen Removal from Aluminum", Metallurgical Transactions B, Vol. 13B, September 1982, pp. 447-460.

[3] Sigworth, G.K., "A Scientific Basis for Degassing Aluminum", 1987 AFS Transactions, pp. 73-78.

[4] Bortor, J., "Kinetics of Hydrogen-degassing of Molten Aluminum by Purge Gases", Communication from Institute of Nonferrous Metals, Gliwice, Poland, Aluminum, vol. 56 (1980), pp. 519-522.

[5] Pehlke, R.D., and Bement, A.L., "Mass Transfer of Hydrogen between Liquid Aluminum and Bubbles of Argon Gas", Trans. of Met. Soc. of AIME, Vol. 224, December 1962, pp. 1237-1942.

[6] Calderbank, P., Chemical Engineer, October 1967, pp. 209-233.

[7] Davies, R.M., and Taylor, G.I., Proc. R. Soc., London, A 200, (1950), pp. 375.

[8] Davidson, L., and Amick, E.H., Jr., A.I.Ch.E.J, Vol. 2, (1956), pp. 337.

[9] Zurecki, Z., Internal Communication, Air Products & Chemicals, Inc.

[10] Nakanishi, K., Fuiji, T., and Szekely, J., Ironmaking Steelmaking Vol. 3, (1975), pp. 193-197.

[11] Themelis, N.J., and Goyal, P., "Gas Injection in Steelmaking: Mechanism and Effects", Canadian Metallurgical Quarterly, Vol. 22, No. 3 (1983), pp. 313-320.

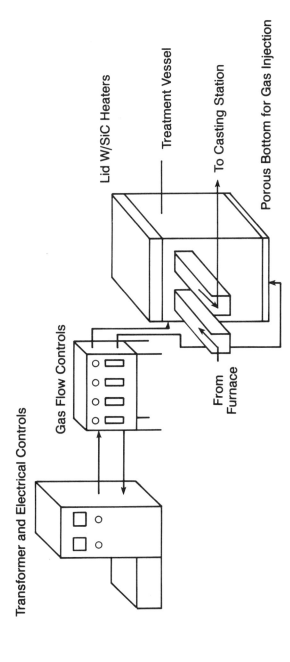

Figure 1 - Schematic of a SIGMA™ Vessel

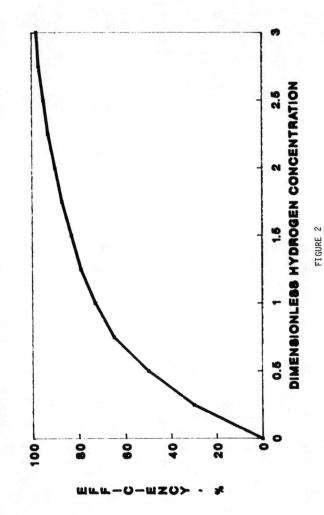

FIGURE 2

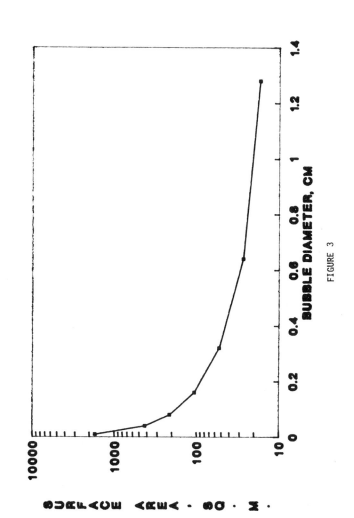

FIGURE 3

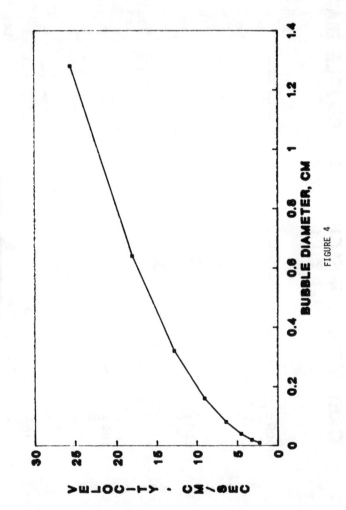

BUBBLE ASCENDING VELOCITY VS. BUBBLE DIAMETER

FIGURE 4

EFFECT OF INERT GAS PURITY ON OXIDE LAYER

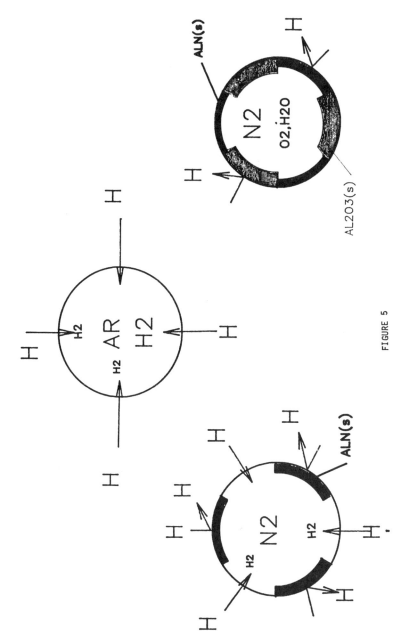

FIGURE 5

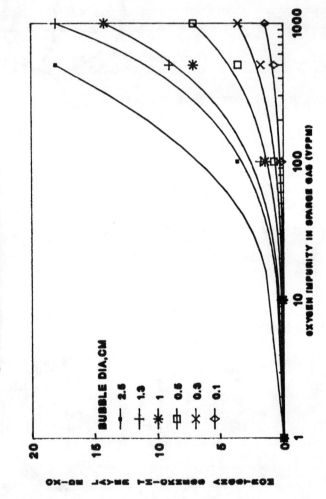

FIGURE 6

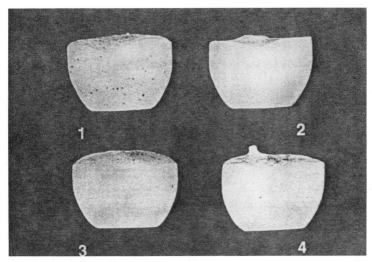

Vacuum solidification test samples (etched, vacuum 25 in. Hg) showing the effect of in-line treatment.
- 1 – Incoming metal from holder (no prior treatment).
- 2 – Immediately after in-line treatment (before filter bowl).
- 3 – Immediately after filter bowl.
- 4 – At the casting table.

Vacuum solidification test samples showing the effect of in-line aluminum treatment (vacuum 25 in. Hg):
Left: (1, 2, 3) – Incoming metal (no prior treatment), taken at the beginning, middle, and end of a casting.
Right: (1A, 2A, 3A) – Corresponding outgoing metal after in-line treatment.

Figure 7 - Vacuum Solidification Test Samples

Figure 8 - SIGMA™ Installation

Figure 9 - Bubble Pattern in a Water Model

MODERNIZATION OF VAW-POTLINES IN THE F.R.G.

Dr. Volker Sparwald
Vereinigte Aluminium-Werke AG,
Aluminiumstraße 2
4048 Grevenbroich 1

The development of the fusion electrolysis has been characterized mainly by rising of current load and pot size decreasing the specific energy consumption at the same time. In the today's modern potlines the comprehensive mechanization and automation of the pot operation is the basis for essential improvements regarding environmental control and working conditions. The erection of new plants corresponding to the actual technical standard needs high investments. As an alternative dry gas cleaning systems with steps to increase the productivity in existing potlines also allows an economically feasible refitting of older plants.

The development of the Hall-Héroult process for the production of aluminium has been characterized during its 100 year history mainly by the increase in amperage and cell size while at the same time decreasing the specific energy consumption. With modern electrolysis systems of our times also significant improvements in environmental protection and working conditions have been achieved. At the Vereinigte Aluminium-Werke the modern state of electrolysis technique is represented by Potline 2 of the Innwerk (Upper Bavaria).

1. New Potrooms for Potline 2 of the Innwerk

In 1980 in the Innwerk 3 potlines with a total of 380 pots operating on an amperage of 36 kA with a combined production capacity of 35.000 tpy were replaced by a new potline having an amperage of 180 kA. In 2 new potroom buildings a total of 120 center worked pots positioned side by side with discontinuous anode carbon blocks were installed in two construction stages. Five of these pots are experimental units for line currents of 240 - 250 kA. The tending of these pots (alumina feeding, anode changing, metal tapping) is carried out with multipurpose cranes supplied by Noell.

Figure 1 - Multipurpose crane in the potroom 2 of the Innwerk

The production capacity has increased to 60.000 tpy aluminium.

The main reason for the increase in amperage and cell size is the improvement of the space-time yield and thus increased efficiency concerning investment and operating costs.

The improvement in space-time yield resulting from increased amperage is presented in Figure 2, which shows the specific annual production of aluminium t Al/m² potroom surface area in relation to the amperage for the VAW smelters.

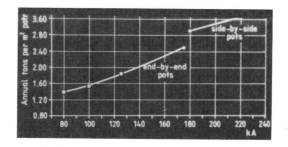

Figure 2 - Production capacity per unit potroom area in dependance on amperage

The diagram in Figure 2 shows a break at 180 kA which results from the change from end-to-end to side-by-side arrangement of the cells in the potrooms.

Considering the comparatively high energy requirements for the production of aluminium the specific consumption of electrical energy represents an important standard in the stage of development of aluminium production technology. During the development of the fusion electrolysis it has been possible at VAW to reduce the DC energy consumption from 25 kWh/kg Al at the beginning to 13,5 kWh/kg Al in the new Potline 2 of the Innwerk. At the same time, the specific man-hour requirements decreased to 1,8 hrs/t Al.

A substantial factor in the reduction of specific DC energy consumption and man-hours is the process control system which regulates the cell resistance and controls the pot tending equipment.

An almost complete direct collection of off-gases from the cells is guaranteed by a closed pot superstructure and center working as well as the control of pot operation by microprocessors and their connection to a central computer system.

The elimination of fluorine from the off-gases from the electrolysis cells is achieved for the first time by two stage dry scrubbing in a circulating fluidized bed of the type VAW/LURGI. By using a fluidized bed reactor and the two stage dust removal the process differs from other dry scrubbing systems that also employ aluminium oxide as means of adsorption. The positioning of an electrostatic precipitator in front of the dry scrubbing system results in the removal of most of the substances that are detrimental to electrolysis and metal quality and are given off during the electrolysis process. The collected dust from the raw gases which amounts to about 40 kg/t of aluminium produced, is subsequently divided into coarse and fine fractions at a ratio of about 7 to 3 by means of a suitable wind sifting process. The separation effect is shown as an example in the following Table 1.

Figure 3 - Wind sifting unit

Table 1 - Results of the Separation of Dust from the Raw Gases by Wind Sifting

		Dust in Raw Gas	Fine Fraction	Coarse Fraction
portion	weight %	100	ca. 30	ca. 70
loss on ignition	weight %	21,8	29,4	16,0
Al_2O_3	weight %	59,9	37,7	72,2
Al_{total}	weight %	36,7	27,1	41,6
F	weight %	15,0	23,1	10,8
C	weight %	8,93	13,0	6,52
Na_2O	weight %	7,56	13,2	4,94
Fe_2O_3	weight %	1,07	3,14	0,39
SiO_2	weight %	0,054	0,068	0,025
P_2O_5	weight %	0,154	0,41	0,049
V_2O_5	weight %	0,063	0,21	0,019
TiO_2	weight %	0,045	0,13	0,018
NiO	weight %	0,043	0,122	0,016
CuO	weight %	0,004	0,009	0,002
ZnO	weight %	0,005	0,009	0,003

The coarse fraction which is rich in aluminium oxide is returned to the electrolysis process. The fine fraction containing the evaporated impurities is carried to the dump. The separation efficiency for the evaporated accompanying elements is about 80% for iron, vanadium and phosphorus and about 70% for titanium, zinc, nickel and copper.

The adsorption stage of the scrubbing system is designed as a venturi reactor to achieve an optimized loading of particulate matter in the gas stream in view of obtaining a high degree of return of alumina to the process. The amount of primary alumina fed into the venturi reactor equivalent to the discharge of fluorine-loaded alumina from the circuit depends on the type of alumina and mainly its degree of calcination. During continuous operation the contents of gaseous fluorine in the stream of purified gas behind the filter system in series can be kept below 1 mg/Nm³.

The construction of the new electrolysis system and the gas scrubbing plant according to the most recent state of development has required an investment effort of DM 180 million or, in relation to production capacity 3.000 DM/tpy Al.

As an alternative, the combination of measures for increasing productivity in existing potlines with the installation of modern off-gas cleaning systems allows an economically justifiable modification of older installations.

2. Modernization of Older Installations

The measures to be considered for modernizing older installations are shown in Table 2.

Table 2 - Measures for the Modernization of Aluminium Electrolysis Plants

1.	Minimizing of fluoride and dust emissions
1.1	Optimizing of gas collection at the cell
1.2	Two-stage off-gas cleaning with alumina as adsorbent in a circulating fluidized bed
2.	Increase in productivity
2.1	Improvement of current efficiency
2.1.1	Optimizing of electrolyte composition
2.1.2	Reduction of influence by magnetic fields
2.2	Increase in amperage
2.3	Increase of the number of operating cells
3.	Lowering of the specific energy consumption
3.1	Improvement of current efficiency
3.2	Lowering of cell voltage
3.2.1	Reduction of current density
3.2.2	Shortening of current paths
3.2.3	Reduction of number of anode effects
4.	Lowering of specific man-hour consumption
4.1	Automation of cell operation
4.1.1	Center working
4.1.2	Process control through microprocessors
4.2	Automation of alumina transport
4.3	Extensive mechanization of anode service

First on the agenda are measures for improvements in environmental protection. The emission limits imposed for the approval by authorities are tied to the modern state of technique. For optimizing gas collection the hooded pots have to be equipped with center working elements. An essential contribution is obtained from the process control with microprocessors that serves for an improvement in productivity at the same time.

In increasing productivity of the electrolysis the most important effect is obtained from an increase in amperage. In addition, a greater number of operating cells has to be taken into consideration.

Finally, the increase in current efficiency gives a substantial contribution to a decrease in specific energy consumption. In addition, the specific energy consumption can be lowered in direct proportion to the cell voltage.

An important part for the economic success of the increase in productivity is, of course, the reduction of man-hours achieved by extensive automation and mechanization of pot operation.

2.1 Modernization of the Rheinwerk

The modernization of the Rheinwerk, the largest aluminium smelter of VAW, started in 1980. The triggering factor for the modernization was the imposition of more stringent measures for environmental protection.

In view of the encouraging experiences in the Innwerk the offgas cleaning in the Rheinwerk has been equipped with a similar installation having double the exhaust volume of 60.000 Nm^3/h each in 4 construction stages. In contrast to Potline 2 of the Innwerk the existing potroom buildings in the Rheinwerk have remained the same as before during the modification. For the cells the principle of the Innwerk has been used as far as possible but the hooded cells with discontinuous anodes and center working were positioned end-to-end. The cathodic part of the cell design as well as the busbar system were left more or less unchanged. The refitting of the cell was carried out while the rest of the potline remained in operation so that losses in production could be avoided. The simultaneous operation of old and new pots can be seen on Figure 4.

Figure 4 - Old/new pots of the Rheinwerk

Because of the design of the existing potroom buildings the heavy multipurpose cranes as in Potline 2 of the Innwerk could not be used. Instead, the changing of the anodes is carried out by means of special anode changing trucks.(Fig. 5)

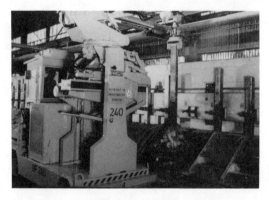

Figure 5 - Changing of the anodes

The filling of the alumina hoppers of the cells is also carried out with special vehicles on most of the cells as shown on Figure 6.

Figure 6 - Alumina feeding by vehicles

For gathering operating experiences part of the cells have been equipped with a pilot installation of a pneumatic conveying system for the alumina supply.

The increase in productivity that has been attained through the modernization of the Rheinwerk is shown in Table 3 in a comparison of characteristic data.

The investment costs for the modernization of the Rheinwerk are at 1.400 DM/tpy production capacity less than half the costs incurred for the new erection of Potline 2 at the Innwerk (3.000 DM/tpy) because the existing potroom buildings were reused. With the same number of operating pots the production capacity of the plant was increased from 150.000 tpy to 210.000 tpy through an increase of line amperage and an improvement of current efficiency. In spite of the considerable rise in current density in the reused busbar system the specific energy consumption decreased by 1 kWh/kg Al.

A substantial contribution to the lowering of the production costs is served by the decrease in specific man-hour requirements from 4,0 hrs/t Al to 2,4 hrs/t Al. The difference to the man-hour requirement of 1,8 hrs/t Al in Potline 2 of the Innwerk can be explained by the fact that rationalization

Table 3

Modernization of Rheinwerk Potlines 1, 2 and 3

	Situation before Change		Present Situation	
	Characteristics	Design Data	Characteristics	Design Data
start-up		62, 69, 73		81, 82, 83
investment costs				150 million DM
specific investment costs (adjusted)				1400 DM/tpy Al
type of anode	"Erftwerk" anodes		discontinuous anodes	
number of operating pots	end-to-end arrangement	470	end-to-end arrangement	470
amperage	three-phase rectifier	124 kA	three-phase rectifier	165 kA
current efficiency		88,0 %		92,5 %
production capacity		150.000 tpy Al		210.000 tpy Al
specific energy consumption		15,8 kWh/kg Al	shorter current paths	14,8 kWh/kg Al
specific man-hour requirement				
- pot operation	side working	4,0 hrs/t Al	center working microprocessors	2,4 hrs/t Al
		2,7 hrs/t Al		1,2 hrs/t Al
- metal tapping	bridge cranes	0,4 hrs/t Al	bridge cranes	0,35 hrs/t Al
- alumina supply	vehicles	0,3 hrs/t Al	vehicles	0,25 hrs/t Al
- anode service	bridge cranes vehicles	0,6 hrs/t Al	vehicles	0,60 hrs/t Al

through the use of multipurpose cranes could not be applied
in the Rheinwerk project.

2.2 Modernization of Potline 1 of the Innwerk

A share of 22.000 tpy of the smelter capacity of the Innwerk
was in the past contributed by pots equipped with anodes of
the Erftwerk type and operating on a line current of 100 kA. A
special feature of this part of the plant is the direct supply
from DC generators at the hydroelectric power plant belonging
to the Innwerk AG, a subsidiary of the VIAG.

Figure 7 - Hydroelectric power plant of the
Innwerk AG

The electrolysis cells of Potline 1 are arranged in several
rows in an old building dating from the year 1924. More stringent requirements for environmental protection here, too,
necessitated operational modifications. Thorough economic considerations led to the decision to retain as much of the
existing equipment as possible and to increase production as
reconstruction advanced in the potline. The measures for modernization have started already and it is planned that while
the positioning of the pots and the paths of the busbar system
will remain more or less unchanged, the existing buildings will
be equipped with center working pots with discontinuous anodes.

Here too, we will have to do without pot tending cranes because of the limited carrying capacity of the crane runways. Also, because of the narrow aisles and passageways the alumina supply can only be accomplished through a pneumatic conveying system.

This concept is considered as a pilot project for the practical demonstration that by using suitable combinations of procedures, older installations can be modified to meet the stringent requirements of the "TA-Luft 1986" (Clean Air Act of 1986).

For Potline 1 the two-stage gas cleaning system VAW-LURGI have been equipped with cloth filter modules at the end of the adsorption stage instead of the electro-filters used on other installations. In comparison to the cleaning systems of the Rheinwerk and Potline 2 of the Innwerk this modification guarantees a further reduction of dust emissions.

Figure 8 shows the installation finished at the end of 1988.

Figure 8

Characteristics and design data of the project are shown in Table 4. (Please the next page)

Table 4

	Situation before Change		Planned Situation	
	Characteristics	Design Data	Characteristics	Design Data
start-up		(24) 54		89, 90
investments costs				60 million DM
specific investment costs (adjusted)				1.800 DM/tpy Al
type of anode	"Erftwerk" anodes		discontinuous anodes	
number of operating pots	end-to-end arrangement	84	end-to-end arrangement	92
amperage	DC generators	100 kA	three-phase rectifier	110-120 kA
current efficiency		90 %		92,5 %
production capacity		22.000 tpy Al		29.000 tpy Al
specific energy consumption		16,0 kWh/kg Al	lower current density shorter current paths	14,0 kWh/kg Al
specific man-hour requirement		4,6 hrs/t Al		2,2 hrs/t Al
- pot operation	side working	2,9 hrs/t Al	center working microprocessors	1,2 hrs/t Al
- metal tapping	bridge cranes	0,4 hrs/t Al	bridge cranes	0,4 hrs/t Al
- alumina supply	vehicles	0,4 hrs/t Al	pneumatic conveyor	-
- anode service	bridge cranes vehicles	0,9 hrs/t Al	vehicles	0,6 hrs/t Al

The start-up of the new pots began in January 1989. The specific investment costs increased to 1.800 DM/tpy annual smelter capacity because of the necessary reinforcement of the busbars in combination with measures for compensation of magnetic fields as well as replacement of the 35-year-old pot shells. In view of the available generator voltage, the reduction in pot voltage is used to increase the number of operating pots.

The raising of the line amperage and the improvement of the current efficiency result in an enhancement of production capacity to 29.000 tpy Al.

The intended lowering of operating costs manifests itself in the reduction of the specific energy consumption by 2 kWh/kg Al and of the specific man-hour requirements from 4,6 hrs/t Al to 2,2 hrs/t Al. The latter figure remains below the result from the Rheinwerk because of the pneumatic alumina supply.

In summary it can be said that the further development of the HALL-HEROULT process as representend by today's modern smelters can, with certain limitations, be applied to the modernization of older installations. By retaining parts and equipment worth reusing, considerably lower investment costs and shorter pay-back periods can be achieved. For all these reasons the described measures serve to assure the further existence of traditional smelter locations and, thus, the continued employment of the experienced personnel.

LITHIUM EXTRACTION FROM SPODUMENE BY METALLO-THERMIC REDUCTION

by

ERNEST MAST RALPH HARRIS
DEPARTMENT OF MINING AND METALLURGICAL ENGINEERING
MCGILL UNIVERSITY, MONTREAL CANADA H3A 2A7

and

J.M. TOGURI,
DEPARTMENT METALLURGY AND MATERIALS SCIENCE
UNIVERSITY OF TORONTO, TORONTO, CANADA M5S 1A4

ABSTRACT

A new metallo-thermic reduction process designated "Melt - Leach - Evaporation " is under development for the extraction of lithium and other group IA and IIA metals. The metal to be recovered is leached out of the ore or concentrate by an excess of liquid metal reductant which is then vacuum refined to recover the metal as a solid or liquid condensate.

The experiments performed contacted spodumene, $LiAlSi_2O_6$, with molten aluminum - magnesium. The molar ratio of the aluminum to spodumene was approximately 86 to 1 and the molar ratio of magnesium to spodumene was increased from zero to ten. The experiments were performed in alumina crucibles, under an argon atmosphere, at 900°C for ninety minutes. Mixing was provided by mechanical stirring.

Assays from the cast ingot confirmed that lithium and silicon contained in the spodumene were reduced and dissolved into the excess molten aluminum reductant. As much as 60 % of the lithium in the spodumene was recovered to the metal ingot cast after the experiment. Increasing the magnesium to spodumene ratio improved the recovery of lithium to the ingot. This was as predicted by a thermodynamic model.

1. INTRODUCTION

1.1. LITHIUM: PROPERTIES AND CURRENT USES AND PRODUCTION.

Lithium is the third element in the Periodic Table and it is the first of the alkali metals. Its most noteworthy chemical property is its density, 0.534 gcm$^{-3(1)}$, which makes it the lightest of all metals. In the pure form, it is too reactive to be used as a structural component, but as an alloying element in aluminum, 1 wt% Li decreases the density by three percent and increases the elastic modulus by 6 percent[2]. Lithium's electrochemical properties such as a standard reduction potential of -3.04 V[3] and an electrochemical equivalence of 3.86 A·hours/g[1] combined with its low density make it an exceptional anode material in batteries. These applications require the use of lithium metal, as opposed to lithium chemicals, which are used in greases, drugs and aluminum production. As a result, the demand of lithium products is changing as shown in Table 1.

Table 1. **USES OF LITHIUM FOR THE U.S. (SHORT TONS)**

	1978[4]	1983[5]	2000[5]
Total Use (Contained Li)	3400	2200	5600
Use in Batteries	0	50	150
Use in Al-Li Alloys	0	0	750
% of Use as Metal	0	2.27	15.1

Two raw materials are presently commercially treated for lithium recovery; sub-surface brines and pegmatite minerals. Currently only a hydrometallurgical route is utilized.

The first product of the hydrometallurgical extraction process for both raw materials is lithium carbonate, Li_2CO_3, which is the most widely used lithium chemical, acting as the raw material for the production of all other lithium chemicals. For lithium metal production at present, LiCl is made from Li_2CO_3, combined with KCl and treated by molten salt electrolysis. The cost of lithium metal is $57.3 U.S. / kg[6] which, it is believed, could be reduced by a more efficient extraction process, such as direct thermal reduction, the subject of this investigation.

The most suitable lithium bearing raw material for reduction would be the pegmatite ores. Brines were discounted as a feed for metallo-thermic reduction since, being aqueous liquids, they could not be treated. Li_2CO_3 was also discounted since lithium recombines with carbon to form extremely stable carbides. Pegmatite reserves are estimated to be 6.1 million tonnes[7], more than enough to supply the demand for lithium metal[8].

The most abundant lithium mineral in pegmatite ores is spodumene. Spodumene concentrates can obtain up to 7.5 wt% Li_2O; equivalent to 3.5 wt% Li. A characteristic of naturally occurring spodumene is an irreversible phase transformation from the α to β phase at about 1000°C. Accompanying this transition is a weakening of the crystal structure and a volume expansion which changes the density from 3.23 g/cm^3 to 2.40 g/cm$^{3(1)}$.

1.2 PREVIOUS STUDIES

Previous work has been performed on the vacuum metallo-thermic reduction of spodumene by Stauffer[9] and on reduction of lithium oxide by Kroll and Schlechten[10]. Aluminum, magnesium, silicon and ferrosilicon powders were used as reductants. The studies found that the reduction was very successful, achieving recoveries in the order of 90%. A lime pretreatment was determined to be essential. Various work has been done on the reaction of lime with lithium materials[11,12,13]. In the most comprehensive study of lime-spodumene sinters, Lanier and Nazirov[14], stated that the first stage of the reaction might proceed according to the equation:

$$Li_2O \cdot Al_2O_3 \cdot 4SiO_2 + 8CaO = Li_2O \cdot Al_2O_3 + 4(2CaO \cdot SiO_2) \quad [1$$

and in a second stage excess lime would break up the lithium aluminate to form lithium oxide.

1.3 THE MELT-LEACH-EVAPORATION (M.L.E.) PROCESS

M.L.E. is a new process being developed at McGill University and the University of Toronto. In the process, a solid, particulate raw material is contacted with an excess of molten metal which extracts a desired species into solution by reducing the compounds of the species. The dissolved species is then recovered by vacuum refining from the bulk metal and condensation[15]. In this paper an investigation of a M.L.E. process for the production of lithium from spodumene, $Li_2O \cdot Al_2O_3 \cdot 4SiO_2$, is discussed.

The advantage of M.L.E. is that it would improve the materials handling and production kinetics compared to a powder process. By using an excess of liquid metal reductant the activity of the lithium is lowered by dilution and more lithium would enter solution to achieve equilibrium.

2. THEORY

The Ellingham Diagram in Figure 1 shows that magnesium and calcium are suitably powerful reducing agents of lithium oxide at atmospheric pressure. However, neither are suitable as bulk molten reductants due to their

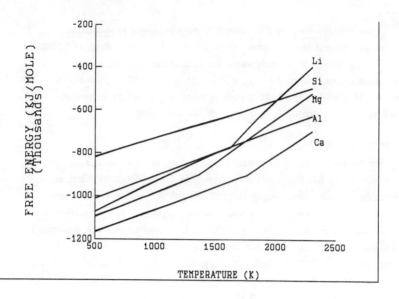

Figure 1. Free Energy of Formation for lithium and potential reductants.

reactivity, vapour pressure and/or cost. Thus aluminum alloyed with either magnesium or calcium was chosen as the reducing agent.

The EQUILIB program of the Facility for the Analysis of Chemical Thermodynamics, F*A*C*T [16] was used to model the system according to the following equations.

$$Li_2O \cdot Al_2O_3 \cdot (SiO_2)_4 + 85.5 \text{ Al} + <A> \text{ Mg} = ? \quad [2$$

$$Li_2O \cdot Al_2O_3 \cdot (SiO_2)_4 + 85.5 \text{ Al} + <A> \text{ Ca} = ? \quad [3$$

Using Gibbs Energy minimization the EQUILIB program determined the most stable mixture of products which could form from the specified reactants. The variable <A> in the equations enabled different alloy compositions to be simulated. The number of moles of aluminum, 85.5, was chosen to simulate the present experiments in which the mole ratio of aluminum to spodumene was 85.5:1. Excess Gibbs Energies of Mixing for Al-Li, Al-Mg and Al-Si were taken into account by the program.

The results of the simulation which are shown in Figure 2 determined that magnesium is the superior reductant. F*A*C*T predicted that there would be complete reduction of silicon at all magnesium concentrations, spinel, $MgO \cdot Al_2O_3$, formation from the product of Mg oxidation and decreasing lithium aluminate, $Li_2O \cdot Al_2O_3$, formation with increased amounts of magnesium addition.

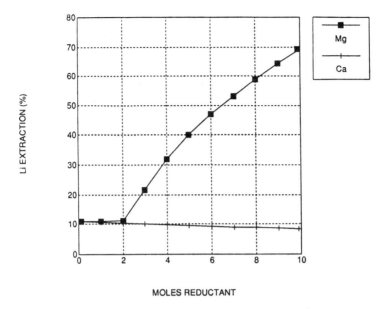

Figure 2. Thermodynamic simulation of the lithium extraction from spodumene by reduction with Al-Mg and Al-Ca alloys. The lithium extracted from the spodumene is plotted against the respective magnesium and calcium to spodumene molar ratios.

3. GOALS

The objectives of the experimental program were twofold; 1) to determine the necessity of pretreating spodumene with lime for the melt-leach process and 2) to determine the effect of magnesium on the recovery of lithium by metallo–thermic reduction from spodumene or a lime pretreated material.

4. LIME-SPODUMENE TESTS
4.1 APPARATUS & REAGENTS

A muffle furnace was used to heat the fire clay crucibles used to contain the reactants in the lime-spodumene tests. Temperature measurement was obtained via Type K thermocouples, placed in alumina sheaths and inserted into the center of the mixture.

The spodumene was a high grade concentrate obtained from The Tantalum Mining Co. of Canada, Bernic Lake, Manitoba containing 7.25 wt% Li_2O. The lime was commercial grade and was calcined at 500°C for 1 hour before use.

4.2 PROCEDURE

Two arbitrarily chosen mixtures containing 38.4 wt% and 68.7 wt% spodumene, respectively, were placed into the fire clay crucibles and heated to temperatures ranging from 900°C to 1150°C in the muffle furnace for four hours. The crucibles were slowly cooled. The mixtures represent lime to spodumene molar ratios of 4:1 and 10.5:1, respectively.

The reaction products were analyzed by X-Ray Diffraction (XRD). The raw data was obtained by a 37 minute scan between 10° and 100° and then analyzed using the **Phillips APD 1700** software analysis system.

4.3 RESULTS

Three distinct reaction products were visually observed upon completion of the experiments. A grey powder which was slightly affected by the $\alpha \rightarrow \beta$ expansion of the spodumene, was observed for both lime - spodumene mixtures at temperatures of 1050°C and below. For trials at 1100°C, a green, friable, sinter that had definitely expanded was obtained. At 1150°C, a white, fused product was obtained. In the experiments at lower temperatures, it is thought that reaction did not occur due to the slow kinetics of solid/solid reactions and the activation energy necessary to commence the reaction which decreases from 753.12 kj/mole below 930°C to 288.7 kj/mole above 900°C[17].

The XRD results listed in Table 2 show that at 1050°C and below only spodumene and the lithium alumino silicate, $LiAlSi_3O_8$, were detected indicating no reaction with lime occurred. At temperatures greater than 1100°C, the products indicating lime reaction were detected. These products were the calcium alumino silicates; gehlenite ($Ca_2Si_2AlO_7$) and larnite (Ca_2SiO_4). Both are greenish in colour and were the cause of the green colour observed. At 1150°C the white mineral, wollastonite ($CaSiO_3$) was detected.

The F*A*C*T EQUILIB program was also used to predict the most thermodynamically stable products of the lime-spodumene pretreatment. The predicted reaction products were the same for all temperatures and are listed in Table 2. The F*A*C*T analysis predicted the formation of Ca_2SiO_4 and $CaSiO_3$, two minerals that were identified in the reaction products by XRD at 1100°C and 1150°C for the 10.5:1 and 4:1 ratios, respectively. In addition F*A*C*T predicted that gehlenite, $Ca_2Al_2SiO_7$, would have a high activity in the system, but not high enough for it to form. F*A*C*T also predicted that the lithium species present would be lithium aluminate, $LiAlO_2$, an extremely stable oxide. This finding agreed with work by Naziroz[15].

Table 2. XRD RESULTS OF LIME SPODUMENE SINTERS

T °C	RATIO	IDENTIFIED PHASES	F*A*C*T
950	10.5:1	SPODUMENE	$LiAlO_2, Ca_2SiO_4, CaO$
	4:1	"	$LiAlO_2, CaSiO_3$
1000	10.5:1		$LiAlO_2, Ca_2SiO_4, CaO$
	4:1	"	$LiAlO_2, CaSiO_3$
1050	10.5:1	"	$LiAlO_2, Ca_2SiO_4, CaO$
	4:1	$LiAlSi_3O_8$	$LiAlO_2, CaSiO_3$
1100	10.5:1	$Ca_2Al_2SiO_7, Ca_2SiO_4$	$LiAlO_2, Ca_2SiO_4, CaO$
	4:1	$Ca_2Al_2SiO_7$	$LiAlO_2, CaSiO_3$
1150	10.5:1	-	$LiAlO_2, Ca_2SiO_4, CaO$
	4:1	$CaSiO_3$	$LiAlO_2, CaSiO_3$

4.4 DISCUSSION

No simple method was available to separate the lithium species from the calcium products, thus it threw into question whether the lime - spodumene sinter product was superior to β spodumene as a feed material for molten metallo-thermic reduction.

The pretreatment did indeed break up the spodumene structure and free lithium oxide species to either form the aluminate or remain as an oxide. However, the lime pretreatment diluted the lithium in the feed. Also, the addition of lime introduces calcium into the system, which has a negative effect on the thermodynamics of the system as shown in Section 2.

Lime's importance in the powder experiments was due to the passive oxide layer which formed on the surface of the reductants. Lime was necessary as a flux to decompose the oxide layer and expose free metal for reaction. Since the reactions were performed in vacuum, no further oxidation, aside from the formation of reaction products, occurred. With the M.L.E. process, molten metals, free of oxide surfaces, would be the reducing agent, so the fluxing properties of lime would not be necessary.

As a result, the lime pretreatment of spodumene was not considered necessary for the M.L.E. process and the reduction of spodumene was attempted without it in this study.

5. METALLO-THERMIC REDUCTION

5.1 APPARATUS

A schematic representation of the apparatus used is shown Figure 3. The equipment consisted of a 100 kW induction furnace with tilting capabilities. The furnace was operated at less than 30% of its rated power. Inside the induction coils, glass wool and silica sand refractory surrounded an alumina crucible 28.8 cm high and 12.8 cm in diameter. A mild steel cap was constructed for the crucible to control the atmosphere within the reactor. The cap had a groove machined into its underside to provide a good fit between itself and the crucible. The cap also had a 2.54 cm hole drilled into its center for an impeller and three 1.59 cm holes for, temperature measurement/viewing, argon gas intake and exhaust gas outlet. Copper tubing was soldered to the cap surface and wound around the exhaust tube for water cooling. The impeller was made from a high-temperature, ceramic cement bonded onto stainless steel reinforcing. It was painted with alumina to provide increased durability and inertness. The impeller had four, 15 cm high, 1 cm thick blades with an edge to edge diameter of 11.8 cm.

5.2 PROCEDURE

Commercial purity aluminum was cut into pieces and melted in the induction furnace. Various amounts of pure magnesium were dissolved into the aluminum via dunking. Spodumene was poured onto the molten surface and the system was heated to 900°C. The cart carrying the impeller and turning drive was moved into place, water cooling and argon gas connections made, and the cap/impeller assembly lowered into the melt. Once the melt reached 900°C, stirring commenced at 36 RPM and argon gas flushing started. Ninety minutes after the commencement of stirring, the power was shut off and agitation stopped. The contents of the crucible were poured into a mold. Drillings were taken from various locations and depths in the ingot and analyzed for lithium and silicon.

5.3 RESULTS

Six experiments were performed with varying magnesium concentrations. They are listed in Table 3. The results of the atomic absorption analysis are shown in Table 4.

To determine the percentage lithium recovery to the ingot, the lithium assay of the ingot was multiplied by the mass of the ingot to obtain the mass of lithium. This was compared to the initial amount of lithium entering the system which was equal to the mass of spodumene multiplied by the lithium concentration (3.38WT.%).

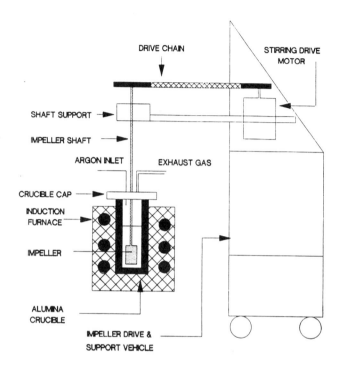

Figure 3: Experimental apparatus used in the present study.

Table 3. **EXPERIMENTS PERFORMED**

EXP.	MASSES (g)			MOLES			Mg/SPOD
	Al	Mg	SPOD	AL	Mg	SPOD	mole ratio
1	2291.0	0.0	343.8	84.8	0.0	0.92	0.0
2	2355.8	104.0	369.0	87.2	4.3	0.99	4.3
3	2410.3	47.6	368.6	89.3	2.0	0.99	2.0
4	2439.8	143.3	370.2	90.3	6.0	1.00	6.0
5	2220.0	191.9	361.8	82.2	8.0	0.97	8.2
6	2269.8	224.8	348.4	84.1	9.4	0.94	10.0

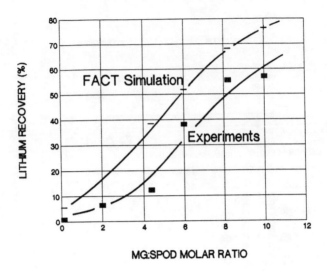

Figure 4. Lithium recoveries to the metal ingot for the experimental and thermodynamic results.

5.4 DISCUSSION
5.4.1 COMPARISON OF RESULTS WITH F*A*C*T SIMULATION

Figure 4 shows that the lithium recovery to the aluminum ingot increased with magnesium addition; which was in agreement with the F*A*C*T simulation. The F*A*C*T simulation predicted that all silicon would be reduced for all magnesium concentrations. Recoveries of silicon to the ingot shown in Table 5 were less than 100%. This may be explained by the fact that pure Si was detected by XRD in the residue powder products. XRD of the powder products also detected spinel, $MgO \cdot Al_2O_3$, which was a Mg compound predicted by F*A*C*T.

5.4.2 REACTION MECHANISM

Various reaction mechanisms were considered possible;
1) reaction between spodumene particles and molten metal throughout the molten metal via thorough mixing of powder and metal.
2) reaction just below the metal surface by temporary submergence of spodumene particles into the molten metal.

Table 4. **RESULTS OF ATOMIC ABSORPTION ANALYSIS**

EXP.	EXCESS METAL INGOT Si(wt%)	Li(wt%)	POWDER Li (wt%)
1	2.72	0.0	2.27
2	3.82	0.13	1.16
3	1.54	0.04	1.93
4	2.32	0.19	1.07
5	3.74	0.30	1.05
6	3.03	0.29	0.82

The experimental lithium recoveries to the ingot versus the magnesium to spodumene molar ratio are compared to the F*A*C*T recoveries in Figure 4 and are listed in Table 5. The silicon recoveries to the ingot are also presented in Table 5. XRD results of the powder products are shown in Table 6.

Table 5. **SILICON AND SILICON RECOVERIES TO INGOT.**

EXP	Mg:SPOD RATIO	SILICON % RECOVERED	LITHIUM % RECOVERED
1	0.0	54.1	0.4
3	2.0	29.1	6.5
2	4.1	53.3	23.6
4	6.0	43.5	38.3
5	8.2	70.3	55.6
6	10.0	63.5	57.1

Table 6. **XRD ANALYSIS**

SPECIES DETECTED
Al
Si
$MgO \cdot Al_2O_3$

3) reaction at the molten metal surface by contact of particle and molten metal surface.
4) reaction via the gas phase by magnesium vapour interacting with the spodumene.

At present, insufficient data are available to distinguish between the various possible mechanisms.

5.4.3 THERMODYNAMIC VARIABLES

The effectiveness of magnesium in increasing lithium recovery can be explained by its large affinity for oxygen; which results in the reduction of the lithium and aluminum oxide species present. Other ways to increase recovery would be to decrease the activity of lithium in the liquid phase. Methods of performing this would be by dilution of lithium in the molten metal by increasing the amount of aluminum solvent, operation in vacuum to remove lithium as a vapour, or increasing the temperature of the system to change the equilibrium conditions.

5.4.4 SURFACE CHEMISTRY

Addition of magnesium to aluminum alloys decreases the surface tension of the molten metal[18] which in turn decreases the contact angle between the solid spodumene and the molten metal and thus improving wetting. However the presence of substantial silicon in the ingot from Experiment 1, where no magnesium was added, indicates that magnesium was not critical to the wetting characteristics.

6. CONCLUSIONS

The lime pretreatment of spodumene ore was not necessary for the reduction of spodumene by metallo-thermic reduction.

The reduction of spodumene by an excess of molten aluminum-magnesium metal was possible at atmospheric pressure. The magnesium reacted with the spodumene to form a magnesium-aluminum oxide, spinel, and silicon and lithium dissolved in excess aluminum alloy.

The recoveries of lithium to the metal ingot mimicked those of a chemical thermodynamic simulation. Using computerized thermodynamic programs, the production of lithium from spodumene could be simulated at different conditions and the best process could be designed.

7. REFERENCES

[1] R. Bach, J.R. Wasson, "Lithium and Lithium Compounds", Kirk-Othmer Encyclopedia of Chemical Technology, Vol. 14, New York, Wiley & Sons, **1981** pp. 449-476.

[2] M. Hunt, " New Frontiers in Superlightweight Alloys ", *Materials Eng.*, Vol. 105, No. 8(**1988**) pp. 29-32.

[3] Robert C. Weast et al eds. CRC Handbook of Chemistry and Physics, Vol. 64, CRC Press Ltd., Boca Raton, Florida, **1984**., p. D-163

[4] J. Searls, "Lithium", Mineral Facts and Problems, U.S. Bureau of Mines Bulletin 671, **1980**,pp 521-534.

[5] J.E. Ferrel, "Lithium", Mineral Facts and Problems, U.S. Bureau of Mines Bulletin 675, **1985**, pp 461-470.

[6] London Metal Exchange, January, **1989**.

[7] R.D. Crozier, "Lithium: Resources and Prospects", *Mining Magazine*, Feb.(**1986**) pp. 148-152.

[8] J.S. Whisnant, J.B. Holman, "Survey of Lithium Resources - Worldwide", *Light Metals 1985*, **1985,** pp. 1479-1482.

[9] R.A. Stauffer, "Vacuum Process for Preparation of Lithium Metal from Spodumene", *Trans* AIME, Vol. 182(**1948**), pp. 275-285.

[10] W.J. Kroll, and A.W. Schlechten, "Laboratory Preparation of Lithium Metal by Vacuum Metallurgy", *Trans* AIME, Vol. 182(**1948**), pp. 266-274.

[11] Mikulinski et al, "Kinetics and Conditions of Condensation During the Preparation of Alkali Metals by a Vacuum-Thermal Method, *Redk. Shchelochnye Elem.,Sb. Dokl. Vses. Soveechch., 2nd. Novosibirsk,* **1964**, pp. 350-360.

[12] A.S. Kozhevnikov, " Reduction of Lithium Oxide With Aluminum".*Redk. Shchelochnye Elem.,Sb. Dokl. Vses. Soveechch., 2nd. Novosibirsk,* **1964**, pp. 343-9.

[13] Mikulinski, A.S. and Efremkin, V.V., "Lithium Ores as Complex Raw Materials", *Redk. Shchelochnye Elem.,Sb. Dokl. Vses. Soveechch., 2nd. Novosibirsk,* **1964**, pp. 339-42.

[14] A.I. Lainer, A. Kh. Nazirov, "Mineral Composition of Lime Spodumene Sinters", *Izv. Akad. Nauk SSSR, Metal.* **1966**(6), pp. 36-39.

[15] Harris, R., Wraith, A.E. and Toguri, J., "Producing Volatile Metals", Canadian Patent Application, Ser.No.: 539,058, June 8, **1987**, USA Patent Application, Ser.No.:,201,446, June 2, **1988**.

[16] F*A*C*T- Facility for the Analysis of Chemical Thermodynamics, (CRCT) Centre for Research in Computational Thermochemistry, W.T. Thompson, A.D. Pelton and C.W. Bale, Ecole Polytechnique, Montréal, **1988**.

[17] Botto, I.L., Cohen Arzi, S., Krenkel, T.G., *Bol. Soc. Esp. Ceram. Vidrio,* Vol. 15(1)**1976**, pp 5-10.

[18] Korol'kov, "Casting Properties of Metals and Alloys" *Consultants Bureau,* New York, **1960**, p.37.

Mineral Processing and Feed Preparation

THE ROUTINE USE OF PROCESS MINERALOGY IN A COST-EFFICIENT

GOLD-SILVER OPERATION

Wolfgang Baum and S. Rick Gilbert

Pittsburgh Mineral & Environmental Technology, Inc.
390 Frankfort Road
Monaca, Pennsylvania 15061 USA

Abstract

In an effort to expedite the development of precious metal ores, process-related mineralogy is too often overlooked. In a similar way, plant modifications are undertaken without a firm understanding of the mineralogical features which ultimately affect the performance of gold-silver ores in the circuit. The end result in either case is the potential loss of revenue, sometimes in the millions of dollars, due to circuit inefficiencies. This paper presents examples in which the routine utilization of mineralogical work in the improvement of gold-silver operations can increase the productivity of a metallurgical plant. Specific processing problems which have been identified during several years of operations support in the precious metals industry could have been avoided, or at least minimized, through early and periodic mineralogical work on the feed material and the metallurgical products.

Introduction

Precious metals development and recovery improvements constitute a major part of present activities in the metallurgical industries. In 1987, 8.95 billion US dollars were spent on gold-silver mine development and plant expansion (1).

Gold recovery operations, more so than with other metals, have adopted risky "fast-track approaches" to process development and plant construction with record completion times of only a few months after the go-ahead decision. Often, process-oriented mineralogical studies are overlooked or ignored in the rush to become operational. The detrimental impact of these short lead times on plant design and process efficiency is documented in numerous operation reports presented in the various mining journals.

A common conclusion of most of these operation updates is the simple but startling fact that many plant and processing problems are due to the mineralogical characteristics of the ore feed. Many operations have recognized the dollar value of process mineralogy and have successfully reaped the benefits. In some operations, these mineralogical process parameters are only haphazardly examined or they are completely neglected during hastened exploration and project development.

Fast-track methods introduce many uncertainties in respect to plant performance and capital costs. As stated in many reports on accelerated developments, the major detriment in implementing a fast-track operation is the lack of recognition of shortcomings and possible problem areas of ore processing before the plant is built. With smaller, cost-sensitive projects, recognition of the problem areas can make the difference between proceeding with the project and a failure. Many of the plant shortcomings and metallurgical problem areas can be predicted and continuously monitored by process mineralogy.

Economic modelling and price predictions alone cannot provide the economic security for a metallurgical operation. A thorough knowledge about the feed material, its response in the process streams and continuous efforts to improve recovery and product quality can lower the operation's cost and ensure survival. Routine plant mineralogy is therefore a vital aspect of this effort.

It has been the authors' experience that the increased revenues gained by applying process mineralogy greatly exceed the cost of such mineralogical evaluations conducted at the front-end of a project. Abortive test work and inexplicable results can often be averted by fundamental mineralogical characterizations.

Typical "side effects" of a metallurgical plant development which disregards or neglects the precious metals mineralogy and other pertinent mineralogical features of the mill or heap feed are:

- Performance of unnecessary test work.
- Biased sample collection.
- Overlooking critical mineralogy-related processing problems.
- Copying existing operations rather than custom-fitting the plant to the specific ore.
- Disregarding the testing of alternative flow sheets.

- Inappropriate equipment selection.
- Tendency to over-size equipment.
- Uncertainties in process design criteria.

Typical Metallurgical Problems in Gold-Silver Operations

The successful utilization of "on-stream" mineralogy as a cost-efficient development and process control tool has been practiced in the past by few operations (2), (3), (4), and (5). However, the farsightedness of the use of process mineralogy in gold-silver plant development as described by Lewis (6) for the Mahd adh Dhahab gold deposit is still found to be an exception. He shows that at the start of the full evaluation of the deposit, a policy decision was made that no further metallurgical test work should be completed until a systematic mineralogical examination of all available sources of ore had been completed. He continues to conclude that in retrospect, the effort spent on prior mineralogical evaluation has been fully justified.

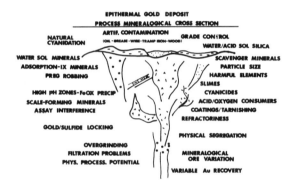

Figure 1 - Graphic presentation showing typical process-related mineralogical parameters in epithermal gold deposits which are in many cases the reasons for metallurgical problems.

The following listing of case histories documents the variety of metallurgical problems which have been identified by the authors during several years of mineralogical support work on precious metals operations in the United States, Canada and South America. Virtually all of these problems could have been avoided, or at least minimized, through early and routine process mineralogy.

- Very fine mine timber material in the mill feed resulted in inefficient screening of a cyclone overflow and in interference with stripping of loaded carbon in the CIP.

- High concentrations of secondary copper minerals caused problems in leaching (high cyanide consumption), copper loading onto the carbon and reduction of doré fineness.

- Very poor strip performance resulted in high soluble losses in the plant. This was caused by a mineralogically complex ore type which had been overlooked during geological core logging and metallurgical test work.

- Carbon fouling occurred due to high concentration of argillaceous alteration minerals and ultrafine sulfides, the significance of which was underestimated during the initial core logging.

- In a gold operation using pressure oxidation, significant fine grained, non-effervescing carbonate minerals from certain portions of the ore body had been overlooked during exploration and were excluded from sampling for process testing. This caused considerable problems during autoclaving (high CO_2 formation) and scale buildup.

- High pH in the flotation circuit caused water-soluble minerals contained in the pulp to precipitate, increasing the viscosity in flotation. This was detected by using optical microscopy in the mill.

- Low recoveries due to slow-floating characteristics of gold and gold-bearing pyrite were aggravated by significant quantities of pyrophyllite (which is hydrophobic) from certain zones of the ore body. Due to routine mineralogy, the mill could respond to these ore types with reagent adjustments, longer residence times and greater cleaner capacity.

- In certain parts of a carbonaceous gold ore body, chlorine pre-treatment was not effective in blinding the preg robbing effects. Mineralogy showed that certain ore types contained significant quantities of clays, zeolites, and chlorites. Chlorine treatment will not eliminate the ion exchange capacities of these minerals.

- Relics of machine oil in the pulp from leaks and careless maintenance (identified by UV-light microscopy) were responsible for frequent reductions of gold extraction ranging from 20% to 30%.

- Problems in a Merrill-Crowe zinc precipitation circuit (zinc stearate fouling) were identified by mineralogical examination of the feed which found considerable amounts of plant roots. Reaction of the organic material with alkaline mill solutions resulted in the formation of stearates.

- Significant amounts of clay slimes were lost during reversed circulation drilling. Metallurgical work performed on these samples was useless since it led to false conclusions regarding adsorption, pulp settling and filtration.

- Frequent changes in silver mineralogy were not noticed during mine exploration work. This resulted in severe problems of gold recovery in the Merrill-Crowe plant since sudden increased silver values reduced gold adsorption on carbon.

The reason that the recoveries for silver are generally much lower than for gold is simply due to a much more complicated mineralogical occurrence of the silver and the fact that very few plants have ever made an attempt to establish, periodically, the reasons for low(er) silver recoveries. There are about 73 silver minerals and probably more than 200 silver-bearing minerals.

- In several operations, ore blending was solely based on gold-silver grades which led to severe recovery problems.

Mill performance and recovery are more affected by the mineral phase composition of the feed and the mineralogical residence of the precious metals than by gold concentrations. Chemical analyses cannot identify the reasons for variable grades, differences in gold-silver particle sizes or the intensity and type of alteration. The following Table I illustrates how a thorough mineralogical characterization can explain differences in gold extraction rates experienced with two ore types in the same deposit.

Table I. Mineralogical Reasons for Variable Gold Extraction Rates.

Type A	Ore Types	Type B
	Typical Cyanide Extraction Rates	
28 - 41 %		0.5 - 20 %
	Mineralogical Reasons for Differences in Gold Extraction	
60-65 %	Gold Particle Size Distribution in -500 Mesh Fraction	36-50 %
22-26 %	Locking of Gold in Coarse Sizes (+200 M)	32-48 %
<1-50 μm	Pyrite/Marcasite Particle Sizes	30 μm->2 mm
<2 %'	Iron Oxide/Hydroxide Content	4-11 %
<2 %	Marcasite Content	2-12 %
>20 %	Clay Content	< 12 %
<1 %	Carbonate Content	5-28 %
<1 %	Formation of Secondary Sulfates	> 5 %

- The alteration mineralogy and related slimes formation during comminution were found to be the primary factors determining reagent consumption, resulting in adsorptive or preg robbing effects and causing slimes interference, i.e., coatings, agglomerations, entrainment, and dissolution.

- Periodic gold particle size analyses were lacking in many operations. Therefore, the plants were not aware of the significance of coarse gold in their feeds. This constituted a major reason for gold losses:

    - Coarse gold did not leach entirely in heap leach operations.
    - Coarse gold-gangue middlings frequently did not respond to cyanidation.
    - Chemical and physical surface contamination of coarse gold, gold-bearing sulfides and middlings resulted in long circuit retention.

- Negligence with regard to feed and gold particle size characteristics and fixed "cyanidation thinking" caused many operations to overlook good potentials for the use of a gravity circuit in their plant.

    A thorough evaluation of the gravity concentration potential should be a standard practice in any precious metals development. Gravity separation can become a key in plant optimization through a beneficial impact on the following processing features:

    - Avoiding overgrinding.
    - Improved middlings recovery.
    - Improved sizing and classification.
    - Coarser tailings and option for retreatment.

- The solubility of various instable and, in part, water-soluble silica phases (opal-CT, volcanic glass, poorly crystalline cristobalite) was increased by high alkalinity in the plant. This caused significant scale formation on particles in the pulp as well as on the equipment. Particularly the formation of colloidal and gelatinuous particle agglomerates resulted in gold-silver entrainment and recovery losses.

    Although calcium salt scaling is the major scale buildup problem in gold processing, there are considerably more scale-forming minerals present. Use of antiscalants can only be effective if the correct phase composition of the scale is known. Mineralogical assessment of all scale buildup (e.g., on leach residue particles, pumps, transfer lines, sprinkler heads, gravity and flotation equipment, filter beds in Merrill-Crowe plants, or activated carbon surfaces in CIP circuits) should become a standard examination.

- Iron hydroxide precipitation caused by local development of high pH zones in the heap, scavenged some of the gold and rendered it refractory. One operation spent over 8 months to determine these problems by metallurgical test work.

    The following Figure 2 shows how early mineralogical work may permit low-cost metallurgical predictions regarding the leachability and refractoriness (locking) of the gold.

LOCKING & LEACHABILITY OF GOLD BASED ON MINERALOGICAL WORK

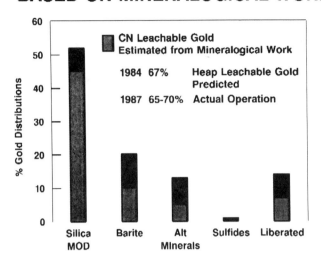

Figure 2 - Estimated predictions for cyanide-leachable gold in the major and minor gold-bearing minerals based on mineralogical work: Silica Mod = various silica modifications, i.e., quartz, chalcedony, opaline silica; Alt Minerals = alteration minerals including goethite, jarosite, alunite; Liberated = liberated native gold at -10 mesh sample size.

- A variety of minor minerals contained in epithermal gold-silver ores became metallurgically and environmentally troublesome as soon as they were exposed to comminution and water with process chemicals. Various thallium and arsenic minerals (e.g., carlinite, orpiment) showed solubility in water. Realgar is appreciably soluble in alkaline solutions, even at low temperatures (7).

- Contaminant wood chips during thermal carbon regeneration became carbonized, entered the circuit as wood carbon with limited adsorption capacity but, due to the brittle nature, disintegrated and caused gold losses to the tailings.

The carbon should be routinely examined for entrained contaminations due to faulty screens, slimes, etc. An eluted carbon sample may contain up to 20% of the ore pulp and/or wood chips comparable in size with the carbon. Wood chips can get into the carbon through slotted screens if the slots are accidentally aligned with the pulp flow direction.

- Certain mineralogical characteristics in gold ores were found to be vital for thiourea leaching. They included:

    - Feed mineralogy.
    - Sulfur formation.
    - Minerals causing high pH.
    - Clay or clay-like minerals (e.g., chlorite).
    - Acid minerals (sulfate alteration).
    - Sulfide mineralogy.

In summary, the majority of gold operations are increasing their capacities by five to sixteen times their initially anticipated milling rate, requiring equipment changes and flow sheet adjustments. That certain mineralogical features may turn into metallurgical problems when milling rates are increased is very seldom taken into consideration.

Computerization of Mineralogical Data

When reviewing the amount of geochemical and chemical data from exploration and processing used for computer evaluation, correlation and process control work, it is astounding to see the lack of mineralogical data in electronic record keeping, plotting of drill holes and ore contour maps, mill circuit monitoring or routine process control of daily metallurgical operations.

In the following areas, tailor-made computer programs can be of considerable help for optimization of the metallurgical operation:

- Isometric mineralogy "stick-model" plots of drill holes.

- Mineralogical 3-D contour and ore type maps.

- Particle size and liberation/locking of gold, silver or selected minerals.

- Slimes characteristics.

- Type and amounts of cyanicides and other reagent consuming minerals.

- Type and amounts of middlings particles.

An automated graphic plotting software package can be created for recording and tracking of process problems related to the mineralogy. Such a program is menu-driven and easy to use. The plant operators can create short text files containing both metallurgical and mineralogical data on typical as well as new process problems and their causes. This enables the metallurgists to automatically keep track of circuit upsets and find a cost-efficient and timely response to treatment problems based on mineralogy.

Ore type blending is probably the most important tool for gold-silver operations to control the harmful effects of alteration minerals. This blending is in the majority of the cases based on grades or visual geological logging. The use of more detailed mineralogical data or the utilization of x-ray mineralogy logging could provide much better metallurgical information which can then be transferred into a computer-compatible format. Thus, the operator will be in a position to refine and "custom-fit" the blending of the ore and/or be on alert for potential process problems resulting from the feed mineralogy. This will save time and reduce the operating costs (Fig. 3).

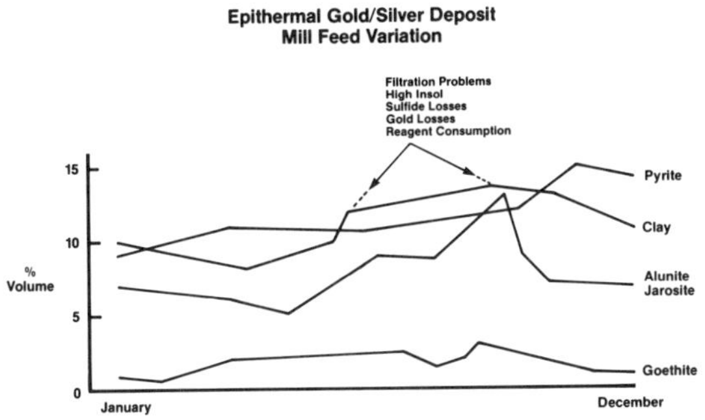

**Epithermal Gold/Silver Deposit
Mill Feed Variation**

Filtration Problems
High Insol
Sulfide Losses
Gold Losses
Reagent Consumption

Pyrite
Clay
Alunite
Jarosite
Goethite

Figure 3 - A continuous monitoring of the mill feed mineralogy will permit the metallurgist to be prepared for potential processing problems and, if necessary, adjust the operation rapidly.

Routine mineralogical work allows the plant metallurgist to store, view, report and plot process-related mineralogical data of the ore and the circuit products. This gives instant access to mill circuit response both in tabular and graphic form.

The use of mineralogical data as a correcting tool for other types of automatic process control (e.g., on-line x-ray fluorescence analysis, particle size analyzers, density and pH measurements) cannot be emphasized enough. Misleading information from conventional on-line control equipment can stem from a variety of mineralogical features such as:

- Minerals soluble in process water.
- Particle coatings.
- Slimes agglomerations.
- Middlings particles.
- "Unconventional" cyanide consumers.
- Artificial contaminants.

The hardware requirements for computer-aided process control may already exist at most plants and the cost for the mineralogical software development is reasonable (less than 5,000 US dollars).

Cost Factors of Process Mineralogy

Not only are quite a large number of individuals in the natural resources industry still under the belief that mineralogy's main application is in museums and collections of precious stones, but there is also a notion that mineralogical work costs too much and is time-consuming.

The authors have performed routine process mineralogical support work for a world-class gold-silver operation in South America on a monthly basis over a period of 3.5 years. The total annual cost of this work was only 2% of the value of the increased revenue gained by applying process mineralogy. The results of this "on-stream" plant mineralogy have been pivotal to optimization of the metallurgical operation and were a key factor in recovery increases of several per cent, thus contributing to revenue improvements in the multimillion dollar range.

Independent mineralogical studies on other operations in the western United States and Canada regarding specific mill circuit and recovery problems ranged from 2,500 to 15,000 US dollars and were completed in three to seven weeks. For comparison, attempts to resolve processing problems via extensive metallurgical test work without mineralogical assistance may require four to six times as much cost and time.

Many operations are profitable and achieve good recoveries. There is, however, an opportunity in precious metals operations to reduce gold and silver losses to the tailings pond (which can amount to several million US dollars) by initiating a thorough mineralogical study.

Conclusions

Lewis and Martin (2) have summed up most distinctly what should be normal practice in any minerals processing operation: without a thorough understanding of the mineralogy, it is difficult to plan a sensible and logical metallurgical test program for any ore deposit.

By reviewing reports of 85 gold operations in North and South America, we found that in up to 70% of the cases, metallurgical and operating problems encountered during the early phases of startup were related to the mineralogical composition of the plant feed.

Not many sectors of the minerals and metallurgical industries have escaped the economic difficulties in the past two decades. The lesson that has been learned by our industry is that, while the prices obtained for products of metallurgical production cannot be controlled, the cost of production can be. The use of periodic process mineralogy in a gold-silver operation is one of the most cost-efficient tools to ensure high precious metals recovery, improved equipment performance and good by-product quality. It is also one of the most useful guides for the metallurgist to maintain a process flow sheet "custom-fitted" to the mineralogical characteristics of the ore.

The value of process mineralogy for plant trouble shooting and optimization lies in its routine and periodic use. A single mineralogical study may not lead to substantial recovery increases or problem solving, just like changing a single metallurgical factor in a plant will not give immediate positive results. It is the continued effort and evaluation that pays the biggest dividends.

References

1. M.P. Sassos, "Mining Investment 1987,"Engineering and Mining Journal, January 1987, 8 p.

2. P.J. Lewis and G.J. Martin, "Mahd adh Dhahab Gold-Silver Deposit, Saudi Arabia: Mineralogical Studies Associated with Metallurgical Process Evaluation," Transactions Inst. Mining and Metallurgy, June 1983, 92, C63-C72.

3. D.M. Hausen, "Process Mineralogy of Select Refractory Carlin-Type Gold Ores," CIM Bulletin, September 1985, vol. 78, No. 881, 83-94.

4. W. Baum, "Optimizing a Gold-Silver Plant Using Mineralogical Data," Randol Gold Forum 1988, Proceedings, Scottsdale, Arizona, 101.

5. W. Baum, J. Sanhueza, E.H. Smith and W. Tufar, "The Use of Process Mineralogy for Plant Optimization at the El Indio Gold-Silver-Copper Operation," in preparation, 15 p., 1989.

6. P.J. Lewis, "Metallurgical Evaluation and Process Development for the Mahd adh Dhahab Gold Deposits, Saudi Arabia," preprint, 1984, 28 p.

7. J.J. Rytuba, "Geochemistry of Hydrothermal Transport and Deposition of Gold and Sulfide Minerals in Carlin-Type Gold Deposits," U.S. Geol. Survey Bulletin, 1646, 1985, 27-34.

STOCHASTIC DYNAMIC SIMULATION OF FLOTATION CIRCUITS

Di Yin, Songren Li and Weibai Hu
Central South University of Technology
Changsha, Hunan
People's Republic of China

Abstract

It was well known that complex flowsheets, many operating variables and many random disturbances keep the flotation process in a highly fluctuating state. Modern automatic control technique has been used as a powerful tool for keeping the process at steady state and significantly improving flotation operation performance. The implementation of process effective automatic control in modern mineral processing plants requires not only good available on-line instrumentation and powerful computer hardware but also good control strategy software based on a profound understanding of the process dynamic responses. In this paper, a general stochastic dynamic simulator for the industrial flotation circuit was developed in a microcomputer IBM-PC/XT by the authors. The parameters of mathematical models were estimated from batch test data in the laboratory. The simulator contained several unit modules, such as conditioner, flotation cell, sump and pipe. The flowsheet topology matrix linked all various unit modules to form the system model of flotation circuit. The mathematical models included a number of independent variables, such as regulator, collector and frother addition rate, air addition rate and water addition rate. The stochastic disturbances were added in the input and independent variables of flotation circuits based on the definite models so as to describe more accurately the dynamic behaviors of practical flotation processes that can be used to study and select alternative control strategies.

SOME MILLING PRACTICES AND TECHNOLOGICAL INNOVATIONS ON BENEFICIATION OF MERCURY SULFIDE ORES IN CHINA

Hsi-keng Hu

Central South University of Technology
Changsha, Hunan 410012
People's Republic of China

Abstract

This paper describes the major technological innovations in the processing of mercury sulfide ores in China. Experiences of the gravity-flotation process and the treatment of the mercurial soot at the three concentrators at Guizhou Mercury Mine are described.

Introduction

China is rich in mercury ores and is one of the major mercury producers in the world.

The mercury sulfide ores in China can be classified into two types: (1) mercury sulfide ore in which cinnabar is the only economic mineral; (2) complex mercury sulfide ore in which cinnabar is associated with stibnite, arsenopyrite and pyrite.

For recovering mercury from the simple cinnabar ores, two principal processes are generally used in China. When the ore grade is higher than 0.2%Hg, the ore is subjected to direct roasting in a blast furnace, rotary furnace, multiple hearth furnace or a fluidized-bed furnace. When the ore grade is lower than 0.2%Hg, gravity concentration and flotation are usually used to obtain higher grade concentrates which are treated by retorting or hydrometallurgical processes. The complex mercury sulfide ores are beneficiated by flotation followed by smelting. The mercurial soot from the smelting process is treated by mechanical process, gravity concentration and flotation.

Milling Practice at the Concentrators of the Guizhou Mercury Mine (1,2)

The Guizhou Mercury Mine is located in the Wanshan County, Guizhou province. The Guizhou Mercury Mine has three concentrators for treating the low grade mercury sulfide ores.

Ore characteristics

The ore from the Guizhou Mercury Mine is an epithermal deposit. The principal economic minerals are cinnabar and native mercury, whereas pyrite, stibnite and sphalerite are present in small amounts in some parts of the deposit. The gangue minerals are dolomite, quartz and calcite. Cinnabar is disseminated in the gangue and is unevenly distributed. The grain size of cinnabar ranges from 12 to 0.002 mm and mostly in the range of 0.1-0.5 mm. The ore grade ranges from 0.1-0.3%Hg. The ore contains very little primary slime.

No.1 Concentrator

No.1 concentrator was built in late of 1950's. At the beginning of operation, a simple flowsheet of gravity concentration was used. Because of high tailing losses, the flowsheet was changed to the present flotation circuit in 1965.

The ore is crushed to -50^{mm} by a jaw crusher, then it is ground in a ball mill operated in closed circuit with a spiral classifier. The overflow from the classifier 30-35% solids and 65-70% minus 200 mesh is fed to flotation circuit consisting of one stage of roughing, one stage of cleaning and two stages of scavening. The flotation concentrate is thickened and filtered and is processed in an electrothermal retorting furnace to obtain a final mercury product. The flotation reagents used are ethyl xanthate 170-180 g/T and pine oil 40-60g/T. The metallurgical summary is as follows:

Feed Grade % Hg	Concentrate % Hg	Tailing % Hg	Recovery %
0.167	16.09	0.0076	95.7

No.2 Concentrator

The No.2 concentrator began operation in July, 1979. In this mill, both gravity concentration and flotation are used to produce a cinnabar product and a mercury concentrate.

The flowsheet of No.2 concentrator is shown in Fig.1.

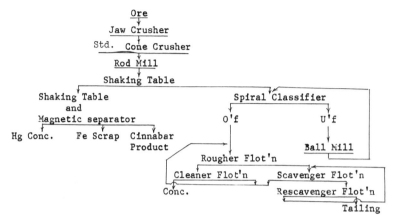

Fig.1 No.2 Concentrator Flowsheet

There are two stages of crushing in an open circuit. The crushed product (-50 mm) is first ground in a rod mill and the product (-3 mm) is fed directly to the shaking tables. Each table produces two products, namely crude cinnabar concentrate and tailing. The crude cinnabar concentrate is further treated by magnetic separation and shaking table to get the fine cinnabar product, iron scrap and the mercury concentrate. The fine cinnabar product is used as the chinese herbal medicine. The tailing from the shaking table is fed to a spiral classifier. The underflow is ground in a ball mill and returned to the spiral classifier. The overflow is fed to the flotation circuit consisting of one stage of roughing, one stage of cleaning and two stages of scavenging to get the mercury concentrate. The Flotation reagents used are $CuSO_4$ (added to ball mill as the activator of cinnabar) 200-240 g/T, EtX 180-200 g/T and pine oil 50-70 g/T.

The metallurgical results are as follows:

Feed Grade	Cinnabar product	Concentrate	Tailing	Recovery
% Hg	% HgS	% Hg	% Hg	%
0.3	99.8	15-20	0.01	95-97

No.3 Concentrator

No.3 concentrator was completed in early 1981 and operations started in October 1981.

The flowsheet of No.3 concentrator is presented in Fig.2
There are two stages of crushing. The secondary crusher is operated in closed circuit with a 16 mm screen. The crushed product is screened . The -16+3 mm product is fed to a rod mill. The rod mill discharge and the undersize product (-3mm) from the screen are treated separately with a shaking table to get a crude cinnabar product and a tailing. The crude cinnabar product is further treated with the same method as No.2 concentrator. The tailing is ground in a ball mill operated with a double spiral classifier. The overflow is fed to a flotation circuit consisting of one stage of roughing, two stages

of cleaning and two stages of scavenging to get the concentrate, which is sent to a retorting furnace to produce mercury product.

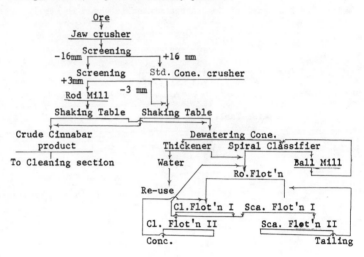

Fig.2 No.3 Concentrator Flowsheet

The flotation reagents used are $CuSO_4$ as the activator (added to the ball mill) 120-140 g/T, ethyl xanthate 120 g/T and pine oil 50 g/T.

The metallurgical summary is as follows:

Feed Grade % Hg	Cinnabar product % HgS	Concentrate %Hg	Tailing % Hg	Recovery %
0.08-0.15	99.5	12-25	0.003-0.008	95

Treatment of the Mercurial Soot (1,2)

The mercurial soot is a product collected from the mercury fume, dust, sludge, etc. of the roasting processes in extracting mercury from the mercury ore and the concentrates.

The mercurial soot usually consists of metallic mercury and mercury sulfide that can be recovered by the following processes.

1. **Mechanical Process** The mercury particles in the mercurial soot usually range in size from 0.1-0.001 mm. When soot is strongly agitated, the small mercury particles may be coalesced to granules or even balls of liquid mercury that may be squeezed out by mechanical machines such as centrifugal machine, mechanical agitator, etc. to separate liquid mercury and the residue. The residue is sent to the retorting furnace for further treatment.

2. **Gravity Concentration** shaking table and hydrocylones are commonly used to separate mercury from the residue. The metallurgical results are as follows:

Process of Treatment	Feed Grade % Hg	Concentrate % Hg	Tailing % Hg	Recovery %
Shaking Table	1.889	99.99	0.068	97.5
Hydrocyclone	0.78-2.551		0.104-0.478*	81-93

Hydrocyclone	39.18	0.084*	99.87

* Overflow of the Hydrocyclone

3. **Flotation** Extensive testwork has been done for recovering mercury from the low-grade mercurial soot by flotation and good results were achieved. The flotation circuit is simple and consists of one stage of roughing, two stages of cleaning and three stages of scavenging. The flotation reagents used are SN-9 (as collector) 240 g/T, pine oil 300 g/T and water glass (as dispersant) 960 g/T. The metallurgical summary is as follows:

Feed Grade	Concentrate	Recovery
% Hg	% Hg	%

(Table supplied during presentation)

4. **Combined Flosheet of Gravity Concentration and Flotation** A great deal of research works have been done by the combined process of gravity concentration and flotation. The low-grade mercurial soot after strong agitation is treated by hydrocyclone to get mercury product, spigot product and overflow. The overflow from the hydrocyclone is fed to a flotation circuit to get the mercury concentrate. The flotation reagents are SN-9 and pine oil. The metallurgical summary is as follows:

Feed Grade % Hg	Concentrate % Hg	Tailing % Hg	Recovery %
0.6-1	99	0.012	98.9

References

Book   Hsi-keng Hu, *Mineral Processing of Non-ferrous Sulfide Ores* (Beijing: The Metallurgical Industry Press, 1987), 321-347

Unpublished Book   Hsi-keng Hu and Yuan-xin Chen, *Mineral Processing of Mercury Ores* (Beijing: The Metallurgical Industry Press, to be published in 1989).

	Book	Hsi-keng Hu, <u>Mineral Processing of Non-ferrous Sulfide Ores</u> (Beijing:The Metallurgical Industry Press, 1987), 321-347
Unpublished	Book	Hsi-keng Hu and Yuan-xin Chen, Mineral Processing of Mercury Ores (Beijing: The Metallurgical Industry Press, to be published in 1989).

ADVANCES IN TECHNOLOGY FOR COMPLEX CUSTOM

FEED MATERIAL TREATMENT AT NORANDA

M. Chapados[1], M. Bédard[2] and G. Kachaniwsky[3]

NORANDA MINERALS INC., - HORNE DIVISION
ROUYN-NORANDA, QUEBEC   CANADA

ABSTRACT

The Noranda Continuous Smelting Process at the Horne Smelter was commissioned in March 1973. The ability of the process to handle a wide variety of feed materials and a steady stream of technical innovations in recent decades has served to maintain the Horne as an efficient and versatile custom smelter.

This paper reviews recent technological development and capabilities in areas such as sampling and preparation of secondary and complex materials, as well as the smelting and related environmental control issues. Future trends and improvements are also addressed.

---

[1] Business Development Superintendent

[2] Materials Handling and Laboratories Superintendent

[3] Technical Smelter Superintendent

INTRODUCTION

The Noranda Horne smelter commenced operations in 1927. It was originally designed to treat 910 tonnes per day of copper concentrates and ores coming from the new Horne mine site. In the 1930's, the company opened its doors to other mines in Northwestern Quebec. Treatment of feed materials on a custom basis has therefore always been an important factor in the operation of the Horne smelter. Throughout the years, the Horne gradually increased its custom smelting capacity, while maintaining its production from the Horne mine. By the early 1950's, 65 percent of the materials processed was from outside customers, and by 1974, this figure had increased to more than 85 percent. With the closure of the Horne mine in 1976, the company extended its smelting operations beyond ore and concentrate, and began to treat more scrap metals and residues rich in copper and precious metal content.

Corresponding with the ever-changing pattern of feed sources, the Horne smelter has adopted technological improvements to meet new and challenging business conditions. The technology of the original smelter of 1927 has been completely replaced over the years in the quest for improvement. One of these major changes occurred in 1973 with the start-up of the Noranda Process Reactor. This extremely efficient smelting system has been exploited to treat a wide variety of copper and precious metal (PM) containing custom materials. This paper reviews recent developments and improvements achieved at the Horne smelter, in the receiving, sampling and treatment of copper and PM scrap and secondaries, with particular emphasis on the development of new technology and equipment in complex feed materials treatment.

MAJOR SMELTER DEVELOPMENTS AND PRESENT OPERATIONS

The original Horne smelter in 1927 consisted of eight roasters, two coal-fired reverberatory furnaces, two Peirce-Smith converters, along with the necessary auxiliary equipment. Plant throughput was gradually increased and two additional roasters were built; the reverberatory furnaces were extended and converter diameters increased. By the 1950's, the proportion of custom feed had increased, typical roaster feed of this period being illustrated in Table I.

In 1957, a third reverberatory furnace based on wet-charge smelting was commissioned to handle additional receipts of custom material from Northern Quebec and Ontario. To cope with increasing matte tonnages, the converters were lengthened, and a new 4.27 m x 9.75 m unit, the largest converter in the world at that time, was commissioned in 1965[1]. The anode plant was upgraded with the start-up of new rotary anode furnaces and the introduction of gaseous reduction of anode copper based on Noranda-developed technology.

While technological changes continued to be recorded at the plant in the nineteen sixties, a new era in smelting operations at the Horne began with the start-up of the Noranda Process on March 1st, 1973. The Noranda Process Reactor has been described in previous papers[2-6]. The ability of the process to readily smelt a heterogenous mixed charge was recognized at an early stage of process development. The reactor has always handled a certain proportion of copper scrap, metallics, purchased secondaries and complex concentrates. This proportion has increased over the years, and typical smelter feed for 1988 is shown in Table I.

Major developments and changes at the Horne smelter since 1973 are summarized in Table II. As shown, modern equipment for handling, sampling and sample preparation of complex feed materials has been installed in the 1980's.

The present operations at the Horne consist of the Noranda Process, a reverberatory furnace with oxy-fuel burners, two oxygen plants with a total capacity of about 580 tonnes per day, five converters (two to three converters operating), three anode furnaces and two casting wheels. An acid plant with a 1,950 tonnes per day capacity is presently under construction, and will be in operation at the end of 1989. A smelter flowsheet is shown in Figure 1.

Table I - Typical Horne Smelter Feed - 1950's and 1988

Material	1950's		1988	
	Feed Rate Tonnes/Day	Analysis Cu % (Average)	Feed Rate Tonnes/Day	Analysis Cu % (Average)
Noranda Horne Concentrate	340	11.6	0	-
Noranda Horne Ore	1,460	1.7	0	-
Noranda Group Concentrate	0	-	620	26
Custom Concentrate	620	21.0	1,520	23
Copper Scrap	0	-	195	35
Secondaries & Misc. Materials	30	20.0	185	25
Flux	310	-	420	0.2
Total Feed to Smelter	2,760		2,940	

On a daily basis, the individual feed components treated in the Noranda Process Horne smelter can vary widely. For example, feed can include conventional chalcopyrite-type concentrates, silver-bearing residues with little or no copper or sulphur, gold-bearing pyritic concentrates with no copper and mixed metallic scrap. Typical material flows and operating data are shown in Table III. The Noranda Process feed system and reactor are capable of comfortably handling feed sizes up to 100 mm and moisture levels up to about 14 percent. In all cases, the Noranda Process Reactor has proven itself to be versatile in handling a wide range of materials varying in both physical and chemical quality. The role of minor elements in the Noranda Process has been discussed elsewhere[7].

The smelting rate at the reactor has increased year after year. Figure 2 shows the net increase in tonnages treated over the years. In November 1988, a new record of 56,065 tonnes of material treated was achieved. The Noranda Process has not yet reached its limit in terms of variety of materials and tonnage treated.

Table II - Major Developments - Horne Smelter
1973 - 1991

Year	Development
1973	- Noranda Process Reactor commissioned on March 1st, 1973 on direct copper production
1975	- Commence ladle cooling of Noranda Process slag
1976	- Rebuild one hot-charge furnace as wet-charge furnace and close five of the ten roasters - Horne mine closed
1977	- Close remaining roasters and last hot-charge reverberatory furnace
1980	- Installation of first converter waffle hood
1982	- 465 tonnes/day oxygen plant on-line and upgrade of Noranda Process feed system - Commence oxy-fuel firing on one reverberatory furnace and shutdown of second reverberatory furnace
1984	- Expanded facilities for scrap receiving and sampling, including the shredding system
1985	- Installation of two 1.3 tonne induction furnaces for scrap sample preparation
1986	- Shipment of lead concentrate commenced from Noranda Process cottrell to Brunswick Smelting at Belledune, New Brunswick, Canada
1988	- Record Noranda Process production achieved in a calendar month period - 56,065 tonnes in November 1988 - Commence construction of 1,950 tonnes/day capacity sulphuric acid plant - Installation of Noranda Sampler for high grade electronic scrap - Installation of Noranda Sampler for P.M. scrap
1989	- Partial closure of last remaining reverberatory furnace and commencement of Noranda Process as sole smelting vessel - Scheduled acid plant start-up (November 1989) - New equipment and facilities for scrap sampling and sample preparations and installation of two additional induction furnaces
1990-1991	- Installation of a pebble mill for slag treatment - Replace the last reverberatory furnace by Peirce-Smith converter smelting technology

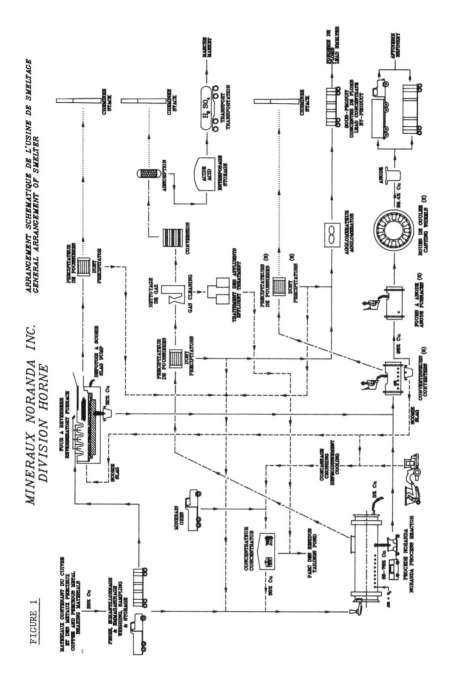

FIGURE 1

MINERAUX NORANDA INC. DIVISION HORNE

ARRANGEMENT SCHEMATIQUE DE L'USINE DE SMELTAGE
GENERAL ARRANGEMENT OF SMELTER

Table III - Typical Material Flow and Operating Data - Noranda Process (1988)

	Instant. Rate Tonnes/Day	Analysis, Wt. (%)					
		Cu	Fe	S	SiO$_2$	Pb	Zn
Input							
Copper Concentrate	1,800	20-23	29	33	5	1.3	2.7
Gold Concentrate	200	1-2	30	33	15	0.6	1.3
Copper Scrap	100	20-60	5	2	12	3.0	5.0
U.P.M. Scrap	30	15-35	6	0	17	3.0	1.0
Custom Misc.,Residues,Slag	190	20-35	8	2	15	2.0	2.0
Slag Concentrate	330	35-40	19	11	13	2.0	2.0
Smelter Reverts	80	20-40	12.1	15.5	2.9	6.0	2.0
Dust Recycle	60	10-20	4	12	2	29.0	9.0
Coal	60	-	-	-	-	-	-
Flux	330	0.6	8	4	65	0.1	0.1
Output							
Matte	580	70-75	3.5	20	0.6	1.8	0.7
Slag	1,600	4-6	38	1.5	22	1.3	3.9
Reverts	90	20-40	12	15	2.8	6.0	2.4
Dust Bleed	40	6	2	11	1.5	34.0	9.0
Dust Recycle	60	12-20	4	12	2	29.0	9.0

Operating Data:
- Average Blowing Rate: 78,000 Nm^3/h
- Average Oxygen Enrichment: 35 %
- Oxygen Rate (inst.): 520 Tonne/Day
- Average Fuel Ratio: 1.0 GJ/Tonne

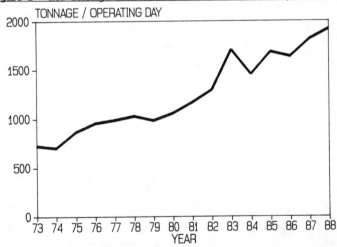

Figure 2 - New Tonnage Treated in the Noranda Process (tonnes/day)

Copper and PM concentrates have always been handled, sampled and treated according to set procedures. With the increasing proportion of scrap and secondaries as feed materials, the Horne smelter has had to develop new procedures and install more sophisticated equipment. Over the last few years, important improvements in the handling, sampling and treatment of these materials have been achieved.

SCRAP AND SECONDARIES FEED TO THE SMELTER

Noranda purchases a variety of scrap and secondaries from suppliers across North America, Europe and Asia. Typical scrap processed at the Horne smelter includes slag, ashes, residues, wire cables, PM containing ingots, jewelry, telephone scrap, automobile parts and PM containing computer and electronic scrap.

Figure 3 shows the increased amounts of scrap and secondaries treated at the Horne smelter. From 23,000 tonnes per year in 1970, the tonnage has increased to 115,000 tonnes per year in 1988. This now represents approximately 15% of total feed to the smelter as illustrated in Figure 4.

Secondary materials are smelted in the reactor and in the converters. Because of increasing matte grades, opportunities to feed scrap to the converters have been substantially reduced. In order to treat the increasing quantities of scrap, a method of rendering more scrap suitable for treatment by the Noranda Process was necessary.

In 1984, a shredding plant was installed to pre-process scrap to maximize Noranda Process feed. Figure 5 shows the layout of the shredding circuit which reduces the scrap to a size of 2 or 4 inches (5 or 10 cm). Operating characteristics have been described in an earlier publication[8]. The successful operation of the shredder has enabled the Horne smelter to significantly increase the total scrap tonnage treated. In 1988, approximately 50% of the scrap was shredded before being smelted in the Noranda Process.

Receiving

The scrap receiving department can accept both trucks and railcars, and unload the material, whether it be loose, in barrels, drums, bags or cardboard boxes. All incoming material and samples are accurately weighed on stationary scales.

Material received in 1988 was typically 50% copper-bearing scrap, 25% slag and residues, 25% PM scrap and secondaries. Upon receipt, the material is unloaded, weighed, sampled and shredded if necessary, before being trucked to the reactor or the converters for smelting.

Depending upon the material classification as regards hazardous wastes, dangerous goods, etc..., the empty trucks or drums containing the material are cleaned at a decontamination station. All residual materials and washings are treated in the slag concentrate dewatering circuit from which slag concentrate containing solids from the decontamination is recycled to the smelting vessel. This efficient decontamination station has enabled the Horne smelter to accept a wide variety of feed, while ensuring all safety, health, environmental and transportation regulations are met.

Sampling

As a custom smelter, the Horne must determine the metal content of all scrap for settlement with its suppliers. Sampling of some PM scrap is performed at other Noranda owned companies, however, the Horne smelter is itself equipped to directly sample most varieties of copper and PM secondary materials.

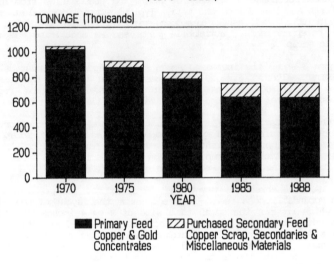

FIGURE 3 CHANGE IN THE HORNE SMELTER FEED (1970 - 1988)

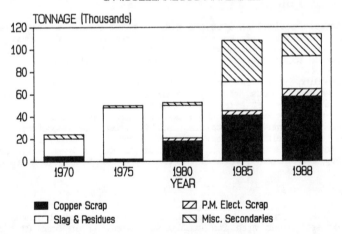

FIGURE 4 PURCHASED COPPER SCRAP SECONDARIES & MISCELLANEOUS MATERIALS

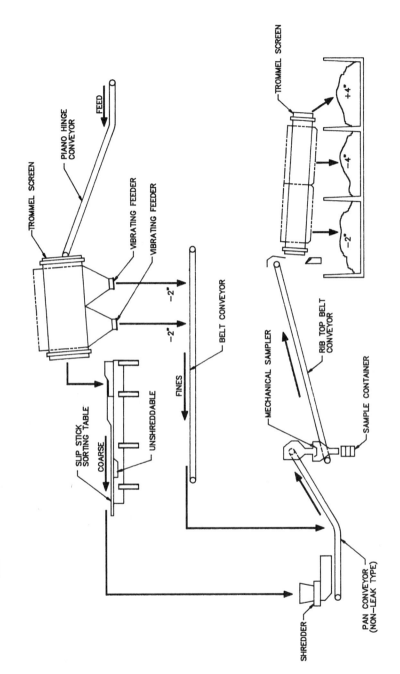

FIGURE 5

Shredding System Flowsheet

Several methods and procedures have been developed for obtaining representative primary samples from various types of scrap. The method used and also the size of the primary sample taken depends largely on PM grade, physical characteristics, chemical homogeneity and lot size of the scrap. In all cases, duplicate samples are taken and one sample is kept in reserve until final settlement with the supplier. The most common primary sampling methods employed at the Horne smelter have been previously published[9], and may be summarized as follows.

Copper concentrates, fine materials such as slags, crushed matte, and fine carbon or residues are pipe sampled. The pipe or the thief sampler is inserted into the drum or bag containers until it reaches the bottom. A minimum of five samples is taken. The aggregate weight must be at least 1 percent of the lot weight, and must total a minimum of 35 kg. Large shipments of coarse slag are processed through the mine crusher and sampled using an in-line Vezin sampler.

Homogeneous low grade ingots are sawed, and the ingot sawings are collected, bagged and sent to the laboratory for assaying. High grade ingots or ingots of poor quality are remelted in the induction furnace, and samples are taken by pin-tube sampling, which has been shown to be precise and accurate[10].

Grab sampling is used for sampling copper scrap (bales and sheets), motors, alternators, radiators and other automobile parts. The method consists of randomly hand-picking pieces to be representative of the entire lot for subsequent melting in the induction furnace. An automatic belt sampler is presently being installed at the discharge end of the shredder to replace the grab sampling practice for copper and low PM electronic scrap.

Until the end of 1988, PM electronic scrap was sampled using the increment sampling method. This consists of spreading the scrap onto a clean, flat metal sheet until a uniform maximum height of 15 cm is reached. Equi-spaced scoop samples are then taken. The number of scoop samples depends on the grade of the material being sampled.

The Noranda Sampler, Figure 6, is used for materials such as PC boards and electronics, chopped silver wires, diodes and computer chips. It has replaced the Horne increment sampling method. The Noranda Sampler was developed by the Horne smelter and was first installed at Noranda Sampling, Rhode Island, U.S. to replace the Gilson splitter. Following successful testing at Noranda Sampling[11], a sampler was installed at the Horne division and commissioned in January 1989. The design is based on the principle of taking a full-cut of a falling stream part of the time. It has 40 chutes positioned on the perimeter of a horizontal carousel which direct material to sample, reserve or bulk containers. The Noranda Sampler, as installed at the Horne, has the flexibility of selecting two inter-penetrating samples in increments of 2.5 percent. This apparatus is being patented by Noranda.

Sample Preparation

After a primary sample is taken, the particle size and sample mass are reduced to obtain a representative sample for laboratory assaying. The Horne employs several methods and procedures for preparing samples for assaying. These methods are outlined in the overall scrap sampling and sample preparation scheme shown in Figure 7.

Figure 6 - Noranda Sampler

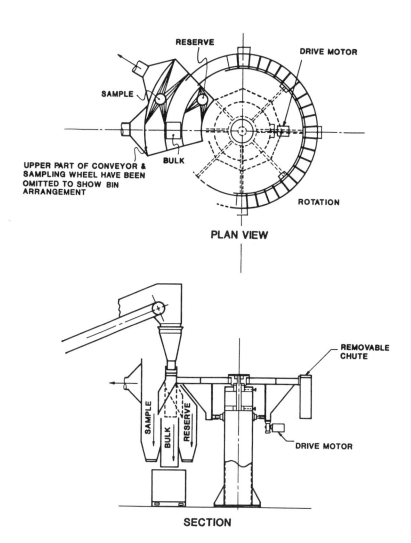

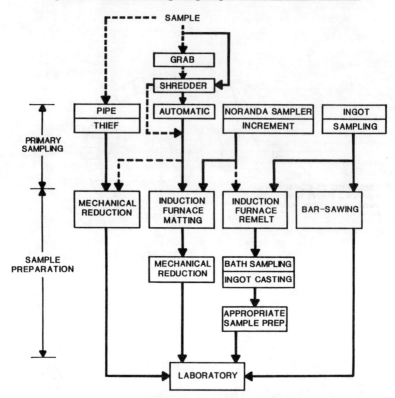

Figure 7 - Overall Scrap Sampling Scheme at Horne Division

The mechanical reduction method is used for materials that are easily crushed to a fine powder such as ashes, mattes and slags, and consists of conventional crushing, screening, mixing and dividing procedures.

Bar-sawing is used for ingots of uniform shape, size and composition, containing less than 2,000 g/tonne Au and 7,000 g/tonne Ag. The bars or ingots are sawed according to set patterns. The sawings are collected, bagged and sent to the laboratory for assaying.

The induction furnace melt/remelt method is used for materials such as high PM ingots, bales of copper, motors, generators and radiators. This method was described in a previous publication[10].

For preparing samples of electronic scrap such as printed boards, the Horne matteing process is used. The process flowsheet is shown in Figure 8. A detailed description of the process has been given elsewhere[9].

The process consists of heating sub-lots of electronic scrap in a 1.3 tonne induction furnace. Pyrite is then added to the baked scrap and melted, leaving a molten bath of iron sulphide (FeS). The metals contained in the scrap (PM, Cu, Pb, Al, etc...) dissolve in the molten FeS forming a solution of metal sulphides.

The resultant matte is crushed to 6 mm in a jaw crusher, and then to -10 mesh in a pan muller, followed by screening. Any +10 mesh material that contains metallics is re-matted. The -10 mesh material is divided in a rotary divider to a subsample of 20 kg. The latter is then processed through a Gilson spinning riffler until a 2 kg subsample is obtained. This subsample is ground to -115 mesh in a ring mill and sent to the laboratory for assaying.

The matteing process reliability has been tested and compared to the conventional method of preparing electronic scrap samples. The conventional method consists of baking the sample to make it friable, followed by ball milling and screening to separate the oversized metallics from the fine powders (sweeps). The sweeps portion is then resampled, while the metallics portion is homogenized by melting with copper before being resampled.

Typical results for duplicate electronic scrap lots containing 300 gr per tonne Au or less are shown in Figure 9. The differences in Au assays were found not to be statistically significant at the 95% confidence level.

The advantage of the matteing process over the conventional sample preparation method is that only one sample is obtained. In addition, less equipment is used in the matteing process, material handling is easier and chances of cross-contamination are less. Tests have also shown that the FeS matte produced in the matteing process has a high capacity to digest aluminum. Table IV shows that the matteing process can accommodate up to 20% aluminium, with a pyrite/PC board sample ratio of 2.

Table IV - Pyrite Consumption Ratio

PC Board Wt. (kg)	Al Added (kg)	Al Total (%)	Pyrite / Sample Ratio
70	8	14	1.75
70	18	24	2.25
70	33	33	2.75

Analysis

Demand for better quality control, greater productivity and faster turnaround time for analysis has increased over the last few years. The Horne laboratory has turned to automation, computerization and intensive use of instrumental techniques. Inductively-coupled plasma (ICP) emission spectroscopy together with fire assay has proven to be a highly successful technique for the analysis of precious metal scrap samples[12]. The recent introduction of Separate Sampling and Excitation Analysis (SSEA) coupled with ICP spectroscopy offers additional flexibility[13].

Figure 8 - Horne Matteing Process Flowsheet

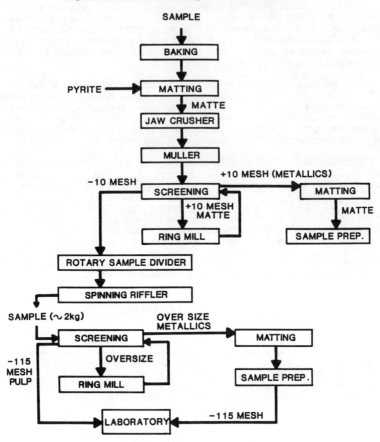

Figure 9 - Comparison of Sample Preparation Methods Matteing Versus Conventional

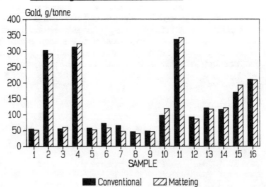

THE EFFECT OF SCRAP FEED ON THE SMELTER

Feeding

The wide variety of materials received is classified, shredded if required to a nominal size of -4 inches (10 cm) and stored in the reactor storage building in different areas based on several criteria: copper content, iron content, reactivity and on impurity or minor element content.

The scrap is then blended with conventional copper concentrates and fed through a 15 cm grizzly. Even reasonably fibrous scrap materials can be mixed in this fashion to provide acceptable Noranda Process feed. As long as the material can be handled on a conveyor system, and pass through chutes and transfer points, it can generally be treated.

The mix of scrap, copper concentrates, revert materials, fluxes and coal are stored in the Noranda Process feed bins.

Smelting

With Noranda Process instantaneous scrap treatment rates increasing in recent years to levels frequently exceeding 20% of the traditional feed, the effect of secondaries on Noranda Process metallurgy has been closely monitored.

Originally heat and mass balance calculations used for process control, ignored the effect of scrap on the process. Operating results were generally not unduly affected by this omission, and acceptable matte grade control, slag quality and operating temperature were maintained. Good feed blending, as described above, and the lower proportions of secondaries fed were the major cause of this.

Higher levels of secondary treatment indicated a need for a re-evaluation of this thesis, and the following secondary material parameters are now monitored in order to establish efficient Noranda Process operating control.

- The copper content and its effect on matte grade and quantity;

- The iron content and the flux requirements to slag off this iron;

- The effect of aluminum and refractory oxides on slag viscosity;

- The heat generated by scrap combustion and its effect on heat balance, or heat required by inert scrap. The use of the tuyere pyrometer[14] is also an invaluable tool in monitoring the Noranda Process heat balance;

- The overall oxygen demand per tonne of secondary material;

These parameters are presently being monitored by Horne Division technical staff. Results to date have shown that good control is possible when treating large and varied tonnages of secondary materials.

Off-Gases

The impact of secondaries on particulate loading of Noranda Process off-gases has been monitored and well controlled with high efficiency ESP's. Of more interest has been the impact of the organic content of scrap treated on the quality of Noranda Process off-gas.

Intensive testing has been conducted to determine if toxic organics, such as PCB's dioxins or furans, were present in Noranda Process off-gas and recycled dusts. These investigations were carried out while the Noranda Process treated unusually large quantities of organic, including PVC, containing secondaries. The testwork[15] did not detect the presence of these compounds.

It is felt that the extremely intensive smelting conditions in the reactor; the highly oxidizing, turbulent, high temperature conditions, and relatively long residence time, all contribute to the total destruction of these compounds.

Further measurements will be conducted at lower detection levels. This program will be undertaken as part of the future acid plant operating checks.

As noted above, a 1,950 tonnes per day acid plant is presently under construction. This plant, based on Chemetics' technology, will handle the entire flow of gases from the Noranda Process. Considering the complex nature of the materials treated in the Noranda Process Reactor, particular emphasis has been placed on designing an efficient and versatile gas cleaning section of the acid plant. Included in this part of the acid plant will be a conditioning tower, a Venturi scrubber, ten electrostatic mist precipitators, ten shell and tube coolers and two Norzink mercury removal towers. The contact section of the acid plant is based on a simple three-pass, single absorption design.

FUTURE CONSIDERATIONS

Although Noranda Minerals is actively engaged in a major exploration program, the present trend is for copper concentrate feed sources to decrease in the north eastern region of Canada. Treatment of complex materials provides a partial buffer against uncertain concentrate supplies. The Horne smelter's objective is to be recognized world-wide for its integrity, flexibility and technical excellence in sampling and processing a broad range of copper and PM bearing materials. An extensive program of testwork to continually improve expertise and treatment efficiency is thus underway. It may be expected that in an environmentally responsable society, this ability to recycle our primary resources will be of increasing value.

REFERENCES

1. G.C. McKerrow, "A 14 ft. x 32 ft. Converter at the Noranda Smelter", Pyrometallurgical Process in Non-Ferrous Metallurgy, edited by J.N. Anderson and P.E. Queneau, (The Metallurgical Society of AIME Conferences, Volume 39, Gorden and Breach Science Publisher, New York, N.Y., 1967, pp 247-258).

2. L.A. Mills, G.D. Hallett and C.J. Newman, "Design and Operation of the Noranda Continuous Smelting Process", Extractive Metallurgy of Copper, edited by J.C. Yannopoulos and J.C. Agarwal, (The Metallurgical Society of AIME, New York, 1976, pp. 458 - 487).

3. J.B.W. Bailey, G.D. Hallett and L.A. Mills, "The Noranda Smelter - 1965 - 1983", Advances in Sulfide Smelting, edited by H.Y. Sohn, D.B. George and A.D. Zunkel, (TMS-AIME, New York, N.Y., 1983, pp. 691-707).

4. P.J. Mackey, J.B.W. Bailey and G.D. Hallett, "The Noranda Process - An Update", Copper Smelting - An Update, edited by D.B. George and J.C. Taylor, (AIME, New York, 1981, pp. 213-236).

5. P. Tarassoff, Process R. & D. - The Noranda Process, (Metallurgical Transactions B, Vol. 15B, Sept. 1984, pp. 411-432).

6. J.B.W. Bailey and W.A. Dutton, Computer Control of the Noranda Continuous Smelting Process, (Paper presented at the 107th AIME annual meeting, Denver, Colorado, February 26th - March 2nd, 1978).

7. H. Persson et al., "The Noranda Process at Different Matte Grades", Journal of Metals, (Vol. 38, No. 9, 1986).

8. D.A. Rawnsley, A. Crépeau and F. Marcil, "Shredded Scrap at the Horne Smelter", Proceedings of International Symposium on Recycled and Secondary Metals, (AIME, Fort Lauderdale, Florida, December 1985, pp. 563 - 573).

9. E. Palumbo et al., Sampling and Sample Preparation of PM Scrap at the Horne Smelter, (13th International Precious Metal Conference, IPMI, Montreal, Canada, June 11 - 15, 1989).

10. E. Palumbo, M. Bédard and C.H. Teh, Induction Furnace Sampling and Sample Preparation Methods, (Precious Metals Sampling and Analysis Seminar, IPMI, Oak Brook, Illinois, October 3-5, 1988).

11. C.H. Teh, J.B.W. Bailey and E. Palumbo, New Sampling Method for PM Scrap - Noranda Sampler, (13th International Precious Metals Conference, IPMI, Montreal, Canada, June 11-15, 1989).

12. M. Bédard, "Determination of Precious Metals in Scrap Samples by Fire Assay ICP Spectroscopy", Proceedings of the Tenth International Precious Metal Institute Conference, (Lake Tahoe, Nevada, pp. 119 - 136, 1986).

13. P. Giasson, D. Pinard and M. Bédard, Recent Improvements in Precious Metal Analysis, (13th International Precious Metal Conference, IPMI, Montreal, Canada, June 11 - 15, 1989).

14. A. Pelletier, J.M. Lucas and P.J. Mackey, The Noranda Tuyère Pyrometer - A New Approach to Furnace Temperature Measurement, (Paper presented at Copper '87, Vina del Mar, Chile, November 30th - December 3rd, 1987, ICM and CIM 1987).

15. F.W. Karasek et al, Analysis of Dust and Stack Samples from Noranda for Dioxins, Furans, PCB's and Other Organic Compounds, (University of Waterloo, Ontario, Canada, April 1986).

16. D.G. Pannell and P.J. Mackey, Noranda Process Operations 1988 and Future Trends, (Paper presented at the Copper Committee Meeting of the GDMB, Antwerk, Belgium, April 27-29, 1988).

THE DEVELOPMENT OF A NEW TYPE OF INDUSTRIAL VERTICAL RING AND PULSATING HGMS SEPARATOR

Xiong Dahe, Liu Shuyi, Liu Yongzhi and Chen Jin

Department of Mineral Engineering, Central South University of Technology, Changsha, Hunan, People's Republic of China

Abstract

This new type industrial vertical ring and pulsating high gradient magnetic separator utilizes the combined force field of magnetic force, pulsating fluid and gravity force to treat fine weakly magnetic particles. It flushes concentrate in the opposite direction of feeding and possesses a slurry pulsating mechanism. Commercial test was carried out in Gushan Iron Mine of Ma Anshan Iron and Steel Company, Anhui, China, to treat fine and refractory hematite ore. It had been successfully running for 3,000 hours by the end of March, 1988, and keeps continuously running up to now in the production line. Compared with centrifugal separator to be substituted, it increases the iron concentrate grade over 4% and the recovery by 10-20%. Moreover, under a new contract another larger separator of the same type is under construction. In this paper, the structure, working principles and commercial test results of the separator will be described.

Introduction

High gradient magnetic separation (HGMS) features the advantage of high recovery of fine weakly magnetic particles (1-4). Cyclic HGMS separators have been successfully applied to kaoline purification and wastewater treatment since 1970s. However, continuous HGMS separators, designed to treat ores containing large portion of weakly magnetic minerals such as hematite ore, are not so competitive in metal mineral processing industry because of their relatively low grade of magnetics and frequent blockage.

We began to study pulsating high gradient magnetic separation(PHGMS) in 1981 (5). Great efforts have been done on preventing magnetic matrix from blockage and improving the quality of magnetic product. The first SLon-1000 vertical ring and pulsating high gradient magnetic separator was worked out under the cooperation of Central South University of Technology and Ganzhou Research Institute of Non-ferrous metallurgy in september,1987. Then it was installed into the hematite processing flowsheet of Gushan Iron Mine to carry out commercial test. After a running period of 3,000 hours test, it still keeps running to meet the industrial needs and up to now (November, 1988) it has been running over 8,000 hours. Commercial test verified that this kind of PHGMS separator possesses a series of advantages for industrial application.

Structure and Working Principles of SLon-1000 Separator

SLon-1000 separator consists of mainly thirteen parts, as shown in Fig.1. The pulsating mechanism, magnetic yoke, energizing coil and working ring are the key parts. Pulsating frequency and ring rotating speed are controlled by speed variable motors. Although pulsating stroke is adjustable, we usually take 13 mm as applied to the slurry in the working zone. The energizing coil is made of hollow copper tube and cooled internally by water. Along the periphery of the ring there are a number of rectangular rooms in which expanded nets sized 0.3:3.2:8 mm made of magnetic conductive stainless steel are filled as matrix. The height of matrix pile is 7 cm. Other kind of matrix may also be used depending on the requirement of different ores. While the separator is working, the ring rotates clockwise. Slurry fed from the feeding box enters the ring from the gaps in the upper yoke. The average speed of slurry in the working zone is about 6 cm per second. The matrix in the working zone is magnetized. Magnetic particles are attracted from slurry onto matrix surface, brought to the top by the ring, where magnetic field is negligible, and flushed out to the concentrate box. Non-magnetic particles enter the tailing box along the gaps in lower yoke under the combined action of slurry pulsation, gravity and drag force. The pulsating mechanism drives the rubber diaphragm on the tailing box to move back and forth. As long as slurry level is adjusted above the level line, the kinetic energy due to pulsation can be effectively transmitted to the working zone.

As the flushing direction of magnetic concentrate is opposite to that of feeding relative to each matrix pile, coarse particles can be washed out without the need of passing through

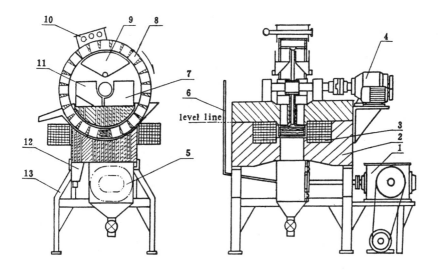

Fig.1-SLon-1000 separator
1-pulsating mechanism; 2-magnetic yoke; 3-energizing coil; 4-ring driver; 5-tailing box; 6-slurry level tube; 7-feeding box; 8-working ring; 9-concentrate box; 10-concentrate flusher; 11-washing water box; 12- draining box; 13-support frame.

matrix pile. Slurry pulsation can keep the particles in matrix pile in a loose state all the time. Evidently, discharging concentrate in an opposite way and slurry pulsating up and down help to prevent matrix from blockage, and magnetic separation can enhance the quality of magnetic concentrate. Moreover, these measures not only ensure the effective recovery of size down to about 10 microns, but also extend the feeding size limit up to 800 microns, thus enlarging particle size range to be treated and simplifying classification unit.

The major parameters and electromagnetic features are shown in table 1 and table 2 respectively. The highest background magnetic induction in the working zone is 1.2 T, though 1.0 T is usually enough for most oxidized iron ores.

Table 1. Parameters of SLon-1000 separator

Ring diameters (outer; inner),	mm	1,000; 800
Ring width,	mm	300
Ring rotating speed,	r/min	0.5-5
Feed size,	mm	0.8-0
Throughput,	t/hr	4-7
Driving power,	Kw	1.1; 1.5
Pulsating stroke,	mm	0-20
Pulsating frequency,	min^{-1}	0-400
Separator weight,	t	5.34

Table 2. Electromagnetic features of SLon-1000 separator

D.C. current, A	300	390	510	720	930	1100	1200	1480
D.C. voltage, V	4.5	6.0	8.0	11.5	15.0	18.0	20.0	24.5
Energizing power, Kw	1.35	2.34	4.08	8.28	14.0	19.8	24.0	36.3
Magnetic field, T	0.30	0.38	0.50	0.70	0.90	1.00	1.08	1.20

Commercial Test

Gushan Iron Mine is one of the major raw material bases of Ma Anshan Iron and Steel Company. The ore is of finely disseminated and colloidal hematite quartzite. The size range of most hematite grain is 30-5 microns. This kind of ore is hard to grind and difficult to beneficiate. The existing flowsheet is that the ore is ground to 50% -200 mesh, then classified by hydrocyclones of D500 and D350 mm. Their underflows are treated by spiral concentrator and the overflow of D350 mm hydrocyclone is treated by centrifugal separator. The particle size of the overflow of D350 mm hydrocyclone is about 90% -200 mesh. It contains a large amount of slime and light particles. Hence it is very difficult for various separators to get good results. For example, the results of flotation in 1982 were: feed 34.14%Fe, concentrate 49.23%Fe, tailing 24.26%Fe, recovery 49.09%; and the results of centrifuge in 1987 were: feed 34.12%Fe, concentrate 52.36%Fe, tailing 26.79%Fe, recovery 43.99%. Because the concentrate grade could not meet metallurgical requirement (Fe=55% or more), the overflow of D350 mm hydrocyclone was mostly discarded as tailing and the mine suffered a heavy loss.

SLon-1000 separator was installed in line with the existing flowsheet to treat the overflow of D350 mm hydrocyclone in parallel with centrifuger. The following are some of the results it achieved during the 3,000hr site test.

Effect of Background Magnetic Induction

The results of background magnetic induction test are listed in table 3. When magnetic induction is increased from 0.53 T to 0.92 T, iron concentrate grade slightly decreases from 56.66% to 56.16%, while recovery increases from 56.57% to 69.44%. When magnetic induction is higher than 0.92 T, the results keep almost the same. The suitable magnetic induction is 0.76-0.92 T.

Table 3. Results of magnetic induction test (Fe%)

Induction, T	0.53	0.76	0.92	1.10
Feed grade	28.70	28.95	30.20	28.45
Concen. grade	56.66	56.28	56.16	55.79
Tailing grade	17.47	15.23	14.73	13.98
Recovery	56.57	64.98	69.44	67.87

Effect of Pulsating Frequency

The results of pulsating frequency test are listed in table 4. When no pulsation (N=0), the concentrate grade is only 48.80% due to serious mechanical trap of non-magnetic particles. As frequency is increased, mechanical trap is greatly reduced and concentrate grade is increased rapidly. When N=350/min, concentrate grade reaches 58.41%, gained by about 10% compared with that in the case of no pulsation. Moreover, because pulsation improves operating condition in the working zone, in a certain range, tailing grade slightly decreases and recovery remaines almost unchanged. When N=400/min, concentrate grade continues rising, but tailing grade also rises and recovery drops due to rapid increase of competing force. The proper frequency is 300-400/min at 13 mm slurry stroke.

Table 4. Results of pulsating frequency test (Fe%)*

N/min	0	200	300	350	400
Feed grade	34.45	33.22	34.45	33.95	33.70
Concen. grade	48.80	53.42	56.78	58.41	59.53
Tailing grade	19.97	16.72	16.47	15.97	16.97
Recovery	71.15	72.52	73.52	72.89	69.44

*Magnetic induction is fixed at 0.9 T and matrix volume package at 6%.

Effect of Feed Size

The statistic results according to feed size are listed in table 5. Although feed size fluctuates in the range of 60-98% -200 mesh, concentrate grade keeps at about 57% and tailing grade remains at about 20%. These results demonstrated that SLon-1000 separator possesses strong adaptability to wide range of particle size.

Table 5. Statistic results related to feed size (Fe%)

-200 mesh in feed, %	60-70	70-80	80-90	90-98
Feed grade	34.57	37.92	36.49	35.18
Concentrate grade	58.88	57.60	57.54	57.09
Tailing grade	19.87	21.90	20.01	19.16
Recovery	64.18	68.16	69.24	68.54

Effect of Feeding Slurry Density

The statistic results related to feeding slurry density are shown in table 6. When density (solid weight percentage in feeding slurry) is 25-30%, the concentrate grade and recovery are fairly good, 58.54% and 67.74% respectively. However, the effect of feeding slurry density seems small. In practice, higher feeding slurry density is always preferred so that the capacity of the separator can be fully utilized.

Table 6. Statistic results related to feed density (Fe%)

Feed density, %	10-20	20-25	25-30	30-42
Feed grade	35.07	35.75	37.22	36.75
Concentrate grade	56.76	57.38	58.54	57.04
Tailing grade	19.21	19.43	21.09	22.41
Recovery	68.36	69.02	67.74	64.27

Effect of Feed Grade

The statistic results according to feed grade are shown in table 7. Although feed grade changes from 23% to 40%, concentrate grade is kept stable at 56.07-58.19%. Recovery increases as feed grade increases. It is very important for the mine to maintain concentrate quality. The results show that SLon-1000 separator possesses the ability to adapt large fluctuations of feed grade.

Table 7. Statistic results related to feed grade (Fe%)

Feed grade	23-27	27-30	30-35	35-38	38-40
Concen. grade	56.07	56.45	56.81	57.40	58.19
Tailing grade	15.16	16.53	18.32	20.72	21.22
Recovery	52.35	60.78	68.46	67.90	71.56

One Roughing and One Cleaning

In order to further improve the quality of iron concentrate, the overflow of D350 mm hydrocyclone was tested by one roughing and one cleaning with SLon-1000 separator. The results are shown in table 8. The grade and recovery of final concentrate are 61.90% and 62.49% respectively. The test shows a practical way for the mine to improve the quality of iron concentrate.

Table 8. Results of one roughing and one cleaning (Fe%)

Product	Feed	Rough concen.	Clean concen.	Middling	Tailing
Grade	38.28	57.61	61.90	42.65	19.16
Weight, %	100	49.73	38.65	11.08	50.27
Recovery	100	74.84	62.49	12.35	25.16

Conclusions

Commercial test of 3,000 hours demonstrated that SLon-1000 vertical ring and pulsating high gradient magnetic separator possesses the advantages of high beneficiation ratio, not easy to be blocked, and strong adaptability to large fluctuations of feeding size, feed grade and slurry density. This type of separator can be widely applied to process weakly magnetic minerals.

References

1. J.A.Oberteuffer,"High gradient magnetic separation",IEEE Trans.on Mag.,1973,No.3,303-306.
2. D.R.Kelland,"HGMS applied to mineral beneficiation",IEEE Trans.on Mag.,1973,No.3,307-309
3. J.A.Oberteuffer,"HGMS of steel mill process and waste water", IEEE Trans.on Mag.,1975,No.5,1591-1593.
4. J.Iannicelli,"Development of high extraction magnetic filtration by the koaline industry of Georgia",IEEE Trans.on Mag., 1976,No.5,489.
5. Liu Shuyi,et al,"Pulsating HGMS of the mixed materials of fine grain tungsten and cassiterite from Shanhu Tin Mine", Mining and Metallurgical Engineering,P.R.China,1983,No.2, 30-34.

Subject Index

Acid rain, 149
Activity coefficient, 248
Alloys
 niobium, titanium & zirconium, 829
Alumina
 VAW technology, 844
Aluminum, 46, 86
 cells, 48, 97
 processing, 855
Arsenic, 99, 545
 behavior, 758
 compounds, 549
 disposal, 779
 in solution, 775
 vapor pressure, 743
 world production, 736
Ashio Smelter, 789
Autoclave leaching, 632

Battery scrap, 483, 495, 501
Bayer process, 844
Birnessite
 randomly stacked, 562
 structure, 566
Black liquor, 243
Blast furnace
 dust, 653
 nonferrous, 187, 498
Boilers, 206
BP Minerals, 75
BSB Smelter, 483
 process, 486
Bulk density
 ferroalloy powders, 287

Carbothermic reduction
 tin, 215
Casting
 continuous, 302
 copper rod, 316
 electromagnetic, 313
 stainless steel, 309
Catalysts
 automotive, 524

 petrochemical, 726
Caustic leach
 zinc, 655
Cementation
 silver, 639
 on lead, 706-707
CollaMat, 717
Consumption
 metals, 159, 736
Contibat process, 501
CONTOP process, 88, 379
Copper
 concentrate, 368, 795
 industry, 28, 69, 87
  Canada, 76
  processing, 28, 87
  tankhouse practice, 90, 717, 783
  United States, 71
Copper Cliff Smelter, 413
Costs
 copper, 71, 73, 75
 nickel, 81
 zinc, 79, 80
Crystal field
 activation energy, 577
 stabilization energy, 566-567, 578

Direct smelting
 tin, 215
 zinc, 241
Drucker
 uncoupling theory, 160-161, 165, 182

Econometric metal demand
 modeling, 167-170
Electric arc furnace dust
 analysis, 653, 654
 processing, 652, 656
Electrical conductivity
 concentrates, 684
Electrolyte, 718
Emissions
 arsenic, 737
Energy, 103

conservation, 93, 150, 459, 473
costs, 387
crisis, 146, 148-149
efficiency, 151, 541
for secondary processing, 153, 473
metal production, 146-153
nuclear, 151
potlines, 52, 87, 881
solar, 151
Environmental protection, 11, 473, 852
Espanola del Zinc, 737
Extraction
zinc, 241

Ferric chloride
leach, 706
Ferroalloy powders
analysis, 287
Ferroalloying, 281, 287
Ferrochrome, 541
Fire refining, 772
Flash smelting
pyrite, 341
FLUXFLOW, 210

Gas cleaning, 209, 500, 744, 792
Glue, 717
Gold, 43, 350, 627, 911
processing, 798, 913
Greenhouse effect, 150
Gushan Iron Mine, 950

Hazelett caster, 319
Heat losses, 415
Hematite, 588
zinc, 603
HOM-TEC, 303, 309
Hudson Bay
Flin Flon, 381

Industrial production index, 180-182
Intensity of use, 165-167
U.S. forecast, 166

Jarosite compounds, 588, 590

Kidd Creek, 77
Krupp Stahl AG, 310

Leach residue, 24, 551, 590, 780
Leachability
zinc, 673, 678
Leaching, 798
catalysts, 727
continuous, 92, 705, 713
galena, 705, 706
manganese oxyhydroxides, 573-581
mechanisms, 576-579, 580-581
sulfuric acid, 573
sulfurous acid, 579
of galena in brine solution, 705, 706
pressure, 383, 659, 798
Lead
blast furnace, 188, 359
chloride, 641
emissions, 360, 493
industry, 15
processing, 90
secondary, 16, 473, 483
Lithium, 894

Magnesium, 57
Magnetite
balance, 370
Manganese
oxyhydroxides, 562
pure and doped, 562, 563
structure, 567-573
synthesis, 563
nodules, 561, 562, 581
genesis, 582, 583
internal structure, 583, 584
minerals, 562
processing, 581-583
Mathematical models, 639, 921
Matte
grade, 88, 368
impurities, 99, 763
Mercury, 924
Metallo
thermic reduction, 900
Metallurgie Hoboken Overpelt (MHO)
plant, 358, 674, 753
Metals
consumption
and economic growth, 160-170, 176-182
causality, 163-164
elasticity, 167-169

ratchet effect, 175-176
structural changes, 169, 172-176
intensity of use, 15, 165
marketing, 183
substitution, 14
Mineral processing, 73, 923, 947
Mineralogy, 909
Mineralogical data, 916, 917

Naoshima Smelter, 366

Nickel industry, 36, 69
nickel/tungsten, 725
niobium alloys, 838
North America, 80
Noranda
Horne, 930
reactor, 89
Flin Flon, 391

Overpelt
zinc plant, 674
Oxy-fuel burners, 473
Oxygen enrichment, 263, 266, 269, 508, 934

Petroleum resources, 149
Phelps Dodge, 75
Platinum metals, 616
Powder metallurgy, 546
Precious metal
consumption, 44
production, 44
Preoxidation
temperature effect, 728-730
Pressure cyanidation, 627
Process mineralogy, 909, 910, 918
gold-silver operation, 909, 911, 912, 916, 918
Productivity, 70, 73, 81, 103, 376, 413
Pyrite, 342
composition, 348

Recovery-platinum metals, 377
Refining processes comparison
copper, 100
zinc, 102
Roasting
arsenic behavior, 748
fluid bed, 257
gold concentrate, 752
pyrite, 754
Rönnskär Smelter, 399

Scrap metal, 935
Scrapped catalysts
nickel/tungsten petrochemical, 726, 727
platinum/palladium automotive, 726
Secondary metals, 152-153
Sherritt zinc pressure leach, 381
Short rotary furnace, 490, 508
Shredding, 937
SIGMA process, 856
plant data, 867
Silver, 43, 350, 632, 639, 911
Slag
composition, 233, 373, 551, 771
losses, 99, 227, 473
Smelting
continuous, 365
processes comparison
copper, 34, 98, 148
lead, 101
zinc, 241
SO_2 emissions, 360, 382, 413, 493
Solvent extraction
precious metals, 613
zinc, 696
South Africa, 160-161, 183
Specific surface area
concentrates, 688
Speisses, 552, 792
Steel
consumption, 172
model, 170-176
Sulphur
acid leaching
nitric acid addition, 728-731
non-preoxidized scrapped catalyst, 728
preoxidized scrapped catalyst, 728-732
selectivity, 726, 728, 730, 732
temperature effect, 730, 731
tubular reactor, 726, 727
production, 345

Tantalum, 470
Tertiary metal, 153
Tin, 60
Titanium alloys, 837

VAW potlines
  Innwerk, 880, 889
  Rheinwerk, 885

Waste heat
  boilers, 206
  utilization, 209, 418
Waste processing, 153, 779

Zinc
  condensation, 246, 254
  direct smelting, 241
  extraction, 241
  industry, 3, 69, 92
    Canada, 78
    processes, 21, 93, 696
    United States, 79
  leachability, 673
  oxide reduction, 254
  roasting, 10, 259
  secondary, 651
  smelting, 93, 241
  solvent extraction, 696
Zirconium alloys, 834

Author Index

Alruiz, O.M., 627
Anfruns, J.P., 627
Argyropoulos, S.A., 281
Arnswald, W., 843
Asteljoki, J.A., 341

Barlin, B., 381
Baum, W., 909
Becker, J.S., 257, 855
Bédard, M., 929
Berendes, H., 315
Biswas, A.K., 561
Blanco, J., 413
Broman, P.G., 399
Buch, E., 315

Canic, N., 651
Castle, J., 3
Chapados, M., 929
Chatfield, P.C., 313
Condliff, A.F., 829
Craigen, W.J.S., 381
Crauwels, D., 357
Crawford, G.A., 69
Cross, M., 187

Dahe, X., 947
Davies, H., 413
Davis, G., 159
de Juan Garcia, D., 695
Distin, P.A., 725
Dutrizac, J.E., 587

Farge, Y., 155
Fischer, R., 501
Floyd, J.M., 241

Garritsen, P., 413
Geskin, E.S., 435
Gilbert, S.R., 909
Girardi, S.C., 627
Gluns, L., 257, 855
Grote, M., 613
Gülbas, M., 815
Gupta, S.K., 241

Hanniala, T.P.T., 341
Harris, G.B., 545
Harris, R., 893
Hu, H.-K., 923
Hu, W., 921

Jacob-Dulière, M., 673
Janssen, W., 541
Jin, C., 947
Jochem, H.-O., 415
Jung, V., 523

Kachaniwsky, G., 929
Kammel, R., 815
Kanamori, K., 365
Kang, T., 639, 705
Kelebek, S., 725
Kellogg, H.H., 145
Kettrup, A., 613
Kim, K.H., 639
Koch, M., 495
Krysa, B., 381
Kumar, R., 561
König, C., 511

Lamm, K.F., 473, 483
Langner, B.E., 717
Lee, C.-K., 705
Leroy, M., 3
Lewis, B.G., 313
Li, S., 921
Lieber, H.-W., 815
Lindgren, P.-O., 399
Lofkvist, B., 399

Maes, R., 3
Mast, E., 893
McLean, A., 281
Melcher, G., 379
Melin, A.E., 483, 735
Mihovilovic, L., 651
Milosevic, S., 651
Misra, V.N., 215, 659
Monette, S., 545

Niklas, H., 495

Paschen, P., 459
Peacey, J.G., 85
Peippo, R., 205
Pickard, F.G.T., 69
Pickert, G., 613
Piret, N.L., 735

Rath, G., 511
Ray, R.K., 561
Roller, E., 301
Ryu, J.-B., 705

Saha, D., 257, 855
Schade, J., 281
Selke, A., 695
Shibasaki, T., 365
Shuyi, L., 947
Siebel, K., 315
Smith, R., 187
Sohn, H.-J., 639, 705
Sparwald, V., 879
Stamatovic, M., 651
Stantke, P., 717

Sterckx, J., 673
Szekely, J., 105

Takeda, Y., 227
Toguri, J.M., 893
Traulsen, H., 3

Ulrich, K.H., 541

Van Camp, M., 357
Vlajcic, T., 511

Wade, K.C., 187
Westerlund, K., 205
Winand, R., 103

Xiong, X.Y., 673

Yarwood, J.C., 313
Yazawa, A., 227
Yin, D., 921
Yongzhi, L., 947
Youssef, M.A., 435

Zivanovic, T., 651